Bioinformatics

Sequence and Genome Analysis

Bioinformatics

Sequence and Genome Analysis

David W. Mount

University of Arizona, Tucson

COLD SPRING HARBOR LABORATORY PRESS
Cold Spring Harbor, New York

Bioinformatics

Sequence and Genome Analysis

Developmental Editor Judy Cuddihy   **Interior Designer** Denise Weiss
Project Coordinator Joan Ebert   **Cover Designer** Ed Atkeson, Berg Design
Production Editor Patricia Barker

Library of Congress Cataloging-in-Publication Data

Mount, David W.
    Bioinformatics : sequence and genome analysis / David W. Mount.
      p. cm.
    Includes bibliographical references and index.
    ISBN 0-87969-597-8 (hard cover : alk. paper)—ISBN 0-87969-608-7 (paperback : alk. paper)
   1. Genetics–Data processing.   2. Bioinformatics.   3. Nucleotide
sequence.   4. Amino acid sequence.   I. Title.
    QH441.2 .M68 2000
    572.8'633–dc21

                 00-060252

10 9 8 7 6 5 4

Front cover: Illustration inspired by the relationship of the 6000 genes in yeast to each other (Lisa Mount and Adam Sherman).

*This book is dedicated to the following individuals who have
contributed much to the field of sequence analysis:
David Lipman,
Bill Pearson,
Temple Smith,
and Michael Waterman
and to the memory of
Margaret Dayhoff and Walter Goad*

The *Bioinformatics* Web Site
Access to the On-line Text and Associated Resources

This print edition of *Bioinformatics* is associated with a Web site (www.bioinformaticsonline.org) that will add to and extend the contents of the book.

When the site is launched, registered purchasers of the book will be able to (at no extra charge):

- Access Web sites referred to in the text.

- Access problem sets for classroom use and other useful material not included in the print edition.

- Receive E-mail alerts about peer-reviewed, new, and updated information that extends the scope and content of the book.

To register at www.bioinformaticsonline.org:

1. Open the home page of the site.

2. Follow the registration procedure that begins on that page.

3. When prompted, enter the unique access code that is printed on the inside front cover of this book.

4. When prompted, enter your E-mail address as your user name and a password of your choice.

5. Complete the registration procedure as requested.

The Web site contains answers to FAQs about the registration procedure and a demonstration of the functions available to registered users. For additional assistance with registration, and for all other inquiries about the *Bioinformatics* Web site, please use the E-mail addresses provided at the site.

Contents

Preface

THIS BOOK IS WRITTEN MAINLY for biologists who want to understand the methods of sequence and structure analysis. I strongly believe that a person using a computer program should understand how it works. Accordingly, one of my main objectives is to help biologists appreciate the underlying algorithms used and assumptions made, as well as limitations of the methods used and strategies for their use. To this end, I have tried to avoid complex formulas and notations and to give instead simple numerical examples whenever possible. I hope that the book will also be of interest to computational biologists who want to learn a little more about the biological questions related to the field of bioinformatics. This book is intended to be a laboratory reference text, as well as a textbook for a course in bioinformatics, rather than a user guide for a specific set of sequence analysis programs.

Most of the chapters include a flowchart that is designed to propose an orderly use of the methods that are discussed in the chapter. There are very few examples of these types of charts and they are quite difficult to produce, requiring assumptions and over-simplifications that may not always be justified. I hope that these charts will be useful for the less experienced in this field, but I expect that the more-experienced practitioners in the field will have other, probably better, ways of achieving the same goal.

There are many references to Web sites and FTP locations where these methods may be applied or programs obtained. In some cases, as for the commonly used and important BLAST and CLUSTALW programs, I have provided a great deal of information about using the program and analyzing the results. However, there are many other important tools and approaches available for biological sequence and genome analysis and I have tried to cover as many of them as possible, given time and space limitations. I have not paid particular attention to simpler types of sequence analyses, e.g., searching for restriction sites, translating sequences, and compositional analysis. There are many commercial and noncommercial packages for performing these tasks, and commercial packages for genome analysis are now appearing.

In writing this book, my first, I found that the amount of information available in the published literature was far more than I could include. I have tried to be thorough and to cover the most significant problems in sequence and genome analysis, but there are also many excellent papers that have not been cited for reasons of time and space, and I apologize to colleagues whose valuable contributions are not mentioned. Because of the space limitations of a printed text, and the ever-changing nature of bioinformatics, material not included in the book, as well as links to all of the Web sites cited, examples, and problems, will appear on a special Web site for the book, which can be found at http://www.bioinformaticsonline.org.

One aspect of this discipline that has been quite remarkable to me is the willingness of most investigators, especially the pioneers in the field, to share their results with colleagues. I have had the privilege of personally knowing several of these early investigators, especially David Lipman, Hugo Martinez (with whom I spent a sabbatical year), and Temple Smith. The tremendous accomplishments of these people became even more meritorious because they freely shared the results of their efforts with colleagues. In doing so, they were very much responsible for the eventual success of the sequence analysis field in both the academic and commercial areas.

This large project has required much support and help. Part of this book was derived from class notes for a course in "Bioinformatics and Genome Analysis" at the University of Arizona in the 1999 and 2000 academic years. Many students made very useful suggestions and were helpful in finding errors; I want to particularly thank Bryan Zeitler for providing many corrections. Any remaining errors will be corrected on the book's Web site. I am grateful to Bill Pearson for information about the FASTA suite of programs, to Julie Thompson and John Kececioglu for comments on Chapter 4, to Steve Henikoff for reading Chapter 3, and to Michael Zuker for helpful comments on the writing of Chapter 5. Bill Montfort provided information about PDB files for Chapter 9, and Roger Miesfeld provided the example of complex gene regulation in Chapter 8. Jun Zhu was very kind in answering my questions about the Bayes block aligner for Chapter 3. My department has been most patient and supportive as I skipped meetings and seminars to complete or revise another chapter, over a period of three years. During this time, Rob Han and Juwon Kim provided the very large number of papers and book chapters that I needed on a regular basis with a very short turnaround time, allowing me more time to digest the information. My editor, Judy Cuddihy of Cold Spring Harbor Laboratory Press, guided me through the process of writing with great skill and was very patient as she tried to keep me to a reasonable writing schedule, providing needed encouragement for completing the project. Elisabeth Cuddihy checked most of the Web sites, carefully went through formulas and numerical examples, and helped to write parts of the glossary. I also thank Joan Ebert and Jan Argentine in the Development Department and Pat Barker and Denise Weiss in the Production Department at the Press.

Last, but not least, I thank my wife Jennifer Hall for her patience and understanding during the many times that book-writing took precedence over family matters.

David W. Mount

Historical Introduction and Overview

THE DEVELOPMENT OF SEQUENCE ANALYSIS METHODS has depended on the contributions of many individuals from varied scientific backgrounds. This chapter provides a brief historical account of the more significant advances that have taken place, as well as an overview of the chapters of this book. Because many contributors cannot be mentioned due to space constraints, additional references to earlier and current reference books, articles, reviews, and journals provide a broader view of the field and are included in the reference lists to this chapter.

THE FIRST SEQUENCES TO BE COLLECTED WERE THOSE OF PROTEINS

Margaret Dayhoff

The development of protein-sequencing methods (Sanger and Tuppy 1951) led to the sequencing of representatives of several of the more common protein families such as cytochromes from a variety of organisms. Margaret Dayhoff (1972, 1978) and her collaborators at the National Biomedical Research Foundation (NBRF), Washington, DC, were the first to assemble databases of these sequences into a protein sequence atlas in the 1960s, and their collection center eventually became known as the Protein Information Resource (PIR, formerly Protein Identification Resource; http://watson.gmu.edu:8080/pirwww/index.html). The NBRF maintained the database from 1984, and in 1988, the PIR-International Protein Sequence Database (http://www-nbrf.georgetown.edu/pir) was established as a collaboration of NBRF, the Munich Center for Protein Sequences (MIPS), and the Japan International Protein Information Database (JIPID).

Dayhoff and her coworkers organized the proteins into families and superfamilies based on the degree of sequence similarity. Tables that reflected the frequency of changes observed in the sequences of a group of closely related proteins were then derived. Proteins that were less than 15% different were chosen to avoid the chance that the observed amino acid changes reflected two sequential amino acid changes instead of only one. From aligned sequences, a phylogenetic tree was derived showing graphically which sequences were most related and therefore shared a common branch on the tree. Once these trees were made, they were used to score the amino acid changes that occurred during evolution of the genes for these proteins in the various organisms from which they originated (Fig. 1.1).

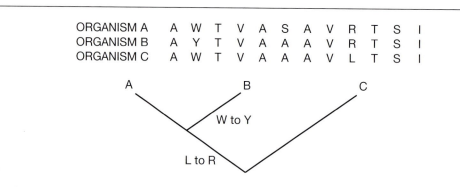

Figure 1.1. Method of predicting phylogenetic relationships and probable amino acid changes during the evolution of related protein sequences. Shown are three highly conserved sequences (A, B, and C) of the same protein from three different organisms. The sequences are so similar that each position should only have changed once during evolution. The proteins differ by one or two substitutions, allowing the construction of the tree shown. Once this tree is obtained, the indicated amino acid changes can be determined. The particular changes shown are examples of two that occur much more often than expected by a random replacement process.

Subsequently, a set of matrices (tables)—the percent amino acid mutations accepted by evolutionary selection or PAM tables—which showed the probability that one amino acid changed into any other in these trees was constructed, thus showing which amino acids are most conserved at the corresponding position in two sequences. These tables are still used to measure similarity between protein sequences and in database searches to find sequences that match a query sequence. The rule used is that the more identical and conserved amino acids that there are in two sequences, the more likely they are to have been derived from a common ancestor gene during evolution. If the sequences are very much alike, the proteins probably have the same biochemical function and three-dimensional structural folds. Thus, Dayhoff and her colleagues contributed in several ways to modern biological sequence analysis by providing the first protein sequence database as well as PAM tables for performing protein sequence comparisons. Amino acid substitution tables are routinely used in performing sequence alignments and database similarity searches, and their use for this purpose is discussed in Chapters 3 and 7.

DNA SEQUENCE DATABASES

Walter Goad

Many types of sequence databases are described in the first annual issue of the journal Nucleic Acids Research.

The growth of the number of sequences in GenBank can be tracked at http://www. ncbi.nlm.nih.gov/Gen Bank/genebankstats. html.

DNA sequence databases were first assembled at Los Alamos National Laboratory (LANL), New Mexico, by Walter Goad and colleagues in the GenBank database and at the European Molecular Biology Laboratory (EMBL) in Heidelberg, Germany. Translated DNA sequences were also included in the Protein Information Resource (PIR) database at the National Biomedical Research Foundation in Washington, DC. Goad had conceived of the GenBank prototype in 1979; LANL collected GenBank data from 1982 to 1992. GenBank is now under the auspices of the National Center for Biotechnology Information (NCBI) (http://www.ncbi.nlm.nih.gov). The EMBL Data Library was founded in 1980 (http://www.ebi.ac.uk). In 1984 the DNA DataBank of Japan (DDBJ), Mishima, Japan, came into existence (http://www.ddbj.nig.ac.jp). GenBank, EMBL, and DDBJ have now formed the International Nucleotide Sequence Database Collaboration (http://www. ncbi.nlm.nih.gov/collab), which acts to facilitate exchange of data on a daily basis. PIR has made similar arrangements.

Initially, a sequence entry included a computer filename and DNA or protein sequence files. These were eventually expanded to include much more information about the sequence, such as function, mutations, encoded proteins, regulatory sites, and references. This information was then placed along with the sequence into a database format that could be readily searched for many types of information. There are many such databases and formats, which are discussed in Chapter 2.

The number of entries in the nucleic acid sequence databases GenBank and EMBL has continued to increase enormously from the daily updates. Annotating all of these new sequences is a time-consuming, painstaking, and sometimes error-prone process. As time passes, the process is becoming more automated, creating additional problems of accuracy and reliability. In December 1997, there were 1.26×10^9 bases in GenBank; this number increased to 2.57×10^9 bases as of April 1999, and 1.0×10^{10} as of September 2000. Despite the exponentially increasing numbers of sequences stored, the implementation of efficient search methods has provided ready public access to these sequences.

To decrease the number of matches to a database search, non-redundant databases that list only a single representative of identical sequences have been prepared. However, many sequence databases still include a large number of entries of the same gene or protein sequences originating from sequence fragments, patents, replica entries from different databases, and other such sequences.

SEQUENCE RETRIEVAL FROM PUBLIC DATABASES

David Lipman

An important step in providing sequence database access was the development of Web pages that allow queries to be made of the major sequence databases (GenBank, EMBL, etc.). An early example of this technology at NCBI was a menu-driven program called GEN-INFO developed by D. Benson, D. Lipman, and colleagues. This program searched rapidly through previously indexed sequence databases for entries that matched a biologist's query. Subsequently, a derivative program called ENTREZ (http://www.ncbi.nlm.nih.gov/Entrez) with a simple window-based interface, and eventually a Web-based interface, was developed at NCBI. The idea behind these programs was to provide an easy-to-use interface with a flexible search procedure to the sequence databases.

Sequence entries in the major databases have additional information about the sequence included with the sequence entry, such as accession or index number, name and alternative names for the sequence, names of relevant genes, types of regulatory sequences, the source organism, references, and known mutations. ENTREZ accesses this information, thus allowing rapid searches of entire sequence databases for matches to one or more specified search terms. These programs also can locate similar sequences (called "neighbors" by ENTREZ) on the basis of previous similarity comparisons. When asked to perform a search for one or more terms in a database, simple pattern search programs will only find exact matches to a query. In contrast, ENTREZ searches for similar or related terms, or complex searches composed of several choices, with great ease and lists the found items in the order of likelihood that they matched the original query. ENTREZ originally allowed straightforward access to databases of both DNA and protein sequences and their supporting references, and even to an index of related entries or similar sequences in separate or the same databases. More recently, ENTREZ has provided access to all of Medline, the full bibliographic database of the National Library of Medicine (NLM), Washington, DC. Access to a number of other databases, such as a phylogenetic database of organisms and a protein structure database, is also provided. This access is provided without cost to any user—private, government, industry, or research—a decision by the staff of NCBI that has provided a stimulus to biomedical research that cannot be underestimated. NCBI presently handles several million independent accesses to their system each day.

A note of caution is in order. Database query programs such as ENTREZ greatly facilitate keeping up with the increasing number of sequences and biomedical journals. However, as with any automated method, one should be wary that a requested database search may not retrieve all of the relevant material, and important entries may be missed. Bear in mind that each database entry has required manual editing at some stage, giving rise to a low frequency of inescapable spelling errors and other problems. On occasion, a particular reference that should be in the database is not found because the search terms may be misspelled in the relevant database entry, the entry may not be present in the database, or there may be some more complicated problem. If exhaustive and careful attempts fail, reporting such problems to the program manager or system administrator should correct the problem.

SEQUENCE ANALYSIS PROGRAMS

Methods for DNA sequencing were developed in 1977 by Maxam and Gilbert (1977) and Sanger et al. (1977). They are described in greater detail at the beginning of Chapter 2.

Because DNA sequencing involves ordering a set of peaks (A, G, C, or T) on a sequencing gel, the process can be quite error-prone, depending on the quality of the data.

As more DNA sequences became available in the late 1970s, interest also increased in developing computer programs to analyze these sequences in various ways. In 1982 and 1984, *Nucleic Acids Research* published two special issues devoted to the application of computers for sequence analysis, including programs for large mainframe computers down to the then-new microcomputers. Shortly after, the Genetics Computer Group (GCG) was started at the University of Wisconsin by J. Devereux, offering a set of programs for analysis that ran on a VAX computer. Eventually GCG became commercial (http://www.gcg.com/). Other companies offering microcomputer programs for sequence analysis, including Intelligenetics, DNAStar, and others, also appeared at approximately the same time. Laboratories also developed and shared computer programs on a no-cost or low-cost basis. For example, to facilitate the collection of data, the programs PHRED (Ewing and Green 1998; Ewing et al. 1998) and PHRAP were developed by Phil Green and colleagues at the University of Washington to assist with reading and processing sequencing data. PHRED and PHRAP are now distributed by CodonCode Corporation (http://www.codoncode.com).

These commercial and noncommercial programs are still widely used. In addition, Web sites are available to perform many types of sequence analyses; they are free to academic institutions or are available at moderate cost to commercial users. Following is a brief review of the development of methods for sequence analysis.

THE DOT MATRIX OR DIAGRAM METHOD FOR COMPARING SEQUENCES

In 1970, A.J. Gibbs and G.A. McIntyre (1970) described a new method for comparing two amino acid and nucleotide sequences in which a graph was drawn with one sequence written across the page and the other down the left-hand side. Whenever the same letter appeared in both sequences, a dot was placed at the intersection of the corresponding sequence positions on the graph (Fig. 1.2). The resulting graph was then scanned for a series of dots that formed a diagonal, which revealed similarity, or a string of the same characters, between the sequences. Long sequences can also be compared in this manner on a single page by using smaller dots.

The dot matrix method quite readily reveals the presence of insertions or deletions between sequences because they shift the diagonal horizontally or vertically by the amount of change. Comparing a single sequence to itself can reveal the presence of a repeat of the same sequence in the same (direct repeat) or reverse (inverted repeat or palindrome) orientation. This method of self-comparison can reveal several features, such as similarity between chromosomes, tandem genes, repeated domains in a protein sequence, regions of low sequence complexity where the same characters are often repeated, or self-complementary sequences in RNA that can potentially base-pair to give a double-stranded structure. Because diagonals may not always be apparent on the graph due to weak similarity, Gibbs and McIntyre counted all possible diagonals and these counts were compared to those of random sequences to identify the most significant alignments.

Maizel and Lenk (1981) later developed various filtering and color display schemes that greatly increased the usefulness of the dot matrix method. This dot matrix representation of sequence comparisons continues to play an important role in analysis of DNA and protein sequence similarity, as well as repeats in genes and very long chromosomal sequences, as described in Chapter 3 (p. 59).

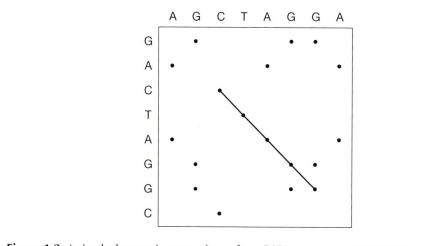

Figure 1.2. A simple dot matrix comparison of two DNA sequences, AGCTAGGA and GACTAG-GC. The diagonal of dots reveals a run of similar sequence CTAGG in the two sequences.

ALIGNMENT OF SEQUENCES BY DYNAMIC PROGRAMMING

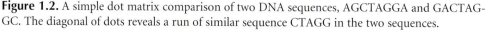

Although the dot matrix method can be used to detect sequence similarity, it does not readily resolve similarity that is interrupted by regions that do not match very well or that are present in only one of the sequences (e.g., insertions or deletions). Therefore, one would like to devise a method that can find what might be a tortuous path through a dot matrix, providing the very best possible alignment, called an optimal alignment, between the two sequences. Such an alignment can be represented by writing the sequences on successive lines across the page, with matching characters placed in the same column and unmatched characters placed in the same column as a mismatch or next to a gap as an insertion (or deletion in the other sequence), as shown in Figure 1.3. To find an optimal alignment in which all possible matches, insertions, and deletions have been considered to find the best one is computationally so difficult that for proteins of length 300, 10^{88} comparisons will have to be made (Waterman 1989).

To simplify the task, Needleman and Wunsch (1970) broke the problem down into a progressive building of an alignment by comparing two amino acids at a time. They started at the end of each sequence and then moved ahead one amino acid pair at a time, allowing for various combinations of matched pairs, mismatched pairs, or extra amino acids in one sequence (insertion or deletion). In computer science, this approach is called dynamic programming. The Needleman and Wunsch approach generated (1) every possible alignment, each one including every possible combination of match, mismatch, and single insertion or deletion, and (2) a scoring system to score the alignment. The object was to determine which was the best alignment of all by determining the highest score. Thus, every match in a trial alignment was given a score of 1, every mismatch a score of 0, and individual gaps a penalty score. These numbers were then added across the alignment to

SEQUENCE A   A   G   Δ   Δ   C   D   E   V   I   G
SEQUENCE B   A   G   E   Y   C   D   Δ   I   I   G

Figure 1.3. An alignment of two sequences showing matches, mismatches, and gaps (Δ). The best or optimal alignment requires that all three types of changes be allowed.

obtain a total score for the alignment. The alignment with the highest possible score was defined as the optimal alignment.

The procedure for generating all of the possible alignments is to move sequentially through all of the matched positions within a matrix, much like the dot matrix graph (see above), starting at those positions that correspond to the end of one of the sequences, as shown in Figure 1.4. At each position in the matrix, the highest possible score that can be achieved up to that point is placed in that position, allowing for all possible starting points in either sequence and any combination of matches, mismatches, insertions, and deletions. The best alignment is found by finding the highest-scoring position in the graph, and then tracing back through the graph through the path that generated the highest-scoring positions. The sequences are then aligned so that the sequence characters corresponding to this path are matched.

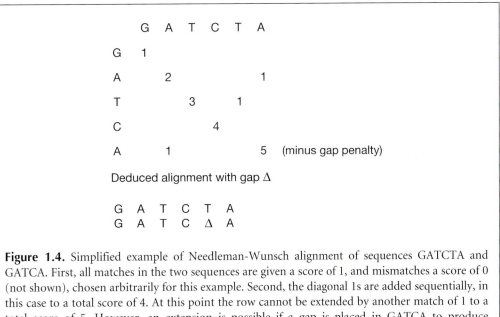

Figure 1.4. Simplified example of Needleman-Wunsch alignment of sequences GATCTA and GATCA. First, all matches in the two sequences are given a score of 1, and mismatches a score of 0 (not shown), chosen arbitrarily for this example. Second, the diagonal 1s are added sequentially, in this case to a total score of 4. At this point the row cannot be extended by another match of 1 to a total score of 5. However, an extension is possible if a gap is placed in GATCA to produce GATC Δ A, where Δ is the gap. To add the gap, a penalty score is subtracted from the total match score of 5 now appearing in the last row and column. The best alignment is found starting with the sequence characters that correspond to the highest number and tracing back through the positions that contributed to this highest score.

FINDING LOCAL ALIGNMENTS BETWEEN SEQUENCES

Mike Waterman

Temple Smith

The above method finds the optimal alignment between two sequences, including the entirety of each of the sequences. Such an alignment is called a global alignment. Smith and Waterman (1981a,b) recognized that the most biologically significant regions in DNA and protein sequences were subregions that align well and that the remaining regions made up of less-related sequences were less significant. Therefore, they developed an important modification of the Needleman-Wunsch algorithm, called the local alignment or Smith-Waterman (or the Waterman-Smith) algorithm, to locate such regions. They also recognized that insertions or deletions of any size are likely to be found as evolutionary changes in sequences, and therefore adjusted their method to accommodate such changes. Finally, they provided mathematical proof that the dynamic programming method is guaranteed to provide an optimal alignment between sequences. The algorithm is discussed in detail in Chapter 3 (p. 64).

Two complementary measurements had been devised for scoring an alignment of two sequences, a similarity score and a distance score. As shown in Figure 1.3, there are three types of aligned pairs of characters in each column of an alignment—identical matches, mismatches, and a gap opposite an unmatched character. Using as an example a simple scoring system of 1 for each type of match, the similarity score adds up all of the matches in the aligned sequences, and divides by the sum of the number of matches and mismatches (gaps are usually ignored). This method of scoring sequence similarity is the one most familiar to biologists and was devised by Needleman and Wunsch and used by Smith and Waterman. The other scoring method is a distance score that adds up the number of substitutions required to change one sequence into the other. This score is most useful for making predictions of evolutionary distances between genes or proteins to be used for phylogenetic (evolutionary) predictions, and the method was the work of mathematicians, notably P. Sellers. The distance score is usually calculated by summing the number of mismatches in an alignment divided by the total number of matches and mismatches. The calculation represents the number of changes required to change one sequence into the other, ignoring gaps. Thus, in the example shown in Figure 1.3, there are 6 matches and 1 mismatch in an alignment. The similarity score for the alignment is 6/7 = 0.86 and the distance score is 1/7 = 0.14, if the required condition is given a simple score of 1. With this simple scoring scheme, the similarity and distance scores add up to 1. Note also the equivalence that the sum of the sequence lengths is equal to twice the number of matches plus mismatches plus the number of deletions or insertions. Thus, in our example, the calculation is 8 + 9 = 2 × (6 + 1) + 3 = 17. Usually more complex systems of scoring are used to produce meaningful alignments, and alignments are evaluated by likelihood or odds scores (Chapter 3), but an inverse relationship between similarity and distance scores for the alignment still holds.

A difficult problem encountered in aligning sequences is deciding whether or not a particular alignment is significant. Does a particular alignment score reveal similarity between two sequences, or would the score be just as easily found between two unrelated sequences (or random sequence of similar composition generated by the computer)? This problem was addressed by S. Karlin and S. Altschul (1990, 1993) and is addressed in detail in Chapter 3 (p. 96).

An analysis of scores of unrelated or random sequences revealed that the scores could frequently achieve a value much higher than expected in a normal distribution. Rather, the scores followed a distribution with a positively skewed tail, known as the extreme value distribution. This analysis provided a way to assess the probability that a score found between two sequences could also be found in an alignment of unrelated or random sequences of the same length. This discovery was particularly useful for assessing matches between a query sequence and a sequence database discussed in Chapter 7. In this case, the evaluation of a particular alignment score must take into account the number of sequence comparisons made in searching the database. Thus, if a score between a query protein sequence and a database protein sequence is achieved with a probability of 10^{-7} of being between unrelated sequences, and 80,000 sequences were compared, then the highest expected score (called the EXPECT score) is $10^{-7} \times 8 \times 10^4 = 8 \times 10^{-3} = 0.008$. A value of 0.02–0.05 is considered significant. Even when such a score is found, the alignment must be carefully examined for shortness of the alignment, unrealistic amino acid matches, and runs of repeated amino acids, the presence of which decreases confidence in an alignment.

MULTIPLE SEQUENCE ALIGNMENT

In addition to aligning a pair of sequences, methods have been developed for aligning three or more sequences at the same time (for an early example, see Johnson and Doolittle 1986). These methods are computer-intensive and usually are based on a sequential aligning of the most-alike pairs of sequences. The programs commonly used are the GCG program PILEUP (http://www.gcg. com/) and CLUSTALW (Thompson et al. 1994) (Baylor College of Medicine, http://dot.imgen.bcm.tmc.edu:9331/multi-align/multi-align.html). Once the alignment of a related set of molecular sequences (a family) has been produced, highly conserved regions (Gribskov et al. 1987) can be identified that may be common to that particular family and may be used to identify other members of the same family. Two matrix representations of the multiple sequence alignment called a PROFILE and a POSITION-SPECIFIC SCORING MATRIX (PSSM) are important computational tools for this purpose.

Multiple sequence alignments can also be the starting point for evolutionary modeling. Each column of aligned sequence characters is examined, and then the most probable phylogenetic relationship or tree that would give rise to the observed changes is identified.

Another form of multiple sequence alignment is to search for a pattern that a set of DNA or protein sequences has in common without first aligning the sequences (Stormo et al. 1982; Stormo and Hartzell 1989; Staden 1984, 1989; Lawrence and Reilly 1990). For proteins, these patterns may define a conserved component of a structural or functional domain. For DNA sequences, the patterns may specify the binding site for a regulatory protein in a promoter region or a processing signal in an RNA molecule. Both statistical and nonstatistical methods have been widely used for this purpose. In effect, these methods sort through the sequences trying to locate a series of adjacent characters in each of the sequences that, when aligned, provides the highest number of matches. Neural networks, hidden Markov models, and the expectation maximization and Gibbs sampling methods (Stormo et al. 1982; Lawrence et al. 1993; Krogh et al. 1994; Eddy et al. 1995) are examples of methods that are used. Explanations and examples of these methods are described in Chapter 4.

PREDICTION OF RNA SECONDARY STRUCTURE

In addition to methods for predicting protein structure, other methods for predicting RNA secondary structure on computers were also developed at an early time. If the complement of a sequence on an RNA molecule is repeated down the sequence in the opposite chemical direction, the regions may base-pair and form a hairpin structure, as illustrated in Figure 1.5.

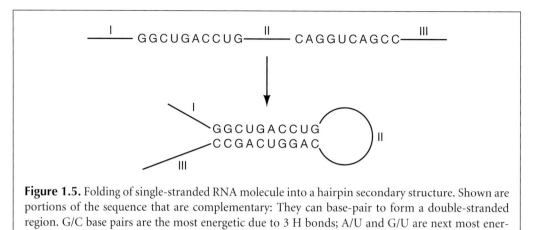

Figure 1.5. Folding of single-stranded RNA molecule into a hairpin secondary structure. Shown are portions of the sequence that are complementary: They can base-pair to form a double-stranded region. G/C base pairs are the most energetic due to 3 H bonds; A/U and G/U are next most energetic with two and one H bonds, respectively.

Tinoco et al. (1971) generated these symmetrical regions in small oligonucleotide molecules and tried to predict their stability based on estimates of the free energy associated with stacked base pairs in the model and of the destabilizing effects of loops, using a table of energy values (Tinoco et al. 1971; Salser 1978). Single-stranded loops and other unpaired regions decreased the predicted energy. Subsequently, Nussinov and Jacobson (1980) devised a fast computer method for predicting an RNA molecule with the highest possible number of base pairs based on the same dynamic programming algorithm used for aligning sequences. This method was improved by Zuker and Stiegler (1981), who added molecular constraints and thermodynamic information to predict the most energetically stable structure.

Another important use of RNA structure modeling is in the construction of databases of RNA molecules. One of the most significant of these is the ribosomal RNA database prepared by the laboratory of C. Woese (1987) (http://www.cme.msu.edu/RDP html/index.html). RNA secondary structure prediction is discussed in Chapter 5. Alignment, structural modeling, and phylogenetic analysis based on these RNA sequences have made possible the discovery of evolutionary relationships among organisms that would not have been possible otherwise.

DISCOVERY OF EVOLUTIONARY RELATIONSHIPS USING SEQUENCES

Variations within a family of related nucleic acid or protein sequences provide an invaluable source of information for evolutionary biology. With the wealth of sequence information becoming available, it is possible to track ancient genes, such as ribosomal RNA and some proteins, back through the tree of life and to discover new organisms based on their sequence (Barns et al. 1996). Diverse genes may follow different evolutionary histories, reflecting transfers of genetic material between species. Other types of phylogenetic analyses can be used to identify genes within a family that are related by evolutionary descent, called orthologs. Gene duplication events create two copies of a gene, called paralogs, and many such events can create a family of genes, each with a slightly altered, or possibly new, function. Once alignments have been produced and alignment scores found, the most closely related sequence pairs become apparent and may be placed in the outer branches of an evolutionary tree, as shown for sequences A and B in Figure 1.1 (p. 2). The next most-alike sequence, sequence C in Figure 1.1, will be represented by the next branch down on the tree. Continuing this process generates a predicted pattern of evolution for that particular gene. Once a tree has been found, the sequence changes that have taken place in the tree branches can be inferred.

The starting point for making a phylogenetic tree is a sequence alignment. For each pair of sequences, the sequence similarity score gives an indication as to which sequences are most closely related. A tree that best accounts for the numbers of changes (distances) between the sequences (Fitch and Margoliash 1987) of these scores may then be derived. The method most commonly used for this purpose is the neighbor-joining method (Saitou and Nei 1987) described in Chapter 6. Alternatively, if a reliable multiple sequence alignment is available, the tree that is most consistent with the observed variation found in each column of the sequence alignment may be used. The tree that imposes the minimum number of changes (the maximum parsimony tree) is the one chosen (Felsenstein 1988).

In making phylogenetic predictions, one must consider the possibility that several trees may give almost the same results. Tests of significance have therefore been derived to determine how well the sequence variation supports the existence of a particular tree branch (Felsenstein 1988). These developments are also discussed in Chapter 6.

IMPORTANCE OF DATABASE SEARCHES FOR SIMILAR SEQUENCES

As DNA sequencing became a common laboratory activity, genes with an important biological function could be sequenced with the hope of learning something about the biochemical nature of the gene product. An example was the retrovirus-encoded v-*sis* and v-*src* oncogenes, genes that cause cancer in animals. By comparing the predicted sequences of the viral products with all of the known protein sequences at the time, R. Doolittle and colleagues (1983) and W. Barker and M. Dayhoff (1982) both made the startling discovery that these genes appeared to be derived from cellular genes. The Sis protein had a sequence very similar to that of the platelet-derived growth factor (PDGF) from mammalian cells, and Src to the catalytic chain of mammalian cAMP-dependent kinases. Thus, it appeared likely that the retrovirus had acquired the gene from the host cell as some kind of genetic exchange event and then had produced a mutant form of the protein that could compromise the function of the normal protein when the virus infected another animal. Subsequently, as molecular biologists analyzed more and more gene sequences, they discovered that many organisms share similar genes that can be identified by their sequence similarity.

These searches have been greatly facilitated by having genetic and biochemical information from model organisms, such as the bacterium *Escherichia coli* and the budding yeast *Saccharomyces cerevisiae*. In these organisms, extensive genetic analysis has revealed the function of genes, and the sequences of these genes have also been determined. Finding a gene in a new organism (e.g., a crop plant) with a sequence similar to a model organism gene (e.g., yeast) provides a prediction that the new gene has the same function as in the model organism. Such searches are becoming quite commonplace and are greatly facilitated by programs such as FASTA (Pearson and Lipman 1988) and BLAST (Altschul et al. 1990).

The methods used by BLAST and other additional powerful methods to perform sequence similarity searching are described further in the next section and in Chapter 7.

THE FASTA AND BLAST METHODS FOR DATABASE SEARCHES

As the number of new sequences collected in the laboratory increased, there was also an increased need for computer programs that provided a way to compare these new sequences sequentially to each sequence in the existing database of sequences, as was done

```
PORTION OF SEQUENCE A      –   –   W   I   V   –   –
PORTION OF SEQUENCE B      –   –   W   I   V   –   –
```

Figure 1.6. Rapid identification of sequence similarity by FASTA and BLAST. FASTA looks for short regions in these two amino acid sequences that match and then tries to extend the alignment to the right and left. In this case, the program found by a quick and simple indexing method that W, I, and then V occurred in the same order in both sequences, providing a good starting point for an alignment. BLAST works similarly, but only examines matched patterns of length 3 of the more significant amino acid substitutions that are expected to align less frequently by chance alone.

Bill Pearson

to identify successfully the function of viral oncogenes. The dynamic programming method of Needleman and Wunsch would not work because it was much too slow for the computers of the time; today, however, with much faster computers available, this method can be used. W. Pearson and D. Lipman (1988) developed a program called FASTA, which performed a database scan for similarity in a short enough time to make such scans routinely possible. FASTA provides a rapid way to find short stretches of similar sequence between a new sequence and any sequence in a database. Each sequence is broken down into short words a few sequence characters long, and these words are organized into a table indicating where they are in the sequence. If one or more words are present in both sequences, and especially if several words can be joined, the sequences must be similar in those regions. Pearson (1990, 1996) has continued to improve the FASTA method for similarity searches in sequence databases.

An even faster program for similarity searching in sequence databases, called BLAST, was developed by S. Altschul et al. (1990). This method is widely used from the Web site of the National Center for Biotechnology Information at the National Library of Medicine in Washington, DC (http://www.ncbi.nlm.nih.gov/BLAST). The BLAST server is probably the most widely used sequence analysis facility in the world and provides similarity searching to all currently available sequences. Like FASTA, BLAST prepares a table of short sequence words in each sequence, but it also determines which of these words are most significant such that they are a good indicator of similarity in two sequences, and then confines the search to these words (and related ones), as described in Figure 1.6. There are versions of BLAST for searching nucleic acid and protein databases, which can be used to translate DNA sequences prior to comparing them to protein sequence databases (Altschul et al. 1997). Recent improvements in BLAST include GAPPED-BLAST, which is threefold faster than the original BLAST, but which appears to find as many matches in databases, and PSI-BLAST (position-specific-iterated BLAST), which can find more distant matches to a test protein sequence by repeatedly searching for additional sequences that match an alignment of the query and initially matched sequences. These methods are discussed in Chapter 7.

PREDICTING THE SEQUENCE OF A PROTEIN BY TRANSLATION OF DNA SEQUENCES

Protein sequences are predicted by translating DNA sequences that are cDNA copies of mRNA sequences from a predicted start and end of an open reading frame. Unfortunately, cDNA sequences are much less prevalent than genomic sequences in the databases. Partial sequence (expressed sequence tags, or ESTs) libraries for many organisms are available, but these only provide a fraction of the carboxy-terminal end of the protein sequence and usually only have about 99% accuracy. For organisms that have few or no introns in their genomic DNA (such as bacterial genomes), the genomic DNA may be translated. For most eukaryotic organisms with introns in their genes, the protein-encoding exons must be predicted and then translated by methods described in Chapter 8. These genome-based predictions are not always accurate, and thus it remains important to have cDNA sequences of protein-encoding genes. Promoter sequences in genomes may also be analyzed for common patterns that reflect common regulatory features. These types of analyses require sophisticated approaches that are also discussed in Chapter 8 (Hertz et al. 1990).

PREDICTING PROTEIN SECONDARY STRUCTURE

There are a large number of proteins whose sequences are known, but very few whose structures have been solved. Solving protein structures involves the time-consuming and highly specialized procedures of X-ray crystallography and nuclear magnetic resonance (NMR). Consequently, there is much interest in trying to predict the structure of a protein, given its sequence. Proteins are synthesized as linear chains of amino acids; they then form secondary structures along the chain, such as α helices, as a result of interactions between side chains of nearby amino acids. The region of the molecule with these secondary structures then folds back and forth on itself to form tertiary structures that include α helices, β sheets comprising interacting β strands, and loops (Fig. 1.7). This folding often leaves amino acids with hydrophobic side chains facing into the interior of the folded molecule and polar amino acids that can interact with water and the molecular environment facing outside in loops. The amino acid sequence of the protein directs the folding pathway, sometimes assisted by proteins called chaperonins. Chou and Fasman (1978) and Garnier et al. (1978) searched the small structural database of proteins for the amino acids associated with each of the secondary structure types—α helices, turns, and β strands. Sequences of proteins whose structures were not known were then scanned to determine whether the amino acids in each region were those often associated with one type of structure. For example, the amino acid proline is not often found in α helices because its side chain is not compatible with forming a helix. This method predicted the structure of some proteins well but, in general, was about as likely to predict a correct as an incorrect structure.

As more protein structures were solved experimentally, computational methods were used to find those that had a similar structural fold (the same arrangement of secondary structures connected by similar loops). These methods led to the discovery that as new protein structures were being solved, they often had a structural fold that was already known in a group of sequences. Thus, proteins are found to have a limited number of ~500 folds (Chothia 1992), perhaps due to chemical restraints on protein folding or to the exis-

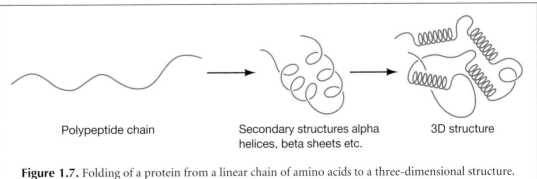

Polypeptide chain   Secondary structures alpha   3D structure
helices, beta sheets etc.

Figure 1.7. Folding of a protein from a linear chain of amino acids to a three-dimensional structure. The folding pathway involves amino acid interactions. Many different amino acid patterns are found in the same types of folds, thus making structure prediction from amino acid sequence a difficult undertaking.

tence of a single evolutionary pathway for protein structure (Gibrat et al. 1996). Furthermore, proteins without any sequence similarity could adopt the same fold, thus greatly complicating the prediction of structure from sequence. Methods for finding whether or not a given protein sequence can occupy the same three-dimensional conformation as another based on the properties of the amino acids have been devised (Bowie et al. 1991). Databases of structural families of proteins are available on the Web and are described in Chapter 9.

Amos Bairoch (Bairoch et al. 1997) developed another method for predicting the biochemical activity of an unknown protein, given its sequence. He collected sequences of proteins that had a common biochemical activity, for example an ATP-binding site, and deduced the pattern of amino acids that was responsible for that activity, allowing for some variability. These patterns were collected into the PROSITE database (http://www.expasy.ch/prosite). Unknown sequences were scanned for the same patterns. Subsequently, Steve and Jorga Henikoff (Henikoff and Henikoff 1992) examined alignments of the protein sequences that make up each MOTIF and discovered additional patterns in the aligned sequences called BLOCKS (see http://www.blocks.fhcrc.org/). These patterns offered an expanded ability to determine whether or not an unknown protein possessed a particular biochemical activity. The changes that were in each column of these aligned patterns were counted and a new set of amino acid substitution matrices, called BLOSUM matrices, similar to the PAM matrices of Margaret Dayhoff, were produced. One of these matrices, BLOSUM62, is most often used for aligning protein sequences and searching databases for similar sequences (Henikoff and Henikoff 1992) (see Chapter 7).

Sophisticated statistical and machine-training techniques have been used in more recent protein structure prediction programs, and the success rate has increased. A recent advance in this now active field of research is to organize proteins into groups or families on the basis of sequence similarity, and to find consensus patterns of amino acid domains characteristic of these families using the statistical methods described in Chapters 4 and 9. There are many publicly accessible Web sites described in Chapter 9 that provide the latest methods for identifying proteins and predicting their structures.

THE FIRST COMPLETE GENOME SEQUENCE

Although many viruses had already been sequenced, the first planned attempt to sequence a free-living organism was by Fred Blattner and colleagues (Blattner et al. 1997) using the bacterium *E. coli*. However, there was some concern over whether such a large sequence, about 4×10^6 bp, could be obtained by the then-current sequencing technology. The first published genome sequence was that of the single, circular chromosome of another bacterium, *Hemophilus influenzae* (Fleischmann et al. 1995), by The Institute of Genetics Research (TIGR, at http://www.tigr.org/), which had been started by researcher Craig Venter. The project was assisted by microbiologist Hamilton Smith, who had worked with this organism for many years. The speedup in sequencing involved using automated reading of DNA sequencing gels through dye-labeling of bases, and breaking down the chromosome into random fragments and sequencing these fragments as rapidly as possible without knowledge of their location in the whole chromosome. Computer analysis of such shotgun cloning and sequencing techniques had been developed much earlier by R. Staden at Cambridge University and other workers, but the TIGR undertaking was much more ambitious. In this genome project, newly read sequences were immediately entered into a computer database and compared with each other to find overlaps and produce contigs of two or more sequences with the assistance of computer programs. This procedure circumvented the need to grow and keep track of large numbers of subclones. Although the same sequence was often obtained up to 10 times, the sequence of the entire chromosome (2×10^9 bp), less a few gaps, was rapidly assembled in the computer over a 9-month period at a cost of about 10^6.

This success heralded a large number of other sequencing projects of various prokaryotic and eukaryotic microorganisms, with a tremendous potential payoff in terms of utilizable gene products and evolutionary information about these organisms. To date, completed projects include more than 30 prokaryotes, yeast *S. cerevisiae* (see Cherry et al. 1997), the nematode *Caenorhabditis elegans* (see *C. elegans* Sequencing Consortium 1998), and the fruit fly *Drosophila* (see Adams et al. 2000). The plant *Arabidopsis thaliana* and the human genome sequencing projects are ongoing and will be completed during 2000 or shortly thereafter.

The Human Genome Project, a large, federally funded collaborative project, will complete sequencing of the entire human genome by 2003. The project was developed from an idea discussed at scientific meetings in 1984 and 1985, and a pilot project, the Human Genome Initiative, was begun by the Department of Energy (DOE) in 1986. National Institutes of Health funding of the project began in 1987 under the Office of Genome Research. Currently, the project is constituted as the National Human Genome Research Initiative. In 1998, a new commercial venture under the leadership of Craig Venter was formed to sequence the majority of the human genome by 2001. This group, which uses a whole genome shotgun cloning approach and intensive computer processing of data, has already completed the Drosophila sequence and will sequence the mouse genome following completion of the human genome. Both groups simultaneously announced completion of the sequencing of the human genome in 2000.

ACEDB, THE FIRST GENOME DATABASE

As more genetic and sequence information became available for the model organisms, interest arose in generating specific genome databases that could be queried to retrieve this information. Such an enterprise required a new level of sharing of data and resources between laboratories. Although there were initial concerns about copyright issues, credits, accuracy, editorial review, and curating, eventually these concerns disappeared or became resolved as resources on the Internet developed. The first genome database, called ACEDB (a *C. elegans* database), and the methods to access this database were developed by Mike Cherry and colleagues (Cherry and Cartinhour 1993). This database was accessible through the internet and allowed retrieval of sequences, information about genes and mutants, investigator addresses, and references. Similar databases were subsequently developed using the same methods for *A. thaliana* and *S. cerevisiae*. Presently, there is a large number of such publicly available databases. Web access to these databases is discussed in Chapter 10 (Table 10.1, p. 482).

REFERENCES

Adams M.D., Celniker S.E., Holt R.A., Evans C.A., Gocayne J.D., Amanatides P.G., Scherer S.E., Li P.W., Hoskins R.A., Galle R.F., et al. 2000. The genome sequence of *Drosophila melanogaster*. *Science* **287:** 2185–2195.

Altschul S.F., Gish W., Miller W., Myers E.W., and Lipman D.J. 1990. Basic local alignment search tool. *J. Mol. Biol.* **215:** 403–410.

Altschul S.F., Madden T.L., Schaffer A.A., Zhang J., Zhang Z., Miller W., and Lipman D.J. 1997. Gapped BLAST and PSI-BLAST: A new generation of protein database search programs. *Nucleic Acids Res.* **25:** 3389–3402.

Bairoch A., Bucher P., and Hofmann K. 1997. The PROSITE database, its status in 1997. *Nucleic Acids Res.* **25:** 217–221.

Barker W.C. and Dayhoff M.O. 1982. Viral *src* gene products are related to the catalytic chain of mammalian cAMP-dependent protein kinase. *Proc. Natl. Acad. Sci.* **79:** 2836–2839.

Barns S.M., Delwiche C.F., Palmer J.D., and Pace N.R. 1996. Perspectives on archaeal diversity, thermophily and monophyly from environmental rRNA sequences. *Proc. Natl. Acad. Sci.* **93:** 9188–9193.

Blattner F.R., Plunkett III, G., Bloch C.A., Perna N.T., Burland V., Riley M., Collado-Vides J., Glasner J.D., Rode C.K., Mayhew G.F., Gregor J., Davis N.W., Kirkpatrick H.A., Goeden M.A., Rose D.J., Mau B., and Shao Y. 1997. The complete genome sequence of *Escherichia coli* K-12. *Science* **277:** 1453–1474.

Bowie J.U., Luthy R., and Eisenberg D. 1991. A method to identify protein sequences that fold into a known three-dimensional structure. *Science* **253:** 164–170.

C. elegans Sequencing Consortium. 1998. Genome sequence of the nematode *C. elegans:* A platform for investigating biology. *Science* **282:** 2012–2018.

Cherry J.M. and Cartinhour S.W. 1993. ACEDB, a tool for biological information. In *Automated DNA sequencing and analysis* (ed. M. Adams et al.). Academic Press, New York.

Cherry J.M., Ball C., Weng S., Juvik G., Schmidt R., Adler C., Dunn B., Dwight S., Riles L., Mortimer R. K., and Botstein D. 1997. Genetic and physical maps of *Saccharomyces cerevisiae*. *Nature* (suppl. 6632) **387:** 67–73.

Chothia C. 1992. Proteins. One thousand families for the molecular biologist. *Nature* **357:** 543–544.

Chou P.Y. and Fasman G.D. 1978. Prediction of the secondary structure of proteins from their amino acid sequence. *Adv. Enzymol. Relat. Areas Mol. Biol.* **47:** 45–147.

Dayhoff M.O., Ed. 1972. *Atlas of protein sequence and structure*, vol. 5. National Biomedical Research Foundation, Georgetown University, Washington, D.C.

———. 1978. Survey of new data and computer methods of analysis. In *Atlas of protein sequence and structure*, vol. 5, suppl. 3. National Biomedical Research Foundation, Georgetown University, Washington, D.C.

Doolittle R.F., Hunkapiller M.W., Hood L.E., Devare S.G., Robbins K.C., Aaronson S.A., and Antoniades H.N. 1983. Simian sarcoma *onc* gene v-*sis* is derived from the gene (or genes) encoding a platelet-derived growth factor. *Science* **221:** 275–277.

Eddy S.R., Mitchison G., and Durbin R. 1995. Maximum discrimination hidden Markov models of sequence consensus. *J. Comput. Biol.* **2:** 9–23.

Ewing B. and Green P. 1998. Base-calling of automated sequence traces using phred. II. Error probabilities. *Genome Res.* **8:** 186–194.

Ewing B., Hillier L., Wendl, M.C., and Green P. 1998. Base-calling of automated sequence traces using phred. I. Accuracy assessment. *Genome Res.* **8:** 175–185.

Felsenstein J. 1988. Phylogenies from molecular sequences: Inferences and reliability. *Annu. Rev. Genet.* **22:** 521–565.

Fitch W.M. and Margoliash E. 1987. Construction of phylogenetic trees. *Science* **155:** 279–284.

Fleischmann R.D., Adams M.D., White O., Clayton R.A., Kirkness E.F., Kerlavage A.R., Bult C.J., Tomb J.F., Dougherty B.A., Merrick J.M., et al. 1995. Whole-genome random sequencing and assembly of *Haemophilus influenzae* Rd. *Science* **269:** 496–512.

Garnier J., Osguthorpe D.J., and Robson B. 1978. Analysis of the accuracy and implications of simple methods for predicting the secondary structure of globular proteins. *J. Mol. Biol.* **120:** 97–120.

Gibbs A.J. and McIntyre G.A. 1970. The diagram, a method for comparing sequences. Its use with amino acid and nucleotide sequences. *Eur. J. Biochem.* **16:** 1–11.

Gibrat J.F., Madej T., and Bryant S.H. 1996. Surprising similarity in structure comparison. *Curr. Opin. Struct. Biol.* **6:** 377–385.

Gribskov M., McLachlan A.D., and Eisenberg D. 1987. Profile analysis: Detection of distantly related proteins. *Proc. Natl. Acad. Sci.* **84:** 4355–4358.

Henikoff S. and Henikoff J.G. 1992. Amino acid substitution matrices from protein blocks. *Proc. Natl. Acad. Sci.* **89:** 10915–10919.

Hertz G.Z., Hartzell III, G.W., and Stormo G.D. 1990. Identification of consensus patterns in unaligned DNA sequences known to be functionally related. *Comput. Appl. Biosci.* **6**: 81–92.

Johnson M.S. and Doolittle R.F. 1986. A method for the simultaneous alignment of three or more amino acid sequences. *J. Mol. Evol.* **23**: 267–268.

Karlin S. and Altschul S.F. 1990. Methods for assessing the statistical significance of molecular sequence features by using general scoring schemes. *Proc. Natl. Acad. Sci.* **87**: 2264–2268.

———. 1993. Applications and statistics for multiple high-scoring segments in molecular sequences. *Proc. Natl. Acad. Sci.* **90**: 5873–5877.

Krogh A., Brown M., Mian I.S., Sjölander K., and Haussler D. 1994. Hidden Markov models in computational biology. Applications to protein modeling. *J. Mol. Biol.* **235**: 1501–1531.

Lawrence C.E. and Reilly A.A. 1990. An expectation maximization (EM) algorithm for the identification and characterization of common sites in unaligned biopolymer sequences. *Proteins Struct. Funct. Genet.* **7**: 41–51.

Lawrence C.E., Altschul S.F., Boguski M.S., Liu J.S., Neuwald A.F., and Wootton J.C. 1993. Detecting subtle sequence signals: A Gibbs sampling strategy for multiple alignment. *Science* **262**: 208–214.

Maizel Jr., J.V. and Lenk R.P. 1981. Enhanced graphic matrix analyses of nucleic acid and protein synthesis. *Proc. Natl. Acad. Sci.* **78**: 7665–7669.

Maxam A.M. and Gilbert W. 1977. A new method for sequencing DNA. *Proc. Natl. Acad. Sci.* **74**: 560–564.

Needleman S.B. and Wunsch C.D. 1970. A general method applicable to the search for similarities in the amino acid sequence of two proteins. *J. Mol. Biol.* **48**: 443–453.

Nussinov R. and Jacobson A.B. 1980. Fast algorithm for predicting the secondary structure of single-stranded RNA. *Proc. Natl. Acad. Sci.* **77**: 6903–6913.

Pearson W.R. 1990. Rapid and sensitive sequence comparison with FASTP and FASTA. *Methods Enzymol.* **183**: 63–98.

———. 1996. Effective protein sequence comparison. *Methods Enzymol.* **266**: 227–258.

Pearson W.R. and Lipman D.J. 1988. Improved tools for biological sequence comparison. *Proc. Natl. Acad. Sci.* **85**: 2444–2448.

Saitou N. and Nei M. 1987. The neighbor-joining method: A new method for reconstructing phylogenetic trees. *Mol. Biol. Evol.* **4**: 406–425.

Salser W. 1978. Globin mRNA sequences: Analysis of base pairing and evolutionary implications. *Cold Spring Harbor Symp. Quant. Biol.* **42**: 985–1002.

Sanger F. and Tuppy H. 1951. The amino acid sequence of the phenylalanyl chain of insulin. *Biochem. J.* **49**: 481–490.

Sanger F., Nicklen S., and Coulson A.R. 1977. DNA sequencing with chain terminating inhibitors. *Proc. Natl. Acad. Sci.* **74**: 5463–5467.

Smith T.F. and Waterman M.S. 1981a. Identification of common molecular subsequences. *J. Mol. Biol.* **147**: 195–197.

———. 1981b. Comparison of biosequences. *Adv. Appl. Math.* **2**: 482–489.

Staden R. 1984. Computer methods to locate signals in nucleic acid sequences. *Nucleic Acids Res.* **12**: 505–519.

———. 1989. Methods for calculating the probabilities of finding patterns in sequences. *Comput. Appl. Biosci.* **5**: 89–96.

Stormo G.D. and Hartzell III, G.W. 1989. Identifying protein-binding sites from unaligned DNA fragments. *Proc. Natl. Acad. Sci.* **86**: 1183–1187.

Stormo G.D., Schneider T.D., Gold L., and Ehrenfeucht A. 1982. Use of the 'Perceptron' algorithm to distinguish translational initiation sites in *E. coli. Nucleic Acids Res.* **10**: 2997–3011.

Thompson J.D., Higgins D.G., and Gibson T.J. 1994. CLUSTAL W: Improving the sensitivity of progressive multiple sequence alignment through sequence weighting, position-specific gap penalties and weight matrix choice. *Nucleic Acids Res.* **22**: 4673–4680.

Tinoco Jr., I., Uhlenbeck O.C., and Levine M.D. 1971. Estimation of secondary structure in ribonucleic acids. *Nature* **230**: 362–367.

Waterman M.S., Ed. 1989. Sequence alignments. In *Mathematical methods for DNA sequences*. CRC Press, Boca Raton, Florida.

Woese C.R. 1987. Bacterial evolution. *Microbiol. Rev.* **51**: 221–271.

Zuker M. and Stiegler P. 1981. Optimal computer folding of large RNA sequences using thermodynamics and auxiliary information. *Nucleic Acids Res.* **9**: 133–148.

Additional Reading

Reference Books and Special Journal Editions

Baldi P. and Brunck S. 1998. *Bioinformatics: The machine learning approach.* MIT Press, Cambridge, Massachusetts.

Baxevanis A.D. and Ouellette B.F., Eds. 1998. *Bioinformatics: A practical guide to the analysis of genes and proteins.* John Wiley & Sons, New York.

Doolittle R.F. 1986. *Of URFS and ORFS: A primer on how to analyze derived amino acid sequences.* University Science Books, Mill Valley, California.

———, Ed. 1990. Molecular evolution: Computer analysis of protein and nucleic acid sequences. *Methods Enzymol.*, vol. 183. Academic Press, San Diego.

———, Ed. 1996. Computer methods for macromolecular sequence analysis. *Methods Enzymol.*, vol. 266. Academic Press, San Diego, California.

Durbin R., Eddy S., Krogh A., and Mitchison G., Eds. 1998. *Biological sequence analysis. Probabilistic models of proteins and nucleic acids.* Cambridge University Press, Cambridge, United Kingdom.

Gribskov M. and Devereux J., Eds. 1991. *Sequence analysis primer.* University of Wisconsin Biotechnology Center Biotechnical Resource Ser. (ser. ed. R.R. Burgess). Stockton Press, New York.

Gusfield D. 1997. *Algorithms on strings, trees, and sequences: Computer science and computational biology.* Cambridge University Press, Cambridge, United Kingdom.

Martinez H., Ed. 1984. Mathematical and computational problems in the analysis of molecular sequences (special commemorative issue honoring Margaret Oakley Dayhoff). *Bull. Math. Biol.* Pergamon Press, New York.

Nucleic Acids Research. 1996–2000. Special database issues published in the January issues of volumes 22–26. Oxford University Press, Oxford, United Kingdom.

Salzberg S.L., Searls D.B., and Kasif S., Eds. 1999. Computational methods in molecular biology. *New Compr. Biochem.*, vol. 32. Elsevier, Amsterdam, The Netherlands.

Sankoff D. and Kruskal J.R., Eds. 1983. *Time warps, string edits, and macromolecules: The theory and practice of sequence comparison.* Addison-Wesley, Don Mills, Ontario.

Söll D. and Roberts R.J., Eds. 1982. The application of computers to research on nucleic acids. I. *Nucleic Acids Res.*, vol. 10. Oxford University Press, Oxford, United Kingdom.

———. 1984. The application of computers to research on nucleic acids. II. *Nucleic Acids Res.*, vol. 12. Oxford University Press, Oxford, United Kingdom.

von Heijne G. 1987. *Sequence analysis in molecular biology — Treasure trove or trivial pursuit.* Academic Press, San Diego, California.

Waterman M.S., Ed. 1989. Mathematical analysis of molecular sequences (special issue). *Bull. Math. Biol.* Pergamon Press, New York.

———. 1995. *Introduction to computational biology: Maps, sequences, and genomes.* Chapman and Hall, London, United Kingdom.

Yap, T.K., Frieder O., and Martino R.L. 1996. *High performance computational methods for biological sequence analysis.* Kluwer Academic, Norwell, Massachusetts.

Journals That Routinely Publish Papers on Sequence Analysis

Bioinformatics (formerly *Comput. Appl. Biosci.* [CABIOS]). Oxford University Press, Oxford, United Kingdom. http://bioinformatics.oupjournals.org/cabios/.

Journal of Computational Biology. Mary Ann Liebert, Larchmont, New York. http://www-hto.usc.edu/jcb/.

Journal of Molecular Biology. Academic Press, London, United Kingdom. http://www.hbuk.co.uk/jmb.

Nucleic Acids Research (sections on Genomics and Computational Biology). Oxford University Press, Oxford, United Kingdom. http://nar.oupjournals.org.

## CHAPTER 2

Collecting and Storing Sequences in the Laboratory

THIS CHAPTER SUMMARIZES METHODS used to collect sequences of DNA molecules and store them in computer files. Once in the computer, the sequences can be analyzed by a variety of methods. Additionally, assembly of the sequences of large molecules from short sequence fragments can readily be undertaken. Assembled sequences are stored in a computer file along with identifying features, such as DNA source (organism), gene name, and investigator. Sequences and accessory information are then entered into a database. This procedure organizes them so that specific ones can be retrieved by a database query program for subsequent use. Unfortunately, most sequence analysis programs require that the information in a sequence file be stored in a particular format. To use these programs, it is necessary to be aware of these formats and to be able to convert one format to another. These programs are outlined in greater detail in Chapter 3.

DNA SEQUENCING

Sequencing DNA has become a routine task in the molecular biology laboratory. Purified fragments of DNA cut from plasmid/phage clones or amplified by polymerase chain reaction (PCR) are denatured to single strands, and one of the strands is hybridized to an oligonucleotide primer. In an automated procedure, new strands of DNA are synthesized from the end of the primer by heat-resistant *Taq* polymerase from a pool of deoxyribonucleotide triphosphates (dNTPs) that includes a small amount of one of four chain-terminating nucleotides (ddNTPs). For example, using ddATP, the resulting synthesis creates a set of nested DNA fragments, each one ending at one of the As in the sequence through the substitution of a fluorescent-labeled ddATP, as shown in Figure 2.1. A similar set of fragments is made for each of the other three bases, but each is labeled with a different fluorescent ddNTP.

The combined mixture of all labeled DNA fragments is electrophoresed to separate the fragments by size, and the ladder of fragments is scanned for the presence of each of the four labels, producing data similar to those shown in Figure 2.2. A computer program then determines the probable order of the bands and predicts the sequence. Depending on the actual procedure being used, one run may generate a reliable sequence of as many as 500 nucleotides. For accurate work, a printout of the scan is usually examined for abnormali-

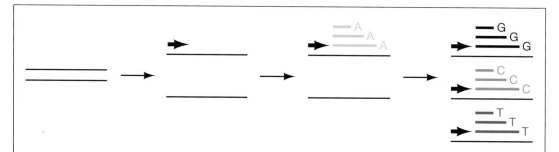

Figure 2.1. Method used to synthesize a nested set of DNA fragments, each ending at a base position complementary to one of the bases in the template sequence. To the left is a double-stranded DNA molecule several kilobases in length. After denaturation, the DNA is annealed to a short primer oligonucleotide primer (*black arrow*), which is complementary to an already sequenced region on the molecule. New DNA is then synthesized in the presence of a fluorescently labeled chain-terminating ddNTP or one of the four bases. The reactions produce a nested set of labeled molecules. The resulting fragments are separated in order by length to give the sequence display shown in Fig. 2.2.

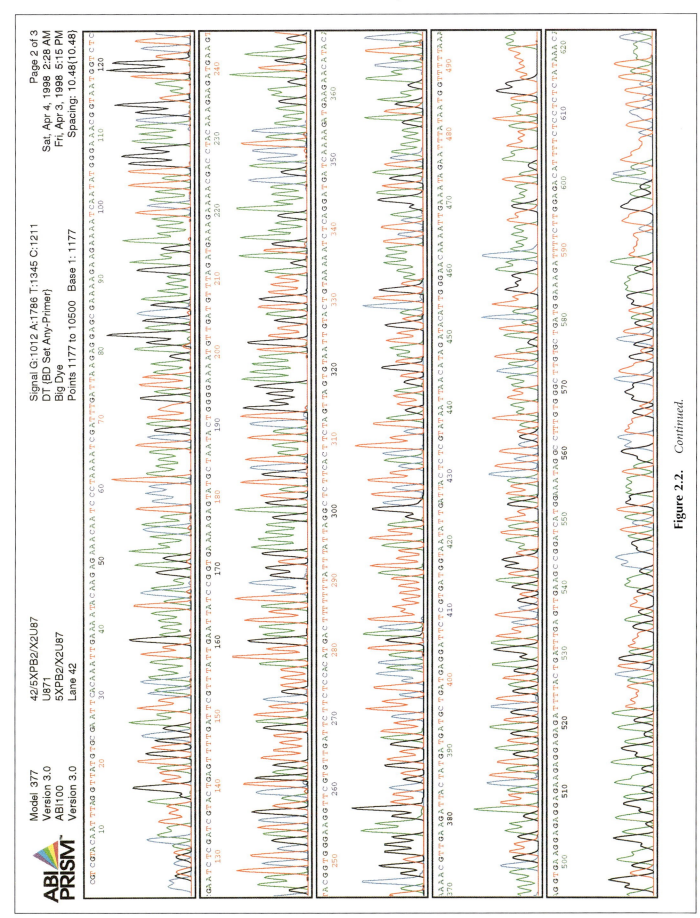

Figure 2.2. *Continued.*

Figure 2.2. *Continued.*

```
  1   CGTCGTACAA TTTAGGTTAT GTGCCGAATTC ACAAATTGAA AATACAAGAG AAACAATCCC TAAAATCGAT TTGATTAAGA GGAGCGAAAA   90
 91   GAAGAAAATC AATATGGGAA ACGGTAATGG  TCTCGAATCT CGATCGTACT GAGTTTTGAT TCGTTTATTG AATTATCCGG TGAAAAGAGT  180
181   ATGCTAATAC TGGGGAAAAT GTTGATGTTT  AGATGAAAGA AAACGACCTA CAAAGAAGAT GGGAAGTTC  GTGTTGATTC            270
271   TTCTCCACAT GACTTTTTTT ATTTATTAGG  CTCTTCACTT CTAGTTAGTG GTAAAAATCT CAGGATGATC AAAAGATGAA            360
361   GAACATACAA AACGTTGAAG ATTACTATGA  TGATGCTGAT GAGGATTCTC GTGATGGTAA TATTGATTAC TCTCGTATAA TTAACATAGA  450
451   TACATTGGGA ACAAAATTGA AATAGAATTT  ATAATGGTTT TTAAAGGTGA AGGAGGAGAG ATTTTACTGA TTTTACTGA  TTTGAGTTGA  540
541   AGCCGGATCA TGGAAATAGG CCTTTGTGGG  CTTGTGCTGA TTCTTGGAGA CATTTTCTCC TCTCTATAAA CAAGCTTATG            630
631   ACTTTCTTAT CGCCATTGCT GAACCCGTTT  GCAGGTTTGA TTTTGATTTG ATTATTATAT CAATGTNAAG TTATGATTTT TGGTGGTGGA  720
721   TTTTCATTCA TTTGGTACTA TTTCAGGCCT  GANTCAATGC CCNAGTTTAA TTAACCCCAC ACTCGTTGNA TGCTGCTGTT TCCTTTGGTC  810
811   TTGANACAGA AACTATCATC TCTGGTTTGA  ATAANCTNTC TAAGAACCAG CTTNCCCGGG GAGATCATTG GATTNAATNC ATGCTTNTAC  900
901   TGNTAATTTT NGGNAAATGA AATTGGGNTT  TGAAAAAAAA TCGGN                                                   990
```

Figure 2.2. Example of a DNA sequence obtained on an ABI-Prism 377 automated sequencer. The target DNA is denatured by heating and then annealed to a specific primer. Sequencing reactions are carried out in a single tube containing Amplitaq (Perkin-Elmer), dNTPs, and four ddNTPs, each base labeled with a different fluorescent dichloro-rhodamine dye. The polymerase extends synthesis from the primer, until a ddNTP is incorporated instead of dNTP, terminating the molecule. The denaturing, reannealing, and synthesis steps are recycled up to 25 times, excess labeled ddNTPs are removed, and the remaining products are electrophoresed on one lane of a polyacrylamide gel. As the bands move down the gel, the rhodamine dyes are excited by a laser within the sequencer. Each of the four ddNTP types emits light at a different wavelength band that is detected by a digital camera. The sequence of changes is plotted as shown in the figure and the sequence is read by a base-calling algorithm. More recently developed machines allow sequencing of 96 samples at a time by capillary electrophoresis using more automated procedures. The accuracy and reliability of high-throughput sequencing have been much improved by the development of the PHRED, PHRAP, and CONSED system for base-calling, sequence assembly, and assembled sequence editing (Ewing and Green 1998; Gordon et al. 1998).

Figure 2.3. Sequential sequencing of a DNA molecule using oligonucleotide primers. One of the denatured template DNA strands is primed for sequencing by an oligonucleotide (*yellow*) complementary to a known sequence on the molecule. The resulting sequence may then be used to produce two more oligonucleotide primers downstream in the sequence, one to sequence more of the same strand (*purple*) and a second (*turquoise*) that hybridizes to the complementary strand and produces a sequence running backward on this strand, thus providing a way to confirm the first sequence obtained.

ties that decrease the quality of the sequence, and the sequence may then be edited manually. The sequence can also be verified by making an oligonucleotide primer complementary to the distal part of the readable sequence and using it to obtain the sequence of the complementary strand on the original DNA template. The first sequence can also be extended by making a second oligonucleotide matching the distal end of the readable sequence and using this primer to read more of the original template. When the process is fully automated, a number of priming sites may be used to obtain sequencing results that give optimal separation of bands in each region of the sequence. By repeating this procedure, both strands of a DNA fragment several kilobases in length can be sequenced (Fig. 2.3).

GENOMIC SEQUENCING

To sequence larger molecules, such as human chromosomes, individual chromosomes are purified and broken into 100-kb or larger random fragments, which are cloned into vectors designed for large molecules, such as artificial yeast (YAC) or bacterial (BAC) chromosomes. In a laborious procedure, the resulting library is screened for fragments called contigs, which have overlapping or common sequences, to produce an integrated map of the chromosome. Many levels of clone redundancy may be required to build a consensus map because individual clones can have rearrangements, deletions, or two separate fragments. These do not reflect the correct map and have to be eliminated. Once the correct map has been obtained, unique overlapping clones are chosen for sequencing. However, these molecules are too large for direct sequencing. One procedure for sequencing these clones is to subclone them further into smaller fragments that are of sizes suitable for sequencing, make a map of these clones, and then sequence overlapping clones (Fig. 2.4). However, this method is expensive because it requires a great deal of time to keep track of all the subclones.

An alternative method is to sequence all the subclones, produce a computer database of the sequences, and then have the computer assemble the sequences from the overlaps that are found. Up to 10 levels of redundancy are used to get around the problem of a small fraction of abnormal clones. This procedure was first used to obtain the sequence of the 4-Mb chromosome of the bacterium *Haemophilus influenzae* by The Institute of Genetics Research (TIGR) team (Fleischmann et al. 1995). Only a few regions could not be joined because of a problem subcloning those regions into plasmids, requiring manual sequencing of these regions from another library of phage subclones.

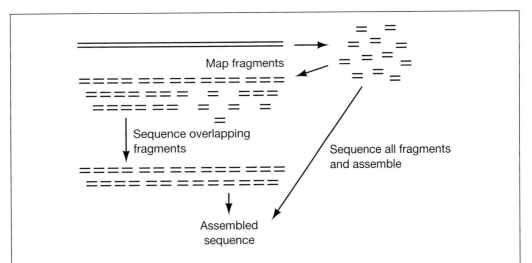

Figure 2.4. Methods for large-scale sequencing. A large DNA molecule 100 kb to several megabases in size is randomly sheared and cloned into a cloning vector. In one method, a map of various-sized fragments is first made, overlapping fragments are identified, and these are sequenced. In a faster method that is computationally intense, fragments in different size ranges are placed in vectors, and their ends are sequenced. Fragments are sequenced without knowledge of their chromosomal location, and the sequence of the large parent molecule is assembled from any overlaps found. As more and more fragments are sequenced, there are enough overlaps to cover most of the sequence.

Shotgun Sequencing

A controversy has arisen as to whether or not the above shotgun sequencing strategy can be applied to genomes with repetitive sequences such as those likely to be encountered in sequencing the human genome (Green 1997; Myers 1997). When DNA fragments derived from different chromosomal regions have repeats of the same sequence, they will appear to overlap. In a new whole shotgun approach, Celera Genomics is sequencing both ends of DNA fragments of short (2 kb), medium (10 kb), and long (BAC or ~100 kb) lengths. A large number of reads are then assembled by computer. This method has been used to assemble the genome of the fruit fly *Drosophila melanogaster* after removal of the most highly repetitive regions (Myers et al. 2000) and also to assemble a significant proportion of the human genome.

SEQUENCING cDNA LIBRARIES OF EXPRESSED GENES

Two common goals in sequence analysis are to identify sequences that encode proteins, which determine all cellular metabolism, and to discover sequences that regulate the expression of genes or other cellular processes. Genomic sequencing as described above meets both goals. However, only a small percentage of the genomic sequence of many organisms actually encodes proteins because of the presence of introns within coding regions and other noncoding regions in the genome. Although there has been a great deal of progress in developing computational methods for analyzing genomic sequences and finding these protein-encoding regions (see Chapter 8), these methods are not completely reliable and, furthermore, such genomic sequences are often not available. Therefore, cDNA libraries have been prepared that have the same sequences as the mRNA molecules produced by organisms, or else cDNA copies are sequenced directly by RT-PCR (copying of mRNA by reverse transcriptase followed by sequencing of the cDNA copy by the polymerase chain reaction). By using cDNA sequence with the introns removed, it is much simpler to locate protein-encoding sequences in these molecules. The only possible difficulty is that a gene of interest may be developmentally expressed or regulated in such a way that the mRNA is not present. This problem has been circumvented by pooling mRNA preparations from tissues that express a large proportion of the genome, from a variety of tissues and developing organs or from organisms subjected to several environmental influences. An important development for computational purposes was the decision by Craig Venter to prepare databases of partial sequences of the expressed genes, called expressed sequence tags or ESTs, which have just enough DNA sequence to give a pretty good idea of the protein sequence. The translated sequence can then be compared to a database of protein sequences with the hope of finding a strong similarity to a protein of known function, and hence to identify the function of the cloned EST. The corresponding cDNA clone of the gene of interest can then be obtained and the gene completely sequenced.

SUBMISSION OF SEQUENCES TO THE DATABASES

Investigators are encouraged to submit their newly obtained sequences directly to a member of the International Nucleotide Sequence Database Collaboration, such as the National Center for Biotechnology Information (NCBI), which manages GenBank (http://www.ncbi.nlm.nih.gov); the DNA Databank of Japan (DDBJ; http://www.ddbj.nig.ac.jp); or the European Molecular Biology Laboratory (EMBL)/EBI Nucleotide Sequence Database (http://www.embl-heidelberg.de). NCBI reviews new entries and updates existing ones, as requested. A database accession number, which is required to publish the sequence, is provided. New sequences are exchanged daily by the GenBank, EMBL, and DDBJ databases.

The simplest and newest way of submitting sequences is through the Web site http://www.ncbi.nlm.nih.gov/ on a Web form page called BankIt. The sequence can also be annotated with information about the sequence, such as mRNA start and coding regions. The submitted form is transformed into GenBank format and returned to the submitter for review before being added to GenBank. The other method of submission is to use Sequin (formerly called Authorin), which runs on personal computers and UNIX machines. The program provides an easy-to-use graphic interface and can manage large submissions such as genomic sequence information. It is described and demonstrated on http://www.ncbi.nlm.nih.gov/Sequin/index.html and may be obtained by anonymous FTP from ncbi.nlm.nih.gov/sequin/. Completed files can also be E-mailed to gb-sub@ncbi.nlm.nih.gov or can be mailed on diskette to GenBank Submissions, National Center for Biotechnology Information, National Library of Medicine, Bldg. 38A, Room 8N-803, Bethesda, Maryland 20894.

SEQUENCE ACCURACY

It should be apparent from the above description of sequencing projects that the higher the level of accuracy required in DNA sequences, the more time-consuming and expensive the procedure. There is no detailed check of sequence accuracy prior to submission to GenBank and other databases. Often, a sequence is submitted at the time of publication of the sequence in a journal article, providing a certain level of checking by the editorial peer-review process. However, many sequences are submitted without being published or prior to publication. In laboratories performing large sequencing projects, such as those engaged in the Human Genome Project or the genome projects of model organisms, the granting agency requires a certain level of accuracy of the order of 1 possible error per 10 kb. This level of accuracy should be sufficient for most sequence analysis applications such as sequence comparisons, pattern searching, and translation. In other laboratories, such as those performing a single-attempt sequencing of ESTs, the error rate may be much higher, approximately 1 in 100, including incorrectly identified bases and inserted or deleted bases. Thus, in translating EST sequences in GenBank and other databases, incorrect bases may translate to the wrong amino acid. The worst problem, however, is that base insertions/deletions will cause frameshifts in the sequence, thus making alignment with a protein sequence very difficult. Another type of database sequence that is error-prone is a fragment of sequence from the immunological variant of a pathogenic organism, such as the regions in the protein coat of the human immunodeficiency virus (HIV). Although this low level of accuracy may be suitable for some purposes such as identification, for more detailed analyses, e.g., evolutionary analyses, the accuracy of such sequence fragments should be verified.

COMPUTER STORAGE OF SEQUENCES

Before using a sequence file in a sequence analysis program, it is important to ensure that computer sequence files contain only sequence characters and not special characters used by text editors. Editing a sequence file with a word processor can introduce such changes if one is not careful to work only with text or so-called ASCII files (those on the typewriter keyboard). Most text editors normally create text files that include control characters in addition to standard ASCII characters. These control characters will only be recognized correctly by the text editor program. Sequence files that contain such control characters may not be analyzed correctly, depending on whether or not the sequence analysis program filters them out. Editors usually provide a way to save files with only standard ASCII characters, and these files will be suitable for most sequence analysis programs.

ASCII and Hexadecimal

Computers store sequence information as simple rows of sequence characters called strings, which are similar to the sequences shown on the computer terminal. Each character is stored in binary code in the smallest unit of memory, called a byte. Each byte comprises 8 bits, with each bit having a possible value of 0 or 1, producing 255 possible combinations. By convention, many of these combinations have a specific definition, called their ASCII equivalent. Some ASCII values are defined as keyboard characters, others as special control characters, such as signaling the end of a line (a line feed and a carriage return), or the end of a file full of text (end-of-file character). A file with only ASCII characters is called an ASCII file. For convenience, all binary values may be written in a hexadecimal format, which corresponds to our decimal format 0, 1, ... 9 plus the letters A, B, ... F. Thus, hexadecimal 0F corresponds to binary 0000 1111 and decimal 15, and FF corresponds to binary 1111 1111 and decimal 255. A DNA sequence is usually stored and read in the computer as a series of 8-bit words in this binary format. A protein sequence appears as a series of 8-bit words comprising the corresponding binary form of the amino acid letters.

Sequence and other data files that contain non-ASCII characters also may not be transferred correctly from one machine to another and may cause unpredictable behavior of the communications software. Some communications software can be set to ignore such control characters. For example, the file transfer program (FTP) has ASCII and binary modes, which may be set by the user. The ASCII mode is useful for transferring text files, and the binary mode is useful for transferring compressed data files, which also contain non-ASCII characters.

Most sequence analysis programs also require not only that a DNA or protein sequence file be a standard ASCII file, but also that the file be in a particular format such as the FASTA format (see below). The use of windows on a computer has simplified such problems, since one merely has to copy a sequence from one window, for example, a window that is running a Web browser on the ENTREZ Web site, and paste it into another, for example, that of a translation program.

In addition to the standard four base symbols, A, T, G, and C, the Nomenclature Committee of the International Union of Biochemistry has established a standard code to represent bases in a nucleic acid sequence that are uncertain or ambiguous. The codes are listed in Table 2.1.

For computer analysis of proteins, it is more convenient to use single-letter than three-letter amino acid codes. For example, GenBank DNA sequence entries contain a translated sequence in single-letter code. The standard, single-letter amino acid code was established by a joint international committee, and is shown in Table 2.2. When the name of only one amino acid starts with a particular letter, then that letter is used, e.g., C, cysteine. In other cases, the letter chosen is phonetically similar (R, arginine) or close by in the alphabet (K, lysine).

Table 2.1. *Base–nucleic acid codes*

Symbol	Meaning	Explanation
G	G	Guanine
A	A	Adenine
T	T	Thymine
C	C	Cytosine
R	A or G	puRine
Y	C or T	pYrimidine
M	A or C	aMino
K	G or T	Keto
S	C or G	Strong interactions 3 h bonds
W	A or T	Weak interactions 2 h bonds
H	A, C or T not G	H follows G in alphabet
B	C, G or T not A	B follows A in alphabet
V	A, C or G not T (not U)	V follows U in alphabet
D	A, G or T not C	D follows C in alphabet
N	A,C,G or T	Any base

Adapted from NC-IUB (1984).

Table 2.2. *Table of standard amino acid code letters*

1-letter code	3-letter code	Amino acid
A[a]	Ala	alanine
C	Cys	cysteine
D	Asp	aspartic acid
E	Glu	glutamic acid
F	Phe	phenylalanine
G	Gly	glycine
H	His	histidine
I	Ile	isoleucine
K	Lys	lysine
L	Leu	leucine
M	Met	methionine
N	Asn	asparagine
P	Pro	proline
Q	Gln	glutamine
R	Arg	arginine
S	Ser	serine
T	Thr	threonine
V	Val	valine
W	Trp	tryptophan
X	Xxx	undetermined amino acid
Y	Tyr	tyrosine
Z[b]	Glx	either glutamic acid or glutamine

Adapted from IUPAC-IUB (1969, 1972, 1983).

[a] Letters not shown are not commonly used.

[b] Note that sometimes when computer programs translate DNA sequences, they will put a "Z" at the end to indicate the termination codon. This character should be deleted from the sequence.

SEQUENCE FORMATS

One major difficulty encountered in running sequence analysis software is the use of differing sequence formats by different programs. These formats all are standard ASCII files, but they may differ in the presence of certain characters and words that indicate where different types of information and the sequence itself are to be found. The more commonly used sequence formats are discussed below.

GenBank DNA Sequence Entry

The format of a database entry in GenBank, the NCBI nucleic acid and protein sequence database, is as follows: Information describing each sequence entry is given, including literature references, information about the function of the sequence, locations of mRNAs and coding regions, and positions of important mutations. This information is organized into fields, each with an identifier, shown as the first text on each line. In some entries, these identifiers may be abbreviated to two letters, e.g., RF for reference, and some identifiers may have additional subfields. The information provided in these fields is described in Figure 2.5 and the database organization is described in Figure 2.6. The CDS subfield in the field FEATURES gives the amino acid sequence, obtained by translation of known and

```
LOCUS           name of locus, length and type of sequence,
                classification of organism, data of entry
DEFINITION      description of entry
ACCESSION       accession numbers of original source
KEYWORDS        key words for cross referencing this entry
SOURCE          source organism of DNA
ORGANISM        description of organism
REFERENCE
COMMENT         biological function or database information
FEATURES        information about sequence by base position or range of positions
     source              range of sequence, source organism
     misc_signal         range of sequence, type of function or signal
     mRNA                range of sequence, mRNA
     CDS                 range of sequence, protein coding region
     intron              range of sequence, position of intron
     mutation            sequence position, change in sequence for mutation
BASE COUNT      count of A, C, G, T and other symbols
ORIGIN          text indicating start of sequence
      1 gaattcgata aatctctggt ttattgtgca gtttatggtt ccaaaatcgc
     51 atatactcac agcataactg tatatacacc caggggcgg aatgaaagcg
//              database symbol for end of sequence
```

Figure 2.5. GenBank DNA sequence entry.

potential open reading frames, i.e., a consecutive set of three-letter words that could be codons specifying the amino acid sequence of a protein. The sequence entry is assumed by computer programs to lie between the identifiers "ORIGIN" and "//".

The sequence includes numbers on each line so that sequence positions can be located by eye. Because the sequence count or a sequence checksum value may be used by the computer program to verify the sequence composition, the sequence count should not be modified except by programs that also modify the count. The GenBank sequence format often has to be changed for use with sequence analysis software.

Accession no	Organism	Reference	Name	Keywords	Sequence
..123	Escherichia. coli	Medline1,.	LexA protein	SOS regulon, repressor, transcriptional regulator,..	ATG..
..124	Escherichia coli	Medline2,.	UmuD protein	SOS regulon,..	GTA..
..125	Saccharomyces. cerevisiae	Medline3,.	GAL4 protein	transcriptional regulator,..	CAT..
..125	Homo. sapiens	Medline4,.	gluco-corticoid receptor	transcriptional regulator,..	TGT..

Figure 2.6. Organization of the GenBank database and the search procedure used by ENTREZ. In this database format, each row is another sequence entry and each column another GenBank field. When one sequence entry is retrieved, all of these fields will be displayed, as in Fig. 2.5. Only a few fields and simple examples are shown for illustration. A search for the term "SOS regulon and coli" in all fields will find two matching sequences. Finding these sequences is simple because indexes have been made listing all of the sequences that have any given term, one index for each field. Similarly, a search for transcriptional regulator will find three sequences.

European Molecular Biology Laboratory Data Library Format

The European Molecular Biology Laboratory (EMBL) maintains DNA and protein sequence databases. The format for each entry in these databases is shown in Figure 2.7. As with GenBank entries, a large amount of information describing each sequence entry is given, including literature references, information about the function of the sequence, locations of mRNAs and coding regions, and positions of important mutations. This information is organized into fields, each with an identifier, shown as the first text on each line. The meaning of each of these fields is explained in Figure 2.7. These identifiers are abbreviated to two letters, e.g., RF for reference, and some identifiers may have additional subfields. The sequence entry is assumed by computer programs to lie between the identifiers "SEQUENCE" and "//" and includes numbers on each line to locate parts of the sequence visually. The sequence count or a checksum value for the sequence may be used by computer programs to make sure that the sequence is complete and accurate. For this reason, the sequence part of the entry should usually not be modified except with programs that also modify this count. This EMBL sequence format is very similar to the GenBank format. The main differences are in the use of the term ORIGIN in the GenBank format to indicate the start of sequence; also, the EMBL entry does not include the sequence of any translation products, which are shown instead as a different entry in the database. This sequence format often has to be changed for use with sequence analysis software.

The output of a DDBJ DNA sequence entry is almost identical to that of GenBank.

SwissProt Sequence Format

The format of an entry in the SwissProt protein sequence database is very similar to the EMBL format, except that considerably more information about the physical and biochemical properties of the protein is provided.

FASTA Sequence Format

The FASTA sequence format includes three parts shown in Figure 2.8: (1) a comment line identified by a ">" character in the first column followed by the name and origin of the

```
      ID              identification code for sequence in the database
      AC              accession number giving origin of sequence
      DT              dates of entry and modification
      KW              key cross-reference words for lookup up this entry
      OS, OC          source organism
      RN, RP, RX, RA, RT, RL  literature reference or source
      DR              i.d. in other databases
      CC              description of biological function
      FH, FT          information about sequence by base position or range of positions
                      source range of sequence, source organism
                      misc_signal range of sequence, type of function or signal
                      mRNA range of sequence, mRNA
                      CDS range of sequence, protein coding region
                      intron range of sequence, position of intron
                      mutation sequence position, change in sequence for mutation
      SQ              count of A, C, G, T and other symbols
      gaattcgata aatctctggt ttattgtgca gtttatggtt ccaaaatcgc cttttgctgt 60
      atatactcac agcataactg tatatacacc caggggcgg aatgaaagcg ttaacggcca 120
      .
      .
      // symbol to indicate end of sequence
```

Figure 2.7. EMBL sequence entry format.

```
>YCZ2_YEAST protein in HMR 3' region
MKAVVIEDGKAVVKEGVPIPELEEGFV
GNPTDWAHIDYKVGPQGSILGCDAAGQ
IVKLGPAVDPKDFSIGDYIYGFIHGSS
VRFPSNGAFAEYSAISTVVAYKSPNEL
KFLGEDVLPAGPVRSLEGAATIPVSLT*
```

Figure 2.8. FASTA sequence entry format.

sequence; (2) the sequence in standard one-letter symbols; and (3) an optional "*" which indicates end of sequence and which may or may not be present. The presence of "*" may be essential for reading the sequence correctly by some sequence analysis programs. The FASTA format is the one most often used by sequence analysis software. This format provides a very convenient way to copy just the sequence part from one window to another because there are no numbers or other nonsequence characters within the sequence. The FASTA sequence format is similar to the protein information resource (NBRF) format except that the NBRF format includes a first line with a ">" character in the first column followed by information about the sequence, a second line containing an identification name for the sequence, and the third to last lines containing the sequence, as described below.

National Biomedical Research Foundation/Protein Information Resource Sequence Format

This sequence format, which is sometimes also called the PIR format, has been used by the National Biomedical Research Foundation/Protein Information Resource (NBRF) and also by other sequence analysis programs. Note that sequences retrieved from the PIR database on their Web site (http://www-nbrf.georgetown.edu) are not in this compact format, but in an expanded format with much more information about the sequence, as shown below. The NBRF format is similar to the FASTA sequence format but with significant differences. An example of a PIR sequence format is given in Figure 2.9. The first line includes an initial ">" character followed by a two-letter code such as P for complete sequence or F for fragment, followed by a 1 or 2 to indicate type of sequence, then a semicolon, then a four- to six-character unique name for the entry. There is also an essential second line with the full name of the sequence, a hyphen, then the species of origin. In FASTA format, the second line is the start of the sequence and the first line gives the sequence identifier after a ">" sign. The sequence terminates with an asterisk.

```
>P1;ILEC
lexA repressor - Escherichia coli
MKALTARQQEVFDLIRDHISQTGMPPTRAE
IAQRLGFRSPNAAEEHLKALARKGVIEIVS
GASRGIRLLQEEEEGLPLVGRVAAGEPLLA
QQHIEGHYQVDPSLFKPNADFLLRVSGMSM
KDIGIMDGDLLAVHKTQDVRNGQVVVARID
DEVTVKRLKKQGNKVELLPENSEFKPIVVD
LRQQSFTIEGLAVGVIRNGDWL
```

Figure 2.9. NBRF sequence entry format.

```
;YEAST protein in HMR 3' region
YCZ2
MKAVVIEDGKAVVKEGVPIPELEEGFV
GNPTDWAHIDYKVGPQGSILGCDAAGQ
IVKLGPAVDPKDFSIGDYIYGFIHGSS
VRFPSNGAFAEYSAISTVVAYKSPNEL
KFLGEDVLPAGPVRSLEGAATIPVSLT1
```

Figure 2.10. Intelligenetics sequence entry format.

Stanford University/Intelligenetics Sequence Format

Started by a molecular genetics group at Stanford University, and subsequently continued by a company, Intelligenetics, the IG format is similar to the PIR format (Fig. 2.10), except that a semicolon is usually placed before the comment line. The identifier on the second line is also present. At the end of the sequence, a 1 is placed if the sequence is linear, and a 2 if the sequence is circular.

Genetics Computer Group Sequence Format

Earlier versions of the Genetics Computer Group (GCG) programs require a unique sequence format and include programs that convert other sequence formats into GCG format. Later versions of GCG accept several sequence formats. A converted GenBank file is illustrated in Figure 2.11. Information about the sequence in the GenBank entry is first included, followed by a line of information about the sequence and a checksum value. This value (not shown) is provided as a check on the accuracy of the sequence by the addition of the ASCII values of the sequence. If the sequence has not been changed, this value should stay the same. If one or more sequence characters become changed through error, a program reading the sequence will be able to determine that the change has occurred because the checksum value in the sequence entry will no longer be correct. Lines of information are terminated by two periods, which mark the end of information and the start of the sequence on the next line. The rest of the text in the entry is treated as sequence. Note the presence of line numbers. Since there is no symbol to indicate end of sequence, no text other than sequence should be added beyond this point. The sequence should not be altered except by programs that will also adjust the checksum score for the sequence. The GCG sequence format may have to be changed for use with other sequence analysis software. GCG also includes programs for reformatting sequence files.

```
BASE COUNT 215 A 224 C 263 G 250 T
ORIGIN
Filename, Length of sequence, Date, Checksum value, ..
 1  GAATTCGATA AATCTCTGGT TTATTGTGCA GTTTATGGTT CCAAAATCGC
51  CTTTTGCTGT ATATACTCAC AGCATAACTG TATATACACC CAGGGGGCGG
```

Figure 2.11. GCG sequence entry format.

Format of Sequence File Retrieved from the National Biomedical Research Foundation/Protein Information Resource

The file format has approximately the same information as a GenBank or EMBL sequence file but is formatted slightly differently, as in Figure 2.12. This format is presently called the PIR/CODATA format.

Plain/ASCII.Staden Sequence Format

This sequence format is a computer file that includes only the sequence with no other accessory information. This particular format is used by the Staden Sequence Analysis programs (http://www/.mrc-lmb.com.ac.uk/pubseq) produced by Roger Staden at Cambridge University (Staden et al. 2000). The sequence must be further formatted to be used for most sequence analysis programs.

```
ENTRY ILEC
      #type complete
TITLE lexA repressor - Escherichia coli
ORGANISM
      #formal_name Escherichia coli
DATE 29-Jul-1981
      #sequence_revision 01-Sep-1981
      #text_change 14-Nov-1997
ACCESSIONS A90808; A93734; S11945; B65212; A03569
REFERENCE A90808
      #authors Horii, T.; Ogawa, T.; Ogawa, H.
      #journal Cell (1981) 23:689-697
      #title Nucleotide sequence of the lexA gene of Escherichia coli.
      #cross-references MUID:81186269
      #contents lexA
      #accession A90808
          ##molecule_type DNA
          ##residues 1-202
          ##label HOR
REFERENCE
   .

   .
COMMENTS
GENETICS
      #gene lexA
      #map_position 92 min
CLASSIFICATION
      #superfamily lexa repressor
KEYWORDS DNA binding, repressor, transcription regulator
SUMMARY
      #length 202
      #molecular_weight 22358
SEQUENCE
                5        10        15        20        25        30
      1 M K A L T A R Q Q E V F D L I R D H I S Q T G M P P T R A E
```

Figure 2.12. Protein Information Resource sequence format.

Abstract Syntax Notation Sequence Format

Abstract Syntax Notation (ASN.1) is a formal data description language that has been developed by the computer industry. ASN.1 (http://www-sop.inria.fr/rodeo/personnel/hoschka/asn1.html; NCBI 1993) has been adopted by the National Center for Biotechnology Information (NCBI) to encode data such as sequences, maps, taxonomic information, molecular structures, and bibliographic information. These data sets may then be easily connected and accessed by computers. The ASN.1 sequence format is a highly structured and detailed format especially designed for computer access to the data. All the information found in other forms of sequence storage, e.g., the GenBank format, is present. For example, sequences can be retrieved in this format by ENTREZ (see below). However, the information is much more difficult to read by eye than a GenBank formatted sequence. One would normally not need to use the ASN.1 format except when running a computer program that uses this format as input.

Genetic Data Environment Sequence Format

Genetic Data Environment (GDE) format is used by a sequence analysis system called the Genetic Data Environment, which was designed by Steven Smith and collaborators (Smith et al. 1994) around a multiple sequence alignment editor that runs on UNIX machines. The GDE features are incorporated into the SEQLAB interface of the GCG software, version 9. GDE format is a tagged-field format similar to ASN.1 that is used for storing all available information about a sequence, including residue color. The file consists of various fields (Fig. 2.13), each enclosed by brackets, and each field has specific lines, each with a given name tag. The information following each tag is placed in double quotes or follows the tag name by one or more spaces.

```
{
name              "Short name for sequence"
longname          "Long (more descriptive) name for sequence"
sequence-ID       "Unique ID number"
creation-date     "mm/dd/yy hh:mm:ss"
direction         [-1|1]
strandedness      [1|2]
type              [DNA|RNA|PROTEIN|TEXT|MASK]
offset            (-999999,999999)
group-ID          (0,999)
creator           "Author's name"
descrip           "Verbose description"
comments          "Lines of comments about a sequence"
sequence          "gctagctagctagctagctcttagctgtagtcgtagctgatgctag
                  ctgatgctagctagctagctagctgatcgatgctagctgatcgtag
                  ctgacggactgatgctagctagctagctagctgtctagtgtcgtag
                  tgcttattgc"
}
```

Figure 2.13. The Genetic Data Environment format.

CONVERSIONS OF ONE SEQUENCE FORMAT TO ANOTHER

READSEQ to Switch between Sequence Formats

READSEQ is an extremely useful sequence formatting program developed by D. G. Gilbert at Indiana University, Bloomington (gilbertd@bio.indiana.edu). READSEQ can recognize a DNA or protein sequence file in any of the formats shown in Table 2.3, identify the format, and write a new file with an alternative format. Some of these formats are used for special types of analyses such as multiple sequence alignment and phylogenetic analysis. The appearance of these formats for two sample DNA sequences, seq1 and seq2, is shown in Table 2.4. READSEQ may be reached at the Baylor College of Medicine site at http://dot.imgen.bcm.tmc.edu:9331/seq-util/readseq.html and also by anonymous FTP from ftp.bio.indiana.edu/molbio/readseq or ftp.bioindiana.edu/molbio/mac to obtain the appropriate files.

Data files that have multiple sequences, such as those required for multiple sequence alignment and phylogenetic analysis using parsimony (PAUP), are also converted. Examples of the types of files produced are shown in Table 2.4. Options to reverse-complement and to remove gaps from sequences are included. SEQIO, another sequence conversion program for a UNIX machine, is described at http://bioweb.pasteur.fr/docs/seqio/seqio.html and is available for download at http://www.cs.ucdavis.edu/~gusfield/seqio.html.

Table 2.3. *Sequence formats recognized by format conversion program READSEQ*

1. Abstract Syntax Notation (ASN.1)
2. DNA Strider
3. European Molecular Biology Laboratory (EMBL)
4. Fasta/Pearson
5. Fitch (for phylogenetic analysis)
6. GenBank
7. Genetics Computer Group (GCG)[a]
8. Intelligenetics/Stanford
9. Multiple sequence format (MSF)
10. National Biomedical Research Foundation (NBRF)
11. Olsen (in only)
12. Phylogenetic Analysis Using Parsimony (PAUP) NEXUS format
13. Phylogenetic Inference package (Phylip v3.3, v3.4)
14. Phylogenetic Inference package (Phylip v3.2)
15. Plain text/Staden[a]
16. Pretty format for publication (output only)
17. Protein Information Resource (PIR or CODATA)
18. Zuker for RNA analysis (in only)

[a] For conversion of single sequence files only. The other conversions can be performed on files with single or multiple sequences.

Table 2.4. *Multiple sequence format conversions by READSEQ*

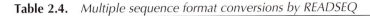

1.   Fasta/Pearson format

```
>seq1
agctagct agct agct
>seq2
aactaact aact aact
```

2.   Intelligenetics format

```
;seq1, 16 bases, 2688 checksum.
seq1
agctagctagctagct1
;seq2, 16 bases, 25C8 checksum.
seq2
aactaactaactaact1
```

3.   GenBank format

```
LOCUS       seq1        16 bp
DEFINITION seq1, 16 bases, 2688 checksum.
ORIGIN
        1 agctagctag ctagct
//
LOCUS       seq2        16 bp
DEFINITION seq2, 16 bases, 25C8 checksum.
ORIGIN
 1   aactaactaa ctaact
//
```

4.   NBRF format

```
>DL;seq1
seq1, 16 bases, 2688 checksum.
 agctagctag ctagct*

>DL;seq2
seq2, 16 bases, 25C8 checksum.
 aactaactaa ctaact*
```

5.   EMBL format

```
ID seq1
DE seq1, 16 bases, 2688 checksum.
SQ          16 BP
   agctagctag ctagct
//
ID seq2
DE seq2, 16 bases, 25C8 checksum.
SQ          16 BP
   aactaactaa ctaact
//
```

Continued.

Table 2.4. *Continued.*

6. **GCG format**

```
seq1
     seq1  Length: 16  Check: 9864 ..
   1  agctagctag ctagct

seq2
     seq2  Length: 16  Check: 9672 ..
   1  aactaactaa ctaact
```

7. **Format for the Macintosh sequence analysis program DNA Strider**

```
; ### from DNA Strider ;-)
; DNA sequence  seq1, 16 bases, 2688 checksum.
;
agctagctagctagct
//
; ### from DNA Strider ;-)
; DNA sequence  seq2, 16 bases, 25C8 checksum.
;
aactaactaactaact
//
```

8. **Format for phylogenetic analysis programs of Walter Fitch**

```
seq1,  16 bases, 2688 checksum.
 agc tag cta gct agc t
seq2,  16 bases, 25C8 checksum.
 aac taa cta act aac t
```

9. **Format for phylogenetic analysis programs PHYLIP of J. Felsenstein v 3.3 and 3.4.**

```
2 16
seq1       agctagctag ctagct
seq2       aactaactaa ctaact
```

10. **Protein International Resource PIR/CODATA format**

```
\\\
ENTRY          seq1
TITLE          seq1, 16 bases, 2688 checksum.
SEQUENCE
                   5        10       15       20
25       30
     1 a g c t a g c t a g c t a g c t
///
ENTRY          seq2
TITLE          seq2, 16 bases, 25C8 checksum.
SEQUENCE
                   5        10       15       20
25       30
     1 a a c t a a c t a a c t a a c t
///
```

Table 2.4. *Continued.*

11. **GCG multiple sequence format (MSF)**

```
/tmp/readseq.in.2449  MSF: 16 Type: N January 01,
1776  12:00  Check: 9536 ..

 Name: seq1              Len:    16 Check:  9864
Weight:  1.00
 Name: seq2              Len:    16 Check:  9672
Weight:  1.00

//

           seq1  agctagctag ctagct
           seq2  aactaactaa ctaact
```

12. **Abstract Syntax Notation (ASN.1) format**

```
Bioseq-set ::= {
seq-set {
  seq {
    id { local id 1 },
    descr { title "seq1" },
    inst {
      repr raw, mol dna, length 16, topology linear,
      seq-data
        iupacna "agctagctagctagct"
      } } ,
  seq {
    id { local id 2 },
    descr { title "seq2" },
    inst {
      repr raw, mol dna, length 16, topology linear,
      seq-data
        iupacna "aactaactaactaact"
      } } ,
} }
```

13. **NEXUS format used by the phylogenetic analysis program PAUP by David Swofford**

```
#NEXUS
[/tmp/readseq.in.2506 -- data title]

[Name: seq1              Len: 16 Check: 2688]
[Name: seq2              Len: 16 Check: 25C8]


begin data;
 dimensions ntax=2 nchar=16;
 format datatype=dna interleave missing=-;
  matrix
     seq1  agctagctagctagct
     seq2  aactaactaactaact
```

Two sequences in FASTA multiple sequence format (1) were used as input for the remainder of the format options (2–14).

GCG Programs for Conversion of Sequence Formats

The "from" programs convert sequence files from GCG format into the named format, and the "to" programs convert the alternative format into GCG format. Shown are the actual program names, no spaces included. There are no programs to convert to GenBank and EMBL formats.

FROMEMBL
FROMFASTA
FROMGENBANK
FROMIG
FROMPIR
FROMSTADEN
TOFASTA
TOIG
TOPIR
TOSTADEN

In addition, the GCG programs include the following sequence formatting programs: (1) GETSEQ, which converts a simple ASCII file being received from a remote PC to GCG format; (2) REFORMAT, which will format a GCG file that has been edited, and will also perform other functions; and (3) SPEW, which sends a GCG sequence file as an ASCII file to a remote PC.

MULTIPLE SEQUENCE FORMATS

Most of the sequence formats listed above can be used to store multiple sequences in tandem in the same computer file. Exceptions are the GCG and raw sequence formats, which are designed only for single sequences. GCG has an alternative multiple sequence format, which is described below. In addition, there are formats especially designed for multiple sequences that can also be used to show their alignments or to perform types of multiple sequence analyses such as phylogenetic analysis. In the case of PAUP, the program will accept MSA format and convert to the NEXUS format. These formats are illustrated below using the same two short sequences.

1. Aligned sequences in FASTA format. The aligned sequence characters occupy the same line and column, and gaps are indicated by a dash.

```
>gi|730305|
MATHHTLWMGLALLGVLGDLQAAPEAQVSVQPNFQQDKFL
RTQTPRAELKEKFTAFCKAQGFTEDTIVFLPQTDKCMTEQ
>gi|404390|
----------------------APEAQVSVQPNFQPDKFL
RTQTPRAELKEKFTAFCKAQGFTEDSIVFLPQTDKCMTEQ
>gi|895868
MAALRMLWMGLVLLGLLGFPQTPAQGHDTVQPNFQQDKFL
RTQTLKDELKEKFTTFSKAQGLTEEDIVFLPQPDKCIQE-

represents the same alignment as:

MATHHTLWMGLALLGVLGDLQAAPEAQVSVQPNFQQDKFL
----------------------APEAQVSVQPNFQPDKFL

RTQTPRAELKEKFTAFCKAQGFTEDTIVFLPQTDKCMTEQ
RTQTLKDELKEKFTTFSKAQGLTEEDIVFLPQPDKCIQE-
```

2. GCG multiple sequence format (MSF) produced by the GCG multiple sequence alignment program PILEUP. The gap symbol is "~". The length indicated is the length of the alignment, which is the length of the longest sequence including gaps.

```
PileUp of: @list4

 Symbol comparison table: GenRunData:blosum62.cmp CompCheck: 6430

                GapWeight: 12
          GapLengthWeight: 4

 list4.msf  MSF: 883  Type: P  February 28, 1997 16:42  Check: 482

 Name: haywire         Len:    883  Check: 3979  Weight:  1.00
 Name: xpb-human       Len:    883  Check: 9129  Weight:  1.00
 Name: rad25           Len:    883  Check: 5359  Weight:  1.00
 Name: xpb-ara         Len:    883  Check: 2015  Weight:  1.00

//

                1                                                  50
    haywire     ~~~~~~~~~~ ~~~~~~~~~~ ~~~~~~~~~~ ~~~~~~~~~~ ~~~~~MGPPK
  xpb-human     ~~~~~~~~~~ ~~~~~~~~~~ ~~~~~~~~~~ ~~~~~~~~~~ ~~~~~~~~~~
      rad25     MTDVEGYQPK SKGKIFPDMG ESFFSSDEDS PATDAEIDEN YDDNRETSEG
    xpb-ara     ~~~~~~~~~~ ~~~~~~~~~~ ~~~~~~~~~~ ~~~~~~~~~~ ~~~~~~~~~~

                51                                                100
    haywire     KSRKDRSG.. GDKFGKKRRA EDEAFTQLVD DNDSLDATES EGIPGAASKN
  xpb-human     MGKRDRAD.. RDKKKSRKRH YED...EEDD EEDAPGNDPQ EAVPSAAGKQ
      rad25     RGERDTGAMV TGLKKPRKKT KSSRHTAADS SMNQMDAKDK ALLQDTNSDI
    xpb-ara     ~~~~~~~~~~ ~~~~~~~~~~ ~~~~~~~~~~M KYGGKDDQKM KNIQNAEDYY

    .
    .
    .
```

3. ALN form produced by multiple sequence alignment program CLUSTALW (Thompson et al. 1994). In addition to the alignment position, the program also shows the current sequence position at the end of each row.

```
Page 1.1
                      1              15 16            30 31           45
   1 gi|730305|  MATHHTLWMGLALLG VLGDLQAAPEAQVSV QPNFQQDKFLGRWFS
                                                                   23
   2 gi|404390|  --------------- -------APEAQVSV QPNFQPDKFLGRWFS
                                                                   45
   3 gi|895868   MAALRMLWMGLVLLG LLGFPQTPAQGHDTV QPNFQQDKFLGRWYS
```

4. Blocked alignment used by GDE and GCG SEQLAB (Fig. 2.14). Unlike the other examples shown, which are all simple text files of an alignment, the following figure is a screen display of an alignment, using GDE and SEQLAB display programs. The underlying alignment in text format would be similar to the GCG multiple sequence alignment file shown above.

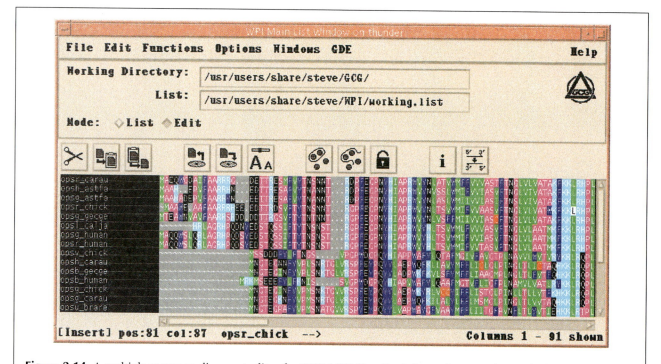

Figure 2.14. A multiple sequence alignment editor for GCG MSF files. For information on using multiple sequence alignment editors and for examples of other editors, see Chapter 4.

5. Format used by Fitch phylogenetic analysis programs.

```
seq1, 16 bases, 2688 checksum.
 agc tag cta gct agc t
seq2, 16 bases, 25C8 checksum.
 aac taa cta act aac t
```

6. Formats used by Felsenstein phylogenetic analysis programs PHYLIP (phylogenetic inference package): 2 for two sequences, 16 for length of alignment.

```
a. version 3.2

2 16 YF
seq1        agctagctag ctagct
seq2        aactaactaa ctaact

b. versions 3.3 and 3.4

2 16
seq1        agctagctag ctagct
seq2        aactaactaa ctaact
```

7. Format used by phylogenetic analysis program PAUP (phylogenetic analysis using parsimony). ntax is number of taxa, nchar is the length of the alignment, and interleave allows the alignment to be shown in readable blocks. The other terms describe the type of sequence and the character used to indicate gaps.

```
#NEXUS

[ comments ]

begin data;
        dimensions ntax=4 nchar=100;
        format datatype=protein interleave gap=-;
        matrix
[          1                                                            50]
      haywire  ---------- ---------- ---------- ---------- ----- MGPPK
    xpb-human  ---------- ---------- ---------- ---------- ---------- -
        rad25  MTDVEGYQPK SKGKIFPDMG ESFFSSDEDS PATDAEIDEN YDDNRETSEG
      xpb-ara  ---------- ---------- ---------- ---------- ---------- -


[         51                                                          100]
      haywire  KSRKDRSG-- GDKFGKKRRA EDEAFTQLVD DNDSLDATES EGIPGAASKN
    xpb-human  MGKRDRAD-- RDKKKSRKRH YED---EEDD EEDAPGNDPQ EAVPSAAGKQ
        rad25  RGERDTGAMV TGLKKPRKKT KSSRHTAADS SMNQMDAKDK ALLQDTNSDI
      xpb-ara  ---------- ---------- --------M KYGGKDDQKM KNIQNAEDYY


        ;
endblock;
```

8. The Selex format used by hidden Markov program HMMER by Sean Eddy has been used to keep track of the alignment of small RNA molecules.

```
# Example selex file

seq1    ACGACGACGACG.
seq2    ..GGGAAAGG.GA
seq3    UUU..AAAUUU.A

seq1    ..ACG
seq2    AAGGG
seq3    AA...UUU
```

Each line contains a name, followed by the aligned sequence. A space, dash, underscore, or period denotes a gap. Long alignments are split into multiple blocks and interleaved or separated by blank lines. The number of sequences, their order, and their names must be the same in every block, and every sequence must be represented even though there are no residues present.

9. The block multiple sequence alignment format (see http://www.blocks.fhcrc.org/).

Identification starts contain a short identifier for the group of sequences from which the block was made and often is the original Prosite group ID. The identifier is terminated by a semicolon, and "BLOCK" indicates the entry type.

AC contains the block number, a seven-character group number for sequences from which the block was made, followed by a letter (A–Z) indicating the order of the block in the sequences. The block number is a 5-digit number preceded by BL (BLOCKS database) or PR (PRINTS database). min,max is the minimum,maximum number of amino acids from the previous block or from the sequence start. DE describes sequences from which the block was made. BL contains information about the block: xxx is the amino acids in the spaced triplet found by MOTIF upon which the block is based. w is the width of the sequence segments (columns) in the block. s is the number of sequence segments (rows) in the block. Other values (n1, n2) describe statistical features of the block. Sequence_id is a list of sequences. Each sequence line contains a sequence identifier, the offset from the beginning of the sequence to the block in parentheses, the sequence segment, and a weight for the segment.

```
ID    short_identifier; BLOCK
AC    block_number; distance from previous block = (min,max)
DE    description
BL    xxx motif; width=w; seqs=s; 99.5%=n1; strength=n2
sequence_id  (offset) sequence_segment  sequence_weight.

//

ID    GLU_CARBOXYLATION; BLOCK
      AC    BL00011; distance from previous block=(1,64)
      DE    Vitamin K-dependent carboxylation domain proteins.
      BL    ECA motif; width=40; seqs=34; 99.5%=1833; strength=1412
      FA10_BOVIN (   45) LEEVKQGNLERECLEEACSLEEAREVFEDAEQTDEFWSKY 31
      FA10_CHICK (   45) LEEMKQGNIERECNEERCSKEEAREAFEDNEKTEEFWNIY 46
      FA10_HUMAN (   45) LEEMKKGHLERECMEETCSYEEAREVFEDSDKTNEFWNKY 33
       FA7_BOVIN (    5) LEELLPGSLERECREELCSFEEAHEIFRNEERTRQFWVSY 57
       FA7_HUMAN (   65) LEELRPGSLERECKEEQCSFEEAREIFKDAERTKLFWISY 42
      OSTC_CHICK (    6) SGVAGAPPNPIEAQREVCELSPDCNELADELGFQEAYQRR 94
//
```

STORAGE OF INFORMATION IN A SEQUENCE DATABASE

As shown by the above examples, each DNA or protein sequence database entry has much information, including an assigned accession number(s); source organism; name of locus; reference(s); keywords that apply to sequence; features in the sequence such as coding regions, intron splice sites, and mutations; and finally the sequence itself. The above information is organized into a tabular form very much like that found in a relational database. (Additional information about databases is given in the box "Database Types.") If one imagines a large table with each sequence entry occupying one row, then each column will include one of the above types of information for each sequence, and each column is called a FIELD (see Fig. 2.6). The last column contains the sequences themselves. It is very easy to make an index of the information in each of these fields so that a search query can locate all the occurrences through the index. Even related sequences are cross-referenced. In addition, the information in one database can be cross-referenced to that in another database. The DNA, protein, and reference databases have all been cross-referenced so that moving between them is readily accomplished (see ENTREZ section below, p. 45).

Database Types

There are several types of databases; the two principal types are the relational and object-oriented databases. The relational database orders data in tables made up of rows giving specific items in the database, and columns giving the features as attributes of those items. These tables are carefully indexed and cross-referenced with each other, sometimes using additional tables, so that each item in the database has a unique set of identifying features. A relational model for the GenBank sequence database has been devised at the National Center for Genome Resources (http://www.ncgr.org/research/sequence/schema.html).

The object-oriented database structure has been useful in the development of biological databases. The objects, such as genetic maps, genes, or proteins, each have an associated set of utilities for analysis and display of the object and a set of attributes such as identifying name or references. In developing the database, relationships among these objects are identified. To standardize some commonly arising objects in biological databases, e.g., maps, the Object Management Group (http://www.omg.org) has formed a Life Science Research Group. The Life Science Research Group is a consortium of commercial companies, academic institutions, and software vendors that is trying to establish standards for displaying biological information from bioinformatics and genomics analyses (http://www.omg.org/home pages/lsr). The Common Object Request Broker Architecture (CORBA) is the Object Management Group's interface for objects that allows different computer applications to communicate with each other through a common language, Interface Definition Language (IDL). To plan an object-oriented database by defining the classes of objects and the relationships among these objects, a specific set of procedures called the Unified Modeling Language (UML) has been devised by the OMG group.

DNA sequence analysis software packages often include sequence databases that are updated regularly. The organizations that manage sequence databases also provide public access through the internet. Using a browser such as Netscape or Explorer on a local personal computer, these sites may be visited through the internet and a form can be filled out with the sequence name. Once the correct sequence has been identified, the sequence is delivered to the browser and may be saved as a local computer file, cut-and-pasted from the browser window into another window of an analysis program or editor, or even pasted into another browser page for analysis at a second Web site. A useful feature of browser programs for sequence analysis is the capability of having more than one browser window running at a time. Hence, one browser window may retrieve sequences from a database and a second may analyze these sequences. At the time of retrieving the sequence, several sequence formats may be available. The FASTA format, which is readily converted into other formats and also is smaller and simpler, containing just a line of sequence identifiers followed by the sequence without numbers, is very useful for this purpose. A list of sequence databases accessible through the internet is provided in Table 2.5.

USING THE DATABASE ACCESS PROGRAM ENTREZ

One straightforward way to access the sequence databases is through ENTREZ, a resource prepared by the staff of the National Center for Biotechnology Information, National Library of Medicine, Bethesda, Maryland, and available through their web site at http://ncbi.nlm.nih.gov/Entrez. ENTREZ provides a series of forms that can be filled out to retrieve a DNA or protein sequence, or a Medline reference related to the molecular biology sequence databases. After search for either a protein or a DNA sequence is chosen at the above address, another Web page is provided with a form to fill out for the search, as shown in Figure 2.15.

Table 2.5. *Major sequence databases accessible through the internet*

1. GenBank at the National Center of Biotechnology Information, National Library of Medicine, Washington, DC accessible from:
   http://www.ncbi.nlm.nih.gov/Entrez

2. European Molecular Biology Laboratory (EMBL) Outstation at Hixton, England
   http://www.ebi.ac.uk/embl/index.html

3. DNA DataBank of Japan (DDBJ) at Mishima, Japan
   http://www.ddbj.nig.ac.jp/

4. Protein International Resource (PIR) database at the National Biomedical Research Foundation in Washington, DC (see Barker et al. 1998)
   http://www-nbrf.georgetown.edu/pirwww/

5. The SwissProt protein sequence database at ISREC, Swiss Institute for Experimental Cancer Research in Epalinges/Lausanne
   http://www.expasy.ch/cgi-bin/sprot-search-de

6. The Sequence Retrieval System (SRS) at the European Bioinformatics Institute allows both simple and complex concurrent searches of one or more sequence databases. The SRS system may also be used on a local machine to assist in the preparation of local sequence databases.
   http://srs6.ebi.ac.uk

The databases are available at the indicated addresses and return sequence files through an internet browser. Many of the sites shown provide access to multiple databases. The first three database centers are updated daily and exchange new sequences daily, so that it is only necessary to access one of them. Additional Web addresses of databases of protein families and structure, and genomic databases, are given in Chapter 9. These databases can also provide access to sequence of a protein family or organism.

On the ENTREZ form, make a selection in the data entry window after the term "Search," then enter search terms in the longer data entry window after "for." The database will be searched for sequence database entries that contain all of these terms or related ones. Using boolean logic, the search looks for database entries that include the first term AND the second, and subsequent terms repeated until the last term. The "Limits" link on the ENTREZ form page is used to limit the GenBank field to be searched, and various logical combinations of search terms may be designed by this method. These fields refer to the GenBank fields described above in Figure 2.5. When searching for terms in a particular field, some knowledge of the terms that are in the database can be helpful. To assist in finding suitable terms, for each field, ENTREZ provides a list of index entries.

Biological databases are beginning to use "controlled vocabularies" for entering data so that these defined terms can confidently be used for database subsequent searches.

For a protein search, for example, current choices for fields include accession (number), all fields, author name, E. C. number, issue, journal name, keyword, modification date, organism, page number, primary accession (number), properties, protein name, publication date (of reference), seqID string, sequence length, substance name, text word, title word, volume, and sequence ID. Similar fields are shown for the DNA database search. Later, the results of searches in separate fields may be combined to narrow down the choices. The number of terms to be searched for and the field to be searched are the main decisions to be made. In doing so, keep in mind that it is important to be as specific as possible, or else there may be a great many possibilities. Thus, knowing accession number, protein name, or name of gene should be enough to find the required entry quickly. If the same protein has been sequenced in several organisms, providing an organism name is also helpful. When the chosen search terms and fields have been decided and submitted, a database comprising all of the currently available sequences (called the nonredundant or NR database) will be searched. Other database selections may also be made.

The program returns the number of matches found and provides an opportunity to narrow this list by including more terms. When the number of matching sequences has been narrowed to a reasonable number, the sequence may be retrieved in a chosen format in

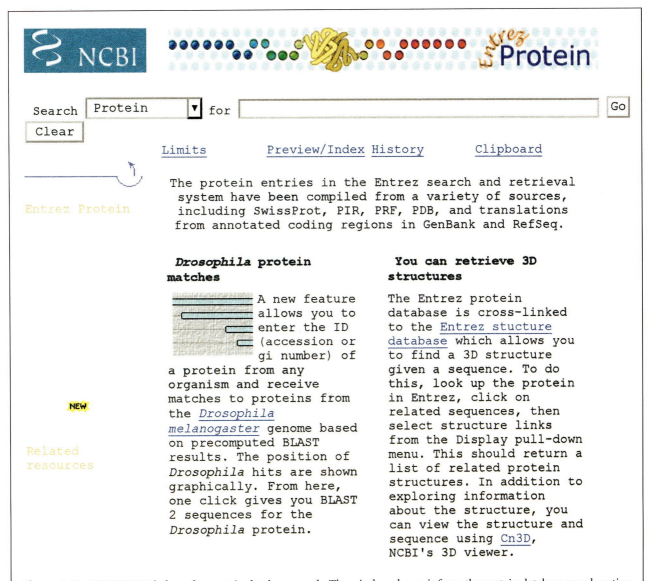

Figure 2.15. ENTREZ Web form for protein database search. The window shown is from the protein database search option at http://www.ncbi.nlm.nih.gov/Entrez/. The search term input window is activated by clicking, one or more search terms are typed, and the "Go" button is clicked (top window). Batch ENTREZ, available from the main ENTREZ Web page, provides a method for retrieving large numbers of sequences at the same time. A particular field (e.g., gene name, organism, protein name) in the GenBank entry can also be searched, by using the "Limits" option. The request is then sent to a server in which all key words in the sequence entries have been indexed, as in looking up a word in the index of a book. GenBank entries with all of the requested terms can be readily identified because the index will indicate in which entry they are all found. The machine returns the number of matches found. Clicking on the retrieve button leads to a list of the found items. Those items chosen are retrieved in a new window format.

several straightforward steps. It is important to look through the sequences to locate the one intended. There may be several different copies of the sequence because it may have been sequenced from more than one organism, or the sequence may be a mutant sequence, a particular clone, or a fragment. There is no simple way to find the correct sequence without manually checking the information provided in each sequence, but this usually takes only a short time. Before leaving ENTREZ, it is often useful to check for sequence database entries that are similar to the one of interest, called "neighbors" by ENTREZ. The expanded query searches other database entries of interest, such as the same protein in another organism, a large chromosomal sequence that includes the gene, or members of the same gene family. While visiting the site, note that ENTREZ has been adapted to search through a number of other biological databases, and also through Medline, and these searches are available from the initial ENTREZ Web page.

Retrieving a Specific Sequence

Even following the above instructions, it can be difficult to retrieve the sequence of a specific gene or protein simply because of the sheer number of sequences in the Gen-Bank database and the complex problem of indexing them. For projects that require the most currently available sequences, the NR databases should be searched. Other projects may benefit from the availability of better curated and annotated protein sequence databases, including PIR and SwissProt. The genomic databases described in Chapter 10 can also provide the sequence of a particular gene or protein. Protein sequences in the Genpro database are generated by automatic translation of DNA sequences. When read from cDNA copies of mRNA sequences, they provide a reliable sequence, given a certain amount of uncertainty as to the translational start site. Many protein sequences are now predicted by translation of genomic sequences, requiring a prediction of exons, a somewhat error-prone step described in more detail in Chapter 8. The origin of protein sequence entries thus needs to be determined, and if they are not from a cDNA sequence, it may be necessary to obtain and sequence a cDNA copy of the gene.

REFERENCES

Barker W.C., Garavelli J.S., Haft D.H., Hunt L.T., Marzec C.R., Orcutt B.C., Srinivasarao G.Y., Yeh L.-S.L., Ledley R.S., Mewes H.-W., Pfeiffer F., and Tsugita A. 1998. The PIR-International Protein Sequence Database. *Nucleic Acids Res.* **26:** 27–32.

Ewing B. and Green P. 1998. Base-calling of automated sequencer traces using phred. II. Error probabilities. *Genome Res.* **8:** 186–194.

Fleischmann R.D., Adams M.D., White O., Clayton R.A., Kirkness E.F., Kerlavage A.R., Bult C.J., Tomb J.F., Dougherty B.A., Merrick J.M., et al. 1995. Whole-genome random sequencing and assembly of *Haemophilus influenzae* Rd. *Science* **269:** 496–512.

Gordon D., Abajian C., and Green P. 1998. Consed: A graphical tool for sequence finishing. *Genome Res.* **8:** 195–202.

Green P. 1997. Against a whole-genome shotgun. *Genome Res.* **7:** 410–417.

IUPAC-IUB: Commission on Biochemical Nomenclature. 1969. A one-letter notation for amino acid sequences. Tentative rules. *Biochem. J.* **113:** 1–4.

———. 1972. Symbols for amino-acid derivatives and peptides. Recommendations 1971. *J. Biol. Chem.* **247:** 977–983.

IUPAC-IUB: Joint Commission on Biochemical Nomenclature (JCBN). 1983. Nomenclature and symbolism for amino acids and peptides. Corrections to recommendations. *Eur. J. Biochem.* **213:** 2.

Myers E.W. 1997. Is whole genome sequencing feasible? In *Computational methods in genome research* (ed. S. Suhai). Plenum Press, New York.

Myers E.W., Sutton G.G., Delcher A.L., Dew I.M., Fasulo D.P., Flanigan M.J., Kravitz S.A., Mobarry C.M., Reinert K.H.J., Remington K.A., et al. 2000. A whole-genome assembly of *Drosophila. Science* **287:** 2196–2204.

NCBI: National Center for Biotechnology Information. 1993. Manual for NCBI Software Development Tool Kit Version 1.8. August 1, 1993. National Library of Medicine, National Institutes of Health.

NC-IUB: Nomenclature Committee of the International Union of Biochemistry. 1984. Nomenclature for incompletely specified bases in nucleic acid sequences. Recommendations. *Eur. J. Biochem.* **150:** 1–5.

Smith S.W., Overbeek R., Woese C.R., Gilbert W., and Gillevet P.M. 1994. The genetic data environment: An expandable GUI for multiple sequence analysis. *Comput. Appl. Biosci.* **10:** 671–675.

Staden R., Beal K.F., and Bonfield J.K. 2000. The Staden package, 1998. *Methods Mol. Biol.* **132:** 115–130.

Thompson J.D., Higgins D.G., and Gibson T.J. 1994. CLUSTAL W: Improving the sensitivity of progressive multiple sequence alignment through sequence weighting, positions-specific gap penalties and weight matrix choice. *Nucleic Acids Res.* **22:** 4673–4680.

Alignment of Pairs of Sequences

INTRODUCTION

Pair-wise sequence alignment is a very large topic to cover as one chapter. Thus, starting with this chapter, more detailed discussions of topics, and information on subjects of more peripheral interest, will be available from the Web site for this book. This site is organized according to the same subject headings as this chapter and can be found at http://www.bioinformaticsonline.org. In addition, starting with this chapter, procedural flowcharts will appear at the beginning of the Methods section of most chapters to provide an overview of the methods of analysis. This chapter discusses pair-wise sequence alignment. Multiple sequence alignment is discussed in Chapter 4.

DEFINITION OF SEQUENCE ALIGNMENT

Sequence alignment is the procedure of comparing two (pair-wise alignment) or more (multiple sequence alignment) sequences by searching for a series of individual characters or character patterns that are in the same order in the sequences. Two sequences are aligned by writing them across a page in two rows. Identical or similar characters are placed in the same column, and nonidentical characters can either be placed in the same column as a mismatch or opposite a gap in the other sequence. In an optimal alignment, nonidentical characters and gaps are placed to bring as many identical or similar characters as possible into vertical register. Sequences that can be readily aligned in this manner are said to be similar.

There are two types of sequence alignment, global and local, and they are illustrated below in Figure 3.1. In global alignment, an attempt is made to align the entire sequence, using as many characters as possible, up to both ends of each sequence. Sequences that are quite similar and approximately the same length are suitable candidates for global alignment. In local alignment, stretches of sequence with the highest density of matches are aligned, thus generating one or more islands of matches or subalignments in the aligned sequences. Local alignments are more suitable for aligning sequences that are similar along some of their lengths but dissimilar in others, sequences that differ in length, or sequences that share a conserved region or domain.

Global Alignment

For the two hypothetical protein sequence fragments in Figure 3.1, the global alignment is stretched over the entire sequence length to include as many matching amino acids as possible up to and including the sequence ends. Vertical bars between the sequences indicate

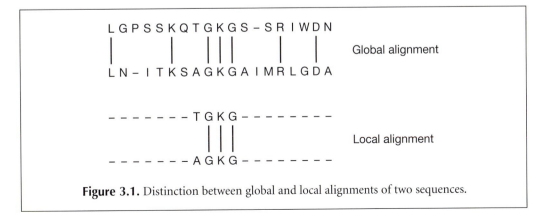

Figure 3.1. Distinction between global and local alignments of two sequences.

the presence of identical amino acids. Although there is an obvious region of identity in this example (the sequence GKG preceded by a commonly observed substitution of T for A), a global alignment may not align such regions so that more amino acids along the entire sequence lengths can be matched.

Local Alignment

In a local alignment, the alignment stops at the ends of regions of identity or strong similarity, and a much higher priority is given to finding these local regions (Fig. 3.1) than to extending the alignment to include more neighboring amino acid pairs. Dashes indicate sequence not included in the alignment. This type of alignment favors finding conserved nucleotide patterns, DNA sequences, or amino acid patterns in protein sequences.

SIGNIFICANCE OF SEQUENCE ALIGNMENT

Sequence alignment is useful for discovering functional, structural, and evolutionary information in biological sequences. It is important to obtain the best possible or so-called "optimal" alignment to discover this information. Sequences that are very much alike, or "similar" in the parlance of sequence analysis, probably have the same function, be it a regulatory role in the case of similar DNA molecules, or a similar biochemical function and three-dimensional structure in the case of proteins. Additionally, if two sequences from different organisms are similar, there may have been a common ancestor sequence, and the sequences are then defined as being homologous. The alignment indicates the changes that could have occurred between the two homologous sequences and a common ancestor sequence during evolution, as shown in Figure 3.2.

With the advent of genome analysis and large-scale sequence comparisons, it becomes important to recognize that sequence similarity may be an indicator of several possible

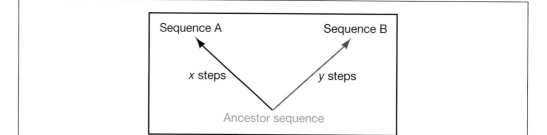

Figure 3.2. The evolutionary relationship between two similar sequences and a possible common ancestor sequence that would make the sequences homologous. The number of steps required to change one sequence to the other is the evolutionary distance between the sequences, and is also the sum of the number of steps to change the common ancestor sequence into one of the sequences (x) plus the number of steps required to change the common ancestor into the other (y). The common ancestor sequence is not available, such that x and y cannot be calculated; only $x + y$ is known. By the simplest definition, the distance $x + y$ is the number of mismatches in the alignment (gaps are not usually counted), as illustrated in Fig. 1.3. In a phylogenetic analysis of three or more similar sequences, the separate distances from the ancestor can be estimated, as discussed in Chapter 6.

types of ancestor relationships, or there may be no ancestor relationship at all, as illustrated in Figure 3.3. For example, new gene evolution is often thought to occur by gene duplication, creating two tandem copies of the gene, followed by mutations in these copies. In rare cases, new mutations in one of the copies provide an advantageous change in function. The two copies may then evolve along separate pathways. Although the resulting separation of function will generate two related sequence families, sequences among both families will still be similar due to the single gene ancestor. In addition, genetic rearrange-

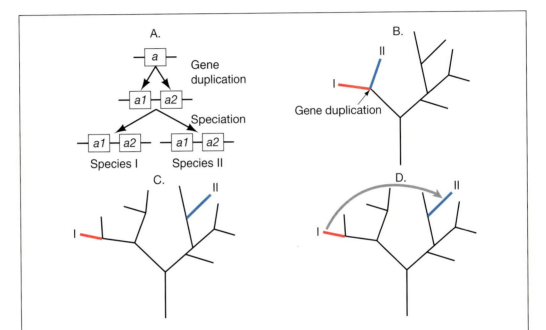

Figure 3.3. Origins of genes having a similar sequence. Shown are illustrative examples of gene evolution. In *A*, a duplication of gene *a* to produce tandem genes *a1* and *a2* in an ancestor of species I and II has occurred. Separation of the duplicated region by speciation gives rise to two separate branches, shown in *B* as blue and red. *a1* in species I and *a1* in species II are orthologous because they share a common ancestor. Similarly, *a2* in species I and *a2* in species II are orthologous. However, the *a1* genes are paralogous to the *a2* genes because they arose from a gene duplication event, indicated in *A*. If two or more copies of a gene family have been separated by speciation in this fashion, they tend to all undergo change as a group, due to gene conversion-type mechanisms (Li and Graur 1991). In *C*, a gene in species I and a different gene in species II have converged on the same function by separate evolutionary paths. Such analogous genes, or genes that result from convergent evolution, include proteins that have a similar active site but within a different backbone sequence. In *D*, genes in species I and II are related through the transfer of genetic material between species, even though the two species are separated by a long evolutionary distance. Although the transfer is shown between outer branches of the evolutionary tree, it could also have occurred in lower-down branches, thus giving rise to a group of organisms with the transferred gene. Such genes are known as xenologous or horizontally transferred genes. Transfer of the P transposable elements between *Drosophila* species is a prime example of such horizontal transfer (Kidwell 1983). Horizontal transfer also is found in bacterial genomes and can be traced as a regional variation in base composition within chromosomes. A similar type of transfer is that of the small ribosomal RNA subunits of mitochondria and chloroplasts, which originated from early prokaryotic organisms. Symbiotic relationships between organisms may be a precursor event leading to such exchanges. Other rearrangements within the genome (not shown) may produce chimeric genes comprising domains of genes that were evolving separately.

Genes that are descended from a common ancestor are called homologs.

ments can reassort domains in proteins, leading to more complex proteins with an evolutionary history that is difficult to reconstruct (Henikoff et al. 1997).

Evolutionary theory provides terms that may be used to describe sequence relationships. Homologous genes that share a common ancestry and function in the absence of any evidence of gene duplication are called orthologs. When there is evidence for gene duplication, the genes in an evolutionary lineage derived from one of the copies and with the same function are also referred to as orthologs. The two copies of the duplicated gene and their progeny in the evolutionary lineage are referred to as paralogs. In other cases, similar regions in sequences may not have a common ancestor but may have arisen independently by two evolutionary pathways converging on the same function, called convergent evolution. There are some remarkable examples in protein structures. For instance, although the enzymes chymotrypsin and subtilisin have totally different three-dimensional structures and folds, the active sites show similar structural features, including histidine (H), serine (S), and aspartic acid (D) in the catalytic sites of the enzymes (for discussion, see Branden and Tooze 1991). Additional examples are given in Chapter 10 (p. 509). In such cases, the similarity will be highly localized. Such sequences are referred to as analogous (Fitch 1970). A closer examination of alignments can help to sort out possible evolutionary origins among similar sequences (Tatusov et al. 1997).

It is important to describe these relationships accurately in publications. A common error in the molecular biology literature is to refer to sequence "homology" when one means sequence similarity. Sequence "similarity" is a measure of the matching characters in an alignment, whereas homology is a statement of common evolutionary origin.

As pointed out by Fitch and Smith (1983), sequences can be either homologous or nonhomologous, but not in between. The genetic rearrangements referred to above can give rise to chimeric genes, in which some regions are homologous and others are not. Referring to the entire sequences as homologous in such situations leads to an inaccurate and incomplete description of the sequence lineage.

Another complication in tracing the origins of similar sequences is that individual genes may not share the same evolutionary origin as the rest of the genome in which they presently reside. Genetic events such as symbioses and viral-induced transduction can cause horizontal transfer of genetic material between unrelated organisms. In such cases, the evolutionary history of the transferred sequences and that of the organisms will be different. Again, with the capability of detecting such events in the genomes of organisms comes the responsibility to describe these changes with the correct evolutionary terminology. In this case, the sequences are xenologous (Gray and Fitch 1983). Recently, Lawrence and Ochman (1997) have shown that horizontal transfer of genes between species is as common in enteric bacteria, if not more common, than mutation. Describing such changes requires a careful description of sequence origins. As discussed in Chapters 6 and 10, phylogenetic and other types of sequence analyses help to uncover such events.

OVERVIEW OF METHODS OF SEQUENCE ALIGNMENT

Alignment of Pairs of Sequences

Alignment of two sequences is performed using the following methods:

1. Dot matrix analysis
2. The dynamic programming (or DP) algorithm
3. Word or *k*-tuple methods, such as used by the programs FASTA and BLAST, described in Chapter 7.

Unless the sequences are known to be very much alike, the dot matrix method should be used first, because this method displays any possible sequence alignments as diagonals on the matrix. Dot matrix analysis can readily reveal the presence of insertions/deletions and direct and inverted repeats that are more difficult to find by the other, more automated methods. The major limitation of the method is that most dot matrix computer programs do not show an actual alignment.

The dynamic programming method, first used for global alignment of sequences by Needleman and Wunsch (1970) and for local alignment by Smith and Waterman (1981a), provides one or more alignments of the sequences. An alignment is generated by starting at the ends of the two sequences and attempting to match all possible pairs of characters between the sequences and by following a scoring scheme for matches, mismatches, and gaps. This procedure generates a matrix of numbers that represents all possible alignments between the sequences. The highest set of sequential scores in the matrix defines an optimal alignment. For proteins, an amino acid substitution matrix, such as the Dayhoff percent accepted mutation matrix 250 (PAM250) or blosum substitution matrix 62 (BLOSUM62) is used to score matches and mismatches. Similar matrices are available for aligning DNA sequences.

The dynamic programming method is guaranteed in a mathematical sense to provide the optimal (very best or highest-scoring) alignment for a given set of user-defined variables, including choice of scoring matrix and gap penalties. Fortunately, experience with the dynamic programming method has provided much help for making the best choices, and dynamic programming has become widely used. The dynamic programming method can also be slow due to the very large number of computational steps, which increase approximately as the square or cube of the sequence lengths. The computer memory requirement also increases as the square of the sequence lengths. Thus, it is difficult to use the method for very long sequences. Fortunately, computer scientists have greatly reduced these time and space requirements to near-linear relationships without compromising the reliability of the dynamic programming method, and these methods are widely used in the available dynamic programming applications to sequence alignment. Other shortcuts have been developed to speed up the early phases of finding an alignment.

The word or k-tuple methods are used by the FASTA and BLAST algorithms (see Chapter 7). They align two sequences very quickly, by first searching for identical short stretches of sequences (called words or k-tuples) and by then joining these words into an alignment by the dynamic programming method. These methods are fast enough to be suitable for searching an entire database for the sequences that align best with an input test sequence. The FASTA and BLAST methods are heuristic; i.e., an empirical method of computer programming in which rules of thumb are used to find solutions and feedback is used to improve performance. However, these methods are reliable in a statistical sense, and usually provide a reliable alignment.

Multiple Sequence Alignment

From a multiple alignment of three or more protein sequences, the highly conserved residues that define structural and functional domains in protein families can be identified. New members of such families can then be found by searching sequence databases for other sequences with these same domains. Alignment of DNA sequences can assist in finding conserved regulatory patterns in DNA sequences. Despite the great value of multiple sequence alignments, obtaining one presents a very difficult algorithmic problem. The methods that have been devised are discussed in Chapter 4.

METHODS

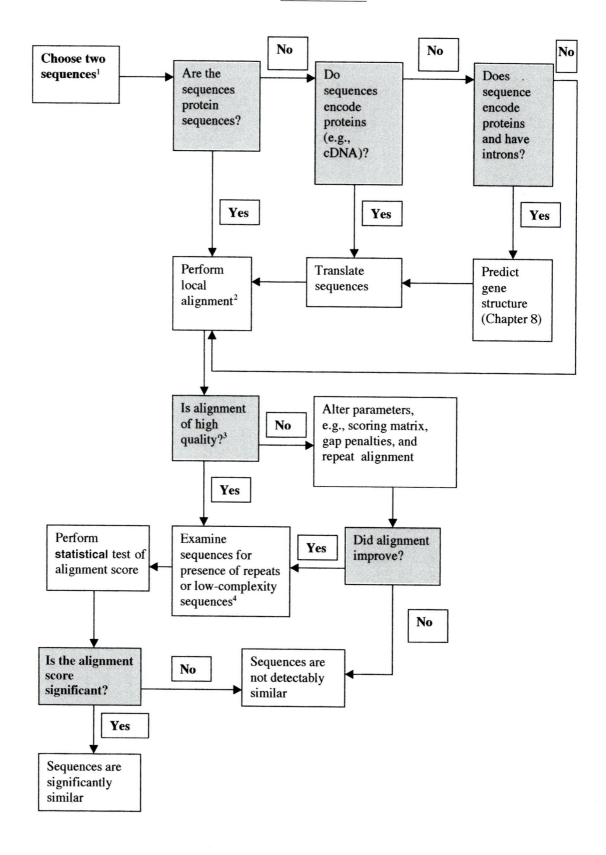

DOT MATRIX SEQUENCE COMPARISON

A dot matrix analysis is primarily a method for comparing two sequences to look for possible alignment of characters between the sequences, first described by Gibbs and McIntyre (1970). The method is also used for finding direct or inverted repeats in protein and DNA sequences, and for predicting regions in RNA that are self-complementary and that, therefore, have the potential of forming secondary structure. Every laboratory that does sequence analysis should have at least one dot matrix program available. In choosing a program, look for as many of the features described below as possible. The dot matrix should be visible on the computer terminal, thus providing an interactive environment so that different types of analyses may be tried. Use of colored dots can enhance the detection of regions of similarity (Maizel and Lenk 1981). Additional descriptions of the dot matrix method have appeared elsewhere (Doolittle 1986; States and Boguski 1991). The examples given below use the dot matrix module of DNA Strider (version 1.3) on a Macintosh computer. The program DOTTER has interactive features for the UNIX X-Windows environment (Sonnhammer and Durbin 1995; http://www.cgr.ki.se/cgr/groups/sonnhammer/Dotter.html). The Genetics Computer Group programs COMPARE and DOTPLOT also perform a dot matrix analysis. Although not a dot matrix method, the program PLALIGN in the FASTA suite may be used to display the alignments found by the dynamic programming method between two sequences on a graph (http://fasta.bioch.virginia.edu/fasta/fasta_list.html; Pearson 1990). A dot matrix program that may be used with a Web browser is described in Junier and Pagni (2000) (http://www.isrec.isb-sib.ch/java/ dotlet/Dotlet.html).

1. This chart assumes that both sequences are protein sequences or that both are DNA sequences. If one is a DNA sequence, that sequence should be translated and then aligned with the second, protein sequence.

2. The local alignment program, e.g., LALIGN or BESTFIT, usually has a recommended scoring matrix and gap penalty combination. It is important to make sure that the combination is one that is known to produce a confined, local alignment with random (or scrambled) sequences. A global alignment program may also be used with sequences of approximately the same length.

3. For protein sequences, a high-quality alignment is one that includes most of each sequence, a significant proportion (e.g., 25%) of identities throughout the alignment, multiple examples of conservative substitutions (chemically and structurally similar amino acids), and relatively few gaps confined to specific regions of the alignment. A poor-quality alignment includes only a portion of the sequences, has few and widely dispersed identities and conservative substitutions, tends to include regions of low complexity (repeats of same amino acid), and includes gaps that are obviously necessary to obtain the alignment. For DNA sequences, a significant alignment must include long runs of identities and few gaps. For two random or unrelated DNA sequences of length 100 and normal composition (0.25 of each base), the longest run of matches that can be expected is 6 or 7 (see text). A clue as to the significance of an alignment may also be obtained by using an alignment program that gives multiple alternative alignments, e.g., LALIGN. The first alignment found, which will be the highest scoring, should have a much higher score than the following ones, which are designed so that the same sequence positions will not be aligned a second time. Hence, these subsequent alignments should usually be random.

4. The result of this analysis can be a guide for the test of significance that follows. In the test described in this chapter, the second sequence is scrambled and realigned with the first sequence. Scrambling can be done at the level of the individual nucleotide or amino acid, or at the level of words by keeping the composition of short stretches of sequence intact.

Pair-wise Sequence Comparison

The major advantage of the dot matrix method for finding sequence alignments is that all possible matches of residues between two sequences are found, leaving the investigator the choice of identifying the most significant ones. Then, sequences of the actual regions that align can be detected by using one of two other methods for performing sequence alignments, e.g., dynamic programming. These methods are automatic and usually show one best or optimal alignment, even though there may be several different, nearly alike alignments. Alignments generated by these programs can be compared to the dot matrix alignment to determine whether the longest regions are being matched and whether insertions and deletions are located in the most reasonable places.

In the dot matrix method of sequence comparison, one sequence (A) is listed across the top of a page and the other sequence (B) is listed down the left side, as illustrated in Figures 3.4 and 3.5. Starting with the first character in B, one then moves across the page keeping in the first row and placing a dot in any column where the character in A is the same. The second character in B is then compared to the entire A sequence, and a dot is placed in row 2 wherever a match occurs. This process is continued until the page is filled with dots representing all the possible matches of A characters with B characters. Any region of similar sequence is revealed by a diagonal row of dots. Isolated dots not on the diagonal represent random matches that are probably not related to any significant alignment.

Detection of matching regions may be improved by filtering out random matches in a dot matrix. Filtering is achieved by using a sliding window to compare the two sequences. Instead of comparing single sequence positions, a window of adjacent positions in the two

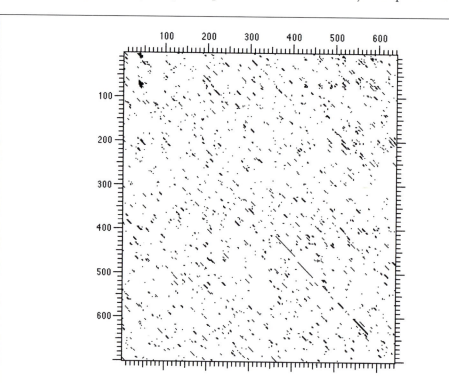

Figure 3.4. Dot matrix analysis of DNA sequences encoding phage λ *cI* (*vertical sequence*) and phage P22 *c2* (*horizontal sequence*) repressors. This analysis was performed using the dot matrix display of the Macintosh DNA sequence analysis program DNA Strider, vers. 1.3. The window size was 11 and the stringency 7, meaning that a dot is printed at a matrix position only if 7 out of the next 11 positions in the sequences are identical.

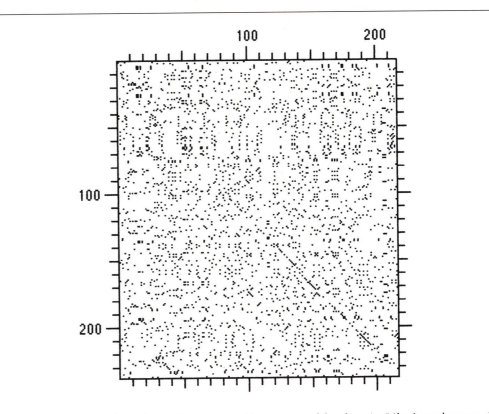

Figure 3.5. Dot matrix analysis of the amino acid sequences of the phage λ *cI* (*horizontal sequence*) and phage P22 *c2* (*vertical sequence*) repressors performed as described in Fig. 3.4. The window size and stringency were both 1.

sequences is compared at the same time, and a dot is printed on the page only if a certain minimal number of matches occur. The window starts at the positions in A and B to be compared and includes characters in a diagonal line going down and to the right, comparing each pair in turn, as in making an alignment. A larger window size is generally used for DNA sequences than for protein sequences because the number of random matches is much greater due to the use of only four DNA symbols as compared to 20 amino acid symbols. A typical window size for DNA sequences is 15 and a suitable match requirement in this window is 10. For protein sequences, the matrix is often not filtered, but a window size of 2 or 3 and a match requirement of 2 will highlight matching regions. If two proteins are expected to be related but to have long regions of dissimilar sequence with only a small proportion of identities, such as similar active sites, a large window, e.g., 20, and small stringency, e.g., 5, should be useful for seeing any similarity. Identification of sequence alignments by the dot matrix method can be aided by performing a count of dots in all possible diagonal lines through the matrix to determine statistically which diagonals have the most matches, and by comparing these match scores with the results of random sequence comparisons (Gibbs and McIntyre 1970; Argos 1987).

An example of a dot matrix analysis between the DNA sequences that encode the *Escherichia coli* phage λ *cI* and phage P22 *c2* repressor proteins is shown in Figure 3.4. With a window of 1 and stringency of 1, there is so much noise that no diagonals can be seen, but, as shown in the figure, with a window of 11 and a stringency of 7, diagonals appear in the lower right. The analysis reveals that there are regions of similarity in the 3' ends of the coding regions, which, in turn, suggests similarity in the carboxy-terminal domains of the encoded repressors. Note that sequential diagonals in matrix C do not line up exactly, indicating the presence of extra nucleotides in one sequence (the lambda *cI* gene on the vertical scale). The diagonals shown in the lower part of the matrix reveal a region of sequence similarity in the carboxy-terminal domains of the proteins. A small insertion in the cI protein that is approximately in the middle of this region and shifts the diagonal slightly downward accounts for this pattern.

An example of a dot matrix analysis between the amino acid sequences of the same two *E. coli* phage lambda cI and phage P22 c2 repressor proteins is shown in Figure 3.5. This matrix was filtered by a window of 1 and a stringency of 1. As found with the DNA sequence alignment of the corresponding genes, diagonals shown in the lower part of the matrix reveal a region of sequence similarity in the carboxy-terminal domains of the proteins. The small insertion in the cI protein approximately in the middle of this region which shifts the diagonal slightly downward and which is also observed in the DNA alignment of these corresponding genes is also visible. Note that these windows are much smaller than required for DNA sequence comparisons due to the greater number of possible symbols (20 amino acids) and therefore fewer random matches.

In conclusion, for DNA sequence dot matrix comparisons, use long windows and high stringencies, e.g., 7 and 11, 11 and 15. For protein sequences, use short windows, e.g., 1 and 1, for window and stringency, respectively, except when looking for a short domain of partial similarity in otherwise not-similar sequences. In this case, use a longer window and a small stringency, e.g., 15 and 5, for window and stringency, respectively.

There are three types of variations in the analysis of two protein sequences by the dot matrix method. First, chemical similarity of the amino acid R group or some other feature for distinguishing amino acids may be used to score similarity. Second, a symbol comparison table such as the PAM250 or BLOSUM62 tables may be used (States and Boguski 1991). These tables provide scores for matches based on their occurrence in aligned protein families. These tables are discussed later in this chapter (pages 78 and 85, respectively). When these tables are used, a dot is placed in the matrix only if a minimum similarity score is found. These table values may also be used in a sliding window option, which averages the score within the window and prints a dot only above a certain average score. Finally, several different matrices can be made, each with a different scoring system, and the scores can be averaged. This method should be useful for aligning more distantly related proteins. The scores of each possible diagonal through the matrix are then calculated, and the most significant ones are identified and shown on a computer screen (Argos 1987).

Sequence Repeats

Dot matrix analysis can also be used to find direct and inverted repeats within sequences. Repeated regions in whole chromosomes may be detected by a dot matrix analysis, and an interactive Web-based program has been designed for showing these regions at increasing levels of detail (http://genome-www.stanford.edu/Saccharomyces/SSV/viewer_start.html). Direct repeats may also be found by performing sequence alignments with dynamic programming methods (see next section). When used to align a sequence with itself, the program LALIGN will show alternative possible alignments between the repeated regions; PLALIGN will plot these alignments on a graph similar in appearance to a dot matrix (see http://fasta.bioch.virginia.edu/fasta/fasta-list.html; Pearson 1990). Here, the sequence is analyzed against itself and the presence of repeats is revealed by diagonal rows of dots. A Bayesian method for finding direct repeats is described on page 122. Inverted repeats require special handling and are discussed in Chapters 5 and 8. In Figure 3.6, an example of such an analysis for direct repeats in the amino acid sequence of the human low-density lipoprotein (LDL) receptor is shown. A list of additional proteins with direct repeats is

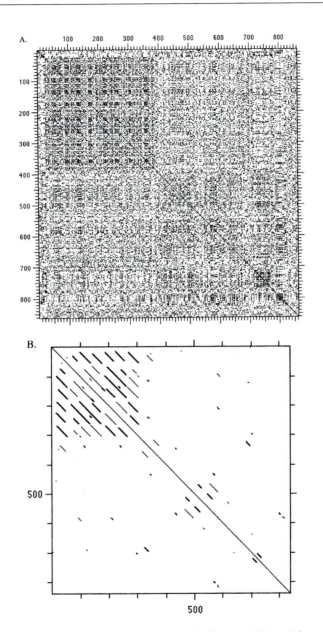

Figure 3.6. Dot matrix analysis of the human LDL receptor against itself using DNA Strider, vers. 1.3, on a Macintosh computer. (*A*) Window 1, Stringency 1. There is a diagonal line from upper left to lower right due to the fact that the same sequence is being compared to itself. The rest of the graph is symmetrical about this line. Other (quite hard to see) lines on either side of this diagonal are also present. These lines indicate repeated sequences perhaps 50 or so long. Patches of high-density dots, e.g., at the position corresponding to position 800 in both sequences representing short repeats of the same amino acid, are also seen. (*B*) Window 23, Stringency 7. The occurrence of longer repeats may be found by using this sliding window. In this example, a dot is placed on the graph at a given position only if 7/23 of the residues are the same. These choices are arbitrary and several combinations may need to be tried. Many repeats are seen in the first 300 positions. A pattern of approximate length 20 and at position 30 is repeated at least six times at positions 70, 100, 140, 180, 230, and 270. Two longer, overlapping repeats of length 70 are also found in this same region starting at positions 70 and 100, and repeated at position 200. Since few of these diagonals remain in new analyses at 11/23 (stringency/window) and all disappear at 15/23, they are not repeats of exactly the same sequence but they do represent an average of about 7/23 matches with no deletions or insertions. The information from the above dot matrix may be used as a basis for listing the actual amino acid repeats themselves by one of the other methods for sequence alignment described below.

given in Doolittle (1986, p. 50), and repeats are also discussed in States and Boguski (1991, p.109). As discussed in Chapters 9 and 10, there are many examples of proteins composed of multiple copies of a single domain.

Repeats of a Single Sequence Symbol

A dot matrix analysis can also reveal the presence of repeats of the same sequence character many times. These repeats become apparent on the dot matrix of a protein sequence against itself as horizontal or vertical rows of dots that sometimes merge into rectangular or square patterns. Such patterns are particularly apparent in the right and lower regions of the dot matrix of the human LDL receptor shown in Figure 3.6 but are also seen throughout the rest of the matrix. The occurrence of such repeats of the same sequence character increases the difficulty of aligning sequences because they create alignments with artificially high scores. A similar problem occurs with regions in which only a few sequence characters are found, called low-complexity regions. Programs that automatically detect and remove such regions from the analysis so that they do not interfere with database similarity searches are discussed in Chapter 7.

DYNAMIC PROGRAMMING ALGORITHM FOR SEQUENCE ALIGNMENT

Dynamic programming is a computational method that is used to align two protein or nucleic acid sequences. The method is very important for sequence analysis because it provides the very best or optimal alignment between sequences. Programs that perform this analysis on sequences are readily available, and there are Web sites that will perform the analysis. However, the method requires the intelligent use of several variables in the program. Thus, it is important to understand how the program works in order to make informed choices of these variables.

The method compares every pair of characters in the two sequences and generates an alignment. This alignment will include matched and mismatched characters and gaps in the two sequences that are positioned so that the number of matches between identical or related characters is the maximum possible. The dynamic programming algorithm provides a reliable computational method for aligning DNA and protein sequences. The method has been proven mathematically to produce the best or optimal alignment between two sequences under a given set of match conditions. Optimal alignments provide useful information to biologists concerning sequence relationships by giving the best possible information as to which characters in a sequence should be in the same column in an alignment, and which are insertions in one of the sequences (or deletions on the other). This information is important for making functional, structural, and evolutionary predictions on the basis of sequence alignments.

Both global and local types of alignments may be made by simple changes in the basic dynamic programming algorithm. A global alignment program is based on the Needleman-Wunsch algorithm, and a local alignment program on the Smith-Waterman algorithm, described below (p. 72). The predicted alignment will be given a score that gives the odds of obtaining the score between sequences known to be related to that obtained by chance alignment of unrelated sequences. There is a method to calculate whether or not an alignment obtained this way is statistically significant. One of the sequences may be scrambled many times and each randomly generated sequence may be realigned with the second sequence to demonstrate that the original alignment is unique. The statistical significance of alignment scores is discussed in detail below (p. 96).

Another feature of the dynamic programming algorithm is that the alignments obtained depend on the choice of a scoring system for comparing character pairs and penalty scores for gaps. For protein sequences, the simplest system of comparison is one based on identity. A match in an alignment is only scored if the two aligned amino acids are identical. However, one can also examine related protein sequences that can be aligned easily and find which amino acids are commonly substituted for each other. The probability of a substitution between any pair of the 20 amino acids may then be used to produce alignments. Recent improvements and experience with the dynamic programming programs and the scoring systems have greatly simplified their use. These enhancements are discussed below and at http://www.bioinformaticsonline.org.

It is important to recognize that several different alignments may provide approximately the same alignment score; i.e., there are alignments almost as good as the highest-scoring one reported by the alignment program. Some programs, e.g., LALIGN, provide several entirely different alignments with different sequence positions matched that can be compared to improve confidence in the best-scoring one. Alignment programs have also been greatly improved in algorithmic design and performance. With the advent of faster machines, it is possible to do a dynamic programming alignment between a query sequence and an entire sequence database and to find the similar sequences in several minutes. Dynamic programming has also been used to perform multiple sequence alignment, but only for a small number of sequences because the complexity of the calculations increases substantially for more than two sequences. Sequence alignment programs are available as a part of most sequence analysis packages, such as the widely used Genetics Computer Group GAP (global alignment) and BESTFIT (local alignment) programs. Sequences can also be pasted into a text area on a guest Web page on a remote host machine that will perform a dynamic programming alignment, and there are also versions of alignment programs that will run on a microcomputer (Table 3.1).

In deciding to perform a sequence alignment, it is important to keep the goal of the analysis in mind. Is the investigator interested in trying to find out whether two proteins have similar domains or structural features, whether they are in the same family with a related biological function, or whether they share a common ancestor relationship? The desired objective will influence the way the analysis is done. There are several decisions to be made along the way, including the type of program, whether to produce a global or local alignment, the type of scoring matrix, and the value of the gap penalties to be used. There are a very large number of amino acid scoring matrices in use (see book Web site), some much more popular than others, and these scoring matrices are designed for different purposes. Some, such as the Dayhoff PAM matrices, are based on an evolutionary model of protein change, whereas others, such as the BLOSUM matrices, are designed to identify members of the same family. Alignments between DNA sequences require similar kinds of considerations. It is often worth the effort to try several approaches to find out which choice of scoring system and gap penalty give the most reasonable result. Fortunately, most alignment programs come with a recommended scoring matrix and gap penalties that are useful for most situations. A more recent development (see Bayesian methods discussed on p. 124) is the simultaneous use of a set of scoring matrices and gap penalties by a method that generates the most probable alignments (see Table 3.1). The final choice as to the most believable alignment is up to the investigator, subject to the condition that reasonable decisions have been made regarding the methods used.

For sequences that are very similar, e.g., >95%, the sequence alignment is usually quite obvious, and a computer program may not even be needed to produce the alignment. As the sequences become less and less similar, the alignment becomes more difficult to produce and one is less confident of the result. For protein sequences, similarity can still be recognized down to a level of approximately 25% amino acid identity. At this level of iden-

Table 3.1. *Web sites for alignment of sequence pairs*

Name of site	Web address	Reference
Bayes block aligner	http://www.wadsworth.org/res&res/bioinfo	Zhu et al. (1998)
BCM Search Launcher:		
Pairwise sequence alignment[a]	http://dot.imgen.bcm.tmc.edu:9331/seq-search/alignment.html	see Web site
SIM—Local similarity program for finding alternative alignments	http://www.expasy.ch/tools/sim.html	Huang et al. (1990); Huang and Miller (1991); Pearson and Miller (1992)
Global alignment programs (GAP, NAP)	http://genome.cs.mtu.edu/align/align.html	Huang (1994)
FASTA program suite[b]	http://fasta.bioch.virginia.edu/fasta/fasta_list.html	Pearson and Miller (1992); Pearson (1996)
BLAST 2 sequence alignment (BLASTN, BLASTP)[c]	http://www.ncbi.nlm.nih.gov/gorf/bl2.html	Altschul et al. (1990)
Likelihood-weighted sequence alignment (lwa)[d]	http://www.ibc.wustl.edu/servive/lwa.html	see Web site

[a] This server provides access to a number of Web sites offering pair-wise alignments between nucleic acid sequences, protein sequences, or between a nucleic acid and a protein sequence.

[b] The FASTA algorithm normally used for sequence database searches (see Chapter 7) provides an alternative method to dynamic programming for producing an alignment between sequences. Briefly, all short patterns of a certain length are located in both sequences. If multiple patterns are found in the same order in both sequences, these provide the starting point for an alignment by the dynamic programming algorithm. Older versions of FASTA performed a global alignment, but more recent versions perform a local alignment with statistical evaluations of the scores. The program PLFASTA in the FASTA program suite provides a plot of the best matching regions, much like a dot matrix analysis, and thus gives an indication of alternative alignments. The FASTA suite is also available from Genestream at http://vega.igh.cnrs.fr/. Programs include ALIGN (global, Needleman-Wunsch alignment), LALIGN (local, Smith-Waterman alignment), LALIGNO (Smith-Waterman alignment, no end gap penalty), FASTA (local alignment, FASTA method), and PRSS (local alignment with scrambled copies of second sequence to do statistical analysis). Versions of these programs that run with a command-line interface on MS-DOS and Macintosh microcomputers are available by anonymous FTP from ftp.virginia.edu/pub/fasta.

[c] The BLAST algorithm normally used for database similarity searches (Chapter 7) can also be used to align two sequences.

[d] A description of the probabilistic method of aligning two sequences is described in Durbin et al. (1998) and Chapter 4. A related topic, hidden Markov models for multiple sequence alignments, is discussed in Chapter 4.

tity, the relative numbers of mismatched amino acids and gaps in the alignment have to be decided empirically and a decision made as to which gap penalties work the best for a given scoring matrix. Alignment of sequences at this level of identity is called the "twilight zone" of sequence alignment by Doolittle (1981). The alignment program may provide a quite convincing alignment, which suggests that the two sequences are homologous. The statistical significance of the alignment score may then be evaluated, as described later in this chapter.

Description of the Algorithm

Alignment of two sequences without allowing gaps requires an algorithm that performs a number of comparisons roughly proportional to the square of the average sequence length, as in a dot matrix comparison. If the alignment is to include gaps of any length at any position in either sequence, the number of comparisons that must be made becomes astronomical and is not achievable by direct comparison methods. Dynamic programming is a method of sequence alignment that can take gaps into account but that requires a manageable number of comparisons.

The method of sequence alignment by dynamic programming and the proof that the method provides an optimal (highest scoring) alignment are illustrated in Figures 3.7 and 3.8. To understand how the method works, we must first recall what is meant by an align-

```
sequence 1   V   D   S   -    C   Y
sequence 2   V   E   S   L    C   Y
SCORE        4   2   4  -11   9   7    SCORE = SUM OF AMINO ACID PAIR SCORES
(26)                                   MINUS SINGLE GAP PENALTY (11) = 15
```

Figure 3.7. Example of scoring a sequence alignment with a gap penalty. The individual alignment scores are taken from an amino acid substitution matrix.

ment, using the two protein sequences shown in Figure 3.7 as an example. The two sequences will be written across the page, one under the other, the object being to bring as many amino acids as possible into register. In some regions, amino acids in one sequence will be placed directly below identical amino acids in the second. In other regions, this process may not be possible and nonidentical amino acids may have to be placed next to each other, or else gaps must be introduced into one of the sequences. Gaps are added to the alignment in a manner that increases the matching of identical or similar amino acids at subsequent portions in the alignment. Ideally, when two similar protein sequences are aligned, the alignment should have long regions of identical or related amino acid pairs and very few gaps. As the sequences become more distant, more mismatched amino acid pairs and gaps should appear.

The quality of the alignment between two sequences is calculated using a scoring system that favors the matching of related or identical amino acids and penalizes for poorly matched amino acids and gaps. To decide how to score these regions, information on the types of changes found in related protein sequences is needed. These changes may be expressed by the following probabilities: (1) that a particular amino acid pair is found in alignments of related proteins; (2) that the same amino acid pair is aligned by chance in the sequences, given that some amino acids are abundant in proteins and others rare; and (3) that the insertion of a gap of one or more residues in one of the sequences (the same as an insertion of the same length in the other sequence), thus forcing the alignment of each partner of the amino acid pair with another amino acid, would be a better choice. The ratio of the first two probabilities is usually provided in an amino acid substitution matrix. Each

```
1.  SCORE OF NEW      =    SCORE OF PREVIOUS +   SCORE OF NEW
    ALIGNMENT              ALIGNMENT (A)         ALIGNED PAIR

V  D  S  -  C  Y           V  D  S  -  C              Y
V  E  S  L  C  Y           V  E  S  L  C              Y

      15            =            8          +         7


II. SCORE OF         =    SCORE OF PREVIOUS +   SCORE OF NEW
    ALIGNMENT (A)         ALIGNMENT (B)         ALIGNED PAIR

V  D  S  -  C              V  D  S  -                 C
V  E  S  L  C              V  E  S  L                 C

       8            =           -1          +         9

III. REPEAT REMOVING ALIGNED PAIRS UNTIL END OF ALIGNMENT IS REACHED.
```

Figure 3.8. Derivation of the dynamic programming algorithm.

table entry gives the ratio of the observed frequency of substitution between each possible amino acid pair in related proteins to that expected by chance, given the frequencies of the amino acids in proteins. These ratios are called odds scores. The ratios are transformed to logarithms of odds scores, called log odds scores, so that scores of sequential pairs may be added to reflect the overall odds of a real to chance alignment of an alignment. Examples are the Dayhoff PAM250 and BLOSUM62 substitution matrices described below (p. 76). These matrices contain positive and negative values, reflecting the likelihood of each amino acid substitution in related proteins. Using these tables, an alignment of a sequential set of amino acid pairs with no gaps receives an overall score that is the sum of the positive and negative log odds scores for each individual amino acid pair in the alignment. The higher this score, the more significant is the alignment, or the more it resembles alignments in related proteins. The score given for gaps in aligned sequences is negative, because such misaligned regions should be uncommon in sequences of related proteins. Such a score will reduce the score obtained from an adjacent, matching region upstream in the sequences. The score of the alignment in Figure 3.7, using values from the BLOSUM62 amino acid substitution matrix and a gap penalty score of -11 for a gap of length 1, is 26 (the sum of amino acid pair scores) $-11 = 15$. The value of -11 as a penalty for a gap of length 1 is used because this value is already known from experience to favor the alignment of similar regions when the BLOSUM62 comparison matrix is used. Choice of the gap penalty is discussed further below where a table giving suitable choices is presented (see Table 3.10 on p. 113). As shown in the example, the presence of the gap decreases significantly the overall score of the alignment.

Calculating the Odds Score of an Alignment from the Odds Scores of Individual Amino Acid Pairs

Sequence alignment scores are based on the individual scores of all amino acid pairs in the alignment. The odds score for an amino acid pair is the ratio of the observed frequency of occurrence of that pair in alignments of related proteins over the expected frequency based on the proportion of amino acids in proteins. Alignments are built by making possible lists of amino acid pairs and by finding the most likely list using odds scores. To calculate the odds score for an alignment, the odds scores for the individual pairs are multiplied. This calculation is similar to finding the probability of one event AND also a second independent event by multiplying the probabilities (if one event OR another is the choice, then the probabilities are added). Thus, if the odds score of C/C is 7/1 and that of W/W is 50/1, then the probability of C/C and W/W being in the alignment is 7/1 × 50/1 = 350/1 (note that the order or position in the alignment does not matter). Usually, log odds scores are used in these calculations, and these scores are added to produce an overall log odds score for the alignment. To perform this optimal alignment using odds scores, the method assumes that the odds score for matching a given pair of sequence positions is not influenced by the odds score of any other matching pair; i.e., that there are no correlations expected among the amino acids found at various sequence positions. Another way of describing this assumption is that the sequences are each being modeled as a Markov chain, with the amino acid found at each position not being influenced by other amino acids in the sequence. Although correlations among sequence positions are expected, since they give rise to structure and function in molecules, this simplifying assumption allows the determination of a reasonable alignment between the sequences.

Although one may be able to align the two short sequences in Figure 3.7 by eye and to place the gap where shown, the dynamic programming algorithm will automatically place gaps in much longer sequence alignments so as to achieve the best possible alignment. The derivation of the dynamic programming algorithm is illustrated in Figure 3.8, using the above alignment as an example. Consider building this alignment in steps, starting with an initial matching aligned pair of characters from the sequences (V/V) and then sequentially adding a new pair until the alignment is complete, at each stage choosing a pair from all the possible matches that provides the highest score for the alignment up to that point. If the full alignment finally reached on the left side of Figure 3.8 (I) has the highest possible or optimal score, then the old alignment from which it was derived (A) by addition of the aligned Y/Y pair must also have been optimal up to that point in the alignment. If this were incorrect, and a different preceding alignment other than A was the highest scoring one, then the alignment on the left would also not be the highest scoring alignment, and we started with that as a known condition. Similarly, in Figure 3.8 (II), alignment A must also have been derived from an optimal alignment (B) by addition of a C/C pair. In this manner, the alignment can be traced back sequentially to the first aligned pair that was also an optimal alignment. One concludes that the building of an optimal alignment in this stepwise fashion can provide an optimal alignment of the entire sequences.

The example in Figure 3.8 also illustrates two of the three choices that can be made in adding to an alignment between two sequences: Match the next two characters in the next positions in each sequence, or match the next character to a gap in the upper sequence. The last possibility, not illustrated, is to add a gap to the lower sequence. This situation is analogous to performing a dot matrix analysis of the sequences, and of either continuing a diagonal or of shifting the diagonal sideway or downward to produce a gap in one of the sequences. An example of using the dynamic programming algorithm to align two short protein sequences is illustrated in Figure 3.9.

Formal Description of the Dynamic Programming Algorithm

The algorithm (Fig. 3.9) may be written in mathematical form, as shown in Figure 3.10. The diagram indicates the moves that are possible to reach a certain matrix position (i,j) starting from the previous row and column at position $(i-1, j-1)$ or from any position in the same row and column.

The following equation describes the algorithm that was illustrated in Figure 3.9. There are three paths in the scoring matrix for reaching a particular position, a diagonal move from position $i-1, j-1$ to position i, j with no gap penalties, or a move from any other position from column j or row i, with a gap penalty that depends on the size of the gap. For two sequences $\mathbf{a} = a_1 a_2 \ldots a_n$ and $\mathbf{b} = b_1 b_2 \ldots b_n$, where $S_{ij} = S(a_1 a_2 \ldots a_i, b_1 b_2 .. b_j)$ then (Smith and Waterman 1981a,b)

$$S_{ij} = \max \left\{ \begin{array}{l} S_{i-1, j-1} + s(a_i b_j), \\[1em] \max_{x \geq 1} (S_{i-x, j} - w_x), \\[1em] \max_{y \geq 1} (S_{ij-y} - w_y) \end{array} \right. \tag{1}$$

where S_{ij} is the score at position i in sequence $\mathbf{a}$ and position j in sequence $\mathbf{b}$, $s(a_i b_j)$ is the score for aligning the characters at positions i and j, w_x is the penalty for a gap of length x

in sequence **a**, and w_y is the penalty for a gap of length y in sequence **b**. Note that S_{ij} is a type of running best score as the algorithm moves through every position in the matrix. Eventually, when all of the matrix positions (all S_{ij}) have been filled, the best score of the alignment will be found as the highest scoring position in the last row and column (for a global alignment), after correcting for any remaining gap penalties to align the sequence ends, if applicable. To determine an optimal alignment of the sequences from the scoring matrix, a second matrix called the trace-back matrix is used (Fig. 3.9). The trace-back matrix keeps track of the positions in the scoring matrix that contributed to the highest overall score found. The sequence characters corresponding to these high scoring positions may align or may be next to a gap, depending on the information in the trace-back matrix. An example of this procedure can be found on the book Web site.

Use of the dynamic programming method requires a scoring system for the comparison of symbol pairs (nucleotides for DNA sequences and amino acids for protein sequences), and a scheme for insertion/deletion (GAP) penalties. Once those parameters have been set, the resulting alignment for two sequences should always be the same. Scoring matrices are

---

Figure 3.9. Example of using the dynamic programming algorithm to align sequences a1 a2 a3 a4 and b1 b2 b3 b4.

1. The sequences are written across the top and down the left side of a matrix, respectively, similar to that done in the dot matrix analysis, except that an extra row and column labeled "gap" are added to allow the alignment to begin with a gap of any length in either sequence. The gap rows are filled with penalty scores for gaps of increasing lengths, as indicated. A zero is placed in the upper right box corresponding to no gaps in either sequence.

2. Maximum possible values are calculated for all other boxes below and to the right of the top row and left column, taking into account any sized gap or no gap, using the steps listed in a through d below. The scores for individual matches a1-b1, a1-b2, etc., are obtained from a scoring matrix (symbol comparison table). To calculate the value for a particular matrix position, trial values are calculated from all moves into that position allowed by the algorithm. The allowed moves are from any position above or to the left of the current position, in the same column or row, or from the upper left diagonal position. The diagonal move attempts to align the sequence characters without introducing a gap. Thus, there is no gap penalty in this case. However, moves from above and to the left will introduce gaps, and thus will require one or more gap penalties to be used. (a) s11 is the score for an a1-b1 match added to 0 in the upper left position. According to the algorithm, there are two other possible paths to this position shown by the vertical and horizontal arrows, but they would probably have to give a lower score because they start at a gap penalty and must include an additional gap penalty. (b) Trial values for s12 are calculated and the maximum score is chosen. Trial 1 is to add the score for the a1-b2 match to s11 and subtract a penalty for a gap of size 1. The other three trials shown by arrows include gap penalties and so likely cannot yield a higher score than trial 1. (c) All possible scores for s21 are calculated by the trial moves indicated. The best score should be obtained by adding the score of an a2-b1 match to s11 since all other moves include gap penalties. (d) Trial values of s22 are calculated by considering moves from s11, s21, and s12, and from the top row and left end column. s22 will be the best score of several possible choices, including adding the score for an a2-b2 match to s11, or to s21 less a single gap penalty. Other trials will normally be attempted from other positions above and to the left of this position, but in this case, they will probably not provide a higher score for s22 because they include multiple gap penalties.

3. As the maximum scores for each matrix position are calculated, a record of the paths that produced the highest scores to reach each matrix position is kept. These short paths, which represent extending the alignment to another matching pair, with or without gaps, are recorded in another matrix called the trace-back matrix, illustrated below. For example, if moving from s11 to s21 gave the highest score of all moves to s21, then the corresponding region of the matrix will appear as shown.

4. The paths in the trace-back matrix are joined to produce an alignment. In the example shown, the highest-scoring matrix position in the sequence comparison matrix is located, in this case s44, and the arrows are then traced back as far as possible, generating the path shown. The corresponding alignment A is shown below the matrix. More than one alignment may be possible if there is more than one path from the highest scoring matrix position. As an example, s43 could also be a high-scoring position, generating trace-back alignment B, an alignment that includes a gap opposite a2. Another gap may also be placed opposite b4, which has no matching symbol. Scoring end gaps is optional in the alignment programs. If

Legend continues.

1.

	gap	a1	a2	a3	a4
gap	0	1 gap	2 gaps	3 gaps	4 gaps
b1	1 gap				
b2	2 gaps				
b3	3 gaps				
b4	4 gaps				

2c.

	gap	a1	a2	a3	a4
gap	0	1 gap	2 gaps	3 gaps	4 gaps
b1	1 gap	s11	s21		
b2	2 gaps	s12			
b3	3 gaps				
b4	4 gaps				

2a.

	gap	a1	a2	a3	a4
gap	0	1 gap	2 gaps	3 gaps	4 gaps
b1	1 gap	s11			
b2	2 gaps				
b3	3 gaps				
b4	4 gaps				

2d.

	gap	a1	a2	a3	a4
gap	0	1 gap	2 gaps	3 gaps	4 gaps
b1	1 gap	s11	s21		
b2	2 gaps	s12	s22		
b3	3 gaps				
b4	4 gaps				

2b.

	gap	a1	a2	a3	a4
gap	0	1 gap	2 gaps	3 gaps	4 gaps
b1	1 gap	s11			
b2	2 gaps	s12			
b3	3 gaps				
b4	4 gaps				

3. Part of trace back matrix

	gap	a1	a2	a3	a4
gap	0	1 gap	2 gaps	3 gaps	4 gaps
b1	1 gap	s11	s21	s31	s41
b2	2 gaps	s12	s22	s32	s42
b3	3 gaps	s13	s23	s33	s43
b4	4 gaps	s14	s24	s34	s44

4. Trace back matrix

	gap	a1	a2	a3	a4
gap	0	1 gap	2 gaps	3 gaps	4 gaps
b1	1 gap	s11	s21 **B**	s31	s41
b2	2 gaps	s12	s22	s32	s42
b3	3 gaps	s13	s23 **A**	s33	s43
b4	4 gaps	s14	s24	s34	s44

Alignment A: a1 a2 a3 a4
              b1 b2 b3 b4

Alignment B: a1 a2 a3 a4 –
              b1 – b2 b3 b4

included in this case, alignment B would be disfavored by an additional gap penalty. In addition to this series of alignments, or so-called clump of alignments starting from the highest scoring position, there will be other possible alignments starting from other high-scoring matrix positions, and these may also have multiple pathways through the scoring matrix, each representing a different alignment. Note that these alignments are global alignments because they include the entire sequences.

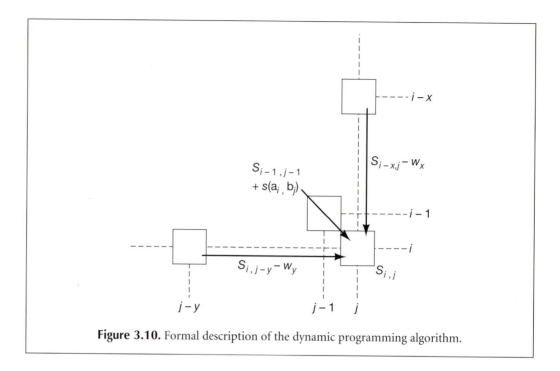

Figure 3.10. Formal description of the dynamic programming algorithm.

described below. The most commonly used ones for protein sequence alignments are the log odds form of the PAM250 matrix and the BLOSUM62 matrix. However, a number of other choices are available.

Dynamic Programming Can Provide Global or Local Sequence Alignments

Global Alignment: Needleman-Wunsch Algorithm

The dynamic programming method as described above gives a global alignment of sequences, as described by Needleman and Wunsch (1970), but was also proven mathematically and extended to include an improved scoring system by Smith and Waterman (1981a,b). The optimal score at each matrix position is calculated by adding the current match score to previously scored positions and subtracting gap penalties, if applicable. Each matrix position may have a positive or negative score, or 0. The Needleman-Wunsch algorithm will maximize the number of matches between the sequences along the entire length of the sequences. Gaps may also be present at the ends of sequences, in case there is extra sequence left over after the alignment. These end gaps are often, but not always, given a gap penalty. The effect of these penalties is illustrated below. An example of a global alignment of two short sequences calculated by hand using the algorithm is shown on the book Web site. The example also reveals that more than one alignment may be equally as likely.

Local Alignment: Smith-Waterman Algorithm

A modification of the dynamic programming algorithm for sequence alignment provides a local sequence alignment giving the highest-scoring local match between two sequences (Smith and Waterman 1981a,b). Local alignments are usually more meaningful than global matches because they include patterns that are conserved in the sequences. They can also be used instead of the Needleman-Wunsch algorithm to match two sequences that may have a matched region that is only a fraction of their lengths, that have different lengths, that overlap, or where one sequence is a fragment or subsequence of the other. The rules for calculating scoring matrix values are slightly different, the most important differences being (1) the scoring system must include negative scores for mismatches, and (2) when a dynamic programming scoring matrix value becomes negative, that value is set to zero, which has the effect of terminating any alignment up to that point. The alignments are produced by starting at the highest-scoring positions in the scoring matrix and following a trace path from those positions up to a box that scores zero. The mathematical formulation of the dynamic programming algorithm is revised to include a choice of zero as the minimum value at any matrix position. For two sequences $\mathbf{a} = a_1 a_2 \ldots a_n$ and $\mathbf{b} = b_1 b_2 \ldots b_n$, where $H_{ij} = H(a_1 a_2 \ldots a_i, b_1 b_2 .. b_j)$, then (Smith and Waterman 1981a)

$$
H_{ij} = \max \left\{
\begin{array}{l}
H_{i-1, j-1} + s(a_i b_j), \\[6pt]
\max_{x \geq 1} (H_{i-x, j} - w_x), \\[6pt]
\max_{y \geq 1} (H_{i, j-y} - w_y), \\[6pt]
0
\end{array}
\right.
\tag{2}
$$

where H_{ij} is the score at position i in sequence $\mathbf{a}$ and position j in sequence $\mathbf{b}$, $s(a_i b_j)$ is the score for aligning the characters at positions i and j, w_x is the penalty for a gap of length x in sequence $\mathbf{a}$, and w_y is the penalty for a gap of length y in sequence $\mathbf{b}$.

To illustrate the difference between the Needleman-Wunsch and Smith-Waterman methods, a local alignment of the same two sequences is shown on the book Web site.

Does a Local Alignment Program Always Produce a Local Alignment and a Global Alignment Program Always Produce a Global Alignment?

Although a computer program that is based on the above Smith-Waterman local alignment algorithm is used for producing an optimal alignment, this feature alone does not assure that a local alignment will be produced. The scoring matrix or match and mismatch scores and the gap penalties chosen also influence whether or not a local alignment is obtained. Similarly, a program based on the Needleman-Wunsch algorithm can also return a local alignment depending on the weighting of end gaps and on other scoring parameters. Often, one can simply inspect the alignment obtained to see how many gaps are present. If the matched regions are long and cover most of the sequences and obviously depend on the presence of many gaps, the alignment is global. A local alignment, on the other hand, will tend to be shorter and not include many gaps, just as in the example given on the book Web site. However, these tests are quite subjective, and a more precise method of knowing whether a given program and set of scoring parameters will provide a local or global alignment is required. Looking ahead in the chapter for a moment, the best way of knowing is by looking at what happens when many random or completely unrelated sequences are aligned under the chosen conditions. As the length of the random sequences being aligned increases, the score of a global alignment will just increase proportionally.

This is easy to see. Because a global alignment matches most of the sequence, and the negative mismatch score and gap penalties are deliberately chosen to be small in comparison to match scores in order to provide a long alignment, only matches count and the score has to be proportional to the length.

If using a scoring matrix, a matrix that gives on the average a positive score to each aligned position, combined with a small enough gap penalty to allow extension of the alignment through poorly matched regions, will give a global alignment. Conversely, for the local alignment, a negative mismatch score and gap penalties are chosen to balance the positive score of a match and to prevent the alignment growing into regions that do not match very well. The scoring matrix in this case will on the average give a negative value to the matched positions, and the gap penalty will be large enough to prevent gaps from extending the alignment. The local alignment score of random sequences does not increase proportionally to sequence length, because the positive score of matches is offset by the mismatch and penalty scores. In this case, it may be shown by theory and experiment that the score of local random alignments increases much more slowly, and proportionally to the logarithm of the product of the sequence lengths. It is this different behavior of the alignment score of random sequences with length that distinguishes global and local alignments.

One may well ask, Does it really matter whether I use a sequence alignment program based on the global alignment algorithm or one based on the local alignment algorithm? The answer is that sometimes both methods will provide the same alignment with the same scoring system and sometimes they will not. The most reasonable approach is to use a program based on the appropriate algorithm for the analysis at hand, and then to choose the scoring system carefully. Small changes in the scoring system can abruptly change an alignment from a local to a global one. There are even examples in the bioinformatics literature where this feature of alignment scoring systems has been overlooked. The rest of this chapter is designed to provide a suitable guide for making the right choices.

Additional Development and Use of the Dynamic Programming Algorithm for Sequence Alignments

Use of Distance Scores for Sequence Alignment

As originally designed by Needleman and Wunsch and Smith and Waterman, the dynamic programming algorithm was used for sequence alignments scored on the basis of the similarity or identity of sequence characters. An alternative method is to score alignments based on differences between sequences and sequence characters; i.e., how many changes are required to change one sequence into another. Using this measure, the greater the distance between sequences, the greater the evolutionary time that has elapsed since the sequences diverged from a common ancestor. Hence, distance scores provide a more biologically natural way to compare sequences than do similarity scores. Using a distance scoring scheme, Sellers (1974, 1980) showed that the dynamic programming method could be used to provide an alignment that highlighted the evolutionary changes. Smith et al. (1981) and Smith and Waterman (1981b) showed that alignments based on a similarity scoring scheme could give a similar alignment. This analysis is discussed further on the book Web site. Conversion between distance and similarity scores is discussed in Chapter 6.

Improvement in Speed and Memory Requirement for the Dynamic Programming Algorithm

The dynamic programming methods for sequence alignments originally required between $n \times m$ and $n \times m^2$ steps and storage in several matrices of size $n \times m$, where n is the length of the shorter sequence (Needleman and Wunsch 1970; Waterman et al. 1976; Smith and Waterman 1981a). On the book Web site, a series of improvements in this algorithm that reduced the number of steps and amount of memory required are described. These steps include: (1) a decreased number of steps in the alignment algorithm by Gotoh (1982); (2) a reduction in the amount of memory required to a linear function of sequence length (Myers and Miller 1988); (3) ability to find near-optimal alignments (Chao et al. 1994) and to align long sequences (Schwartz et al. 1991); and (4) ability to find the best-scoring alternative alignments that do not include alignments of the same sequence positions (Waterman and Eggert 1987; Huang et al. 1990; Huang and Miller 1991).

The alignment programs listed in Table 3.1 include these features.

An alternative global alignment is found by giving the matrix position that begins with an alignment score of zero, and then all matrix positions that are affected by this change are recalculated. The next highest matrix score and the path leading to it provide an alternative alignment of the sequences that does not include the same sequence matches as were present in the original alignment (Waterman and Eggert 1987). Alternative local alignments are found by a more complex algorithm (the SIM algorithm) that includes the improvements listed above (Huang et al. 1990; Huang and Miller 1991).

Examples of Global and Local Alignments

An example of global and local alignments between two phage repressor proteins using the Genetics Computer Group (GCG) programs GAP (Needleman-Wunsch algorithm) and BESTFIT (Smith-Waterman algorithm) is shown in Figure 3.11. Note that the proteins are 58% similar in the carboxy-terminal domain, which is the region required for protein–protein interactions and a self-cleavage function that leads to phage induction. In these GCG implementations of the Needleman-Wunsch and Smith-Waterman algorithms, the alignments found in the carboxy-terminal domain are identical. However, the Smith-Waterman method (B) only reports the most alike regions, as expected by the focus on a local alignment strategy. In contrast, the Needleman-Wunsch method shows the entire alignment of the sequences but reports a lower score of similarity due to the longer alignment.

LALIGN (Fig. 3.12) is an implementation of the SIM algorithm for finding multiple unique (nonintersecting) alignments in DNA and protein sequences (Huang and Miller 1991) distributed in the FASTA package from W. Pearson. The program is also available on Web sites (see Table 3.1). Two features of these alignments are noteworthy: First, the highest-scoring alignment is similar to that found by the GAP program using a different amino acid substitution matrix and different gap penalties, with some minor variations in the more dissimilar regions and extension of the alignment farther into the amino-terminal domains. Second, by design, the alternative alignments never align the same amino acids and, in this example, the second and third alignments score much lower than the first one. These observations that strongly aligning regions are not significantly influenced by the scoring system, and that alternative high-scoring alignments are not possible, add convincing support that the initial alignment represents true similarity between these sequences. Another example of an alignment of these same sequences using ALIGN with a different scoring system is given on page 116.

A. GAP (Needleman-Wunsch algorithm)

Percent Similarity: 44.651   Percent Identity: 36.279

```
     1 MSTKKKPLTQEQLEDARRLKAIYEKKKNELGLSQESVADKMGMGQSGVGA 50
       | · |        | |       |     | · | · |   ·   |   · | · · · · ·
     1 MNT........QLMGER....IRARRKK.LKIRQAALGKMVGVSNVAISQ 37

    51 LFNGINALNAYNAALLAKILKVSVEEFSPSIAREIYEMYEAVSMQPSLRS 100
          | · |     | · | · |    | · : |   ·       |     · |
    38 WERSETEPNGENLLALSKALQCSPDYLLKGDLSQTNVAYHS...RHEPRG 84

   101 EYEYPVFSHVQAGMFSPELRTFTKGDAERWVSTTKKASDSAFWLEVEGNS 150
       | | · | | | | |      ·     | | | | | |:  · | | | : | · | |
    85 ..SYPLISWVSAGQWMEAVEPYHKRAIENWHDTTVDCSEDSFWLDVQGDS 132

   151 MTAPTGSKPSFPDGMLILVDPEQAVEPGDFCIARLGGD.EFTFKKLIRDS 199
       | | | | · |     | : | | · | | | | |    · | : | | · | | | | | | |  · | |
   133 MTAPAG..LSIPEGMIILVDPEVEPRNGKLVVAKLEGENEATFKKLVMDA 180

   200 GQVFLQPLNPQYPMIPCNESCSVVGKVIASQWPEETFG 237
       | · | | | | | | | | | |   | · | : : | | :  ·
   181 GRKFLKPLNPQYPMIEINGNCKIIGVVVDAKLAN..LP 216
```

B. BESTFIT (Smith-Waterman algorithm)

Percent Similarity: 58.871   Percent Identity: 48.387

```
   104 YPVFSHVQAGMFSPELRTFTKGDAERWVSTTKKASDSAFWLEVEGNSMTA 153
       | | · | | | |      · | | | | | |:  · | | | : | : | | | |
    86 YPLISWVSAGQWMEAVEPYHKRAIENWHDTTVDCSEDSFWLDVQGDSMTA 135

   154 PTGSKPSFPDGMLILVDPEQAVEPGDFCIARLGGD.EFTFKKLIRDSGQV 202
       | · |     | | : | | · | | | | |    | : | : | | : | | | | | |  | · | ·
   136 PAG..LSIPEGMIILVDPEVEPRNGKLVVAKLEGENEATFKKLVMDAGRK 183

   203 FLQPLNPQYPMIPCNESCSVVGKVIAS 229
       | | | | | | | | | | |   | · | : : | | :  ·
   184 FLKPLNPQYPMIEINGNCKIIGVVVDA 210
```

Figure 3.11. Example of local alignment of phage λ *cI* and phage P22 *c2* repressors by dynamic programming using the GCG GAP (Needleman-Wunsch algorithm) and BESTFIT (Smith-Waterman algorithm) programs. The log odds form of the PAM120 amino acid substitution matrix was used. PAM120 is optimal for proteins that are ~40% similar. The alignment reveals that the proteins are similar in the carboxy-terminal domain. The penalty for opening a gap in one of the sequences is 11 and for extending the gap 8; these were the default values assigned by the programs. Gaps at the unaligned ends of sequences were also weighted. In the program output, percent identity indicates the number of identical amino acids in the alignment, and percent similarity, the number of similar amino acids. Similar amino acids are defined by high-scoring matches between the amino acid pairs in the substitution matrix, and were defined at the time the program was run. The most similar pairs were indicated by a ':', less similar pairs by a '.' and unrelated pairs by a space, ' ', between the amino acid pairs. Although these dynamic programming programs provide a single optimal alignment, it is important to realize that a series of alignments are usually possible. Other programs, such as ALIGN in the FASTA set (Table 3.1 ALIGN-SITES), provide a user-specified number of alignments (see Fig. 3.12). Additionally, the alignments depend on the method used by the program to convert the traceback matrix into an alignment. GCG programs GAP and BESTFIT provide a method for printing two extremes of alignment, depending on whether gaps are favored in one sequence or the other. These options are called high road and low road.

USE OF SCORING MATRICES AND GAP PENALTIES IN SEQUENCE ALIGNMENTS

Amino Acid Substitution Matrices

Protein chemists discovered early on that certain amino acid substitutions commonly occur in related proteins from different species. Because the protein still functions with these substitutions, the substituted amino acids are compatible with protein structure and function. Often, these substitutions are to a chemically similar amino acid, but other changes also occur. Yet other substitutions are relatively rare. Knowing the types of changes that are most and least common in a large number of proteins can assist with predicting alignments for any set of protein sequences, as illustrated in Figure 3.13. If related

```
            LALIGN finds the best local alignments between two sequences
            version 2.0u64 March 1998
          Please cite:
            X. Huang and W. Miller (1991) Adv. Appl. Math. 12:373-381

          Comparison of:
          (A) lamc1.pro  LAMC1  REFORMAT of: cipro.pro  from: 1 - 237
          (B) p22c2.pro  P22C2  REFORMAT of: p22c2.pro  from: 1 - 216
          using matrix file: pam250.mat, gap penalties: -12/-2

            34.0% identity in 206 aa overlap; score:  338

                  30        40        50        60        70        80
          LAMC1  KKNELGLSQESVADKMGMGQSGVGALFNGINALNAYNAALLAKILKVSVEEFSPSIAREI
                 ....: .:.:..........:... ....  .. .. :. :  :.: :. : . .  . ...
          P22C2  RRKKLKIRQAALGKMVGVSNVAISQWERSETEPNGENLLALSKALQCSPDYLLKGDLSQT
                    20        30        40        50        60        70

                  90       100       110       120       130       140
          LAMC1  YEMYEAVSMQPSLRSEYEYPVFSHVQAGMFSPELRTFTKGDAERWVSTTKKASDSAFWLE
                 :..    ...    .:..: : :: .  .. .. : . :.: .:: . :...:::.
          P22C2  NVAYHSRHEPRG-----SYPLISWVSAGQWMEAVEPYHKRAIENWHDTTVDCSEDSFWLD
                   80        90       100       110       120

                 150       160       170       180       190       200
          LAMC1  VEGNSMTAPTGSKPSFPDGMLILVDPEQAVEPGDFCIARLGGD-EFTFKKLIRDSGQVFL
                 :.:.:::::.: :.:.::.:::::: .  :.. .:.:.:. : :::::::.::.:. ::
          P22C2  VQGDSMTAPAG--LSIPEGMIILVDPEVEPRNGKLVVAKLEGENEATFKKLVMDAGRKFL
                   130       140       150       160       170       180

                 210       220       230
          LAMC1  QPLNPQYPMIPCNESCSVVGKVIASQ
                 .:::::::::: :..:...: :....
          P22C2  KPLNPQYPMIEINGNCKIIGVVVDAK
                   190       200       210


          ----------

            17.8% identity in 90 aa overlap; score:   37

                  20        30        40        50        60        70
          LAMC1  RRLKAIYEKKKNELGLSQESVAD-KMGMGQSGVGALFNGINALNAYNAALLAKILKVSVE
                 ..::   . ..:.: .... .... ..:. .::.   . :: :..
          P22C2  KKLKIRQAALGKMVGVSNVAISQWERSETEPNGENLLALSKALQCSPDYLLKGDLSQTNV
                   20        30        40        50        60        70

                  80        90       100
          LAMC1  EF-SPSIAREIYEMYEAVSMQPSLRSEYEY
                 .. :.  .:. : .. ::   . . .  :
          P22C2  AYHSRHEPRGSYPLISWVSAGQWMEAVEPY
                   80        90       100


          ----------

            40.0% identity in 15 aa overlap; score:   36

                 220       230
          LAMC1  SCSVVGKVIASQWPE
                 :...... : :.:: :
          P22C2  SYPLISWVSAGQWME
                   90
```

Figure 3.12. Example of LALIGN program for finding multiple local alignments of two protein sequences. Three independent alignments of the phage λ and P22 repressors are shown. The amino acid substitution matrix used was the log odds form of the Dayhoff PAM250 matrix provided with the program, with a gap opening penalty of −12 and a gap extension penalty of −2.

protein sequences are quite similar, they are easy to align, and one can readily determine the single-step amino acid changes. If ancestor relationships among a group of proteins are assessed, the most likely amino acid changes that occurred during evolution can be predicted. This type of analysis was pioneered by Margaret Dayhoff (1978).

Amino acid substitution matrices or symbol comparison tables, as they are sometimes called, are used for such purposes. Although the most common use of such tables is for comparison of protein sequences, other tables of nucleic acid symbols are also used for comparison of nucleic acid sequences in order to accommodate ambiguous nucleotide

	Alignment			
sequence A	Tyr	Cys	Asp	Ala
sequence B	Phe	Met	Glu	Gly
BLOSUM62 matrix value	3	−1	2	0

Total score for alignment of sequence A with sequence B
= 3 − 1 + 2 + 0 = 4

Figure 3.13. Use of amino acid substitution matrix to evaluate an alignment of two protein sequences. The score for each amino acid pair (Tyr/Phe, etc.) is looked up in the BLOSUM62 matrix. Each value represents an odds score, the likelihood that the two amino acids will be aligned in alignments of similar proteins divided by the likelihood that they will be aligned by chance in an alignment of unrelated proteins. In a series of individual matches in an alignment, these odds scores are multiplied to give an overall odds score for the alignment itself. For convenience, odds scores are converted to log odds scores so that the values for amino acid pairs in an alignment may be summed to obtain the log odds score of the alignment. In this case, the logarithms are calculated to the base 2 and multiplied by 2 to give values designated as half-bits (a bit is the unit of an odds score that has been converted to a logarithm to the base 2). The value of 4 indicates that the 4 amino acid alignment is $2^{(4/2)} = 4$-fold more likely than expected by chance.

characters or models of expected sequence changes during different periods of evolutionary time that vary scoring of transitions and transversions.

In the amino acid substitution matrices, amino acids are listed both across the top of a matrix and down the side, and each matrix position is filled with a score that reflects how often one amino acid would have been paired with the other in an alignment of related protein sequences. The probability of changing amino acid A into B is always assumed to be identical to the reverse probability of changing B into A. This assumption is made because, for any two sequences, the ancestor amino acid in the phylogenetic tree is usually not known. Additionally, the likelihood of replacement should depend on the product of the frequency of occurrence of the two amino acids and on their chemical and physical similarities. A prediction of this model is that amino acid frequencies will not change over evolutionary time (Dayhoff 1978).

Dayhoff Amino Acid Substitution Matrices (Percent Accepted Mutation or PAM Matrices)

This family of matrices lists the likelihood of change from one amino acid to another in homologous protein sequences during evolution. There is presently no other type of scoring matrix that is based on such sound evolutionary principles as are these matrices. Even though they were originally based on a relatively small data set, the PAM matrices remain a useful tool for sequence alignment. Each matrix gives the changes expected for a given period of evolutionary time, evidenced by decreased sequence similarity as genes encoding the same protein diverge with increased evolutionary time. Thus, one matrix gives the changes expected in homologous proteins that have diverged only a small amount from each other in a relatively short period of time, so that they are still 50% or more similar. Another gives the changes expected of proteins that have diverged over a much longer period, leaving only 20% similarity. These predicted changes are used to produce optimal alignments between two protein sequences and to score the alignment. The assumption in this evolutionary model is that the amino acid substitutions observed over short periods of evolutionary history can be extrapolated to longer distances. The BLOSUM matrices (see below) are based on scoring substitutions found over a range of evolutionary periods and reveal that substitutions are not always as predicted by the PAM model.

In deriving the PAM matrices, each change in the current amino acid at a particular site is assumed to be independent of previous mutational events at that site (Dayhoff 1978). Thus, the probability of change of any amino acid **a** to amino acid **b** is the same, regardless of the previous changes at that site and also regardless of the position of amino acid **a** in a protein sequence. Amino acid substitutions in a protein sequence are thus viewed as a Markov model (see also hidden Markov models in Chapter 4), characterized by a series of changes of state in a system such that a change from one state to another does not depend on the previous history of the state. Use of this model makes possible the extrapolation of amino acid substitutions observed over a relatively short period of evolutionary time to longer periods of evolutionary time.

To prepare the Dayhoff PAM matrices, amino acid substitutions that occur in a group of evolving proteins were estimated using 1572 changes in 71 groups of protein sequences that were at least 85% similar. Because these changes are observed in closely related proteins, they represent amino acid substitutions that do not significantly change the function of the protein. Hence they are called "accepted mutations," defined as amino acid changes "accepted" by natural selection. Similar sequences were first organized into a phylogenetic tree, as illustrated in Figure 1.1 in Chapter 1. The number of changes of each amino acid into every other amino acid was then counted. To make these numbers useful for sequence analysis, information on the relative amount of change for each amino acid was needed.

Relative mutabilities were evaluated by counting, in each group of related sequences, the number of changes of each amino acid and by dividing this number by a factor, called the exposure to mutation of the amino acid. This factor is the product of the frequency of occurrence of the amino acid in that group of sequences being analyzed and the total number of all amino acid changes that occurred in that group per 100 sites. This factor normalizes the data for variations in amino acid composition, mutation rate, and sequence length. The normalized frequencies were then summed for all sequence groups. By these scores, Asn, Ser, Asp, and Glu were the most mutable amino acids, and Cys and Trp were the least mutable.

The above amino acid exchange counts and mutability values were then used to generate a 20 × 20 mutation probability matrix representing all possible amino acid changes. Because amino acid change was modeled by a Markov model, the mutation at each site being independent of the previous mutations, the changes predicted for more distantly related proteins that have undergone N mutations could be calculated. By this model, the PAM1 matrix could be multiplied by itself N times, to give transition matrices for comparing sequences with lower and lower levels of similarity due to separation of longer periods of evolutionary history. Thus, the commonly used PAM250 matrix represents a level of 250% of change expected in 2500 my. Although this amount of change seems very large, sequences at this level of divergence still have about 20% similarity. For example, alanine will be matched with alanine 13% of the time and with another amino acid 87% of the time.

Do not confuse this mutation probability form of the PAM250 matrix with the log odds form of the matrix described below.

The percentage of remaining similarity for any PAM matrix can be calculated by summing the percentages for amino acids not changing (Ala versus Ala, etc.) after multiplying each by the frequency of that amino acid pair in the database (e.g., 0.089 for Ala) (Dayhoff 1978). The PAM120, PAM80, and PAM60 matrices should be used for aligning sequences that are 40%, 50%, and 60% similar, respectively. Simulations by George et al. (1990) have shown that, as predicted, the PAM250 matrix provides a better-scoring alignment than lower-numbered PAM matrices for distantly related proteins of 14–27% similarity.

PAM matrices are usually converted into another form, called log odds matrices. The odds score represents the ratio of the chance of amino acid substitution by two different hypotheses—one that the change actually represents an authentic evolutionary variation at that site (the numerator), and the other that the change occurred because of random sequence variation of no biological significance (the denominator). Odds ratios are converted to logarithms to give log odds scores for convenience in multiplying odds scores of amino acid pairs in an alignment by adding the logarithms (Fig. 3.13).

Example: Calculations for obtaining the log odds score for changes between Phe and Tyr at an evolutionary distance of 250 PAMs

1. Of 1572 observed amino acid changes, there were 260 changes between Phe and Tyr. These numbers were multiplied by (1) the relative mutability of Phe (see text), and (2) the fraction of Phe to Tyr changes over all changes of Phe to any other amino acid (since Phe to Tyr and Tyr to Phe changes are not distinguished in the original mutation counts, sums of changes are used to calculate the fraction) to obtain a mutation probability score of Phe to Tyr. A similar score was obtained for changes of Phe to each of the other 18 amino acids, and also for the calculated probability of not changing at all. The resulting 20 scores were summed and divided by a normalizing factor such that their sum represented a probability of change of 1%, as illustrated in Table 3.2.

   In this matrix, the score for changing Phe to Tyr was 0.0021, as opposed to a score of Phe not changing at all of 0.9946, as shown in Table 3.2. These calculations were repeated for Tyr changing to any other amino acid. The score for changing Tyr to Phe was 0.0028, and that of not changing Tyr was 0.9946 (not shown). These scores were placed in the PAM1 matrix, in which the overall probability of each amino acid changing to another is ~1%, and that of each not changing is ~99%.

2. The above PAM1 matrix was multiplied by itself 250 times to obtain the distribution of changes expected for 250 PAMs of evolutionary change. These changes can include both forward changes to another amino acid and reverse changes to a former one. At this distance, the probability of change of Phe to Tyr was 0.15 as opposed to a probability of 0.32 of no change in Phe. The corresponding probabilities for Tyr to Phe at 250 PAMs were 0.20 and 0.31 for no change.

3. The log odds values for changes between Phe and Tyr were then calculated. The Phe-Tyr score in the 250 PAM matrix, 0.15, was divided by the frequency of Phe in the sequence data, 0.040, to give the relative frequency of change. This ratio, $0.15/0.04 = 3.75$, was converted to a logarithm to the base 10 ($\log_{10}3.75 = 0.57$) and multiplied by 10 to remove fractional values ($0.57 \times 10 = 5.7$). Similarly, the Tyr to Phe score is $0.20/0.03 = 6.7$, and the logarithm of this number is $\log_{10}6.7 = 0.83$, and multiplied by 10 ($0.83 \times 10 = 8.3$). The average of 5.7 and 8.3 is 7, the number entered in the log odds table for changes between Phe and Tyr at 250 PAMs of evolutionary distance.

   The log odds from the PAM250 matrix, which is sometimes referred to as the mutation data matrix (MDM) at 250 PAMs and also as MDM_{78}, is shown in Figure 3.14. The log odds scores in this table lie within the range of −8 to +17. A value of 0 indicates that the frequency of the substitution between a matched pair of amino acids in related proteins is as expected by chance; a value less than 0 or greater than 0 indicates that the frequency is less than or greater than that expected by chance, respectively. Using such a matrix, a high positive score between two amino acids means that the pair is more likely to be found aligned in sequences that are derived from a common ancestor, i.e., homologous, than in unrelated or nonhomologous sequences. The highest-scoring replacements are for amino acids whose side chains are chemically similar, as might be expected if the amino acid substitution is not to impede function. In the original data, the largest number of observed changes (83) was between Asp (D) and Glu (E). This number is reflected as a log odds score of +3 in the MDM. Many changes were not observed. For example, there were no changes between Gly (G) and Trp (W), resulting in a score of −7 in the table.

Table 3.2. *Normalized probability scores for changing Phe to any other amino acid (or of not changing) at PAM1 and PAM250 evolutionary distances*

Amino acid change	PAM1	PAM250
Phe to Ala	0.0002	0.04
Phe to Arg	0.0001	0.01
Phe to Asn	0.0001	0.02
Phe to Asp	0.0000	0.01
Phe to Cys	0.0000	0.01
Phe to Gln	0.0000	0.01
Phe to Glu	0.0000	0.01
Phe to Gly	0.0001	0.03
Phe to His	0.0002	0.02
Phe to Ile	0.0007	0.05
Phe to Leu	0.0013	0.13
Phe to Lys	0.0000	0.02
Phe to Met	0.0001	0.02
Phe to Phe	0.9946	0.32
Phe to Pro	0.0001	0.02
Phe to Ser	0.0003	0.03
Phe to Thr	0.0001	0.03
Phe to Trp	0.0001	0.01
Phe to Tyr	0.0021	0.15
Phe to Val	0.0001	0.05
SUM[a]	1.0000	1.00

[a]Approximate since scores are rounded off.

The multiplication of two PAM1 matrices to give a PAM2 matrix. Only three rows and columns are shown for illustrative purposes.

$$
\begin{array}{cc|ccc}
 & & aa1 & aa2 & aa3 \rightarrow \\
\hline
aa1 & & a & b & c \\
aa2 & & d & e & f \\
aa3 & & g & h & i \\
\downarrow & & & &
\end{array}
\times
\begin{array}{cc|ccc}
 & & aa1 & aa2 & aa3 \rightarrow \\
\hline
aa1 & & a & b & c \\
aa2 & & d & e & f \\
aa3 & & g & h & i \\
\downarrow & & & &
\end{array}
$$

$$
=
\begin{array}{cc|ccc}
 & & aa1 & aa2 & aa3 \rightarrow \\
\hline
aa1 & & A & B & C \\
aa2 & & D & E & F \\
aa3 & & G & H & I \\
\downarrow & & & &
\end{array}
\qquad
\begin{aligned}
A &= a^2 + bd + cg + \dots \\
B &= ab + be + ch + \dots \\
C &= ac + bf + ci + \dots \\
D &= da + ed + fg + \dots, \text{ etc.}
\end{aligned}
$$

		C	S	T	P	A	G	N	D	E	Q	H	R	K	M	I	L	V	F	Y	W		
C		12																					C
S		0	2																				S
T		-2	1	3																			T
P		-3	1	0	6																		P
A		-2	1	1	1	2																	A
G		-3	1	0	-1	1	5																G
N		-4	1	0	-1	0	0	2															N
D		-5	0	0	-1	0	1	2	4														D
E		-5	0	0	-1	0	0	1	3	4													E
Q		-5	-1	-1	0	0	-1	1	2	2	4												Q
H		-3	-1	-1	0	-1	-2	2	1	1	3	6											H
R		-4	0	-1	0	-2	-3	0	-1	-1	1	2	6										R
K		-5	0	0	-1	-1	-2	1	0	0	1	0	3	5									K
M		-5	-2	-1	-2	-1	-3	-2	-3	-2	-1	-2	0	0	6								M
I		-2	-1	0	-2	-1	-3	-2	-2	-2	-2	-2	-2	-2	2	5							I
L		-6	-3	-2	-3	-2	-4	-3	-4	-3	-2	-2	-3	-3	4	2	6						L
V		-2	-1	0	-1	0	-1	-2	-2	-2	-2	-2	-2	-2	2	4	2	4					V
F		-4	-3	-3	-5	-4	-5	-4	-6	-5	-5	-2	-4	-5	0	1	2	-1	9				F
Y		0	-3	-3	-5	-3	-5	-2	-4	-4	-4	0	-4	-4	-2	-1	-1	-2	7	10			Y
W		-8	-2	-5	-6	-6	-7	-4	-7	-7	-5	-3	2	-3	-4	-5	-2	-6	0	0	17		W
		C	S	T	P	A	G	N	D	E	Q	H	R	K	M	I	L	V	F	Y	W		

Figure 3.14. The log odds form (the mutation data matrix or MDM) of the PAM250 scoring matrix. Amino acids are grouped according to the chemistry of the side group: (C) sulfhydryl, (STPAG) small hydrophilic, (NDEQ) acid, acid amide and hydrophilic, (HRK) basic, (MILV) small hydrophobic, and (FYW) aromatic. Each matrix value is calculated from an odds score, the probability that the amino acid pair will be found in alignments of homologous proteins divided by the probability that the pair will be found in alignments of unrelated proteins by random chance. The logarithm of these ODDS scores to the base 10 is multiplied by 10 and then used as the table value (see text for details). Thus, +10 means the ancestor probability is greater, 0 that the probabilities are equal, and −4 that the alignment is more often a chance one than due to an ancestor relationship. Because these numbers are logarithms, they may be added to give a combined probability of two or more amino acid pairs in an alignment. Thus, the probability of aligning two Ys in an alignment YY/YY is 10 + 10 = 20, a very significant score, whereas that of YY with TP is −2 −5 = − 7, a rare and unexpected alignment between homologous sequences.

At one time, the PAM250 scoring matrix was modified in an attempt to improve the alignment obtained. All scores for matching a particular amino acid were normalized to the same mean and standard deviation, and all amino acid identities were given the same score to provide an equal contribution for each amino acid in a sequence alignment (Gribskov and Burgess 1986). These modifications were included as the default matrices for the GCG sequence alignment programs in versions 8 and earlier and are optional in later versions. They are not recommended because they will not give an optimal alignment that is in accord with the evolutionary model.

Choosing the Best PAM Scoring Matrices for Detecting Sequence Similarity. The ability of PAM scoring matrices to distinguish statistically between chance and biologically meaningful alignments has been analyzed using a recently developed statistical theory for sequences (Altschul 1991) that is discussed later in this chapter. As discussed above, each PAM matrix is designed to score alignments between sequences that have diverged by a particular degree of evolutionary distance. Altschul (1991) has examined how well the PAM matrices actually can distinguish proteins that have diverged to a greater or lesser extent, when these proteins are subjected to a local alignment.

Initially, when using a scoring matrix to produce an alignment, the amount of similarity between sequences may not be known. However, the ungapped alignment scores obtained are maximal when the correct PAM matrix, i.e., the one corresponding to the degree of similarity in the target sequences, is used (Altschul 1991). Altschul (1991) has also examined the ability of PAM matrices to provide a reliable enough indication of an ungapped local alignment score between sequences on an initial attempt of alignment. For sequence alignments, the PAM200 matrix is able to detect a significant ungapped alignment of 16–62 amino acids whose score is within 87% of the optimal one. Alternatively, several combinations, such as PAM80 and PAM250 or PAM120 and PAM350, can also be used. Altschul (1993) has also proposed using a single matrix and adjusting a statistical parameter in the scoring system to reach more distantly related sequences, but this change would primarily be for database searches.

Scoring matrices are also used in database searches for similar sequences. The optimal matrices for these searches have also been determined (see book Web site and Chapter 7). It is important to remember that these predictions assume that the amino acid distributions in the set of protein families used to make the scoring matrix are representative of all families that are likely to be encountered. The original PAM matrices represent only a small number of families. Scoring matrices obtained more recently, such as the BLOSUM matrices, are based on a much larger number of protein families. BLOSUM matrices are not based on a PAM evolutionary model in which changes at large evolutionary distance are predicted by extrapolation of changes found at small distances. Matrix values are based on the observed frequency of change in a large set of diverse proteins. As is discussed on the book Web site, the BLOSUM scoring matrices (especially BLOSUM62) appear to capture more of the distant types of variations found in protein families.

In addition to the aforementioned differences among PAM scoring matrices for scoring alignments of more- or less-related proteins, the ability of each PAM matrix to discriminate real local alignments from chance alignments also varies. To calculate the ability of the entire matrix to discriminate related from unrelated sequences (H, the relative entropy), the score for each amino acid pair s_{ij} (in units of $\log_2$, called bits) is multiplied by the probability of occurrence of that pair in the original dataset, q_{ij} (Altschul 1991). This weighted score is then summed over all of the amino acid pairs to produce a score that represents the ability of the average amino acid pair in the matrix to discriminate actual from chance alignments.

$$H = \sum_{i=1}^{20} \sum_{j=1}^{i} q_{ij} \times s_{ij} \tag{3}$$

In information theory, this score is called the average mutual information content per pair, and the sum over all pairs is the relative entropy of the matrix (termed H). The relative entropy will be a small positive number. For the PAM250 matrix the number is $+0.36$, for PAM120, $+0.98$, and for PAM160, $+0.70$. In general, all other factors being equal, the higher the value of H for a scoring matrix, the more likely it is to be able to distinguish real from chance alignments.

Analysis of the Dayhoff Model of Protein Evolution as Used in PAM Matrices. As outlined above, the Dayhoff model of protein evolution is a Markov process. In this model, each amino acid site in a protein can change at any time to any of the other 20 amino acids with probabilities given by the PAM table, and the changes that occur at each site are independent of the amino acids found at other sites in the protein and depend only on the current amino acid at the site. The assumptions that underlie the method of constructing the Dayhoff scoring matrix have been challenged (for discussion, see George et al. 1990; States and Boguski 1991). First, it is assumed that each amino acid position is equally mutable, whereas, in fact, sites vary considerably in their degree of mutability. Mutagenesis hot spots are well known in molecular genetics, and variations in mutability of different amino acid sites in proteins are well known.

The more conserved amino acids in similar proteins from different species are ones that play an essential role in structure and function and the less conserved are in sites that can vary without having a significant effect on function. Thus, there are many factors that influence both the location and types of amino acid changes that occur in proteins. Wilbur (1985) has tested the Markov model of evolution (see box, below) and has shown that it can be valid if certain changes are made in the way that the PAM matrices are calculated.

Test of Markov Model of Evolution in Proteins

To test the model, Wilbur addressed a major criticism of the PAM scoring matrix, namely that the frequency of amino acid changes that require two nucleotide changes is higher than would be expected by chance. About 20% of the observed amino acid changes require more than a single mutation for the necessary codon changes. This fraction is far greater than would be expected by chance.

To correct for changes that require at least two mutations, Wilbur recalculated the PAM1 matrix using only amino acid substitution data from 150 amino acid pairs that are accountable by single mutations. To accomplish this calculation, he used a refined mathematical model that provided a more precise measure of the rate of substitution. He then estimated frequencies of the other 230 amino acid substitutions reachable only by at least two mutations, and compared these frequencies to the values calculated by Dayhoff, who had assumed these were single-step changes. If these numbers agreed, argued Wilbur, then the PAM model used to produce the Dayhoff matrix is a reliable one. In fact, the Dayhoff values exceeded the two-step model values by a factor of about 117. One source of discrepancy was the assumption that the two-step changes were a linear function of evolutionary time over short evolutionary periods of 1 PAM (average time of 1 PAM = 10 my), whereas, because two mutations are required to make the change, a quadratic function is expected. With this correction made to the Dayhoff calculations for amino acid substitutions requiring two mutations, agreement with the two-step model improved about 10-fold, leaving another 11.7-fold unaccounted for.

Wilbur analyzed the remainder by the covarion hypothesis (Fitch and Markowitz 1970; Miyamoto and Fitch 1995), in which it is assumed that only a certain fraction of amino acid sites in a protein are variable and that one site influences another. Thus, a change in one site may influence the variability of others. This model seems to be reasonable from many biological perspectives. The prediction of this hypothesis is that the frequency of two-step changes would be overestimated because we did not take into account the failure of many sites to be mutable. Using a reasonable estimate of 0.3 for the fraction of the sites that could change, the effect on the Dayhoff calculations for frequencies of two-step changes would be 3.3-fold. The remaining discrepancy in the 11.7-fold ratio between Dayhoff values and two-step values may be attributable to variations in mutation rates from site to site, or to the exclusion of certain amino acids at a particular site. In conclusion, Wilbur (1985) has shown that the Dayhoff model for protein evolution appears to give predictable and consistent

results, but that frequencies of change between amino acids that require two mutational steps must be calculated as a two-step process. Failure to do so generates errors
due to variations in site-to-site mutability. George et al. (1990) have counterargued
that it has never been demonstrated that two independent mutations must occur,
each becoming established in a population before the next appears.

A further criticism of the PAM scoring matrices is that they are not more useful
for sequence alignment than simpler matrices, such as one based on a chemical grouping of amino acid side chains. Although alignment of related proteins is straightforward
and quite independent of the symbol comparison scoring scheme, alignments of less-
related proteins are much more speculative (Feng et al. 1985). These matrices and the
BLOSUM matrices have been very useful for finding more distantly related sequences
(George et al. 1990). There have been recent changes in the way that members of protein
families are identified (see Chapters 4 and 9). Once a family has been identified, family-
specific scoring matrices can be produced, and there is no point in using these general
matrices. As described in Chapter 4, a scoring matrix representing a section of aligned
sequences with no gaps, or a matrix representing a section of aligned sequences with
matches, mismatches, and gaps (a profile), are the best tools to search for more family
members.

Another criticism of the PAM matrix is that constructing phylogenetic relationships
prior to scoring mutations has limitations, due to the difficulty of determining ancestral
relationships among sequences, a topic discussed in Chapter 6. Early on in the Dayhoff
analysis, the evolutionary trees were estimated by a voting scheme for the branches in the
tree, each node being estimated by the most abundant amino acid in distal parts of the tree.
Once available, the PAM matrices were used to estimate the evolutionary distance between
proteins, given the amount of sequence similarity. Such data can be used to produce a tree
based on evolutionary distances (Chapter 6). This circular analysis of using alignments to
score amino acid changes and then to use the matrices to produce new alignments has also
been criticized. However, no method has yet been devised in any type of sequence analysis
for completely circumventing this problem. Evidence that the values in the scoring matrix
are insensitive to changes in the phylogenetic relationships has been provided (George et
al. 1990).

Finally, the Dayhoff PAM matrices have been criticized because they are based on a
small set of closely related proteins. The Dayhoff data set has been augmented to include
the 1991 protein database (Gonnet et al. 1992; Jones et al. 1992). The ability of the Dayhoff
matrices to identify homologous sequences has also been extensively compared to that of
other scoring matrices. These comparisons are discussed on the book Web site.

Blocks Amino Acid Substitution Matrices (BLOSUM)

The BLOSUM62 substitution matrix (Henikoff and Henikoff 1992) is widely used for scoring protein sequence alignments. The matrix values are based on the observed amino acid
substitutions in a large set of ~2000 conserved amino acid patterns, called blocks. These
blocks have been found in a database of protein sequences representing more than 500
families of related proteins (Henikoff and Henikoff 1992) and act as signatures of these
protein families. The BLOSUM matrices are thus based on an entirely different type of
sequence analysis and a much larger data set than the Dayhoff PAM matrices.

These protein families were originally identified by Bairoch in the Prosite catalog. This catalog provides lists of proteins that are in the same family because they have a similar biochemical function. For each family, a pattern of amino acids that are characteristic of that function is provided. Henikoff and Henikoff (1991) examined each Prosite family for the presence of ungapped amino acid patterns (blocks) that were present in each family and that could be used to identify members of that family. To locate these patterns, the sequences of each protein family were searched for similar amino acid patterns by the MOTIF program of H. Smith (Smith et al. 1990), which can find patterns of the type aa1 d1 aa2 d2 aa3, where aa1 and aa2 are conserved amino acids and d1 and d2 are stretches of intervening sequence up to 24 amino acids long located in all sequences. These initial patterns were organized into larger ungapped patterns (blocks) between 3 and 60 amino acids long by the Henikoffs' PROTOMAT program (http://www.blocks.fhcrc.org). Because these blocks were present in all of the sequences in each family, they could be used to identify other members of the same family. Thus, the family collections were enlarged by searching the sequence databases for more proteins with these same conserved blocks.

The blocks that characterized each family provided a type of multiple sequence alignment for that family. The amino acid changes that were observed in each column of the alignment could then be counted. The types of substitutions were then scored for all aligned patterns in the database and used to prepare a scoring matrix, the BLOSUM matrix, indicating the frequency of each type of substitution. As previously described for the PAM matrices, BLOSUM matrix values were given as logarithms of odds scores of the ratio of the observed frequency of amino acid substitutions divided by the frequency expected by chance. An example of the calculations is shown in Figure 3.15.

This procedure of counting all of the amino acid changes in the blocks, however, can lead to an overrepresentation of amino acid substitutions that occur in the most closely related members of each family. To reduce this dominant contribution from the most alike sequences, these sequences were grouped together into one sequence before scoring the amino acid substitutions in the aligned blocks. The amino acid changes within these clustered sequences were then averaged. Patterns that were 60% identical were grouped together to make one substitution matrix called BLOSUM60, and those 80% alike to make another matrix called BLOSUM80, and so on. As with the PAM matrices, these matrices differ in the degree to which the more common amino acid pairs are scored relative to the less common pairs. Thus, when used for aligning protein sequences, they provide a greater or lesser distinction between the more common and less common amino acid pairs. The ability of these different BLOSUM matrices to distinguish real from chance alignments and to identify as many members as possible of a protein family has been determined (Henikoff and Henikoff 1992).

Two types of analyses were performed: (1) an information content analysis of each matrix, as was described above for the PAM matrices, and (2) an actual comparison of the ability of each matrix to find members of the same families in a database search, discussed below. As the clustering percentage was increased, the ability of the resulting matrix to distinguish actual from chance alignments, defined as the relative entropy of the matrix or the average information content per residue pair (see above), also increased. As clustering increased from 45% to 62%, the information content per residue increased from ~0.4 to 0.7 bits per residue, and was ~1.0 bits at 80% clustering. However, at the same time, the number of blocks that contributed information decreased by 25% between no clustering and 62% clustering. BLOSUM62 represents a balance between information content and data size. The BLOSUM62 matrix is shown in Figure 3.16.

```
...A...
...A...
...A...
...A...
...S...
...A...
...A...
...A...
...A...
...A...
```

Figure 3.15. Derivation of the matrix values in the BLOSUM62 scoring matrix. As an example of the calculations, if a column in one of the blocks consisted of 9 A and 1 S amino acids, the following is true for this data set (see Henikoff and Henikoff 1992).

1. Since the original sequence from which the others were derived is not known, each column position has to be considered a possible ancestor of the other nine columns. Hence, there are $8+7+6...+1 = 36$ possible AA pairs (f_{AA}) and 9 possible AS pairs (f_{AS}) to be compared.

2. There are $20+19+18+...+1 = 210$ possible amino acid pairs.

3. The frequency of occurrence of an AA pair, $q_{AA} = f_{AA}/(f_{AA} + f_{AS}) = 36/(36+9) = 0.8$, and that of an AS pair, $q_{AS} = f_{AS}/(f_{AA} + f_{AS}) = 9/(36+9) = 0.2$.

4. The expected frequency of A being in a pair, $p_A = (q_{AA} + q_{AS}/2) = 0.8 + 0.2/2 = 0.9$, and that of $p_S = q_{AS}/2 = 0.1$.

5. The expected frequency of occurrence of AA pairs, $e_{AA} = p_A \times p_A = 0.9 \times 0.9 = 0.81$, and that of AS, $e_{AS} = 2 \times p_S \times p_A = 2 \times 0.9 \times 0.1 = 0.18$.

6. The matrix entry for AA will be calculated from the ratio of the occurrence frequency to the expected frequency. For AA, ratio = $q_{AA}/e_{AA} = 0.8/0.81 = 0.99$, and for AS, ratio = $q_{AS}/e_{AS} = 0.2/0.18 = 1.11$.

7. Both ratios are converted to logarithms to the base 2 and then multiplied by 2 (1/2 bit units). Matrix entry for AA, $s_{AA} = \log_2(q_{AA}/e_{AA}) = -0.04$, and for AS, $s_{AS} = \log_2(q_{AS}/e_{AS}) = 0.30$. These logarithms are both rounded to 1 $^1/_2$ bit unit.

Henikoff and Henikoff (1993) have prepared a set of interval BLOSUM matrices that represent the changes observed between more closely related or more distantly related representatives of each block. Rather than representing the changes observed in very alike sequences up to sequences that were *n*% alike to give a BLOSUM-*n* matrix, the new BLOSUM-*nm* matrix represented the changes observed in sequences that were between *n*% alike and *m*% alike. The idea behind these matrices was to have a set of matrices corresponding to amino acid changes in sequence blocks that are separated by different evolutionary distances.

Comparison of the PAM and BLOSUM Amino Acid Substitution Matrices

There are several important differences in the ways that the PAM and BLOSUM scoring matrices were derived, and these differences should be appreciated in order to interpret the results of protein sequence alignments obtained with these matrices. First, the PAM matrices are based on a mutational model of evolution that assumes amino acid changes occur as a Markov process, each amino acid change at a site being independent of previous changes at that site. Changes are scored in sequences that are 85% similar after predicting

	C	S	T	P	A	G	N	D	E	Q	H	R	K	M	I	L	V	F	Y	W	
C	9																				C
S	-1	4																			S
T	-1	1	5																		T
P	-3	-1	-1	7																	P
A	0	1	0	-1	4																A
G	-3	0	-2	-2	0	6															G
N	-3	1	0	-2	-2	0	6														N
D	-3	0	-1	-1	-2	-1	1	6													D
E	-4	0	-1	-1	-1	-2	0	2	5												E
Q	-3	0	-1	-1	-1	-2	0	0	2	5											Q
H	-3	-1	-2	-2	-2	-2	1	-1	0	0	8										H
R	-3	-1	-1	-2	-1	-2	0	-2	0	1	0	5									R
K	-3	0	-1	-1	-1	-2	0	-1	1	1	-1	2	5								K
M	-1	-1	-1	-2	-1	-3	-2	-3	-2	0	-2	-1	-1	5							M
I	-1	-2	-1	-3	-1	-4	-3	-3	-3	-3	-3	-3	-3	1	4						I
L	-1	-2	-1	-3	-1	-4	-3	-4	-3	-2	-3	-2	-2	2	2	4					L
V	-1	-2	0	-2	0	-3	-3	-3	-2	-2	-3	-3	-2	1	3	1	4				V
F	-2	-2	-2	-4	-2	-3	-3	-3	-3	-3	-1	-3	-3	0	0	0	-1	6			F
Y	-2	-2	-2	-3	-2	-3	-2	-3	-2	-1	2	-2	-2	-1	-1	-1	-1	3	7		Y
W	-2	-3	-2	-4	-3	-2	-4	-4	-3	-2	-2	-3	-3	-1	-3	-2	-3	1	2	11	W
	C	S	T	P	A	G	N	D	E	Q	H	R	K	M	I	L	V	F	Y	W	

Figure 3.16. The BLOSUM62 amino acid substitution matrix. The amino acids in the table are grouped according to the chemistry of the side group: (C) sulfhydryl, (STPAG) small hydrophilic, (NDEQ) acid, acid amide, and hydrophilic, (HRK) basic, (MILV) small hydrophobic, and (FYW) aromatic. Each entry is the logarithm of the odds score, found by dividing the frequency of occurrence of the amino acid pair in the BLOCKS database (after sequences 62% or more in similarity have been clustered) by the likelihood of an alignment of the amino acids by random chance. The denominator in this ratio is calculated from the frequency of occurrence of each of the two individual amino acids in the BLOCKS database and provides a measure of a chance alignment of the two amino acids. The actual/expected ratio is expressed as a log odds score in so-called half-bit units, obtained by converting the odds ratio to a logarithm to the base 2, and then multiplying by 2. A zero score means that the frequency of the amino acid pair in the database is as expected by chance, a positive score that the pair is found more often than by chance, and a negative score that the pair is found less often than by chance. The accumulated score of an alignment of several amino acids in two sequences may be obtained by adding up the respective scores of each individual pair of amino acids. As with the PAM250-derived matrix, the highest-scoring matches are between amino acids that are in the same chemical group, and the very highest-scoring matches are for cysteine–cysteine matches and for matches among the aromatic amino acids. Compared to the PAM160 matrix, however, the BLOSUM62 matrix gives a more positive score to mismatches with the rare amino acids, e.g., cysteine, a more positive score to mismatches with hydrophobic amino acids, but a more negative score to mismatches with hydrophilic amino acids (Henikoff and Henikoff 1992).

a phylogenetic history of the changes in each family. Thus, the PAM matrices are based on prediction of the first changes that occur as proteins diverge from a common ancestor during evolution of a protein family. Matrices that may be used to compare more distantly related proteins are then derived by extrapolation from these short-term changes, assuming that these more distant changes are a reflection of the short-term changes occurring over and over again. For each longer evolutionary interval, each amino acid can change to any other with the same frequency as observed in the short term. In contrast, the BLOSUM matrices are not based on an explicit evolutionary model. They are derived from considering all amino acid changes observed in an aligned region from a related family of proteins, regardless of the overall degree of similarity between the protein sequences. However, these proteins are known to be related biochemically and, hence, should share common ancestry. The evolutionary model implied in such a scheme is that the proteins in each family share a common origin, but closer versus distal relationships are ignored, as if they all were derived equally from the same ancestor, called a starburst model of protein evolution (see Chapter 6). Second, the PAM matrices are based on scoring all amino acid positions in related sequences, whereas the BLOSUM matrices are based on substitutions and conserved positions in blocks, which represent the most alike common regions in related sequences. Thus, the PAM model is designed to track the evolutionary origins of proteins, whereas the BLOSUM model is designed to find their conserved domains.

Other Amino Acid Scoring Matrices

In addition to the Dayhoff PAM, and related Gonnet et al. (1992), Benner et al. (1994), and Jones et al. (1992) matrices and the BLOSUM matrices, a number of other amino acid substitution matrices have been used for producing protein sequence alignments, and several representative ones are listed in Table 3.3. For a more complete list and comparison, see Vogt et al. (1995). These tables vary from a comparison of simple chemical properties of amino acids to a complex analysis of the substitutions found in secondary structural domains of proteins. Because most of these tables are designed to align proteins on the basis of some such feature of the amino acids, and not on an evolutionary model, they are not particularly suitable for evolutionary analysis. They can be very useful, however, for discovering structural and functional relationships, or family relationships among proteins. A sequence alignment program that uses a combination of these tables has been found to be particularly useful for detecting distant protein relationships (Argos 1987; Rechid et al. 1989). There have been extensive comparisons of the usefulness of various amino acid substitution matrices for aligning sequences, for finding similar sequences in a protein sequence database, or for aligning similar sequences based on structure that are described on the book Web site.

Table 3.3. *Criteria used in amino acid scoring matrices for sequence alignments*

1. Simple identity, which scores only identical amino acids as a match and all others as a mismatch.
2. Genetic code changes, which score the minimum number of nucleotide changes to change a codon for one amino acid into a codon for another, due to Fitch (1966), and also with added information based on structural similarity of amino acid side chains (Feng et al. 1985). A similar matrix based on the assumption that genetic code is the only factor influencing amino acid substitutions has been produced (Benner et al. 1994).
3. Matrices based on chemical similarity of amino acid side chains, molecular volume, and polarity and hydrophobicity of amino acid side chains (see Vogt et al. 1995).
4. Amino acid substitutions in structurally aligned three-dimensional structures (Risler et al. 1988; matrix JO93, Johnson and Overington 1993). A similar matrix was described by Henikoff and Henikoff (1993). Sander and Schneider (1991) prepared a similar matrix based on these same substitutions but augmented by substitutions found in proteins which are so similar to the structure-solved group that they undoubtedly have the same three-dimensional structure.
5. Gonnet et al. (1994) have prepared a 400 × 400 dipeptide substitution matrix for aligning proteins based on the possibility that amino acid substitutions at a particular site are influenced by neighboring amino acids, and thus that the environment of an amino acid plays a role in protein evolution.
6. Jones et al. (1994) have prepared a scoring matrix specifically for transmembrane proteins. This matrix was prepared using an analysis similar to that used for preparing the original Dayhoff PAM matrices, and therefore provides an estimate of evolutionary distances among members of this class of proteins.

Nucleic Acid PAM Scoring Matrices

Just as amino acid scoring matrices have been used to score protein sequence alignments, nucleotide scoring matrices for scoring DNA sequence alignments have also been developed. The DNA matrix can incorporate ambiguous DNA symbols (see Table 2.1) and information from mutational analysis, which reveals that transitions (substitutions between the purines A and G or between the pyrimidines C and T) are more probable than transversions (substitutions between purine to pyrimidine or pyrimidine to purine) (Li and Graur 1991). These substitution matrices may be used to produce global or local alignments of DNA sequences.

States et al. (1991) have developed a series of nucleic acid PAM matrices based on a Markov transition model similar to that used to generate the Dayhoff PAM scoring matrices. Although designed to improve the sensitivity of similarity searches of sequence databases, these matrices also may be used to score nucleic acid alignments. The advantage of using these matrices is that they are based on a defined evolutionary model and that the statistical significance of alignment scores obtained by local alignment programs may be evaluated, as described later in this chapter.

To prepare these DNA PAM matrices, a PAM1 mutation matrix representing 99% sequence conservation and one PAM of evolutionary distance (1% mutations) was first calculated. For a model in which all mutations from any nucleotide to any other are equally likely, and in which the four nucleotides are present at equal frequencies, the four diagonal elements of the PAM1 matrix representing no change are 0.99 whereas the six other elements representing change are 0.00333 (Table 3.4). The values are chosen so that the sum of all possible changes for a given nucleotide in the PAM1 matrix is 1% ($3 \times 0.00333 = 0.00999$). For a biased mutation model in which a given transition is threefold more likely than a transversion (Table 3.4), the off-diagonal matrix elements corresponding to the one possible transition for each nucleotide are 0.006 and those for the two possible transversions are 0.002, and the sum for each nucleotide is again 1% ($0.006 + 0.002 + 0.002 = 0.01$).

As with the amino acid matrices, the above matrix values are then used to produce log odds scoring matrices that represent the frequency of substitutions expected at increasing

Table 3.4. *Nucleotide mutation matrix for an evolutionary distance of 1 PAM, which corresponds to a probability of a change at each nucleotide position of 1%*

A. Model of uniform mutation rates among nucleotides			
	A	**G**	**T**	**C**
A	0.99			
G	0.00333	0.99		
T	0.00333	0.00333	0.99	
C	0.00333	0.00333	0.00333	0.99

B. Model of threefold higher transitions than transversions			
	A	**G**	**T**	**C**
A	0.99			
G	0.006	0.99		
T	0.002	0.002	0.99	
C	0.002	0.002	0.006	0.99

Values are frequency of change at each site, or of no change for all base combinations.

evolutionary distances. In terms of an alignment, the probability (s_{ij}) of obtaining a match between nucleotides i and j, divided by the random probability of aligning i and j, is given by

$$s_{ij} = \log (p_i M_{ij} / p_i p_j) \tag{4}$$

where M_{ij} is the value in the mutation matrix given in Table 3.4, and p_i and p_j are the fractional composition of each nucleotide, assumed to be 0.25. The base of the logarithm can be any value, corresponding to multiplying every value in the matrix by the same constant. With such scaling variations, the ability of the matrix to distinguish among significant and chance alignments will not be altered. The resulting tables with s_{ij} expressed in units of bits (logarithm to the base 2) and rounded off to the nearest whole integer are shown in Table 3.5.

From these PAM1 matrices, additional log odds matrices at an evolutionary distance of n PAMs may be obtained by multiplying the PAM1 matrix by itself n times. The ability of each matrix to distinguish real from random nucleotide matches in an alignment, designated H, measured in bit units ($\log_2$) can be calculated using the equation

$$H = \sum_{i,j} p_i p_j s_{ij} 2^{s_{ij}} \tag{5}$$

where the s_{ij} scores are also expressed in bit units. In Table 3.6 are shown the log odds values of the match and mismatch scores for PAM matrices at increasing evolutionary distances, assuming a uniform rate of mutation among all nucleotides. Also shown is the percentage of nucleotides that will be changed at that distance. The identity score will be 100 minus this value. This percentage is not as great as the PAM score due to expected back-mutation over longer time periods. Also shown are the H scores of the matrices at each PAM value.

Table 3.5. *Nucleotide substitution matrix at 1 PAM of evolutionary distance*

A. Model of uniform mutation rates among nucleotides

	A	G	T	C
A	2			
G	−6	2		
T	−6	−6	2	
C	−6	−6	−6	2

B. Model of threefold higher transitions than transversions

	A	G	T	C
A	2			
G	−5	2		
T	−7	−7	2	
C	−7	−7	−5	2

Units are log odds scores obtained as described in the text.

Table 3.6. *Properties of nucleic acid substitution matrices assuming a uniform rate of mutation among nucleotides*

PAM distance	Percentage difference	Match score (bits)	Mismatch score (bits)	Average information per position (bits)
10	9.4	1.86	−3.00	1.40
25	21.3	1.66	−1.82	0.92
50	36.5	1.34	−1.04	0.47
100	55.2	0.84	−0.44	0.13
125	60.8	0.65	−0.30	0.07

The following points may be made:

1. If comparing sequences that are quite similar, it is better to use a lower scoring matrix because the information content of the small PAM matrices is relatively higher. As discussed earlier for lower-numbered Dayhoff PAM matrices for more-alike protein sequences, a more optimal alignment will be obtained.

2. As the PAM distance increases, the mismatch scores in the biased mutational model in Table 3.7 become positive and appear as conservative substitutions. Thus, the bias model can provide considerably more information than the uniform mutation model when aligning sequences that are distantly related (>30% different) and may be used for this purpose (States et al. 1991).

3. The scoring matrices at large evolutionary distances provide very little information per aligned nucleotide pair. When sequences have so little similarity, a much longer alignment is necessary to be significant.

As with amino acid scoring matrices, the average information content shown is only achieved by using the scoring matrix that matches the percentage difference between the sequences. For example, for sequences that are 21% different (79% identical), the matrix at 25 PAM distance should be used. One cannot know ahead of time what the percentage similarity or difference between two sequences actually is until an alignment is done, thus a trial alignment must first be done. States et al. (1991) have calculated how efficient a given scoring matrix is at achieving the highest possible score in aligning two sequences that vary in their levels of similarity. Once the initial similarity score has been obtained with these matrices, a more representative score can be obtained by using another PAM matrix designed specifically for sequences at that level of similarity.

Gap Penalties

The inclusion of gaps and gap penalties is necessary in order to obtain the best possible alignment between two sequences. A gap opening penalty for any gap (g) and a gap exten-

Table 3.7. *Properties of nucleic acid substitution matrices assuming transitions are threefold more frequent than transversions*

PAM distance	Percentage difference	Match score (bits)	Transition score (bits)	Transversion score (bits)	Average information per position (bits)
10	9.3	1.86	−2.19	−3.70	1.42
25	21.0	1.66	−1.06	−2.46	0.96
50	35.8	1.36	−0.37	−1.60	0.54
100	53.7	0.89	0.06	−0.86	0.19
150	62.9	0.57	0.16	−0.52	0.08

sion penalty for each element in the gap (r) is most often used, to give a total gap score w_x, according to the equation

$$w_x = g + rx \qquad (6)$$

where x is the length of the gap. Note that in some formulations of the gap penalty, the equation $w_x = g + r(x - 1)$ is used. Thus, the gap extension penalty is not added to the gap opening penalty until the gap size is 2. Although this difference does not affect the alignment obtained, one needs to distinguish which method is being used by a particular computer program if the correct results are to be obtained. In the former case, the penalty for a gap of size 1 is $g + x$, whereas in the latter case this value is g. The values for these penalties have to be chosen to balance the scores in the scoring matrix that is used. Thus, the Dayhoff log odds matrix at PAM250 is expressed in units of $\log_{10}$, which is approximately 1/3 bits, but if this matrix were converted to 1/2 bits, the same gap penalties would no longer be appropriate.

If too high a gap penalty is used relative to the range of scores in the substitution matrix, gaps will never appear in the alignment. Conversely, if the gap penalty is too low compared to the matrix scores, gaps will appear everywhere in the alignment in order to align as many of the same characters as possible. Fortunately, most alignment programs will suggest gap penalties that are appropriate for a given scoring matrix in most situations. In the GCG and FASTA program suites, the scoring matrix itself is formatted in a way that includes default gap penalties. Examples of the values of g and r used by various alignment programs are shown on the book Web site. When deciding gap penalties for local alignment programs, another consideration is that the penalties should be large enough to provide a local alignment of the sequences. Examples of suitable values are given in Table 3.10 on p. 114. Altschul and Gish (1996) and Pearson (1996, 1998) have found that use of appropriate gap penalties will provide an improved local alignment based on statistical analysis. These studies are described in detail in the following section.

Mathematician Peter Sellers (1974) showed that if sequence alignment was formulated in terms of distances instead of similarity between sequences, a biologically more appealing interpretation of gaps is possible. The distance is the number of changes that must be made to convert one sequence into the other and represents the number of mutations that will have occurred following separation of the genes during evolution; the greater the distance, the more distantly related are the sequences in evolution. In this case, substitution produces a positive score of 1. Notice that the distance score plus the similarity score for an alignment is equal to 1. Sellers proved that this distance formulation of sequence alignment has a desirable mathematical property that also makes evolutionary sense. If three sequences, **a**, **b**, and **c**, are compared using the above scoring scheme, the distance score as defined above is described as a metric that satisfies the triangle inequality relationship

$$d(\mathbf{a},\mathbf{b}) + d(\mathbf{b},\mathbf{c}) \geq d(\mathbf{a},\mathbf{c}) \qquad (7)$$

where $d(\mathbf{a},\mathbf{b})$ is the distance between sequences **a** and **b**, and likewise for the other two d values. Expressed another way, if the three possible distances between three sequences are obtained, then the distance between any first pair plus that for any second pair cannot underscore the third pair. Violating this rule would not be consistent with the expected evolutionary origin of the sequences. To satisfy the metric requirement, the scoring of individual matches, mismatches, and gaps must be such that in an alignment of two identical sequences **a** and **a**′, $d(\mathbf{a},\mathbf{a}')$ must equal 0 and for two totally different sequences **b** and **b**′, $d(\mathbf{b},\mathbf{b}')$ must equal 1. For any other two sequences **a** and **b**, $d(\mathbf{a},\mathbf{b}) = d(\mathbf{b},\mathbf{a})$. Hence, it is important that the distance score for changing one sequence character into a second is the same as the converse score for changing the second into the first, if the distance score of the alignment is to remain a metric and to make evolutionary sense. The above relationships were shown by Sellers to be true for gaps of length 1 in a sequence alignment. He also showed that the smallest number of steps required to change one sequence into the other could be calculated by the dynamic programming algorithm. The method was similar to that discussed above for the Needleman-Wunsch global and Smith-Waterman local alignments, except that these former methods found the maximum similarity between two sequences, as opposed to the minimum distance found by the Sellers analysis.

Subsequently, Smith et al. (1981) and Smith and Waterman (1981a,b) showed that gaps of any length could also be included in an alignment and still provide a distance metric for the alignment score. In this formulation, the gap penalty was required to increase as a function of the gap length. The argument was made that a single mutational event involving a single gap of *n* residues should be more likely to have occurred than *n* single gaps. Thus, to increase the likelihood of such gaps of length >1 being found, the penalty for a gap of length *n* was made smaller than the score for *n* individual gaps. The simplest way of implementing this feature of the gap penalty was to have the gap score w_x be a linear function of gap length by consisting of two parts, a larger gap opening penalty (*g*) and a smaller gap extension penalty (*r*) for each extra position in the gap, or $w_x = g + rx$, where *x* is the length of the gap, as described above. This type of gap penalty is referred to as an affine gap penalty in the literature. Any other formula for scoring gap penalties should also work, provided that the score increases with length of the gap but that the score is less than *x* individual gaps. Scoring of gaps by the above linear function of gap length has now become widely used in sequence alignment. However, more complex gap penalty functions have been used (Miller and Myers 1988).

Penalties for Gaps at the Ends of Alignments

Sequence alignments are often produced that include gaps opposite nonmatching characters at the ends of an alignment. These gaps may be given the same penalty score as gaps inside of the alignment or, alternatively, they may not be given any penalty score. End gaps were an important component in the mathematical formulation of both the similarity and distance methods of sequence alignment for producing both global and local alignments. Failure to include them in distance calculations can result in a failure to obtain distance scores that make evolutionary sense (Smith et al. 1981). Examples of using or of not using end gap penalties in the Needleman-Wunsch alignment are shown on the book Web site. Without scoring end alignments, gaps may be liberally placed at the ends of alignments by the dynamic programming algorithm to increase the matching of internal characters, as opposed to including these gaps as a part of the overall alignment.

If comparing sequences that are homologous and of about the same length, it makes a great deal of sense to include end gap penalties to achieve the best overall alignment. For sequences that are of unknown homology or of different lengths, it may be better to use an alignment that does not include end gap penalties (States and Boguski 1991). If one sequence is expected to be contained within the other, it is reasonable to include end gap penalties only for the shorter sequence. However, for any test alignment, these end penalties should be included in at least one alignment to assure that they do not have an effect. It is also important to use alignment programs that include them as an option.

Parametric Sequence Alignments

Computer methods that find a range of possible alignments in response to varying the scoring system used for matches, mismatches, and gaps, called parametric sequence comparisons (Waterman et al. 1992; Waterman 1994 and references therein), have been developed. There is also an effort to use scores such that the results of global and local types of sequence alignments provide consistent results. For example, if two sequences are similar along their entire lengths, both global and local methods should provide the same alignment. The program Xparal (Gusfield and Stelling 1996), which can perform this type of analysis, is available from http://theory.cs.ucdavis.edu/~stevenk. The program runs on a UNIX environment under X-Windows. When provided with two sequences and some of the alignment parameters, such as gap score, the program displays graphically the types of possible alignments when the remaining parameters are varied. Another sequence alignment program that performs parametric sequence alignment is the Bayes block aligner, discussed below (p. 124).

Effects of Varying Mismatched Gap Penalties on Local Alignment Scores

Vingron and Waterman (1994) have reviewed the effect of varying the parameters of the scoring system on the alignment of random DNA and protein sequences. To simplify the number of parameters, a constant penalty for any size gap was used. If a very high mismatch penalty is used relative to a positive score for a match, with zero gap penalty, the local alignment of these sequences will not include any gaps and is defined as the longest common subsequence. The global alignment with the same scoring parameters will have no mismatches but will have many gaps so placed as to maximize the matches, and the score will be positive. In this case, the score of the local alignment of the sequences is predicted to increase linearly with the length of the sequences being compared.

Another case of varying alignment is penalizing gaps heavily. Then the best scoring local alignment between the sequences will be one that optimizes the score between matches and mismatches, without any gaps. If both mismatches and gaps are heavily penalized, the resulting alignment will also be a local alignment that contains the longest region of exact matches. In the above two cases, the alignment score of the highest-scoring local alignment will increase as the logarithm of the length of the sequences. Under these same conditions, the score of the corresponding global alignment between the sequences will be negative. The transition between a linear and logarithmic dependence of the local similarity score on sequence length occurs when the score of the corresponding global alignment is zero. When both the mismatch and gap penalties are varied between zero and a high negative score, the number of possible alignments of random DNA sequences is very large.

Three general conclusions can be drawn from this theoretical study of random sequence alignments: (1) Use of high mismatch and gap penalties that are greater than a match score will find local alignments, of which there are relatively few in number; (2) when the penalty for a mismatch is greater than twice the score for a match, the gap penalty becomes the decisive parameter in the alignment; and (3) for a mismatch penalty less than twice the score of a gap and a wide range of gap penalties, there are a large number of possible alignments that depend on both the mismatch and gap penalty scores.

Distinguishing local from global alignments has an important practical application. A local alignment is rarely produced between random sequences. Accordingly, the significance of a local alignment between real sequences may be readily calculated, as described below. In contrast, the significance of a global alignment is difficult to determine since a global alignment is readily produced between random sequences.

Optimal Combinations of Scoring Matrices and Gap Penalties for Finding Related Proteins

The usefulness of combinations of scoring matrices and gap penalties for identifying related proteins, including distantly related ones, has been compared (Feng et al. 1985; Doolittle 1986; Henikoff and Henikoff 1993; Pearson 1995, 1996, 1998; Agarwal and States 1998; Brenner et al. 1998). The method generally used is to start with a database of protein sequences organized into families, either based on sequence similarity or structural similarity (described in Chapters 7 and 9, respectively). A member of a family is then selected and used as a query sequence in a search of the entire database from which the sequence came, using a database similarity search method (FASTA, BLAST, SSEARCH), as described in Chapter 7. These methods basically use the dynamic programming algorithm and a choice of scoring matrix and gap penalties to produce alignment scores. Details of these studies are described on the book Web site.

In summary, the following general observations have been made: (1) Some scoring matrices are superior to others at finding related proteins based on either sequence or structure. For example, matrices prepared by examining the full range of amino acid substitutions in families of related proteins, such as the BLOSUM62 matrix, perform better than matrices based on variations in closely related proteins that are extrapolated to produce matrices for more distantly related sequences, such as the Dayhoff PAM250 matrix. (2) Gap penalties that for a given scoring matrix are adjusted to produce a local alignment are the most suitable. (3) To identify related sequences, the significance of the alignment scores should be estimated, as described in the following section.

These methods provide the means to demonstrate sequence similarity in even the most distantly related proteins. For closely related proteins, a PAM-type scoring matrix that matches the evolutionary separation of the sequences may provide a higher-scoring alignment, as described on page 82. Another set of studies has suggested that a global alignment algorithm in combination with scoring matrices that have all positive values and suitable gap penalties can be used to align proteins that have limited sequence similarity (i.e., 25% identity) but that have similar structure (Vogt et al. 1995; Abagyan and Batalov 1997).

ASSESSING THE SIGNIFICANCE OF SEQUENCE ALIGNMENTS

One of the most important recent advances in sequence analysis is the development of methods to assess the significance of an alignment between DNA or protein sequences. For sequences that are quite similar, such as two proteins that are clearly in the same family, such an analysis is not necessary. A significance question arises when comparing two sequences that are not so clearly similar but are shown to align in a promising way. In such a case, a significance test can help the biologist to decide whether an alignment found by the computer program is one that would be expected between related sequences or would just as likely be found if the sequences were not related. The significance test is also needed to evaluate the results of a database search for sequences that are similar to a sequence by the BLAST and FASTA programs (Chapter 7). The test will be applied to every sequence matched so that the most significant matches are reported. Finally, a significance test can also help to identify regions in a single sequence that have an unusual composition suggestive of an interesting function. Our present purpose is to examine the significance of sequence alignment scores obtained by the dynamic programming method.

Originally, the significance of sequence alignment scores was evaluated on the basis of the assumption that alignment scores followed a normal statistical distribution. If sequences are randomly generated in a computer by a Monte Carlo or sequence shuffling method, as in generating a sequence by picking marbles representing four bases or 20 amino acids out of a bag (the number of each type is proportional to the frequency found in sequences), the distribution may look normal at first glance. However, further analysis of the alignment scores of random sequences will reveal that the scores follow a different distribution than the normal distribution called the Gumbel extreme value distribution (see p. 104). In this section, we review some of the earlier methods used for assessing the significance of alignments, then describe the extreme value distribution, and finally discuss some useful programs for this type of analysis with some illustrative examples.

The statistical analysis of alignment scores is much better understood for local alignments than for global alignments. Recall that the Smith-Waterman alignment algorithm and the scoring system used to produce a local alignment are designed to reveal regions of closely matching sequence with a positive alignment score. In random or unrelated sequence alignments, these regions are rarely found. Hence, their presence in real sequence alignments is significant, and the probability of their occurring by chance alignment of unrelated sequences can be readily calculated. The significance of the scores of global alignments, on the other hand, is more difficult to determine. Using the Needleman-Wunsch algorithm and a suitable scoring system, there are many ways to produce a global alignment between any pair of sequences, and the scores of many different alignments may be quite similar. When random or unrelated sequences are compared using a global alignment method, they can have very high scores, reflecting the tendency of the global algorithm to match as many characters as possible. Thus, assessment of the statistical significance of a global alignment is a much more difficult task. Rather than being used as a strict test for sequence homology, a global alignment is more appropriately used to align sequences that are of approximately the same length and already known to be related. The method will conveniently show which sequence characters align. One can then use this information to perform other types of analyses, such as structural modeling or an evolutionary analysis.

Significance of Global Alignments

In general, global alignment programs use the Needleman-Wunsch alignment algorithm and a scoring system that scores the average match of an aligned nucleotide or amino acid pair as a positive number. Hence, the score of the alignment of random or unrelated sequences grows proportionally to the length of the sequences. In addition, there are many possible different global alignments depending on the scoring system chosen, and small changes in the scoring system can produce a different alignment. Thus, finding the best global alignment and knowing how to assess its significance is not a simple task, as reflected by the absence of studies in the literature.

Waterman (1989) provided a set of means and standard deviations of global alignment scores between random DNA sequences, using mismatch and gap penalties that produce a linear increase in score with sequence length, a distinguishing feature of global alignments. However, these values are of limited use because they are based on a simple gap scoring system. Abagyan and Batalov (1997) suggested that global alignment scores between unrelated protein sequences followed the extreme value distribution, similar to local alignment scores. However, since the scoring system that they used favored local alignments, these alignments they produced may not be global but local (see below). Unfortunately, there is no equivalent theory on which to base an analysis of global alignment scores as there is for local alignment scores. For zero mismatch and gap penalties, which is the most extreme condition for a global alignment giving the longest subsequence common to two sequences, the score between two random or unrelated sequences P is proportional to sequence length n, such that $P \simeq cn$ (Chvátal and Sankoff 1975), but it has not proven possible to calculate the proportionality constant c (Waterman and Vingron 1994a).

To evaluate the significance of a Needleman-Wunsch global alignment score, Dayhoff (1978) and Dayhoff et al. (1983) evaluated Needleman-Wunsch alignment scores for a large number of randomized and unrelated but real protein sequences, using their log odds scoring matrix at 250 PAMs and a constant gap penalty. The distribution of the resulting random scores matched a normal distribution. On the basis of this analysis, the significance of an alignment score between two apparently related sequences A and B was determined by obtaining a mean and standard deviation of the alignment scores of 100 random permutations or shufflings of A with 100 of B, conserving the length and amino acid composition of each. If the score between A and B is significant, the authors specify that the real score should be at least 3–5 standard deviations greater than the mean of the random scores. This level of significance means that the probability that two unrelated sequences would give such a high score is 1.35×10^{-3} (3 s.d.s) and 2.87×10^{-6} (5 s.d.s). In evaluating an alignment, two parameters were varied to maximize the alignment score: First, a constant called the matrix bias was added to each value in the scoring matrix and, second, the gap penalty was varied. The statistical analysis was then performed after the score between A and B had been maximized. Recall that the log odds PAM250 matrix values vary from -7 to 17 in units of 1/3 bits. The bias varied from 2 to 20 and had the effect of increasing the score by the bias times the number of alignment positions where one amino acid is matched to another. As a result, the alignment frequently decreases in length because there are fewer gaps, assuming the gap penalty is not also changed. It was these optimized alignments on which the significance test was performed. Feng et al. (1985) used the same method to compare the significance of alignment scores obtained by using different scoring matrices. They used 25–100 pairs of randomized sequences for each test of an alignment.

There are several potential problems with this approach, some of which apply to other methods as well. First, the method is expensive in terms of the number of computational steps, which increase at least as much as the square of sequence length because many Needleman-Wunsch alignments must be done. However, this problem is much reduced with the faster computers and more efficient algorithms of today. Second, if the amino acid composition is unusual, and if there is a region of low complexity (for example, many occurrences of one or two amino acids), the analysis will be oversimplified. Third, when natural sequences were compared more closely, the patterns found did not conform to a random set of the basic building blocks of sequences but rather to a random set of sequence segments that were varying. Consider use of the 26-letter alphabet in English sentences. Alphabet letters do not appear in any random order in these sentences but rather in a vocabulary of meaningful words. What happens if sentences, which are made up of words, are compared? On the one hand, if just the alphabet composition of many sentences is compared, not much variation is seen. On the other hand, if words are compared, much greater variation is found because there are many more words than alphabet characters. If random sequences are produced from segments of sequences, rather than from individual residues, more variation is observed, more like that observed when unrelated natural sequences are compared. The increased variation found among natural sequences is not surprising when one thinks of DNA and proteins as sources of information. For example, protein-encoding regions of DNA sequences are constrained by the genetic code and by amino acid patterns that produce functional domains in proteins.

Lipman et al. (1984) analyzed the distribution of scores among 100 vertebrate nucleic acid sequences and compared these scores with randomized sequences prepared in different ways. When the randomized sequences were prepared by shuffling the sequence to conserve base composition, as was done by Dayhoff and others, the standard deviation was approximately one-third less than the distribution of scores of the natural sequences. Thus, natural sequences are more variable than randomized ones, and using such randomized sequences for a significance test may lead to an overestimation of the significance. If, instead, the random sequences were prepared in a way that maintained the local base composition by producing them from overlapping fragments of sequence, the distribution of scores has a higher standard deviation that is closer to the distribution of the natural sequences. The conclusion is that the presence of conserved local patterns can influence the score in statistical tests such that an alignment can appear to be more significant than it actually is. Although this study was done using the Smith-Waterman algorithm with nucleic acids, the same cautionary note applies for other types of alignments. The final problem with the above methods is that the correct statistical model for alignment scores was not used. However, these earlier types of statistical analysis methods set the stage for later ones.

The GCG alignment programs have a RANDOMIZATION option, which shuffles the second sequence and calculates similarity scores between the unshuffled sequence and each of the shuffled copies. If the new similarity scores are significantly smaller than the real alignment score, the alignment is considered significant. This analysis is only useful for providing a rough approximation of the significance of an alignment score and can easily be misleading.

Dayhoff (1978) and Dayhoff et al. (1983) devised a second method for testing the relatedness of two protein sequences that can accommodate some local variation. This method is useful for finding repeated regions within a sequence, similar regions that are in a different order in two sequences, or a small conserved region such as an active site. As used in a computer program called RELATE (Dayhoff 1978), all possible segments of a given length of one sequence are compared with all segments of the same length from another. An alignment score using a scoring matrix is obtained for each comparison to give a score distribution among all of the segments. A segment comparison score in standard deviation units is calculated as the difference between the value for real sequences minus the average value for random sequences divided by the standard deviation of the scores from the random sequences. A version of the program RELATE that runs on many computer platforms is included with the FASTA distribution package by W. Pearson. An example of the output of the RELATE program for the phage λ and P22 repressor sequences is shown in Table 3.8. This program also calculates a distribution based on the normal distribution, thus it provides only an approximate indication of the significance of an alignment.

Modeling a Random DNA Sequence Alignment

The above types of analyses assume that alignment scores between random sequences follow a normal distribution that can be used to test the significance of a score between two test sequences. For a number of reasons, mathematicians were concerned that this statistical model might not be correct. Let's start by creating two aligned random DNA sequences by drawing pairs of marbles from a large bag filled with four kinds of labeled marbles. The marbles are in equal proportions and labeled A, T, G, and C to represent an assumed equal representation of the four nucleotides in DNA. Now consider the probability of removing 10 identical pairs representing 10 columns in an alignment between two random sequences. The probability of removing an identical pair (an A and another A) is $1/4 \times 1/4$, but there are 4 possible identical pairs (A/A, C/C, G/G, and T/T), so that the probability of removing any identical pair is $4 \times 1/4 \times 1/4 = 1/4$ and that for removing 6 identical pairs is $(1/4)^6 = 2.4 \times 10^{-4}$. The probability of drawing a mismatched pair is $1 - 1/4 = 3/4$, and that of drawing 6/6 mismatched pairs $(3/4)^6 = 0.178$. Most random alignments produced in this manner will have a mixture of a few matches and many mismatches.

The calculations are a little more complex if the four nucleotides are not equally represented, but the results will be approximately the same. The probability of drawing the same

Table 3.8. *Distribution of alignment scores produced by program RELATE*

```
  < -120      0 :
    -115      0 :
    -110      0 :
    -105      0 :
    -100      0 :
     -95      0 :
     -90      0 :
     -85      0 :
     -80      0 :
     -75      0 :
     -70      0 :
     -65      9 :.
     -60     69 :===
     -55    293 :==============
     -50    932 :================================================
     -45   1868 :==================================================
     -40   3214 :==================================================
     -35   4784 :==================================================
     -30   5858 :==================================================
 0   -25   6091 :==================================================
     -20   5384 :==================================================
     -15   4470 :==================================================
 1   -10   2960 :==================================================
      -5   2076 :==================================================
       0   1131 :==================================================
 2     5    590 :=============================
      10    288 :==============
 3    15    154 :=======
      20     67 :===
      25     34 :=
 4    30     18 :.
      35     10 :.
 5    40      1 :.
      45      0 :
      50      0 :
 6    55      0 :
      60      0 :
      65      0 :
 7    70      0 :
      75      0 :
 8    80      0 :
      85      0 :
      90      0 :
      95      0 :
     100      0 :
     105      0 :
     110      0 :
     115      0 :
     120      0 :
     125      0 :
 >   125      0 :
40301 comparisons of window: 25, mean score: -27.3 (13.34)
matrix file: PAM250
29 segments >= 4 sd above mean
```

The sequences of two phage repressors were broken down into overlapping 25-amino-acid segments, and all 40,301 combinations of these segments were compared. The first column gives the approximate location of the number of standard deviations (13.34) from the mean score of −27.3. The second column is increasing ranges of the alignment score, and the third, the number of segment alignment scores, that fall within the range. Twenty-nine scores were greater than 3 standard deviations from the mean. Thus, these two sequences share segments that are significantly more related than the average segment, and the proteins share strong regions of local similarity. In such cases of strong local similarity, a local alignment program such as LFASTA, PLFASTA, or LALIGN can provide the alignments and a more detailed statistical analysis, as described below. Graph is truncated on right side.

pair is p, where $p = p_A{}^2 + p_C{}^2 + p_G{}^2 + p_T{}^2$, where p_X is the proportion of nucleotide X. p is an important parameter to remember for the discussion below. An even more complicated situation is when the two random sequences to align have different nucleotide distributions. One way would be to use an average p for the two sequences. This example illustrates the difficulty of modeling sequence alignments between two different organisms that have a different base composition.

The above model is not suitable for predicting the number of sequentially matched positions between random sequences of a given length. To estimate this number, a DNA sequence alignment may also be modeled by coin-tossing experiments (Arratia and Waterman 1989; Arratia et al. 1986, 1990). Random alignments will normally comprise mixtures of matches and mismatches, just as a series of coin tosses will produce a mixture of heads and tails. The chance of producing a series of matches in a sequence alignment with no mismatches is similar to the chance of tossing a coin and coming up with a series of only heads. The numbers of interest are the highest possible score that can be obtained and the probability of obtaining such a score in a certain number of trials. In such models, coins are usually considered to be "fair" in that the probability of a head is equal to that of a tail. The coin in this example has a certain probability p of scoring a head (H) and $q = 1 - p$ of scoring a tail (T). The longest run of heads R has been shown by Erdös and Rényi to be given by $\log_{1/p}(n)$. If $p = 0.5$ as for a normal coin, then the base of the logarithm is $1/p = 2$. For the example of $n = 100$ tosses, then $R = \log_2 100 = \log_e 100 / \log_e 2 = 4.605/0.693 = 6.65$.

To use the coin model, an alignment of two random sequences $\mathbf{a} = a_1, a_2, a_3$---$a_n$ and $\mathbf{b} = b_1, b_2, b_3$---$b_n$, each of the same length n is converted to a series of heads and tails. If $a_i = b_i$ then the equivalent toss result is an H, otherwise the result is a T. The following example illustrates the conversion of an alignment to a series of H and T tosses.

$$a_1\ a_2\ a_3 \text{ --- } a_n \text{----> } H\ T\ H \text{ ---}$$
$$b_1\ b_2\ b_3 \text{ --- } b_n \qquad \text{where } a_1 = b_1 \text{ and } a_3 = b_3 \text{ only} \qquad (8)$$

The longest run of matches in the alignment is now equivalent to the longest run of heads in the coin-tossing sequence, and it should be possible to use the Erdös and Rényi law to predict the longest run of matches. This score, however, only applies to one particular alignment of random sequences, such as generated above by the marble draw. In performing a sequence alignment, two sequences are in effect shifted back and forth with respect to each other to find regions that can be aligned. In addition, the sequences may be of different lengths. If two random sequences of length m and n are aligned in this same manner, the same law still applies but the length of the predicted match is $\log_{1/p}(mn)$ (Arratia et al. 1986). If $m = n$, the longest run of matches is doubled. Thus, for DNA sequences of length 100 and $p = 0.25$ (equal representation of each nucleotide), the longest expected run of matches is $2 \times \log_{1/p}(n) = 2 \times \log_4 100 = 2 \times \log_e 100 / \log_e 4 = 2 \times 4.605 / 1.386 = 6.65$, the same number as in the coin-tossing experiment. This number corresponds to the longest subalignment that can be expected between two random sequences of this length and composition.

A more precise formula for the expectation value or mean of the longest match M and its variance has been derived (Arratia et al. 1986; Waterman et al. 1987; Waterman 1989).

$$E\,(M) \simeq \log_{1/p}(mn) + \log_{1/p}(q) + \gamma \log(e) - 1/2 \qquad (9)$$

$$\text{Var}\,[M(n,m)] \simeq [\pi \log_{1/p}(e)]^2/6 + 1/12 \qquad (10)$$

where $\gamma = 0.577$ is Euler's number and $q = 1 - p$. Note that Equation 9 can be simplified

$$E\,(M) \simeq \log_{1/p}(Kmn) \qquad (11)$$

where K is a constant that depends on the base composition.

Equation 11 also applies when there are k mismatches in the alignment, except that another term $= k \log_{1/p} \log_{1/p}(qmn)$ appears in the equation (Arratia et al. 1986). K, the constant in Equation 11, depends on k. The log log term is small and can be replaced by a constant (Mott 1992), and simulations also suggest that it is not important (Altschul and Gish 1996). Altschul and Gish (1996) have found a better match to Equation 11 when the length of each sequence is reduced by the expected length of a match. In the example given above with two sequences of length 100, the expected length of a match was 6.65. As the sequences slide align each other, it is not possible to have overlaps on the ends that are shorter than 7 because there is not enough sequence remaining. Hence, the effective length of the sequences is $100 - 7 = 93$ (Altschul and Gish 1996). This correction is also used for the calculation of statistical significance by the BLAST algorithm discussed in Chapter 7.

Equation 11 is fundamentally important for calculating the statistical significance of alignment scores. Basically, it states that as the lengths of random or unrelated sequences increase, the mean of the highest possible local alignment scores will be proportional to the logarithm of the product of the sequence lengths, or twice the logarithm of the sequence length if the lengths are equal (since $\log\,(nn) = 2 \log n$). Equation 10 also predicts a constant variance among scores of random or unrelated sequences, and this prediction is also borne out by experiment. It is important to emphasize once again that this relationship depends on the use of scoring parameters appropriate for a local alignment algorithm, such as 1 for a match and -0.9 for a mismatch, or a scoring matrix that scores the average aligned position as negative, and also upon the use of sufficiently large gap penalties. This type of scoring system gives rise to positive scoring regions only rarely. The significance of these scores can then be estimated as described herein.

Another way of describing the result in Equation 11 uses a different parameter, λ, where $\lambda = \log_e(1/p)$ (Karlin and Altschul 1990)

$$E\,(M) = [\log_e\,(Kmn)]\,/\,\lambda \qquad (12)$$

Recall that p is the probability of a match between the same two characters, given above as 1/4 for matching a random pair of DNA bases, assuming equal representation of each base in the sequences. p may also be calculated as the probability of a match averaged over scoring matrix and sequence composition values. Instead, it is λ that is more commonly used with scoring matrix values. The calculation of λ and also of K is described below and in more detail on the book Web site.

It is more useful in sequence analysis to use alignment scores instead of lengths for comparing alignments. The expected or mean alignment length between two random sequences given by Equations 11 and 12 can be easily converted to an alignment score just by using match and mismatch or scoring matrix values along with some simple normalization procedures. Thus, in addition to predicting length, these equations can also predict the mean or expected value of the alignment scores $E(S)$ between random sequences of lengths m and n. Assessing statistical significance then boils down to calculating the probability that an alignment score between two random or unrelated sequences will actually go above $E(S)$. Hence, the expected score or mean extreme score is

$$E(S) = [\log_e (Kmn)] / \lambda \qquad (13)$$

Another important mathematical result bearing on this question was that the number of matched regions that exceeds the mean score $E(S)$ in Equation 13 could be predicted by the Poisson distribution where the mean x of the Poisson distribution is given by $E(S)$ (Waterman and Vingron 1994b). The Poisson distribution applies when the probability of success in a single trial is small, but the number of trials is large (as in comparing many pairs of random sequences or a test sequence to many scrambled versions of a second sequence) so that some trials end in success but others do not. Some alignments do not reach the expected score, but others will reach or even exceed that score. The Poisson distribution gives the probability P_n of the number of successes, i.e., 0, 1, 2, 3 . . . when the average number is x and is given by the formula $P_n = e^{-x} x^n / n!$. The probability that no score from many test alignments will exceed x is therefore approximated by ($P_0 = e^{-x}$). The probability that at least one score exceeds x is $1 - P_0$ and is given by $P(S > x) = 1 - e^{-x}$, so that

$$P(S < x) \simeq \exp(-E(S))$$
$$\simeq \exp(-Kmne^{-\lambda x}) \qquad (14)$$

$$P(S > x) \simeq 1 - \exp(-Kmne^{-\lambda x}) \qquad (15)$$

Equation 15 estimates the probability of a score greater than x between two random sequences and is identical to the extreme value distribution described below. The Poisson approximation provides a very convenient way to estimate K and λ from alignment scores between many random or unrelated sequences by using the fraction of alignments that have a score less than value x (see book Web site).

Alignments with Gaps

It was predicted on mathematical grounds and shown experimentally that a similar type of analysis holds for sequence alignments that include gaps (Smith et al. 1985). Thus, when Smith et al. (1985) optimally aligned a large number of unrelated vertebrate and viral DNA sequences of different lengths (n and m) and their complements to each other, using a dynamic programming local alignment method that allowed for a score of +1 for matches, −0.9 for mismatches, and −2 for a single gap penalty (longer gaps were not considered in order to simplify the analysis), a plot of the similarity score (S) versus the $\log_1/p(nm)$ produced a straight line with approximately constant variance. This result is as expected in the above model except that with the inclusion of gaps, the slope was increased and was of the form

$$S_{\text{mean}} = 2.55 \, (\log_l/p(mn)) - 8.99 \tag{16}$$

with constant standard deviation $\sigma = 1.78$. This result was then used to calculate how many standard deviations were between the predicted mean and variance of the local alignment scores for unrelated sequences and the scores for test pairs of sequences. If the actual alignment score exceeded the predicted S_{mean} by several standard deviations, then the alignment score should be significant. For example, the expected score between two unrelated sequences of lengths 2948 and 431, average $p = 0.279$, was $S_{\text{mean}} = 2.55 \times \log_{1/0.279}(2948 \times 431) - 8.99 = 2.55 \times (\log_e(2948 \times 431)/\log_e(1/0.279)) - 8.99 = 2.55 \times 14.1 / 1.28 - 8.99 = 28.1 - 8.99 = 19.1$. The actual optimal alignment score between the two real sequences of these lengths was 37.20, which exceeds the alignment score expected for random sequences by $(37.20 - 19.1) / 1.78 = 10.2\sigma$. Is this number of standard deviations significant? Smith et al. (1985) and Waterman (1989) suggested the use of a conservative statistic known as Chebyshev's inequality, which is valid for many probability distributions: The probability that a random variable exceeds its mean is less than or equal to the square of 1 over the number of standard deviations from the mean. In this example where the actual score is 10 standard deviations above the mean, the probability is $(1/10)^2 = 0.01$.

Waterman (1989) has noted that for low mismatch and gap penalties, e.g., $+1$ for matches, -0.5 for mismatches, and -0.5 for a single gap penalty, the predicted alignment scores between random sequences as estimated above are not accurate because the score will increase linearly with sequence length instead of with the logarithm of the length. The linear relationship arises when the alignment is more global in nature, and the logarithmic relationship when it is local. Waterman (1989) has fitted alignment scores from a large number of randomly generated DNA sequences of varying lengths to either the predicted $\log(n)$ or n linear relationships expected for low- and high-valued mismatch and gap penalties. The results provide the mean and standard deviation of an alignment score for several scoring schemes, assuming a constant gap penalty.

With further mathematical analysis, it became apparent that the expected scores between alignment of random and unrelated sequences follow a distribution called the Gumbel extreme value distribution (Arratia et al. 1986; Karlin and Altschul 1990). This type of distribution is typical of values that are the highest or best score of a variable, such as the number of heads only expected in a coin toss discussed previously. Subsequently, S. Karlin and S. Altschul (1990, 1993) further developed the use of this distribution for evaluating the significance of ungapped segments in comparisons between a test sequence and a sequence database using the BLAST program (for review, see Altschul et al. 1994). The method is also used for evaluating the statistical features of repeats and amino acid patterns and clusters in the same sequence (Karlin and Altschul 1990; Karlin et al. 1991). The program SAPS developed by S. Karlin and colleagues at Stanford University and available at http://ulrec3.unil.ch/software/software.html provides this type of analysis. The extreme value distribution is now widely used for evaluating the significance of the score of local alignments of DNA and protein sequence alignments, especially in the context of database similarity searches.

The Gumbel Extreme Value Distribution

When two sequences have been aligned optimally, the significance of a local alignment score can be tested on the basis of the distribution of scores expected by aligning two random sequences of the same length and composition as the two test sequences (Karlin and

Altschul 1990; Altschul et al. 1994; Altschul and Gish 1996). These random sequence alignment scores follow a distribution called the extreme value distribution, which is somewhat like a normal distribution with a positively skewed tail in the higher score range. When a set of values of a variable are obtained in an experiment, biologists are used to calculating the mean and standard deviation of the entire set assuming that the distribution of values will follow the normal distribution. For sequence alignments, this procedure would be like obtaining many different alignments, both good and bad, and averaging all of the scores. However, biologically interesting alignments are those that give the highest possible scores, and lower scores are not of interest. The experiment, then, is one of obtaining a set of values, and then of using only the highest value and discarding the rest. The focus changes from the statistical approach of wanting to know the average of scores of random sequences, to one of knowing how high a value will be obtained next time another set of alignment scores of random sequences is obtained.

The distribution of alignment scores between random sequences follows the extreme value distribution, not the normal distribution. After many alignments, a probability distribution of highest values will be obtained. The goal is to evaluate the probability that a score between random or unrelated sequences will reach the score found between two real sequences of interest. If that probability is very low, the alignment score between the real sequences is significant and the sequence similarity score is significant.

The probability distribution of highest values in an experiment, the extreme value distribution, is compared to the normal probability distribution in Figure 3.17. The equations giving the respective y coordinate values in these distributions, Y_{ev} and Y_n, are

$$Y_{ev} = \exp\left[-x - e^{-x}\right] \quad \text{for the extreme value distribution} \tag{17}$$

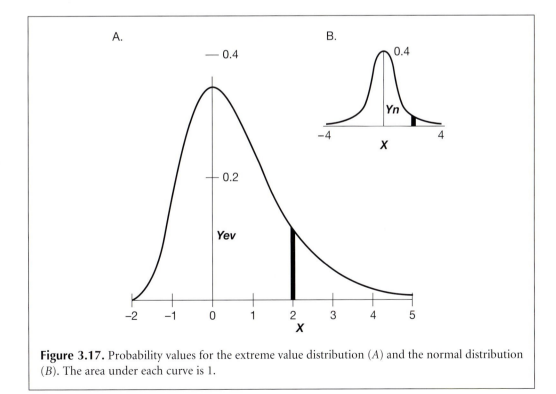

Figure 3.17. Probability values for the extreme value distribution (*A*) and the normal distribution (*B*). The area under each curve is 1.

$$Y_n = 1/\sqrt{(2\pi)} \exp\left[(-x^2)/2\right] \quad \text{for the normal distribution} \tag{18}$$

The area under both curves is 1. The normal curve is symmetrical about the expectation value or mean at $x = 0$, such that the area under the curve below the mean (0.5) is the same as that above the mean (0.5) and the variance σ^2 is 1. The probability of a particular value of x for the normal distribution is obtained by calculating the area under curve B, usually between $-x$ and $+x$. For $x = 2$, often used as an indication of a significant deviation from the mean, the area between -2 and $+2$ is 0.9544. For the extreme value distribution, the expectation value or mean of x is the value of the Euler-Mascheroni constant, 0.57722 . . . and the variance of x, σ^2, is the value of $\pi^2 / 6 = 1.6449$. The probability that score S will be less than value x, $P(S < x)$, is obtained by calculating the area under curve A from $-\infty$ to x, by integration of Equation 17 giving

$$P(S < x) = \exp\left[-e^{-x}\right] \tag{19}$$

and the probability of $S \geq x$ is 1 minus this probability

$$P(S \geq x) = 1 - \exp\left[-e^{-x}\right] \tag{20}$$

For the extreme value distribution, the area below $x = 0$, which represents the peak or mode of the distribution, is $1/e$ or 0.368 of the total area of 1, and the area above the mean is $1 - 0.368 = 0.632$. At a value of $x = 2$, $Yev = 0.118$ and $P(S < 2) = \exp\left[-e^{-2}\right] = 0.873$. Thus, just over 0.87 of the area under the curve is found below $x = 2$. An area of 0.95 is not reached until $x = 3$. The difference between the two distributions becomes even greater for larger values of x. As a result, for a variable whose distribution comes from extreme values, such as random sequence alignment scores, the score must be greater than expected from a normal distribution in order to achieve the same level of significance.

The above equations are modified for use with scores obtained in an analysis. For a variable x that follows the normal distribution, values of x are used to estimate the mean m and standard deviation σ of the distribution, and the probability curve given by Equation 18 then becomes

$$Yn = 1/(\sigma\sqrt{(2\pi)}) \exp\left[-(x-m)^2/2\sigma^2\right] \tag{21}$$

The probability of a particular value of x can be estimated by using m and σ to estimate the number of standard deviations from the mean, Z, where $Z = (x - m)/\sigma$. Similarly, Equations 17 and 20 can be modified to accommodate the extreme values such as sequence alignment scores

$$P(S \geq x) = 1 - \exp\left[-e^{-\lambda(x-u)}\right] \tag{22}$$

where u is the mode, highest point, or characteristic value of the distribution, and λ is the decay or scale parameter. As is apparent in Equation 22, λ converts the experimentally measured values into standard values of x after subtraction of the mode from each score.

It is quite straightforward to calculate u and λ, and several methods using alignment scores are discussed on the book Web site. There is an important relationship between u and λ, and the mean and standard deviation of a set of extreme values. The mean and standard deviation do not only apply to the normal distribution, but in fact are mathematically defined for any probability distribution. The mean of any set of values of a variable may always be calculated as the sum of the values divided by their number. The mean m or expected value of a variable x, $E(x)$, is defined as the first moment of the values of the variable around the mean. From this definition, the mean is that number from which the sum of deviations to all values is zero. The variance σ^2 is the second moment of the values about the mean and is the sum of the squares of the devations from the mean divided by the number of observations less one ($n - 1$). The mean $\underline{x}$ and standard deviation σ of a set of extreme values can be calculated in the same way, and then u and λ can be calculated using the following equations derived by mathematical evaluation of the first and second moments of the extreme value distribution (Gumbel 1962; Altschul and Erickson 1986).

$$\lambda = \pi / (\sigma\sqrt{6}) = 1.2825 / \sigma \tag{23}$$

$$u = \underline{x} - \gamma / \lambda = \underline{x} - 0.4500\,\sigma \tag{24}$$

where γ was already introduced. Equation 23 is derived from the ratio of the variance σ^2 of the two distributions in Figure 3.17, or 1 to $\pi^2 / 6$. Equation 24 is derived from the observation that the mode or the EV distribution (zero in Fig. 3.17) has the value of γ less than the mean. However, the value of γ must be scaled by the ratio of the standard deviations. Hence γ / λ is subtracted from the mean. This method of calculating u and λ from means and standard deviations is called the method of moments.

As with the normal distribution, z scores may be calculated for each extreme value x, where $z = (x - m) / \sigma$ is the number of standard deviations from the mean m to each score. z scores are used by the FASTA, version 3, programs distributed by W. Pearson (1998). Equation 22 may be written in a form that directly uses z scores to evaluate the probability that a particular score Z exceeds a value z,

$$P(Z > z) = 1 - exp(-e^{-1.2825\,z\,-\,0.5772}) \tag{25}$$

For sequence analysis, u and λ depend on the length and composition of the sequences being compared, and also on the particular scoring system being used. They can be calculated directly or estimated by making many alignments of random sequences or shuffled natural sequences, using a scoring system that gives local alignments. The parameters will change when a different scoring system is used. Examples of programs that calculate these values are given below.

For alignments that do not include any gaps, u and λ may be calculated from the scoring matrix. The scaling factor λ is calculated as the value of x, which satisfies the condition

$$\Sigma \, p_i p_j \, e^{s_{ij}x} = 1 \qquad (26)$$

where p_i and p_j are the respective fractional representations of residues i and j in the sequences, and s_{ij} is the score for a match being i and j, taken from a log odds scoring matrix. u, the characteristic value of the distribution, is given by (Altschul and Gish 1996)

$$u = (\ln Kmn) \, / \, \lambda \qquad (27)$$

where m and n are the sequence lengths and K is a constant that can also be calculated from the values of p_i and s_{ij}. Note that this value originates from the coin toss analysis that gave rise to Equation 14. Combining Equations 25 and 27 eliminates u and gives the following relationship

$$P\,(S \geq x) = 1 - \exp\,[-\, e^{-\, \lambda(x \,-\, u)}]$$
$$= 1 - \exp\,[-\, e^{-\, \lambda(x \,-\, (\ln Kmn) \,/\, \lambda)}]$$
$$= 1 - \exp\,[-\, e^{-\, \lambda x \,+\, \ln Kmn}] \qquad (28)$$
$$= 1 - \exp\,[-\, Kmn \, e^{-\lambda x}] \qquad (29)$$

To facilitate calculations, a sequence alignment score S may also be normalized to produce a score S'. The effect of normalization is to change the score distribution into the form shown above in Figure 3.17 with $u = 0$ and $\lambda = 1$. From Equation 28, S' is calculated by

$$S' = \lambda S - \ln Kmn \qquad (30)$$

The probability of $P\,(S' > x)$ is then given by Equation 20 with $S = S'$

$$P\,(S' \geq x) = 1 - \exp\,[-\, e^{-x}] \qquad (31)$$

The probability of a particular normalized score may then be readily calculated. This capability depends on a determination of the λ and K to calculate the normalized scores S' by Equation 30.

The probability function $P(S' \geq x)$ decays exponentially in x as x increases and $P(S' \geq x) = 1 - \exp\,[-\, e^{-x}] -> e^{-x}$. Consequently, an important approximation for Equations 29 and 31 for the significant part of the extreme value distribution where $x > 2$ is shown in Equations 32 and 33. Note that the replacement equations are single and not double exponentials.

$$P\,(S \geq x) \simeq Kmn \, e^{-\lambda x} \qquad (32)$$

$$P\,(S' \geq x) \simeq e^{-x} \qquad (33)$$

Table 3.9. *Approximation of P(S′ ≥ x) by e^{-x}*

x	1−exp [− e^{-x}]	e^{-x}
0	0.63	1
1	0.308	0.368
2	0.127	0.135
3	0.0486	0.0498
4	0.0181	0.0183

A comparison of probability calculations using this approximation instead of that given in Equation 31 is shown in Table 3.9. For $x > 2$, the estimates differ by less than 2%. The estimate given in Equation 32 also provides a quicker method for estimating the significance of an alignment score.

A Quick Determination of the Significance of an Alignment Score

Scoring matrices are most useful for statistical work if they are scaled in logarithms to the base 2 called bits. Scaling the matrices in this fashion does not alter their ability to score sequence similarities, and thereby to distinguish good matches from poor ones, but does allow a simple estimation of the significance of an alignment. The actual alignment may then be calculated by summing the matrix values for each of the aligned pairs, using matrix values in bit units. If the actual alignment score in bits is greater than expected for alignment of random sequences, the alignment is significant.

For a typical amino acid scoring matrix and protein sequence, K = 0.1 and λ depends on the values of the scoring matrix. If the log odds matrix is in units of bits as described above, then $\lambda = \log_e 2 = 0.693$, and the following simplified form of Equation 32 may be derived (Altschul 1991) by taking logarithms to the base 2 and setting p as the probability of the scores of random or unrelated alignments reaching a score of S or greater

$$\begin{aligned} \log_2 p &= \log_2 (Kmn\, e^{-\lambda S}) \\ &= \log_2 (Kmn) + \log_2(e^{-\lambda S}) \\ &= \log_2 (Kmn) + (\log_e(e^{-\lambda S}))/\log_e 2 \\ &= \log_2 (Kmn) - \lambda S/\log_e 2 \\ &= \log_2 (Kmn) - S \end{aligned} \tag{34}$$

then S, the score corresponding to probability P, may be obtained by rearranging terms of Equation 34 as follows

$$\begin{aligned} S &= \log_2 (Kmn) - \log_2 P \\ &= \log_2 (K/P) + \log_2(nm) \end{aligned} \tag{35}$$

Since for most scoring matrices K $\simeq$ 0.1 and choosing $P = 0.05$, the first term is 1, and the second term in Equation 35 becomes the most important one for calculating the score (Altschul 1991), thus giving

$$S \simeq \log_2 (nm) \tag{36}$$

Example: Using the Extreme Value Distribution to Calculate the Significance of a Local Alignment

Suppose that two sequences approximately 250 amino acids long are aligned by the Smith-Waterman local alignment algorithm using the PAM250 matrix and a high gap score to omit gaps from the alignment, and that the following alignment is found.

```
FWLEVEGNSMTAPTG
FWLDVQGDSMTAPAG
```

1. By Equation 36, a significant alignment between unrelated or random sequences will have a score of $S \simeq \log_2(nm) = \log_2(250 \times 250) = 16$ bits.

2. The score of the above actual alignment is 75 using the scores in the Dayhoff mutation data matrix (MDM) that provides log odds scores at 250 PAMs evolutionary distance.

3. A correction to the alignment score must be made because the MDM table at 250 PAMs is not in bit units but in units of logarithm to the base 10, multiplied by 10. These MDM scores actually correspond to units of 1/3 bits ([MDM score in units of $\log_{10}$] $\times$ 10 = [MDM score in bits of $\log_2 \times \log_2 10$] / 10 = [MDM score in units of $\log_{10} \times 10$] $\times$ 0.333). Thus, the score of the alignment in bits is 75/3 = 25 and 9 bits greater than the 16 expected by chance. Therefore, this alignment score is highly significant.

4. Altschul and Gish (1996) have provided estimates of $K = 0.09$ and $\lambda = 0.229$ for the PAM250 scoring matrix, for a typical amino acid distribution and for an alignment score based on using a very high gap penalty. By Equations 3.30 and 3.31, $S' = 0.229 \times 75 - \ln(0.09 \times 250 \times 250) = 17.18 - 8.63 = 8.55$ bits, and $P(S' \geq 8.55) = 1 - \exp[-e^{-8.55}] = 1.9 \times 10^{-4}$. Thus, the chance that an alignment between two random sequences will achieve a score greater than or equal to 75 using the MDM matrix is 1.9×10^{-4}. Note that the calculated S' of 8.55 bits in step 4 is approximately the same as the 9 bits calculated by the simpler method in step 3.

5. The probability may also be calculated by the approximation given in Equation 3.33 $P(S' > x) \simeq e^{-x} = e^{-8.55} = 1.9 \times 10^{-4}$.

The Importance of the Type of Scoring Matrix for Statistical Analyses

Using a log odds matrix in bit units simplifies estimation of the significance of an alignment. The Dayhoff PAM matrices, the BLOSUM matrices, and the nucleic acid PAM scoring matrices are examples of this type. Such matrices are also useful for finding local alignments because the matrix includes both positive and negative values. Another important feature of the log odds form of the scoring matrix is that this design is optimal for assessing statistical significance of alignment scores. A set of matrices, each designed to detect similarity between sequences at a particular level, is best for this purpose. Use of a matrix that is designed for aligning sequences that have a particular level of similarity (or evolutionary distance) assures the highest-scoring alignment and therefore the very best estimate of significance. Thus, lower-numbered PAM matrices are most suitable for aligning sequences that are more similar. In the above example, the Dayhoff PAM250 matrix designed for sequences that are 20% similar was used to align sequences that are approximately 20% identical and 50% similar (identities plus common replacements in the alignment). Using a lower PAM120 matrix produces a slightly higher score for this alignment, and thus increases the significance of the alignment score.

Another important parameter of the scoring matrix for statistical purposes is the expected value of the average amino acid pair, calculated as shown in Equation 37. This value should be negative if alignment scores for the matrix are to be used for statistical tests, as performed in the above example. Otherwise, in any aligned pair of sequences the scores will increase with length faster than the logarithm of the length. Not all scoring matrices will meet this requirement. To calculate the expected score (E), the score for each amino acid pair (s_{ij}) is multiplied by the fractional occurrences of each amino acid (p_i and p_j). This weighted score is then summed over all of the amino acid pairs. The expected values of the log odds matrices such as the Dayhoff PAM, BLOSUM, JTT, JO93, PET91, and Gonnet92 matrices all meet this statistical requirement.

$$E = \sum_{i=1}^{20} \sum_{j=1}^{i} p_i\, p_j\, s_{ij} \tag{37}$$

For example, for the PAM120 matrix in one-half bits $E = -1.64$ and for PAM160 in one-half bits, $E = -1.14$. Thus, scores obtained with these matrices may be used in the above statistical analysis. Ungapped alignment scores obtained using the BLOSUM62 matrix may also be subject to a significance test, as described above for the PAM matrices. The test is valid because the expect score for a random pair of amino acids is negative ($E = -0.52$). Because the matrix is in half-bit units, the alignment is significant when a score exceeds $16/0.52 \approx 32$ half-bits.

To assist in keeping track of information, scoring matrices have appeared in a new format suitable for use by many types of programs. An example is given in Figure 3.18. The matrix includes: (1) the scale of the matrix and the value of the statistical parameter λ; (2) E, the expect score of the average amino acid pair in the matrix, which if negative assures that local alignments will be emphasized (Eq. 37); (3) H, the information content or entropy of the matrix (Eq. 3) giving the ability of the matrix to discriminate related from unrelated sequence alignments, not shown here; and (4) suitable gap penalties. The BLOSUM matrices are also available in this same format.

Significance of Gapped, Local Alignments

When random sequences of varying lengths are optimally aligned with the Smith-Waterman dynamic programming algorithm using an appropriate scoring matrix and gap penalties, the distribution of scores also matches the extreme value distribution (Altschul and Gish 1996). Similarly, in optimally aligning a given sequence to a database of sequences, and after removing the high scores of the closely related sequences, the scores of the unrelated sequences also follow this distribution (Altschul et al. 1994; Pearson 1996, 1998). In these and other cases, optimal scores are found to increase linearly with log (n), where n is the sequence length. Equation 36 predicts that the optimal alignment score (x) expected between two random or unrelated sequences should be proportional to the logarithm of the product of the sequence lengths, $x \simeq \log_2(nm)$. If the sequence lengths are approximately equal, $n \simeq m$, then x should be proportional to $\log_2(n^2) = 2\log_2(n)$, and the predicted score should also increase linearly with $\log(n)$. $\log_2(n)$ is equivalent to $\log(n)$ because, to change the base of a logarithm, one merely multiplies by a constant. In comparing one sequence of length m to a sequence database of length n, m is a constant and

```
#
# This matrix was produced by "pam" Version 1.0.6 [28-Jul-93]
#
# PAM 120 substitution matrix, scale = ln(2)/2 = 0.346574   [1/2 bits]
#
# Expected score = -1.64, Entropy = 0.979 bits
#
# Lowest score = -8, Highest score = 12
#
   A  R  N  D  C  Q  E  G  H  I  L  K  M  F  P  S  T  W  Y  V  B  Z  X  *
A  3 -3 -1  0 -3 -1  0  1 -3 -1 -3 -2 -2 -4  1  1  1 -7 -4  0  0 -1 -1 -8
R -3  6 -1 -3 -4  1 -3 -4  1 -2 -4  2 -1 -5 -1 -1 -2  1 -5 -3 -2 -1 -2 -8
N -1 -1  4  2 -5  0  1  0  2 -2 -4  1 -3 -4 -2  1  0 -4 -2 -3  3  0 -1 -8
D  0 -3  2  5 -7  1  3  0  0 -3 -5 -1 -4 -7 -3  0 -1 -8 -5 -3  4  3 -2 -8
C -3 -4 -5 -7  9 -7 -7 -4 -4 -3 -7 -7 -6 -6 -4  0 -3 -8 -1 -3 -6 -7 -4 -8
Q -1  1  0  1 -7  6  2 -3  3 -3 -2  0 -1 -6  0 -2 -2 -6 -5 -3  0  4 -1 -8
E  0 -3  1  3 -7  2  5 -1 -1 -3 -4 -1 -3 -7 -2 -1 -2 -8 -5 -3  3  4 -1 -8
G  1 -4  0  0 -4 -3 -1  5 -4 -4 -5 -3 -4 -5 -2  1 -1 -8 -6 -2  0 -2 -2 -8
H -3  1  2  0 -4  3 -1 -4  7 -4 -3 -2 -4 -3 -1 -2 -3 -3 -1 -3  1  1 -2 -8
I -1 -2 -2 -3 -3 -3 -3 -4 -4  6  1 -3  1  0 -3 -2  0 -6 -2  3 -3 -3 -1 -8
L -3 -4 -4 -5 -7 -2 -4 -5 -3  1  5 -4  3  0 -3 -4 -3 -2  1 -4 -4 -3 -2 -8
K -2  2  1 -1 -7  0 -1 -3 -2 -3 -4  5  0 -7 -2 -1 -1 -5 -5 -4  0 -1 -2 -8
M -2 -1 -3 -4 -6 -1 -3 -4 -4  1  3  0  8 -1 -3 -2 -1 -6 -4  1 -4 -2 -2 -8
F -4 -5 -4 -7 -6 -6 -7 -5 -3  0  0 -7 -1  8 -5 -3 -4 -1  4 -3 -5 -6 -3 -8
P  1 -1 -2 -3 -4  0 -2 -2 -1 -3 -3 -2 -3 -5  6  1 -1 -7 -6 -2 -2 -1 -2 -8
S  1 -1  1  0  0 -2 -1  1 -2 -2 -4 -1 -2 -3  1  3  2 -2 -3 -2  0 -1 -1 -8
T  1 -2  0 -1 -3 -2 -2 -1 -3  0 -3 -1 -1 -4 -1  2  4 -6 -3  0  0 -2 -1 -8
W -7  1 -4 -8 -8 -6 -8 -8 -3 -6 -3 -5 -6 -1 -7 -2 -6 12 -2 -8 -6 -7 -5 -8
Y -4 -5 -2 -5 -1 -5 -5 -6 -1 -2 -2 -5 -4  4 -6 -3 -3 -2  8 -3 -3 -5 -3 -8
V  0 -3 -3 -3 -3 -3 -3 -2 -3  3  1 -4  1 -3 -2 -2  0 -8 -3  5 -3 -3 -1 -8
B  0 -2  3  4 -6  0  3  0  1 -3 -4  0 -4 -5 -2  0  0 -6 -3 -3  4  2 -1 -8
Z -1 -1  0  3 -7  4  4 -2  1 -3 -3 -1 -2 -6 -1 -1 -2 -7 -5 -3  2  4 -1 -8
X -1 -2 -1 -2 -4 -1 -1 -2 -2 -1 -2 -2 -2 -3 -2 -1 -1 -5 -3 -1 -1 -1 -2 -8
* -8 -8 -8 -8 -8 -8 -8 -8 -8 -8 -8 -8 -8 -8 -8 -8 -8 -8 -8 -8 -8 -8 -8  1
```

Figure 3.18. Example of BLASTP format of the Dayhoff MDM giving log odds scores at 120 PAMs. Note that the matrix has mirror-image copies of the same score on each side of the main diagonal. Besides the standard single-letter amino acid symbols, there are four new symbols, B, Z, X, *. B is the frequency-weighted average of entries for D and N pairs, Z similarly for Q and E entries, X similarly for all pairs in each row, and * is the lowest score in the matrix for matches with any other sequence character that may be present.

the predicted score should increase linearly as log(n). This log(n) relationship has been found in several studies of the distribution of optimal local alignment scores that have included gap penalties (Smith et al. 1985; Arratia et al. 1986; Collins et al. 1988; Pearson 1996, 1998; for additional references, see Altschul et al. 1994). Thus, the same statistical methods described above for assessing the significance of ungapped alignment scores may also be used for gapped alignment scores. Methods for calculating the parameters K and λ for a given combination of scoring matrix methods and gap penalties are described on the book Web site.

Methods for Calculating the Parameters of the Extreme Value Distribution

In the analysis by Altschul and Gish (1996), 10,000 random amino acid sequences of variable lengths were aligned using the Smith-Waterman method and a combination of the scoring matrix and a reasonable set of gap penalties for the matrix. The scores found by this method followed the same extreme value distribution predicted by the underlying statistical theory. Values of K and λ were then estimated for each combination by fitting the data to the predicted extreme value distribution. Some representative results are shown in

Table 3.10. Readers should consult Tables V–VII in Altschul and Gish (1996) for a more detailed list of the gap penalties tested.

Altschul and Gish (1996) have cautioned users of these statistical parameters. First, the parameters were generated by alignment of random sequences that were produced assuming a particular amino acid distribution, which may be a poor model for some proteins. Second, the accuracy of λ and K cannot be estimated easily. Finally, for gap costs that give values of $H < 0.15$, the optimal alignment length is a significant fraction of the sequence lengths and produces a source of error called the edge effect. The effect occurs when the expected length of an alignment is a significant fraction of the sequence length, and, as discussed earlier, alignments between sequences that overlap at their ends cannot be completed. The expected length is then subtracted from the sequence length before λ is estimated. If no such correction is done, λ may be overestimated.

These values for gap penalties should also not be construed to represent the best choice for a given pair of sequences or the only choices, simply because the statistical parameters are available. The process of choosing a gap penalty remains a matter of reasoned choice. In trying the effects of varying the gap penalty, it is important to recognize that as the gap penalty is lowered, the alignments produced will have more gaps and will eventually change from a local to a global type of alignment, even though a local alignment program is being used. In contrast, higher H values are generated by a very large gap penalty and produce alignments with no gaps (Table 3.10), thus suggesting an increased ability to discriminate between related and unrelated sequences. In this respect, Altschul and Gish (1996) note that beyond a certain point increasing the gap

Table 3.10. *Statistical parameters for combination of scoring matrices and affine gap penalties*

Scoring matrix	Gap opening penalty[b]	Gap extension penalty[b]	K	λ	H^c
BLOSUM50	∞^a	0-∞	0.232	0.11	0.34
BLOSUM50	15	8–15	0.09	0.222	0.31
BLOSUM50	11	8–11	0.05	0.197	0.21
BLOSUM50	11	1	—	—	—
BLOSUM62	∞^a	0-∞	0.318	0.13	0.40
BLOSUM62	12	3–12	0.1	0.305	0.38
BLOSUM62	8	7–8	0.06	0.270	0.25
BLOSUM62	7	1	—	—	—
PAM250	∞^a	0-∞	0.229	0.09	0.23
PAM250	15	5–15	0.06	0.215	0.20
PAM250	10	8-10	0.031	0.175	0.11
PAM250	11	1	—	—	—

Dashes indicate that no value can be calculated because the relationship between alignment score and sequence length is linear and not logarithmic, indicating that the alignment is global, not local, in character. Statistical significance may not be calculated for these gap penalty-scoring matrix combinations. The corresponding values for gap penalties define approximate lower limits that should be used.

[a] A value of ∞ for gap penalty will produce alignments with no gaps.

[b] The penalty for a gap opening of length 1 is the value of the gap opening penalty shown. The gap extension penalty is not added until the gap length is 2. Make sure that the alignment program uses this same scheme for scoring gaps. The extension penalty is shown over a range of values; values within this range did not change K and λ.

[c] The entropy in units of the natural logarithm.

extension penalty does not change the parameters, indicating that most gaps in their simulations are probably of length 1. However, reducing the gap penalty can also allow an alignment to be extended and create a higher scoring alignment. Eventually, however, the optimal local alignment score between unrelated sequences will lose the log length relationship with sequence length and become a linear function. At this point, gap penalties are no longer useful for obtaining local alignments and the above statistical relationships are no longer valid.

The higher the H value, the better the matrix can distinguish related from unrelated sequences. The lower the value of H, the longer the expected alignment. These conditions may be better if a longer alignment region is required, such as testing a structural or functional model of a sequence by producing an alignment. Conversely, scoring parameters giving higher values of H should produce shorter, more compact alignments. If $H < 0.15$, the alignments may be very long. In this case, the sequences have a shorter effective length since alignments starting near the ends of the sequences may not be completed. This edge effect can lead to an overestimation of λ but was corrected for in the above table (Altschul and Gish 1996).

Unfortunately, the above method for calculating the significance of an alignment score may not be used to test the significance of a global alignment score. The theory does not apply when these same substitution matrices are used for global alignments. Transformation of these matrices by adding a fixed constant value to each entry or by multiplying each value by a constant has no effect on the relative scores of a series of global alignments. Hence, there is no theoretical basis for a statistical analysis of such scores as there is for local alignments (Altschul 1991).

As discussed in Chapter 7, two programs are commonly used for database similarity searches: FASTA and BLAST. These programs both calculate the statistical significance of the higher scores found with similar sequences, but the types of analyses used to determine the statistical significance of these scores are somewhat different. BLAST uses the value of K and λ found by aligning random sequences and Equation 29, where n and m are shortened to compensate for inability of ends to align. FASTA calculates the statistical significance using the distribution of scores with unrelated sequences found during the database search. In effect, the mean and standard deviation of the low scores found in a given length range are calculated. These scores represent the expected range of scores of unrelated sequences for that sequence length (recall that the local alignment scores increase as the logarithm of the sequence length). The number of standard deviations to the high scores of related sequences in the same length range (z score) is then determined. The significance of this z score is then calculated according to the extreme value distribution expected of the z scores, given in Equation 25. This method is discussed in greater detail in Chapter 7. Pearson (1996) showed that these two methods are equally useful in database similarity searches for detecting sequences more distantly related to the input query sequence.

Pearson (1996) has also determined the influence of scoring matrices and gap penalties on alignment scores of moderately related and distantly related protein sequences in the same family. For two examples of moderately related sequences, the choice of scoring matrix and gap penalties (gap opening penalty followed by penalty for each additional gap position) did not matter, i.e., BLOSUM50 $-12/-2$, BLOSUM62 $-8/-2$, Gonnet93 $-10/-2$, and PAM250 -12, -2 all produced statistically significant scores. The scores of distantly related proteins in the same family depended more on the choice of scoring matrix and gap penalty, and some scores were significant and others were not. Pearson recommends using caution in evaluating alignment scores using only one particular combination of scoring matrix and gap penalties. He also suggests that using a larger gap penalty, e.g., -14, -2 with BLOSUM50, can increase the selectivity of a database search for similarity (fewer sequences known to be unrelated will receive a significant alignment score).

A difficulty encountered by FASTA in calculating statistical parameters during a database search is that of distinguishing unrelated from related sequences, because only scores of unrelated sequences must be used. As score and sequence length information is accumulated during the search, the scores will include high, intermediate, and sometimes low scores of sequences that are related to the query sequence, as well as low scores and sometimes intermediate and even high scores of unrelated sequences. As an example, a high score with an unrelated database sequence can occur because the database sequence has a region of low complexity, such as a high proportion of one amino acid. Regardless of the reason, these high scores must be pruned from the search if accurate statistical estimates are to be made. Pearson (1998) has devised several such pruning schemes, and then determined the influence of the scheme on the success of a database search at demonstrating statistically significant alignment scores among members of the same protein family or superfamily. However, no particular scheme proved to be better than another.

Example: Use of the Above Principles to Estimate the Significance of a Smith-Waterman Local Alignment Score

The alignment shown in step 1 in the next example box is a local alignment between the phage λ and P22 repressor protein sequences used previously. The alignment is followed by a statistical analysis of the score in steps 2 and 3. To perform this analysis, the second sequence (the P22 repressor sequence) was shuffled 1000 times and realigned with the first sequence to create a set of random alignments. Two types of shuffling are available: first, a global type of shuffling in which random sequences are assembled based on amino acid composition and, second, a local one in which the random sequences are assembled by random selection of an amino acid from a sliding window of length n in the original sequence in order to preserve local amino acid composition as described on page 98 (an example of a global analysis is shown in step 2). The distribution of scores in each case was fitted to the extreme value distribution (Altschul and Gish 1996) to obtain estimates of λ and K to be used in the estimation of significance.

The program and parameters used were LALIGN (see Table 3.1 , p. 66), which produces the highest-scoring n independent alignments and which was described previously (p. 75), and the scoring matrix BLOSUM50 with a gap opening penalty of -12 and -2 for extra positions in the gap, with end gaps weighted. These programs do not presently have windows or Web page interfaces, and must be run using command line options.

The program PRSS performs a statistical analysis based on the correct statistical distribution of alignment scores, as shown below. PRSS version 3 (PRSS3) gives the results as z scores.

Example: Estimation of Statistical Significance of a Local Alignment Score

1. Optimal alignment of phage λ and P22 repressor sequences using the program LALIGN. The command line used was lalign -f -12 -g -2 lamc1.pro p22c2.pro 3> results.doc. The -f and -g flags indicate the gap opening and extension parameters to be used, and are followed by the sequence files in FASTA format, then a request for 3 alignments. No scoring matrix was specified and the default BLOSUM50 matrix was therefore used. Program output is directed to the file results.doc, as indicated by the symbol >. The alignment shown is the highest-scoring or optimal one using this scoring matrix and these gap penalties. The next two alignments reported were only 9 and 15 amino acids long and each one had a score of 35 (not shown). As discussed in the text, these alignments are produced by repeatedly erasing the previous alignment from the dynamic programming matrix and then rescoring the matrix to find the next best alignment. The fact that the first alignment has a much higher score than the next two is an indication that (1) there are no other reasonable alignments of these sequences and (2) the first alignment score is highly significant.

```
LALIGN finds the best local alignments between two sequences
version 2.0u64 March 1998
Please cite:
X. Huang and W. Miller (1991) Adv. Appl. Math. 12:373-381

Comparison of:
(A) lamc1.pro  LAMC1   REFORMAT of: cipro.pro  check: -1  from: 1  - 237 aa
(B) p22c2.pro  P22C2   REFORMAT of: p22  check: 4729  from: 1  to  - 216 aa
 using matrix file: blosum50.mat, gap penalties: -12/-2

   36.1% identity in 208 aa overlap; score:  401 [1/2 bits]

         30        40        50        60        70        80
LAMC1  KKNELGLSQESVADKMGMGQSGVGALFNGINALNAYNAALLAKILKVSVEEFSPSIAREI
       ....:  . :  ...  .:.... ..   .  . :.  :. ::.: . ... .
P22C2  RRKKLKIRQAALGKMVGVSNVAISQWERSETEPNGENLLALSKALQCSPDYLLKGDLSQT
          20        30        40        50        60        70

         90        100       110       120       130       140
LAMC1  YEMYEAVSMQPSLRSEYEYPVFSHVQAGMFSPELRTFTKGDAERWVSTTKKASDSAFWLE
       :..    .: :. : :.:.:.. .:::. ...  .. :  :  . ::  :...:::.
P22C2  NVAYHS-RHEP--RGSY--PLISWVSAGQWMEAVEPYHKRAIENWHDTTVDCSEDSFWLD
           80          90        100       110       120

         150       160       170       180       190       200
LAMC1  VEGNSMTAPTGSKPSFPDGMLILVDPEQAVEP--GDFCIARLGGD-EFTFKKLIRDSGQV
       :.:.:::::: ::  :  :.::.::::::: ::: .:  . .:.: . :: :::::::::::
P22C2  VQGDSMTAPAGL--SIPEGMIILVDPE--VEPRNGKLVVAKLEGENEATFKKLVMDAGRK
       130       140       150         160       170       180

         210       220       230
LAMC1  FLQPLNPQYPMIPCNESCSVVGKVIASQ
       ::.:::::::::  : .:..:: :..  ..
P22C2  FLKPLNPQYPMIEINGNCKIIGVVVDAK
       190       200       210
```

2. Statistical analysis with program PRSS using a global shuffling strategy. The program prompts for input information and requests the name of a file for saving output. The second sequence has been shuffled 1000 times conserving amino acid composition, and realigned to the first sequence. The distribution of scores is shown. Fitting the extreme value distribution to these scores provides an estimate of λ and K needed for performing the statistical estimate by Equation 31. Recent versions of PRSS estimate these parameters by the method of maximum likelihood estimation (Mott 1992; W. Pearson, pers. comm.) described on the book Web site.

```
  lamc1.pro, 237 aa vs p22c2.pro


         s-w  est
< 24     0    0:
  26     0    0:
  28     3    1:*==
  30    13    6:=====*=======
  32    27   21:===================*======
  34    68   50:=================================================*
  36    98   84:=================================================*
  38   128  111:=================================================*
  40   129  123:=================================================*
  42   105  121:=================================================*
  44   110  108:=================================================*
  46    63   91:=================================================*
  48    75   72:=================================================*
  50    35   56:===================================           *
  52    48   42:=========================================*======
  54    30   32:=========================     *
  56    19   23:=================     *
  58    17   16:==============*=
  60     6   13:======        *
  62     7    9:======= *
  64     7    6:=====*=
  66     2    5:==   *
  68     4    3:==*=
  70     0    2: *
  72     1    2:=*
  74     0    1:*
  76     1    1:*
  78     2    1:*=
  80     0    0:
  82     0    0:
  84     0    0:
  86     1    0:=
  88     1    0:=
  90     0    0:
  92     0    0:
  94     0    0:
> 96     0    0: O
 216000 residues in  1000 sequences,
 BLOSUM50 matrix, gap penalties: -12,-2
 unshuffled s-w score: 401; shuffled score range: 30 - 89
Lambda: 0.16931 K: 0.020441; P(401)= 3.7198e-27
For 1000 sequences, a score >=401 is expected 3.72e-24 times
```

The above method does not necessarily ensure that the choice of scoring matrix and gap penalties provides a realistic set of local alignment scores. In the comparable situation of matching a test sequence to a database of sequences, the scores also follow the extreme value distribution. For this situation, Mott (1992) has explained that for local alignments the end point of the alignment should on the average be half-way along the query sequence, and for global alignments, the end point should be beyond that half-way point. Pearson (1996) has pointed out that the presence of known, unrelated sequences in the upper part of the curve where $E > 1$ (see Chapter 7) can be an indication of an inappropriate scoring system.

The Statistical Significance of Individual Alignment Scores between Sequences and the Significance of Scores Found in a Database Search Are Calculated Differently

In performing a database search between a query sequence and a sequence database, a new comparison is made for each sequence in the database. Alignment scores between unrelated sequences are employed by FASTA to calculate the parameters of the extreme value distribution. The probability that scores between unrelated sequences could reach as high as those found for matched sequences can then be calculated (Pearson 1998). Similarly, in the database similarity search program BLAST, estimates of the statistical parameters are calculated based on the scoring matrix and sequence composition. The parameters are then used to calculate the probability of finding conserved patterns by chance alignment of unrelated sequences (Altschul et al. 1994). When performing such database searches, many trials are made in order to find the most strongly matching sequences.

As more and more comparisons between unrelated sequences are made, the chance that one of the alignment scores will be the highest one yet found increases. The probability of finding a match therefore has to be higher than the value calculated for a score of one sequence pair. The length of the query sequence is about the same as it would be in a normal sequence alignment, but the effective database sequence is very large and represents many different sequences, each one a different test alignment. Theory shows that the Poisson distribution should apply (Karlin and Altschul 1990, 1993; Altschul et al. 1994), as it did above for estimating the parameters of the extreme value distribution from many alignments between random sequences.

The probability of observing, in a database of D sequences, no alignments with scores higher than the mean of the highest possible local alignment scores s is given by e^{-Ds}, and that of observing at least one score s is $P \simeq 1 - e^{-Ds}$. For the range of values of P that are of interest, i.e., $P < 0.1$, $P = Ds$. If two sequences are aligned by PRSS as given in the above example, and the significance of the alignment is calculated, two scores must be considered. The probability of the score may first be calculated using the estimates of λ and K. Thus, in the phage repressor alignment, $P(s > 401) = 3.7 \times 10^{-27}$. However, to estimate the EV parameters, 1000 shuffled sequences were compared, and the probability that one of those sequences would score as high as 401 is given by Ds, or $1000 \times 3.7 \times 10^{-27} = 3.7 \times 10^{-24}$. These numbers are also shown in the statistical estimates computed by PRSS. Finally, if the score had arisen from a database search of 50,000 sequences, the probability of a score of 401 among this many sequence alignments is 5×10^{-19}, still a small number, but 50,000 larger than that for a single comparison. These probability calculations are used for reporting the significance of scores with database sequences by FASTA and BLAST, as described in Chapter 7.

SEQUENCE ALIGNMENT AND EVOLUTIONARY DISTANCE ESTIMATION BY BAYESIAN STATISTICAL METHODS

A recent development in sequence alignment methods is the use of Bayesian statistical methods to produce alignments between pairs of sequences (Zhu et al. 1998) and to calculate distances between sequences (Agarwal and States 1996). Before discussing these methods further, we provide some introductory comments about Bayesian probability.

Introduction to Bayesian Statistics

Bayesian statistical methods differ from other types of statistics by the use of conditional probabilities. These probabilities are used to derive the joint probability of two events or conditions. An example of a conditional probability is $P(B|A)$, meaning the probability of B, given A, whereas $P(B)$ is the probability of B, regardless of the value of A. Suppose that A can have two states, A1 and A2, and that B can also have two states, B1 and B2, as shown in Table 3.11. These states might, for instance, correspond to two allelic states of two genes. Then, $P(B) = P(B1) + P(B2) = 1$ and $P(A) = P(A1) + P(A2) = 1$. Suppose, further, that the probability $P(B1) = 0.3$ is known. Hence $P(B2) = 1 - 0.3 = 0.7$. In our genetic example, each probability might correspond to the frequency of an allele, for which p and q are often used. These probabilities $P(B1)$, etc., can be placed along the right margins of the table as the respective sum of each row or column and are referred to as the marginal probabilities.

Interest is now focused on filling in the missing data in the middle two columns of the table. The probability of A1 and B1 occurring together (the value to be entered in row B1 and column A1) is called the joint probability, $P(B1 \text{ and } A1)$ (also denoted $P[B1, A1]$). The marginal probability $P(A1)$ is also missing. The available information up to this point, called the prior information, is not enough to calculate the joint probabilities. With additional data on the co-occurrence of A1 with B1, etc., these joint probabilities may be derived by Bayes' rule. Suppose that the conditional probabilities $P(A1|B1) = 0.8$ and $P(A2|B2) = 0.70$ are known, the first representing, for example, the proportion of a population with allele B1 that also has allele A1. First, note that $P(A1|B1) + P(A2|B1) = 1$, and hence that $P(A2|B1) = 1.0 - 0.8 = 0.2$. Similarly, $P(A1|B2) = 1.0 - 0.70 = 0.3$. Then the joint probabilities and other conditional probabilities may be calculated by Bayes' rule, illustrated using the joint probability for A1 and B1 as an example.

$$P(A1 \text{ and } B1) = P(B1)\, P(A1|B1) \tag{38}$$

$$P(A1 \text{ and } B1) = P(A1)\, P(B1|A1) \tag{39}$$

Thus, $P(A1 \text{ and } B1) = P(B1) \times P(A1|B1) = 0.3 \times 0.8 = 0.24$, and $P(A2 \text{ and } B2) = P(B2) \times P(A2|B2) = 0.7 \times 0.7 = 0.49$. The other joint probabilities may be calculated by subtraction; e.g., $P(A2 \text{ and } B1) = P(B1) - P(A1 \text{ and } B1) = 0.30 - 0.24 = 0.06$. To calculate

Table 3.11. *Prior information for a Bayes analysis*

	A1	A2	
B1			0.3
B2			0.7
			1.0

Table 3.12. *Completed table of joint and marginal probabilities*

	A1	A2	
B1	0.24	0.06	0.3
B2	0.21	0.49	0.7
	0.45	0.55	1.0

$P(A1)$ and $P(A2)$, the joint probabilities in each column may be added, thereby completing the additions to the table, and shown in Table 3.12.

However, note that $P(A1)$ may also be calculated in the following manner,

$$P(A1) = P(A1 \text{ and } B1) + P(A1 \text{ and } B2)$$
$$= P(B1)\,P(A1|B1) + P(B2)\,P(A1|B2) \tag{40}$$

Other conditional probabilities may be calculated from Equations 38 and 39 by rearranging terms and by substituting Equation 40, and the following form of Bayes' rule may be derived,

$$P(B2|A1) = P(A1 \text{ and } B2) / P(A1)$$
$$= P(B2)\,P(A1|B2) / P(A1)$$
$$= P(B2)\,P(A1|B2) / [P(B1)\,P(A1|B1) + P(B2)\,P(A1|B2)] \tag{41}$$

Using Equation 41, $P(B2|A1) = 0.7 \times 0.30 / [0.3 \times 0.80 + 0.7 \times 0.3] = 0.467$, and also $P(B1|A1) = 1.0 - 0.467 = 0.533$. Such calculated probabilities are called posterior probabilities or posteriors, as opposed to the prior probabilities or priors initially available. Thus, based on the priors and additional information, application of Bayes' rule allows the calculation of posterior estimates of probabilities not initially available. This procedure of predicting probability relationships among variables may be repeated as more data are collected, with the existing model providing the prior information and the new data providing the information to derive a new model. The initial beliefs concerning a parameter of interest are expressed as a prior distribution of the parameter, the new data provide a likelihood for the parameter, and the normalized product of the prior and likelihood (Eq. 41) forms the posterior distribution.

Example: Bayesian Analysis

Another illustrative example of a Bayesian analysis is the game played by Monty Hall in the television game show "Let's Make a Deal." Behind one of three doors a prize is placed by the host. A contestant is then asked to choose a door. The host opens one door (one that he knows the prize is not behind) and reveals that the prize is not behind that door. The contestant is then given the choice of changing to the other door of the three to win. The initial or prior probability for each door is 1/3, but after the new information is provided, these probabilities must be revised. The original door chosen still has a probability of 1/3, but the second door that the prize could be behind now has a probability of 2/3. These new estimates are posterior probabilities based on the new information provided.

In the above example, note that the joint probability of A1 and B1 [P(A1 and B1)] is not equal to the product of P(A1) and P(B1); i.e., 0.24 is not equal to $0.3 \times 0.45 = 0.135$. Such would be the case if the states of A and B were completely independent; i.e., if A and B were statistically independent variables as, for example, in a genetic case of two unlinked genes A and B. In the above example, the state of one variable is influencing the state of the other such that they are not independent of each other, as might be expected for linked genes in the genetic example.

A more general application of Bayes' rule is to consider the influence of several variables on the probability of an outcome. The analysis is essentially the same as that outlined above. To see how the method works with three instead of two values of a variable, think first of an example of three genes, each having three alleles, and of deriving the corresponding conditional probabilities. The resulting joint probabilities will depend on the choice made of the three possible values for each variable. To go even farther, instead of a small number of discrete sets of alternative values of a variable, Bayesian statistical methods may also be used with a large number of values of variables or even with continuous variables.

For sequence analysis by Bayesian methods, a slightly different approach is taken. The variables may include combinations of possible alignments, gap scoring systems, and log odds substitution matrices. The most probable alignments may then be identified. The scoring system used for sequence alignments is quite readily adapted to such an analysis. In an earlier discussion, it was pointed out that a sequence alignment score in bits is the logarithm to the base 2 of the likelihood of obtaining the score in alignments of related sequences divided by the likelihood of obtaining the score in alignments of unrelated sequences. It was also indicated that the highest alignment score should be obtained if the scoring matrix is used that best represents the nucleotide or amino acid substitutions expected between sequences at the same level of evolutionary distance. Bayesian methodology carries this analysis one step farther by examining the probabilities of all possible alignments of the sequences using all possible variations of the input parameters and matrices. These selections are the prior information for the Bayesian statistical analysis and provide various estimates of the alignment that allow us to decide on the most probable alignments. The alignment score for each combination of these variables provides an estimate of the probability of the alignment. By using equations of conditional probability such as Equation 41, posterior information on the probability of alignments, gap scoring system, and substitution matrix can be obtained. For further reading, a Bayesian bioinformatics tutorial by C. Lawrence is available at http://www.wadsworth.org/resnres/bioinfo/.

Application of Bayesian Statistics to Sequence Analysis

To use an example from sequence analysis, a local alignment score (s) between two sequences varies with the choice of scoring matrix and a gap scoring system. In the previous sections, an amino acid scoring matrix was chosen on the basis of its performance in identifying related sequences. Gap penalties were then chosen for a particular scoring matrix on the basis of their performance in identifying known sequence relationships and of their keeping a local alignment behavior by the increase in score between unrelated sequences remaining a logarithmic function of sequence length. The alignment score expressed in bit units was the ratio of the alignment score expected between related sequences to that expected between unrelated sequences, expressed as a logarithm to the base 2. The scores may be converted to an odds ratio (r) using the formula $r = 2^s$. The probability of such a score between unrelated or random sequences can then be calculated using the parameters for the extreme value distribution for that combination of scoring matrix and gap penalty. Finally, the above analysis may provide several different alignments, without providing any information as to which is the most likely. With the application of Bayesian statistics, the approach is different.

The application of Bayesian statistics to this problem allows one to examine the effect of prior information, such as the chosen amino acid substitution matrix, on the probability that two sequences are homologous. The method provides a posterior probability distribution of all alignments taking into account all possible scoring systems. Thus, the most likely alignments and their probabilities may be determined. This method circumvents the need to choose a particular scoring matrix and gap scoring system because a range of available choices can be tested. The approach also provides conditional posterior distributions on the gap number and substitution matrix. Another application of Bayes statistics for sequence analysis is to find the PAM DNA substitution matrix that provides the maximum probability of a given level of mismatches in a sequence alignment, and thus to predict the evolutionary distance between the sequences.

Bayesian Evolutionary Distance

Agarwal and States (1996) have applied Bayesian methods to provide the best estimate of the evolutionary distance between two DNA sequences. The examples used are sequences of the same length that have a certain level of mismatches. Consequently, there are no gaps in the alignment between the sequences. Sequences of this type originated from gene duplication events in the yeast and *Caenorhabditis elegans* genomes. When there are multiple mismatches between such repeated sequences, it is difficult to determine the most likely length of the repeats. With the application of Bayesian methods, the most probable repeat length and evolutionary time since the repeat was formed may be derived.

The alignment score in bits between sequences of this type may be calculated from the values for matches and mismatches in the DNA PAM scoring matrices described earlier (Table 3.6). Recall that a PAM1 evolutionary distance represents a change of 1 sequence position in 100 and is thought to correspond roughly to an evolutionary distance of 10^7 years. Higher PAMN tables are calculated by multiplying the PAM1 scoring matrix by itself n times. This Markovian model of evolution assumes that any sequence position can change with equal probability, and subsequent changes at a site are not influenced by preceding changes at that site. In addition, a changed position can revert to the original nucleotide at that position. The problem is to discover which scoring matrix (PAM50, 100, etc.) gives the most likely alignment score between the sequences. This corresponding evolutionary distance will then represent the time at which the sequence duplication event could have occurred.

An approach described earlier was to evaluate the alignment scores using a series of matrices and then to identify the matrix giving the highest similarity score. For example, if there are 60 mismatches between sequences that are 100 nucleotides long, the PAM50 matrix score of the alignment in bits ($\log_2$) is $40 \times 1.34 - 60 \times 1.04 = -8.8$, but the PAM125 matrix score is much higher, $40 \times 0.65 - 60 \times 0.30 = 8$. When these log odds scores in bits are converted to odds scores, the difference is 0.002 versus 256. Thus, the PAM125 matrix provides a much better estimate of the evolutionary distance between sequences that have diverged to this degree. The Bayesian approach continues this type of analysis to discover the probability of the alignment as a function of each evolutionary distance represented by a different PAM matrix. If x is the evolutionary distance represented by the PAMN matrix divided by 100, and k is the number of mismatches in a sequence of length n, then by Bayes' rule and related formulas discussed above

$$P(x|k) = P(k|x)\,P(x)\,/\,P(k)$$
$$= P(k|x)\,P(x)\,/\,\Sigma_x\,P(k|x) \tag{42}$$

$P(x|k)$ is the probability of distance x given the sequence with k mismatches (and $n - k$ matches), $P(k|x)$ is the odds score for the sequence with k mismatches using the log odds scores in the DNA PAM100x matrix, and $P(x)$ is the prior probability of distance x (usually 1 over the number of matrices, thus making each one equally possible). The denominator is the sum of the odds scores over the range of x, which is $0.01 - 4$, representing PAM1 to PAM400 ($\sim$10 million to 4 billion years) times the prior probability of each value of x. Like the conditional probabilities calculated by Equation 42, this sum represents the area under the probability curve and has the effect of normalizing the probability for each individual scoring matrix used. The shape of the probability curve reveals how $P(x|k)$ varies with x. An example is shown in Figure 3.19.

The probability curves have a single mode or highest score for $k < 3n/4$. Because the curves are not symmetrical about this mode but are skewed toward higher distances, the expected value or mean of the distribution and its standard deviation are the best indication of evolutionary distance. For a sequence 100 nucleotides long with 40 mismatches, the expected value of x is 0.60 with $s = 0.11$, representing a distance of $\sim$600 million years. These estimates are different from the earlier method that was described of finding the matrix that gives the highest alignment score, which would correspond to the mode or highest scoring distance. Other methods of calculating evolutionary distances are described in Chapter 6.

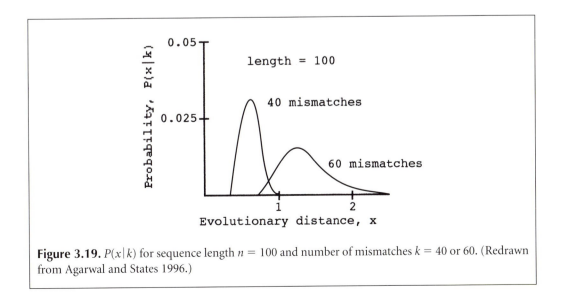

Figure 3.19. $P(x|k)$ for sequence length $n = 100$ and number of mismatches $k = 40$ or 60. (Redrawn from Agarwal and States 1996.)

Working with Odds Scores

Odds scores, and probabilities in general, may be either multiplied or added, depending on the type of analysis. If the purpose is to calculate the probability of one event AND a second event, the odds scores for the events are multiplied. An example is the calculation of the odds of an alignment of two sequences from the alignment scores for each of the matched pairs of bases or amino acids in the alignment. The odds scores for the pairs are multiplied. Usually, the log odds score for the first pair is added to that for the second, etc., until the scores for every pair have been added. An odds score of the alignment in units of logarithm to the base 2 (bits) may then be calculated by the formula odds score = 2 raised to the power of the log odds score. A second type of probability analysis is to calculate the odds score for one event OR a second event, or of a series of events (event 1 OR event 2 OR event 3). In this case, the odds scores are added. An example is the calculation of the odds score for a given sequence alignment using a series of alternative PAM scoring matrices. The alignment scores are calculated in log odds units and then converted into odds scores as described above. The odds scores for the sequences using matrix 1 are added to the odds score using matrix 2, then to the score using matrix 3, and so on, thereby generating the odds score for the set of matrices. From this sum of odds scores, the probability of obtaining one of the odds scores S is S divided by the sum. There are also a number of other uses of this same type of calculation for locating common patterns in a set of sequences by statistical methods that are discussed in Chapter 4.

One difficulty with making such estimations is that the estimate depends on the assumption that the mutation rate in sequences has been constant with time (the molecular clock hypothesis) and that the rate of mutation of all nucleotides is the same. Such problems may be solved by scoring different portions of a sequence with a different scoring matrix, and then using the above Bayesian methods to calculate the best evolutionary distance. Another difficulty is deciding on the length of sequence that was duplicated. In genomes, the presence of repeats may be revealed by long regions of matched sequence positions dispersed among regions of sequence positions that do not match. However, as the frequency of mismatches is increased, it becomes difficult to determine the extent of the repeated region. The application of the above Bayesian analysis allows a determination of the probability distributions as a function of both length of the repeated region and evolutionary distance. A length and distance that gives the highest overall probability may then be determined. Such alignments are initially found using an alignment algorithm and a particular scoring matrix. Analysis of the yeast and *C. elegans* genomes for such repeats has underscored the importance of using a range of DNA scoring matrices such as PAM1 to PAM120 if most repeats are to be found (Agarwal and States 1996). One disadvantage of the Bayesian approach is that a specific mutational model is required, whereas other methods, such as the maximum likelihood approach described in Chapter 6, can be used to estimate the best mutational model as well as the distance. Computationally, however, the Bayesian method is much more practical.

Bayesian Sequence Alignment Algorithms

Zhu et al. (1998) have devised a computer program called the Bayes block aligner which in effect slides two sequences along each other to find the highest scoring ungapped regions or blocks. These blocks are then joined in various combinations to produce alignments. There is no need for gap penalties because only the aligned sequence positions in blocks are scored. Instead of using a given substitution matrix and gap scoring system to find the highest scoring alignment, a Bayesian statistical approach is used. Given a range of substitution matrices and number of blocks expected in an alignment as the prior information, the method provides posterior probability distributions of alignments. The Bayes aligner is available through a licensing agreement from http://www.wadsworth.org/resnres/bioinfo. A graphical interface for X windows in a UNIX environment and a nongraphical interface for PCs running Windows are available. The method may be used for both protein and DNA sequences. An alignment block between two sequences is defined as a run of one or more identical characters in the sequence alignment that can include intervening mismatches but no gaps, as shown in the following example. Only the aligned blocks are identified and scored; regions of unaligned sequence and gaps between these blocks are not scored. The probability of a given alignment is given by the product of the probabilities of the individual alignment scores in the blocks, as indicated in the following example. The Bayes block aligner scores every possible combination of blocks to find the best scoring alignment.

Example: Block Alignment of Two Sequences and of the Scoring of the Alignment as Used in the Bayes Block Aligner (Zhu et al. 1998)

The score of the alignment is obtained by adding the log odds scores of each aligned pair in each block. Sequence not within these blocks is not scored and there is no penalty for gaps. Regions of both sequences that are not aligned can be present within the gap. The sequence alignment score is therefore determined entirely by the placement of block boundaries.

```
                     Block 1              Block 2

Sequence 1      S  G  T  G  K (gap) K  K  R  L  E
Sequence 2      P  G  S  G  K (gap) K  Q  R  L  T


BLOSUM62
score          -1  6  1  6  5        5  1  5  4  -1


Sum of scores               =    31 half bits
                            =    15.5 bits


Odds of alignment score


                            =    2^{15.5} to 1
                            =    4.6 x 10^4 to 1
```

Unlike the commonly used methods for aligning a pair of sequences, the Bayesian method does not depend on using a particular scoring matrix or designated gap penalties. Hence, there is no need to choose a particular scoring system or gap penalty. Instead, a number of different scoring matrices and range of block numbers up to some reasonable maximum are examined, and the most probable alignments are determined. The Bayesian method provides a distribution of alignments weighted according to probability and can also provide an estimate of the evolutionary distance between the sequences that is independent of scoring matrix and gaps.

Like dynamic programming methods and the BLAST and FASTA programs, the Bayes block aligner has been used to find similar sequences in a database search. The most extensive comparisons of database searches have shown that the program SSEARCH based on the Smith-Waterman algorithm, with the BLOSUM50 -12,-2 matrix and gap penalty scoring system, can find the most members of protein families previously identified on the basis of sequence similarity (Pearson 1995, 1996, 1998) or structural homology (Brenner et al. 1998). In a similar comprehensive analysis, Zhu et al. have shown that the Bayes block aligner has a slightly better rate than even SSEARCH of finding structurally related sequences at a 1% false-positive level. Hence, this method may be the best one to date for database similarity searching.

The Bayes block aligner defines blocks by an algorithm due to Sankoff (1972). This algorithm is designed to locate blocks by finding the best alignment between two sequences for any reasonable number of blocks. The example shown in Figure 3.20 illustrates the basic block-finding algorithm.

Following the initial finding of block alignments in protein sequences by the Sankoff method, the Bayes block aligner calculates likelihood scores for these alignments for various block numbers and amino acid or DNA substitution matrices. To be biologically more meaningful by avoiding too many blocks, the number of protein sequence blocks k is limited from zero to 20 or the length of the shorter sequence divided by 10, whichever is smaller. For a set of amino acid substitution matrices such as the Dayhoff PAM or BLOSUM matrices, the only requirement is that they should be in the log odds format in order to provide the appropriate likelihood scores by additions of rows and columns in the V and W matrices (Fig. 3.20). A large number of matrices like the V and W matrices in Figure 3.20 are used, each for a different amino acid substitution matrix and block number. In each of these matrices, a number of alignments of the block regions that are found are possible. The score in the lower right-hand corner of each matrix is the sum of the odds scores of all possible alignments in that particular matrix. The odds scores thus calculated in each matrix are summed to produce a grand total of odds scores. The fraction of this total that is shared by a set of alignments under given conditions (e.g., a given number of blocks or an amino acid substitution matrix) provides the information needed to calculate the most probable scoring matrix, block number, etc., by Bayesian formulas. The joint probabilities equivalent to the interior row and column entries in Tables 3.11 and 3.12 are then calculated. In this case, each joint probability is the likelihood of the alignment given a particular block alignment, number of blocks, and substitution matrix, multiplied by the prior probabilities. These prior probabilities of particular alignment, block number, and scoring matrix are treated as having an equally likely prior probability. Once all joint probabilities have been computed for every combination of the alignment variables, the conditional posterior information can be obtained by Bayes' rule, using equations similar to Equation 41. As in Equation 41, the procedure involves dividing the sum of all alignment likelihoods that apply to a particular value of a particular variable by the sum of all alignment likelihoods found for all variables.

Use of the Bayes Block Aligner for Pair-wise Sequence Alignment

There are several possible uses of the Bayes block aligner for sequence alignment. The overall probability that a given pair of residues should be aligned may be found by several methods. In the first, alignments may be sampled in proportion to their joint posterior probability, as for example, alignments produced by a particular combination of substitution matrix and gap number. A particular substitution matrix and gap number may be chosen based on their posterior probabilities. An alignment may then be obtained from the alignment matrix in much the same manner as the trace-back procedure used to find an alignment by dynamic programming. Once a number of sample alignments has been obtained, these samples may be used to estimate the marginal distribution of all alignments. This distribution then gives the probability that each pair of residues will align. An alternative method of sampling the joint posterior probability distribution is to identify an average alignment for k blocks by sampling the highest peaks in the marginal posterior alignment distribution and by using each successively lower peak as the basis for another alignment block down to a total of k blocks, concatenating any overlaps. These alignments may then be used to obtain the probability of each aligned residue. In the second method, the exact marginal posterior alignment distribution of a specific pair of residues may be obtained by summing over all substitution matrices and possible blocks.

Third, optimal alignment and near-optimal alignments for a given number of blocks can also be obtained. Finally, the Bayes block aligner provides an indication as to whether or not the sequence similarity found is significant. Bayesian statistics examines the posterior probabilities of all alternative models over all possible priors. The Bayesian evidence that two sequences are related is given by the probability that K, the maximum allowed number of blocks, is greater than 0, as calculated in the following example taken from Zhu et al. (1998). The posterior probability of the number of blocks, the substitution matrices, and the aligned residues can all be calculated as described above.

Example: Bayes Block Aligner (Zhu et al. 1998)

The proteins guanylate kinase from yeast (PDB id. 1GKY) and adenylate kinase from beef heart (PDB id. 2AK3, chain A) are known to be structurally related and are from a database of protein sequences that are 26–35% identical. These proteins were aligned with the Bayes block aligner using as prior information an equal chance that the block number k can be any number between 0 and 18, and that the BLOSUM30 to 100 substitution matrices can each equally well predict the aligned positions. The posterior probability distribution of the number of blocks, k, is shown in Figure 3.21A. Values $k > 0$ indicate the possibility of finding one or more blocks. In this example, the probability for values of k is approximately the same for $k > 8$. Below 8, the values decrease gradually to a low value at $k = 1$ and then increase again abruptly for $k = 0$. The total area under the curve from $k = 0$ to $k = 18$ has been set to 1.

The cumulative posterior probability that the block number K is greater than a given value k is shown in Figure 3.21B. The area under the curve for $k \geq 1$ has the value 0.938. Although at first glance this number appears to represent the probability that the sequences are related, i.e., that $K > 0$, the probability is actually higher by Bayesian standards. Instead, the maximum value for $P(k|\text{sequences})$ in Figure 3.21A, i.e., 0.0731 at $k = 8$, is used. This number times the maximum number of blocks $0.0731 \times 18 = 1.316$, represents the accumulated best evidence that the blocks are related or that $K > 0$. This calculation assumes that all block numbers are equally likely or that $p(k|k>0) = 1/K = 1/18$. The value $P(k = 0|\text{sequences}) = 0.0621$ is the corresponding best evidence that the sequences are not related or that $K = 0$. The probability that the sequences are related is then calculated as $1.316 / (1.316 + 0.0621) = 0.955$. This value is the supremum of $P(k > 0)$ taken over all prior distributions on k, where the supremum is a mathematical term that refers to the least upper bound of a set of numbers. This high a Bayesian probability is strong evidence for the hypothesis that the sequences are homologous. Normally, a Bayesian probability of $p > 0.5$ will suffice (Zhu et al. 1998).

The posterior probability distribution for the BLOSUM scoring matrices for alignment of these same two proteins is shown in Table 3.13. Note that the highest probabilities are for BLOSUM tables between BLOSUM50 and BLOSUM 80, and that the highest probability is at BLOSUM62, which is commonly used for protein sequence alignment and database searches. Thus, BLOSUM62 seems best to represent the amino acid substitutions observed in all of the computed alignments between these two proteins. In another alignment of 1GKY and 2AK3-A using the Dayhoff PAM matrices instead of the BLOSUM matrices, the posterior probability distribution of the matrices shown in Figure 3.22 was found. Note that peaks are found at PAM110,

Figure 3.20. The Sankoff algorithm for finding the maximum number of identical residues in two sequences without scoring gaps. The example of two DNA sequences shown is taken from Sankoff (1972). A series of scoring matrices called V and W are made according to the matrix scoring scheme shown in parts A—D. In A, the algorithm first examines the maximum number of bases that can match. The scoring scheme used in this example is that a match between two bases is scored as 1 and a mismatch as 0. This number, 4, is shown in the lower right-hand corner of the matrix. To obtain this number, the method does not consider the number of gapped regions between each group of matched pairs, defined as an unconstrained set of matches by Sankoff. For example, a_1 can pair with b_3, and a_2 with b_4, to comprise a group of two sequential pairs, shown in bold. Then there is an unmatched region followed by a match of a_4 with b_6, unmatched base a_5, and finally a match between a_6 and b_7. Thus, two unmatched (gapped) regions will be included in this alignment. A second such set of matches that gives a maximum number of matches is shown as italicized positions. In this case, there is one unmatched region between the groups of matches. In B–D, a slightly different computational method is used to find the maximum possible number of matches given that there are zero gapped regions, one gapped region, two gapped regions, etc. In B, a matrix V_0, where subscript 0 indicates the number of gapped regions permitted, is first calculated. The bold and italicized positions indicate the scores found for the two groups of matches. To simplify the calculation of higher-level V matrices (V_1, V_2, etc.), another set of matrices (W_1, W_2, etc.) is also calculated. In C, the calculation of W_0 is shown. Using the scores calculated in W_0, matrix position and the algorithm shown in D, V_1 is then produced. V_1 shows the same combinations of matches found in the

A. W matrix

	j	b	C	C	A	G	T	C	T
i		0	1	2	3	4	5	6	7
a	0	0	0	0	0	0	0	0	0
A	1	0	0	0	1	1	1	1	1
G	2	0	0	0	0	2	2	2	2
C	3	0	*1*	1	1	2	2	3	3
C	4	0	0	2	2	2	2	3	3
A	5	0	1	2	3	3	3	3	3
T	6	0	0	1	2	3	4	4	4

$W(i,j)$

$$= \max \begin{cases} W(i-1,j), \\ W(i,j-1), \\ W(i-1,j-1) + s(a_i, b_j) \end{cases}$$

where $s(a_i, b_j)$ is score of match of a_i with b_j.

B. V_0 matrix

	j	b	C	C	A	G	T	C	T
i		0	1	2	3	4	5	6	7
a	0	0	0	0	0	0	0	0	0
A	1	0	0	0	1	0	0	0	0
G	2	0	0	0	0	2	0	0	0
C	3	0	*1*	1	0	0	2	1	0
C	4	0	1	2	1	0	1	3	3
A	5	0	1	1	3	1	0	0	3
T	6	0	0	0	1	3	2	0	1

$V_0(i,j) =$

$$V_0(i-1, j-1) + s(a_i, b_j)$$

C. W_0 matrix

	j	b	C	C	A	G	T	C	T
i		0	1	2	3	4	5	6	7
a	0	0	0	0	0	0	0	0	0
A	1	0	0	0	1	1	1	1	1
G	2	0	0	0	1	2	2	2	2
C	3	0	1	1	1	2	2	2	2
C	4	0	1	2	2	2	2	3	3
A	5	0	1	2	3	3	3	3	3
T	6	0	1	2	3	3	3	3	3

$W_0(i,j)$

$$= \max \begin{cases} W_0(i-1, j), \\ V_0(i,j), \\ W_0(i,j-1) \end{cases}$$

where $V_0(i, j)$ is from the V_0 matrix in part B.

D. V_1 matrix

	j	b	C	C	A	G	T	C	T
i		0	1	2	3	4	5	6	7
a	0	0	0	0	0	0	0	0	0
A	1	0	0	0	1	0	0	0	0
G	2	0	0	0	0	2	1	1	1
C	3	0	*1*	1	0	1	2	3	2
C	4	0	1	2	1	1	2	3	3
A	5	0	0	1	3	2	2	2	3
T	6	0	0	1	2	3	4	3	4

$V_1(i, j)$

$$= \max \begin{cases} V_1(i-1, j-1), \\ W_0(i-1, j-1) \\ \quad + s(a_i, b_j) \end{cases}$$

where $W_0(i, j)$ are obtained from the W_0 matrix in part C.

unconstrained case in A, and, therefore, no further calculation of matrices is necessary. In other cases, q V and W matrices will be calculated so that alignments with an increased number of unmatched or gapped regions may be found according to the formulas:

$$W_q(i, j) = \max \begin{cases} W_q(i-1, j), \\ V_q(i, j), \\ W_q(i, j-1) \end{cases}$$

$$V_q(i, j) = \max \begin{cases} V_q(i-1, j-1), \\ W_{q-1}(i-1, j-1) \\ \quad + s(a_i, b_j) \end{cases}$$

The number of computational steps required is equal to the product of the sequence lengths times the number of cycles needed to reach the unconstrained alignment, as shown in the lower right-hand corner of the matrix (A). The method may also be used for aligning protein sequences (Zhu et al. 1998) that are distantly related, as described below.

Table 3.13. *Posterior probability distribution of BLOSUM scoring matrices for alignment of 1GKY and 2AK3-A*

Matrix	Posterior probability
BLOSUM30	0.0257
BLOSUM35	0.0449
BLOSUM40	0.0825
BLOSUM45	0.1115
BLOSUM50	0.1755
BLOSUM62	0.2867
BLOSUM80	0.2350
BLOSUM100	0.0382

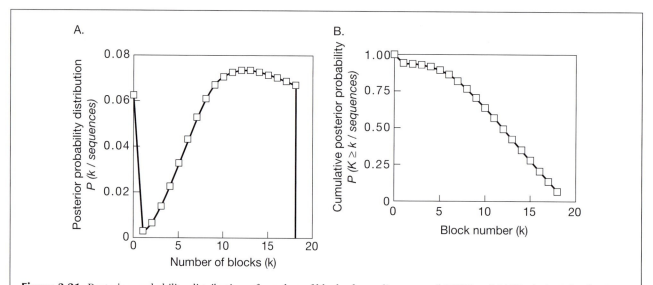

Figure 3.21. Posterior probability distribution of number of blocks from alignment of 1GKY and 2AK3-chain A by the Bayes block aligner (analysis of Zhu et al. 1998). (*A*) Posterior probability distribution of the block number, *k*. (*B*) Cumulative posterior probability distribution. This distribution shows the probability of a block number *K* greater than or equal to the value *k*. Values are derived from the probability distribution of *k* given in A. For example, $P(k{\geq}1) = P(k{\geq}0) - P(k{=}0) = 1 - 0.062 = 0.938$.

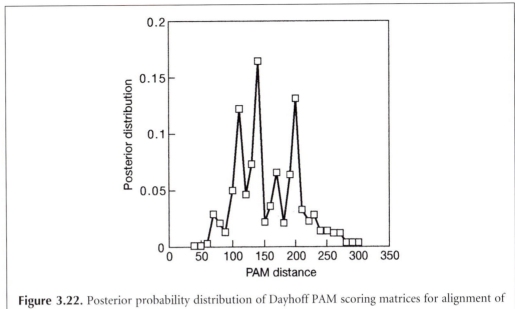

Figure 3.22. Posterior probability distribution of Dayhoff PAM scoring matrices for alignment of 1GKY and 2AK3-A.

140, and 200, thereby suggesting that substitution matrices for different evolutionary distances reflect the observed substitutions in different block alignments. The lower PAM matrix may be recognizing a more conserved domain, for example. This interesting observation implies that the alignment blocks found may be separated by different evolutionary distances, or at least may have undergone increased mutational variation. Thus, this type of analysis can provide information as to the evolutionary history of genes, including the possible involvement of duplications, rearrangements, and genetic events producing chimeras.

Another type of analysis that can be performed with the Bayes block aligner is to examine the probability of the alignments. The procedure is entirely different from other methods of sequence alignment such as dynamic programming. On the one hand, with dynamic programming methodology, a single best alignment is found for a given scoring matrix and gap penalty, and the odds for finding as good a score between random sequences of the same length and complexity is determined. On the other hand, with Bayesian alignment methods, all possible alignments are considered for a reasonable number of blocks and a set of substitution matrices. Rather than a probability of a single alignment, the probabilities of many alignments are provided. Many possible alignments may be examined and compared, and the frequency of certain residues in the sequences in these alignments may be determined.

For 1GKY and 2AK3-A, no highly probable single optimal or near-optimal alignment is found, suggesting these alignments are not representative of the best possible alignment of these sequences. Experience with the method has suggested that a minimum number of blocks that best represents the expected domain structure is the best approach. An average

A. Bayes block aligner

I

```
                              . . . . . . . .
                        : : : : : : : : : : : : : .
                    : : : : : : : : : : : : : : : : : .
                . : : : : : : : : : : : : : : : : : : : : : .
     1  U S R P I V I S G P S G T G K S T L L K K L F A E Y P D S F G  31
     5  R L L R A A I M G A P G S G K G T V S S R I T K H F E L K H L  35
            *   *      *    * *    *
        s s s s s s s s s s s s s s s s s s s s s s s s s s s
```

II

```
              . . . . . . : : . . . . . . . . . .
    54  V S V D E F K S M I K N N E F I E W A Q F  74
    73  L V L H E L K N L T Q Y N W L L D G F P R  93
            *     *                *
```

III

```
        . . . . . : : : :
   126  V E D L K K R L E  134
   117  F E V I K Q R L T  125
            *        * *
   +12  s s s s s s s s s  +2
```

IV

```
                        . . : : : : : : : : . . . : .
                  . . : : : : : : : : : : : : : : : : .
              : : : : : : : : : : : : : : : : : : : : . . .
   135  G R G T E T E E S I N K R L S A A Q A E L A Y A E  159
   159  Q R E D D R P E T V V K R L K A Y E A Q T E P V L  178
            *              *      * * *    *      *
        s s s s s s s s s s s s s s s s s s s s s s s s s  +3
```

B. SSEARCH

```
   123  P P S - - - V E D L K K R - L E G R G T E T E E S I N K R L S A A Q A E  154
   143  P P K T M G I D D L T G E P L V Q R E D D R P E T V V K R L K A Y E A Q  178
            * *          * *        *      *              * * *        *
```

Figure 3.23. The alignment of 1GKY and 2AK3-A obtained with the Bayes aligner (*A*) and by SSEARCH (*B*), a dynamic programming method that provides local alignments (from Zhu et al. 1998). The highest-scoring sequence positions in the marginal posterior alignment distribution for the sequences for a block number of probability greater than 0.9 and the BLOSUM substitution matrices were successively sampled, and are shown in *A*. Neighboring aligned positions with scores greater than 0.25 of the peak value were included. Dots above the sequences indicate the relative probability of the aligned sequence positions. Asterisks are placed to highlight sequence identities. There is a clear correlation between the number of identities and the posterior probabilities. Alignment positions marked with an 's' were also identified by structural alignment using the program VAST (see Chapter 9). In regions III and IV, longer aligned regions were found by VAST than by the Bayes aligner. Three other regions identified by VAST of lengths 7, 7, and 8, two of which include 1–2 identities, were not reported by the Bayes aligner. In *B*, a local alignment of the sequences with SSEARCH is shown. The alignment parameters (BLOSUM50 substitution table and scoring penalties of −12,−2) are optimized for superfamily and family alignments. The center and right end of the alignment shown are approximately the same as that of alignment IV, but gaps are incorrectly predicted in the left end.

alignment for a number of blocks of probability greater than 0.9 has been found to give good agreement with predicted structural alignments. Values of k are obtained from the probability distribution for k such as in Figure 3.21. Using this approach with the Bayes aligner, the alignments between 1GKY and 2AK3-A shown in Figure 3.23 have been predicted. Although most of the predicted alignments correspond to expected structural alignments with the active site of the enzyme, alignment II does not so correspond (Fig. 3.24). Such false-negative predictions of structural alignments are the commonest error of Bayesian methods, probably because of relaxed conditions for scoring alignments in the

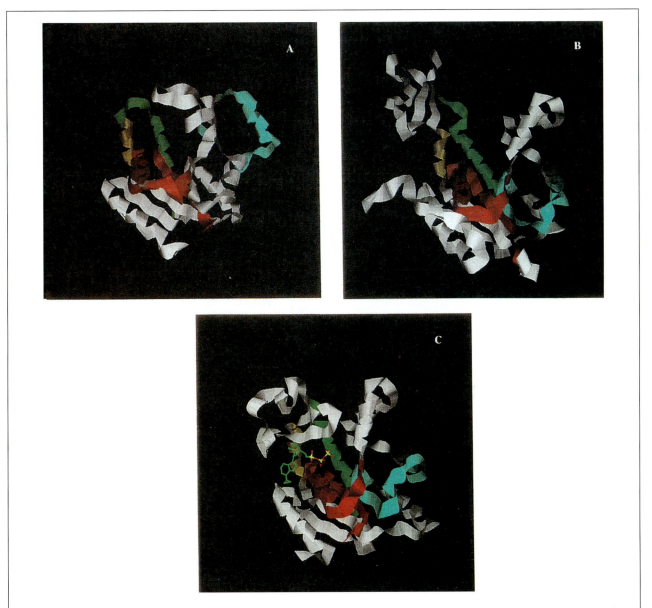

Figure 3.24. The positions of the alignments predicted by the Bayes block aligner. Predicted alignment I is shown in *red*, II in *cyan*, III in *orange*, and IV in *green*. (A) 1GKY, (B) 2AK3-A, and (C) 2AKY, which is similar to 2AK3-A. 2AKY is cocrystallized with an ATP analog. I, III, and IV may be structurally superimposed, but not II. (Reprinted, with permission, from Zhu et al. 1998 [copyright Oxford University Press].)

use of unconstrained prior information (Zhu et al. 1998). For these proteins, which share little sequence identity, the Bayes aligner correctly predicts many, but not all, features of the structural alignment, and does so better than a dynamic programming method that provides local alignments. In other cases, the Bayes aligner may not perform as well as dynamic programming. The prudent choice is to use the Bayes aligner as one of several computer tools for aligning sequences.

REFERENCES

Abagyan R.A. and Batalov S. 1997. Do aligned sequences share the same fold? *J. Mol. Biol.* **273:** 355–368.

Agarwal P. and States D.J. 1996. A Bayesian evolutionary distance for parametrically aligned sequences. *J. Comput. Biol.* **3:** 1–17.

———. 1998. Comparative accuracy of methods for protein sequence similarity search. *Bioinformatics* **14:** 40–47.

Altschul S.F. 1991. Amino acid substitution matrices from an information theoretic perspective. *J. Mol. Biol.* **219:** 555–565.

———. 1993. A protein alignment scoring system sensitive to all evolutionary distances. *J. Mol. Evol.* **36:** 290–300.

Altschul S.F. and Erickson B.W. 1986. A nonlinear measure of subalignment similarity and its significance levels. *Bull. Math. Biol.* **48:** 617–632.

Altschul S.F. and Gish G. 1996. Local alignment statistics. *Methods Enzymol.* **266:** 460–480.

Altschul S.F., Boguski M.S., Gish W., and Wootton J.C. 1994. Issues in searching molecular databases. *Nat. Genet.* **6:** 119–129.

Altschul S.F., Gish W., Miller W., Myers E.W., and Lipman D.J. 1990. Basic local alignment search tool. *J. Mol. Biol.* **215:** 403–410.

Argos P. 1987. A sensitive procedure to compare amino acid sequences. *J. Mol. Biol.* **193:** 385–396.

Arratia R. and Waterman M.S. 1989. The Erdös-Rényi strong law for pattern matching with a given proportion of mismatches. *Ann. Probab.* **17:** 1152–1169.

Arratia R., Gordon L., and Waterman M. 1986. An extreme value theory for sequence matching. *Ann. Stat.* **14:** 971–993.

———. 1990. The Erdös-Rényi law in distribution, for coin tossing and sequence matching. *Ann. Stat.* **18:** 539–570.

Bairoch A. 1991. PROSITE: A dictionary of sites and patterns in proteins. *Nucleic Acids Res.* **19:** 2241–2245.

Benner S.A., Cohen M.A., and Gonnet G.H. 1994. Amino acid substitution during functionally constrained divergent evolution of protein sequences. *Protein Eng.* **7:** 1323–1332.

Branden C. and Tooze J. 1991. *Introduction to protein structure.* Garland Publishing, New York.

Brenner S.E., Chothia C., and Hubbard T. 1998. Assessing sequence comparison methods with reliable structurally identified distant evolutionary relationships. *Proc. Natl. Acad. Sci.* **95:** 6073–6078.

Chao K.-M., Hardison R.C., and Miller W. 1994. Recent developments in linear-space alignment methods: A survey. *J. Comput. Biol.* **1:** 271–291.

Chvátal V. and Sankoff D. 1975. Longest common subsequences of two random sequences. *J. Appl. Probab.* **12:** 306–315.

Collins J.F., Coulson A.F., and Lyall A. 1988. The significance of protein sequence similarities. *Comput. Appl. Biosci.* **4:** 67–71.

Dayhoff M.O. 1978. Survey of new data and computer methods of analysis. In *Atlas of protein sequence and structure,* vol. 5, suppl. 3. National Biomedical Research Foundation, Georgetown University, Washington, D.C.

Dayhoff M.O., Barker W.C., and Hunt L.T. 1983. Establishing homologies in protein sequences. *Methods Enzymol.* **91:** 524–545.

Doolittle R.F. 1981. Similar amino acid sequences: Chance or common ancestry. *Science* **214:** 149–159.

———. 1986. *Of URFs and ORFs: A primer on how to analyze derived amino acid sequences.* University Science Books, Mill Valley, California.

Durbin R., Eddy S., Krogh A., and Mitchison G. 1998. *Biological sequence analysis: Probabilistic models of proteins and nucleic acids.* Cambridge University Press, United Kingdom.

Feng D.F., Johnson M.S., and Doolittle R.F. 1985. Aligning amino acid sequences: Comparison of commonly used methods. *J. Mol. Evol.* **21:** 112–125.

Fitch W.M. 1966. An improved method of testing for evolutionary homology. *J. Mol. Biol.* **16:** 9–16.

———. 1970. Distinguishing homologous from analogous proteins. *Syst. Zool.* **19:** 99–113.

Fitch W.M. and Markowitz E. 1970. An improved method for determining codon variability in a gene and its application to the rate of fixation of mutations in evolution. *Biochem. Genet.* **4:** 579–593.

Fitch W.M. and Smith T.F. 1983. Optimal sequences alignments. *Proc. Natl. Acad. Sci.* **80:** 1382–1386.

George D.G., Barker W.C., and Hunt L.T. 1990. Mutation data matrix and its uses. *Methods Enzymol.* **183:** 333–351.

Gibbs A.J. and McIntyre G.A. 1970. The diagram, a method for comparing sequences. Its use with amino acid and nucleotide sequences. *Eur. J. Biochem.* **16:** 1–11.

Gonnet G.H., Cohen M.A., and Benner S.A. 1992. Exhaustive matching of the entire protein sequence database. *Science* **256:** 1443–1445.

———. 1994. Analysis of amino acid substitution during divergent evolution: The 400 by 400 dipeptide substitution matrix. *Biochem. Biophys. Res. Commun.* **199:** 489–496.

Gotoh O. 1982. An improved algorithm for matching biological sequences. *J. Mol. Biol.* **162:** 705–708.

Gray G.S. and Fitch W.M. 1983. Evolution of antibiotic resistance genes: The DNA sequence of a kanamycin resistance gene from *Staphylococcus aureus. Mol. Biol. Evol.* **1:** 57–66.

Gribskov M. and Burgess R.R. 1986. Sigma factors from *E. coli, B. subtilis,* phage SP01, and phage T4 are homologous proteins. *Nucleic Acids Res.* **14:** 6745–6763.

Gumbel E.J. 1962. Statistical theory of extreme values (main results). In *Contributions to order statistics* (ed A.E. Sarhan and B.G. Greenberg), chap. 6, p. 71. Wiley, New York.

Gusfield D. and Stelling P. 1996. Parametric and inverse-parametric sequence alignment with XPARAL. *Methods Enzymol.* **266:** 481–494.

Henikoff S. and Henikoff J.G. 1991. Automated assembly of protein blocks for database searching. *Nucleic Acids Res.* **19:** 6565–6572.

———. 1992. Amino acid substitution matrices from protein blocks. *Proc. Natl. Acad. Sci.* **89:** 10915–10919.

———. 1993. Performance evaluation of amino acid substitution matrices. *Proteins Struct. Funct. Genet.* **17:** 49–61.

Henikoff S., Greene E.A., Pietrokovski S., Bork P., Attwood T.K., and Hood L. 1997. Gene families: The taxonomy of protein paralogs and chimeras. *Science* **278:** 609–614.

Huang X. 1994. On global sequence alignment. *Comput. Appl. Biosci.* **10:** 227–235.

Huang X. and Miller W. 1991. A time-efficient, linear-space local similarity algorithm. *Adv. Appl. Math.* **12:** 337–357.

Huang X., Hardison R.C., and Miller W. 1990. A space-efficient algorithm for local similarities. *Comput. Appl. Biosci.* **6:** 373–381.

Johnson M.S. and Overington J.P. 1993. A structural basis for sequence comparisons: An evaluation of scoring methodologies. *J. Mol. Biol.* **233:** 716–738.

Jones D.T., Taylor W.R., and Thornton J.M. 1992. The rapid generation of mutation data matrices from protein sequences. *Comput. Appl. Biosci.* **8:** 275–282.

———. 1994. A mutation data matrix for transmembrane proteins. *FEBS Lett.* **339:** 269–275.

Karlin S. and Altschul S.F. 1990. Methods for assessing the statistical significance of molecular sequence features by using general scoring schemes. *Proc. Natl. Acad. Sci.* **87:** 2264–2268.

———. 1993. Applications and statistics for multiple high-scoring segments in molecular sequences. *Proc. Natl. Acad. Sci.* **90:** 5873–5877.

Karlin S., Bucher P., and Brendel P. 1991. Statistical methods and insights for protein and DNA sequences. *Annu. Rev. Biophys. Biophys. Chem.* **20:** 175–203.

Kidwell M.G. 1983. Evolution of hybrid dysgenesis determinants in *Drosophila melanogaster. Proc. Natl. Acad. Sci.* **80:** 1655–1659.

Lawrence J.G. and Ochman H. 1997. Amelioration of bacterial genomes: Rates of change and exchange. *J. Mol. Biol.* **44:** 383–397.

Li W. and Graur D. 1991. *Fundamentals of molecular evolution.* Sinauer Associates, Sunderland, Massachusetts.

Lipman D.J., Wilbur W.J., Smith T.F., and Waterman M.S. 1984. On the statistical significance of nucleic acid similarities. *Nucleic Acids Res.* **12:** 215–226.

Maizel J.V., Jr. and Lenk R.P. 1981. Enhanced graphic matrix analysis of nucleic acid and protein sequences. *Proc. Natl. Acad. Sci.* **78:** 7665–7669.

Miller W. and Myers E.W. 1988. Sequence comparison with concave weighting functions. *Bull. Math. Biol.* **50:** 97-120.

Miyamoto M.M. and Fitch W.M. 1995. Testing the covarion hypothesis of evolution. *Mol. Biol. Evol.* **12:** 503–513.

Mott R. 1992. Maximum-likelihood estimation of the statistical distribution of Smith-Waterman local sequence similarity scores. *Bull. Math. Biol.* **54:** 59–75.

Myers E.W. and Miller W. 1988. Optimal alignments in linear space. *Comput. Appl. Biosci.* **4:** 11–17.

Needleman S.B. and Wunsch C.D. 1970. A general method applicable to the search for similarities in the amino acid sequence of two proteins. *J. Mol. Biol.* **48:** 443–453.

Pearson W.R. 1990. Rapid and sensitive sequence comparison with FASTP and FASTA. *Methods Enzymol.* **183:** 63–98.

———. 1995. Comparison of methods for searching protein sequence databases. *Protein Sci.* **4:** 1150–1160.

———. 1996. Effective protein sequence comparison. *Methods Enzymol.* **266:** 227–258.

———. 1998. Empirical statistical estimates for sequence similarity searches. *J. Mol. Biol.* **276:** 71–84.

Pearson W.R. and Miller W. 1992. Dynamic programming algorithm for biological sequence comparison. *Methods Enzymol.* **210:** 575–601.

Rechid R., Vingron M., and Argos P. 1989. A new interactive protein sequence alignment program and comparison of its results with widely used programs. *Comput. Appl. Biosci.* **5:** 107–113.

Risler J.L., Delorme M.O., Delacroix H., and Henaut A. 1988. Amino acid substitutions in structurally related proteins: A pattern recognition approach. *J. Mol. Biol.* **204:** 1019–1029.

Sander C. and Schneider R. 1991. Database of homology derived protein structures and the structural meaning of sequence alignment. *Proteins* **9:** 56–68.

Sankoff D. 1972. Matching sequences under deletion/insertion constraints. *Proc. Natl. Acad. Sci.* **69:** 4–6.

Schwartz S., Miller W., Yang C.-M., and Hardison R.C. 1991. Software tools for analyzing pairwise alignments of long sequences. *Nucleic Acids Res.* **19:** 4663–4667.

Sellers P.H. 1974. On the theory and computation of evolutionary distances. *SIAM J. Appl. Math.* **26:** 787–793.

———. 1980. The theory and computation of evolutionary distances: Pattern recognition. *J. Algorithms* **1:** 359–373.

Smith H.O., Annau T.M., and Chandrasegaran S. 1990. Finding sequence motifs in groups of functionally related proteins. *Proc. Natl. Acad. Sci.* **87:** 826–830.

Smith T.F. and Waterman M.S. 1981a. Identification of common molecular subsequences. *J. Mol. Biol.* **147:** 195–197.

———. 1981b. Comparison of biosequences. *Adv. Appl. Math.* **2:** 482–489.

Smith T.F., Waterman M.S., and Burks C. 1985. The statistical distribution of nucleic acid similarities. *Nucleic Acids Res.* **13:** 645–656.

Smith T.F., Waterman M.S., and Fitch W.M. 1981. Comparative biosequence metrics. *J. Mol. Evol.* **18:** 38–46.

Sonnhammer E.L. and Durbin R. 1995. A dot-matrix program with dynamic threshold control suited for genomic DNA and protein sequence analysis. *Gene* **167:** GC1–10.

States D.J. and Boguski M.S. 1991. Similarity and homology. In *Sequence analysis primer* (ed. M. Gribskov and J. Devereux), pp. 92–124. Stockton Press, New York.

States D.J., Gish W., and Altschul S.F. 1991. Improved sensitivity of nucleic acid database searches using application-specific scoring matrices. *Methods* **3:** 66–70.

Tatusov R.L., Koonin E.V., and Lipman D.J. 1997. A genomic perspective on protein families. *Science* **278:** 631–637.

Vingron M. and Waterman M.S. 1994. Sequence alignment and penalty choice: Review of concepts, case studies and implications. *J. Mol. Biol.* **235:** 1–12.

Vogt G., Etzold T., and Argos P. 1995. An assessment of amino acid exchange matrices: The twilight zone re-visited. *J. Mol. Biol.* **249:** 816–831.

Waterman M.S., Ed. 1989. Sequence alignments. In *Mathematical methods for DNA sequences.* CRC Press, Boca Raton, Florida.

————. 1994. Parametric and ensemble sequence alignment algorithms. *Bull. Math. Biol.* **56:** 743–767.

Waterman M.S. and Eggert M. 1987. A new algorithm for best subsequence alignments with application to tRNA-tRNA comparisons. *J. Mol. Biol.* **197:** 723–728.

Waterman M.S. and Vingron M. 1994a. Rapid and accurate estimates of statistical significance for sequence database searches. *Proc. Natl. Acad. Sci.* **91:** 4625–4628.

————. 1994b. Sequence comparison significance and Poisson distribution. *Stat. Sci.* **9:** 367–381.

Waterman M.S., Eggert M., and Lander E. 1992. Parametric sequence comparisons. *Proc. Natl. Acad. Sci.* **89:** 6090–6093.

Waterman M.S., Gordon L., and Arratia R. 1987. Phase transitions in sequence matches and nucleic acid structure. *Proc. Natl. Acad. Sci.* **84:** 1239–1243.

Waterman M.S., Smith T.F., and Beyer W.A. 1976. Some biological sequence metrics. *Adv. Math.* **20:** 367–387.

Wilbur W.J. 1985. On the PAM model of protein evolution. *Mol. Biol. Evol.* **2:** 434–447.

Zhu J., Liu J.S., and Lawrence C.E. 1998. Bayesian adaptive sequence alignment algorithms. *Bioinformatics* **14:** 25–39.

Multiple Sequence Alignment

INTRODUCTION

O NE OF THE MOST IMPORTANT CONTRIBUTIONS of molecular biology to evolutionary analysis is the discovery that the DNA sequences of different organisms are often related. Similar genes are conserved across widely divergent species, often performing a similar or even identical function, and at other times, mutating or rearranging to perform an altered function through the forces of natural selection. Thus, many genes are represented in highly conserved forms in organisms. Through simultaneous alignment of the sequences of these genes, sequence patterns that have been subject to alteration may be analyzed.

Because the potential for learning about the structure and function of molecules by multiple sequence alignment (msa) is so great, computational methods have received a great deal of attention. In msa, sequences are aligned optimally by bringing the greatest number of similar characters into register in the same column of the alignment, just as described in Chapter 3 for the alignment of two sequences. Computationally, msa presents several difficult challenges. First, finding an optimal alignment of more than two sequences that includes matches, mismatches, and gaps, and that takes into account the degree of variation in all of the sequences at the same time poses a very difficult challenge. The dynamic programming algorithm used for optimal alignment of pairs of sequences can be extended to three sequences, but for more than three sequences, only a small number of relatively short sequences may be analyzed. Thus, approximate methods are used, including (1) a progressive global alignment of the sequences starting with an alignment of the most alike sequences and then building an alignment by adding more sequences, (2) iterative methods that make an initial alignment of groups of sequences and then revise the alignment to achieve a more reasonable result, (3) alignments based on locally conserved patterns found in the same order in the sequences, and (4) use of statistical methods and probabilistic models of the sequences. A second computational challenge is identifying a reasonable method of obtaining a cumulative score for the substitutions in the column of an msa. Finally, the placement and scoring of gaps in the various sequences of an msa presents an additional challenge.

The msa of a set of sequences may also be viewed as an evolutionary history of the sequences. If the sequences in the msa align very well, they are likely to be recently derived from a common ancestor sequence. Conversely, a group of poorly aligned sequences share a more complex and distant evolutionary relationship. The task of aligning a set of sequences, some more closely and others less closely related, is identical to that of discovering the evolutionary relationships among the sequences.

As with aligning a pair of sequences, the difficulty in aligning a group of sequences varies considerably with sequence similarity. On the one hand, if the amount of sequence variation is minimal, it is quite straightforward to align the sequences, even without the assistance of a computer program. On the other hand, if the amount of sequence variation is great, it may be very difficult to find an optimal alignment of the sequences because so many combinations of substitutions, insertions, and deletions, each predicting a different alignment, are possible.

The availability of a subset of the many multiple sequence alignment programs is shown in Table 4.1. A flowchart illustrating the considerations to be made in choosing an alignment method is shown on page 144.

When dealing with a sequence of unknown function, the presence of similar domains in several similar sequences implies a similar biochemical function or structural fold that may become the basis of further experimental investigation. A group of similar sequences may

Table 4.1. *Web sites and program sources for multiple sequence alignment*

Name	Source	Reference
Global alignments including progressive		
CLUSTALW or CLUSTALX (latter has graphical interface)	FTP to ftp.ebi.ac.uk/pub/software[a,d]	Thompson et al. (1994a, 1997); Higgins et al. (1996)
MSA	http://www.psc.edu/[b]	Lipman et al. (1989); Gupta et al. (1995)
	http://www.ibc.wustl.edu/ibc/msa.html[c]	
	FTP to fastlink.nih.gov/pub/msa	
PRALINE	http://mathbio.nimr.mrc.ac.uk/~jhering/praline	Heringa (1999)
Iterative and other methods		
DIALIGN segment alignment	http://www.gsf.de/biodv/dialign.html	Morgenstern et al. (1996)
MultAlin	http://protein.toulouse.inra.fr/multalin.html	Corpet (1988)
PRRP progressive global alignment (randomly or doubly nested)	ftp.genome.ad.jp/pub/genome/saitama-cc	Gotoh (1996)
SAGA genetic algorithm	http://igs-server.cnrs-mrs.fr/~cnotred/Projects_home_page/saga_home_page.html	Notredame and Higgins (1996)
Local alignments of proteins		
Aligned Segment Statistical Evaluation Tool (Asset)	FTP to ncbi.nlm.nih.gov/pub/neuwald/asset	Neuwald and Green (1994)
BLOCKS Web site	http://blocks.fhcrc.org/blocks/	Henikoff and Henikoff (1991, 1992)
eMOTIF Web server	http://dna.Stanford.EDU/emotif/	Nevill-Manning et al. (1998)
GIBBS, the Gibbs sampler statistical method	FTP to ncbi.nlm.nih.gov/pub/neuwald/gibbs9_95/	Lawrence et al. (1993); Liu et al. (1995); Neuwald et al. (1995)
HMMER hidden Markov model software	http://hmmer.wustl.edu/	Eddy (1998)
MACAW, a workbench for multiple alignment construction and analysis	FTP to ncbi.nlm.nih.gov/pub/macaw	Schuler et al. (1991)
MEME Web site, expectation maximization method	http://meme.sdsc.edu/meme/website/	Bailey and Elkan (1995); Grundy et al. (1996, 1997); Bailey and Gribskov (1998)
Profile analysis at UCSD[a,e]	http://www.sdsc.edu/projects/profile/	Gribskov and Veretnik (1996)
SAM hidden Markov model Web site	http://www.cse.ucsc.edu/research/compbio/sam.html	Krogh et al. (1994); Hughey and Krogh (1996)

[a] Lists of additional Web sites for msa are maintained at: http://www.ebi.ac.uk/biocat/, http://www.hgmp.mrc.ac.uk/Registered/Menu/prot-mult.html, http://www.hum-molgen.de/BioLinks/Biocomp.html, http://biocenter.helsinki.fi/bi/rnd/biocomp/. Reviews on the performance of msa software are given in McClure et al. (1994; progressive alignment methods), Gotoh (1996) and Thompson et al. (1999), a review of Web sites is given in Briffeuil et al. (1998) and a review on iterative algorithms is given in Hirosawa et al. (1995) and Gotoh (1999). The performance of msa programs is commonly assessed by comparing the computed msa with a structural alignment of the proteins and by other objective methods (Notredame et al. 1998). Many of these programs are computationally complex and must be set up on a local site.

[b] The Biomedical Supercomputing facility at the University of Pittsburgh Supercomputing Facility provides accounts (see http://www.psc.edu/biomed/seqanal/grants.html) that provide access to several different versions of MSA and profile analysis. MSA 50 150 will align no more than 50 sequences each less than 150 residues long, MSA 25 500 will align no more than 25 sequences each less than 200 residues long, and MSA10 1000 will align no more than 10 sequences each less than 1000 residues long.

[c] The MSA server at the University of Washington will take up to 8 sequences, each less than 500 long.

[d] CLUSTALW is also available as freeware that runs on PCs and Macintosh computers from the same FTP site.

[e] Profile generating programs are available by FTP from ftp.sdsc.edu/pub/sdsc/biology and are included in the Genetics Computer Group suite of programs (http://www.gcg.com/), although the most recent features of Gribskov and Veretnik (1996) are not included.

define a protein family that may share a common biochemical function or evolutionary origin. Similar proteins have been organized into databases of protein families that are described in Chapter 9.

GENOME SEQUENCING

One application of multiple sequence alignment algorithms is in genome sequencing projects discussed in Chapter 2. Instead of cloning and arranging a very large number of fragments of a large DNA molecule, and then moving along the molecule and sequencing the fragments in order, random fragments of the large molecule are sequenced, and those that overlap are found by a msa program. This approach enables automated assembly of large sequences. Bacterial genomes have been quite readily sequenced by this method, and it has also been used to assemble portions of the *Drosophila* and human genomes at Celera Genomics (Weber and Myers 1997 and see Chapter 10).

The requirements for a msa program for genome projects differ in several respects from those for general sequence analysis. First, the sequences are fragments of the same large sequence molecule, and the sequences of overlapping fragments should be the same except for sequence copying and reading errors, which may introduce the equivalent of substitutions and insertions/deletions between the compared fragments. Thus, there should be one correct alignment that corresponds to that of the genome sequence instead of a range of possibilities. Second, the sequences may be from one DNA strand or the other and hence the complements of each sequence must also be compared. Third, sequence fragments will usually overlap, but by an unknown amount, and, in some cases, one sequence may be included within another. Finally, all of the overlapping pairs of sequence fragments must be assembled into a large, composite genome sequence, taking into account any redundant or inconsistent information. Interested readers may wish to consult a description of the type of methodology (Myers 1995 and see Chapter 10) and a comparison of the methods, including several commercial packages that are useful for managing the sequence data from laboratory sequencing projects (Miller and Powell 1994). The Institutue of Genome Research (http://www.tigr.org/) has also developed and made available software and methods for genome assembly and analysis.

USES OF MULTIPLE SEQUENCE ALIGNMENTS

Just as the alignment of a pair of nucleic acid or protein sequences can reveal whether or not there is an evolutionary relationship between the sequences, so can the alignment of three or more sequences reveal relationships among multiple sequences. Multiple sequence alignment of a set of sequences can provide information as to the most alike regions in the set. In proteins, such regions may represent conserved functional or structural domains.

If the structure of one or more members of the alignment is known, it may be possible to predict which amino acids occupy the same spatial relationship in other proteins in the alignment. In nucleic acids, such alignments also reveal structural and functional relationships. For example, aligned promoters of a set of similarly regulated genes may reveal consensus binding sites for regulatory proteins. Methods for finding such sites in nucleic acid sequences are discussed in Chapter 8.

Another use for consensus information retrieved from a multiple sequence alignment is for the prediction of specific probes for other members of the same group or family of similar sequences in the same or other organisms. There are both computer and molecular biology applications. Once a consensus pattern has been found, database searching programs (Chapter 7) may be used to find other sequences with a similar pattern. In the laboratory, a reasonable consensus of such patterns may be used to design polymerase chain reaction (PCR) primers for amplification of related sequences.

RELATIONSHIP OF MULTIPLE SEQUENCE ALIGNMENT TO PHYLOGENETIC ANALYSIS

Once the msa has been found, the number or types of changes in the aligned sequence residues may be used for a phylogenetic analysis. The alignment provides a prediction as to which sequence characters correspond. Each column in the alignment predicts the mutations that occurred at one site during the evolution of the sequence family, as illustrated in Figure 4.1. Within the column are original characters that were present early, as well as other derived characters that appeared later in evolutionary time. In some cases, the position is so important for function that mutational changes are not observed. It is these conserved positions that are useful for producing an alignment. In other cases, the position is less important, and substitutions are observed. Deletions and insertions may also be present in some regions of the alignment. Thus, starting with the alignment, one can hope to dissect the order of appearance of the sequences during evolution.

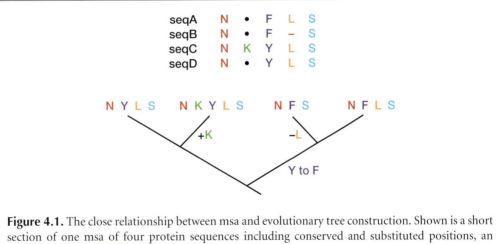

Figure 4.1. The close relationship between msa and evolutionary tree construction. Shown is a short section of one msa of four protein sequences including conserved and substituted positions, an insertion (of K) and a deletion (of L). Below is a hypothetical evolutionary tree that could have generated these sequence changes. Each outer "branch" in the tree represents one of the sequences. The outer branches are also referred to as "leaves." The deepest, oldest branch is that of sequence D, followed by A, then by B and C. The optimal alignment of several sequences can thereby be thought of as minimizing the number of mutational steps in an evolutionary tree for which the sequences are the outer branches or leaves. The mathematical solution to this problem was first outlined by Sankoff (1975). Fast multiple sequence alignment programs that are tree-based have since been developed (Ravi and Kececioglu 1998). However, such an approach depends on knowing the evolutionary tree to perform an alignment, and often this is not the case. Usually, pair-wise alignments are generated first and then used to predict the tree. In this example, the alignment could be explained by several different trees, including the one shown, following one of several types of analyses described in Chapter 6. The sequences then become the outer leaves of the tree, and the inner branches are constructed by this analysis.

METHODS

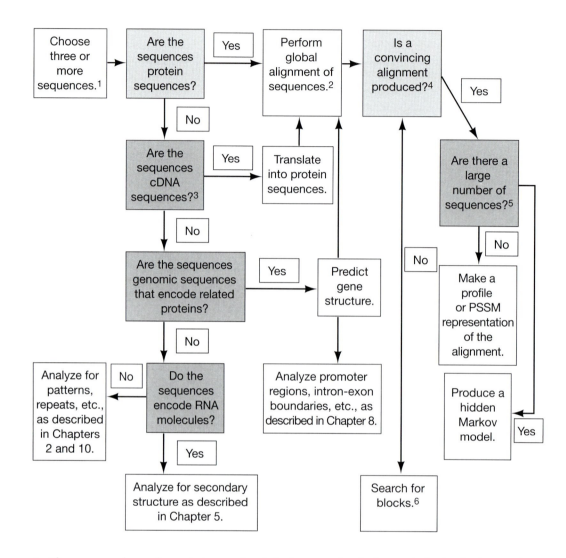

1. The sequence chosen for analysis may already be known to be similar on the basis of pair-wise alignments (Chapter 2), but sequences related by other criteria may also be used. Complex features of the sequences, including repeated or low-complexity regions that interfere with alignments, can be analyzed as described in Chapters 2 and 7. The flowchart describes the production of four classes of multiple sequence alignment.

    a. A global alignment includes the entire range of each sequence in the alignment, and is usually produced by extensions to the dynamic programming global alignment algorithm that is used for aligning pairs of sequences, but other methods are also used.

    b. A sequence block is an alignment of common patterns in protein sequences that includes matches and mismatches in each column found by using pattern-finding algorithms, but no gaps (insertions and deletions) are present.

    c. An alignment of common patterns in protein sequences that includes matches, mismatches, insertions, and deletions may be used to make a type of scoring matrix called a profile.

    d. A hidden Markov model is a probabilistic model of a global alignment of protein sequences or of a conserved local region (similar to a sequence profile) in those sequences that includes matches, mismatches, insertions, and deletions. The model is "trained" to represent the set of sequences.

Methods for finding common patterns in DNA sequences are discussed in Chapter 8.

2. Examples of global alignment, as well as other programs from which to choose, are given in the global alignments and iterative and other methods sections of Table 4.1.

3. cDNA sequences of the same gene from a group of organisms may be multiply aligned by a global method so that synonymous (i.e., change the amino acid) and nonsynonymous (i.e., do not change the amino acid) sequences may be analyzed, as described in Chapter 6 (see also note 2).

4. A convincing alignment should include a series of columns in which a majority of the sequences have the same amino acid or an amino acid that is a conservative substitution for that amino acid, with relatively few examples of other substitutions or gaps in these columns. These columns of alike amino acids should be found throughout the alignment, often clustered into domains. There may also be variable regions in the alignment that represent sequences that diverged more during the evolution of the protein family.

5. This decision rests on whether or not there are enough sequences on which to build a hidden Markov model of the entire alignment or of a well-defined region in the alignment (a profile hidden Markov model). For sequences that are related but show considerable variations in many columns, as many as 100 sequences may be needed to produce a hidden Markov model of the alignment. This number is reduced to approximately 25–50 if there is less variation among the sequences. A scoring matrix representing the sequence variation found in each column of the alignment may also be made. These matrices may accommodate gaps in the alignment (a profile or HMM profile) or may not include gaps (position-specific scoring matrix).

6. For finding patterns common to the sequences, pattern-searching algorithms and statistical methods are used. The former search for a set of matched sequence characters that are present in the sequences. The latter perform an exhaustive analysis of sequence "windows" in the sequences to find the most alike amino acid patterns by the expectation maximization (EM) or Gibbs sampling algorithms. These methods are described in the text.

MULTIPLE SEQUENCE ALIGNMENT AS AN EXTENSION OF SEQUENCE PAIR ALIGNMENT BY DYNAMIC PROGRAMMING

The dynamic programming algorithm described in Chapter 2 provides an optimal alignment of two sequences. In the program MSA (Lipman et al. 1989), application of the global alignment algorithm has been extended to provide an optimal alignment of a small number of sequences greater than two. Gupta et al. (1995) have shown, however, that MSA rarely produces a provable optimal alignment. The number of sequences that can be aligned is limited because the number of computational steps and the amount of memory required grow exponentially with the number of sequences to be analyzed. This limitation means that the program has somewhat limited application to a small number of sequences.

Recall that the dynamic programming method of sequence alignment between two sequences builds a scoring matrix where each position provides the best alignment up to that point in the sequence comparison. The number of comparisons that must be made to fill this matrix without using any short cuts and excluding gaps is the product of the length of the two sequences. Imagine extending this analysis to three or more sequences. For three sequences, instead of the two-dimensional matrix for two sequences, think of the lattice of a cube that is to be filled with calculated dynamic programming scores. Scoring positions on three surfaces of the cube will represent the alignment values between a pair of the sequences, ignoring the third sequence, as illustrated in Figure 4.2. In MSA, positions inside the lattice of the cube are given values based on the sum of the initial scores of the three pairs of sequences.

For three protein sequences each 300 amino acids in length and excluding gaps, the number of comparisons to be made by dynamic programming is equal to $300^3 = 2.7 \times 10^7$, whereas only $300^2 = 9 \times 10^4$ is required for two sequences of this length. This number is sufficiently small that alignment of three sequences by this method is practical. For alignment of more than three sequences, one has to imagine filling an N-dimensional space or hypercube. The number of steps and memory required for a 300-amino-acid sequence (300^N, where N is the number of sequences) then becomes too large for most practical purposes, and it is necessary to find a way to reduce the number of comparisons that must be made without compromising the attempt to find an optimal alignment. Fortunately, Carrillo and Lipman (1988) found such a method, called the sum of pairs, or SP method. Since the publication of the MSA program, Gupta et al. (1995) have substantially reduced the memory requirements and number of steps required. The enhanced version of MSA is available by anonymous FTP from fastlink.nih.gov/pub/msa.

The basic idea is that a multiple sequence alignment imposes an alignment on each of the pairs of sequences. The heavy arrow in Figure 4.2 represents the path followed in the cube to find a msa for three sequences, but the msa can be projected on to the sides of the cube, thus defining an alignment for each pair of sequences. The alignments found for each pair of sequences likewise impose bounds on the location of the msa within the cube, and thus defines the number of positions within the cube that have to be evaluated. Pair-wise alignments are first computed between each pair of sequences. Next, a trial msa is produced by first predicting a phylogenetic tree for the sequences (Saitou and Nei 1987; see Chapter 6 for the neighbor-joining method of tree construction), and the sequences are

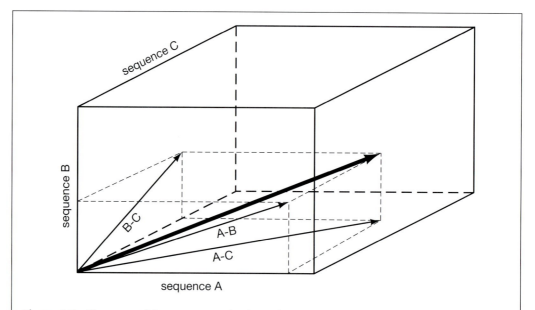

Figure 4.2. Alignment of three sequences by dynamic programming. Arrows on the surfaces of the cube indicate the direction for filling in the scoring matrix for pairs of sequences, A with B, etc., performed as previously described. The alignment of all three sequences requires filling in the lattice of the cube space with optimal alignment scores following the same algorithm. The best score at each interior position requires a consideration of all possible moves within the cube up to that point in the alignment. The trace-back matrix will align positions in all three sequences including gaps.

then multiply aligned in the order of their relationship on the tree. This method is used by other programs described below (e.g., PILEUP, CLUSTALW) and provides a heuristic alignment that is not guaranteed to be optimal. However, the alignment serves to provide a limit to the space within the cube within which optimal alignments are likely to be found. In Figure 4.3, the green area on the left surface of the cube is bounded by the optimal alignment of sequences B and C and a projection of the heuristic alignment for all three sequences. The orange and blue areas are similarly defined for other sequence pairs. The dark gray volume within the cube is bounded by projections from each of the three surface areas. For more sequences, a similar type of analysis of bounds may be performed in the corresponding higher-order space.

In practice, MSA calculates the multiple alignment score within the cube lattice by adding the scores of the corresponding pair-wise alignments in the msa. This measure is known as the SP measure (for sum of pairs), and the optimal alignment is based on obtaining the best SP score. These scores may or may not be weighted so as to reduce the influence of more closely related sequences in the msa. The Dayhoff PAM250 matrix and an associated gap penalty are used by MSA for aligning protein sequences. MSA uses a constant penalty for any size of gap and scores gaps according to the scheme illustrated in Figure 4.4 (Altschul 1989; Lipman et al. 1989). MSA calculates a value ϵ for each pair of sequences that provides an idea of how much of a role the alignment of those two sequences plays in the msa. ϵ for a given sequence pair is the difference between the score of the alignment of that pair in the msa and the score of the optimal pair-wise alignment. The bigger the value of ϵ, the more divergent the msa from the pair-wise alignment and the smaller the contribution of that alignment to the msa. For example, if an extra copy of one

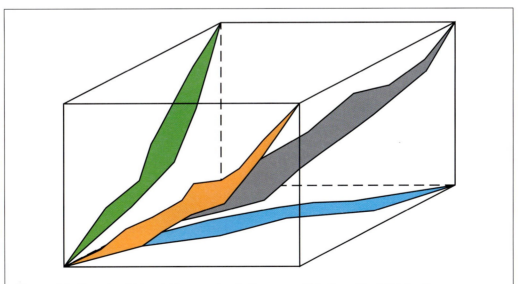

Figure 4.3. Bounds within which an optimal alignment will be found by MSA for three sequences. For MSA to find an optimal alignment among three sequences by the DP algorithm, it is only necessary to calculate optimal alignment scores within the gray volume. This volume is bounded on the one side by the optimal alignments found for each pair of sequences, and on the other by a heuristic multiple alignment of the sequences. The colored areas on each cube surface are two-dimensional projections of the gray volume.

		Natural gap cost	Quasi-natural gap cost
sequence 1	x – – – x		
sequence 2	x x – x x	3	4
sequence 3	x x x x x		

Figure 4.4. Method of scoring gap penalties by the msa program MSA. *x* indicates aligned residues, which may be a match or a mismatch, and – indicates a gap. In this example, each gap cost is 1, regardless of length. The "natural" gap cost is the sum of the number of gaps in all pair-wise combinations (sequences 1 and 2, 1 and 3, and 2 and 3). Note that the alignment of a gap of three in sequence 1 with a gap of length one in sequence 2 scores as gap of 1 because the gap in sequence 1 is longer. The quasi-natural gap cost is the natural cost for the gap plus an additional value for any gap that begins and ends within another. In this example, there is an additional penalty score for the presence of a single gap in sequence 2 that falls within a larger gap in sequence 1. The inclusion of this extra cost for a gap has little effect on the alignments produced but provides an enormous reduction in the amount of information that must be maintained in the DP scoring matrix (Altschul 1989), thus making possible the simultaneous alignment of more sequences by MSA.

of the sequences is added to the alignment project, then ε for sequence pairs that do not include that sequence will increase, indicating a lesser role because the contributions of that pair have been out-voted by the alike sequences (Altschul et al. 1989). Weighting the sequence pairs is designed to get around the common difficulty that some pairs in most sets of sequences are similar. Another score δ is the sum of the εs and gives an indication of the degree of divergence among the sequences—closely related sequences will have low εs and δs and distantly related sequences will have high εs and δs.

The MSA program avoids the bias in an alignment due to alike sequences by weighting the pair-wise scores before they are added to give the SP score. These weights are determined by using the predicted tree of the sequences discussed above. The pair-wise scores between all sequence pairs are adjusted to reduce the influence of the more unlike sequence pairs that occupy more distant "leaves" on the evolutionary tree (i.e., by sequences that are joined by more branches) based on the argument that these sequence pairs provide less useful information for computing the msa. This scheme is different from that used by other msa programs (see below), which generally increase the weight of scores from more distant sequences because these sequences represent greater divergence in the evolutionary tree (see Vingron and Sibbald 1993).

In using MSA, several additional practical considerations should be considered (described on MSA Web sites given in Table 4.1): (1) MSA is a heavy user of machine resources and is limited to a small number of sequences of relatively short lengths. (2) In the UNIX command line mode of the program, there are options that allow users to specify gap costs, force the alignment of certain residues, specify maximum values for ε, and tune the program in other ways. (3) When the output shows that some ε are greater than the respective maximum ε, a better alignment usually can be found by increasing the maximum ε in question. However, increasing ε also increases the computational time. (4) If the program bogs down, try dividing the problem into several smaller ones.

Below is an example from http://www.psc.edu of using MSA to align a group of phospholipase a2 proteins. Note that the program uses the FASTA sequence format. The following steps are used:

1. Calculate all pair-wise alignment scores (alignment costs).
2. Use the scores (costs) to predict a tree.
3. Calculate pair weights based on the tree.

4. Produce a heuristic msa based on the tree.

5. Calculate the maximum ϵ for each sequence pair.

6. Determine the spatial positions that must be calculated to obtain the optimal alignment.

7. Perform the optimal alignment.

8. Report the ϵ found compared to the maximum ϵ.

Example of MSA

```
MSA release 2.1  (PSC revision b) started on Thu Jun 19 14:55:31 1997
Sequence file format is Fasta.
Calculating pairwise alignments.
..........
**********
Calculating weights.


---------------- Tree given from ancestor ----------------
On the left:    Internal Node  Distance to parent =  278.83
On the left:    Internal Node  Distance to parent =   23.63
On the left:    Internal Node  Distance to parent =  118.62
On the left:    SEQ#01         Distance to parent =  230.50
On the right:   SEQ#04         Distance to parent =  205.50
On the right:   SEQ#05         Distance to parent =  238.37
On the right:   SEQ#02         Distance to parent =  256.17
On the right:   SEQ#03         Distance to parent =    0.00


Calculating epsilons.


Sequence      ID         Description
   1        SEQ#01       P1;1POA Phospholipase a2 (EC 3.1.1.4) - Chinese cobra
   2        SEQ#02       P1;1POD Phospholipase a2 (EC 3.1.1.4) - human
   3        SEQ#03       P1;1PPA Phospholipase a2 (EC 3.1.1.4) lys 49 variant
   4        SEQ#04       P1;1BPQ phospholipase A2 (EC 3.1.1.4) mutant (K56M) -
   5        SEQ#05       P1;1PP2R phospholipase A2 (EC 3.1.1.4) (calcium-free)



              ***  Heuristic Multiple Alignment  ***


**********23541        ****************************35214 **35214       ***
NLYQFKNMIQCTVPSR-SWWDFADYGCYCGRGGSGTPVDDLDRCCQVHDNCYNEAEKISGC-----WPYFKTYSY
NLVNFHRMIK-LTTGKEAALSYGFYGCHCGVGGRGSPKDATDRCCVTHDCCYKRLEK-RGC-----GTKFLSYKF
SVLELGKMIL-QETGKNAITSYGSYGCNCGWGHRGQPKDATDRCCFVHKCCYKKLT---DC-----NHKTDRYSY
ALWQFNGMIKCKIPSSEPLLDFNNYGCYCGLGGSGTPVDDLDRCCQTHDNCYKQAMKLDSCKVLVDNPYTNNYSY
SLVQFETLIM-KIAGRSGLLWYSAYGCYCGWGGHGLPQDATDRCCFVHDCCYGKAT---DC-----NPKTVSYTY
```

```
*********35214 *****************14325
ECSQGTLTCKGGNNACAAAVCDCDRLAAICFAG--APYNDNDYNINLKARC-------
SNSGSRITC-AKQDSCRSQLCECDKAAATCFARNKTTYNKKYQYYS-NKHCRGSTPRC
SWKNKAIIC-EEKNPCLKEMCECDKAVAICLRENLDTYNKKYKAYF-KLKCKKPDT-C
SCSNNEITCSSENNACEAFICNCDRNAAICFSK--VPYNKEHKNLD-KKNC-------
SEENGEIIC-GGDDPCGTQICECDKAAAICFRDNIPSYDNKYWLFP-PKDCREEPEPC


Calculating pairwise projection costs.
..........
*********


Calculating multiple alignment.
....1....2....3....4....5....6....7....8....9....0
**************************************************

                *** Optimal Multiple Alignment ***


NLYQFKNMIQCTVPSR-SWWDFADYGCYCGRGGSGTPVDDLDRCCQVHDNCYNEAEKISGC-----WPYFKTYSY
NLVNFHRMIK-LTTGKEAALSYGFYGCHCGVGGRGSPKDATDRCCVTHDCCYKRLEK-RGC-----GTKFLSYKF
SVLELGKMIL-QETGKNAITSYGSYGCNCGWGHRGQPKDATDRCCFVHKCCYKKL---TDC-----NHKTDRYSY
ALWQFNGMIKCKIPSSEPLLDFNNYGCYCGLGGSGTPVDDLDRCCQTHDNCYKQAMKLDSCKVLVDNPYTNNYSY
SLVQFETLIM-KIAGRSGLLWYSAYGCYCGWGGHGLPQDATDRCCFVHDCCYGKA---TDC-----NPKTVSYTY


ECSQGTLTCKGGNNACAAAVCDCDRLAAICFAG--APYNDNDYNINLKARC-------
SNSGSRITC-AKQDSCRSQLCECDKAAATCFARNKTTYNKKYQYYS-NKHCRGSTPRC
SWKNKAIIC-EEKNPCLKEMCECDKAVAICLRENLDTYNKKYKAYF-KLKCK-KPDTC
SCSNNEITCSSENNACEAFICNCDRNAAICFSK--VPYNKEHKNLD-KKNC-------
SEENGEIIC-GGDDPCGTQICECDKAAAICFRDNIPSYDNKYWLFP-PKDCREEPEPC


End gaps not penalized.
Costfile:                pam250
Alignment cost:   35132     Lower bound:    34945
Delta:              187     Max. Delta:       285
```

Sequences		Proj. Cost	Pair. Cost	Epsilon	Max. Epsi.	Weight	Weight*Cost
1	2	1864	1825	39	39	1	1864
1	3	1891	1843	48	57	1	1891
1	4	1654	1653	1	5	4	6616
1	5	1814	1787	27	28	2	3628
2	3	1735	1733	2	8	4	6940
2	4	1876	1866	10	10	1	1876
2	5	1713	1712	1	8	2	3426
3	4	1901	1889	12	21	1	1901
3	5	1648	1648	0	11	2	3296
4	5	1847	1842	5	6	2	3694

```
Elapsed time =   0.895


Tree
     A tree is given for the heuristic alignment (not shown).
```

SCORING MULTIPLE SEQUENCE ALIGNMENTS

As discussed above, the SP method provides a way to score the msa by summing the scores of all possible combinations of amino acid pairs in a column of a msa. The method assumes a model for evolutionary change in which any of the sequences could be the ancestor of the others, as illustrated in Figure 4.5. This figure also illustrates a difficulty with the SP method when a substitution table of log odds scores such as BLOSUM62 is used for protein sequences (see Durbin et al. 1998, pp. 139–140). Shown is the effect of adding a small number of amino acid subsitutions to a column that initially has all matching amino acids. Scores in the msa column decrease rapidly as the number of mismatched residue pairs increases. For a larger number of sequences than five with all N, or with one or two C substitutions, these decreases should be greater because there will be more N-N matched pairs relative to mismatched N-C pairs. However, the reverse is true with the SP method of scoring. For n sequences, the number of combinations of pairs in a column is

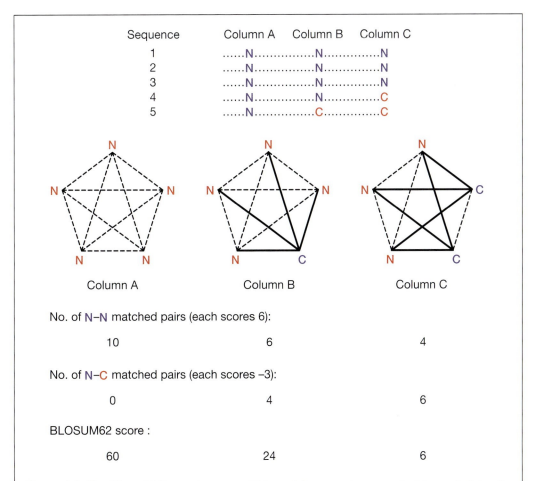

Figure 4.5. The SP model for scoring a msa. This model represents one method for optimizing the msa by maximizing the number of matched pairs (or minimizing the cost or number of mismatched pairs) summed over all columns in the msa. Shown first are three columns of a five-sequence msa with all matched (A), four matched and one mismatched (B), or three matched and two mismatched (C) sequence characters. The SP method of calculating the cumulative scores for columns of a msa is then illustrated by a graph with the five sequences as vertices and representing the ten possible sequence pair-wise sequence comparisons. Solid lines represent a matched pair and dotted lines a mismatched pair. Shown are the BLOSUM62 scores for each column calculated by the SP method. (Adapted from Altschul 1989.)

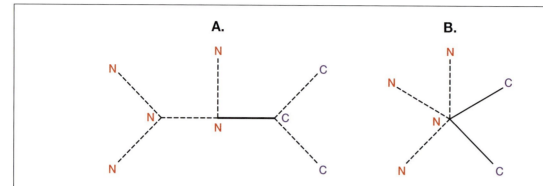

Figure 4.6. Alternative methods for scoring a column in the msa (Altschul 1989b). The variations in column C of Fig. 4.5 are shown modeled by a phylogenetic tree (*A*) and a simplified phylogenetic tree called a star phylogeny (*B*) where one of the sequences is treated as the ancestor of all the others (instead of treating them as all equally possible ancestors as in the original sum of pairs scoring method).

$n(n - 1)/2$. If all are amino acid N, as in column A, then the BLOSUM62 score for the column is $6 \times n(n - 1)/2$. If there is one C in the column, as in column B, then $n - 1$ matched N-N pairs will be replaced by $n-1$ mismatched N-C pairs, giving a score of $9(n - 1)$ less. The score for one C in the column divided by that for zero Cs is $9(n - 1)/[6n(n - 1)/2] = 3/n$. For three sequences, the relative difference is 1, whereas for six sequences, the relative difference is 2. As more sequences are present in the column, the relative difference increases, not in agreement with expectation. Hence, the SP method is not providing a reasonable result when this type of scoring matrix is used. Two other methods for scoring a msa (Altschul 1989) have been described and are illustrated in Figure 4.6. The first is a tree-based method. Because a phylogenetic tree describing the relationships among the sequences is found by the MSA program, the sum of the lengths of the tree branches can be calculated using the substitutions in the column of the msa. Alternatively, a simplified tree with one of the sequences as the ancestor of all of the others (a star phylogeny) can also be used (see Chapter 6). msa programs using these methods have not been implemented. Other scoring methods include information content (see p. 195) and a graph-based method called the trace method (Kececioglu 1993). A novel branch-and-cut algorithm for msa has been developed based on the trace method (Kececioglu et al. 2000). Other methods of scoring and producing an alignment guided by a tree are described below.

PROGRESSIVE METHODS OF MULTIPLE SEQUENCE ALIGNMENT

The MSA program described above for obtaining an optimal alignment of multiple sequences is limited to three sequences or to a small number (six to eight) of relatively short sequences. Progressive alignment methods use the dynamic programming method to build a msa starting with the most related sequences and then progressively adding less-related sequences or groups of sequences to the initial alignment (Waterman and Perlwitz 1984; Feng and Doolittle 1987, 1996; Thompson et al. 1994a; Higgins et al. 1996). Relationships among the sequences are modeled by an evolutionary tree in which the outer branches or leaves are the sequences (Fig. 4.7). The tree is based on pair-wise comparisons of the sequences using one of the phylogenetic methods described in Chapter 6. Progenitor sequences represented by the inner branches of the tree are derived by alignment of the outermost sequences. These inner branches will have uncertainties where positions in the

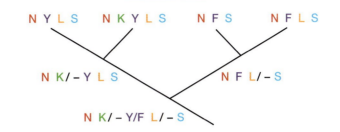

Figure 4.7. Progressive sequence alignment. Sequences are represented as the outermost branches (leaves) on an evolutionary tree. The most closely related sequences are first aligned by dynamic programming, providing a representation of ancestor sequences in deeper branches with uncertainties where amino acids have been substituted or positioned opposite a gap. These sequences are the same as those shown in EVMSA. The challenge to the msa method is to utilize an appropriate combination of sequence weighting, scoring matrix, and gap penalties so that the correct series of evolutionary changes may be found.

outermost sequences are dissimilar, as illustrated in Figure 4.7. Two examples of programs that use progressive methods are CLUSTALW and the Genetics Computer Group program PILEUP.

CLUSTALW

CLUSTAL has been around for more than 10 years, and the authors have done much to support and improve the program (Higgins and Sharp 1988; Thompson et al. 1994a; Higgins et al. 1996). CLUSTALW is a more recent version of CLUSTAL with the W standing for "weighting" to represent the ability of the program to provide weights to the sequence and program parameters, and CLUSTALX provides a graphic interface (see Table 4.1). These changes provide more realistic alignments that should reflect the evolutionary changes in the aligned sequences and the more appropriate distribution of gaps between conserved domains.

CLUSTAL performs a global-multiple sequence alignment by a different method than MSA, although the initial heuristic alignment obtained by MSA is calculated the same way. The steps include: (1) Perform pair-wise alignments of all of the sequences; (2) use the alignment scores to produce a phylogenetic tree (for an explanation of the neighbor-joining method that is used, see Chapter 6); and (3) align the sequences sequentially, guided by the phylogenetic relationships indicated by the tree. Thus, the most closely related sequences are aligned first, and then additional sequences and groups of sequences are added, guided by the initial alignments to produce a msa showing in each column the sequence variations among the sequences. The initial alignments used to produce the guide tree may be obtained by a fast k-tuple or pattern-finding approach similar to FASTA that is useful for many sequences, or a slower, full dynamic programming method may be used. An enhanced dynamic programming alignment algorithm (Myers and Miller 1988; see book Web site) is used to obtain optimal alignment scores. For producing a phylogenetic tree, genetic distances between the sequences are required. The genetic distance is the number of mismatched positions in an alignment divided by the total number of matched positions (positions opposite a gap are not scored).

As with MSA, sequence contributions to the msa are weighted according to their relationships on the predicted evolutionary tree. A rooted tree with known branch lengths of which the sequences are outer branches (leaves) is examined (see Chapter 6). Weights are based on the distance of each sequence from the root, as illustrated in Figure 4.8. The alignment scores between two positions in the msa are then calculated using the resulting weights as multiplication factors.

The scoring of gaps in a msa has to be performed in a different manner from scoring gaps in a pair-wise alignment. As more sequences are added to a profile of an existing msa, gaps accumulate and influence the alignment of further sequences (Thompson et al. 1994b; Taylor 1996). CLUSTALW calculates gaps in a novel way designed to place them between conserved domains. When Pascarella and Argos (1992; see book Web site) aligned sequences of structurally related proteins, the gaps were preferentially found between secondary structural elements. These authors also prepared a table of the observed frequency

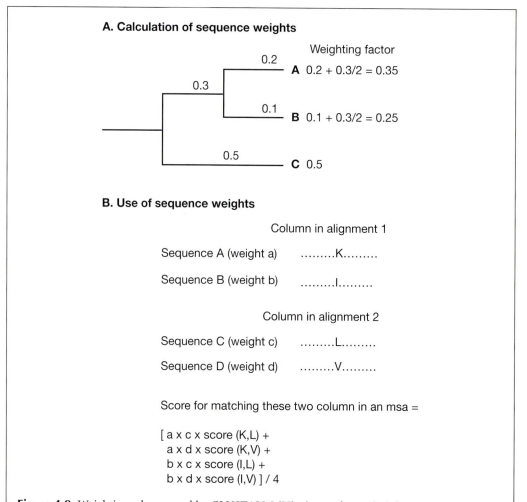

Figure 4.8. Weighting scheme used by CLUSTALW (Higgins et al. 1996). (*A*) Sequences that arise from a unique branch deep in the tree receive a weighting factor equal to the distance from the root. Other sequences that arise from branches shared with other sequences receive a weighting factor that is less than the sum of the branch lengths from the root. For example, the length of a branch common to two sequences will only contribute one-half of that length to each sequence. Once the specific weighting factors for each sequence have been calculated, they are normalized so that the largest weight is 1. As CLUSTALW aligns sequences or groups of sequences, these fractional weights are used as multiplication factors in the calculation of alignment scores. (*B*) Illustration of using sequence weights for aligning two columns in two separate alignments. Note that this sequence weighting scheme is the opposite to that used by MSA, because the more distant a sequence from the others, the higher the weight given. For a comparison of additional weighting schemes, see Vingron and Sibbald (1993).

of gaps next to each amino acid in these regions. CLUSTALW uses the information in this table and also attempts to locate what may be the corresponding domains by appropriate gap placement in the msa. Like other alignment programs, CLUSTAL uses a penalty for opening a gap in a sequence alignment and an additional penalty for extending the gap by one residue. These penalties are user-defined (defaults are available). Gaps found in the initial alignments remain fixed. New gaps introduced as more sequences are added also receive this same gap penalty, even when they occur within an existing gap, but the gap penalties for an alignment are then modified according to the average match value in the substitution matrix, the percent identity between the sequences, and the sequence lengths (Higgins et al. 1996). These changes are attempts to compensate for the scoring matrix, expected number of gaps (alignment with more identities should have fewer gaps), and differences in sequence length (should limit placement of gaps if one sequence shorter). Tables of gaps are then calculated for each group of sequences to be aligned to confine them to less conserved regions in the alignment. Gap penalties are decreased where gaps already occur (another method for achieving this same result is to enhance the scores of more closely matching regions on the alignment as described in Taylor 1996), increased in regions adjacent to already gapped regions, decreased within stretches of hydrophilic regions (amino acids DEGKNQPRS), and increased or decreased according to the table in Pascarella and Argos (1992). These rules are most useful when a correct alignment of some of the sequences is already known. The CLUSTALW algorithm and the results of using the above sequence weighting gap adjustment method are illustrated in Figure 4.9.

CLUSTALW also has options for adding one or more additional sequences with weights or an alignment to a existing alignment (Higgins et al. 1996). Once an alignment has been made, a phylogenetic tree may be made by the neighbor-joining method, with corrections for possible multiple changes at each counted position in the alignment (see Chapter 6). The predicted trees may also be displayed by various programs described in Chapter 6.

PILEUP

PILEUP is the msa program that is a part of the Genetics Computer Group package of sequence analysis programs, owned since 1997 by Oxford Communications, and is widely used due to the popularity and availability of this package. PILEUP uses a method for msa that is very similar to CLUSTALW. The sequences are aligned pair-wise using the Needleman-Wunsch dynamic programming algorithm, and the scores are used to produce a tree by the unweighted pair-group method using arithmetic averages (UPGMA; Sneath and Sokal 1973 and see Chapter 6). The resulting tree is then used to guide the alignment of the most closely related sequences and groups of sequences. The resulting alignment is a global alignment produced by the Needleman-Wunsch algorithm. Standard scoring matrices and gap opening/extension penalties are used. Unfortunately, there have not been any recent enhancements of this program such as gap modifications or sequence weighting comparable to those introduced for CLUSTALW. As with other progressive alignment msa programs, PILEUP does not guarantee an optimal alignment.

Problems with Progressive Alignment

The major problem with progressive alignment programs such as CLUSTAL and PILEUP is the dependence of the ultimate msa on the initial pair-wise sequence alignments. The very first sequences to be aligned are the most closely related on the sequence tree. If these sequences align very well, there will be few errors in the initial alignments. However, the more distantly related these sequences, the more errors will be made, and these errors will

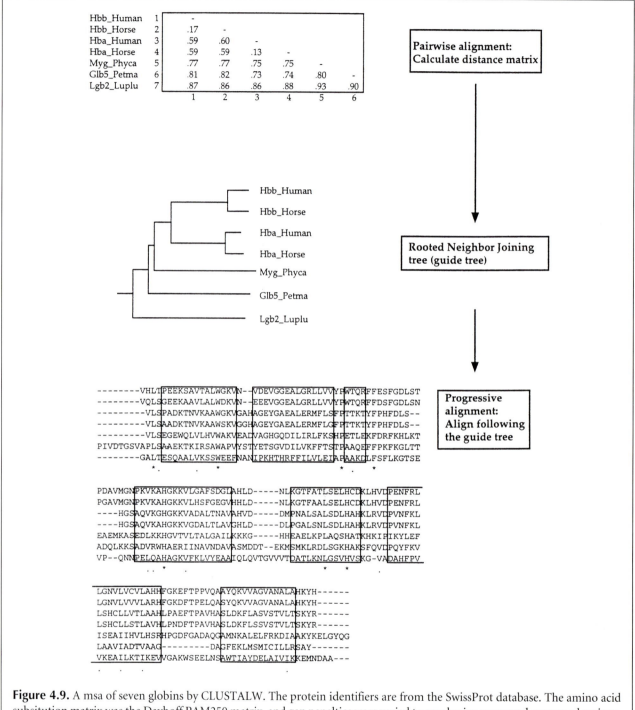

Figure 4.9. A msa of seven globins by CLUSTALW. The protein identifiers are from the SwissProt database. The amino acid subsitution matrix was the Dayhoff PAM250 matrix, and gap penalties were varied to emphasize conserved ungapped regions. The approximate and known locations of seven α-helices in the structure of this group are shown in boxes. (Reprinted, with permission, from Higgins et al. 1996 [copyright Academic Press].)

be propagated to the msa. There is no simple way to circumvent this problem. A second problem with the progressive alignment method is the choice of suitable scoring matrices and gap penalties that apply to the set of sequences (Higgins et al. 1996).

For the difficult task of aligning more distantly related sequences, using Bayesian methods such as hidden Markov models (HMMs) may be useful. For more closely related sequences, CLUSTALW is designed to provide an adequate alignment of a large number of sequences and provide a very good indication of the domain structure of those sequences.

ITERATIVE METHODS OF MULTIPLE SEQUENCE ALIGNMENT

The major problem with the progressive alignment method described above is that errors in the initial alignments of the most closely related sequences are propagated to the msa. This problem is more acute when the starting alignments are between more distantly related sequences. Iterative methods attempt to correct for this problem by repeatedly realigning subgroups of the sequences and then by aligning these subgroups into a global alignment of all of the sequences. The objective is to improve the overall alignment score, such as a sum of pairs score. Selection of these groups may be based on the ordering of the sequences on a phylogenetic tree predicted in a manner similar to that of progressive alignment, separation of one or two of the sequences from the rest, or a random selection of the groups. These methods are compared in Hirosawa et al. (1995).

MultAlin (Corpet 1988) recalculates pair-wise scores during the production of a progressive alignment and uses these scores to recalculate the tree, which is then used to refine the alignment in an effort to improve the score. The program PRRP (Table 4.1) uses iterative methods to produce an alignment. An initial pair-wise alignment is made to predict a tree, the tree is used to produce weights for making alignments in the same manner as MSA except that the sequences are analyzed for the presence of aligned regions that include gaps rather than being globally aligned, and these regions are iteratively recalculated to improve the alignment score. The best scoring alignment is then used in a new cycle of calculations to predict a new tree, new weights, and new alignments, as illustrated in Figure 4.10. The process is repeated until there is no further increase in the alignment score (Gotoh 1994, 1995, 1996).

The program DIALIGN (see Table 4.1) finds an alignment by a different iterative method. Pairs of sequences are aligned to locate aligned regions that do not include gaps, much like continuous diagonals in a dot matrix plot. Diagonals of various lengths are identified. A consistent collection of weighted diagonals that provides an alignment which is a maximum sum of weights is then found.

Additional methods that use iterative procedures are described below.

Genetic Algorithm

The genetic algorithm is a general type of machine-learning algorithm that has no direct relationship to biology and that was invented by computer scientists. The method has been recently adapted for msa by Notredame and Higgins (1996) in a computer program package called SAGA (Sequence Alignment by Genetic Algorithm; see Table 4.1). Zhang and Wong (1997) have developed a similar program. The method is of considerable interest because the algorithm can find high-scoring alignments as good as those found by other methods. Similar genetic algorithms have been used for RNA sequence alignment (Notredame et al. 1997) and for prediction of RNA secondary structure (Shapiro and Navetta 1994). Although the method is relatively new and not used extensively, it likely represents the first of a series of sequence analysis programs that produce alignments by attempted simulation of the evolutionary changes in sequences.

The basic idea behind this method is to try to generate many different msas by rearrangements that simulate gap insertion and recombination events during replication in

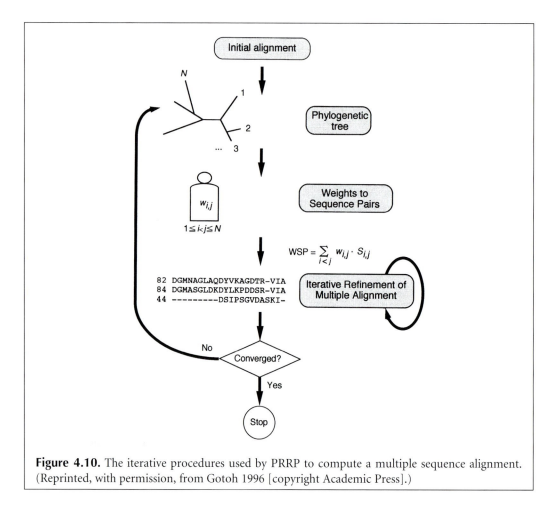

Figure 4.10. The iterative procedures used by PRRP to compute a multiple sequence alignment. (Reprinted, with permission, from Gotoh 1996 [copyright Academic Press].)

order to generate a higher and higher score for the msa. The alignments are not guaranteed to be optimal or to be the highest scoring that is achievable (optimal alignment). Although SAGA can generate alignments for many sequences, the program is slow for more than about 20 sequences.

A similar approach for obtaining a higher-scoring msa by rearranging an existing alignment uses a probability approach called simulated annealing (Kim et al. 1994). The program MSASA (Multiple Sequence Alignment by Simulated Annealing) starts with a heuristic msa and then changes the alignment by following an algorithm designed to identify changes that increase the alignment score.

The success of the genetic algorithm may be attributed to the steps used to rearrange sequences, many of which might be expected to have occurred during the evolution of the protein family. The steps in the algorithm are as follows:

1. The sequences to be aligned (up to ~20 in number) are written in rows, as on a page, except that they are made to overlap by a random amount of sequence, up to 50 residues long for sequences about 200 in length. The ends are then padded with gaps. A typical population of 100 of these msas is made, although other numbers may be set.

```
XXXXXXXXXX----
---XXXXXXXXXXX
-XXXXXXXXXX---
```

Shown is an initial msa for the genetic algorithm (1 of ~100 in number).

2. The 100 initial msas are scored by the sum of pairs method, except that both natural and quasi-natural gap-scoring schemes (Fig. 4.4) are used. Recall that the best SSP score for a msa is the minimum one and the one that is closest to the sum of the pair-wise sequence alignment. Standard amino acid scoring matrices and gap opening and extension penalties are used.

3. These initial msas are now replicated to give another generation of msas. The half with the lowest SSP scores are sent to the next generation unchanged. The remaining half for the next generation are selectively chosen by lot, like picking marbles from a bag, except that the chance for a particular choice is inversely proportional to the msa score (the lower the score, the better the msa, therefore gives that one a greater chance of replicating). These latter one-half of the choices for the next generation are now subject to mutation, as described in step 4 below, to produce the children of the next generation. All members of the next-generation msas undergo recombination to make new child msas derived from the two parents, as described in step 5 below. The relative probabilities of these separate events are governed by program parameters. These parameters are also adjusted dynamically as the program is running to favor those processes that have been most useful for improving msa scores.

4. In the mutation process, the sequence is not changed (else it would no longer be an alignment), but gaps are inserted and rearranged in an attempt to create a better-scoring msa. In the gap insertion process, the sequences in a given msa are divided into two groups based on an estimated phylogenetic tree, and gaps of random length are inserted into random positions in the alignment. Alternatively, in a "hill-climbing" version of the procedure, the position is so chosen as to provide the best possible score following the change.

```
XXXXXXXXX                        XXX - - XXXXXXX
XXXXXXXXX                        XXX - - XXXXXXX
XXXXXXXXX         - - - - >       XXXXXXXXX - - X
XXXXXXXXX                        XXXXXXXXX - - X
XXXXXXXXX                        XXXXXXXXX - - X
```

Shown above are random gap insertions into phylogenetically related sequences. The first two and last three sequences comprise the two related groups in this example. x indicates any sequence character.

Another mutational process is to move common blocks of sequence (overlapping ungapped regions) delineated by a gap, or blocks of gaps (overlapping gaps). Some of the possible moves are illustrated below. These moves may also be tailored to improve the alignment score.

```
XXX - - XXXXX      XX - - XXXXXX      XXX - - XXXXX      XXXXX - - XXX
XXXXXXXXXX        XXXXXXXXXX        XXXXXXXXXX        XXXXXXXXXX
XX - - XXXXX       X - - XXXXXXX      XXX - - XXXXX      XX - XX - XXXX
XXXXXXXXXX        XXXXXXXXXX        XXXXXXXXXX        XXXXXXXXXX
```

| Starting block | Whole block move | Split block horizontally (guided by phylogenetic grouping) | Split block vertically |

5. Recombination among next-generation parent msas is accomplished by one of two mechanisms. The first is not homology-driven. One msa is cut vertically through, and the other msa is cut in a staggered manner that does not lose any sequence after the fragments are spliced. The higher scoring of the two reciprocal recombinants is kept. The second, illustrated below, is recombination between msas driven by conserved sequence positions. It is driven by homology expressed as a vertical column of the same residue and is very like standard homologous recombination.

```
xxGxxxxDxx      xxGxx-xDxx      xxGxx-xDxx
xxGx-xxDxx      xxGxxxxDxx      xxGxxxxDxx
xxGxx-xDxx      xxGxxxxDxx      xxGxxxxDxx
xxGxxxxDxx      xxGx-xxDxx      xxGx-xxDxx
```

| **Parent A** | **Parent B** | **Child** |
| alignment | alignment | alignment |

6. The next generation, an overlapping one of the previous one-half of the best-scoring parental msas and the mutated children, is now evaluated as in step 2, and the cycle of steps 2–5 is typically repeated as much as 100 times, although as many as 1000 generations can be run. The best-scoring msa is then obtained.

7. The entire process of producing a set of msas for replication and mutation is repeated several times to obtain several possible msas, and the best scoring is chosen.

Hidden Markov Models of Multiple Sequence Alignment

The HMM is a statistical model that considers all possible combinations of matches, mismatches, and gaps to generate an alignment of a set of sequences. A localized region of similarity, including insertions and deletions, may also be modeled by an HMM. Analysis of sequences by an HMM is discussed on page 185 along with other statistical methods.

OTHER PROGRAMS AND METHODS FOR MULTIPLE SEQUENCE ALIGNMENT

The msa method often used, especially for 10 or more sequences, is to first determine sequence similarity between all pairs of sequences in the set. On the basis of these similarities, various methods are used to cluster the sequences into the most related groups or into a phylogenetic tree.

In the group approach, a consensus is produced for each group and then used to make further alignments between groups. Two examples of programs using the group approach are the program PIMA (Smith and Smith 1992), which uses several novel alignment techniques, and the program MULTAL described by Taylor (1990, 1996; see Table 4.1).

The tree method uses the distance method of phylogenetic analysis to arrange the sequences. The two closest sequences are then aligned, and the resulting consensus alignment is aligned with the next best sequence or cluster of sequences, and so on, until an alignment is obtained that includes all of the sequences. The programs PILEUP and CLUSTALW discussed above are examples. The ALIGN set of programs (Feng and Doolittle 1996) and the MS-DOS program by Corpet (1988) use this method. Additional programs for msa are also described in Barton (1994), Kim et al. (1994), and Morgenstern et al. (1996).

Another program (Vingron and Argos 1991) aligns all possible pairs of sequences to create a set of dot matrices, and the matrices are then filtered sequentially to find motifs that provide a starting point for sequence alignment. A set of programs for interactive msa by dot matrix analysis and other alignment techniques has also been developed (Boguski et al. 1992).

The program TREEALIGN takes the approach that multiple sequence alignments should be done in a fashion that simultaneously minimizes the number of changes needed during evolution to generate the observed sequence variation (Hein 1990). TREEALIGN (also named ALIGN in the program versions) has a method for performing the alignment and the most parsimonious tree construction at the same time. The initial steps are similar to other multiple sequence alignment methods, except for the use of a distance scale: i.e., the sequences are aligned pair-wise and the resulting distance scores are used sequentially to produce a tree, which is rearranged as more sequences are added. The sequences are then realigned so that the same tree can be produced by maximum parsimony. Finally, the tree is rearranged to maximize parsimony. The advantage to this method is the increased use of phylogenetic analysis to improve the multiple sequence alignment.

LOCALIZED ALIGNMENTS IN SEQUENCES

Multiple sequence alignment programs based on the methods discussed above report a global alignment of the sequences, including all parts of all sequences. A portion of the alignment that is highly conserved may then be identified and a type of scoring matrix called a profile may be produced. A profile includes scores for amino acid substitutions and gaps in each column of the conserved region so that an alignment of the region to a new sequence can be determined. Alternatively, the alignment may be scanned for regions that include only substituted regions without gaps, called blocks, and these blocks may then be used in sequence alignments.

There is also a third method for finding a localized region of sequence similarity in a set of sequences without first having to produce an alignment. In this method, the sequences are analyzed by pattern-searching or statistical methods. All of these methods for finding localized sequence similarity are discussed below.

Profile Analysis

Profiles are found by performing the global msa of a group of sequences and then removing the more highly conserved regions in the alignment into a smaller msa. A scoring matrix for the msa, called a profile, is then made. The profile is composed of columns much like a mini-msa and may include matches, mismatches, insertions, and deletions. A tutorial on preparing profiles by the first method, prepared by M. Gribskov, is at Web address http://www.sdsc.edu/projects/profile/profile_tutorial.html, and the Web site at http://www.sdsc.edu/projects/profile/ will perform a motif analysis on the University of California at San Diego Supercomputer Center. The program Profilemake can be used to produce a profile from a msa (Gribskov et al. 1987, 1990; Gribskov and Veretnik 1996). A version of the Profilesearch program, which performs a database search for matches to a profile, is available at the University of Pittsburgh Supercomputer Center (http://www.psc.edu/general/software/packages/profiless/profiless.html). A special grant application may be needed to use this facility. Profile-generating programs are available by FTP from ftp.sdsc.edu/pub/sdsc/biology and are included in the Genetics Computer Group suite of programs (http://www.gcg.com/), although the more recent features (Gribskov and Veretnik 1996) are not included in GCG, v. 9.1.

Once produced, the profile is used to search a target sequence for possible matches to the profile using the scores in the table to evaluate the likelihood at each position. For example, the table value for a profile that is 25 amino acids long will have 25 rows of 20 scores, each score in a row for matching one of the amino acids at the corresponding position in the profile. If a sequence 100 amino acids in length is to be searched, each 25-amino-acid-long stretch of sequence will be examined, 1–25, 2–26, ... 76–100. The first 25-amino-acid-long stretch will be evaluated using the profile scores for the amino acids in that sequence, then the next 25-long stretch, and so on. The highest-scoring sections will be the most similar to the profile.

The disadvantage of this method of profile extraction from an msa is that the profile produced is only as representative of the variation in the family of sequences as the msa itself. If several sequences in the msa are similar, the msa and the derived profile will be biased in favor of those sequences. Methods have been devised for partially circumventing this problem with the profile (Gribskov and Veretnik 1996), but the difficulty with the msa itself is not easily reconciled, as discussed at the beginning of this section. Sequence weighting is based on the production of a simple phylogenetic tree by distance methods; more closely related sequences then receive a reduced weight in the profile. Another problem is that some amino acids may not be represented in a particular column because not enough sequences have been included. Athough absence of an amino acid may mean that the amino acid may not occur at that position in the protein family, adding counts to such positions generally increases the usefulness of the profile. This feature is built into the profile method discussed below.

An example of the generation of a profile and the matrix representation of this profile for a set of heat shock proteins is illustrated in Figure 4.11. The profile is similar to the log odds form of the amino acid substitution table, such as the PAM250 and BLOSUM62

Cons	A	B	C	D	E	F	G	H	I	K	L	M	N	P	Q	R	S	T	V	W	Y	Z	Gap	Len
I	8	3	-2	5	4	5	5	-4	<u>24</u>	0	15	13	1	1	1	-7	2	22	21	-18	-6	4	100	100
T	13	19	-5	24	18	-18	19	7	1	7	-7	-4	14	11	10	-1	9	<u>29</u>	3	-28	-14	15	100	100
L	5	5	-5	3	4	13	4	2	8	-4	<u>14</u>	12	8	-5	0	-10	0	10	10	-1	5	2	22	22
S	17	14	17	13	10	-12	29	-5	-5	6	-14	-9	12	10	0	-2	<u>34</u>	19	1	-8	-15	4	100	100
T	15	3	22	0	-1	-5	12	-2	7	-3	-8	-6	5	7	-8	-7	16	<u>29</u>	9	-22	6	-4	100	100
T	8	-1	12	-2	0	5	6	-4	19	-4	8	5	-1	2	-8	-8	7	<u>22</u>	19	-15	4	-3	100	100
C	17	0	<u>24</u>	-1	-3	11	8	-1	7	-10	1	-2	1	-3	-8	-14	8	5	9	-5	14	-7	100	100
V	11	0	18	-1	-2	2	14	-10	26	-4	9	7	-3	7	-7	-7	21	10	<u>31</u>	-19	-5	-5	100	100
C	10	-8	<u>15</u>	-11	-11	6	8	-7	11	-10	4	3	-7	0	-11	-4	11	5	15	-22	14	-11	100	100
V	7	7	-3	8	8	-3	11	1	20	-1	14	10	4	2	8	-5	0	5	<u>26</u>	-24	-6	8	100	100

Figure 4.11. Pattern identification by the profile method. A set of heat shock 70 (hsp70) proteins from a diverse group of organisms were aligned by the Genetics Computer Group msa program PILEUP. A profile was then made from one region in the alignment with the Genetics Computer Group program Profilemake. The profile represents the specific motif pattern found for the chosen location shown for this set of hsp70 proteins. The first column gives the consensus amino acid at each position in the profile. Thus, the consensus pattern is ITLSTTCVCV. This profile is used to search a target sequence for matches to the profile. The table values are a log odds score of giving the probability of finding the amino acid in the target sequence at that position in the profile divided by the probability of aligning the two amino acids by random chance. If a gap must be placed in the target sequence to align the sequence with the profile, then the penalties for opening a gap and extending the gap, respectively, are subtracted. The profile itself may include gaps, in which case the penalty is reduced, as seen for example in the row 3 of the profile table. The method of producing the substitution scores shown in the table is described in the text.

matrices used for sequence alignments. The matrix is 23 columns wide, one column for each of the 20 amino acids, plus one column for an unknown amino acid z and two columns for a gap opening and extension penalty. There is one row for each column in the msa. The consensus sequence, derived from the most common amino acid in each column of the msa, is listed down the left-hand column. The scores on each row reflect the number of occurrences of each amino acid in the aligned sequences. For example, in the first row, I, T, and V were found, with I being the majority amino acid. The highest positive score on each row (underlined) is in the column corresponding to the consensus amino acid, the most negative score for an amino acid not expected at that position. These values are derived from the log odds amino acid substitution matrix that was used to produce the alignment, such as the log odds form of the Dayhoff PAM250 matrix. Two methods are used to produce profile tables, the average method and the evolutionary method. The evolutionary method seems somewhat better for finding family members.

In the average method, the profile matrix values are weighted by the proportion of each amino acid in each column of the msa. For example, if column 1 in the msa has 5 Ile (I), 3 Thr (T), and 2 Val (V), then the frequency of each amino acid in this column is 0.5 I, 0.3 T, and 0.2 V. These amino acids are considered to have arisen with equal probability from any of the 20 amino acids as ancestors. In the example in Figure 4.11, the I, T, and V in column 1 could have arisen from any of the 20 amino acids by mutation. Suppose that they arose from an Ile (I). The profile values in the Ile (I) column of the corresponding row in the profile matrix would then use the amino acid scoring matrix values for I-I, I-T, and I-V, which are log odds scores of 5, 0, and 4 in the Dayhoff PAM250 matrix. Then the profile value for the I column is the frequency-weighted value, or $0.5 \times 5 + 0.3 \times 0 + 0.2 \times 4 = 3.3$.

The profile table also includes penalties for matching a gap in the target sequence, shown in the two right columns. All of these table values are multiplied by a constant for convenience so that only the value of a score with one sequence relative to the score with another sequence matters. Once a profile table has been obtained, the table may be used in database searches for additional sequences with the same pattern (program Profilesearch) or as a scoring matrix for aligning sequences (program Profilegap). If several profiles characteristic of a protein family can be identified, the chance of a positive identification of additional family members is greatly increased (Bailey and Gribskov 1998; also see http://www.sdsc.edu/MEME).

The evolutionary method for producing a profile table is based on the Dayhoff model of protein evolution (Chapter 2) (Gribskov and Veretnik 1996). The amino acids in each column of the msa are assumed to be evolving at a different rate, as reflected in the amount of amino acid variation that is observed. As with the average model, the object is to consider each of the 20 amino acids as a possible ancestor of the pattern of each column. In the evolutionary model, the evolutionary distance in PAM units that would be required to give the observed amino acid distribution in each column is determined. Recall that each PAM unit represents an overall probability of 1% change in a sequence position. For example, in the original Dayhoff PAM1 matrix for an evolutionary distance of 1 PAM unit (very roughly 10 my), the probability of an I not changing is 0.9872, and the probabilities for changing to a T or a V are 0.0011 and 0.0057, respectively. All of the probabilities of changing I to any other amino acid add up to 1.0000, for a combined probability of change of 1% for I. For an evolutionary distance of n PAM, the PAM1 matrix is multiplied by itself n times to give the expected changes at that distance. At a distance of 250 PAMs, the above three probabilities of an I not changing or of changing to a T or V are 0.10, 0.06, and 0.15, respectively, representing a much greater degree of change than for a shorter time, as might be expected (Dayhoff 1978).

Do not confuse these probabilities of one amino acid changing to another in the original Dayoff PAM250 matrix with scores from the log odds form of the PAM250 matrix, which have been used up to now. The log odds scores are derived from the original Dayhoff matrix by dividing each probability of change with the probability of a chance matching of the amino acids in a sequence alignment; i.e., that the one amino acid is not an ancestor of the other. These ratios are then converted to logarithms.

Thus, for the example of the msa column 1 with 5 Ile (I), 3 Thr (T), and 2 Val (V), the object is to find what amount of PAM distance from each of the 20 amino acids as possible ancestors will generate this much diversity. This amount can be found by a formula giving the amount of information (entropy) of the observed column variation given the expected variation in the evolutionary model,

$$H = -\sum_{\text{all } a\text{'s}} f_a \log(p_a) \tag{1}$$

where f_a is the observed proportion of each amino acid a in the msa column and p_a is the expected frequency of the amino acid when derived from a given ancestor amino acid. For a given column in the msa, H is calculated for each 20 ancestor amino acids and for a large number of evolutionary distances (PAM1, PAM2, PAM4, ...). The distance that gives the minimum value for H for each column-possible ancestor combination is the best estimate of the distance that generates the column diversity from that ancestor. This analysis provides 20 possible models (M_a for $a = 1,2,3, . . . 20$) as to how the amino acid frequencies in a column (F) may have originated. The next step in the evolutionary profile construction determines the extent to which each M_a predicts F by the now-familiar Bayes conditional probability analysis.

$$P(M_a|F) = P(M_a) \times P(F|M_a) / \sum_{\text{all } a\text{'s}} P(M_a) \times P(F|M_a) \tag{2}$$

where the prior distribution $P(M_a)$ is the given by the background amino acid frequencies and

$$P(F|M_a) = p_{aa1}^{faa1} \times p_{aa2}^{faa2} \times p_{aa3}^{faa3} \ldots \ldots p_{aa20}^{faa20} \tag{3}$$

i.e., the product of the expected amino acid frequencies in M_a raised to the power of the fraction observed for each amino acid in the msa column, as defined above. From $P(M_a | F)$, the weights for each of the 20 possible distributions that give rise to the msa column diversity are calculated as follows:

$$W_a = P(M_a | F) - P(M_{\text{random}} | F) \tag{4}$$

where W_a is the weight given to M_a and $P(M_{\text{random}} | F)$ is calculated as above using the background amino acid distribution.

The log odds scores for the profile (Profile$_{ij}$) are given by:

$$\text{Profile}_{ij} = \log \left[\sum_{\text{all } a\text{'s}} (W_{ai} \times p_{aij}) / p_{\text{random } j} \right] \tag{5}$$

where W_{ai} is the weight of an ancestral amino acid a at row i in the profile, p_{aij} is the frequency of amino acid j in the PAM amino acid distribution that best matches at row i, and

$p_{\text{random } j}$ is the background frequency of amino acid j. An example of a profile matrix for the ATP-dependent RNA helicase ("DEAD" box family) from the M. Gribskov laboratory is given in Figure 4.12.

The usefulness of the evolutionary profile is demonstrated by the following: A profile for the 4Fe-4S ferredoxin family was prepared from six sequences. This profile was then used to search the SwissProt database for family members. Success was measured by the so-called receiver operating characteristic test (ROC) plot. The fraction of scores equal to or greater than a certain value is plotted for the true positive matches (a correct family member identified) on the y axis and for the true negatives (unrelated sequences) on the x axis. The area under the curve and the x axis gives the probability of correct identification. The ROC_{50} is the area under the curve when it is truncated to the first 50 incorrect sequences, and can be used as a standard for success in a database search (Gribskov and Veretnik 1996). For the ferredoxin family search, the ROC_{50}, $95.6 \pm 0.6\%$ of the known family members, was identified in a search of SwissProt by an evolutionary profile, whereas $93.0 \pm 2.0\%$ was identified by the average profile method (Gribskov and Veretnik 1996). The success rate was increased 0.4–0.6% by using 12 training sequences and 2–3% by using 134 training sequences.

Block Analysis

Like profiles, blocks represent a conserved region in the msa. Blocks differ from profiles in lacking insert and delete positions in the sequences. Instead, every column includes only matches and mismatches. Like profiles, blocks may be made by searching for a section of an msa alignment that is highly conserved. However, aligned regions may also be found by searching each sequence in turn for similar patterns of the same length. These patterns may include a region with one or a few matching characters followed by a short spacer region of unmatched characters and then by another set of a few matching characters, and so on, until the sequences start to be different. These patterns are all of the same length, and when they are aligned, the matching sequence characters will appear in columns. The first alignments of this type were performed by computer programs that searched for patterns in sequences (Henikoff and Henikoff 1991; Neuwald and Green 1994). Several blocks located in different regions in a set of sequences may be used to produce a msa (Zhang et al. 1994), and blocks may be constructed from a set of aligned sequence pairs (Miller et al. 1994). Statistical and Bayesian statistical methods are also used to locate the most alike regions of sequences (Lawrence et al. 1993; Lawrence and Reilly 1990). Web sites that perform some of these types of analyses are discussed below and also given in Table 4.1. Finally, the information content of these tables can be displayed by a sequence logo (see p. 195). Note that few of these types of analyses presently provide a method for phylogenetic estimates of the sequence relationships so that sequence weighting can be used to make the changes more reflective of the phylogenetic histories among the sequences. Additionally, except where noted, these methods do not use substitution matrices such as the PAM and BLOSUM matrices to score matches. Rather, they are based on finding exact matches that have the same spacing in at least some of the input sequences, and that may be repeated in a given sequence.

Extraction of Blocks from a Global or Local Multiple Sequence Alignment

A global msa of related protein sequences usually includes regions that have been aligned without gaps in any of the sequences. These ungapped patterns may be extracted from these aligned regions and used to produce blocks. Blocks found in this manner are

A. The multiple sequence alignment.

```
rhle_ecoli    GVDVLVATPG RLLDLEHQNA ....VKLDQV EILVLDEADR MLDMGFIHDI
dbp2_schpo    GVEICIATPG RLLDMLDSNK ....TNLRRV TYLVLDEADR MLDMGFEPQI
dbp2_yeast    GSEIVIATPG RLIDMLEIGK ....TNLKRV TYLVLDEADR MLDMGFEPQI
dbpa_ecoli    APHIIVATPG RLLDHLQKGT ....VSLDAL NTLVMDEADR MLDMGFSDAI
rm62_drome    GCEIVIATPG RLIDFLSAGS ....TNLKRC TYLVLDEADR MLDMGFEPQI
 p68_human    GVEICIATPG RLIDFLECGK ....TNLRRT TYLVLDEADR MLDMGFEPQI
rhlb_ecoli    GVDILIGTTG RLIDYAKQNH ....INLGAI QVVVLDEADR MYDLGFIKDI
yn21_caeel    RPHIIVATPG RLVDHLENTK ...GFNLKAL KFLIMDEADR ILNMDFEVEL
yhm5_yeast    KPHIIIATPG RLMDHLENTK ...GFSLRKL KFLVMDEADR LLDMEFGPVL
me31_drome    KVQLIIATPG RILDLMDKKV ....ADMSHC RILVLDEADK LLSLDFQGML
drs1_yeast    RPDIVIATPG RFIDHIRNSA ...SFNVDSV EILVMDEADR MLEEGFQDEL
if4a_rabit    APHIIVGTPG RVFDMLNRRY ....LSPKYI KMFVLDEADE MLSRGFKDQI
if41_human    APHIIVGTPG RVFDMLNRRY ....LSPKYI KMFVLDEADE MLSRGFKDQI
vasa_drome    GCHVVIATPG RLLDFVDRTF ....ITFEDT RFVVLDEADR MLDMGFSEDM
srmb_ecoli    NQDIVVATTG RLLQYIKEEN ....FDCRAV ETLILDEADR MLDMGFAQDI
DEAD_ecoli    GPQIVVGTPG RLLDHLKRGT ....LDLSKL SGLVLDEADE MLRMGFIEDV
if4a_orysa    GVHVVVGTPG RVFDMLRRQS ....LRPDYI KMFVLDEADE MLSRGFKDQI
DEAD_klepn    GPQIVVGTPG RLLDHLKRGT ....LDLSKL SGLVLDEADE MLRMGFIEDV
pl10_mouse    GCHLLVATPG RLVDMMERGK ....IGLDFC KYLVLDEADR MLDMGFEPQI
 p54_human    TVHVVIATPG RILDLIKKGV ....AKVDHV QMIVLDEADK LLSQDFVQIM
if4a_drome    GCHVVVGTPG RVYDMINRKL .....RTQYI KLFVLDEADE MLSRGFKDQI
ded1_yeast    GCDLLVATPG RLNDLLERGK ....ISLANV KYLVLDEADR MLDMGFEPQI
ms16_yeast    RPNIVIATPG RLIDVLEKYS ..NKFFRFV DYKVLDEADR LLEIGFRDDL
pr28_yeast    GCDILVATPG RLIDSLENHL ....LVMKQV ETLVLDEADK MYDLGFEDQV
if4n_human    GQHVVAGTPG RVFDMIRRRS ....LRTRAI KMLVLDEADE MLNKGFKEQI
an3_xenla     GCHLLVATPG RLVDMMERGK ....IGLDFC KYLVLDEADR MLDMGFEPQI
dbp1_yeast    GCDLLVATPG RLNDLLERGK ....VSLANI KYLVLDEADR MLDMGFEPQI
if4a_yeast    DAQIVVGTPG RVFDNIQRRR ....FRTDKI KMFILDEADE MLSSGFKEQI
spb4_yeast    RPQILIGTPG RVLDFLQMPA ....VKTSAC SMVVMDEADR LLDMSFIKDT
if4a_caeel    GIHVVVGTPG RVGDMINRNA ....LDTSRI KMFVLDEADE MLSRGFKDQI
pr05_yeast    GTEIVVATPG RFIDILTLND .GKLLSTKRI TFVVMDEADR LFDLGFEPQI
if42_mouse    APHIVVGTPG RVFDMLNRRY ....LSPKWI KMFVLDEADE MLSRGFKDQI
dhh1_yeast    TVHILVGTPG RVLDLASRKV ....ADLSDC SLFIMDEADK MLSRDFKTII
db73_drome    KADIVVTTPG RLVDHLHATK ...GFCLKSL KFLVIDEADR IMDAVFQNWL
yk04_yeast    GCNFIIGTPG RVLDHLQNTK VIKEQLSQSL RYIVLDEGDK LMELGFDETI
ybz2_yeast    SGQIVIATPG RFLELLEKDN .TLIKRFSKV NTLILDEADR LLQDGHFDEF
yhw9_yeast    KPHFIIATPG RLAHHIMSSG DDTVGGLMRA KYLVLDEADI LLTSTFADHL
glh1_caeel    GATIIVGTVG RIKHFCEEGT ....IKLDKC RFFVLDEADR MIDAMGFGTD
```

Figure 4.12. msa and the derived evolutionary profile.

B. The evolutionary profile. Note the location of red conserved regions in the alignment in the corresponding profile of these sequences.

Cons	A	B	C	D	E	F	G	H	I	K	L	M	N	P	Q	R	S	T	V	W	Y	Z	Gap	Len
G	17	18	0	19	14	-22	31	0	-9	12	-15	-5	15	10	9	6	18	14	1	-15	-22	11	100	100
P	18	0	13	0	0	-12	13	0	8	-3	-3	-1	-2	23	2	-2	12	11	17	-31	-8	1	100	100
H	5	24	-12	29	25	-20	8	32	-9	9	-10	-9	22	7	30	10	0	4	-8	-20	-7	27	100	100
I	-1	-12	6	-13	-11	33	-12	-13	63	-11	40	29	-15	-9	-14	-15	-6	7	50	-17	8	-11	100	100
V	3	-11	1	-11	-9	22	-3	-11	46	-9	37	30	-3	-9	-13	-6	6	50	-19	2	-8	100	100	100
V	5	-9	9	-9	-9	19	-1	-13	57	-9	35	26	-2	-11	-13	-4	9	58	-29	0	-9	100	100	100
A	54	15	12	20	17	-24	44	-6	-4	-1	-11	12	19	9	-13	21	19	9	-39	-20	10	100	100	100
T	40	20	20	20	20	-30	40	-10	20	20	-10	0	30	-10	-10	30	150	20	-60	-30	10	100	100	100
P	31	6	7	6	6	-41	19	11	-9	6	-16	-11	89	17	17	24	22	9	-50	-48	12	100	100	100
G	70	60	20	70	50	-60	150	-20	-30	-10	-50	-30	30	20	-30	60	40	20	-100	-70	30	100	100	100
! 11																								
R	-30	10	-30	0	0	-50	-30	50	-30	80	-40	20	30	40	150	10	-10	-30	140	-60	20	100	100	100
L	-2	-17	-15	-18	-12	38	-13	-9	38	-12	49	39	-9	-9	-15	-11	0	38	6	12	-10	100	100	100
L	0	-12	-15	-14	-9	32	-12	-7	32	-7	41	35	-9	-6	-12	-9	0	29	6	9	-7	100	100	100
D	15	58	-27	78	54	-52	35	27	-12	16	-26	-21	6	41	3	9	10	-12	-57	-25	50	100	100	100
L	-5	-5	-7	-8	-4	24	-12	13	13	-6	25	17	-7	0	-2	-8	-3	10	11	17	-2	100	100	100
L	3	-13	-13	-13	-8	31	-11	-8	34	-9	41	36	-7	-5	-13	-8	2	31	-1	8	-6	100	100	100
E	6	19	-15	23	27	-21	9	15	-6	18	-8	-1	6	23	12	6	5	-6	-15	-16	25	100	100	100
K	3	14	-12	11	12	-16	2	10	-5	23	-7	4	6	15	22	8	3	-5	7	-15	14	100	100	100
G	11	17	0	16	14	-16	19	5	-6	11	-11	-5	9	8	4	14	15	-1	-13	-14	11	100	100	100
T	12	9	-1	7	7	-8	9	2	4	12	0	4	5	4	3	9	12	7	-8	-8	5	100	100	100

Figure 4.12. *Continued.*

↓21																									
D	1	1	0	2	1	-1	1	0	1	0	0	1	0	1	0	1	1	0	2	-3	-1	1	22	22
T	2	2	0	3	2	-2	3	0	2	0	1	1	1	4	1	1	4	1	2	-5	-2	2	22	22
K	0	1	-3	0	1	0	0	4	1	0	4	0	0	3	0	1	3	1	1	0	-2	1	22	22
G	3	3	0	4	4	-1	6	-1	3	1	0	4	1	3	1	3	5	2	3	-6	-3	2	22	22
L	5	-6	-4	-7	-4	16	-2	-4	21	23	17	-4	-4	-5	-4	-2	4	19	-4	0	6	-4	22	22
B	5	16	-6	15	11	-15	10	6	-3	-8	-1	16	9	4	-2	7	7	-2	-3	-3	-11	10	100	100
L	1	-13	-12	-14	-9	27	-8	-7	24	36	30	-8	-5	-7	-4	-10	23	6	6	-3	9	-8	100	100
D	7	19	-7	22	17	-22	13	-6	-6	-11	-3	19	14	15	8	14	6	-5	-5	3	-18	16	100	100
K	11	10	-3	10	9	-12	5	-4	-4	-6	0	16	10	11	6	10	4	-4	3	-8	-8	10	100	100
V	7	-10	11	7	-10	14	0	-8	-11	-11	19	16	-10	0	-10	-12	2	8	34	-22	9	-10	100	100
↓31																								
K	4	16	-11	16	17	-21	6	-7	27	-12	-7	0	16	7	20	12	8	-7	-18	0	17	100	100
F	-3	-10	3	-14	-10	29	-3	20	-9	26	19	-10	-7	-12	-6	-6	5	14	20	10	-11	100	100
L	-5	-18	-19	-21	-14	42	-8	-16	34	51	40	-8	-12	-15	-12	-3	-3	32	16	17	-11	100	100
V	10	-11	11	-11	-11	15	8	-17	66	46	34	-17	-12	4	-6	11	83	-43	-4	-4	100	100
L	-4	-23	-37	-24	-14	53	-23	-11	39	73	67	-19	-14	-14	-19	-4	39	17	11	-9	100	100
D	30	110	-50	150	100	-100	70	40	-20	30	-50	-40	70	10	70	0	20	-20	-110	-50	90	100	100
E	30	70	-60	100	150	-70	50	40	-20	30	-30	-20	50	10	70	0	20	-20	-110	-50	110	100	100
A	121	17	24	25	25	-41	58	-8	0	0	-9	-24	16	41	16	33	33	16	-66	-25	16	100	100
D	30	110	-50	150	100	-100	70	40	-20	30	-50	-40	70	10	70	0	20	-20	-110	-50	90	100	100
R	-7	9	-16	7	10	-23	-7	17	-9	36	-14	49	18	10	8	5	0	-10	35	-25	14	100	

Figure 4.12. *Continued.*

| 41																							
M	-1	-16	-19	-10	33	-16	-12	31	1	60	62	-15	-10	-2	-1	-14	-1	31	-2	2	-6	100	100
L	-6	-25	-26	-16	61	-26	-10	41	-15	74	64	-20	-17	-7	-20	-20	-5	39	24	18	-12	100	100
D	12	32	41	32	-30	23	11	-7	12	-17	-12	23	7	20	3	19	10	-7	-27	-19	27	100	100
M	5	-2	-3	1	8	-3	-3	13	6	25	30	-3	0	5	6	-1	2	13	-1	-6	3	100	100
G	26	26	31	23	-25	54	-5	-8	-1	-16	-9	17	11	10	-10	23	17	8	-41	-27	15	100	100
F	-31	-44	-63	-44	96	-36	-4	44	-45	77	31	-31	-45	-50	-32	-19	-19	13	82	90	-44	100	100
E	7	8	10	13	-7	5	4	7	10	3	5	6	2	10	3	6	5	5	-9	-7	12	100	100
D	12	24	31	27	-27	19	12	-6	12	-12	-8	17	12	21	4	8	9	-3	-30	-19	25	100	100
D	8	18	25	22	-18	11	13	0	9	-3	-1	13	5	24	5	3	8	0	-24	-12	24	100	100
I	0	-8	-8	-7	27	-10	-9	43	-7	36	29	-10	-8	-7	-11	-6	6	35	-9	5	-6	100	100
| 51																							
E	6	15	18	21	-11	7	11	-2	8	-7	-3	12	2	12	6	6	6	-3	-8	-7	16	100	100
N	3	17	20	21	-19	5	17	-7	18	-9	-3	15	5	20	15	4	8	-7	-10	-12	21	100	100
I	2	-9	-10	-9	27	-9	-13	60	-9	34	26	-13	-6	-13	-13	-3	11	49	-23	4	-9	100	100
L	1	-10	-11	-8	26	-9	-8	33	-7	34	28	-10	-7	-8	-9	-5	2	28	-2	-7	-7	100	100
K	8	9	9	9	-13	11	1	0	16	-4	4	7	8	11	13	12	3	-2	-15	8	8	100	100
L	3	4	3	6	3	-2	8	9	7	10	10	0	8	3	0	5	7	-2	0	7	7	100	100
L	1	-13	-13	-9	32	-11	-7	32	-9	42	36	-12	-6	-13	-9	-9	33	2	8	-7	-7	100	100
*	99	0	25	120	94	137	44	181	105	256	94	41	62	64	144	59	99	162	3	35	0		

Figure 4.12. *Continued.*

only as good as the msa from which they are derived. Using the BLOCKS (http://www.blocks.fhcrc.org/blocks/process_blocks.html), blocks of width 10–55 are extracted from a protein msa of up to 400 sequences (Henikoff and Henikoff 1991, 1992). The program accepts FASTA, CLUSTAL, or MSF formats, or manually reformatted msas. Several types of analyses may be performed with such extracted blocks. The BLOCKS server primarily generates blocks from unaligned sequences. The eMOTIFs server at http://dna.stanford.edu/emotif/ (Nevill-Manning et al. 1998) similarly extracts motifs from msas in several msa formats and provides a formatter for additional msa formats. These types of analyses are discussed below in greater detail.

Pattern Searching

This type of analysis was performed on groups of related proteins, and the amino acid patterns that were located may be found in the Prosite catalog (Bairoch 1991). This catalog groups proteins that have similar biochemical functions on the basis of amino acid patterns such as those in the active site. Subsequently, these families were searched for amino acid patterns by the MOTIF program (Smith et al. 1990), which finds patterns of the type aa1 d1 aa2 d2 aa3, where aa1 and aa2 are conserved amino acids and d1 and d2 are stretches of intervening sequence up to 24 amino acids long. These initial patterns are then organized into blocks between 3 and 60 amino acids long by the Henikoff PROTOMAT program (Henikoff and Henikoff 1991, 1992). The BLOCKS database can be accessed at http://www.blocks.fhcrc.org/, and the server may also be used to produce new blocks by the original pattern-finding method or other methods described below.

Although used successfully for making the BLOCKS database, the MOTIF program is limited in the pattern sizes that can be found. The MOTIF program distinguishes true motifs from random background patterns by requiring that motifs occur in a number of the input sequences and tend not to be internally repeated in any one sequence. As the length of the motif increases, there are many possible combinations of patterns of a given length where only a few characters match, e.g., $>10^9$ possible patterns for a 15-amino-acid-long pattern with only five matches. The MOTIF program always provides a motif, even for random sequences, thus making it difficult to decide how significant the found motif really is. This problem has been circumvented by combining the analysis performed by MOTIF with that of the Gibbs sampler (discussed on p. 177), which is based on sound statistical principles. A rigorous searching algorithm called Aligned Segment Statistical Evaluation Tool (ASSET) has been devised (Neuwald and Green 1994) that can find patterns in sequence up to 50 amino acids long, group them, and provide a measure of the statistical significance of the patterns. These patterns may also include certain pairs, the 26 positive scoring pairs in the BLOSUM62 scoring matrix. Consideration of all BLOSUM pairs is not possible because this would greatly increase the complexity of the analysis.

The efficiency of ASSET is achieved by a combination of an efficient pattern search strategy called the depth-first method, which assures searching for the same patterns only once, and the use of formulas for efficiently organizing the patterns. Low-complexity regions with high proportions of the same residue and use of sequences, some of which are more similar than the others, can interfere with the ability of the method to find a range of patterns. ASSET removes low-complexity regions and redundant sequences from consideration. The program was easily able to find subtle motifs in the DNA methylase, reverse transcriptase, and tRNA ligase families, and previously identified by the MOTIF program. In addition, however, ASSET gave these motifs an expect score, the probability that these are random matches of unrelated sequences, of <0.001. The program also found motifs in families with only a fraction of the sequences sharing a motif (the acyltransferase family) and in a set of distantly related sequences sharing the helix-turn-helix motif. Finally, the program found several repeat sequences in a prenyltransferase and ankyrin-like repeats in an *E. coli* protein. This source code of the program is available by anonymous FTP from ncbi.nlm.nih.gov/pub/neuwald/asset. The European Bioinformatics Institute has a Web page for another complex pattern-finding program (PRATT) at http://www2.ebi.ac.uk/pratt/ (Jonassen et al. 1995).

Blocks Produced by the BLOCKS Server from Unaligned Sequences

As described above, the BLOCKS server can extract a conserved, ungapped region from a msa to produce a sequence block. This same server can also find blocks in a set of unaligned, input sequences and maintains a large database of blocks based on an analysis of proteins in the Prosite catalog. Blocks are found by the Protomat program (Henikoff and Henikoff 1991). Blocks are found in two steps: First, the program MOTIF (Smith et al. 1990) described on the previous page is used to locate spaced patterns. The second step takes the best and most consistent patterns found in step 1 and uses the program MOTOMAT to merge overlapping triplets and extend them, orders the resulting blocks, and chooses those that are in the largest subset of sequences. Since 1993, the Gibbs sampler (see below) has been used as an additional tool for finding the initial set of short patterns also by specifying that the sampler search for short motifs. This program is based on a statistical analysis of the sequences and can identify the most significant common patterns in a set of sequences.

An example of BlockMaker output using an example from Lawrence et al. (1993) is shown below. The program first searches for blocks using either the MOTIFS or Gibbs sampler program to identify patterns, then the Protomat program to consolidate the patterns into meaningful blocks. The results of both types of analyses are reported.

```
A.  Motif analysis


                LipocalA, width = 15        LipocalB, width = 11
   BBP_PIEBR    16 NFDWSNYHGKWWEVA (  70)   101 VLSTDNKNYII
   ICYA_MANSE   17 DFDLSAFAGAWHEIA (  73)   105 VLATDYKNYAI
   LACB_BOVIN   25 GLDIQKVAGTWYSLA (  70)   110 VLDTDYKKYLL
   MUP2_MOUSE   27 NFNVEKINGEWHTII ( 101)   143 DLSSDIKERFA
   RETB_BOVIN   14 NFDKARFAGTWYAMA (  77)   106 IIDTDYETFAV



B.  Gibbs sampler analysis


                LipocalA, width = 15        LipocalB, width = 11
   BBP_PIEBR    16 NFDWSNYHGKWWEVA (  70)   101 VLSTDNKNYII
   ICYA_MANSE   17 DFDLSAFAGAWHEIA (  73)   105 VLATDYKNYAI
   LACB_BOVIN   25 GLDIQKVAGTWYSLA (  70)   110 VLDTDYKKYLL
   MUP2_MOUSE   27 NFNVEKINGEWHTII (  68)   110 IPKTDYDNFLM
   RETB_BOVIN   14 NFDKARFAGTWYAMA (  77)   106 IIDTDYETFAV
```

In the above example, two blocks identified as Lipocal A and B are reported using both the MOTIF and Gibbs sampler programs for step 1, the initial pattern-finding step. The MOTIF program is based on a heuristic method that will always find motifs, even in random sequences, whereas the Gibbs sampler discriminates found motifs based on sound statistical methods. These blocks are identical to those determined from analysis of three-dimensional structures. Note that MOTIF aligned MUP2_MOUSE incorrectly in the B

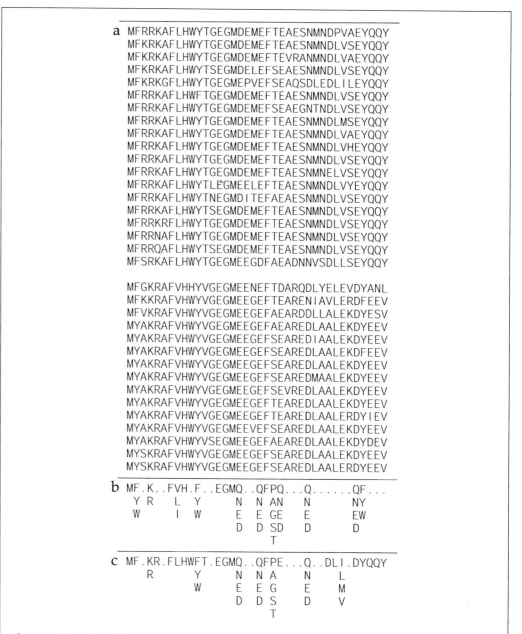

Figure 4.13. Aligned block of 34 tubulin proteins. (*a*) The sequences are divided into two groups based on the occurrence of R or L in the fourth position and Y in the last position. (*b*) Specific substitution groups found in the columns of the block. If a group cannot be found, then the position is ambiguous and a dot is printed at the position. (*c*) If only the first group of sequences is used, a more specific motif may be found because sequences in this group are more closely related to each other. (Reprinted, with permission, from Nevill-Manning et al. 1998 [copyright National Academy of Sciences].)

block. The Gibbs sampler results may differ when the same sequences are submitted repeatedly with a different initial alignment (see below).

The eMOTIF Method of Motif Analysis

Another somewhat different but extemely useful method of identifying motifs in protein sequences has been described (Nevill-Manning et al. 1998). From the BLOCKS database (derived from msa of proteins in the Prosite catalog) and the HSSP database (derived from msa of proteins based on predicted structural similarities), a set of amino acid substitution groups characteristic of each column in all of the alignments was found. These patterns reflect the higher log odds scores in the amino acid substitution matrices. A statistical analysis was performed to identify amino acids that are found together in the same msa column as opposed to amino acids that are found in different columns at the 0.01 level of significance. Thirty and 51 substitution groups that met this criterion were found in the BLOCKS and HSSP msas, respectively. For example, the chemically aromatic group of amino acids F, W, and Y were found to define a group often located in the same column of the msa.

From the msa for a particular group of proteins, each column is examined to see whether these groups are represented in the column, as illustrated in Figure 4.13. In column 1, M is always present, and because M is one group, M is used in column 1 of the motif, as shown in part *b*. Similarly for column 2, Y and F, which are members of the group FYW, are found, and hence this group is used as column 2 in the motif. The final motif shown in *b* describes the variation in all the sequences. Instead, a motif may be made for only the first group of 19 sequences, and is shown in *c*. This second motif (*c*) has less variability and greater specificity for the first 19 sequences and thus would be more likely to find those sequences in a database search (i.e., it is a more sensitive motif for those sequences) than motif *b*.

The probability of each motif is estimated from the frequencies of the individual amino acids in the SwissProt database. The probability of the motif *b* above is given by the product of the probability sums in each column, or $p(\text{Motif}) = p(M) \times 1 \times [p(F)+p(W)+p(y)] \times [p(Y)+p(R)] \times \ldots$ This value has been found to provide a good estimate of false positives, or of the selectivity of the motif, in a database search. Both the sensitivity and selectivity of a given motif must be taken into account in using the motif for a database search. Ideally, a motif can find all of the sequences used to generate the motif but none other. In practice, eMOTIF produces a large set of motifs, some more and some less sensitive for the set of aligned sequences. The more sensitive ones, which are also the most selective based on the value of $p(\text{Motif})$, are then chosen. Some are useful for specifying subfamilies of a protein superfamily. A database of such motifs called Identify is a useful resource for discovering the function of a gene (Nevill-Manning et al. 1998; http://dna.stanford.edu/emotif/).

STATISTICAL METHODS FOR AIDING ALIGNMENT

Expectation Maximization Algorithm

This algorithm has been used to identify both conserved domains in unaligned proteins and protein-binding sites in unaligned DNA sequences (Lawrence and Reilly 1990), including sites that may include gaps (Cardon and Stormo 1992). Given are a set of sequences that are expected to have a common sequence pattern and may not be easily recognizable by eye. An initial guess is made as to the location and size of the site of interest in each of the sequences, and these parts of the sequence are aligned. The alignment provides an estimate of the base or amino acid composition of each column in the site. The EM algorithm then consists of two steps, which are repeated consecutively. In step 1, the expectation step, the column-by-column composition of the site already available is used to estimate the probability of finding the site at any position in each of the sequences. These probabilities are used in turn to provide new information as to the expected base or amino acid distribution for each column in the site. In step 2, the maximization step, the new counts of bases or amino acids for each position in the site found in step 1 are substituted for the previous set. Step 1 is then repeated using these new counts. The cycle is repeated until the algorithm converges on a solution and does not change with further cycles. At that time, the best location of the site in each sequence and the best estimate of the residue composition of each column in the site will be available.

As an example, suppose that there are 10 DNA sequences having very little similarity with each other, each about 100 nucleotides long and thought to contain a binding site near the middle 20 residues, based on biochemical and genetic evidence. As we will later see when examining the EM program MEME, the size and number of binding sites, the location in each sequence, and whether or not the site is present in each sequence do not necessarily have to be known. For the present example, the following steps would be used by the EM algorithm to find the most probable location of the binding sites in each of the 10 sequences.

The Initial Setup of the Algorithm

The 20-residue-long binding motif patterns in each sequence are aligned as an initial guess of the motif. The base composition of each column in the aligned patterns is then determined. The composition of the flanking sequence on each side of the site provides the surrounding base or amino acid composition for comparison, as illustrated below. For illustration purposes, each sequence is assumed to be the same length and to be aligned by the ends, and each character in the alignment represents five sequence positions (o, not in motif; x, in motif).

OOOOOOOOXXXXOOOOOOOO
OOOOOOOOXXXXOOOOOOOO
OOOOOOOOXXXXOOOOOOOO
OOOOOOOOXXXXOOOOOOOO
OOOOOOOOXXXXOOOOOOOO
OOOOOOOOXXXXOOOOOOOO
OOOOOOOOXXXXOOOOOOOO
OOOOOOOOXXXXOOOOOOOO
OOOOOOOOXXXXOOOOOOOO
OOOOOOOOXXXXOOOOOOOO

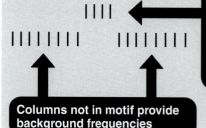

Columns defined by a preliminary alignment of the sequences provide initial estimates of frequencies of amino acids in each motif column

Columns not in motif provide background frequencies

The number of each base in each column is determined and then converted to fractions. Suppose, for example, that there are four Gs in the first column of the 10 sequences, then the frequency of G in the first column of the site, $fs_G = 4/10 = 0.4$. This procedure is repeated for each base and each column. For the rest of the sequences not included in the sites, the background frequency of each base is calculated. For example, let one of these four values for the background frequency, the frequency of G, be $fb_G = 224/800 = 0.28$. These values are now placed in a 5×20 matrix of values, the first column for the background frequencies, and the next 20 columns for the base frequencies in each successive column in the sites. Thus, the counts in the first three columns of the matrix may appear as shown in Table 4.2.

The following calculations are performed in the expectation step of the EM algorithm:

1. The above estimates provide an initial estimate of the composition of the site and the location in each sequence. The object of this step is to improve this estimate by discriminating to the greatest possible extent between sequence within and sequence not within the site. Using the above estimates of base frequencies for (1) background sequences that are not within the site and (2) each column within the site, each sequence is scanned for all possible locations for the site to find the most probable location of the site. For the 10-residue DNA sequence example, there are $100 - 20 + 1$ possible starting sites for a 20-residue-long site, the first one being at position 1 in the sequence ending at 20 and the last beginnning at position 81 and ending at 100 (there is not enough sequence for a 20-residue-long site beyond position 81).

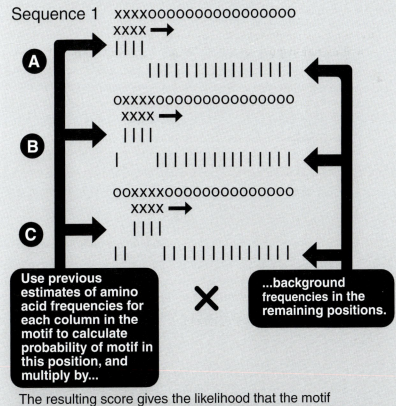

The resulting score gives the likelihood that the motif matches positions (a) 1-20, (b) 6-25, or (c) 11-30 in sequence 1. Repeat for all other positions and find most likely locator. Then repeat for the remaining sequences.

Table 4.2. *Column frequencies of each base in the example given*

	Background	Site column 1	Site column 2	...
G	0.27	0.4	0.1	...
C	0.25	0.4	0.1	...
A	0.25	0.2	0.1	...
T	0.23	0.2	0.7	...
	1.00	1.0	1.0	

The first column gives the background frequencies in the flanking sequence. Subsequent columns give base frequencies within the site given in the above example.

For each possible site location, the probability that the site starts is just the product of the probabilities given by Table 4.2. For example, suppose that the site starts in column 1 and that the first two positions in sequence 1 are A and T, respectively. The site will then end at position 20 and the first two nonsites, flanking background sequence positions, are 21 and 22. Suppose that these positions have an A and a T, respectively. Then the probability of this location of the site in sequence 1 is given by $P_{site1,sequence1} = 0.2$ (for A in position 1) $\times$ 0.7 (for T in position 2) $\times$ Ps for next 18 positions in site $\times$ 0.25 (for A in first flanking position) $\times$ 0.23 (for T in second flanking position) $\times$ Ps for next 78 flanking positions. Similar probabilities for $P_{site2,\ sequence1}$ to $P_{site78,\ sequence1}$ are then calculated, thus providing a comparative set of probabilities for the site location. The probability of this best location in sequence 1, say at site k, is the ratio of the site probability at k divided by the sum of all the other site probabilities $P(site\ k\ in\ sequence\ 1) = P_{site\ k,\ sequence\ 1} / (P_{site\ 1,\ sequence\ 1} + P_{site\ 2,\ sequence\ 1} + \ldots + P_{site\ 78,\ sequence\ 1})$. The probability of the site location in each sequence is then calculated in this manner.

2. The above site probabilities for each sequence are then used to provide a new table of expected values for base counts for each of the site positions using the site probabilities as weights. For example, suppose that P (site 1 in sequence 1) = 0.01 and that P (site 2 in sequence 1) = 0.02. In the above example, the first base in site 1 is an A and the first base for site 2 is a T. Then 0.01 As and 0.02 Ts are added to the accumulated list of bases at site column 1. This procedure is repeated for every other 76 possible first columns in sequence 1. Similarly, site column 2 in the new table of expected values is augmented by counts from the 78 possible column 2 positions in sequence 1, the first, for example, being 0.01 Ts. The weighted sequence data from the remaining sequences are also added to the new table, resulting finally in a new estimate of the expected number of each base at each site position and providing a new version of Table 4.2.

In this maximization step, the base frequencies found in the expectation step are used as an updated estimate of the site residue composition. In this case, the data are more complete than the initial estimate because all possible sites in each of the sequences have been evaluated. The expectation and maximization steps are repeated until the estimates of the base frequencies do not change.

An Alternative Method of Calculating Site Probabilities by the EM Algorithm

The example shown above uses the frequencies of each base in the trial alignment and background base frequencies to calculate the probabilities of each possible location in each sequence. An alternative method is to produce an odds scoring matrix calculated by dividing each base frequency by the background frequency of that base. The probability of each location is then found by multiplying the odds scores from each column. An even simpler method is to use log odds scores in the matrix. The column scores are then simply added. In this case, the log odds scores must be converted to odds scores before position probabilities are calculated.

Multiple EM for Motif Elicitation (MEME)

A Web resource for performing local msas by the above expectation maximization method is the program Multiple EM for Motif Elicitation (MEME) developed at the University of California at San Diego Supercomputing Center. The Web page for two versions of MEME, ParaMEME, a Web program that searches for blocks by an EM algorithm (described below), and a similar program MetaMEME (which searches for profiles using HMMs, described below) is found at http://www.sdsc.edu/MEME/meme/website/meme.html. The Motif Alignment and Search Tool (MAST) for searching through databases for matches to motifs may also be found at http://www.sdsc.edu/MEME/meme/website/mast.html.

MEME will locate one or more ungapped patterns in a single DNA or protein sequence or in a series of DNA or protein sequences. A search is conducted for a range of possible motif widths, and the most likely width for each profile is chosen on the basis of the log-likelihood score after one iteration of the EM algorithm. The EM algorithm then iterates to find the best EM estimate for that width. Three types of possible motif models may be chosen. The OOPS model is for one expected occurrence of a motif per sequence, the ZOOPS model is for zero or one occurrence per sequence, and the TCM model is for a motif to appear any number of times in a sequence. These models are reflected in the choices on the Web page (Fig. 4.14). The current version of MEME can use prior knowledge about a motif being present in all or only some of the sequences, the length of the motif and whether it is a palindrome (DNA sequences), and the expected patterns in individual motif positions (Dirichlet mixtures, see section on HMMs, p. 189) that provide information as to which amino acids are likely to be interchangeable in a motif (Bailey and Elkan 1995). Once a motif has been found, the motif and its position are effectively erased to prevent finding the same one twice. An example of the output from a ParaMEME analysis is given in Figure 4.15.

The Gibbs Sampler

Another statistical method for finding motifs in sequences is the Gibbs sampler. The method is similar in principle to the EM method described above, but the algorithm is different. Like the EM method, given a set of sequences, the Gibbs sampler searches for the statistically most probable motifs and can find the optimal width and number of these motifs in each sequence (Lawrence et al. 1993; Liu et al. 1995; Neuwald et al. 1995). The source code of the program code is available by anonymous FTP from ncbi.nlm.nih.gov/pub/neuwald/gibbs9-95. A combinatorial approach of the Gibbs sampler and MOTIF may be used to make blocks at the BLOCKS Web site (http://www.blocks.fhcrc.org/). The expected number of blocks in the search is one block for approximately each 40 residues of sequence. The Gibbs sampler is also an option of the msa block-alignment and editing program MACAW (Schuler et al. 1991), which runs on MS-DOS, Macintosh, and other computer platforms and is available by anonymous FTP from ncbi.nlm.nih.gov/pub/schuler/macaw.

MEME -- Multiple EM for Motif Elicitation: Version 2.2

Motif discovery tool

Data Submission Form - Advanced Version

| Basic MEME |

Your data will be processed on the Cray-T3E supercomputer at the **San Diego Supercomputer Center** and the results will be sent to you by e-mail.

Please enter the **e-mail** address where you would like your results sent:

[Optional] Please enter a brief **description** of your sequences.

Please enter the **sequences** which you believe share one or more motifs. The sequences may contain no more than **100,000 characters** total in any of a large number of **formats**. Please enter either:

1. the **name of a file** containing the sequences here:
2. or the **actual sequences** here:

How do you think the occurrences of a single motif are **distributed** among the sequences?

⬤ **One** per sequence
⬤ **Zero or one** per sequence
⬤ **Any number** of repetitions

How many different motifs would you like to look for?

| 3 | ⬍ |

MEME can choose the **width** of each motif favoring **short** or **wide** motifs. **Wide** is recommended if there are fewer than 10 occurrences of any motif in your sequences. Choosing a number will cause all motifs reported to have that width. Select the width you want with the select button below, or enter a width in the text window. Legal choices are "short", "wide" or any number from 2 to 300. (If you enter something in the text window, it will override what is shown on the select button.)

| short ⬍ | or | |

Brief output format:

ADVANCED OPTIONS

Shuffle letters in input sequences:

DNA-ONLY OPTIONS

DNA **palindromes:**	ignore allow force
	complementary strand, 5' to 3' (inverse complement)
Additional **strands/directions** to search:	**main strand, 3' to 5'**
	complementary strand, 3' to 5'
Strength of the prior (enter a positive number):	

Click here for **more information** on MEME.
Return to **MEME SYSTEM introduction**.

You might be interested in trying other motif-making programs such as BLOCK MAKER at the Fred Hutchinson Cancer Research Center.

Please send comments and questions to: tbailey@sdsc.edu .

Figure 4.14. The MEME Web page. The MEME program finds ungapped motifs (blocks) in unaligned protein or DNA sequences. As indicated, the program can be directed to search for the size and expected number of motifs or can predict motifs based on a statistical analysis based on the EM algorithm described in the text.

A. Summary line

MOTIF 1   width = 9   sites = 29.5

B. Letter-probability matrix

```
Simplified        A  ::1::::8:
motif letter-     C  :::::::::
probability       D  :8:::::::
matrix            E  :::::::::
                  F  :::::::::
                  G  ::1:::::9
                  H  :::::::::
                  I  2:212::::
                  K  :::::::::
                  L  3:18:::::
                  M  :::::::::
                  N  :::::89::
                  P  :::::::::
                  Q  :::::::::
                  R  :::::::::
                  S  :::::::::
                  T  :::::::::
                  V  3:3:7::::
                  W  :::::::::
                  Y  :::::::::
```


C. Information content of the profile

```
Information   bits 6.2
content            5.6
( 22.0 bits )      5.0
                   4.4
                   3.7
                   3.1   *      * *
                   2.5   *      * *  *
                   1.9   *  * * * * *
                   1.2  * *  * * * * *
                   0.6  * * * * * * * *
                   0.0  - - - - - - - - -
```


D. The multilevel consensus sequence

Multilevel   VDVLVNNAG
consensus
sequence   L

Figure 4.15. Results produced by a MEME analysis of sequences for motifs. The output diagrams are discussed in the text. (*A*) Summary line giving the number of the next motif found in order of statistical significance, width, and expected number of occurrences in the given sequences. (*B*) Simplified motif letter-probability matrix showing the frequency of each amino acid in each column of the matrix. The columns are the columns of the motif. For easier reading, the numbers shown are frequencies rounded to the nearest one-tenth and multiplied by 10, and zeros are shown as colons. (*C*) The information content of the profile is given in a diagram. Basically, the diagram shows the degree of amino acid variation in each column of the profile: the lower the value, the greater the variation. The scale is logarithmic to the base 2 (bits). The total of all columns is also shown. The subject of information content is discussed in greater detail below under position-specific scoring matrices. (*D*) The multilevel consensus sequence shows all letters in each column of the motif that occur with a frequency of >0.2. *Continued.*

E. The next motif

Motif 1 in BLOCKS format

BL MOTIF 1; width = 9; seqs = 33

2BHD_STREX	(81)	VDGLVNNAG	1
3BHD_COMTE	(81)	LNVLVNNAG	1
ADH_DROME	(86)	VDVLINGAG	1
AP27_MOUSE	(77)	VDLLVNNAA	1
BA72_EUBSP	(86)	LDVMINNAG	1
BDH_HUMAN	(138)	MWGLVNNAG	1
BPHB_PSEPS	(79)	IDTLIPNAG	1
BUCD_KLETE	(80)	FNVIVNNAG	1
DHES_HUMAN	(84)	VDVLVCNAG	1
DHGB_BACME	(87)	LDVMINNAG	1
DHMA_FLAS1	(198)	VDVTGNNTG	1
ENTA_ECOLI	(73)	LDALVNAAG	1
FIXR_BRAJA	(112)	LHALVNNAG	1
GUTD_ECOLI	(82)	VDLLVYSAG	1
HDE_CANTR	(396)	IDILVNNAG	1
HDHA_ECOLI	(89)	VDILVNNAG	1
NODG_RHIME	(81)	VDILVNNAG	1
RIDH_KLEAE	(89)	LDIFHANAG	1
YINL_LISMO	(83)	VDAIFLNAG	1
YRTP_BACSU	(84)	IDILINNAG	1
CSGA_MTXXA	(13)	VDVLINNAG	1
DHB2_HUMAN	(161)	LWAVINNAG	1
DHB3_HUMAN	(125)	IGILVNNVG	1
DHCA_HUMAN	(83)	LDVLVNNAG	1
FVT1_HUMAN	(115)	VDMLVNCAG	1
HMTR_LEIMA	(103)	CDVLVNNAS	1
MAS1_AGRRA	(320)	IDGLVNNAG	1
PCR_PEA	(165)	LDVLINNAA	1
YURA_MYXXA	(90)	LDLVVANAG	1
//			

Figure 4.15. *Continued.* (*E*) Possible examples of the motif in the training set are shown. This list is based on using a position-dependent scoring matrix (log-odds matrix) to search each sequence. The threshold score for displaying a site is chosen such that the expected number of incorrect assignments will equal the expected number of missed but correct assignments. Positions before and after the motif are also shown. *Continued.*

F. Possible examples of motif 1 in the training set

Sequence name	Start	Score		Site	
-------------------	------	------		------	
2BHD_STREX	81	28.80	VAYAREEFGS	VDGLVNNAG	ISTGMFLETE
3BHD_COMTE	81	25.99	MAAVQRRLGT	LNVLVNNAG	ILLPGDMETG
ADH_DROME	86	22.33	LKTIFAQLKT	VDVLINGAG	ILDDHQIERT
AP27_MOUSE	77	24.36	TEKALGGIGP	VDLLVNNAA	LVIMQPFLEV
BA72_EUBSP	86	26.39	VGQVAQKYGR	LDVMINNAG	ITSNNVFSRV
BDH_HUMAN	138	23.46	PFEPEGPEKG	MWGLVNNAG	ISTFGEVEFT
BPHB_PSEPS	79	18.60	ASRCVARFGK	IDTLIPNAG	IWDYSTALVD
BUDC_KLETE	80	20.97	VEQARKALGG	FNVIVNNAG	IAPSTPIESI
DHES_HUMAN	84	25.67	AARERVTEGR	VDVLVCNAG	LGLLGPLEAL
DHGB_BACME	87	26.39	VQSAIKEFGK	LDVMINNAG	MENPVSSHEM
DHMA_FLAS1	198	16.36	ILVNMIAPGP	VDVTGNNTG	YSEPRLAEQV
ENTA_ECOLI	73	21.90	CQRLLAETER	LDALVNAAG	ILRMGATDQL
FIXR_BRAJA	112	23.67	EVKKRLAGAP	LHALVNNAG	VSPKTPTGDR
GUTD_ECOLI	82	17.17	SRGVDEIFGR	VDLLVYSAG	IAKAAFISDF
HDE_CANTR	92	20.90	VETAVKNFGT	VHVIINNAG	ILRDASMKKM
HDE_CANTR	396	29.32	IKNVIDKYGT	IDILVNNAG	ILRDRSFAKN
HDHA_ECOLI	89	30.18	ADFAISKLGK	VDILVNNAG	GGGPKPFDMP
NODG_RHIME	81	30.18	GQRAEADLEG	VDILVNNAG	ITKDGLFLHM
RIDH_KLEAE	89	16.02	LQGILQLTGR	LDIFHANAG	AYIGGPVAEG
YINL_LISMO	83	14.65	VELAIERYGK	VDAIFLNAG	IMPNSPLSAL
YRTP_BACSU	84	27.41	VAQVKEQLGD	IDILINNAG	ISKFGGFLDL
CSGA_MYXXA	13	28.94	AFATNVCTGP	VDVLINNAG	VSGLWCALGD
DHB2_HUMAN	161	19.62	KVAAMLQDRG	LWAVINNAG	VLGFPTDGEL
DHB3_HUMAN	125	18.63	HIKEKLAGLE	IGILVNNVG	MLPNLLPSHF
DHCA_HUMAN	83	30.23	RDFLRKEYGG	LDVLVNNAG	IAFKVADPTP
FVT1_HUMAN	115	24.21	IKQAQEKLGP	VDMLVNCAG	MAVSGKFEDL
HMTR_LEIMA	103	24.02	VAACYTHWGR	CDVLVNNAS	SFYPTPLLRN
MAS1_AGRRA	320	27.93	VTAAVEKFGR	IDGLVNNAG	YGEPVNLDKH
PCR_PEA	165	23.97	VDNFRRSEMP	LDVLINNAA	VYFPTAKEPS
YURA_MYXXA	90	18.59	IRALDAEAGG	LDLVVANAG	VGGTTNAKRL

Figure 4.15. *Continued.* (*F*) The next motif is given in the format used for the BLOCKS database (http://www. blocks.fhcrc.org/blocks). The predicted locations of this motif in each sequence and the probability that the motif starts at that location are shown. The sites reported depend on the motif search model used: (1) OOPS, the most probable location in each sequence is given; (2) ZOOPS, the most probable location in each sequence is reported but only probabilities greater than 0.5 (a significant level for Bayesian statistics); TCM, all positions in each sequence with probabilities > 0.5 are shown. *Continued.*

G. Position-specific scoring matrix

Log-odds matrix: alength = 20 w = 9 n = 9732 bayes = 8.36118

```
-2.725   0.818  -5.204  -4.539  -0.082  -4.432  -3.515   1.560  -4.218   1.814   0.701  -4.126  -3.146  -3.848  .
-3.441  -3.841  -4.023  -1.204  -4.313  -2.395  -0.889  -4.226  -4.009  -4.571  -3.882  -0.220  -4.682  -3.547  ·
-0.768  -2.342  -4.756  -4.189  -2.319   0.376  -3.154   1.757  -3.870   0.288   0.918  -3.149  -4.229  -3.492  .
-3.379  -2.600  -5.066  -4.331  -0.586  -5.089  -3.668  -0.081  -4.098   3.045   1.107  -4.393  -4.287  -3.383  ·
-1.373  -1.895  -3.823  -3.574  -1.086  -1.952  -0.466   1.480  -3.565  -2.234  -1.834  -3.701  -3.612  -3.536  ·
-1.879  -0.980  -2.231  -4.187  -3.807  -3.562  -0.892  -3.306  -3.238  -2.753  -3.337   4.193  -2.276  -2.750  ·
-2.460  -0.912  -2.252 -4.176  -3.833  -2.391  -0.968  -3.339  -3.262  -4.256  -3.364   4.217  -4.026  -2.768  ·
-3.475  -1.137  -3.874  -3.535  -3.304  -2.080  -2.080  -2.826  -3.544  -3.127  -2.263  -3.592  -4.599  -3.533  ·
-0.693  -3.833  -3.137  -3.879  -4.963   3.663  -3.647  -3.364  -3.716  -5.287  -4.212  -2.849  -4.518  -4.155  .
```

H. Motif letter-frequency matrix

Letter-probability matrix: alength = 20 w = 9 n = 9732

```
0.011063  0.032022  0.001403  0.002682  0.038055  0.003212  0.001962  0.165990  0.003143  0.322510  0.037503  0.011063
0.006738  0.001268  0.841023  0.027061  0.002026  0.013178  0.012108  0.003008  0.003632  0.003860  0.001564  0.011063
0.124630  0.003583  0.001915  0.003418  0.008070  0.089951  0.002520  0.190255  0.004000  0.112000  0.043590  0.011063
0.007032  0.002996  0.001544  0.003098  0.026845  0.002037  0.001765  0.053213  0.003415  0.756853  0.049683  0.011063
0.028238  0.004883  0.003655  0.005236  0.018977  0.017917  0.016240  0.156947  0.004942  0.019499  0.006470  0.011063
0.019895  0.009211  0.011023  0.003422  0.002878  0.005871  0.012089  0.005691  0.006199  0.013606  0.002282  0.011063
0.013301  0.009656  0.010865  0.003449  0.002827  0.013217  0.011467  0.005564  0.006098  0.004800  0.002240  0.011063
0.813801  0.008259  0.003529  0.005378  0.004079  0.016396  0.005304  0.007937  0.005014  0.010499  0.004806  0.011063
0.045249  0.001275  0.005879  0.004237  0.001291  0.878064  0.001790  0.005467  0.004450  0.002354  0.001244  0.011063
```

Figure 4.15. *Continued.* (*G*) Position-specific scoring matrix. This matrix is a log-odds matrix calculated by taking the log (base 2) of the ratio of the observed to expected counts for each amino acid in each column of the profile. Columns and rows in the matrix correspond to the amino acids in each column and positions of the motif, respectively. The counts for each column may have additional pseudocounts added to compensate for zero occurrences of an amino acid in a column or for a small number of sequences, as discussed below for this type of matrix. (*H*) Motif letter-frequency matrix is given, showing the frequency of amino acid found in each column of the profile. Columns and rows correspond to the amino acids in each column and rows to columns in the motif, respectively. Shown also are the numbers of types of residues, the width of the motif, and number of characters in the sequences. Only portions of the output are shown.

To understand the algorithm, consider a simple example using the Gibbs sampler algorithm to locate a single 20-residue-long motif in 10 sequences, each 200 residues long, as was done above to illustrate the EM algorithm. The method iterates through two steps. In the first step, the predictive update step, a random start position for the motif is chosen for all sequences but for one that is chosen at random or in a specified order. So let us choose sequence 1 as the outlier and use the other 9 to find an initial guess of the motif. These other 9 sequences are aligned with random overlaps. The following figure illustrates how this initial motif is located (an x equals 20 sequence positions, M indicates the random location of the motif chosen for each sequence, and − the 20 initially aligned motif positions).

The objective is to find the most probable pattern common to all of the sequences by sliding them back and forth until the ratio of the motif probability to the background probability is a maximum. This is accomplished by first using the initial alignment shown above to estimate the residue frequencies in each column of the motif, and the sequence residues

Steps of the Gibbs sampler algorithm.

A. Estimate the amino acid frequencies in the motif columns of all but
    1 sequence. Also obtain background

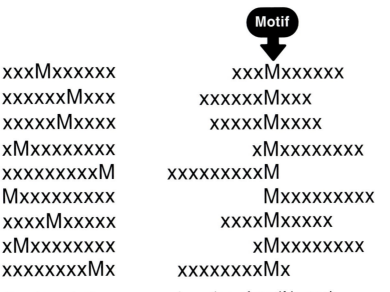

xxxMxxxxxx
xxxxxxMxxx
xxxxxMxxxx
xMxxxxxxxx
xxxxxxxxxM
Mxxxxxxxxx
xxxxMxxxxx
xMxxxxxxxx
xxxxxxxxMx

Random start
positions chosen

xxxMxxxxxx
xxxxxxMxxx
xxxxxMxxxx
xMxxxxxxxx
xxxxxxxxxM
Mxxxxxxxxx
xxxxMxxxxx
xMxxxxxxxx
xxxxxxxxMx

Location of motif in each
sequence provides first
estimate of motif composition

B. Use the estimates from A to calculate the ratio of probability of
    motif to background score at each position in the left out sequence.
    This ratio for each possible location in the sequence is the weight
    of the position.

xxxxxxxxxxx   xxxxxxxxxxx   xxxxxxxxxxx   xxxxxxxxxxx   xxxxxxxxxxx
M ->   M ->   M ->   M ->   M ->

C. choose a new location for the motif in the left out sequence by a
    random selection using the weights to bias the choice.

xxxxxxxxMxx   Estimated location of the motif in left out sequence

D. Repeat steps A to C >100 times.

that are not included in the motif to estimate the background residue frequencies. For
example, if these sequences are DNA sequences and the first column of the estimated motif
in the 10 sequences includes 3 Gs, then the value for $f_{g,\,\text{column1}} = 3/9 = 0.33$. Similarly, let
$f_{t,\,\text{column2}} = 1/9 = 0.11$ for illustration. These frequencies are determined for each of the 20
columns in our example. Similarly, if there are 240 Gs among the $10 \times 80 = 800$ sequence
positions not within the estimated motif, then $f_{g,\,\text{background}} = 240/800 = 0.30$. Also let
$f_{t,\,\text{background}} = 180/800 = 0.225$. If the first two positions in sequence 1 are G and T in that
order, then the probability of the motif starting at position 1, Q_1, is calculated as $0.33 \times
0.11 \times \ldots \ldots x\, f_{\text{last base, column20}}$. The background probability of this first possible motif,
P_1, is also calculated as $0.30 \times 0.225 \times \ldots .. x\, f_{\text{last base, background}}$.

Note the difference between the Gibbs sampler method and the EM method, which calculates the probability of the entire sequence using the motif column frequencies within the motif and the background frequencies elsewhere.

The ratio Q_1/P_1 is designated as weight A_1 for motif position 1 in sequence 1. A_1s are then calculated for all other $100 - 20 + 1 = 81$ possible locations of the 20-residue-long motif in sequence 1. These weights are then normalized by dividing each weight by their sum to give a probability for each motif position. From this probability distribution, a random start position is chosen for position 1. In so doing, the chance of choosing a particular position is proportional to the weight of that position so that a higher scoring position is more likely to be chosen. (You can think of a bag with 81 kinds of balls, with the number of each ball proportional to the weight or probability of that kind. Drawing a random ball will favor the more prevalent ones.) This position in the left-out sequence is then used as an estimate of the location for the motif in sequence 1. The procedure is then repeated. Select the next sequence to be scanned, align the motifs in the other 9 sequences with sequence 1 now using the estimated location found above, and so on. This process is repeated until the residue frequencies in each column of the motif do not change. For different starting alignments, the number of iterations needed may range from several hundred to several thousand.

As the above cycles are repeated, the more accurate the initial estimate of the motif in the aligned sequences, the more accurate the pattern location in the outlier sequence. The second step in the algorithm tends to move the sequence alignments in a direction that favors a better score but also has a random element to search for other possible better locations. When correct start positions have been selected in several sequences by chance, the compositions of the motif columns begin to reflect a pattern that the algorithm can search for in the other sequences, and the method converges on the optimal motif and the probability distribution of the motif location in each sequence.

Several additional procedures are used to improve the performance of the algorithm.

1. For a correct Bayesian statistical analysis, the amino acid counts in the motif and the background in the outlier sequence are estimated and added to the counts in the remaining aligned sequences. This step is the equivalent of combining prior and updated information to improve the estimation of the motif. These counts may be estimated by Dirichlet mixtures (see discussion of HMMs, p. 189), which give frequencies expected based on prior information from amino acid distributions (Liu et al. 1995). The missing background counts for each residue b_i are estimated by the formula $b_i = f_i x, B$ where B is chosen based on experience with the method as $\sqrt{N}$, the number of sequences in the motif, and f_i is the frequency of residue i in the sequences (Lawrence et al. 1993).

2. Another feature is a procedure to prevent the algorithm from getting locked in a suboptimal solution. In the HMM method (see below), noise is introduced for this purpose. In the Gibbs sampler, after a certain number of iterations, the current alignments are shifted a certain number of positions to the right and left, and the scores from these shifted positions are found. A probability distribution of these scores is then used as a basis for choosing a new random alignment.

3. The results of a range of motif widths can be investigated. The major difficulty in exploring motif width is to arrive at a criterion for comparing the resulting scores. One suitable measure is to optimize the average information (see below) per free parameter in the motif, a value that can be calculated (Lawrence et al. 1993; Liu et al. 1995). The number of free parameters for proteins is $20 - 1 = 19$, and for DNA, $4 - 1 = 3$, times the model width.

4. The method can be readily extended to search for multiple motifs in the same set of sequences.

5. The method has been extended to seek a pattern in only a fraction of the input sequences.

The Gibbs sampler was used to align 30 helix-turn-helix DNA-binding domains showing very little sequence similarity. The information per parameter criterion was used to find the best motif width. Multiple motifs were found in lipocalins, a family with quite dissimilar motif sequences separated by variable spacer regions, and also in protein isoprenyltransferase subunits, which have very large numbers of repeats of several kinds (Lawrence et al. 1993). Thus, the method is widely applicable for discovering complex and variable motifs in proteins.

Hidden Markov Models

The HMM is a statistical model that considers all possible combinations of matches, mismatches, and gaps to generate an alignment of a set of sequences (Fig. 4.16). A model of a sequence family is first produced and initialized with prior information about the sequences. A set of 20–100 sequences or more is then used as data to train the model. The trained model may then be used to produce the most probable msa as posterior information. Alternatively, the model may be used to search sequence databases to identify additional members of a sequence family. A different HMM is produced for each set of sequences. HMMs have been previously used very successfully for speech recognition, and an excellent review of the methodology is available (Rabiner 1989). In addition to their use in producing multiple sequence alignments (Baldi et al. 1994; Krogh et al. 1994; Eddy 1995, 1996), HMMs have also been used in sequence analysis to produce an HMM that represents a sequence profile (a profile HMM), to analyze sequence composition and patterns (Churchill 1989), to locate genes by predicting open reading frames (Chapter 8), and to produce protein structure predictions (Chapter 9). Pfam, a database of profiles that represent protein families, is based on profile HMMs (Sonhammer et al. 1997).

HMMs often provide a msa as good as, if not better than, other methods. The approach also has a number of other strong features: It is well grounded in probability theory, no sequence ordering is required, insertion/deletion penalties are not needed, and experimentally derived information can be used. Two disadvantages to using HMMs are that at least 20 sequences and sometimes many more are required to accommodate the evolutionary history (see Mitchison and Durbin 1995). The HMM can be used to improve an existing heuristic alignment. The two HMM programs in common use are Sequence Alignment and Modeling Software System, or SAM (Krogh et al. 1994; Hughey and Krogh 1996), and HMMER (see Eddy 1998). The software is available at http://www.cse.ucsd.edu/research/compbio/sam.html and http://hmmer.wustl.edu/. The algorithms used for producing HMMs are extensively discussed in Durbin et al. (1998). A comparison of HMMs with other methods is given at the end of this section.

The HMM representation of a section of multiple sequence alignment that includes deletions and insertions was devised by Krogh et al. (1994) and is shown in Figure 4.6. This HMM generates sequences with various combinations of matches, mismatches, insertions, and deletions, and gives these a probability, depending on the values of the various parameters in the model. The object is to adjust the parameters so that the model represents the observed variation in a group of related protein sequences. A model trained in this manner will provide a statistically probable msa of the sequences.

As illustrated in Figure 4.6, the object is to calculate the best HMM for a group of sequences by optimizing the transition probabilities between states and the amino acid compositions of each match state in the model. The sequences do not have to be aligned to use the method. Once a reasonable model length reflecting the expected length of the sequence alignment is chosen, the model is adjusted incrementally to predict the sequences. Several methods for training the model in this fashion have been described (Baldi et al. 1994; Krogh et al. 1994; Eddy et al. 1995; Eddy 1996; Hughey and Krogh 1996;

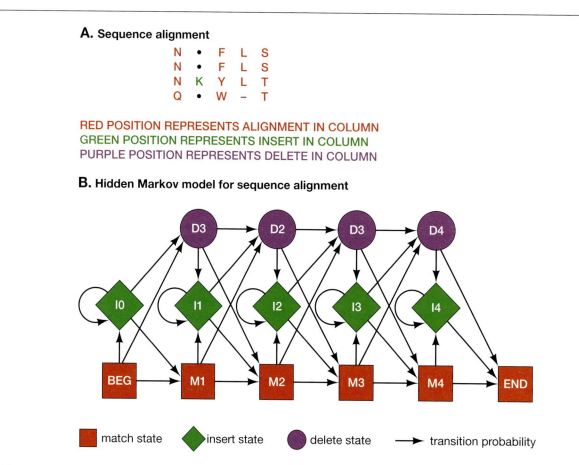

A. Sequence alignment

```
N   •   F   L   S
N   •   F   L   S
N   K   Y   L   T
Q   •   W   -   T
```

RED POSITION REPRESENTS ALIGNMENT IN COLUMN
GREEN POSITION REPRESENTS INSERT IN COLUMN
PURPLE POSITION REPRESENTS DELETE IN COLUMN

B. Hidden Markov model for sequence alignment

■ match state　◆ insert state　● delete state　→ transition probability

Figure 4.16. Relationship between the sequence alignment and the hidden Markov model of the alignment (Krogh et al. 1994). This particular form for the HMM was chosen to represent the sequence, structural, and functional variation expected in proteins. The model accommodates the identities, mismatches, insertions, and deletions expected in a group of related proteins. (*A*) A section of a multiple sequences alignment. The illustration shows the columns generated in a multiple sequence alignment. Each column may include matches and mismatches (*red* positions), insertions (*green* positions), and deletions (*purple* position). (*B*) The HMM. Each column in the model represents the possibility of a match, insert, or delete in each column of the alignment in *A*. The HMM is a probabilistic representation of a section of a msa. Sequences can be generated from the HMM by starting at the beginning state labeled BEG and then by following any one of many pathways from one type of sequence variation to another (states) along the state transition arrows and terminating in the ending state labeled END. Any sequence can be generated by the model and each pathway has a probability associated with it. Each square match state stores an amino acid distribution such that the probability of finding an amino acid depends on the frequency of that amino acid within that match state. Each diamond-shaped insert state produces random amino acid letters for insertions between aligned columns and each circular delete state produces a deletion in the alignment with probability 1. For example, one of many ways of generating the sequence **N K Y L T** in the above profile is by the sequence BEG→M1→I1→M2→M3→M4→END. Each transition has an associated probability, and the sum of the probabilities of transitions leaving each state is 1. The average value of a transition would thus be 0.33, since there are three transitions from most states (there are only two from M4 and D4, hence the average from them is 0.5). For example, if a match state contains a uniform distribution across the 20 amino acids, the probability of any amino acid is 0.05. Using these average values of 0.33 or 0.5 for the transition values and 0.05 for the probability of each amino acid in each state, the probability of the above sequence **N K Y L T** is the product of all of the transition probabilities in the path BEG→M1→I1→M2→M3→M4→END, and the probability that each state will produce the corresponding amino acid in the sequences, or $0.33 \times 0.05 \times 0.33 \times 0.05 \times 0.33 \times 0.05 \times 0.33 \times 0.05 \times 0.33 \times 0.05 \times 0.5 = 6.1 \times 10^{-10}$. Since these probabilities are very small numbers, amino acid distributions and transition probabilities are converted to log odds scores, as done in other statistical methods (see pp. 176–177), and the logarithms are added to give the overall probability score. The secret of the HMM is to adjust the transition values and the distributions in each state by training the model with the sequences. The training involves finding every possible pathway through the model that can produce the sequences, counting the number of times each transition is used

Continued.

Durbin et al. 1998). For example, an EM algorithm from speech recognition methods known as the Baum-Welch algorithm is used as follows:

1. The model is initialized with estimates of transition probabilities and amino acid composition for each match and insert date. If an initial alignment of the sequences is known, or some other kinds of data suggest which sequence positions are the same, then these data may be used in the model. For other cases, the initial distribution of amino acids to be used in each state is described below. The initial transition probabilities generally favor transitions from one match state to the next rather than favoring insert and delete states, which build more uncertainty into a sequence motif.

2. All possible paths through the model for generating each sequence in turn are examined. There are many possible such paths for each sequence. This procedure would normally require a huge amount of time computationally. Fortunately, an algorithm, the forward-backward algorithm, reduces the number of computations to the number of steps in the model times the total length of the training sequences. This calculation provides a probability of the sequence, given all possible paths through the model, and, from this value, the probability of any particular path may be found. Another algorithm, the Baum-Welch algorithm, then counts the number of times a particular state-to-state transition is used and a particular amino acid is required by a particular match state to generate the corresponding sequence position.

3. A new version of the HMM is produced that uses the results found in step 2 to generate new transition probabilities and match-insert state compositions.

4. Steps 3 and 4 are repeated up to 10 more times until the parameters do not change significantly.

5. The trained model is used to provide the most likely path for each sequence, as described in Figure 4.16. The algorithm used for this purpose, the Viterbi algorithm, does not have to go through all of the possible alignments of a given sequence to the HMM to find the most probable alignment, but instead can find the alignment by a dynamic programming technique very much like that used for the alignment of two sequences, discussed in Chapter 3. The collection of paths for the sequences provides a msa of the sequences with the corresponding match, insert, and delete states for each sequence. The columns in the msa are defined by the match states in the HMM such that amino acids from a particular match state are placed in the same column. For columns that do not correspond to a match state, a gap is added.

6. The HMM may be used to search a sequence database for additional sequences that share the same sequence variation. In this case, the sum of the probabilities of all possible sequence alignments to the model is obtained. This probability is calculated by the forward component of the forward-backward algorithm described above. This analysis

and which amino acids were required by each match and insert state to produce the sequences. This training procedure leaves a memory of the sequences in the model. As a consequence, the model will be able to give a better prediction of the sequences. Once the model has been adequately trained, of all the possible paths through the model that can generate the sequence N K Y L T, the most probable should be the match-insert-3 match combination (as opposed to any other combination of matches, inserts, and deletions). Likewise, the other sequences in the alignment would also be predicted with highest probability as they appear in the alignment; i.e., the last sequence would be predicted with highest probability by the path match-match-delete-match. In this fashion, the trained HMM provides a multiple sequence alignment, such as shown in A. For each sequence, the objective is to infer the sequence of states in the model that generate the sequences. The generated sequence is a Markov chain because the next state is dependent on the current one. Because the actual sequence information is hidden within the model, the model is described as a hidden Markov model.

gives a type of distance score of the sequence from the model, thus providing an indication of how well a new sequence fits the model and whether the sequence may be related to the sequences used to train the model. In later derivations of HMMs, the score was divided by the length of the sequence because it was found to be length-dependent. A *z* score giving the number of standard deviations of the sequence length-corrected score from the mean length-corrected score is therefore used (Durbin et al. 1998).

Recall that for the Bayes block aligner, the initial or prior conditions were amino acid substitution matrices, block numbers, and alignments of the sequences. The sequences were then used as new data to examine the model by producing scores for every possible combination of prior conditions. By using Bayes' rule, these data provided posterior probability distributions for all combinations of prior information. Similarly, the prior conditions of the HMM are the initial values given to the transition values and amino acid compositions. The sequences then provide new data for improving the model. Finally, the model provides a posterior probability distribution for the sequences and the maximum posterior probability for each sequence represented by a particular path through the model. This path provides the alignment of the sequence in the msa; i.e., the sequence plus matches, inserts, and deletes, as described in Figure 4.16.

The success of the HMM method depends on having appropriate initial or prior conditions, i.e., a good prior model for the sequences and a sufficient number of sequences to train the model. The prior model should attempt to capture, for example, the expected amino acid frequencies found in various types of structural and functional domains in proteins. As the distributions are modified by adding amino acid counts from the training sequences, new distributions should begin to reflect common patterns as one moves through the model and along the sequences. It is important that the model reflect not only the patterns in the training sequences, but also pattern variations that might be present in other members of the same protein family. Otherwise, the model will only recognize the training sequences but not other family members. Thus, some smoothing of the amino acid frequencies is desirable, but not to the extent of suppressing highly conserved pattern information from the training sequences. Such problems are avoided by using a method called regularization to avoid overfitting the data to the model. Basically, the method involves using a carefully designed amino acid distribution as the prior condition and then modifying this distribution in a manner that uses training information in a complementary manner.

Rather than using simple amino acid composition as a prior condition for the match states in the HMM, amino acid patterns that capture some of the important features of protein structure and function have been used with considerable success (Sjölander et al. 1996). Other prior conditions include using Dayhoff PAM or BLOSUM amino acid substitution matrices modified by adding additional counts (pseudocounts) to smooth the distributions (Tatusov et al. 1994; Eddy 1996; Henikoff and Henikoff 1996; Sonnhammer et al. 1997; and see Chapter 2). Sjölander et al. (1996) have prepared particularly useful amino acid distributions called Dirichlet mixtures to use as prior information in the match states of the HMM. These mixtures provide amino acid compositions that have proven to be useful for the detection of weak but significant sequence similarity. As an example, the amino acid frequencies that are characteristic of a particular set of nine blocks in the BLOCKS database have been determined. These blocks represent amino acid frequencies that are favored in certain chemical environments such as aromatic, neutral, and polar residues and are useful for detecting such environments in test sequences. The nine-component system has been used successfully for producing an HMM for globin sequences (Hughey and Krogh 1996). To use these frequencies as prior information, they are treated as possible posterior distributions that could have generated the given amino acid frequencies as posterior probabilities. The probability of a particular amino acid distribution given a known frequency distribution, i.e., 100 A, 67 G, 5 C, etc., where pA is the probability of A given by the frequency of A, pG the probability of G, etc., and n is the total number of amino acids given by the multinomial distribution

$$P\,(100A, 67G, 5C\ldots) = n!\,pA^{100}pG^{67}\,pC^{5}\ldots/\,100!\,67!\,5!\ldots \qquad (6)$$

The prior distribution for the multinomial distribution is the Dirichlet distribution (Carlin and Louis 1996), whose formulation is similar to that given in Equation 6 with a similar set of parameters but with factorial and powers reduced by 1. The idea behind using this particular distribution is that if additional sequence data with a related pattern are added, then by the Bayesian procedure of multiplying prior probabilities with the likelihood of the new data to obtain the posterior distribution, the probability of finding the correct frequency of amino acids is favored statistically. Because the amino acid frequencies in the test sequences could be any one of several alternatives, a prior distribution that reflects these several choices is necessary. There is a method for weighting the prior distributions expected for several different multinomial distributions into a combined frequency distribution, the Dirichlet mixture. Calculation of these mixtures is a complex mathematical procedure (Sjölander et al. 1996). Dirichlet mixtures recommended for use in aligning proteins by the HMM method have been described previously (Karplus 1995) and are available from http://www.cse.ucsc.edu/research/compbio/. After the prior amino acid frequencies are in place in the match states of the model, these are modified by training the HMM with the sequences, as described in steps 2 and 3 above. For each match state in the model, a new frequency for each amino acid is calculated by dividing the sum of all new and prior counts for that amino acid by the new total of all amino acids. In this fashion, the new HMM (step 4 above) reflects a combination of expected distributions averaged over patterns in the Dirichlet mixture and patterns exhibited in the training sequences. A similar method is used to refashion the transition probabilities in the HMM during training following manual insertion of initial values.

Another consideration in using HMMs is the number of sequences. If a good prior model such as the above Dirichlet distribution is used, it should be possible to train the HMM with as few as 20 sequences (SAM manual; Eddy 1996; Hughey and Krogh 1996). In general, the smaller the sequence number, the more important the prior conditions. If the number of sequences is >50, the initial conditions play a lesser role because the training step is more effective. As with any msa method, the more sequence diversity, the more challenging the task of aligning sequences with HMMs. HMMs are also more effective if methods to inject statistical noise into the model are used during the training procedure. As the model is refashioned to fit the sequence data, it sometimes goes into a form that provides locally optimal instead of globally optimal alignments of the sequences. One of several noise injection methods (Baldi et al. 1994; Krogh et al. 1994; Eddy et al. 1995; Eddy 1996; Hughey and Krogh 1996) may be used in the training procedure. One method called simulated annealing is used by SAM (Hughey and Krogh 1996). A user-defined number of sequences are generated from the model at each cycle and the counts so generated are added to those from the training sequences. The noise generated in this way is reduced as the cycle number is increased. Finally, the HMM program SAM has a built-in feature of model surgery during training. If a match state is used by fewer than half of the sequences, it is deleted. These same sequences then have to use an insert state in the revised model. Similarly, if an insert state is used by more than half of the sequences, a number of additional match states equal to the average number of insertions is added, and the model has to be revised accordingly. These fractions may be varied in SAM to test the effect on the type of HMM model produced (Hughey and Krogh 1996).

In trying to produce an HMM for a set of related sequences, the recommended procedure is to produce several models by varying the prior conditions. Using regularization by adding prior Dirichlet mixtures to the match states produces models that are more representative of the protein family from which the training sequences are derived. Varying the noise and model surgery levels is another way to vary the training procedure and the HMM model. The best HMM model is the one that predicts a family of related sequences with the lowest and most narrow distribution of NLL scores. An example of a portion of an HMM trained on a set of globin sequences is shown in Figure 4.17.

Motif-based Hidden Markov Models

The program Meta-MEME uses the HMM method to find motifs (conserved sequence domains) in a set of related protein sequences and the spacer regions between them (Grundy et al. 1997) and is built in part on the HMM program HMMER (Eddy et al. 1995). A similar method was originally used to analyze prokaryotic promoters with two conserved patterns separated by a variable spacer region (Cardon and Stormo 1992). A Meta-MEME analysis may be performed at http://www.sdsc.edu/MEME using the University of California at San Diego Supercomputing Center. The use of hidden Markov models for producing a global msa is described in the above section. A problem with HMMs is that the training set has to be quite large (50 or more sequences) to produce a useful model for the sequences. For a smaller number of sequences, it is possible to obtain a model if suitable prior data are used, and an amino acid frequency that is a mixture of frequencies characteristic of certain structural domains (the Dirichlet mixture) is used as prior information of the match states of the model. This mixture is a reasonable guess of combinations of amino acid patterns that are likely to be found. A difficulty in training the HMM residues is that many different parameters must be found (the amino acid distributions, the number and positions of insert and delete states, and the state transition frequencies add up to thousands of parameters) to obtain a suitable model, and the purpose of the prior and training data is to find a suitable estimate for all of these parameters. When trying to make an alignment of short sequence fragments to produce a profile HMM, this problem is worsened because the amount of data for training the model is even further reduced.

Two methods are used by Meta-MEME to circumvent this problem. First, another pattern-finding algorithm, the EM algorithm (discussed on p. 173), is used to locate ungapped regions that match in the majority of the sequences. Second, a simplified HMM with a much reduced number of parameters is produced. The model includes a series of match states that model the patterns located by MEME with transition probabilities of 1 between them and a single insert state between each of these patterns, as illustrated in Figure 4.18. As a result, fewer parameters need to be used, mostly for the amino acid frequencies in the match states.

The most probable order and spacing of the patterns is next found. Another program (Motif Alignment and Search Tool, or MAST; Bailey and Gribskov 1997) is used for this purpose. MAST searches a sequence database for the patterns and reports the database sequences that have the statistically most significant matches. The order and spacing of the patterns found in the highest-scoring database sequences are then used by Meta-MEME as a basis for designing the number of match and insert states and the transition probabilities for the insert states. The match states are filled with modified Dirichlet mixtures (Baylor

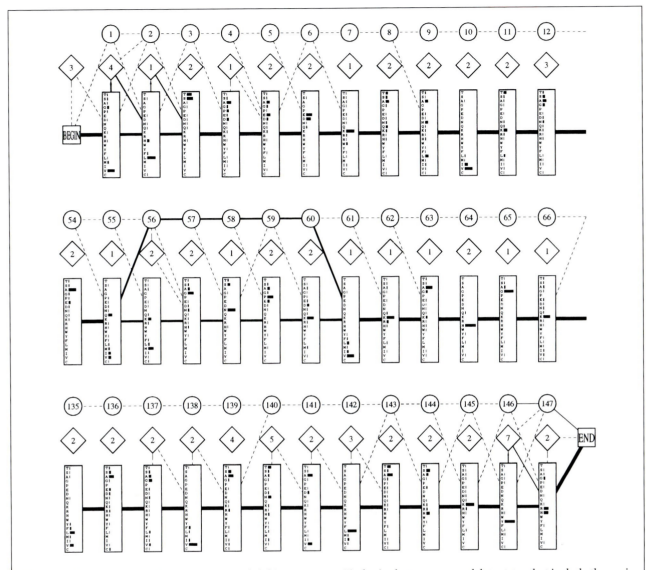

Figure 4.17. HMM trained for recognition of globin sequences. Circles in the top row are delete states that include the position in the alignment; the diamonds in the second row are insert states showing the average length of the insertion, and the rectangles in the bottom row show the amino acid distribution in the match states: V is common at match position 1, L at 2, and so on. The width of each transition line joining these various states indicates the extent of use of that path in the training procedure, and dotted lines indicate a rarely used path. The most used paths are between the match states, but about one-half of the sequences use the delete states at model positions 56–60. Thus, for most of the sequences, the msa or profile will show the first two columns aligned with a V followed by an L, but at 56–60, about one-half of the sequences will have a 5-amino-acid deletion. (Reprinted, with permission, from Krogh et al. 1994 [copyright Academic Press].)

and Gribskov 1996), and the model is trained by the motif models found by MEME. For the 4Fe-4S ferredoxins, a measure of the success of the HMM for database search, the ROC_{50} score (see p. 165), was approximately 0.6–0.8 for 4 to 8 training sequences, compared to 0.95–0.96 using an evolutionary profile of 6 to 12 sequences. However, this family was one of the most difficult ones to model, and other families produced an ROC_{50} of 0.9 or better when trained by 20 or more sequences.

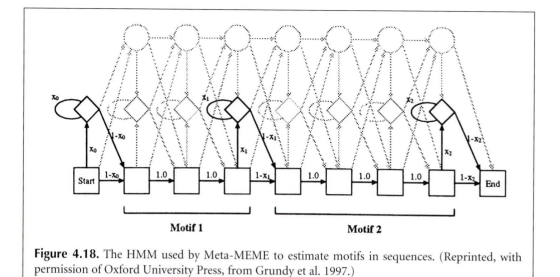

Figure 4.18. The HMM used by Meta-MEME to estimate motifs in sequences. (Reprinted, with permission of Oxford University Press, from Grundy et al. 1997.)

POSITION-SPECIFIC SCORING MATRICES

Analysis of msas for conserved blocks of sequence leads to production of the position-specific scoring matrix, or PSSM. An example of a PSSM produced by the MEME Web site is shown in Figure 4.15G. The PSSM may be used to search a sequence to obtain the most probable location or locations of the motif represented by the PSSM. Alternatively, the PSSM may be used to search an entire database to identify additional sequences that also have the same motif. Consequently, it is important to make the PSSM as representative of the expected sites as possible. The quality and quantity of information provided by the PSSM also varies for each column in the motif, and this variation profoundly influences the matches found with sequences. This situation can be accurately described by information theory, and the results can be displayed by a colored graph called a sequence logo (see Fig. 4.19).

The PSSM is constructed by a simple logarithmic transformation of a matrix giving the frequency of each amino acid in the motif. Two considerations arise in trying to tune the PSSM so that it adequately represents the training sequences. First, if the number of sequences with the found motif is large and reasonably diverse, the sequences represent a good statistical sampling of all sequences that are ever likely to be found with that same motif. If a given column in 20 sequences has only isoleucine, it is not very likely that a different amino acid will be found in other sequences with that motif because the residue is probably important for function. In contrast, another column in the motif from the 20 sequences may have several amino acids, and some amino acids may not be represented at all. Even more variation may be expected at that position in other sequences, although the more abundant amino acids already found in that column would probably be favored. Thus, if a good sampling of sequences is available, the number of sequences is sufficiently large, and the motif structure is not too complex, it should, in principle, be possible to obtain frequencies highly representative of the same motif in other sequences also (Henikoff and Henikoff 1996; Sjölander et al. 1996).

However, the number of sequences for producing the motif may be small, highly diverse, or complex, giving rise to a second level of consideration. If the data set is small, then unless the motif has almost identical amino acids in each column, the column frequencies in the motif may not be highly representative of all other occurrences of the motif. In such cases, it is desirable to improve the estimates of the amino acid frequencies by adding extra amino acid counts, called pseudocounts, to obtain a more reasonable distribution of amino acid frequencies in the column. Knowing how many counts to add is a difficult but fortunately solvable problem. On the one hand, if too many pseudocounts are added in comparison to real sequence counts, the pseudocounts will become the dominant influence in the amino acid frequencies, and searches using the motif will not work. On the other hand, if there are relatively few real counts, many amino acid variations may not be present because of the small sample of sequences. The resulting matrix would then only be useful for finding the sequences used to produce the motif. In such a case, the pseudocounts will broaden the evolutionary reach of the profile to variations in other sequences. Even in this case, the pseudocounts should not drown out but serve to augment the influence of the real counts. In summary, relatively few pseudocounts should be added when there is a good sampling of sequences, and more should be added when the data are more sparse.

The goal of adding pseudocounts is to obtain an improved estimate of the probability p_{ca} that amino acid a is in column c in all occurrences of the blocks, and not just the ones in the present sample. The current estimate of p_{ca} is f_{ca}, the frequency of counts in the data. A simplified Bayesian prediction improves the estimate of p_{ca} by adding prior information in the form of pseudocounts (Henikoff and Henikoff 1996):

$$p_{ca} = (n_{ca} + b_{ca}) \, / \, (N_c + B_c) \tag{7}$$

where n_{ca} and b_{ca} are the real counts and pseudocounts, respectively, of amino acid a in column c, N_c and B_c are the total number of real counts and pseudocounts, respectively, in the column, and $f_{ca} = n_{ca} / N_c$. It is obvious that as b_{ca} becomes larger, the pseudocounts will have a greater infuence on p_{ca}. Furthermore, not only the types of pseudocounts but also the total number added to the column (B_c) will influence p_{ca}. Finally, fractions such as p_{ca} are used to produce the log odds form of the motif matrix, the PSSM, which is the most suitable representation of the data for sequence comparisons. A count and probability of zero for an amino acid a in a given column, which is quite common in blocks, may not be converted to logarithms. Addition of a small number of b_{ca} will correct this problem without producing a major change in the PSSM values. An equation similar to Equation 7 is used in the Gibbs sampler (p. 177), except that the number of sequences is $N - 1$.

Pseudocounts are added based on simple formulas or on the previous variations seen in aligned sequences. The amino acid substitution matrices, including the Dayhoff PAM and BLOSUM matrices, provide one source of information on amino acid variation. Another source is the Dirichlet mixtures derived as a posterior probability distribution from the amino acid substitutions observed in the BLOCKS database (see HMMs; Sjölander et al. 1996).

One simple formula that has worked well in some studies is to make B in Equation 7 equal to $\sqrt{N}$, where N is the number of sequences, and to allot these counts to the amino acids in proportion to their frequencies in the sequences (Lawrence et al. 1993; Tatusov et al. 1997). As N increases, the influence of pseudocounts will decrease because $\sqrt{N}$ will increase more slowly. The main difficulties with this method are that it does not take into account known substitutions of amino acids in alignments and the observed amino acid variations from one column in the motif to the next, and it does not add enough pseudocounts when the number of sequences is small.

The information in scoring matrices may be used to produce an average sequence profile, as illustrated in Figure 4.12. Rather than count amino acids, the scoring table values are averaged between each possible 20 amino acids and those amino acids found in the column of the scoring matrix. Zero counts in a column are not a problem because amino acids not present are not used in the calculations. Because these averaging methods do not take into account the number of sequences in the block, they do not have the desirable effect of a reduced influence when there is a large number of sequences.

Another method of using the information from amino acid substitution matrices is to base pseudocounts on these matrices. Recall the log odds form of the matrices is derived by taking the logarithm of the frequency of substitution q_{ia} of amino acid i for amino acid a divided by the frequency of occurrence of amino acid a, p_a. Then, b_{ca} may be estimated from the total number of pseudocounts in the column by (Henikoff and Henikoff 1996),

$$b_{ca} = B_c Q_i \text{ where } Q_i = \sum_{\text{all } i} q_{ia} \tag{8}$$

b_{ca} in column c can also be made to depend on the observed data in that column (Tatusov et al. 1997), which is given by multiplying B_c by the following conditional probabilities.

$$b_{ca} = B_c \sum_{\text{all } i} \text{prob (amino acid } i|\text{column } c) \times \text{prob (amino acid } a|i)$$

$$= B_c \sum_{\text{all } i} (n_{ci}/N_c \times q_{ia}/Q_i) \tag{9}$$

where n_{ci} is the real count of amino acid i in column c.

The total number of pseudocounts in each column needs also to be estimated. As described above, one estimate is to make B_c for each column equal to $\sqrt{N}$, where N is the number of sequences, but this method does not take into account the differences between columns and, for a small number of sequences, the total number of pseudocounts is not sufficient. Allowing B_c to be a constant that can exceed N_c overcomes this limitation but still does not take into account variations in amino acid frequencies between columns, such that a column with conserved amino acids should receive fewer pseudocounts. Using the number of different amino acids in column c, R_c, as an indicator, B_c has been estimated by the formula (Henikoff and Henikoff 1996)

$$B_c = m \times R_c \tag{10}$$

where m is a positive number derived from trial database searches and $m \leq m \times B_c \leq \min$ ($m \times N_c$, $m/20$) (the latter term meaning the minimum of the two given values). By this formula and a given value of m, when $N_c \leq m \times 20$, the total number of pseudocounts B_c is greater, and when $N_c > m \times 20$, B_c is smaller than the total number of real counts, N_c, regardless of the value of R_c. The number of pseudocounts is also reduced when $R_c = 1$. In a test search of the SwissProt and Prosite catalogs with various values of m, a value of 5–6 for m produced the most efficient PSSMs for finding known family members. Of the several methods for making PSSMs discussed above, the one with pseudocounts derived by Equations 9 and 10 was most successful. This search was performed with PSSMs derived from blocks with amino acid counts also weighted to account for redundancy (Henikoff and Henikoff 1996). However, pseudocounts added from Dirichlet mixtures, which also vary in each column of the scoring matrix, are also very effective (Henikoff and Henikoff 1996; Tatusov et al. 1997).

Once pseudocounts have been added to real counts of amino acids in each column of the motif, the PSSM may be calculated. The PSSM has one column (or row) for each position in the motif and one row (or column) for each amino acid, and the entries are log odds entries. Each entry is derived by taking the logarithm to the base 2 (bit units, but sometimes also natural logarithms in nat units are used) of the total of the real counts plus pseudocounts for each amino acid, divided by the probability of that amino acid (b_{ca} / N_c). An example of a PSSM produced by MEME is shown in Figure 4.15G.

As a sequence is searched with the PSSM, the value of the first amino acid in the sequence is looked up in the first column of the PSSM, then the value of the second amino acid in the matrix, and so on, until the length scanned is the same as the motif width represented by the matrix. All the log odds scores are added to produce a summed score for start position 1 in the sequence. The process is repeated starting at the second position in the sequence, and so on, until there is not enough sequence left. The highest log odds scoring sequence positions have the closest match statistically to the PSSM. Adding logarithms in this manner is the equivalent of mutiplying the probabilities of the amino acids at each sequence position. To convert each summed log odds score (S) to a likelihood or odds score of the sequence matching the PSSM, use the formula odds score = 2^S. These odds scores may be summed and each individual score divided by the sum to normalize them and to thereby produce a probability of the motif at each sequence location.

The above description and example are of using a PSSM to define motifs in protein families. PSSM are also used to define DNA sequence patterns that define regulatory sites, such as promoters or exon–intron junctions in genomic sequences. These topics are discussed in Chapter 8.

Information Content of the PSSM

The usefulness of a PSSM in distinguishing real sequence patterns from background may be measured. The unit of measure is the information content in bits. The PSSM described above gives the log odds score for finding a particular matching amino acid in a target sequence corresponding to each motif position. Variations in the scores found in each column of the table are an indication of the amino acid variation in the original training sequences that were used to produce the motif. In some columns, only one amino acid may have been present, whereas in others several may have been present. The columns with highly conserved positions have more information than do the variable columns and will be more definitive for locating matches in target sequences. There is a formal method known as information theory for describing the amount of information in each column that is useful for evaluating each PSSM. The information content of a given amino acid substitution matrix was previously introduced (p. 83) and is discussed in greater detail here. T. Schneider has prepared a Web site that gives excellent tutorials and a review on the topic of information theory, along with methods to produce sequence logos (Schneider and Stephens 1990) at http://www-lmmb.ncifcrf.gov/~toms/sequencelogo. html.

To illustrate the concepts of information and uncertainty (see above Web site), consider 64 cups in a row with an object hidden under one of them. The goal is to find the object with as few questions as possible. The solution is quite simple. First, ask whether the object is hidden under the first or second half of the cups. If the answer is the first 32, then ask which half of that 32, the first 16 or the second 16, and so on. The sequential questions reduce the possibilities from 64-32-16-8-4-2-1, and six questions will therefore suffice to locate the object. This number is also a measure of the amount of uncertainty in the data because this number of questions must be asked to find the object. After the first question has been asked, uncertainty has been reduced by 1, so that only five questions then need to be asked to find the object. The uncertainty is zero when the object is found.

A method to calculate uncertainty (the number of questions to be asked) may be derived from the probability of finding the object under a given cup [p(object) = 1/64]. Uncertainty is found by taking the negative logarithm to the base 2 of 1/64 [$-\log_2(1/64) = 6$ bits]. A situation similar to the hidden object example is found with amino acids in the columns of a PSSM. Here, the interest is to find which amino acid belongs at a particular column in the motif. When we have no information at all, since there are 20 possible choices in all, the amount of uncertainty is $\log_2 20 = 4.32$.

The data from the PSSM provide information that reduces this uncertainty. If only one amino acid is observed in a column of the PSSM, the uncertainty is zero because there are no other possibilities. If two amino acids are observed with equal frequency, there is still uncertainty as to which one it is, and one question must be asked to find the answer, or uncertainty = 1. The formula for finding the uncertainty in this example is the sum of the fractional information provided by each amino acid, or $-$ [0.5 $\times$ $\log_2 0.5$ + 0.5 $\times$ $\log_2 0.5$] = 1. In general, the average amount of uncertainty (H_c) in bits per symbol for column c of the PSSM is given by

$$H_c = -\sum_{\text{all } i} p_{ic} \log_2(p_{ic}) \tag{11}$$

where p_{ic} is the frequency of amino acid i in column c and is estimated by the frequency of occurrence of each amino acid (b_{ca}/N_c) and $\log_2(p_{ic})$ is the log odds score for each amino acid in column c. Uncertainty for the entire PSSM may then be calculated as

$$H = \sum_{\text{all columns}} H_c \tag{12}$$

H is also known as the entropy of the PSSM position in information theory because the higher the value, the greater the uncertainty. The lower the value of the uncertainty H for the PSSM, the greater the ability of the PSSM to distinguish real occurrences of the motif from random matches. Conversely, the higher the information content, calculated as shown below, the more useful the PSSM.

Sequence Logos

Sequence logos are graphs that illustrate the amount of information in each column of a motif. The logo is derived from sequence information in the PSSM described above. Conserved patterns in both protein and DNA sequences can be represented by sequence logos. A program for producing logos, along with several examples, is available from http://www-lmmb.ncifcrf.gov/~toms/sequencelogo.html. The Web site of S.E. Brenner at http://www.bio.cam.ac.uk/seqlogo/ will produce sequence logos from an input alignment using the Gibbs sampler method, and an implementation of an extension of the logo method for structural RNA alignment (Gorodkin et al. 1997) is at http://www.cbs.dtu.dk/gorodkin/appl/plogo.html. A logo representation for the BLOCKS database has been implemented (Henikoff et al. 1995) and may be viewed when the information on a particular block is retrieved from the BLOCKS Web server (http://www.blocks.fhcrc.org/). An example of a Block logo is shown in Figure 4.19. Another example of a simple graph of information content is given in Figure 4.15C. In this case, the information for the entire motif has been calculated by the MEME server by summing the values in each column to a total value of 22 bits. Although logos are primarily used with ungapped motifs and sequence patterns, logos of alignments that include gaps in some sequence positions may also be made. If such is the case, then the height of the column with gaps is reduced by the proportion of sequence positions that are not gaps.

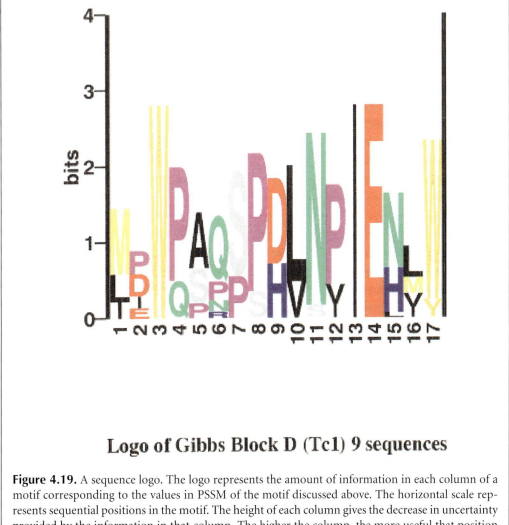

Logo of Gibbs Block D (Tc1) 9 sequences

Figure 4.19. A sequence logo. The logo represents the amount of information in each column of a motif corresponding to the values in PSSM of the motif discussed above. The horizontal scale represents sequential positions in the motif. The height of each column gives the decrease in uncertainty provided by the information in that column. The higher the column, the more useful that position for finding matches in sequences. In each column are shown symbols of the amino acids found at the corresponding position in the motif, with the height of the amino acid proportional to the frequency of that amino acid in the column, and the amino acids shown in decreasing order of abundance from the top of the column. From each logo, the following information may thus be found: The consensus may be read across the columns as the top amino acid in each column, the relative frequency of each amino acid in each column of the motif is given by the size of the letters in each column, and the total height of the column provides a measure of how useful that column is for reducing the level of uncertainty in a sequence matching experiment. Note that the highest values are for columns with less diversity.

The height of each logo position is calculated as the amount by which uncertainty has been decreased by the available data; in this case, the amino acid frequencies in each column of the motif. The relative heights of each amino acid within each column are calculated by determining how much each amino acid has contributed to that decrease. The uncertainty at column c is given by Equation 11. Because the maximum uncertainty at a position/column when no information is available is $\log_2 20 = 4.32$, as more information about the motif is obtained by new data, the decrease in uncertainty (or increase in the amount of information) R_c is

$$R_c = \log_2 20 - (H_c + \epsilon_n) \tag{13}$$

where H_c is given by Equation 11 and ϵ_n is a correction factor for a small sequence number n. R_c is used as the total height of the logo column. The height of amino acid a at position c in the motif logo is then given by $f_{ac} \times R_c$.

The above description applies to protein sequences. Sequence logos are also produced for DNA sequences. The methodology is very similar to the above except that there are only four possible choices for each logo location. Hence, the maximum amount of uncertainty is $\log_2 4 = 2$. The above method assumes that the sequence pattern is less random than the background or expected sequence variation, and this assumption limits the ability of the method to locate subtle patterns in sequences.

An improved method for finding more subtle patterns in sequences is called the relative entropy method (Durbin et al. 1998). In this case, differences between the observed frequencies and background frequencies are used (Gorodkin et al. 1997), and the decrease in uncertainty from background to observed (or amount of information) in bits is given by

$$R_c = \sum_{\text{all } i} p_{ic} \log_2(p_{ic}/b_i) \tag{14}$$

where b_i is the background frequency of residue i in the organism and the maximum uncertainty in column c is given by $-\sum_{\text{all } i} [p_{ic} \log_2(1/b_i)]$. When background frequencies are taken into account, and the column frequency is less than the background frequency, it is possible for the information given by a particular residue in a logo column to be negative. To accommodate this change, the corresponding sequence character is inverted in the logo to indicate a less than expected frequency. There are also two ways used to illustrate the contribution of each character through the height of the symbol. The first method is described above. The second method is to display symbol heights in proportion to the ratio of the observed to the expected frequency, i.e., by the fraction $(p_{ic}/b_i) / (\sum_{\text{all } i} p_{ic}/b_i)$ for each symbol i. Gaps are included in the analysis by using $p_{\text{gap}} = 1$ and, as a result, will always give a negative contribution to the information (Gorodkin et al. 1997).

MULTIPLE SEQUENCE ALIGNMENT EDITORS AND FORMATTERS

Once a multiple sequence alignment has been obtained by the global msa program, it may be necessary to edit the sequence manually to obtain a more reasonable or expected alignment. Several considerations must be kept in mind when choosing a sequence editor, which should include as many of the following features as possible: (1) provision for displaying the sequence on a color monitor with residue colors to aid in a clear visual representation of the alignment, (2) recognition of the multiple sequence format that was output by the msa program and maintenance of the alignment in a suitable format when the editing is completed, (3) provision of a suitable windows interface, allowing use of the mouse to add, delete, or move sequence followed by an updated display of the alignment. In addition, there are other types of editing that are commonly performed on msas such as, for example, shading conserved residues in the alignment.

The large number of multiple sequence alignment formats that are in use were discussed in Chapter 2. Two commonly encountered examples are the Genetics Computer Group's MSF format and the CLUSTALW ALN format. Because these formats follow a precise outline, one may be readily converted to another by computer programs. READSEQ by D.G. Gilbert at Indiana University at Bloomington is one such program. This program will run on almost any computer platform and may be obtained by anonymous FTP from ftp.bio.indiana.edu/molbio/readseq. There is also a Web-based interface for READSEQ from Baylor College of Medicine at http://dot.imgen.bcm.tmc.edu:9331/seq-util/seq-util.html/. A software package SEQIO, which provides C program modules for conversion of sequence files from one format to another, is available by anonymous FTP from ftp.pasteur.fr/pub/GenSoft/unix/programming/seqio-1.2.tar.gz; documentation is available at http://bioweb.pasteur.fr/docs/doc-gensoft/seqio/.

A short list of the many available programs that have or exceed the above-listed features is discussed below. For a more comprehensive list, visit the catalog of software page at Web address http://www.ebi.ac.uk/biocat/.

Sequence Editors

1. CINEMA (Colour Interactive Editor for Multiple Alignments) at http://www.biochem.ucl.ac.uk/bsm/dbbrowser/CINEMA2.02/kit.html is a broadly functional program for sequence editing and analysis, including dot matrix analysis. It features drag-and-drop editing, sequence shifting to left or right, viewing of different parts of an alignment using the split-screen option, multiple motif selection and manipulation, and a number of added features such as viewing of protein structures. CINEMA was developed by A.W.R. Payne, D.J. Parry-Smith, A.D. Michie, and T.K. Attwood. CINEMA is an applet that runs under a Web browser and therefore will run on almost any computer platform.

2. GDE (Genetic Data Environment) provides a general interface on UNIX machines for sequence analysis, sequence alignment editing, and display (Smith et al. 1994) and is available from several anonymous FTP sites including ftp.ebi.ac.uk/pub/software/unix. GDE is described at http://bimas.dcrt.nih.gov/gde_sw.html, and http://www.tigr.org/~jeisen/GDE/GDE.html. GDE features are incorporated into the Seqlab interface for the GCG software, vers. 9. This interface requires communication with a host UNIX machine running the Genetics Computer Group software. Interface with MS-DOS or Macintosh is possible if the computer is equipped with the appropriate X-Windows client software.

3. GeneDoc is an alignment editing and display editor by K. Nicholas and H. Nicholas of the Pittsburgh Supercomputing Center for MSF-formatted msas. It can also import files in other formats. GeneDoc can move residues by inserting or deleting gap, and features drag-and-drop editing. As the alignment is edited, a new alignment score is calculated by sum of pairs method or based on a phylogenetic tree. GeneDoc is available from http://www.psc.edu/biomed/genedoc/ and runs under MS Windows.

4. MACAW is both a local multiple sequence alignment program and a sequence editing tool (Schuler et al. 1991). Given a set of sequences, the program finds ungapped blocks in the sequences and gives their statistical significance. Later versions of the program

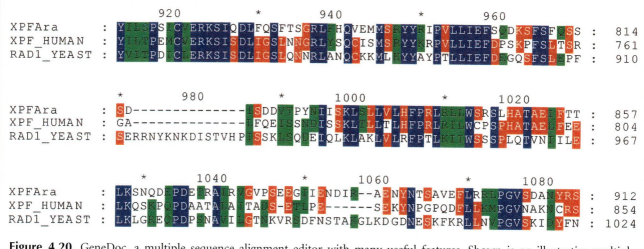

Figure 4.20. GeneDoc, a multiple sequence alignment editor with many useful features. Shown is an illustrative multiple sequence alignment of three DNA repair genes similar to the *S. cerevisiae Rad1* gene. The sequences were aligned with *CLUSTALW*, and the FASTA-formatted alignment (Chapter 2) was imported into GeneDoc on a PC.

find blocks by one of three user-chosen methods: by searching for maximum segment pairs or common patterns present in the sequences scored by a scoring matrix such as PAM250 or BLOSUM matrices (the methods used by the BLAST algorithm), by using the Gibbs sampling strategy, a statistical method, or by searching for user-provided patterns provided in a particular format called a regular expression. Executable programs that run under MS-DOS Windows, Macintosh, and other computer platforms are available by anonymous FTP from ncbi.nlm.nih.gov/pub/schuler/macaw.

Sequence Formatters

1. Boxshade is a formatting program by K. Hofmann for marking identical or similar residues in msas with shaded boxes, and is available by anonymous FTP from http://www.isrec.isb-sib.ch/sib-isrec/boxshade. The Web server at http://www.ch.embnet.org/software/BOX_form.html takes a multiple-alignment file in either the Genetics Computer Group MSF format or CLUSTAL ALN format and can output a file in many forms including Postscript/EPS and PICT for editing on Macintosh and MS-DOS machines.

2. CLUSTALX is a sequence formatting tool that provides a Windows interface for a CLUSTALW msa and is available for many computer platforms, including MS-DOS and Macintosh machines by anonymous FTP from ftp-igbmc.u-strasbg.fr/pub/ClustalX/ (Thompson et al. 1997).

REFERENCES

Altschul S.F. 1989. Gap costs for multiple sequence alignment. *J. Theor. Biol.* **138:** 297–309.

Altschul S.F., Carroll R.J., and Lipman D.J. 1989. Weights for data related by a tree. *J. Mol. Biol.* **207:** 647–653.

Bailey T.L. and Elkan C. 1995. The value of prior knowledge in discovering motifs with MEME. In *Pro-*

ceedings of the 3rd International Conference on Intelligent Systems for Molecular Biology (ed. C. Rawlings et al.), pp. 21–29. AAAI Press, Menlo Park, California.

Bailey T.L. and Gribskov M. 1997. Score distributions for simultaneous matching to multiple motifs. *J. Comput. Biol.* **4:** 45–59.

———. 1998. Methods and statistics for combining motif match searches. *J. Comput. Biol.* **5:** 211–221.

Bairoch A. 1991. PROSITE: A dictionary of sites and patterns in proteins. *Nucleic Acids Res.* (suppl.) **19:** 2241–2245.

Baldi P., Chauvin Y., Hunkapillar T., and McClure M.A. 1994. Hidden Markov models of biological primary sequence information. *Proc. Natl. Acad. Sci.* **91:** 1059-1063.

Barton G.J. 1994. The AMPS package for multiple protein sequence alignment. Computer analysis of sequence data, part II. *Methods Mol. Biol.* **25:** 327–347.

Baylor T.L. and Gribskov M. 1996. The megaprior heuristic for discovering protein sequence patterns. In *Proceedings of the 4th International Conference on Intelligent Systems for Molecular Biology* (ed. D.J. States et al.), pp. 15–24. AAAI Press, Menlo Park, California.

Boguski M., Hardison R.C., Schwartz S., and Miller W. 1992. Analysis of conserved domains and sequence motifs in cellular regulatory proteins and locus control regions using software tools for multiple alignment and visualization. *New Biol.* **4:** 247–260.

Briffeuil P., Baudoux G., Reginster I., Debolle X., Depiereux E., and Feytmans E. 1998. Comparative analysis of seven multiple protein sequence alignment servers: Clues to enhance reliability of predictions. *Bioinformatics* **14:** 357–366.

Cardon L.R. and Stormo G.D. 1992. Expectation maximization algorithm for identifying protein-binding sites with variable lengths from unaligned DNA fragments. *J. Mol. Biol.* **223:** 159–170.

Carlin B.P. and Louis T.A. 1996. *Bayes and empirical Bayes methods for data analysis (Monographs on statistics and applied probability* [ed. D.R. Cox et al.]). Chapman and Hall, New York.

Carrillo H. and Lipman D. 1988. The multiple sequence alignment problem in biology. *SIAM J. Appl. Math.* **48:** 197–209.

Churchill G.A. 1989. Stochastic models for heterogeneous DNA sequences. *Bull. Math. Biol.* **51:** 79–94.

Corpet F. 1988. Multiple sequence alignment with hierarchical clustering. *Nucleic Acids Res.* **16:** 10881–10890.

Dayhoff M.O. 1978. Survey of new data and computer methods of analysis. In *Atlas of protein sequence and structure,* vol. 5, suppl. 3. National Biomedical Research Foundation, Georgetown University, Washington, D.C.

Durbin R., Eddy S., Krogh A., and Mitchison G. 1998. *Biological sequence analysis: Probabilistic models of proteins and nucleic acids.* Cambridge University Press, United Kingdom.

Eddy S.R. 1995. Multiple alignment using hidden Markov models. *Ismb* **3:** 114–120.

———. 1996. Hidden Markov models. *Curr. Opin. Struct. Biol.* **6:** 361–365.

———. 1998. Profile hidden Markov models. *Bioinformatics* **14:** 755–763.

Eddy S.R., Mitchison G., Durbin R. 1995. Maximum discrimination hidden Markov models of sequence consensus. *J. Comput. Biol.* **2:** 9–23.

Feng D.F. and Doolittle R.F. 1987. Progressive sequence alignment as a prerequisite to correct phylogenetic trees. *J. Mol. Evol.* **25:** 351–360.

———. 1996. Progressive alignment of amino acid sequences and construction of phylogenetic trees from them. *Methods Enzymol.* **266:** 368–382.

Gorodkin J., Heyer L.J., Brunak S., and Stormo G.D. 1997. Displaying the information contents of structural RNA alignments: The structure logos. *Comput. Appl. Biosci.* **13:** 583–586.

Gotoh O. 1994. Further improvement in methods of group-to-group sequence alignment with generalized profile operations. *Comput. Appl. Biosci.* **10:** 379–387.

———. 1995. A weighting system and algorithm for aligning many phylogenetically related sequences. *Comput. Appl. Biosci.* **11:** 543–551.

———. 1996. Significant improvement in accuracy of multiple protein sequence alignments by iterative refinement as assessed by reference to structural alignments. *J. Mol. Biol.* **264:** 823–838.

———. 1999. Multiple sequence alignment: Algorithms and applications. *Adv. Biophys.* **36:** 159–206.

Gribskov M. and Veretnik S. 1996. Identification of sequence patterns with profile analysis. *Methods Enzymol.* **266:** 198–212.

Gribskov M., Luethy R., and Eisenberg D. 1990. Profile analysis. *Methods Enzymol.* **183:** 146–159.

Gribskov M., McLachlan A.D., and Eisenberg D. 1987. Profile analysis: Detection of distantly related proteins. *Proc. Natl. Acad. Sci.* **84:** 4355–4358.

Grundy W.N., Bailey T.L., and Elkan C.P. 1996. Para-MEME: A parallel implementation and a web interface for a DNA and protein motif discovery tool. *Comput. Appl. Biosci.* **12:** 303–310.

Grundy W.N., Bailey T.L., Elkan C.P., and Baker M.E. 1997. Meta-MEME: Motif-based hidden Markov models of protein families. *Comput. Appl. Biosci.* **13:** 397–406.

Gupta S.K., Kececioglu J.D., and Schaffer A.A. 1995. Improving the practical space and time efficiency of the shortest-paths approach to sum-of-pairs multiple sequence alignment. *J. Comput. Biol.* **2:** 459–472.

Hein J. 1990. Unified approach to alignment and phylogenies. *Methods Enzymol.* **183:** 626–645.

Henikoff J.G. and Henikoff S. 1996. Using substitution probabilities to improve position-specific scoring matrices. *Comput. Appl. Biosci.* **12:** 135–143.

Henikoff S. and Henikoff J.G. 1991. Automated assembly of protein blocks for database searching. *Nucleic Acids Res.* **19:** 6565–6572.

———. 1992. Amino acid substitution matrices from protein blocks. *Proc. Natl. Acad. Sci.* **89:** 10915–10919.

Henikoff S., Henikoff J.G., Alford W.J., and Pietrokovski S. 1995. Automated construction and graphical presentation of protein blocks from unaligned sequences. *Gene* **163:** GC17–GC26.

Heringa J. 1999. Two strategies for sequence comparison: Profile-preprocessed and secondary structure-induced multiple alignment. *Comput. Chem.* **23:** 341–364.

Higgins D.G. and Sharp P.M. 1988. CLUSTAL: A package for performing multiple sequence alignment on a microcomputer. *Gene* **73:** 237–244.

Higgins D.G., Thompson J.D., and Gibson T.J. 1996. Using CLUSTAL for multiple sequence alignments. *Methods Enzymol.* **266:** 383–402.

Hirosawa M., Totoki Y., Hoshida M., and Ishikawa M. 1995. Comprehensive study on iterative algorithms of multiple sequence alignment. *Comput. Appl. Biosci.* **11:** 13–18.

Hughey R. and Krogh A. 1996. Hidden Markov models for sequence analysis: Extension and analysis of the basic method. *Comput. Appl. Biosci.* **12:** 95–107.

Jonassen I., Collins J.F., and Higgins D. 1995. Finding flexible patterns in unaligned protein sequences. *Protein Sci.* **4:** 1587–1595.

Karplus K. 1995. Regularizers for estimating the distributions of amino acids from small samples. In *UCSC Technical Report* (UCSC-CRL-95-11). University of California, Santa Cruz.

Kececioglu J. 1993. The maximum weight trace problem in multiple sequence alignment. In *Proceedings of the 4th Symposium on Combinatorial Pattern Matching: Lecture notes in computer science*, no. 684, pp. 106–119. Springer Verlag, New York.

Kececioglu J., Lehof H.-P., Mehlhorn K., Mutzel P., Reinert K., and Vingron M. 2000. A polyhedral approach to sequence alignment problems. *Discrete Appl. Math.* **104:** 143–186.

Kim J., Pramanik S., and Chung M.J. 1994. Multiple sequence alignment by simulated annealing. *Comput. Appl. Biosci.* **10:** 419–426.

Krogh A., Brown M., Mian I.S., Sjölander K., and Haussler D. 1994. Hidden Markov models in computational biology. Applications to protein modeling. *J. Mol. Biol.* **235:** 1501–1531.

Lawrence C.E. and Reilly A.A. 1990. An expectation maximization (EM) algorithm for the identification and characterization of common sites in unaligned biopolymer sequences. *Proteins Struct. Funct. Genet.* **7:** 41–51.

Lawrence C.E., Altschul S.F., Boguski M.S., Liu J.S., Neuwald A.F., and Wootton J.C. 1993. Detecting subtle sequence signals: A Gibbs sampling strategy for multiple alignment. *Science* **262:** 208–214.

Lipman D.J., Altschul S.F., and Kececioglu J.D. 1989. A tool for multiple sequence alignment. *Proc. Natl. Acad. Sci.* **86:** 4412–4415.

Liu J.S., Neuwald A.F., and Lawrence C.E. 1995. Alignment and Gibbs sampling strategies. *J. Am. Stat. Assoc.* **90:** 1156–1170.

McClure M.A., Vasi T.K., and Fitch W.M. 1994. Comparative analysis of multiple protein-sequence alignment methods. *Mol. Biol. Evol.* **11:** 571-592.

Miller M.J. and Powell J.I. 1994. A quantitative comparison of DNA sequence assembly programs. *J. Comput. Biol.* **1:** 257–269.

Miller W., Boguski M., Raghavachari B., Zhang Z., and Hardison R.C. 1994. Constructing aligned sequence blocks. *J. Comput. Biol.* **1:** 51–64.

Mitchison G.J. and Durbin R.M. 1995. Tree-based maximal likelihood substitution matrices and hidden Markov models. *J. Mol. Evol.* **41:** 1139–1151.

Morgenstern B., Dress A., and Werner T. 1996. Multiple DNA and protein sequence alignment based on segment-to-segment comparison. *Proc. Natl. Acad. Sci.* **93:** 12098–12103.

Morgenstern B., Frech K., Dress A., and Werner T. 1998. DIALIGN: Finding local similarities by multiple sequence alignment. *Bioinformatics* **14:** 290–294.

Myers E.W. 1995. Toward simplifying and accurately formulating fragment assembly. *J. Comput. Biol.* **2:** 275–290.

Myers E.W. and Miller W. 1988. Optimal alignments in linear space. *Comput. Appl. Biosci.* **4:** 11–17.

Neuwald A.F. and Green P. 1994. Detecting patterns in protein sequences. *J. Mol. Biol.* **239:** 698–712.

Neuwald A.F., Liu J.S., and Lawrence C.E. 1995. Gibbs motif sampling: Detection of bacterial outer membrane protein repeats. *Protein Sci.* **4:** 1618–1632.

Nevill-Manning C.G., Wu T.D., and Brutlag D.L. 1998. Highly specific protein sequence motifs for genome analysis. *Proc. Natl. Acad. Sci.* **95:** 5865–5871.

Notredame C. and Higgins D.G. 1996. SAGA: Sequence alignment by genetic algorithm. *Nucleic Acids Res.* **24:** 1515–1524.

Notredame C., Holme L., and Higgins D.G. 1998. COFFEE: A new objective function for multiple sequence alignment. *Bioinformatics* **14:** 407–422.

Notredame C., O'Brien E.A., and Higgins D.G. 1997. RAGA: RNA sequence alignment by genetic algorithm. *Nucleic Acids Res.* **25:** 4570–4580.

Pascarella S. and Argos P. 1992. Analysis of insertions/deletions in protein sequences. *J. Mol. Biol.* **224:** 461–471.

Rabiner L.R. 1989. A tutorial on hidden Markov models and selected applications in speech recognition. *Proc. IEEE* **77:** 257–1531.

Ravi R. and Kececioglu J. 1998. Approximation algorithms for multiple sequence alignment under a fixed evolutionary tree. *Discrete Appl. Math.* **88:** 355–366.

Saitou N. and Nei M. 1987. The neighbor-joining method: A new method for reconstructing phylogenetic trees. *Mol. Biol. Evol.* **4:** 406–425.

Sankoff D. 1975. Minimal mutation trees of sequences. *SIAM J. Appl. Math.* **78:** 35–42.

Schneider T.D. and Stephens R.M. 1990. Sequence logos: A new way to display consensus sequences. *Nucleic Acids Res.* **18:** 6097–6100.

Schuler G.D., Altschul S.F., and Lipman D.J. 1991. A workbench for multiple alignment construction and analysis. *Proteins* **9:** 180–190.

Shapiro B. and Navetta J. 1994. A massively parallel genetic algorithm for RNA secondary structure prediction. *J. Supercomput.* **8:** 195–207.

Sjölander K., Karplus K., Brown M., Hughey R., Krogh A., Mian I.S., and Haussler D. 1996. Dirichlet mixtures: A method for improved detection of weak but significant protein sequence homology. *Comput. Appl. Biosci.* **12:** 327–345.

Smith H.O., Annau T.M., and Chandrasegaran S. 1990. Finding sequence motifs in groups of functionally related proteins. *Proc. Natl. Acad. Sci.* **87:** 826–830.

Smith R.F. and Smith T.F. 1992. Pattern-induced multi-sequence alignment (PIMA) algorithm employing secondary structure-dependent gap penalties for use in comparative protein modelling. *Protein Eng.* **5:** 35–41.

Smith S.W., Overbeek R., Woese C.R., Gilbert W., and Gillevet P.M. 1994. The genetic data environment and expandable GUI for multiple sequence analysis. *Comput. Appl. Biosci.* **10:** 671–675.

Sneath P.H.A. and Sokal R.R. 1973. *Numerical taxonomy.* W.H. Freeman, San Francisco, California.

Sonnhammer E.L., Eddy S.R., and Durbin R. 1997. Pfam: A comprehensive database of protein domain families based on seed alignments. *Proteins* **28:** 405–420.

Tatusov R.L., Altschul S.F., and Koonin E.V. 1994. Detection of conserved segments in proteins: Iterative scanning of sequence databases with alignment blocks. *Proc. Natl. Acad. Sci.* **91:** 12091–12095.

Tatusov R.L., Koonin E.V., and Lipman D.J. 1997. A genomic perspective on protein families. *Science* **278:** 631–637.

Taylor W.R. 1990. Hierarchical method to align large numbers of biological sequences. *Methods Enzymol.* **183:** 456–474.

———. 1996. Multiple protein sequence alignment: Algorithms and gap insertion. *Methods Enzymol.* **266:** 343–367.

Thompson J.D., Higgins D.G., and Gibson T.J. 1994a. CLUSTAL W: Improving the sensitivity of progressive multiple sequence alignment through sequence weighting, position-specific gap penalties and weight matrix choice. *Nucleic Acids Res.* **22:** 4673–4680.

———. 1994b. Improved sensitivity of profile searches through the use of sequence weights and gap excision. *Comput. Appl. Biosci.* **10:** 19–29.

Thompson J.D., Plewniak F., and Poch O. 1999. A comprehensive comparison of multiple sequence alignment programs. *Nucleic Acids Res.* **27:** 2682–2690.

Thompson J.D., Gibson T.J., Plewniak F., Jeanmougin F., and Higgins D.G. 1997. The CLUSTAL X windows interface: Flexible strategies for multiple sequence alignment aided by quality analysis tools. *Nucleic Acids Res.* **25:** 4876–4882.

Vingron M. and Argos P. 1991. Motif recognition and alignment for many sequences by comparison of dot matrices. *J. Mol. Biol.* **218:** 33–43.

Vingron M. and Sibbald P.R. 1993. Weighting in sequence space: A comparison of methods in terms of generalized sequences. *Proc. Natl. Acad. Sci.* **90:** 8777–8781.

Waterman M. and Perlwitz M.D. 1984. Line geometries for sequence comparisons. *Bull. Math. Biol.* **46:** 567–577.

Weber J.L. and Myers E.W. 1997. Human whole-genome shotgun sequencing. *Genome Res.* **7:** 401–409.

Zhang C. and Wong A.K. 1997. A genetic algorithm for multiple molecular sequence alignment. *Comput. Appl. Biosci.* **13:** 565–581.

Zhang Z., Raghavachari B., Hardison R.C., and Miller W. 1994. Chaining multiple-alignment blocks. *J. Comput. Biol.* **1:** 217–226.

Prediction of RNA Secondary Structure

THE PREVIOUS TWO CHAPTERS DISCUSS the alignment of protein and nucleic acid sequences. The methods used either align entire sequences or search for common patterns in the sequences. In either case, the objective is to locate a set of sequence characters in the same order in the sequences. Nucleic acid sequences that specify RNA molecules have to be compared differently. Sequence variations in RNA sequences maintain base-pairing patterns that give rise to double-stranded regions (secondary structure) in the molecule. Thus, alignments of two sequences that specify the same RNA molecules will show covariation at interacting base-pair positions, as illustrated in Figure 5.1. In addition to these covariable positions, sequences of RNA-specifying genes may also have rows of similar sequence characters that reflect the common ancestry of the genes.

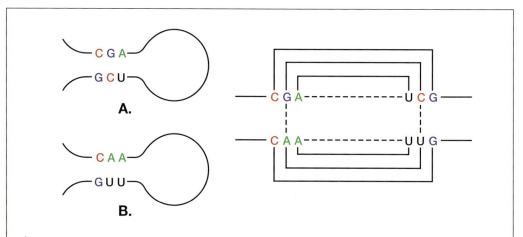

Figure 5.1. Complementary sequences in RNA molecules maintain RNA secondary structure. Shown is a simple stem-and-loop structure formed by the RNA strand folding back on itself. Molecule *A* depends on the presence of two complementary sequences CGA and UCG that are base-paired in the structure. In *B*, two sequence changes, G → A and C → U, which maintain the same structure, are present. Aligning RNA sequences required locating such regions of sequence covariation that are capable of maintaining base-pairing in the corresponding structure.

INTRODUCTION

As genomic sequences of organisms become available, it is important to be able to identify the various classes of genes, including the major class of genes that encodes RNA molecules. There are a large number of Web sites listed in Table 5.1 that provide programs

Table 5.1. *RNA databases and RNA analysis Web sites*

Site or resource	Web address	Reference
5S Ribosomal RNA data bank	http://rose.man.poznan.pl/5SData/ and mirrored at http://userpage.chemie.fu-berlin. de/fb_chemie/ibc/agerdmann/5S_rRNA.html	Szymanski et al. (1999)
5S rRNA database	http://www.bchs.uh.edu/~nzhou/temp/5snew.html	Shumyatsky and Reddy (1993)
Comparative RNA Web site	http://www.rna.icmb.utexas. edu/	see Web site
GenLang linguistic sequence analyzer	http://www.cbil.upenn.edu/	Dong and Searls (1994)
Gobase for mitochondrial sequences	http://alice.bch.umontreal.ca/genera/gobase/ gobase.html	Korab-Laskowska et al. (1998)

Site or resource	Web address	Reference
Intron analysis—*Saccharomyces cerevisiae*	http://www.cse.ucsc.edu/research/compbio/yeast_introns.html	Spingola et al. (1999)
tRNA genes, higher plant mitochondria	ftp://ftp.ebi.ac.uk/pub/databases/plmitrna/	Ceci et al. (1999)
MFOLD minimum energy RNA configuration	http://bioinfo.math.rpi.edu/~zukerm/rna/	Zuker et al. (1991)
Nucleic acid database and structure resource	http://ndbserver.rutgers.edu/	Berman et al. (1998)
Pseudobase–pseudoknot database maintained by E. van Batenburg, Leiden University	http://wwwbio.leidenuniv.nl/~batenburg/pkb.html	see Web page
Ribonuclease P database Web site	http://jwbrown.mbio.ncsu.edu/RNaseP/home.html	Brown (1999)
Ribosomal RNA database project (RDP II)	http://www.cme.msu.edu/RDP/	Maidak et al. (1999)
Ribosomal RNA mutation databases	http://www.fandm.edu/Departments/Biology/Databases/RNA.html	Triman and Adams (1997)
RiboWeb Project–3D models of *E. coli* 30S ribosomal subunit and 16s rRNA	http://www-smi.stanford.edu/projects/helix/ribo3dmodels/index.html	Chen et al. (1997)
RNA aptamer sequence database (University of Texas)	http://speak.icmb.utexas.edu/ellington/aptamers.html	see Web site
RNA editing Web site, UCLA	http://www.lifesci.ucla.edu/RNA/index.html	Simpson et al. (1998)
RNA editing, uridine insertion/deletion	http://www.lifesci.ucla.edu/RNA/trypanosome/	Simpson et al. (1998)
RNA modification database	http://medlib.med.utah.edu/RNAmods/	Limbach et al. (1994); Rozenski et al. (1999)
RNA secondary structures, Group I introns, 16S rRNA, 23S rRNA	http://www.rna.icmb.utexas.edu	Gutell (1994); Schnare et al. (1996 and references therein)
RNA structure database	http://www.rnabase.org/	see Web page
RNA world at IMB Jena	http://www.imb-jena.de/RNA.html	Sühnel (1997)
rRNA–Database of ribosomal subunit sequences	http://rrna.uia.ac.be/	De Rijk et al. (1992, 1999)
Signal recognition particle database	http://psyche.uthct.edu/dbs/SRPDB/SRPDB.html	Samuelsson and Zwieb (2000)
Small RNA database	http://mbcr.bcm.tmc.edu/smallRNA/smallrna.html	see Web page
snoRNA database for *S. cerevisiae*	http://rna.wustl.edu/snoRNAdb/	Lowe and Eddy (1999)
tmRNA[a] database	http://psyche.uthct.edu/dbs/tmRDB/tmRDB.html	Wower and Zwieb (1999)
tmRNA[a] Web site	http://www.indiana.edu/~tmrna/	Williams (1999)
tRNAscan-SE search server	http://www.genetics.wustl.edu/eddy/tRNAscan-SE/	Lowe and Eddy (1997)
tRNA and tRNA gene sequences	http://www.uni-bayreuth.de/departments/biochemie/sprinzl/trna/	Sprinzl et al. (1998)
u RNA database	http://psyche.uthct.edu/dbs/uRNADB/uRNADB.html	Zwieb (1997)
Vienna RNA package for RNA secondary structure prediction and comparison	http://www.tbi.univie.ac.at/~ivo/RNA/	Hofacker et al. (1998); Wuchty et al. (1999)
Viroid and viroid-like RNA sequences	http://www.callisto.si.usherb.ca/~jpperra	Lafontaine et al. (1999)

[a] tmRNA adds a carboxy-terminal peptide tag to the incomplete protein product from a broken mRNA molecule and thereby targets the protein for proteolysis.

A list of RNA Web sites and databases is available at http://bioinfo.math.rpi.edu/~zukerm/ and at http://pundit.colorado.edu:8080/.

and guest sites for RNA analysis or for access to databases of RNA molecules and sequences. These molecules perform a variety of important biochemical functions, including translation; RNA splicing, processing, and editing; and cellular localization. As with proteins, RNA-specifying genes may be identified by using the unknown gene as a query sequence for DNA sequence similarity searches, as described in Chapter 7. If a significant match to the sequence of an RNA molecule of known structure and function is found, then the query molecule should have a similar role. For some small molecules, the amount of sequence variation necessitates the use of more complex search methods, described later in this chapter.

RNA STRUCTURE PREDICTION BASICS

A computational method for predicting the most likely regions of base-pairing in an RNA molecule has been designed, just given the sequence, thus providing an ab initio prediction of secondary structure. From the many possible choices of complementary sequences that can potentially base-pair, the compatible sets that provide the most energetically stable molecules are chosen. Structures with energies almost as stable as the most stable one may also be produced, and regions whose predictions are the most reliable can be identified from such an analysis. Sequence variations found in related sequences may also be used to predict which base pairs are likely to be found in each of the molecules. One variation of RNA structure prediction methods will predict a set of sequences that are able to form a particular structure. Methods for predicting three-dimensional structures from sequence are also being developed (see http://bioinfo.math.rpi.edu/~zuker/rna/).

Another type of RNA secondary structure prediction method takes into account conserved patterns of base-pairing that are conserved during evolution of a given class of RNA molecules. Sequence positions that base-pair are found to vary at the same time during evolution of RNA molecules so that structural integrity is maintained. For example, if two positions G and C form a base pair in a given type of molecule, then sequences that have C and G reversed, or A and U or U and A at the corresponding positions, would be considered reasonable matches. These patterns of covariation in RNA molecules are a manifestation of secondary structure that lead to a structural prediction. The computational challenge is to discover these covariable positions against the background of other sequence changes.

FEATURES OF RNA SECONDARY STRUCTURE

Like protein secondary structure, RNA secondary structure can be conveniently viewed as an intermediate step in the formation of a three-dimensional structure. RNA secondary structure is composed primarily of double-stranded RNA regions formed by folding the single-stranded molecule back on itself. To produce such double-stranded regions, a run of bases downstream in the RNA sequence must be complementary to another upstream run so that Watson–Crick base-pairing between the complementary nucleotides G/C and A/U (analogous to the G/C and A/T base pairs in DNA) can occur. In addition, however, G/U wobble pairs may be produced in these double-stranded regions. As in DNA, the G/C base pairs contribute the greatest energetic stability to the molecule, with A/U base pairs contributing less stability than G/C, and G/U wobble base pairs contributing the least. From the RNA structures that have been solved, these base pairs and a number of addi-

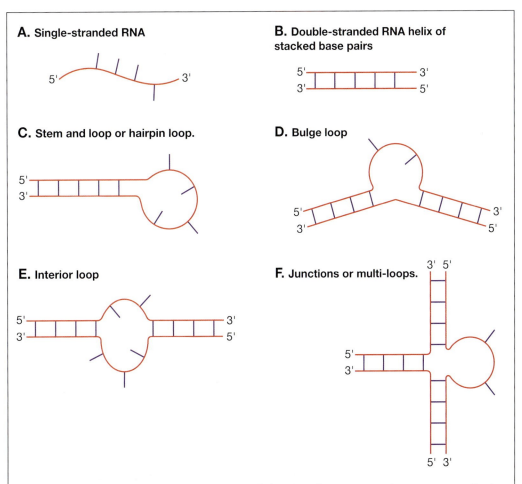

A. Single-stranded RNA

B. Double-stranded RNA helix of stacked base pairs

C. Stem and loop or hairpin loop.

D. Bulge loop

E. Interior loop

F. Junctions or multi-loops.

Figure 5.2. Types of single- and double-stranded regions in RNA secondary structures. Single-stranded RNA molecules fold back on themselves and produce double-stranded helices where complementary sequences are present. A particular base may either not be paired, as in *A*, or paired with another base, as in *B*. The double-stranded regions will most likely form where a series of bases in the sequence can pair with a complementary set elsewhere in the sequence. The stacking energy of the base pairs provides increased energetic stability. Combinations of double-stranded and single-stranded regions produce the types of structures shown in *C–F*, with the single-stranded regions destabilizing neighboring double-stranded regions. The loop of the stem and loop in *C* must generally be at least four bases long to avoid steric hindrance with base-pairing in the stem part of the structure. The stem and loop reverses the chemical direction of the RNA molecule. Interior loops, as in *D*, form when the bases in a double-stranded region cannot form base pairs, and may be asymmetric with a different number of base pairs on each side of the loop, as shown in *E*, or symmetric with the same number on each side. Junctions, as in *F*, may include two or more double-stranded regions converging to form a closed structure. The RNA backbone is red, and both unpaired and paired bases are blue. The types of loop structures can be represented mathematically, thereby aiding in the prediction of secondary structure (Sankoff et al. 1983; Zuker and Sankoff 1984). (Adapted from Burkhard et al. 1999b.)

tional ones (see Burkhard et al. 1999a,b) have been identified. RNA structure predictions comprise base-paired and non-base-paired regions in various types of loop and junction arrangements, as shown in Figure 5.2.

In addition to secondary structural interactions in RNA, there are also tertiary interactions, illustrated by the examples in Figure 5.3. These kinds of structures are not predictable by secondary structure prediction programs. They can be found by careful covariance analysis.

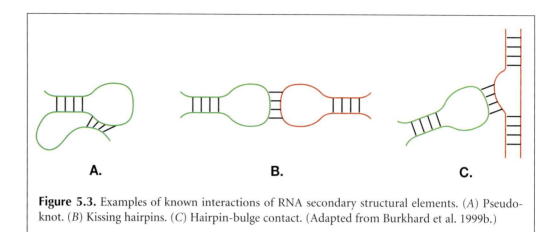

Figure 5.3. Examples of known interactions of RNA secondary structural elements. (*A*) Pseudo-knot. (*B*) Kissing hairpins. (*C*) Hairpin-bulge contact. (Adapted from Burkhard et al. 1999b.)

LIMITATIONS OF PREDICTION

In predicting RNA secondary structure, some simplifying assumptions are usually made. First, the most likely structure is similar to the energetically most stable structure. Second, the energy associated with any position in the structure is only influenced by local sequence and structure. Thus, the energy associated with a particular base pair in a double-stranded region is assumed to be influenced only by the previous base pair and not by the base pairs farther down the double-stranded region or anywhere else in the structure. These energies can be reliably estimated by experimentation with small, synthetic RNA oligonucleotides (Tinoco et al. 1971, 1973; Freier et al. 1986; Turner and Sugimoto 1988; SantaLucia 1998) recently improved to include sequence dependence (Mathews et al. 1999). They are most reliable when used for standard Watson–Crick base pairs and single G-U pairs surrounded

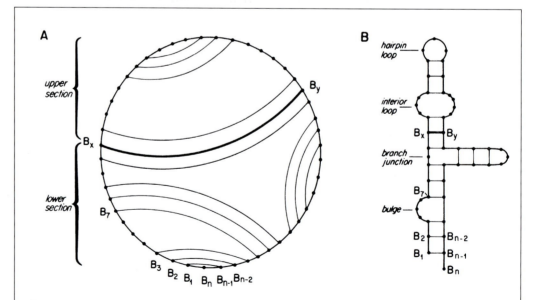

Figure 5.4. Display of base pairs in an RNA secondary structure by a circle plot. The predicted minimum free-energy structure shown in *B* is represented by a plot of the predicted base pairs as arcs connecting the bases in the sequence, which is drawn around the circumference of a circle, as shown in *A* (see Nussinov and Jacobson 1980). Note that none of the lines cross, a representation that the structure does not include any knots. (Reprinted from Nussinov and Jacobson 1980.)

by Watson–Crick pairs. Finally, the structure is assumed to be formed by folding of the chain back on itself in a manner that does not produce any knots. The best way of representing this requirement is to draw the sequence in a circular form. The paired bases are then joined by arcs. If the total structure with all predicted base pairs is to be free of knots, none of the arcs must cross (Fig. 5.4). Note, however, that if a pseudoknot (Fig. 5.3) is represented on such a diagram, the lines will cross.

DEVELOPMENT OF RNA PREDICTION METHODS

The development of methods for predicting RNA secondary structure has been reviewed by von Heijne (1987). Tinoco et al. (1971) first estimated the energy associated with regions of secondary structure by extrapolation from studies with small molecules and then attempted to predict which configurations of larger molecules were the most energetically stable. Energy estimates included the stabilizing energy associated with stacking base pairs in a double-stranded region and the destabilizing influence of regions that were not paired. Pipas and McMahon (1975) developed computer programs that listed all possible helical regions in tRNA sequences; using modified Watson–Crick base-pairing rules, they created all possible secondary structures by forming permutations of compatible helical regions, and evaluated each possible structure for total free energy. Studnicka et al. (1978) designed a method for adding compatible double-stranded regions together to produce the energetically most favorable structure. Martinez (1984) made a list of possible double-stranded regions, and these regions were then given weights in proportion to their equilibrium constants, calculated by the Boltzmann function [$\exp(-\Delta G/RT)$], where $-\Delta G$ is the free energy of the regions, R is the gas constant, and T is the temperature. The RNA molecule is folded by a Monte Carlo method in which one initial region is chosen at random from a weighted pool, similar to the method used in Gibbs sampling (see p. 177).

In the Monte Carlo method, a random drawing is made from a pool of all possible double-stranded regions, with the number of each type weighted in proportion to energetic stability.

Imagine each possible double-stranded region being represented by a marble in a bag. The number of each type of marble is weighted by the Boltzmann probability so that marbles corresponding to more energetically stable regions are more likely to be chosen. Additional compatible regions are then added sequentially by further selections from the weighted pool until no more can be added. This method generates a set of possible structures weighted by energy, but it does not take into account the destabilizing effect of unpaired regions. The Boltzmann probability function is used in more recent applications (described below) to find the most probable secondary structures (Hofacker et al. 1998; Wuchty et al. 1999).

Nussinov and Jacobson (1980) were the first to design a precise and efficient algorithm for predicting secondary structure. The algorithm generates two scoring matrices—one $M(i,j)$ to keep track of the maximum number of base pairs that can be formed in any interval i to j in the sequence and a second $K(i,j)$ to keep track of the base position k that is paired with j. From these matrices, a structure with the maximum possible number of base pairs could be deduced by a trace-back procedure similar to that used in performing sequence alignments by dynamic programming. Zuker and Stiegler (1981) used the dynamic programming algorithm and energy rules for producing the most energetically favorable structure. Their method assumes that the most energetic, and usually longest, predicted dsRNA regions are present in the molecule. Because many double-stranded regions are predictable for most RNA sequences, the number of predictions is reduced by including known biochemical or structural information to indicate which bases should be paired or not paired, by enforcing topological restraints and by requiring that the structure be in an energetically stable configuration.

MFOLD, written by Dr. Michael Zuker and colleagues, is commonly used to predict the energetically most stable structures of an RNA molecule (Jaeger et al. 1989, 1990; Zuker 1989, 1994). MFOLD provides a set of possible structures within a given energy range and provides an indication of their reliability. The program also uses covariance information from phylogenetically related sequences (Zuker et al. 1991). MFOLD includes methods for graphic display of the predicted molecules. This program is one of the most demanding on computer resources that is currently used because the algorithm is of N^3 complexity, where N is the sequence length. For each doubling of sequence length, the time taken to compute a structure increases eightfold. The program also requires a large amount of memory for storing intermediate calculations of structure energies in multiple scoring matrices. As a result, MFOLD is most often used to predict the structure of sequences less than 1000 nucleotides in length. This method is most reliable for small molecules and becomes less reliable as the length of the molecule increases.

MFOLD and many other types of useful information on RNA are found at the Web site of Dr. Michael Zuker, at http://bioinfo.math.rpi.edu/~zuker/rna/. Details of running MFOLD are not given here because the user manual for MFOLD is widely available (Jaeger et al. 1990). Recently, a new method called the partition function method for finding the most probable secondary structural configuration of an RNA molecule and the most probable base pairs has been reported by the Vienna RNA group (Wuchty et al. 1999) and is discussed below (p. 219).

One advance in the prediction of RNA structure has come from the recognition that certain RNA sequences form specific structures and that the presence of these sequences is strongly predictive of such a structure. For example, the hairpin CUUCGG occurs in different genetic contexts and forms a very stable structure (Tuerk et al. 1988). Databases of such RNA structures and RNA sequences can greatly assist in RNA structure prediction (Table 5.1).

The genetic algorithm (see Chapter 4, p. 157) has also been used to predict secondary structure (Shapiro and Navetta 1994); for aligning RNA sequences, taking into account both sequence and secondary structure and including pseudoknots (Notredame et al. 1997); and for simulation of RNA-folding pathways (Gultyaev et al. 1995). The program FOLDALIGN uses a dynamic programming algorithm to align RNAs based on sequence and secondary structure and locates the most significant motifs (Gorodkin et al. 1997). Chan et al. (1991) have described another algorithm for the same purpose, and Chetouani et al. (1997) have developed ESSA, a method for viewing and analyzing RNA secondary structure.

METHODS

SELF-COMPLEMENTARY REGIONS IN RNA SEQUENCES PREDICT SECONDARY STRUCTURE

One of the simplest types of analyses that can be performed to find stretches of sequence in RNA that are self-complementary is a dot matrix sequence comparison for self-complementary regions. For single-stranded RNA molecules, these repeats represent regions that can potentially self-hybridize to form RNA double strands (von Heijne 1987; Rice et al. 1991). All types of RNA secondary structure analysis begin by the identification of these regions, and, once identified, the compatible regions may be used to predict a minimum free-energy structure. A more advanced type of dot matrix can be used to show the most energetic parts of the molecule (see Fig. 5.8, below).

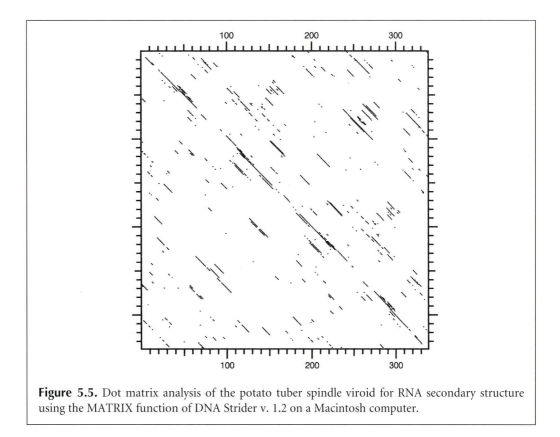

Figure 5.5. Dot matrix analysis of the potato tuber spindle viroid for RNA secondary structure using the MATRIX function of DNA Strider v. 1.2 on a Macintosh computer.

Self-complementary regions in RNA may be found by performing a dot matrix analysis with the sequence to be analyzed listed in both the horizontal and vertical axes. In one method for finding such regions, the sequence is listed in the 5′→3′ direction across the top of the page and the sequence of the complementary strand is listed down the side of the page, also in the 5′→3′ direction. The matrix is then scored for identities. Self-complementary regions appear as rows of dots going from upper left to lower right. For RNA, these regions represent sequences that can potentially form A/U and G/C base pairs. G/U base pairs will not usually be included in this simple type of analysis. As with matching DNA sequences, there are many random matches between the four bases in RNA, and the diagonals are difficult to visualize. A long window and a requirement for a large number of matches within this window are used to filter out these random matches.

An example of the RNA secondary structure analysis using a DNA matrix option of DNA Strider is shown in Figure 5.5. An analysis of the potato spindle tuber viroid is shown, using a window of 15 and a required match of 11. Note the appearance of a diagonal running from the center of the matrix to the upper left, and a mirror image of this diagonal running to the lower right. The presence of this diagonal indicates the occurrence of a large self-complementary sequence such that the entire molecule can potentially fold into a hairpin structure. An alternative dot matrix method for finding RNA secondary structure is to list the given RNA sequence across the top of the page and also down the side of the page and then to score matches of complementary bases (G/C, A/U, and G /U). Diagonals indicating complementary regions will go from upper right to lower left in this type of matrix. This is the kind of matrix used to produce an energy matrix (see Fig. 5.8, below).

MINIMUM FREE-ENERGY METHOD FOR RNA SECONDARY STRUCTURE PREDICTION

To predict RNA secondary structure, every base is first compared to every other base by a type of analysis very similar to the dot matrix analysis. The sequence is listed across the top and down the side of the page, and G/C, A/U, and G/U base pairs are scored (for an example using a dot matrix method to find hairpins, see Fig. 5.5). Just as a diagonal in a two-sequence comparison indicates a range of sequence similarity, a row of matches in the RNA matrix indicates a succession of complementary nucleotides that can potentially form a double-stranded region. The energy of each predicted structure is estimated by the nearest-neighbor rule by summing the negative base-stacking energies for each pair of bases in double-stranded regions and by adding the estimated positive energies of destabilizing regions such as loops at the end of hairpins, bulges within hairpins, internal bulges, and other unpaired regions. Representative examples of the energy values that are currently used are given in Table 5.2. To evaluate all the different possible configurations and to find the most energetically favorable, several types of scoring matrices are used. The complementary regions are evaluated by a dynamic programming algorithm to predict the most energetically stable molecule. The method is similar to the dynamic programming method used for sequence alignment (see Chapter 3).

To calculate the stacking energy of a row of base pairs in the molecule, the stacking energies similar to those shown in Table 5.2 are used. An illustrative example for evaluation of energy in a double-stranded region is shown in Figure 5.6. The sequence is listed down the side of the matrix, and a portion of the same sequence is also listed across the top of the matrix; matching base pairs have been identified within the matrix. The object is to find a diagonal row of matches that goes from upper right to lower left, and such a row is shown in the example. In Figure 5.6, a match of four complementary bases in a row produces a molecule of free energy −6.4 kcal/mole. In general, each matrix value is obtained by considering the minimum energy values obtained by all previous complementary pairs

Table 5.2. *Predicted free-energy values (kcal/mole at 37°C) for base pairs and other features of predicted RNA secondary structures*

| | **A. Stacking energies for base pairs** | | | | | |
	A/U	**C/G**	**G/C**	**U/A**	**G/U**	**U/G**
A/U	−0.9	−1.8	−2.3	−1.1	−1.1	−0.8
C/G	−1.7	−2.9	−3.4	−2.3	−2.1	−1.4
G/C	−2.1	−2.0	−2.9	−1.8	−1.9	−1.2
U/A	−0.9	−1.7	−2.1	−0.9	−1.0	−0.5
G/U	−0.5	−1.2	−1.4	−0.8	−0.4	−0.2
U/G	−1.0	−1.9	−2.1	−1.1	−1.5	−0.4

| | **B. Destabilizing energies for loops** | | | | |
Number of bases	**1**	**5**	**10**	**20**	**30**
Internal	–	5.3	6.6	7.0	7.4
Bulge	3.9	4.8	5.5	6.3	6.7
Hairpin	–	4.4	5.3	6.1	6.5

(*Upper*) Stacking energy in double-stranded region when base pair listed in left column is followed by base pair listed in top row. C/G followed by U/A is therefore the dinucleotide 5′ CU 3′ paired to 5′ AG 3′. (*Lower*) Destabilizing energies associated with loops. Hairpin loops occur at the end of a double-stranded region, internal loops are unpaired regions flanked by paired regions, and a bulge loop is a bulge of one strand in an otherwise paired region (Fig. 5.2). An updated and more detailed list of energy parameters may be found at the Web site of M. Zuker (http://bioinfo.math.rpi.edu/~zuker/rna/energy/).
From Turner and Sugimoto (1988); Serra and Turner (1995).

A. Base comparisons

5'	A	C	G	U	3'
A					
C					
G					
U					
–					
–					
G		C/G		U/G	
C			G/C		
G		C/G		U/G	
U	A/U	C/U	G/U		

3'

B. Free energy calculations

5'	A	C	G	U	3'
A					
C					
G					
U					
–					
–					
G				−6.4	
C			−5.2		
G		−1.8			
U					

3'

Figure 5.6. Evaluation of secondary structure in RNA sequence by the method described in the text. The sequence is listed down the first column of *A* and *B* in the 5'→3' orientation, and the first four bases of the sequence are also listed in the first row of the tables in the 5'→3' direction. Several complementary base pairs between the first and last four bases that could lead to secondary structure are shown in *A*. The most 5' base is listed first in each pair. The diagonal set of base pairs A/U, C/G, G/C, and U/G reveals the presence of a potential double-stranded region between the first and last four bases. The free energy associated with such a row of base pairs is shown in *B*. A C/G base pair following an A/U base pair has a base stacking energy of −1.8 kcal/mole (Turner and Sugimoto 1988). This value is placed in the corresponding position in *B*. Similarly, a C/G base pair followed by a G/C provides energy of −3.4, and a G/C followed by a U/G, −1.2 kcal/mole. Hence, the energy accumulated after stacking of these additional two base pairs is −5.2 and −6.4. The energy of this double-stranded structure will continue to decrease (become more stable) as more base pairs are added, but will be increased if the structure is interrupted by noncomplementary base pairs.

decreased by the stacking energy of any additional complementary base pairs or increased by the destabilizing energy associated with noncomplementary bases. The increase depends on the type and length of loop that is introduced by the noncomplementary base pair, whether internal loop, bulge loop, or hairpin loop, as shown in Table 5.2. This comparison of all possible matches and energy values is continued until all nucleotides have been compared. The pattern followed in comparing bases within the RNA molecule is illustrated in Figure 5.7.

SUBOPTIMAL STRUCTURE PREDICTIONS BY MFOLD AND THE USE OF ENERGY PLOTS

Originally, the FOLD program of M. Zuker predicted only one structure having the minimum free energy. However, changes in a single nucleotide can result in drastic changes in the predicted structure. A later version, called MFOLD, has improved prediction of non-base-paired interactions and predicts several structures having energies close to the minimum free energy. These predictions accurately reflect structures of related RNA molecules derived from comparative sequence analysis (Jaeger et al. 1989; Zuker 1989, 1994; Zuker et al. 1991; Zuker and Jacobson 1995). To find these suboptimal structures, the dynamic programming method was modified (Zuker 1989, 1991) to evaluate parts of a new scoring matrix in which the

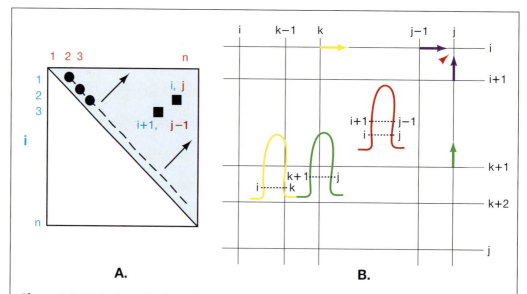

A.   **B.**

Figure 5.7. Method used in dynamic programming analysis for identifying the most energetically favorable configuration of a linear RNA molecule. (*A*) The sequence of an RNA molecule of length *n* bases is listed across the top of the page and down the side. The index of the sequence across the top is *j* and that down the side is *i*. The search only includes the upper right part of the matrix shown in gray and begins at the first diagonal line for matching base pairs. First positions $i = 1$ and $j = 2$ are compared for potential base-pairing, and if pairing can occur, an energy value is placed in an energy matrix *W* at position 1,2. Then, $i = 2$ and $j = 3$ base are compared, and so on, until all base combinations along the dashed diagonal have been made. Then, comparisons are made along the next upper right diagonal. As each pair of bases is compared, an energy calculation is made that is the optimal one up to that point in the comparison. In the simplest case, if $i + 1$ pairs with $j - 1$, and *i* pairs with *j*, and if this structure is the most favorable up to that point, the energy of the *i/j* base pair will be added to that of the $i + 1/j - 1$ base pair. Other cases are illustrated in *B*. The process of obtaining the most stable energy value at each matrix position is repeated following the direction of the arrows until the last position, $i = 1$ and $j = n$, has been compared and the energy value placed at this position in matrix *W*, the value entered in $W(1,n)$, will be the energy of the most energetically stable structure. The structure is then found by a trace-back procedure through the matrices similar to that used for sequence alignments. The method used is a combination of a search for all possible double-stranded regions and an energy calculation based on energy values similar to those in Table 5.2. The search for the most energetic structure uses an algorithm (Zuker and Stiegler 1981) similar to that for finding the structure with maximum base-pairing (Nussinov and Jacobson 1980). These authors recognized that there are three possible ways, illustrated here by the colored arrows, of choosing the best energy value at position *i,j* in an energy matrix *W*. The simplest calculation (*red arrow*) is to use the energy value found up to position $i - 1$, $j - 1$ diagonally below *i,j*. If *i* and *j* can form a base pair (and if there are at least four bases between them in order to allow enough sequence for a hairpin) and $i + 1$ and $j - 1$ also pair, then the stacking energy of *i/j* upon $i + 1/j - 1$ will reduce the energy value at $i + 1$, $j - 1$, producing a more stable structure, and the new value can be considered a candidate for the energy value entered at position *i,j*. If *i* and *j* do not pair, then another choice for the energy at *i,j* is to use the values at positions *i*, $j - 1$ or $i + 1$, *j* illustrated by the blue arrows. *i* and *j* then become parts of loop structures. Finally, *i* and *j* may each be paired with two other bases, *i* with *k* and *j* with $k + 1$, where *k* is between *i* and *j* ($i < k < j$), illustrated by the structure shown in yellow and green, reflecting the location of the paired bases. The minimum free-energy value for all values of *k* must be considered to locate the best choice as a candidate value at *i,j*. Finally, of the three possible choices for the minimum free-energy value at *i,j* indicated by the four colored arrows, the best energy value is placed at position $W(i,j)$. The procedure is repeated for all values of *i* and *j*, as illustrated in *A*. Besides the main energy scoring matrix *W*, additional scoring matrices are used to keep track of auxiliary information such as the best energy up to *i,j* where *i* and *j* form a pair, and the influence of bulge loops, interior loops, and other destabilizing energies. An essential second matrix is $V(i,j)$, which keeps track of all substructures in the interval *i,j* in which *i* forms a base pair with *j*. Some values in the *W* matrix are derived from values in the *V* matrix and vice versa (Zuker and Stiegler 1981).

sequence is represented in two tandem copies on both the vertical and horizontal axes. The regions from $i = 1$ to n and $j = 1$ to n are used to calculate an energy $V(i,j)$ for the best structure that includes an i,j base pair and is called the included region. A second region, the excluded region, is used to calculate the energy of the best structure that includes i,j but is not derived from the structure at $i+1, j-1$ (Fig. 5.7). After certain corrections are made, the difference between the included and excluded values is the most energetic structure that includes the base pair i,j. All complementary base pairs can be sampled in this fashion to determine which are present in a suboptimal structure that is within a certain range of the optimal one.

An energy dot plot is produced showing the locations of alternative base pairs that produce the most stable or suboptimally stable structures, as illustrated in Figure 5.8. The program may be instructed to find structures within a certain percentage of the minimum free energy. Parameter d provides a measure of similarity between two structures. When MFOLD is established on a suitable local host machine, the window is interactive, and clicking a part of the display will lead to program output of the corresponding structure. The dot plot may be filtered so that only suboptimal regions with helices of a certain minimal length are shown. One of the predicted structures is shown in Figure 5.9.

Reliability of Secondary Structure Prediction

Three scores, Pnum (i), Hnum (i,j), and Ssum, have been derived to assist with a determination of the reliability of a secondary structure prediction for a particular base i or a base pair i,j. Pnum(i) is the total number of energy dots regardless of color in the ith row and ith column of the energy dot plot, and represents in an unfiltered dot plot the number of base pairs that the ith base can form with all other base pairs in structures within the defined energy range. The lower this value, the more well defined or "well determined" the local structure because there are few competitive foldings. Hnum(i,j) is the sum of Pnum(i) and Pnum(j) less 1 and is the total number of dots in the ith row and jth column and represents the total number of base pairs with the ith or jth base in the predicted structures. The Hnum for a double-stranded region is the average Hnum value for the base pairs in that helix. The lower this number, the more well determined the double-stranded region. In an analysis of tRNAs, 5S RNAs, ribosomal RNAs, and other published secondary structure models based on sequence variation (Jaeger et al. 1990; Zuker and Jacobson 1995), these methods correctly predict about 70% of the double-stranded regions. Snum, also called ss-count, is the number of foldings in which base i is single-stranded divided by m, the number of foldings, and gives the probability that base i is single-stranded. If Snum is approximately 1, then base i is probably in a single-stranded region, and if Snum is approximately 0, then base i is probably not in such a region. This reliability information has been used to annotate output files of MFOLD and other RNA display programs (Zuker and Jacobsen 1998). Plots of these values against sequence position are given by the MFOLD program and the Zuker Web site.

OTHER ALGORITHMS FOR SUBOPTIMAL FOLDING OF RNA MOLECULES

A limitation of the Zuker method and other methods (Nakaya et al. 1995) for computing suboptimal RNA structures is that they do not compute all the structures within a given energy range of the minimum free-energy structure. For example, no alternative structures

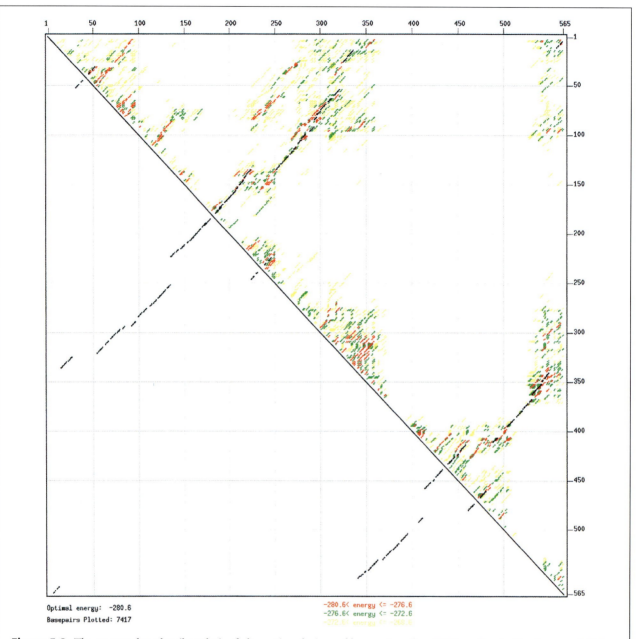

Figure 5.8. The energy dot plot (boxplot) of alternative choices of base pairs of an RNA molecule (Jacobson and Zuker 1993). The sequence is that of a human adenovirus pre-terminal protein (GenBank U52533) that is given by M. Zuker as an example on his Web site at http://bioinfo.math.rpi.edu/~zukerm. Foldings were computed using the default parameters of the MFOLD program at http://bioinfo.math.edu/~mfold/rna/form1.cgi (Mathews et al. 1999) using the thermodynamic values of SantaLucia (1998). The minimum energy of the molecule is −280.6 kcal/mole and the maximum energy increment is 12 kcal/mole. Black dots indicate base pairs in the minimum free-energy structure and are shown both above and the mirror image below the main diagonal. Red, blue, and yellow dots are base pairs in foldings of increasing 4, 8, and 12 kcal/mole energies greater than the minimum energy, respectively. A region with very few alternative base pairs such as the pairing of 370–395 with 530–505 is considered to be strongly predictive, whereas regions with many alternative base pairs such as the base-pairing in the region of 340–370 with 570–530 are much less predictive.

are produced that have the absence of base pairs in the best structure, and, if two sub-structures are joined by a stretch of unpaired bases, no structures are produced that are suboptimal for both structures. These factors limit the number of alternative structures predicted compared to known variations based on sequence variations in tRNAs (Wuchty et al. 1999).

These limitations have been largely overcome by using an algorithm originally described by Waterman and Byers (1985) for finding sequence alignments within a certain range of the optimal one by modifications of the trace-back procedure used in dynamic programming. This method efficiently calculates a large number of alternative structures, up to a very large number, within a given energy range of the minimum free-energy structure (see Fig. 5.10). The method has been used to demonstrate that natural tRNA sequences can form many alternative structures which are close to the minimum free-energy structure and that base modification plays a major role in this energetic stability (Wuchty et al. 1999). The method may also be used to assess the thermodynamic stability of RNA structures given expected changes in energies associated with base pairs and loops as a function of temperature. The RNA secondary structure prediction and comparison Web site at http://www.tbi.univie.ac.at/~ivo/RNA/ will fold molecules of length > 300 bases, and the Vienna RNA Package software for folding larger molecules on a local machine is available from this site.

PREDICTION OF MOST PROBABLE RNA SECONDARY STRUCTURE

In the above types of analyses, the energy associated with predicted double-stranded regions in RNA is used to produce a secondary structure. Stabilizing energies associated with base-paired regions and destabilizing energies associated with loops are summed to produce the most stable structure or suboptimal RNA secondary structure. A different way of predicting the structures is to consider the probability that each base-paired region will form based on principles of thermodynamics and statistical mechanics. The probability of forming a region with free energy ΔG is expressed by the Boltzmann distribution, which states that the likelihood of finding a structure with free energy $-\Delta G$ is proportional to [exp $(-\Delta G/kT)$] where k is the Boltzmann gas constant and T is the absolute temperature.

The Boltzmann constant k is 8.314510 J/mole/degree K.

Note that the more stable a structure, the lower the value of ΔG. Since ΔG is a negative number, the value of $\exp(-\Delta G/kT)$ increases for more stable structures and also grows exponentially with a decrease in energy. The probability of these regions forming increases in the same manner. Conversely, the effect of destabilizing loops that have a positive ΔG is to decrease the probability of formation. By using these probability calculations and a dynamic programming method similar to that used in MFOLD, it is possible to predict the most probable RNA secondary structures and to assess the probability of the base pairs that contribute energetic stability to this structure.

For a set of possible structural states, the likelihood of each may be calculated using this formula, and the sum of these likelihoods provides a partition function that can be used to normalize each individual likelihood, providing a probability that each will occur. Thus, probability of structure A of energy $-\Delta G_a$ is [exp $(-\Delta G_a/kT)$] divided by the partition function Q, where $Q = \Sigma_s$ [exp $(-\Delta G_s/kT)$], the sum of probabilities of all possible struc-

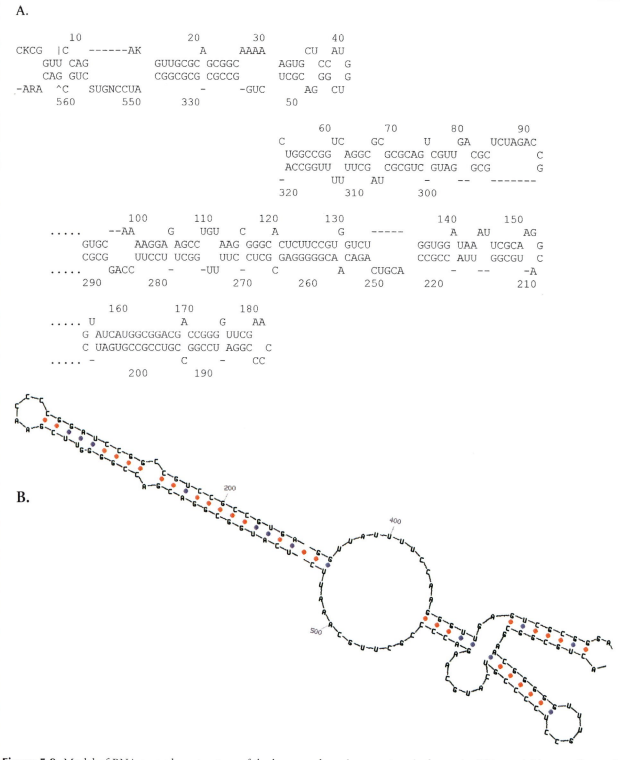

Figure 5.9. Model of RNA secondary structure of the human adenovirus pre-terminal protein. This model is one of several alternative structures represented by the above energy plot and provided as an output by the current versions of MFOLD. (*A*) Simple text representation of one of the predicted structures. Each stem-and-loop structure is shown separately and the left end of each structure is placed below the point of connection to the one above. (*B*) More detailed rendition of one part of the predicted structures. The structure continues beyond the right side of the page.

tures, s. This kind of analysis allows one to calculate the probability of a certain base pair forming.

The key to this analysis is the calculation of the partition function Q. A dynamic programming algorithm for calculating this function exactly for RNA secondary structure has been developed (McCaskill 1990). The algorithm is very similar to that used for computing an optimal folding by MFOLD. Complexity similarly increases as the cube of the sequence length, and the energy values used for base pairs and loops are also the same except that structures with very large interior loops are ignored. Just as the minimum free-energy value is given at $W(1,n)$ in the Zuker MFOLD algorithm, the value of the partition function is given at matrix position $Q(1,n)$ in the corresponding partition matrix.

As indicated above, the partition function is calculated as the sum of the probabilities of each possible secondary structure. Because there are a very large possible number of structures, the calculation is simplified by calculating an auxiliary function, $Q^b(i,j)$, which is the sum of the probabilities of all structures that include the base pair i,j. The partition function $Q(i,j)$ includes both these structures and the additional ones where i is not paired with j. An example illustrating the difference between the minimum free energy and the partition function methods should be instructive. Suppose that the bases at positions $i+1, j-1$ and i,j can both form base pairs. They then form a stack of two base pairs. In the minimum free-energy method, the energy of the i,j pair stacked on the $i+1, j-1$ pair will be added to $V(i+1, j-1)$ to give $V(i,j)$, where V is a scoring matrix that keeps track of the best structure that includes an i,j base pair. In contrast, the value for $Q^b(i,j)$ will be calculated by multiplying the matrix value $Q^b(i+1, j-1)$ by the probability of the base pair i,j given by the Boltzmann probability [exp $(-\Delta G/kT)$], where ΔG is the negative stacking energy of the i,j base pair on the $i+1, j-1$ base pair, and will be a large number reflecting the probability given the stability of the base-paired region.

For a hairpin structure with a row of successive base pairs, the probability will be the product of the Boltzmann factors associated with the stacked pair, giving a high number for the relative likelihood of formation. The procedure followed by the partition function algorithm is to calculate $Q^b(i,j)$ and $Q(i,j)$ iteratively in a scoring matrix similar to that illustrated in Figure 5.7A until $Q(1,n)$ is reached. This matrix position contains the value of the full partition function Q.

Both the partition function and the probabilities of all base pairs are computed by this algorithm, and the most probable structural model is thereby found. Information about intermediate structures, base-pair opening and slippage, and the temperature dependence of the partition function may also be determined. The latter calculation provides information about the melting behavior of the secondary structure.

A suite of RNA-folding programs available from the Vienna RNA secondary structure prediction Web site (http://www.tbi.univie.ac.at/~ivo/RNA/) uses this methodology to predict the most probable and alternative RNA secondary structures. An example of the folding of a 300-base RNA molecule is given in Figure 5.10. The probability of forming each base pair is shown in a dot matrix display in which the dots are squares of increasing size reflecting the probability of the base pair formed by the bases in the horizontal and vertical positions of the matrix. Secondary structure prediction is done by two kinds of dynamic programming algorithms: the minimum free-energy algorithm of Zuker and Stiegler (1981) and the partition function algorithm of McCaskill (1990).

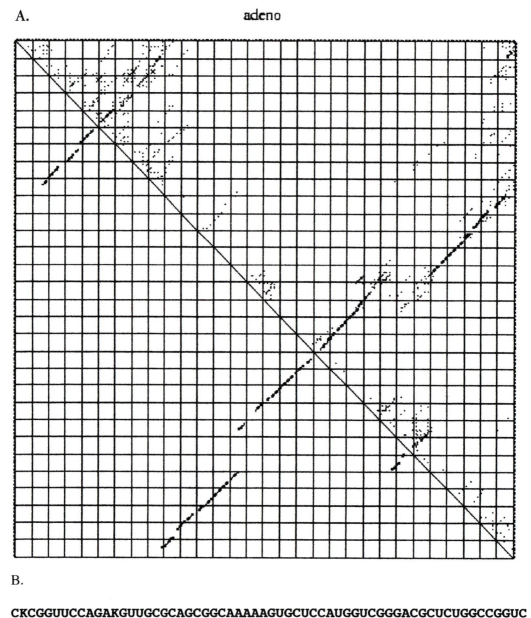

B.

CKCGGUUCCAGAKGUUGCGCAGCGGCAAAAAGUGCUCCAUGGUCGGGACGCUCUGGCCGGUC
AGGCGCGCGCAGUCGUUGACGCUCUAGACCGUGCAAAAGGAGAGCCUGUAAGCGGGCACUCU
UCCGUGGUCUGGUGGAUAAAUUCGCAAGGGUAUCAUGGCGGACGACCGGGGUUCGAACCCCG
GAUCCGGCCGUCCGCCGUGAUCCAUGCGGUUACCGCCCGCGUGUCGAACCCAGGUGUGCGAC
GUCAGACAACGGGGGAGCGCUCCUUUUGGCUUCCUUCCAGGCGCGGCGGAUG

```
. . . . . . . . . . . . . . . . . . ( ( ( ( ( ( ( ( ( ( . . . . ( ( ( ( . ( ( . ( ( ( ( ( ( ( ( . . . . ) ) ) ) ) ) )
. . . . ) ) . ) ) ) ) ) . . . ) ) ) ) ) ) ) ) . ) ) ) . . . . ( . ( ( ( ( ( ( . . ( ( ( ( ( . ( ( ( ( . . . ( ( ( . ( ( (
( . ( ( ( ( ( ( ( ( ( . ( ( ( ( ( ( ( ( ( . . . . . . ( ( ( ( ( . . ( ( . ( ( ( ( ( ( ( ( ( ( ( ( ( . ( ( ( ( ( . . (
( ( . . . . . ) ) ) . ) ) ) ) ) . ) ) ) ) ) ) ) ) ) ) ) ) ) ) . ) ) ) ) ) . . . . . . . ) ) ) . ( ( ( ( ( . ( ( .
. . ) ) . . . ) ) ) ) ) . ) ) ) ) ) . ) ) ) ) ) ) ) ) ) . ) ) ) ) ) ) . . ) ) ) ) ) ) ) ) ) ) . . . . ) ) ) ) ) )
. . . . .
```

Figure 5.10. Suboptimal foldings of an RNA sequence using probability distributions of base-pairings. The first 300 bases of the same adenovirus sequence used in Fig. 5.8 was submitted to the Vienna Web server. (*A*) The region shown represents structures within the range of bases 150–300 and may be compared to the same region in Fig. 5.8. The minimum free energy of this thermodynamic ensemble is −134.85 kcal/mole, compared to a minimum free energy of 125.46 kcal/mole. The size of the square box at highlighted matrix positions indicates the probability of the base pair and decreases in steps of 10-fold; i.e., order of magnitude decreases. The size variations shown in the diagram cover a range of ~4–6 orders of magnitude. Calculations of base-pair probabilities are discussed in the text. (*B*) The minimum free-energy structure representing base pairs as pairs of nested parentheses. A low-resolution picture was also produced (not shown).

USING SEQUENCE COVARIATION TO PREDICT STRUCTURE

The second major method that has been used to make RNA secondary structure predictions (Woese et al. 1983) and also tertiary structure analyses such as those shown in Figure 5.3 (Gutell et al. 1986) is RNA sequence covariation analysis. This method examines sequences of the same RNA molecules from different species for positions that vary together in a manner that would allow them to produce a base pair in all of the molecules. The idea is quite simple. On the one hand, for double-stranded regions in RNA molecules, sequence changes that take place in evolution should maintain the base-pairing. On the other hand, sequence changes in loops and single-stranded regions should not have such a constraint. The method of analysis is to look for sequence positions at which covariation maintains the base-pairing properties. The justification for this method is that these types of joint substitutions or covariations actually are found to occur during evolution of such genes. As shown in Figure 5.11, when one position corresponding to a base pair is changed, another position corresponding to the base-pairing partner will also change. For example, if two positions G and C form a base pair, then sequences that have C and G reversed, or A and T or T and A at the corresponding positions, would also be considered reasonable matches. Sequence covariability has been used to improve thermodynamic structure prediction as described in the above section (Hofacker et al. 1998). An example of using covariation analysis to decipher base-pair interactions in tRNA is shown in Figure 5.12.

One method of covariation analysis also examines which phylogenetic groups exhibit change at a given position. For each position, the base that generally predominates in one particular part of the tree is determined. These methods have required manual examination of sequences and structures for covariation, but automatic methods have also been devised and demonstrated to produce reliable predictions (Winker et al. 1990; Han and Kim 1993; see box below).

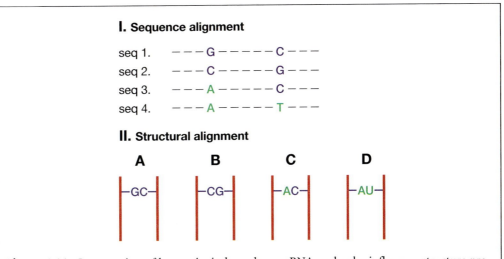

Figure 5.11. Conservation of base pairs in homologous RNA molecules influences structure prediction. The predicted structure takes into account sequence covariation found at aligned sequence positions, and may also use information about conserved positions in components of a phylogenetic tree. In the example shown, sequence covariations in A, B, and D found in sequences 1, 2, and 4, respectively, permit Watson–Crick base and G-U base-pairing in the corresponding structure, but variation C found in sequence 3 is not compatible. Sometimes correlations will be found that suggest other types of base interactions, or the occurrence of a common gap in a multiple sequence alignment may be considered a match. Positions with greater covariation are given greater weight in structure prediction. Molecules with only one of the two sequence changes necessary for conservation of the base-paired position may be functionally deleterious.

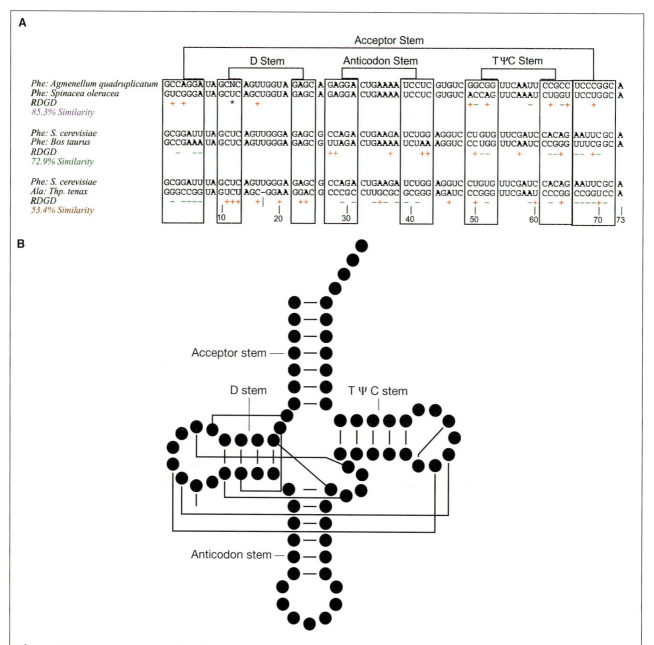

Figure 5.12. Covariation found in tRNA sequences reveals base interactions in tRNA secondary and tertiary structure. (*A*) Alignment of tRNA sequences showing regions of interacting base pairs. (+) Transition; (−) transversions; (|) deletion; (*) ambiguous nucleotide. (*B*) Diagram of tRNA structure illustrating base–base interactions revealed by a covariance analysis. Adapted from the Web site of R. Gutell at http://www.rna.icmb.utexas.edu.

Methods of Covariation Analysis in RNA Sequences

Secondary and tertiary features of RNA structure may be determined by analyzing a group of related sequences for covariation. Two sequence positions that covary in a manner that frequently maintains base-pairing between them provides evidence that the bases interact in the structure. Combinations of the following methods have been used to locate such covarying sites in RNA sequences (see R. Gutell for additional details and at http://www.rna.icmb.utexas.edu/METHODS/menu.html).

1. Optimally align pairs of sequence to locate conserved primary sequence, mark transitions and transversions from a reference sequence, and then visually examine these changes to identify complementary patterns that represent potential secondary structure.

2. Perform a multiple sequence alignment, highlight differences using one of the sequences as a reference, and visually examine for complementary patterns.

3. Mark variable columns in the multiple sequence alignment by numbers that mark changes (e.g., transitions or transversions) from a reference sequence; examine marked columns for a similar or identical number pattern that can represent potential secondary structure.

4. Perform a statistical analysis (Chi-square test) of the number of observations of a particular base pair in columns i and j of the multiple sequence alignment, compared to the expected number based on the frequencies of the two bases.

5. Calculate the mutual information score (mixy) for each pair of columns in the alignment, as described in the text and illustrated in Figure 5.13.

6. Score the number of changes in each pair of columns in the alignment divided by the total number of changes (the ec score), examine the phylogenetic context of these changes to determine the number of times the changes have occurred during evolution, and choose the highest scores that are representative of multiple changes.

7. Measure the covariance of each pair of positions in the alignment by counting the numbers of all 16 possible base-pair combinations and dividing by the expected number of each combination (number of sequence $\times$ frequency of base in first position $\times$ frequency of base in second position), choose the most prevalent pair, and examine remaining combinations for additional covariation; then sum frequency of all independently covarying sites to obtain covary score.

Mutual Information Content

A method used to locate covariant positions in a multiple sequence alignment is the mutual information content of two columns. First, for each column in the alignment, the frequency of each base is calculated. Thus, the frequencies in column m, $f_m(B_1)$, are $f_m(A)$, $f_m(U)$, $f_m(G)$, and $f_m(C)$ and those for column n, $f_n(B_2)$, are $f_n(A)$, $f_n(U)$, $f_n(G)$, and $f_n(C)$. Second, the 16 joint frequencies of two nucleotides, $f_{m,n}(B_1,B_2)$ one base B_1 in column m and the same or another base B_2 in column n are calculated. If the base frequencies in any two columns are independent of each other, then the ratio of $f_{m,n}(B_1,B_2)$ / $[f_m(B_1) \times f_n(B_2)]$ is expected to equal 1, and if the frequencies are correlated, then this ratio will be greater than 1. If they are perfectly covariant, then $f_{m,n}(B_1,B_2) = f_m(B_1) = f_n(B_2)$. To calculate the mutual information content H (m,n) in bits between the two columns m and n, the logarithm of this ratio is calculated and summed over all possible 16 base-pair combinations.

$$H(m,n) = \Sigma_{B1,B2}\, f_{m,n}(B_1,B_2) \times \log_2 \{f_{m,n}(B_1,B_2) / [f_m(B_1)\, f_n(B_2)]\}$$

$H(m,n)$ varies from the value of 0 bits of mutual information representing no correlation to that of 2 bits of mutual information, representing perfect correlation (Eddy and Durbin 1994).

The mutual information content may be plotted on a motif logo (Gorodkin et al. 1997), similar to that described in Chapter 4, page 196, for illustrating a sequence motif. The example shown in Figure 5.13 shows the mutual information content M superimposed on the information content of each sequence position in an RNA alignment.

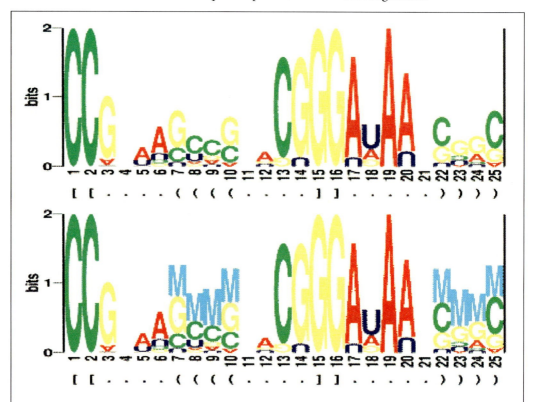

Figure 5.13. RNA structure logo. The top panel is the normal sequence logo showing the size of each base in proportion to the contribution of that base to the amount of information in that column of the multiple sequence alignment. The relative entropy method is used in which the frequency of bases in each column is compared to the background frequency of each base. Inverted sequence characters indicate a less than background frequency (see Chapter 4, page 196). The bottom panel includes the same information plus the mutual information content in pairs of columns. The amount of information is indicated by the letter M, and the matching columns are shown by nested sets of brackets and parentheses. All sequences have a C in column 1 and a matching G in column 16. Similar columns 2 and 15 can form a second base pair stacked upon the first. Columns 7–10 and 25–22 also can form G/C base pairs most of the time. Sequences with a G in column 7 frequently have a C in column 25, and those with a C in column 7 may have a G in column 25. Thus, there is mutual information in these two columns (Gorodkin et al. 1997 [using data of Tuerk and Gold 1990]).

A formal covariance model has been devised by Eddy and Durbin (1994). Although very accurate when used for identifying tRNA genes, the algorithm is extremely slow and unsuitable for searching through large genomes. Instead, the method has been used to screen through putative tRNA genes previously identified by faster methods (Lowe and Eddy 1997). The difficulty that is faced in modeling RNA molecules is to identify the potential base pairs in a set of related RNA molecules based on covariation at two sites. Recall from Chapter 4 that the hidden Markov model is used for capturing the types of variations observed in a sequence profile, including matches, mismatches, insertions, and deletions. This type of model assumes each sequence can be predicted by a series of states in the model, one after the other, as in a series of independent events in a Markov chain. The hidden Markov model does not analyze joint variations at sequence positions such as occur in RNA molecules. The model that is used for analyzing RNA secondary structure (but not tertiary structure) is an ordered tree model. A simplified tree representation of RNA secondary structure is shown in Figure 5.14.

The above assumes that we know which bases are paired in a model of RNA secondary structure, whereas the goal is to build a model that discovers this information. The task is achieved by constructing a more general model, training the model with a set of sequences,

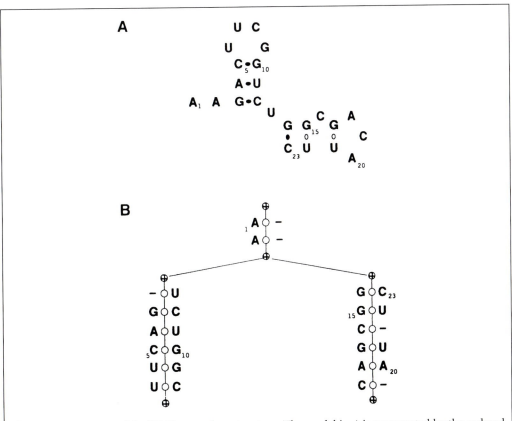

Figure 5.14. Tree model of RNA secondary structure. The model in *A* is represented by the ordered binary tree shown in *B*. This model attempts to capture both the sequence and the secondary structure of the RNA molecule. The tree is read like a sequence starting at the root node at the top of the model, then moving down the main branch to the bifurcation mode. Along the main trunk are nodes that represent matched or unmatched base pairs. Shown are two A's matching a "-," indicating no pairing with these bases. After the bifurcation mode, one then moves down the most leftward branch to the end node. Along the branch are unmatched bases, matched base pairs, and mismatched pairs. After the end node is reached, go back to the previous bifurcation node and follow the right branch. (Reprinted, with permission of Oxford University Press, from Eddy and Durbin 1994.)

and then having the model reveal the most likely base-paired regions. The approach is similar to training a hidden Markov model for proteins to recognize a family of protein sequences, thereby producing the most probable multiple sequence alignment. In the case of RNA secondary structure, a tree model is trained by the sequences, and the model may then be used to predict the most probable secondary structure. In addition, the model may also be used to search a database for sequences that produce a high score when aligned to the model. These sequences are likely to encode a similar type of RNA molecule such as tRNA or 5S RNA. Each model is derived by training a more general tree model with the sequences.

The general tree model needs to represent the types of variations that are found in aligning a series of related sequences, such as insertions, deletions, and mismatches. To allow for such variations, each node in the tree is replaced by a set of states that correspond to all of the possible sequence variations that might be encountered at that position. These states are illustrated in Figure 5.15.

The mutual information content of all sequence positions is used in designing the model, and the expectation maximization method (Chapter 4) is used to optimize the parameters of the model. A dynamic programming method is used to find a model that maximizes the amount of covariation. The structure of the model may subsequently be altered during training. Once a covariance model suitable for an RNA molecule has been established, the model is trained by the sequences. The methodology is similar to that of hidden Markov models and is described in detail in Chapter 4. Basically, the model is initialized by giving starting values to the base and dinucleotide frequencies in each MATCH and INS state and to the transition probabilities. All possible paths through the model are found for each sequence in the training set. The frequencies and transition probabilities are modified each time a particular path in the model is used. The base pairs are found from MATP (see Fig. 5.15), which gives probabilities to the 16 possible dinucleotides.

Once the model has been trained, the most probable path for each sequence provides a consensus structural alignment of the sequences. A dynamic programming algorithm is used that matches subsequence alignments to the nodes of the covariance model. The result is a log odds score of the sequence matching the covariance model. A similar method may be used to find sequences in a genomic database with high matching scores to the covariance model. The method was used to predict the structural alignment of representative sets of tRNA sequences, and it provided alignments that closely matched actual structural alignments based on other methods. The software for the COVELS program is available by request from the authors (Eddy and Durbin 1994).

STOCHASTIC CONTEXT-FREE GRAMMARS FOR MODELING RNA SECONDARY STRUCTURE

In the above section, we discussed the need to have models for RNA secondary structure that reflect the interaction among base pairs. Simpler models of sequence variation treat sequences as simple strings of characters without such interactions and are therefore not suitable for RNA. A general theory for modeling strings of symbols, such as bases in DNA sequences, has been developed by linguists. There is a hierarchy of these so-called transformational grammars that deal with situations of increasing complexity. The application of these grammars to sequence analysis has been extensively discussed elsewhere (Durbin et al. 1998). The context-free grammar is suitable for finding groups of symbols in different parts of the input sequence that thus are not in the same context. Complementary regions in sequences, such as those in RNA that will form secondary structures, are an

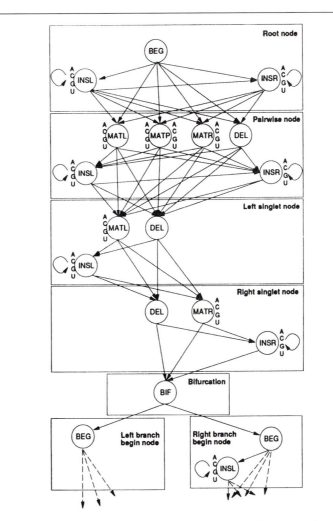

Figure 5.15. Details of tree model for RNA secondary structure. Each type of node in the tree shown in Fig. 5.14 is replaced by a pattern of states corresponding to the types of sequence variations that are expected in a family of related RNA sequences. These states each store a table of frequencies of 4 bases or of 16 possible dinucleotides. The seven different types of nodes are illustrated. BEG node includes insert states for sequence of any length on the right or left side of the node. The pair-wise node includes a state MATP for storing the 16 possible dinucleotide frequencies; MATL and MATR states for storing single base frequencies on either the left or right side of the node, respectively; a DEL state for allowing deletions; and INSL and INSR states that allow for insertions of any length on the left or right of the node. DEL does not store information. The other five node types have the same types of states. Each state is joined to other states by a set of transition probabilities shown by the arrows. These probabilities are similar to those used in hidden Markov models. BIF is a bifurcation state with transition probabilities entering the state from above and then leaving to one or the other of two branches. (Reprinted, with permission of Oxford University Press, from Eddy and Durbin 1994.)

example of such context-free sequences. Stochastic context-free grammars (SCFG) introduce uncertainty into the definition of such regions, allowing them to use alternative symbols as found in the evolution of RNA molecules. Thus, SCFGs can help define both the types of base interactions in specific classes of RNA molecules and the sequence variations at those positions. SCFGs have been used to model tRNA secondary structure (Sakakibara et al. 1994). Although SCFGs are computationally complex (Durbin et al. 1998), they are likely to play an important future role in identifying specific types of RNA molecules.

The application of SCFGs to RNA secondary structure analysis is very similar in form to the probabilistic covariance models described in the above section. For RNA, the symbols of the alphabet are A, C, G, and U. The context-free grammar establishes a set of rules called productions for generating the sequence from the alphabet, in this case an RNA molecule with sections that can base-pair and others that cannot base-pair. In addition to the sequence symbols (named terminal symbols because they end up in the sequence), another set of symbols (nonterminal symbols) designated S_0, S_1, S_2 . . . , determines intermediate production stages. The initial symbol is S_0 by convention. The next terminal symbol S_1 is produced by modifying S_0 in some fashion by productions indicated by an arrow. For example, the productions $S_0 \rightarrow S_1$, $S_1 \rightarrow C\ S_2\ G$ generate the sequence $C\ S_2\ G$ where S_2 has to be defined further by additional productions. The example shown in Figure 5.16 (from Sakakibara et al. 1994) shows a set of productions for generating the sequence CAUCAGGGAAGAUCUCUUG and also the secondary structure of this molecule. The productions chosen describe both features.

In this example of a context-free grammar, only one sequence is produced at each production level. In a SCFG, each production of a nonterminal symbol has an associated probability for giving rise to the resulting product, and there are a set of productions, each giving a different result. For example, the production $S_1 \rightarrow C\ S_2\ G$ could also be represented by 15 other base-pair combinations, and each of these has a corresponding probability. Thus, each production can be considered to be represented by a probability distribution over the possible outcomes. Note the identity of the SCFG representation of the predicted structure to that shown for the tree representation of the covariance model in Figure 5.14. The use of SCFGs in RNA secondary structure production analysis is in fact very similar to that of the covariance model, with the grammatical productions resembling the nodes in the ordered binary tree. As with hidden Markov models, the probability distribution of each production must be derived by training with known sequences. The algorithms used for training the SCFG and for aligning a sequence with the SCFG are somewhat different from those used with hidden Markov models, and the time and memory requirements are greater (Sakakibara et al. 1994: Durbin et al 1998).

SEARCHING GENOMES FOR RNA-SPECIFYING GENES

One goal in RNA research has been to design methods to identify sequences in genomes that encode small RNA molecules. Larger, highly conserved molecules can simply be identified based on their sequence similarity with already-known sequences. For smaller sequences with more sequence variation, this method does not work. A number of methods for finding small RNA genes have been described and are available on the Web (Table 5.1). A major problem with these methods in searches of large genomes is that a small false-positive rate becomes quite unacceptable because there are so many false positives to check out.

One of the first methods used to find tRNA genes was to search for sequences that are self-complementary and can fold into a hairpin like the three found in tRNAs (Staden 1980).

Figure 5.16. A set of transformation rules for generating an RNA sequence and the secondary structure of the sequence from the RNA alphabet (ACGU). (*A*) The set of production rules for producing the sequence and the secondary structure. These rules reveal which bases are paired and which are not paired. (*B*) Derivation of the sequence. (*C*) A parse tree showing another method for displaying the derivation of the sequence in *B*. (*D*) Secondary structure from applying the rules. (Redrawn, with permission of Oxford University Press, from Sakakibara et al. 1994.)

A. Productions

$P = \{\; S_0 \rightarrow S_1,$ $\quad S_7 \rightarrow G\,S_8,$

$\quad S_1 \rightarrow C\,S_2\,G,$ $\quad S_8 \rightarrow G,$

$\quad S_2 \rightarrow A\,S_3\,U,$ $\quad S_9 \rightarrow A\,S_{10}\,U,$

$\quad S_3 \rightarrow S_4 \;\; S_9,$ $\quad S_{10} \rightarrow G\,S_{11}\,C,$

$\quad S_4 \rightarrow U\,S_5\,A,$ $\quad S_{11} \rightarrow A\,S_{12}\,U,$

$\quad S_5 \rightarrow C\,S_6\,G,$ $\quad S_{12} \rightarrow U\,S_{13},$

$\quad S_6 \rightarrow A\,S_7,$ $\quad S_{13} \rightarrow C \qquad \}$

B. Derivation

$S_0 \;\rightarrow\; S_1 \;\rightarrow\; CS_2G \;\rightarrow\; CAS_3UG \;\rightarrow\; CAS_4S_9UG$

$\rightarrow\; CAUS_5AS_9UG \;\rightarrow\; CAUCS_6GAS_9UG$

$\rightarrow\; CAUCAS_7GAS_9UG \;\rightarrow\; CAUCAGS_8GAS_9UG$

$\rightarrow\; CAUCAGGGAS_9UG \;\rightarrow\; CAUCAGGGAAS_{10}UUG$

$\rightarrow\; CAUCAGGGAAGS_{11}CUUG$

$\rightarrow\; CAUCAGGGAAGAS_{12}UCUUG$

$\rightarrow\; CAUCAGGGAAGAUS_{13}UCUUG$

$\rightarrow\; CAUCAGGGAAGAUCUCUUG.$

C. Parse tree

C A U C A G G G A A G A U C U C U U G

D. Secondary structure

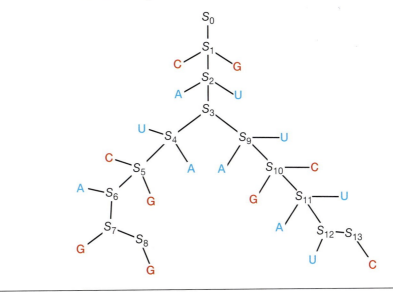

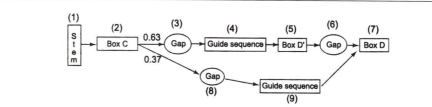

Figure 5.17. Probabilistic model of snoRNAs. The numbered boxes and ovals represent conserved sequence and structural features that have been modeled by training on snoRNAs. Secondary structural features of Stem were modeled with an SCFG. Boxes with ungapped hidden Markov models, the guide sequence with a hidden Markov model, and gapped regions (spacers) are shown by ovals. The guide sequence interacts with methylation sites on rRNA and is targeted in each search to a complementary sequence near one of those sites. The alignment of this model produces a log odds score that provides an indication of the reliability of the match. The transition probabilities are 1, except where the model bifurcates to allow identification of two types of target sequences. The model is highly specific and seldom identifies incorrect matches in random sequences. (Reprinted, with permission, from Lowe and Eddy 1999 [copyright AAAS, Washington, D.C.].)

Fichant and Burks (1991) described a program, tRNAscan, that searches a genomic sequence with a sliding window searching simultaneously for matches to a set of invariant bases and conserved self-complementary regions in tRNAs with an accuracy of 97.5%. Pavesi et al. (1994) derived a method for finding the RNA polymerase III transcriptional control regions of tRNA genes using a scoring matrix derived from known control regions that is also very accurate. Finally, Lowe and Eddy (1997) have devised a search algorithm tRNAscan-SE that uses a combination of three methods to find tRNA genes in genomic sequences—tRNAscan, the Pavesi algorithm, and the COVELS program based on sequence covariance analysis (Eddy and Durbin 1994). This method is reportedly 99–100% accurate with an extremely low rate of false positives.

The probabilistic model shown in Figure 5.17 was used to identify small nucleolar (sno) RNAs in the yeast genome that methylate ribosomal RNA. The model is not used to search genomic sequences directly. Instead, a list of candidate sequences is first found by searching for patterns that match the sequences in the model (Lowe and Eddy 1999). The probability model was a hybrid combination of HMMs and SCFGs trained on snoRNAs. These RNAs vary sufficiently in sequence and structure that they are not found by straightforward similarity searches. The RNAs found were shown to be snoRNAs by insertional mutagenesis.

APPLICATIONS OF RNA STRUCTURE MODELING

In summary, methods for predicting the structure of RNA molecules include (1) an analysis of all possible combinations of potential double-stranded regions by energy minimization methods and (2) identification of base covariation that maintains secondary and tertiary structure of an RNA molecule during evolution. Energy minimization methods have been so well refined that a series of energetically feasible models and the most thermodynamically probable structural models may be computed. Covariation analysis by C. Woese led to his building of detailed structural models for rRNAs. By examining the evolutionary variation in these structures, he was able to predict three domains of life—the Bacteria, the Eukarya, and a newly identified Archaea. Although a large amount of horizontal transfer among evolutionary lineages of other genes has added a great deal of noise to the evolutionary signal, the rRNA-based prediction is supported by other types of genomic analyses. In addition to these uses of rRNA structural analysis, excellent probabilistic models of two small RNA molecules, tRNA and snoRNA, have been built, and these models may be used to search reliably through genomic sequences for genes that encode these RNA molecules. The successful analysis of these types of RNA molecules should be readily extensible to other classes of RNA molecules.

REFERENCES

Berman H.M., Zardecki C., and Westbrook J. 1998. The nucleic acid database: A resource for nucleic acid science. *Acta Crystallogr. D Biol. Crystallogr.* **54:** 1095–1104.

Brown J.W. 1999. The ribonuclease P database. *Nucleic Acids Res.* **27:** 314.

Burkhard M.E., Turner D.H., and Tinoco I., Jr. 1999a. The interactions that shape RNA secondary structure. In *The RNA world*, 2nd edition (ed. R.F. Gesteland et al.), pp. 233–264. Cold Spring Harbor Laboratory Press, Cold Spring Harbor, New York.

———. 1999b. Appendix 2: Schematic diagrams of secondary and tertiary structure elements. In *The RNA world*, 2nd edition (ed. R.F. Gesteland et al.), pp. 681–685. Cold Spring Harbor Laboratory Press, Cold Spring Harbor, New York.

Ceci L.R., Volpicella M., Liuni S., Volpetti V., Licciulli F., and Gallerani R. 1999. PLMItRNA, a database for higher plant mitochondrial tRNAs and tRNA genes. *Nucleic Acids Res.* **27:** 156–157.

Chan L., Zuker M., and Jacobson A.B. 1991. A computer method for finding common base paired helices in aligned sequences: Application to the analysis of random sequences. *Nucleic Acids Res.* **19:** 353–358.

Chen R.O., Felciano R., and Altman R.B. 1997. RIBOWEB: Linking structural computations to a knowledge base of published experimental data. *Ismb* **5:** 84–87.

Chetouani F., Monestié P., Thébault P., Gaspin C., and Michot B. 1997. ESSA: An integrated and interactive computer tool for analysing RNA secondary structure. *Nucleic Acids Res.* **25:** 3514–3522.

De Rijk P., Neefs J.M., Van de Peer Y., and De Wachter R. 1992. Compilation of small ribosomal subunit RNA sequences. *Nucleic Acids Res.* **20:** 2075–2089.

De Rijk P., Robbrecht E., de Hoog S., Caers A., Van de Peer Y., and De Wachter R. 1999. Database on the structure of large subunit ribosomal RNA. *Nucleic Acids Res.* **27:** 174–178.

Dong S. and Searls D.B. 1994. Gene structure prediction by linguistic methods. *Genomics* **23:** 540–551.

Durbin R., Eddy S., Krogh A., and Mitchison G., Eds. 1998. *Biological sequence analysis. Probabilistic models of proteins and nucleic acids*, chapters 9 and 10. Cambridge University Press, Cambridge, United Kingdom.

Eddy S. and Durbin R. 1994. RNA sequence analysis using covariance models. *Nucleic Acids Res.* **22:** 2079–2088.

Fichant G.A. and Burks C. 1991. Identifying potential tRNA genes in genomic DNA sequences. *J. Mol. Biol.* **220:** 659–671.

Freier S.M., Kierzek R., Jaeger J.A., Sugimoto N., Caruthers M.H., Neilson T., and Turner D.H. 1986. Improved free-energy parameters for predictions of RNA duplex stability. *Proc. Natl. Acad. Sci.* **83:** 9373–9377.

Gorodkin J., Heyer L.J., Brunak S., and Stormo G.D. 1997. Displaying the information contents of structural RNA alignments: The structure logos. *Comput. Appl. Biosci.* **13:** 583–586.

Gultyaev A.P., van Batenburg F.H., and Pleij C.W. 1995. The computer simulation of RNA folding pathways using a genetic algorithm. *J. Mol. Biol.* **250:** 37–51.

Gutell R.R. 1994. Collection of small subunit (16S- and 16S-like) ribosomal RNA structures *Nucleic Acids Res.* **22:** 3502–3507.

Gutell R.R., Noller H.F., and Woese C.R. 1986. Higher order structure in ribosomal RNA. *EMBO J.* **5:** 1111–1113.

Han K. and Kim H.-J. 1993. Prediction of common folding structures of homologous RNAs. *Nucleic Acids Res.* **21:** 1251–1257.

Hofacker I.L., Fekete M., Flamm C., Huynen M.A., Rauscher S., Stolorz P.E., and Stadler P.F. 1998. Automatic detection of conserved RNA structure elements in complete RNA virus genomes. *Nucleic Acids Res.* **26:** 3825–3836.

Jacobson A.B. and Zuker M. 1993. Structural analysis by energy dot plot of a large mRNA. *J. Mol. Biol.* **233:** 261–269.

Jaeger J.A., Turner D.H., and Zuker M. 1989. Improved predictions of secondary structures for RNA. *Proc. Natl. Acad. Sci.* **86:** 7706–7710.

———. 1990. Predicting optimal and suboptimal secondary structure for RNA. *Methods Enzymol.* **183:** 281–306.

Korab-Laskowska M., Rioux P., Brossard N., Littlejohn T.G., Gray M.W., Lang B.F., and Burger G. 1998. The organelle genome database project (GOBASE). *Nucleic Acids Res.* **26:** 138–144.

Lafontaine D.A., Deschenes P., Bussiere F., Poisson V., and Perreault J.P. 1999. The viroid and viroid-like RNA database. *Nucleic Acids Res.* **27:** 186–187.

Limbach P.A., Crain P.F., and McCloskey J.A. 1994. Summary: The modified nucleosides of RNA. *Nucleic Acids Res.* **22:** 2183–2196.

Lowe T.M. and Eddy S.R. 1997. tRNAscan-SE: A program for improved detection of transfer RNA genes in genomic sequence. *Nucleic Acids Res.* **25:** 955–964.

———. 1999. A computational screen for methylation guide snoRNAs in yeast. *Science* **283:** 1168–1171.

Maidak B.L., Cole J.R., Parker C.T., Jr., Garrity G.M., Larsen N., Li B., Lilburn T.G., McCaughey M.J., Olsen G.J., Overbeek R., Pramanik S., Schmidt T.M., Tiedje J.M., and Woese C.R. 1999. A new version of the RDP (ribosomal database project). *Nucleic Acids Res.* **27:** 171–173.

Martinez H.M. 1984. An RNA folding rule. *Nucleic Acids Res.* **12:** 323–334.

Mathews D.H., Sabina J., Zuker M., and Turner D.H. 1999. Expanded sequence dependence of thermodynamic parameters provides robust prediction of RNA secondary structure. *J. Mol. Biol.* **288:** 911–940.

McCaskill J.S. 1990. The equilibrium partition function and base pair binding probabilities for RNA secondary structure. *Biopolymers* **29:** 1105–1119.

Nakaya A., Yamamoto K., and Yonezawa A. 1995. RNA secondary structure prediction using highly parallel computers. *Comput. Appl. Biosci.* **11:** 685–692.

Notredame C., O'Brien E.A., and Higgins D.G. 1997. RAGA: RNA sequence alignment by genetic algorithm. *Nucleic Acids Res.* **25:** 4570–4580.

Nussinov R. and Jacobson A.B. 1980. Fast algorithm for predicting the secondary structure of single-stranded RNA. *Proc. Natl. Acad. Sci.* **77:** 6903–6913.

Pavesi A., Conterio F., Bolchi A., Dieci G., and Ottonello S. 1994. Identification of new eukaryotic tRNA genes in genomic DNA databases by a multistep weight matrix analysis of transcriptional control regions. *Nucleic Acids Res.* **122:** 1247–1256.

Pipas J.M. and McMahon J.E. 1975. Method for predicting RNA secondary structure. *Proc. Natl. Acad. Sci.* **72:** 2017–2021.

Rice P.M., Elliston K., and Gribskov M. 1991. DNA. In *Sequence analysis primer* (ed. M. Gribskov and J. Devereux), pp. 51–57. Stockton Press, New York.

Rozenski J., Crain P.F., and McCloskey J.A. 1999. The RNA modification database: 1999 update. *Nucleic Acids Res.* **27:** 196–197.

Sakakibara Y., Brown M., Hughey R., Mian I.S., Sjölander K., Underwood R.C., and Haussler D. 1994. Stochastic context-free grammars for tRNA modeling. *Nucleic Acids Res.* **22:** 5112–5120.

Samuelsson T. and Zwieb C. 2000. SRPDB (signal recognition particle database). *Nucleic Acids Res.* **28:** 171–172.

Sankoff D., Kruskal J.B., Mainville S., and Cedergren R.J. 1983. Fast algorithms to determine RNA secondary structures containing multiple loops. In *Time warps, string edits, and macromolecules: The theory and practice of sequence comparison* (ed. D. Sankoff and J.B. Kruskal), chap. 3, pp. 93–120. Addison-Wesley, Reading, Massachusetts.

SantaLucia J., Jr. 1998. A unified view of polymer, dumbbell, and oligonucleotide DNA nearest-neighbor thermodynamics. *Proc. Natl. Acad. Sci.* **95:** 1460–1465.

Schnare M.N., Damberger S.H., Gray M.W., and Gutell R.R. 1996. Comprehensive comparison of structural characteristics in eukaryotic cytoplasmic large subunit (23 S-like) ribosomal RNA. *J. Mol. Biol.* **256:** 701–719.

Serra M.J. and Turner D.H. 1995. Predicting thermodynamic properties of RNA. *Methods Enzymol.* **259:** 242–261.

Shapiro B.A. and Navetta J. 1994. A massively parallel genetic algorithm for RNA secondary structure prediction. *J. Supercomput.* **8:** 195–207.

Shumyatsky G. and Reddy R. 1993. Compilation of small RNA sequences. *Nucleic Acids Res.* **21:** 3017.

Simpson L., Wang S.H., Thiemann O.H., Alfonzo J.D., Maslov D.A., and Avila H.A. 1998. U-insertion/deletion edited sequence database. *Nucleic Acids Res.* **26:** 170–176.

Souza A.E. and Göringer H.U. 1998. The guide RNA database. *Nucleic Acids Res.* **26:** 168–169.

Spingola M., Grate L., Haussler D., and Ares M., Jr. 1999. Genome-wide bioinformatic and molecular analysis of introns of *Saccharomyces cerevisiae*. *RNA* **5:** 221–234.

Sprinzl M., Horn C., Brown M., Ioudovitch A., and Steinberg S. 1998. Compilation of tRNA sequences and sequences of tRNA genes. *Nucleic Acids Res.* **26:** 148–153.

Staden R. 1980. A computer program to search for tRNA genes. *Nucleic Acids Res.* **8:** 817–825.

Studnicka G.M., Rahn G.M., Cummings I.W., and Salser W.A. 1978. Computer method for predicting the secondary structure of single-stranded RNA. *Nucleic Acids Res.* **5:** 3365–3387.

Sühnel J. 1997. Views of RNA on the world wide web. *Trends Genet.* **13:** 206–207.

Szymanski M., Barciszewska M.Z., Barciszewski J., and Erdmann V.A. 1999. 5S ribosomal RNA Data Bank. *Nucleic Acids Res.* **27:** 158–160.

Tinoco I., Jr., Uhlenbeck O.C., and Levine M.D. 1971. Estimation of secondary structure in ribonucleic acids. *Nature* **230:** 362–367.

Tinoco I., Jr., Borer P.N., Dengler B., Levine M.D., Uhlenbeck O.C., Crothers D.M., and Gralla J. 1973. Improved estimation of secondary structure in ribonucleic acids. *Nat. New Biol.* **246:** 40–41.

Triman K.L. and Adams B.J. 1997. Expansion of the 16S and 23S ribosomal RNA mutation databases (16SMDB and 23SMDB). *Nucleic Acids Res.* **25:** 188–191.

Tuerk C. and Gold L. 1990. Systematic evolution of ligands by exponential enrichment: RNA ligands to bacteriophage T4 DNA polymerase. *Science* **249:** 505–510.

Tuerk C., Gauss P., Thermes C., Groebe D.R., Gayle M., Guild N., Stormo G., d'Aubenton-Carafa Y., Uhlenbeck O.C., Tinoco I., Jr., et al. 1988. CUUCGG hairpins: Extraordinarily stable RNA secondary structures associated with various biochemical processes. *Proc. Natl. Acad. Sci.* **85:** 1364–1368.

Turner D.H. and Sugimoto N. 1988. RNA structure prediction. *Annu. Rev. Biophys. Biophys. Chem.* **17:** 167–192.

von Heijne G. 1987. *Sequence analysis in molecular biology — Treasure trove or trivial pursuit*, pp. 58–72. Academic Press, San Diego, California.

Waterman M.S. and Byers T.H. 1985. A dynamic programming algorithm to find all solutions in a neighborhood of the optimum. *Math. Biosci.* **77:** 179–188.

Williams K.P. 1999. The tmRNA website. *Nucleic Acids Res.* **27:** 165–166.

Winker R., Overbeek R., Woese C., Olsen G.J., and Pfluger N. 1990. Structure detection through automated covariance search. *Comput. Appl. Biosci.* **6:** 365–371.

Woese C.R., Gutell R., Gupta R., and Noller H.F. 1983. Detailed analysis of the higher-order structure of 16S-like ribosomal ribonucleic acids. *Microbiol. Rev.* **47:** 621–669.

Wower J. and Zwieb C. 1999. The tmRNA database (tmRDB). *Nucleic Acids Res.* **27:** 167.

Wuchty S., Fontana W., Hofacker I.L., and Schuster P. 1999. Complete suboptimal folding of RNA and the stability of secondary structures. *Biopolymers* **49:** 145–165.

Zuker M. 1989. On finding all suboptimal foldings of an RNA molecule. *Science* **244:** 48–52.

———. 1991. Suboptimal sequence alignment in molecular biology. Alignment with error analysis. *J. Mol. Biol.* **221:** 403–420.

———. 1994. Predicting optimal and suboptimal secondary structure for RNA. *Methods Mol. Biol.* **25:** 267–294.

Zuker M. and Jacobson A.B. 1995. "Well-determined" regions in RNA secondary structure prediction: Analysis of small subunit ribosomal RNA. *Nucleic Acids Res.* **23:** 2791–2798.

———. 1998. Using reliability information to annotate RNA secondary structures. *RNA* **4:** 669–679.

Zuker M. and Sankoff D. 1984. RNA secondary structures and their prediction. *Bull. Math. Biol.* **46:** 591–621.

Zuker M. and Stiegler P. 1981. Optimal computer folding of large RNA sequences using thermodynamics and auxiliary information. *Nucleic Acids Res.* **9:** 133–148.

Zuker M., Jaeger J.A., and Turner D.H. 1991. A comparison of optimal and suboptimal RNA secondary structures predicted by free energy minimization with structures determined by phylogenetic comparison. *Nucleic Acids Res.* **19:** 2707–2714.

Zwieb C. 1997. The uRNA database. *Nucleic Acids Res.* **25:** 102–103.

Phylogenetic Prediction

INTRODUCTION

A PHYLOGENETIC ANALYSIS OF A FAMILY of related nucleic acid or protein sequences is a determination of how the family might have been derived during evolution. The evolutionary relationships among the sequences are depicted by placing the sequences as outer branches on a tree. The branching relationships on the inner part of the tree then reflect the degree to which different sequences are related. Two sequences that are very much alike will be located as neighboring outside branches and will be joined to a common branch beneath them. The object of phylogenetic analysis is to discover all of the branching relationships in the tree and the branch lengths.

Phylogenetic analysis of nucleic acid and protein sequences is presently and will continue to be an important area of sequence analysis. In addition to analyzing changes that have occurred in the evolution of different organisms, the evolution of a family of sequences may be studied. On the basis of the analysis, sequences that are the most closely related can be identified by their occupying neighboring branches on a tree. When a gene family is found in an organism or group of organisms, phylogenetic relationships among the genes can help to predict which ones might have an equivalent function. These functional predictions can then be tested by genetic experiments. Phylogenetic analysis may also be used to follow the changes occurring in a rapidly changing species, such as a virus. Analysis of the types of changes within a population can reveal, for example, whether or not a particular gene is under selection (McDonald and Kreitman 1991; Comeron and Kreitman 1998; Nielsen and Yang 1998), an important source of information in applications like epidemiology.

Procedures for phylogenetic analysis are strongly linked to those for sequence alignment discussed in Chapters 3 and 4, and similar difficulties are encountered. Just as two very similar sequences can be easily aligned even by eye, a group of sequences that are very similar but with a small level of variation throughout can easily be organized into a tree. Conversely, as sequences become more and more different through evolutionary change, they can be much more difficult to align. A phylogenetic analysis of very different sequences is also difficult to do because there are so many possible evolutionary paths that could have been followed to produce the observed sequence variation. Because of the complexity of this problem, considerable expertise is required for difficult situations.

Phylogenetic analysis programs are widely available at little or no cost. A comprehensive list will not be given here since one has been published previously (Swofford et al. 1996). The main ones in use are PHYLIP (phylogenetic inference package) (Felsenstein 1989 1996) available from Dr. J. Felsenstein at http://evolution.genetics.washington.edu/ phylip.html and PAUP (phylogenetic analysis using parsimony) available from Sinauer Associates, Sunderland, Massachusetts, http://www.lms.si.edu/PAUP/. Current versions of these programs provide the three main methods for phylogenetic analysis—parsimony, distance, and maximum likelihood methods (described below)—and also include many types of evolutionary models for sequence variation. Examples using these programs are given later in the chapter. Each program requires a particular type of input sequence format that is described below and in Chapter 2. Another program, MacClade, is useful for detailed analysis of the predictions made by PHYLIP, PAUP, and other phylogenetic programs and is also available from Sinauer (also see http://phylogeny.arizona.edu/macclade/ macclade.html). MacClade, as the name suggests, runs on a Macintosh computer. PHYLIP and PAUP run on practically any machine, but the user interface for PAUP has been most developed for use on the Macintosh computer.

There are also several Web sites that provide information on phylogenetic relationships among organisms (Table 6.1). There are several excellent descriptions of phylogenetic

Table 6.1. *Phylogenetic relationships among organisms*

Site name	Address	Description	Reference
Entrez	http://www3.ncbi.nlm.nih.gov/ Taxonomy/taxonomyhome.html	taxonomically related structures or group of organisms	see Web page
RDP (Ribosomal database project)	http://www.cme.msu.edu/RDP/	ribosomal RNA-derived trees	Maidak et al. (1999)
Tree of life	http://phylogeny.arizona.edu/tree/ phylogeny.html	information about phylogeny and biodiversity	Maddison and Maddison (1992)

analysis in which the methods are covered in considerable depth (Li and Graur 1991; Miyamoto and Cracraft 1991; Felsenstein 1996; Li and Gu 1996; Saitou 1996; Swofford et al. 1996; Li 1997).

RELATIONSHIP OF PHYLOGENETIC ANALYSIS TO SEQUENCE ALIGNMENT

When the sequences of two nucleic acid or protein molecules found in two different organisms are similar, they are likely to have been derived from a common ancestor sequence. Chapter 3 discusses sequence alignment methods used to determine sequence similarity. Chapter 4 discusses multiple sequence alignment methods that need to be applied to a set of related sequences before a phylogenetic analysis can be performed. Chapter 7 describes methods for searching through a database of sequences to locate sequences that are similar to a query sequence. A sequence alignment reveals which positions in the sequences were conserved and which diverged from a common ancestor sequence, as illustrated in Figure 6.1. When one is quite certain that two sequences share an evolutionary relationship, the sequences are referred to as being homologous.

The commonest method of multiple sequence alignment (the progressive alignment method, p. 152) first aligns the most closely related pair of sequences and then sequentially adds more distantly related sequences or sets of sequences to this initial alignment (see flowchart, p. 144). The alignment so obtained is influenced by the most alike sequences in the group and thus may not represent a reliable history of the evolutionary changes that have occurred. Other methods of multiple sequence alignment attempt to circumvent the influence of alike sequences (see Chapter 4, p. 157). Once a multiple sequence alignment has been obtained, each column is assumed to correspond to an individual site that has

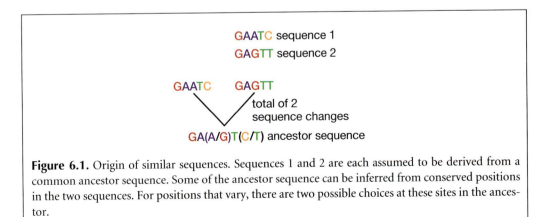

Figure 6.1. Origin of similar sequences. Sequences 1 and 2 are each assumed to be derived from a common ancestor sequence. Some of the ancestor sequence can be inferred from conserved positions in the two sequences. For positions that vary, there are two possible choices at these sites in the ancestor.

been evolving according to the observed sequence variation in the column. Most methods of phylogenetic analysis assume that each position in the protein or nucleic acid sequence changes independently of the others (analysis of RNA sequence evolution is an exception: see Chapter 5).

As indicated above, the analysis of sequences that are strongly similar along their entire lengths is quite straightforward. However, to align most sequences requires the positioning of gaps in the alignment. Gaps represent an insertion or deletion of one or more sequence characters during evolution. Proteins that align well are likely to have the same three-dimensional structure. In general, sequences that lie in the core structure of such proteins are not subject to insertions or deletions because any amino acid substitutions must fit into the packed hydrophobic environment of the core. Gaps should therefore be rare in regions of multiple sequence alignments that represent these core sequences. In contrast, more variation, including insertions and deletions, may be found in the loop regions on the outside of the three-dimensional structure because these regions do not influence the core structure as much. Loop regions interact with the environment of small molecules, membranes, and other proteins (see Chapter 9).

Gaps in alignments can be thought of as representing mutational changes in sequences, including insertions, deletions, or rearrangements of genetic material. The expectation that a gap of virtually any length can occur as a single event introduces the problem of judging how many individual changes have occurred and in what order. Gaps are treated in various ways by phylogenetic programs, but no clear-cut model as to how they should be treated has been devised. Many methods ignore gaps or focus on regions in an alignment that do not have any gaps. Nevertheless, gaps can be useful as phylogenetic markers in some situations.

Another approach for handling gaps is to avoid analysis of individual sites in the sequence alignment and instead to use sequence similarity scores as a basis for phylogenetic analysis. Rather than trying to decide what has happened at each sequence position in an alignment, a similarity score based on a scoring matrix with penalties for gaps is often used. As discussed below, these scores may be converted to distance scores that are suitable for phylogenetic analysis (Feng and Doolittle 1996) by distance methods (p. 254).

GENOME COMPLEXITY AND PHYLOGENETIC ANALYSIS

When performing a phylogenetic analysis, it is important to keep in mind that the genomes of most organisms have a complex origin. Some parts of the genome are passed on by vertical descent through the normal reproductive cycle. Other parts may have arisen by horizontal transfer of genetic material between species through a virus, DNA transformation, symbiosis, or some other horizontal transfer mechanism. Accordingly, when a particular gene is being subjected to phylogenetic analysis, the evolutionary history of that gene may not coincide with the evolutionary history of another.

One of the most significant uses of phylogenetic analysis of sequences is to make predictions concerning the tree of life. For this purpose, a gene should be selected that is universally present in all organisms and easily recognizable by the conservation of sequence in many species. At the same time, there should be enough sequence variation to determine which groups of organisms share the same phylogenetic origin. Ideally, the gene should also not be under selection, meaning that as variation occurs in populations of organisms, certain sequences are not favored with a loss of the more primitive variation.

Two molecules of this type that carry a great deal of evolutionary history in inter-species sequence variations are the small rRNA subunit and mitochondrial sequences. A large number of rRNA sequences from a variety of organisms were aligned and the secondary structure was deduced following methods discussed in Chapter 5. Phylogenetic predictions were then made using the distance method described below (Woese 1987). On the basis of rRNA sequence signatures, or regions within the molecule that are conserved in one group of organisms but different in another (Fig. 6.2), Woese (1987) predicted that early life diverged into three main kingdoms—Archaea, Bacteria, and Eukarya—a view that has been challenged (Mayr 1998). Evidence for the presence of additional organisms in these groups has since been found by PCR amplification of environmental samples of RNA (Barns et al. 1996). A more detailed analysis was used to find relationships among individual species within each group. The types of relationships found among the prokaryotic organisms are illustrated in Figure 6.3. The use of mitochondrial sequences for analysis of primate evolution is given below in the description of the parsimony method of phylogenetic analysis.

Although these studies of rRNA sequences suggest a quite clear-cut model for the evolution of life, phylogenetic analysis of other genes and gene families has revealed that the situation is probably more complex and that a more appropriate model might be the one shown in Figure 6.4. There are now many examples of horizontal or lateral transfer of genes between species (see Fig. 3.3, p. 55) that introduce new genes and sequences into an organism (Brown and Doolittle 1997; Doolittle 1999). These types of transfers are inferred from the finding that the phylogenetic histories of different genes in an organism, such as genes for metabolic functions, are not the same or that codon use in different genes varies (see Chapter 10). Another type of phylogenetic analysis is based on the number of genes shared between genomes and produces a tree that is similar to the rRNA tree (Snel et al. 1999).

To track the evolutionary history of genes, more attention has also been paid to the methodology of phylogenetic analysis and to the inherent errors in many of the assumptions (Doolittle 1999). Problems associated with variations between rates of change in different sites and of analyzing more distantly related sequences are discussed below. Moreover, there is evidence that genomes undergo extensive rearrangements, placing sequences of different evolutionary origin next to each other and even causing rearrangements within protein-encoding genes (Henikoff et al. 1997).

The different regions of independent evolutionary origin in a sequence therefore need to be identified. As discussed in Chapter 9, proteins are modular with functional domains, sometimes repeated within a protein and sometimes shared within a protein family. These regions are identified by their sharing of significant sequence similarity. The remainder of the aligned regions in the group may have variable levels of similarity. In nucleic acid sequences, a given sequence pattern may provide a binding site for a regulatory molecule, leading to promoter function, RNA splicing, or some other function. It may be difficult to decide the extent of these patterns for phylogenetic analysis; however, statistical approaches discussed in Chapter 4 may be used.

Another feature of genome evolution that should be considered in phylogenetic analysis is the occurrence of gene duplication events that create tandem copies of a gene. These two copies may then evolve along separate pathways leading to different functions. However, these copies maintain a certain level of similarity and undergo concerted evolution, a process of acquiring mutations in a coordinated way, probably through gene conversion or recombination events. Speciation events following gene duplications will give rise to two independent sets of genes and sequences, one set for each gene copy. As discussed in Chapter 3 and illustrated in Figure 3.3, two genes in the same lineage can have different relationships. In the example shown in Figure 3.3, genes a1 and a2 have been derived from gene a. The pair is then segregated by speciation such that there is one a1 a2 pair in one species evolving along one path and a second a1 a2 pair in a second species evolving along

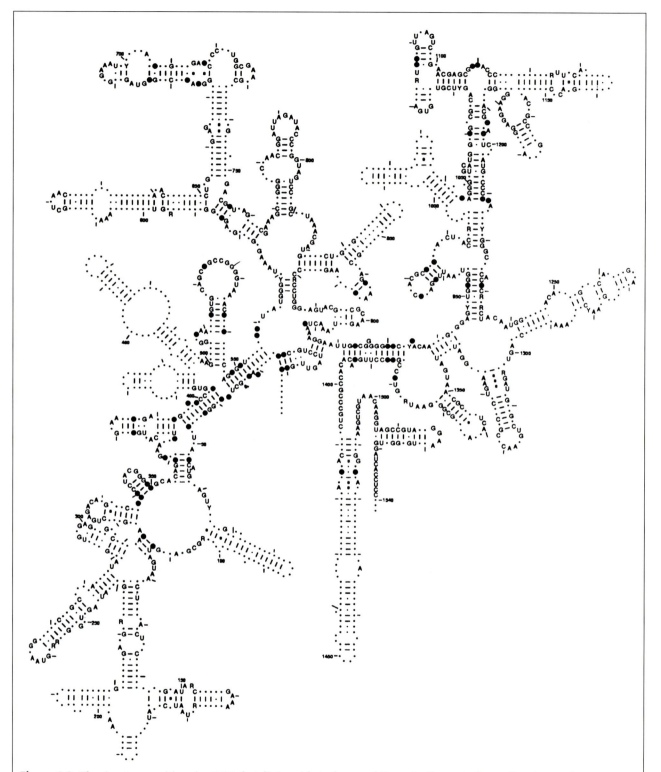

Figure 6.2. The signature positions in rRNA that distinguish Archaea and Bacteria. Shown is the predicted secondary structure for *E. coli* 16S ribosomal RNA with the most highly conserved sequence positions marked by the sequence character and the positions that distinguish Archaea and Bacteria shown by a black dot. Other marker positions in the sequence were used to define the third group, the Eukarya. (Reprinted, with permission, from Woese 1987 [copyright American Society for Microbiology].)

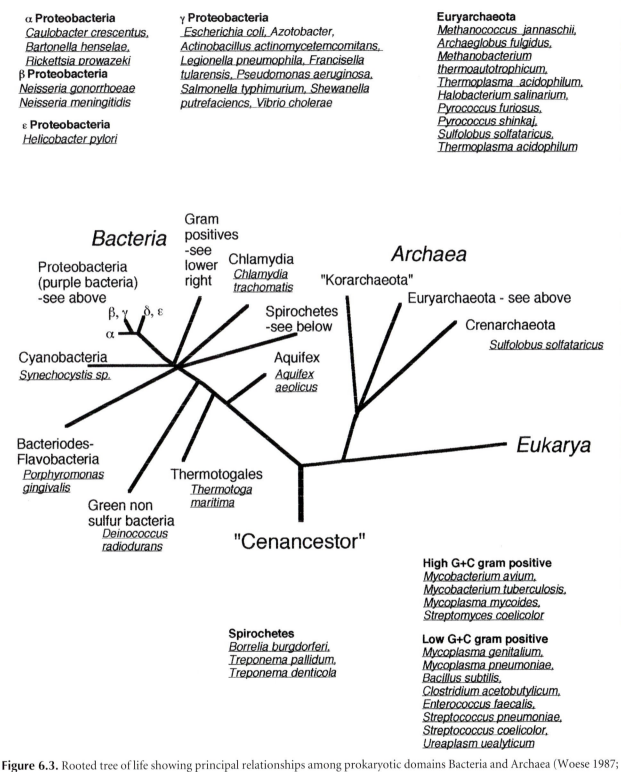

α Proteobacteria
Caulobacter crescentus,
Bartonella henselae,
Rickettsia prowazeki
β Proteobacteria
Neisseria gonorrhoeae
Neisseria meningitidis

ε Proteobacteria
Helicobacter pylori

γ Proteobacteria
 Escherichia coli, Azotobacter,
Actinobacillus actinomycetemcomitans,
Legionella pneumophila, Francisella
tularensis, Pseudomonas aeruginosa,
Salmonella typhimurium, Shewanella
putrefaciencs, Vibrio cholerae

Euryarchaeota
Methanococcus jannaschii,
Archaeglobus fulgidus,
Methanobacterium
thermoautotrophicum,
Thermoplasma acidophilum,
Halobacterium salinarium,
Pyrococcus furiosus,
Pyrococcus shinkaj,
Sulfolobus solfataricus,
Thermoplasma acidophilum

Bacteria

Gram
positives
-see
lower
right

Proteobacteria
(purple bacteria)
-see above

β, γ   δ, ε
α

Chlamydia
Chlamydia
trachomatis

Spirochetes
-see below

Cyanobacteria
Synechocystis sp.

Aquifex
Aquifex
aeolicus

Archaea

"Korarchaeota"

Euryarchaeota - see above

Crenarchaeota

Sulfolobus solfataricus

Bacteriodes-
Flavobacteria
Porphyromonas
gingivalis

Thermotogales
Thermotoga
maritima

Green non
sulfur bacteria
Deinococcus
radiodurans

Eukarya

"Cenancestor"

Spirochetes
Borrelia burgdorferi,
Treponema pallidum,
Treponema denticola

High G+C gram positive
Mycobacterium avium,
Mycobacterium tuberculosis,
Mycoplasma mycoides,
Streptomyces coelicolor

Low G+C gram positive
Mycoplasma genitalium,
Mycoplasma pneumoniae,
Bacillus subtilis,
Clostridium acetobutylicum,
Enterococcus faecalis,
Streptococcus pneumoniae,
Streptococcus coelicolor,
Ureaplasm uealyticum

Figure 6.3. Rooted tree of life showing principal relationships among prokaryotic domains Bacteria and Archaea (Woese 1987; Barns et al. 1996; Brown and Doolittle 1997). Branch lengths are approximate only. Species that have been sequenced or are being sequenced are shown. A comprehensive database of sequenced microbial genomes is maintained at http://www.tigr.org/.

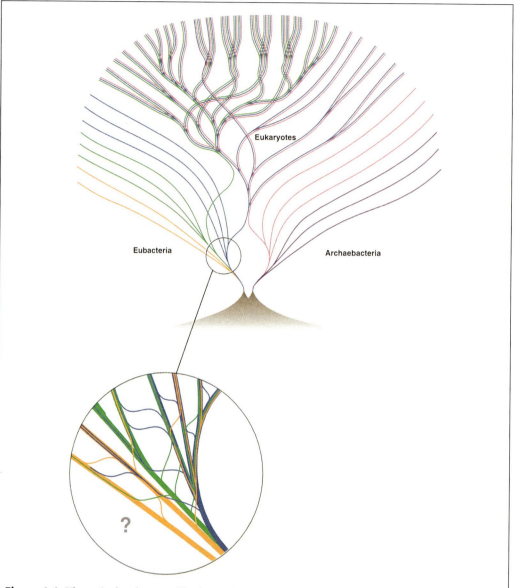

Figure 6.4. The reticulated or net-like form of the tree of life. Analysis of rRNA sequences originally suggested three main branches in the tree of life, Archaea, Bacteria, and Eukarya. Subsequent phylogenetic analysis of genes for some metabolic enzymes is not congruent with the rRNA tree. Hence, for these metabolic genes, the tree has a reticulated form due to horizontal transfer of these genes between species. (Reprinted, with permission, from Martin 1999 [copyright Wiley-Liss, Inc.].)

a second path, reproductively and genetically isolated from each other. The a1 genes in the different species are orthologous to each other, as are the a2 genes, but the a1 and a2 genes are paralogous because they arose from a gene duplication event. These relationships can be determined by a careful analysis of genomes and sequence relationships (Tatusov et al. 1997) that is discussed further in Chapter 10.

THE CONCEPT OF EVOLUTIONARY TREES

An evolutionary tree is a two-dimensional graph showing evolutionary relationships among organisms, or in the case of sequences, in certain genes from separate organisms.

The separate sequences are referred to as taxa (singular taxon), defined as phylogenetically distinct units on the tree. The tree is composed of outer branches (or leaves) representing the taxa and nodes and branches representing relationships among the taxa, illustrated as sequences A–D in Figure 6.5. Thus, sequences A and B are derived from a common ancestor sequence represented by the node below them, and C and D are similarly related. The A/B and C/D common ancestors also share a common ancestor represented by a node at the lowest level of the tree. It is important to recognize that each node in the tree represents a splitting of the evolutionary path of the gene into two different species that are isolated reproductively. Beyond that point, any further evolutionary changes in each new branch are independent of those in the other new branch. The length of each branch to the next node represents the number of sequence changes that occurred prior to the next level of separation. Note that, in this example, the branch length between the A/B node and A is approximately equal to that between the A/B node and B, indicating the species are evolving at the same rate.

The amount of evolutionary time that has transpired since the separation of A and B is usually not known. What is estimated by phylogenetic analysis is the amount of sequence change between the A/B node and A and also between the A/B node and B. Hence, judging by the branch lengths from this node to A and B, the same number of sequence changes has occurred. However, it is also likely that for some biological or environmental reason unique to each species, one taxon may have undergone more mutations since diverging from the ancestor than the other. In this case, different branch lengths would be shown on the tree. Some types of phylogenetic analyses assume that the rates of evolution in the tree branches are the same, whereas others assume that they vary, as discussed below. The assumption of a uniform rate of mutation in the tree branches is known as the molecular clock hypothesis and is usually most suitable for closely related species (Li and Graur 1991; Li 1997). Tests for this hypothesis have been devised as described below. Even if there is a common rate of evolutionary change, statistical variations from one branch to another can influence the analysis. The number of substitutions in each branch is generally assumed to vary according to the Poisson distribution (see Chapter 3, p. 103, for an explanation of the Poisson distribution), and the rate of change is assumed to be equal across all sequence positions (Swofford et al. 1996).

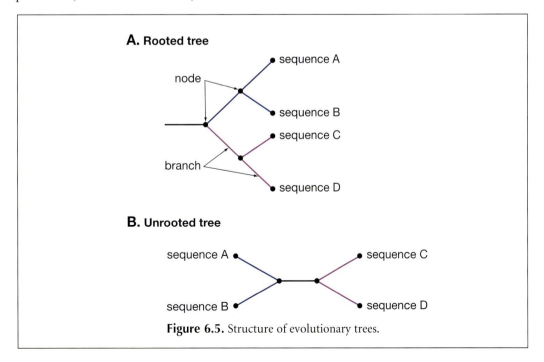

Figure 6.5. Structure of evolutionary trees.

The tree shown is only one of many, each predicting a different evolutionary relationship among the sequences or taxa. The number of possible rooted trees increases very rapidly with the number of sequences or taxa, as shown in Table 6.2. A root has been placed at this position indicating that in this evolutionary model of the sequences this basal node is the common ancestor of all of the other sequences. A unique path leads from the root node to any other node, and the direction of the path indicates the passage of evolutionary time. The root is defined by including a taxon that we are reasonably sure branched off earlier than the other taxa under study but should be related to the remaining taxa. It is also possible to predict a root, assuming that the molecular clock hypothesis holds.

The sum of all the branch lengths in a tree is referred to as the tree length. The tree is also a bifurcating or binary tree, in that only two branches emanate from each node. This situation is what one would expect during evolution—only one splitting away of a new species at a time. Trees can have more than one branch emanating from a node if the events separating taxa are so close that they cannot be resolved, or to simplify the tree.

An alternative representation of the relationships among sequences A–D in Figure 6.5A is shown in Figure 6.5B. The difference between the tree in A and that in B is that the tree in B is unrooted. The unrooted tree also shows the evolutionary relationships among sequences A–D, but it does not reveal the location of the oldest ancestry. B could be converted into A by placing another node and adjoining root to the black line. A root could also be placed anywhere else in the tree. Hence, there are a great many more possibilities for rooted than for unrooted trees for a given number of taxa or sequences, as shown in Table 6.2.

Three methods—maximum parsimony, distance, and maximum likelihood—are generally used to find the evolutionary tree or trees that best account for the observed variation in a group of sequences. Each of these methods uses a different type of analysis as described below. The flowchart on page 247 descibes the types of considerations that need to be made in choosing a method. These methods may find that more than one tree meets the criterion chosen for being the most likely tree. The branching patterns in these trees may be compared to find which branches are shared and therefore are more strongly supported. PAUP provides methods for finding consensus trees, and such trees are also calculated by the CONSENSE program in the PHYLIP package. Trees are stored as a tree file that shows the relationships in nested-parenthesis notation, i.e., a file with the line (A,(B,(C,D))) represents the tree shown below in Table 6.2. Sometimes, branch lengths are

Table 6.2. *Number of possible evolutionary trees to consider as a function of number of sequences*

Taxa or sequence no.	No. of rooted trees	No. of unrooted trees
3	3	1
4	15	3
5	105	15
—	—	—
7	10,395	954

also included next to the names, e.g., A:0.05. From this information, a tree-drawing program may be used to produce a tree representation of the data.

METHODS

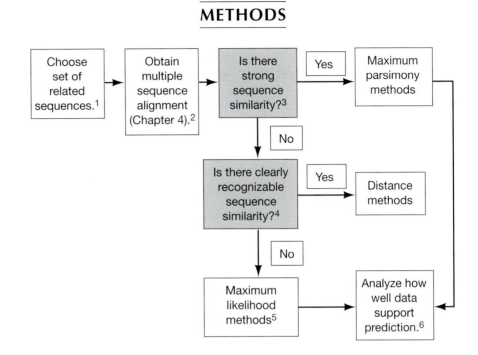

1. The sequences chosen can be either DNA or protein sequence: Different programs and program options are used for each type. RNA sequences are analyzed by covariation methods and by analyzing changes in secondary structure, as outlined in Chapter 5. The selected sequences should align with each other along their entire lengths, or else each should have a common set of patterns or domains that provides a strong indication of evolutionary relatedness.

2. The alignment of the sequence pairs should not have a large number of gaps that are obviously necessary to align identical or related characters (see Chapter 3 flowchart, p. 58). A phylogenetic analysis should only be performed on parts of sequences that can be reasonably aligned. In general, phylogenetic methods analyze conserved regions that are represented in all the sequences. The more similar the sequences are to each other, the better. The simplest evolutionary models assume that the variation in each column of the multiple sequence alignment represents single-step changes and that no reversals (A → T → A) have occurred. As the observed variation increases, more multiple-step changes (A → T → G) and reversions are likely to be present. Corrections may be applied for such variation, thereby increasing the observed amount of change to a more reasonable value. These corrections assume a uniform rate of change at all sequence positions over time. Gaps in the multiple sequence alignment are usually not scored because there is no suitable model for the evolutionary mechanisms that produce them.

3. This question is designed to select sequences suitable for maximum parsimony analysis. Other methods may also be used with these same sequences. For parsimony analysis, the best results are obtained when the amount of variation among all pairs of sequences is similar (no very different sequences are present) and when the amount of variation is small. Some columns in the multiple sequence alignment will have the same residue in all sequences; other columns will include both conserved and nonconserved residues. There should be a clear-cut majority of certain residues in some columns of the alignment but also some variation. These more common residues are taken to represent an earlier group of sequences from which others were derived. If there is too much variation, there will be too many possible ancestral relationships. Because the maximum parsimony method has to attempt to fit all possible trees to the data, the method is not suitable for more than 11 or 12 sequences because there are too many trees to test. More than one tree may be found to be equally parsimonious. A consensus tree representing the conserved features of the different trees may then be produced.

4. The purpose of this question is to select sequences for phylogenetic analysis by distance methods. Distance methods are able to predict an evolutionary tree when variation among the sequences is present (some sequences are more alike than others) and when the amount of variation is intermediate. The number of changed positions in an alignment between two sequences divided by the total number of matched positions is the distance between the sequences. As distances increase, corrections are necessary for deviations from single-step changes between sequences (see note 3). Of course, as distances increase, the uncertainty of alignments also increases (see Chapter 4), and a reassessment of the suitability of the multiple sequence alignment method may be necessary. Sequences with this type of variation may also be suitable for phylogenetic analysis by maximum likelihood methods. Distance methods may be used with a large number of sequences. The program CLUSTALW produces a distance-based tree at the same time as a multiple sequence alignment (Higgins et al. 1996).

5. Maximum likelihood methods may be used for any set of related sequences, but they are particularly useful when the sequences are more variable. These methods are computationally intense, and computational complexity increases with the number of sequences since the probability of every possible tree must be calculated as described in the text. An advantage of these methods is that they provide evolutionary models to account for the variation in the sequences.

6. The data in the multiple sequence alignment columns is resampled to test how well the branches on the evolutionary tree are supported (boot-strapping).

MAXIMUM PARSIMONY METHOD

This method predicts the evolutionary tree (or trees) that minimizes the number of steps required to generate the observed variation in the sequences. For this reason, the method is also sometimes referred to as the minimum evolution method. A multiple sequence alignment is required to predict which sequence positions are likely to correspond. These positions will appear in vertical columns in the multiple sequence alignment. For each aligned position, phylogenetic trees that require the smallest number of evolutionary changes to produce the observed sequence changes are identified. This analysis is continued for every position in the sequence alignment. Finally, those trees that produce the smallest number of changes overall for all sequence positions are identified. This method is used for sequences that are quite similar and for small numbers of sequences, for which it is best suited. The algorithm followed is not particularly complicated, but it is guaranteed to find the best tree, because all possible trees relating a group of sequences are examined. For this reason, the method is quite time-consuming and is not useful for data that include a large number of sequences or sequences with a large amount of variation. One or more unrooted trees are predicted and other assumptions must be made to root the predicted tree.

PAUP offers a number of options and parameter settings for a parsimony analysis in the Macintosh environment. The main programs for maximum parsimony analysis in the PHYLIP package (Felsenstein 1996) are listed below.

For analysis of nucleic acid sequences, programs are:

1. DNAPARS, which treats gaps as a fifth nucleotide state.

2. DNAPENNY, which performs parsimonious phylogenies by branch-and-bound search that can analyze more sequences (up to 11 or 12).

3. DNACOMP, which performs phylogenetic analysis using the compatibility criterion. Rather than searching for overall parsimony at all sites in the multiple sequence alignment, this method finds the tree that supports the largest number of sites. This method is recommended when the rate of evolution varies among sites.

4. DNAMOVE, which performs parsimony and compatibility analysis interactively.

For analysis of protein sequences, the program is:

1. PROTPARS, which counts the minimum number of mutations to change a codon for the first amino acid into a codon for the second amino acid, but only scores those mutations in the mutational path that actually change the amino acid. Silent mutations that do not change the amino acid are not scored on the grounds that they have little evolutionary significance.

The maximum parsimony analysis is illustrated in the following example of four sequences shown in Table 6.3 and Figure 6.6 (adapted from Li and Graur 1991). An example of a parsimony analysis of mitochondrial sequences using PAUP and MacClade is then given. Note that in a multiple sequence alignment, only certain sequence variations at a given site are useful for a parsimony analysis. In the analysis, all of the possible unrooted trees (three trees for four sequences) are considered. The sequence variations at each site in the alignment are placed at the tips of the trees, and the tree that requires the smallest number of changes to produce this variation is determined. This analysis is repeated for each informative site, and the tree (or trees) that supports the smallest number of changes overall is found. The length of the tree, defined as the sum of the number of steps in each branch of the tree, will be a minimum.

Example: Maximum Parsimony Analysis of Sequences

Table 6.3 shows an example of phylogenetic analysis by maximum parsimony. This method finds the tree that changes any sequence into all of the others by the least number of steps.

Rules for analysis by maximum parsimony in this example are:

1. There are four taxa giving three possible unrooted trees.
2. Some sites are informative, i.e., they favor one tree over another (site 5 is informative but sites 1, 6, and 8 are not).
3. To be informative, a site must have the same sequence character in at least two taxa (sites 1, 2, 3, 4, 6, and 8 are not informative; sites 5, 7, and 9 are informative).
4. Only the informative sites need to be analyzed.

The three possible trees are shown in Figure 6.6. The optimal tree is obtained by adding the number of changes at each informative site for each tree, and picking the tree requiring the least number of changes. A scoring matrix may be used instead of scoring a change as 1. Tree 1 is the correct one and the tree length will be 4 (one change at each of positions 5 and 7 and two changes at position 9).

In the above example, because there were only four sequences to consider, it was necessary to consider only three possible unrooted trees. For a larger number of sequences, the number of trees becomes so large that it may not be feasible to examine all possible trees. The example of 12 sequences below took only a few seconds on a Macintosh G3. The exhaustive and branch-and-bound options of the program PAUP will analyze all possible trees, and if the number is too large, the program can keep running for a very long time.

Branch-and-bound is a method that stops analyzing a particular branching pattern in trees when it is not possible to obtain a more parsimonious solution than has been already found.

For large numbers of sequences, PAUP provides a program option called "heuristic," which searches among all possible trees and keeps representative trees that best fit the data. The presence of common branch patterns in these trees reveals some of the broader features of the phylogenetic relationships among the sequences.

Table 6.3. *Example of phylogenetic analysis to find the correct unrooted tree from four aligned sequences by the maximum parsimony method*

Taxa	Sequence position (sites) and character								
	1	2	3	4	5	6	7	8	9
1	A	A	G	A	G	T	G	C	A
2	A	G	C	C	G	T	G	C	G
3	A	G	A	T	A	T	C	C	A
4	A	G	A	G	A	T	C	C	G

Adapted from Li and Graur 1991.

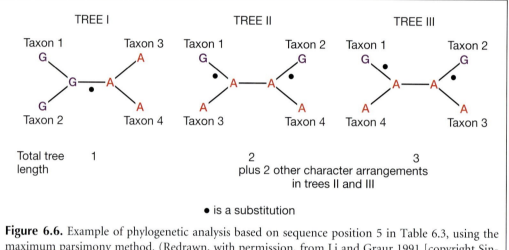

Figure 6.6. Example of phylogenetic analysis based on sequence position 5 in Table 6.3, using the maximum parsimony method. (Redrawn, with permission, from Li and Graur 1991 [copyright Sinauer Associates].)

Analysis of Mitochondrial Sequences by PAUP

To search for this tree, which best fits all the sequence data, the trees that best fit each vertical column of sequence characters in Figure 6.7A were first determined. In some columns, the data are not informative, as in the case of all nucleotides being the same. For a nucleotide position to be informative, at least two different nucleotides must be present in at least two of the sequences. A tree that provides the least number of evolutionary steps to satisfy the data in all columns, the most parsimonious tree, is then found.

Parsimony can give misleading information when rates of sequence change vary in the different branches of a tree that are represented by the sequence data. These variations produce a range of branch lengths, long ones representing more extended periods of time and short ones representing shorter times. For example, the real tree shown below in Figure 6.8A includes two long branches in which G has turned to A independently, probably with a number of intermediate changes that are not observed in the sequence data. Because in a parsimony analysis rates of change along all branches of the tree are assumed to be equal, the tree predicted by parsimony and shown in Figure 6.8B will not be correct.

Although other columns in the sequence alignment that show less variation may provide the correct tree, the columns representing greater variation dominate the analysis

(Swofford et al. 1996). Such long branches may be broken down if additional taxa are present that are more closely related to taxa 1 and 4, thereby providing branches that intersect the long branches and give a better resolution of the changes.

Another method for identifying such long branches is called Lake's method of invariants or evolutionary parsimony, available in PAUP. In this method, four of the sequences are chosen at a time, and only transversions in the aligned positions are scored as changes on the grounds that transversions are the most significant base changes during evolution. Transversions of any base to each possible derivative, e.g., A → C or T, are assumed to change at the same rate to create a balanced distribution, and the changes in each column of the alignment (each sequence position) are assumed to occur independently of each other. Suppose that there are two long branches as in the case discussed immediately above. The correct tree is shown in Figure 6.9A, and one of the sites has changed multiply but ends up as the same base A by chance. Traditional parsimony will identify this tree incorrectly, as indicated above. If these long branches do indeed exist, then other sites should give the type of transversion events shown in Figure 6.9B. The greater the number of B-type sites, the less one can depend on the A-type sites revealed in A. The evolutionary parsimony method subtracts the number of type B from the number of type A. If, on the one hand, long branches are not present in the quartet of sequences, there will be very few type B, and type A will be taken as evidence for the correct tree. On the other hand, if many examples of type B are present, the A type will carry little weight. These calculations are performed for all three possible unrooted trees and all possible types of transversions for the four sequences, and the tree receiving the most support is chosen. These methods and other more sophisticated methods for correcting uneven branch lengths are discussed in detail in Swofford et al. (1996). The PHYLIP program DNAINVAR computes Lake's and other phylogenetic invariants for nucleic acid sequences. PAUP also includes an option for Lake's invariant.

Compared to the above methods, maximum likelihood and distance methods provide more reliable predictions when corrections are made for multiple substitutions. Distance methods such as neighbor joining discussed below have been shown generally to be better predictors than both standard and evolutionary parsimony methods when branch lengths are varying (Jin and Nei 1990; Swofford et al. 1996).

There are options in PAUP and MacClade for selecting among the most parsimonious trees. With MacClade it is possible to view the changes in sequence characters in each branch of the tree to arrive at the current base in each sequence or taxon, as shown below. As these characters are traced from positions lower in the tree to upper positions, some nodes in the tree may be assigned an unambiguous character (shown in color, Fig. 6.10). For other nodes, the assignment may be ambiguous because the node is leading to two different characters above (thin black line). It is possible to arrange these ambiguities optionally in two ways: one is to delay them going as far up the tree away from the root as possible (the Deltran option; not shown in figure); a second is to introduce them as soon as possible and as close to the root as possible (the Acctran option; not shown in figure). The effect of using Deltran is to force parallel changes in the upper branches of the tree, that of Acctran is to force reversals in the upper branches. Using these options is not recommended unless such variations are expected, as in analysis of more divergent sequences (Maddison and Maddison 1992).

Homoplasy refers to the occurrence of the same sequence change in more than one branch of the tree. If all the sequence character changes support the same tree, there is no homoplasy. In reality, homoplasy is usually found for some characters for any tree. MacClade allows changing of the tree to avoid homoplasy at a sequence position, but the new tree length will often increase, thus making the tree a less parsimonious choice than the

A. Mitochondrial sequences.

```
#NEXUS

begin taxa;
      dimensions ntax=12;
end;

begin characters;
      dimensions nchar=898;
      format missing=? gap=- matchchar=. interleave datatype=dna;
      options gapmode=missing;
      matrix

Lemur_catta        AAGCTTCATAGGAGCAACCATTCTAATAATCGCACATGGCCTTACATCATCCATATTATT
Homo_sapiens       AAGCTTCACCGGCGCAGTCATTCTCATAATCGCCCACGGGCTTACATCCTCATTACTATT
Pan                AAGCTTCACCGGCGCAATTATCCTCATAATCGCCCACGGACTTACATCCTCATTATTATT
Gorilla            AAGCTTCACCGGCGCAGTTGTTCTTATAATTGCCCACGGACTTACATCATCATTATTATT
Pongo              AAGCTTCACCGGCGCAACCACCCTCATGATTGCCCATGGACTCACATCCTCCCTACTGTT
Hylobates          AAGCTTTACAGGTGCAACCGTCCTCATAATCGCCCACGGACTAACCTCTTCCCTGCTATT
Macaca_fuscata     AAGCTTTTCCGGCGCAACCATCCTTATGATCGCTCACGGACTCACCTCTTCCATATATTT
M._mulatta         AAGCTTTTCTGGCGCAACCATCCTCATGATTGCTCACGGACTCACCTCTTCCATATATTT
M._fascicularis    AAGCTTCTCCGGCGCAACCACCCTTATAATCGCCCACGGGCTCACCTCTTCCATGTATTT
M._sylvanus        AAGCTTCTCCGGTGCAACTATCCTTATAGTTGCCCATGGACTCACCTCTTCCATATACTT
Saimiri_sciureus   AAGCTTCACCGGCGCAATGATCCTAATAATCGCTCACGGGTTTACTTCGTCTATGCTATT
Tarsius_syrichta   AAGTTTCATTGGAGCCACCACTCTTATAATTGCCCATGGCCTCACCTCCTCCCTATTATT

Lemur_catta        CTGTCTAGCCAACTCTAACTACGAACGAATCCATAGCCGTACAATACTACTAGCACGAGG
Homo_sapiens       CTGCCTAGCAAACTCAAACTACGAACGCACTCACAGTCGCATCATAATCCTCTCTCAAGG
Pan                CTGCCTAGCAAACTCAAATTATGAACGCACCCACAGTCGCATCATAATTCTCTCCCAAGG
Gorilla            CTGCCTAGCAAACTCAAACTACGAACGAACCCACAGCCGCATCATAATTCTCTCTCAAGG
Pongo              CTGCCTAGCAAACTCAAACTACGAACGAACCCACAGCCGCATCATAATCCTCTCTCAAGG
Hylobates          CTGCCTTGCAAACTCAAACTACGAACGAACTCACAGCCGCATCATAATCCTATCTCGAGG
Macaca_fuscata     CTGCCTAGCCAATTCAAACTATGAACGCACTCACAACCGTACCATACTACTGTCCCGAGG
M._mulatta         CTGCCTAGCCAATTCAAACTATGAACGCACTCACAACCGTACCATACTACTGTCCCGGGG
M._fascicularis    CTGCTTGGCCAATTCAAACTATGAGCGCACTCATAACCGTACCATACTACTATCCCGAGG
M._sylvanus        CTGCTTGGCCAACTCAAACTACGAACGCACCCACAGCCGCATCATACTACTATCCCGAGG
Saimiri_sciureus   CTGCCTAGCAAACTCAAATTACGAACGAATTCACAGCCGAACAATAACATTTACTCGAGG
Tarsius_syrichta   TTGCCTAGCAAATACAAACTACGAACGAGTCCACAGTCGAACAATAGCACTAGCCCGTGG
.
.
.

end;
```

B. Phylogenetic tree

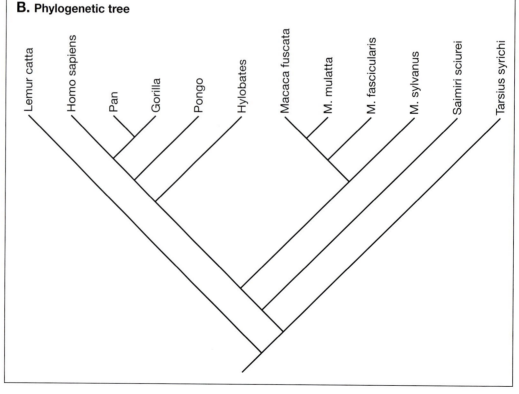

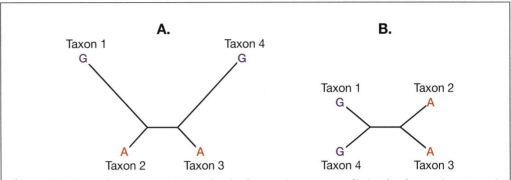

Figure 6.8. Type of sequence variation that leads to an incorrect prediction by the maximum parsimony method.

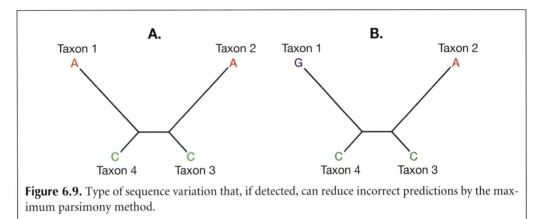

Figure 6.9. Type of sequence variation that, if detected, can reduce incorrect predictions by the maximum parsimony method.

original. Another parameter used is the consistency index (CI), which is the minimum possible tree length divided by the actual tree length. The more homoplasy, the greater the actual tree length, and the smaller the value of CI.

Parsimony methods can use information on the number of changes required or steps to change one residue into another. For example, the number of mutations required to change one amino acid into another in one branch of a tree can be taken into account. The parsimony method then attempts to minimize the number of such steps. This number of steps for interchanging characters can be incorporated into a matrix, called a step or cost matrix for programs such as PAUP and MacClade to use.

A program designated PROTPARS for protein squences in the PHYLIP package scores only those mutations that produce amino acid changes (Felsenstein 1996). This program uses an algorithm similar to one described by Sankoff (1975) for determining the mini-

Figure 6.7. Analysis of mitochondrial sequences using the maximum parsimony method provided by the PAUP program. (*A*) Portion of a multiple sequence alignment of the mitochondrial sequences provided in the PAUP distribution package. PAUP will import sequences in other multiple sequence alignment format and convert them into the NEXUS format. The program READSEQ will reformat multiple sequence alignments into the NEXUS format. This format includes information about type of sequence, coding information, codon positions, differential weights for transitions and transversions, treatment of gaps, and preferred groupings (see Chapter 2). Only a portion of the NEXUS file is shown. In this analysis, branch-and-bound and otherwise default options were used. Gaps are treated as missing information. The number of sequences is indicated as ntaxa, number of alignment columns as nchar, and the interleave command allows the data to be entered in readable blocks of sequence 60 characters long. (*B*) One of the two predicted trees. The tree file of PAUP was edited in MacClade and output as a graphics file.

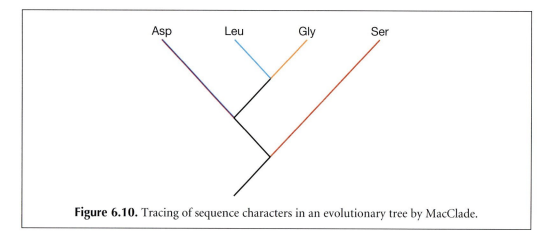

Figure 6.10. Tracing of sequence characters in an evolutionary tree by MacClade.

mum number of mutations in a tree for changing one sequence into another. Similar types of analyses for proteins are also available in PAUP and MacClade. The PAUP program uses a 3+1 option in the stepmatrices option, which is a short cut for analyzing trees that represent the most possible ancestors of an amino acid (PAUP vers 3.1 manual, pp. 124–126).

DISTANCE METHODS

The distance method employs the number of changes between each pair in a group of sequences to produce a phylogenetic tree of the group. The sequence pairs that have the smallest number of sequence changes between them are termed "neighbors." On a tree, these sequences share a node or common ancestor position and are each joined to that node by a branch. The goal of distance methods is to identify a tree that positions the neighbors correctly and that also has branch lengths which reproduce the original data as closely as possible. Finding the closest neighbors among a group of sequences by the distance method is often the first step in producing a multiple sequence alignment, as discussed in Chapter 4.

The distance method was pioneered by Feng and Doolittle, and a collection of programs by these authors will produce both an alignment and tree of a set of protein sequences (Feng and Doolittle 1996). The program CLUSTALW, discussed in Chapter 4, uses the neighbor-joining distance method as a guide to multiple sequence alignments. PAUP version 4 has options for performing a phylogenetic analysis by distance methods. Programs of the PHYLIP package that perform a distance analysis include the following programs, which automatically read in a sequence in the PHYLIP infile format (see Chapter 2) and automatically produce a file called outfile with a distance table.

1. DNADIST computes distances among input nucleic acid sequences. There are choices given for various models of evolution as described below and a choice for the expected ratio of transitions to transversions.

2. PROTDIST computes a distance measure for protein sequences, based on the Dayhoff PAM model (see p. 78) or other models of evolutionary change in proteins (Felsenstein 1996).

Once distance matrices have been produced, they may be used as input to the following distance analysis programs in PHYLIP. The PHYLIP programs all automatically read an input file called infile and produce an output file called outfile. Hence, file names have to be edited when using these programs. In this example, the distance outfile must be edited to include only the distance table and the number of taxa, and then the file is saved under the sequence name infile.

Distance analysis programs in PHYLIP:

1. FITCH estimates a phylogenetic tree assuming additivity of branch lengths using the Fitch-Margoliash method described below and does not assume a molecular clock (allows rates of evolution along branches to vary).

2. KITSCH estimates a phylogenetic tree using the Fitch-Margoliash method but under the assumption of a molecular clock.

3. NEIGHBOR estimates phylogenies using the neighbor-joining or unweighted pair group method with arithmetic mean (UPGMA) described below. The neighbor-joining method does not assume a molecular clock and produces an unrooted tree. The UPGMA method assumes a molecular clock and produces a rooted tree.

Recall that in aligning sequences, we normally calculate a similarity score, defined as the sum of the number of identities and number of conservative substitutions in the alignment of the two sequences, with gaps being ignored. An identity score between the sequences showing just the identities may also be found from the alignment. For phylogenetic analysis, the distance score between two sequences is used. This score between two sequences is the number of mismatched positions in the alignment or the number of sequence positions that must be changed to generate the other sequence. Gaps may be ignored in these calculations or treated like substitutions. When a scoring or substitution matrix is used, the calculation is slightly more complicated, but the principle is the same. These methods are described below.

The success of distance methods depends on the degree to which the distances among a set of sequences can be made additive on a predicted evolutionary tree. Suppose there are four sequences, A–D, as shown below in Figure 6.11A, and that they were derived from evolutionary changes reflected by the tree in Figure 6.11D. The number of changes along the branches of the tree corresponds to distances between the sequences shown in Figure 6.11, B and C. In this tree, each change only occurs once, and there are no examples of the same change occurring twice (homoplasy). Although this pattern of change is idealized and most groups of sequences would have examples of the same change occurring more than once, as well as reversions, this example illustrates the additivity principle for four sequences. The principle is that for four sequences predicted by this tree, $d_{AB} + d_{CD} \leq d_{AC} + d_{BD} = d_{AD} + d_{BC}$. In this example the additivity is $3 + 3 \leq 7 + 7 = 8 + 6$. For any other tree, there would be examples of parallel changes and reversions. The additivity condition can be relaxed such that $d_{AB} + d_{CD} \leq d_{AC} + d_{BD}$ and $d_{AB} + d_{CD} \leq d_{AD} + d_{BC}$ will still hold even for sequences in which the changes in the sequence are not fully additive. For each set of four sequences, the tree for which the above additivity condition among the distances best holds provides information as to which sequences are neighbors. This method may be used to evaluate trees and find the minimum evolution tree for four sequences and for any additional number of sequences by extending the analysis to additional groups of four sequences (Sattath and Tversky 1977; Fitch 1981; for references, see Swofford et al. 1996). In order to calculate branch lengths, distance methods assume additivity in the distances between sequences. However, real sequence data may not fit these idealized conditions. As a result, a small positive, zero, or even a negative value may be calculated for a branch length. This result may be due to errors in the sequences or sequence alignment, statistical variation, or simply a reflection of two or more sequences diverging at approximately the same time from a common ancestor.

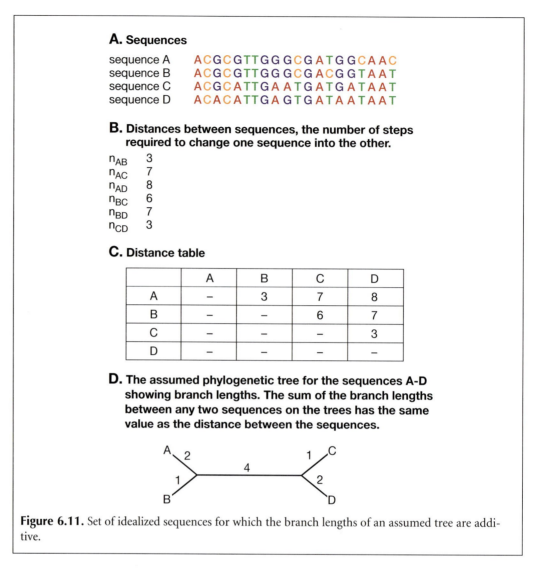

A. Sequences

sequence A	ACGCGTTGGGCGATGGCAAC
sequence B	ACGCGTTGGGCGACGGTAAT
sequence C	ACGCATTGAATGATGATAAT
sequence D	ACACATTGAGTGATAATAAT

B. Distances between sequences, the number of steps required to change one sequence into the other.

n_{AB}	3
n_{AC}	7
n_{AD}	8
n_{BC}	6
n_{BD}	7
n_{CD}	3

C. Distance table

	A	B	C	D
A	–	3	7	8
B	–	–	6	7
C	–	–	–	3
D	–	–	–	–

D. The assumed phylogenetic tree for the sequences A–D showing branch lengths. The sum of the branch lengths between any two sequences on the trees has the same value as the distance between the sequences.

Figure 6.11. Set of idealized sequences for which the branch lengths of an assumed tree are additive.

An even more demanding condition, rarely found in real distance data, is that the distances are ultrametric, meaning that for three taxa, $d_{AC} \leq \max(d_{AB}, d_{BC})$. If the data meet this condition, the distances between two taxa and their common ancestor are equal (Swofford et al. 1996). If the distances follow this relationship, the rates of evolution in the tree branches are approximately the same, thereby meeting the expectations of the molecular clock hypothesis. If these conditions are not met, an analysis based on the assumption of a molecular clock may give misleading results. One method of finding the best tree under such conditions is to transform the sequences after identifying one or more sequences that are least like the rest, called an outgroup (Li and Graur 1991). Some distance methods are based on this assumption and others are not. The overall objective of the distance methods described below is to find this tree by the identification of consecutive sets of neighbors starting with the most alike sequence pair.

Fitch and Margoliash Method and Related Methods

The Fitch and Margoliash (1987) method uses a distance table illustrated in Figure 6.11C. The sequences are combined in threes to define the branches of the predicted tree and to calculate the branch lengths of the tree. The branch lengths are assumed to be additive, as described above. This method of averaging distances is most accurate for trees with short branches. The presence of long branches tends to decrease the reliability of the predictions (Swofford et al. 1996). The following first example describes the use of the algorithm for three sequences, and the second example expands the analysis to more than three sequences.

Example 1: Use of Fitch Margoliash Algorithm for Three Sequences

Steps in algorithm for three sequences:

1. Draw an unrooted tree with three branches emanating from a common node and label ends of branches as shown in Figure 6.12. Given the closer distance between A and B, the branch lengths between these sequences are expected to be shorter, as indicated.

2. Calculate lengths of tree branches algebraically:

   Distances among sequences A, B, and C are shown in the following table.

	A	B	C
A	—	22	39
B		—	41
C			—

   The branch lengths may be calculated algebraically using the branch labels a–c in Figure 6.12:

   distance from A to B = $a + b$ = 22 (1)
   distance from A to C = $a + c$ = 39 (2)
   distance from B to C = $b + c$ = 41 (3)

   subtract (3) from (2), $a - b = -2$ (4)
   add (1) and (4), $2a = 20, a = 10$
   from (1) and (2), $b = 12, c = 29$

   Note that this calculation finds that the branch lengths of A and B from their common ancestor are not the same. Hence, A and B are diverging at different rates of evolution by this calculation and model. For the rates to be the same, these distances would be the same and equal to the distance from A to B divided by 2 = $22/2 = 11$.

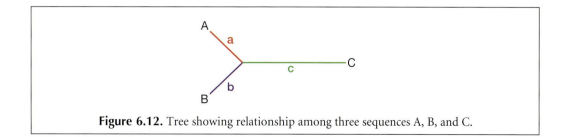

Figure 6.12. Tree showing relationship among three sequences A, B, and C.

Example 2: Use of Fitch-Margoliash Algorithm for Five Sequences

	A	B	C	D	E
A	—	22	39	39	41
B	—	—	41	41	43
C	—	—	—	18	20
D	—	—	—	—	10
E	—	—	—	—	—

These distance data are derived from the unrooted tree shown in Figure 6.13. The Fitch-Margoliash method may be extended from three sequences as shown in example 1 by following the steps shown in the box below, Steps in Fitch-Margoliash algorithm for more than three sequences. The method will find the correct tree and provide the branch lengths a–g, as illustrated below.

1. The most closely related sequences given in the distance table are D and E. A new table is made with the remaining sequences combined.

2. The average distances from D to A, B and C and from B to A, B and C are calculated.

	D	E	ave. ABC
D	—	10	32.7
E	—	—	34.7
average ABC	—	—	—

3. The average distances from D to ABC and from E to ABC can also be found by averaging the sum of the appropriate branch lengths a–g.

   Distance between D and E = $d + e$
   Average distance between D and ABC = $d + m$, $m = g + [(c + 2f + a + b)/3]$
   Average distance between E and ABC = $e + m$

   By subtracting the third from the second equation and adding the result to the first equation, $d = 4$ and $e = 6$.

4. D and E are now treated as a single composite sequence (DE), and a new distance table is made. The distance from A to (DE) is the average of the distance of A to D and of A to E. The other distances to (DE) are calculated accordingly.

	A	B	C	(DE)
A	—	22	39	40
B	—	—	41	42
C	—	—	—	19
(DE)	—	—	—	—

5. The next most closely related sequences are identified, in this case C with the (DE) composite group. The new table is:

	DE	C	Ave. AB
DE	—	19	41
C	—	—	40
Ave. AB	—	—	—

By algebraic manipulations similar to those described above, $c = 9$ and the composite distance of $g + [(e + f)/2] = 10$.

6. Given the above composite distance and the previously calculated values of e and f, then $g = 10 - + [(e + f)/2] = 5$.

The next round of tree-building is that A and B are the next matching pair, giving $a = 10$ and $b = 12$, and a composite distance of $29.7 = [3f + c + 2g + d + e]/3$ giving $f = 29.7 - [(9 + 10 + 10)/3] = 20$. These values are precisely those given in the original tree.

7. Although by design we have generated the correct tree, normally the next step is to repeat the process starting with another sequence pair, such as A and B. We will leave this step as a student exercise to show that the correct tree will again be predicted.

The procedure generally followed is to join all combinations of sequences in pairs to find a tree that best predicts the data in the distance table. The percent change from the actual to the predicted distance is determined for each sequence pair. These values are squared and summed over all possible pairs. This sum divided by the number of pairs = $n(n-1)/2$ less one (the number of degrees of freedom) provides the square of the percent standard deviation of the result.

Steps Followed by Fitch-Margoliash Algorithm for Phylogenetic Analysis of More Than Three Sequences

Steps in algorithm for more than three sequences:

1. Find the most closely related pair of sequences, for example, A and B.
2. Treat the rest of the sequences as a single composite sequence. Calculate the average distance from A to all of the other sequences, and B to all of the other sequences.
3. Use these values to calculate the distances a and b as in the above example with three sequences.

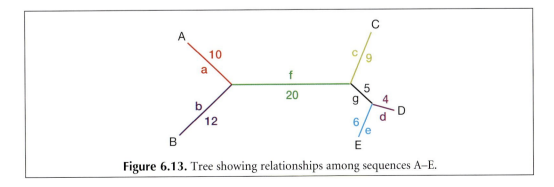

Figure 6.13. Tree showing relationships among sequences A–E.

4. Now treat A and B as a single composite sequence AB, calculate the average distances between AB and each of the other sequences, and make a new distance table from these values.

5. Identify the next pair of most closely related sequences and proceed as in step 1 to calculate the next set of branch lengths.

6. When necessary, subtract extended branch lengths to calculate lengths of intermediate branches.

7. Repeat the entire procedure starting with all possible pairs of sequences A and B, A and C, A and D, etc.

8. Calculate the predicted distances between each pair of sequences for each tree to find the tree that best fits the original data.

The Neighbor-joining Method and Related Neighbor Methods

The neighbor-joining method (Saitou and Nei 1987) is very much like the Fitch-Margoliash method except that the choice as to which sequences to pair is determined by a different algorithm. The neighbor-joining method is especially suitable when the rate of evolution of the separate lineages under consideration varies. When the branch lengths of trees of known topology are allowed to vary in a manner that simulates varying levels of evolutionary change, the neighbor-joining method and the Sattath and Taversky method, described below, are the most reliable in predicting the correct tree (Saitou and Nei 1987). Pearson et al. (1999) have enhanced the neighbor-joining method so that a set of trees that fit the data, rather than just a single tree, may be determined. The general neighbor joining (GNJ) is available from ftp.virginia.edu/pub/fasta/GNJ.

Neighbor-joining chooses the sequences that should be joined to give the best least-squares estimates of the branch lengths that most closely reflect the actual distances between the sequences. It is not necessary to compare all possible trees to find the least-squares fit as in the Fitch-Margoliash method. The method pairs sequences based on the effect of the pairing on the sum of the branch lengths of the tree. To start, the distances between the sequences are used to calculate the sum of the branch lengths for a tree that has no preferred pairing of sequences. The star-like appearance of such a tree and the calculation of the length of the tree using the data in Example 2 above are shown in Figure 6.14.

The next step in the neighbor-joining algorithm is to decompose or modify the star-like tree in Figure 6.14 by combining pairs of sequences. When this step is performed for sequences A and B in Example 2, the new tree shown in Figure 6.15 will be produced. The tree has A and B paired from a common node that is joined by a new branch j to a second node to which C, D, and E are joined. The sum of the branch lengths of this new tree is calculated as shown in Figure 6.15.

In the neighbor-joining algorithm, each possible sequence pair is chosen and the sum of the branch lengths of the corresponding tree is calculated. For example, using the data of Example 2, $S_{AB} = 67.7$, $S_{BC} = 81$, $S_{CD} = 76$, and $S_{DE} = 70$, plus six other possible combinations. Of these, S_{AB} has the lowest value. Hence, A and B are chosen as neighbors on the grounds that they reduce the total branch length to the largest extent. Once the choice of neighbors has been made, the branch lengths a and b and the average distance from AB to CDE may be calculated by the FM method, as described in the last section. a is calculated by $a = [d_{AB}+(d_{AC}+d_{AD}+d_{AE})/3-(d_{BC}+d_{BD}+d_{DE})/3]/2 = (22+39.7-41.70)/2 = 10$, and b is calculated by $b = [d_{AB}+(d_{BC}+d_{BD}+d_{BE})/3-(d_{AC}+d_{AD}+d_{AE})/3]/2 = (22+41.7-39.7)/2 = 12$.

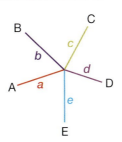

Figure 6.14. Tree for five sequences with no pairing of sequences. In the neighbor-joining method, the sum of the branch lengths $S_0 = a + b + c + d + e$ is calculated. The known distances from (1) A to B, $D_{AB} = a + b$; (2) A to C $= D_{AC} = a + c$; (3) B to C $= D_{BC} = b + c$; and finally (4) D to E, $D_{DE} = d + e$ for a total of 4 + 3 + 2 + 1 = 10 combinations. In summing the 10 distances = 22 + 39 + . . . + 10 = 314, each branch a, b, c, etc., is counted four times. Hence, the sum of branch lengths is 314/4 = 78.5. In general, for N sequences, $S_0 = \Sigma D_{ij} /(N - 1)$, where D_{ij} represents the distances between sequences i and j, $i < j$.

The next step of the neighbor-joining algorithm is like that of the Fitch-Margoliash method: a new distance table with A and B forming a single composite sequence is produced. The neighbor-joining algorithm is then used to find the next sequence pair and Fitch-Margoliash is then used to find the next branch lengths. The cycle is repeated until the correctly branched tree and the branch distances on that tree have been identified.

The neighbors relation method (Sattath and Tversky 1977; Li and Graur 1991) also is a reliable predictor of trees when the rate of evolution varies. In this method, the sequences are divided into all possible groups of four. The sum of the pair-wise distances for the three possible neighbor groupings (AB/CD, AC/BD, AD/BC) for each group are then compared to find which grouping of the three gives the lowest sum of pairs. This procedure is repeated for all possible groups of four. The pair that appears most often in the lowest sum of pairs is selected as neighbors. An example of this method is shown in Table 6.4. The pair is then treated as a composite grouping and the entire process is repeated to find the next closest neighbor until all of the sequences have been included.

The Unweighted Pair Group Method with Arithmetic Mean

The above distance methods provide a good estimate of an evolutionary tree and are not influenced by variations in the rates of change along the branches of the tree. The UPGMA

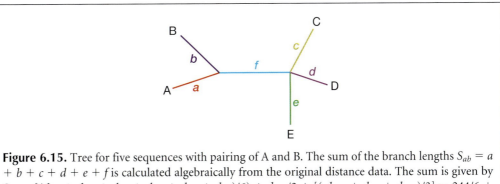

Figure 6.15. Tree for five sequences with pairing of A and B. The sum of the branch lengths $S_{ab} = a + b + c + d + e + f$ is calculated algebraically from the original distance data. The sum is given by $S_{ab} = [(d_{AC} + d_{AD} + d_{CE} + d_{BC} + d_{BD} + d_{BE})/6] + d_{AB} /2 + [(d_{CD} + d_{CE} + d_{DE})/3] = 244/6 + 22/2 + 48/3 = 67.7$. In general, the formula for N sequences when m and n are paired is $S_{mn} = [(\Sigma d_{im} + d_{in})/2(N - 2)] + d_{mn}/2 + \Sigma d_{ij}/N - 2$ where i and j represent all sequences except m and n, and $i < j$.

Table 6.4. *The Sattath and Tversky (1977) method for finding repeated neighbors*

Chosen set of 4	Sum of distances	Pairs chosen
ABCD	$n_{AB} + n_{CD} = 22 + 18 = 40$	AB, CD
	$n_{AC} + n_{BD} = 39 + 41 = 80$	
	$n_{AD} + n_{BC} = 39 + 41 = 80$	
	$n_{AB} + n_{CE} = 22 + 20 = 42$	
	$n_{AC} + n_{BE} = 39 + 43 = 82$	
ABCE	$n_{AE} + n_{BC} = 39 + 41 = 82$	AB, CE
	$n_{AB} + n_{DE} = 22 + 10 = 32$	
ABDE	$n_{AD} + n_{BE} = 39 + 43 = 82$	AB, DE
	$n_{AE} + n_{BD} = 41 + 41 = 82$	
	$n_{AC} + n_{DE} = 39 + 10 = 49$	
ACDE	$n_{AD} + n_{CE} = 39 + 20 = 59$	AC, DE
	$n_{AE} + n_{CD} = 41 + 18 = 59$	
	$n_{BC} + n_{DE} = 41 + 10 = 51$	
BCDE	$n_{BD} + n_{CE} = 41 + 20 = 61$	BC, DE
	$n_{BE} + n_{CD} = 43 + 18 = 61$	

Totals from Column 3 giving the number of times a pair gives the lowest score: AB (3), DE (3), CD (1), CE (1), and BC (1). AB and DE are therefore closest neighbors.

The five sequences used in the above example (see Fig. 6.13) are divided into the five possible groups of four. The sums of distances for each set of sequence pairs for the three possible groupings are then determined and the closest pairs in each grouping are determined. The closest neighbors overall are those that appear as neighbors most often. In this example, AB and DE appear most often as neighbors. These sequences are then chosen as neighbors to calculate the branch lengths on the phylogenetic tree by the method of Fitch and Margoliash.

method is a simple method for tree construction that assumes the rate of change along the branches of the tree is a constant and the distances are approximately ultrametric (see above). There are also a number of variations of this method for pairing or clustering sequences. The UPGMA method starts by calculating branch lengths between the most closely related sequences, then averages the distance between this pair or sequence cluster and the next sequence or sequence cluster, and continues until all the sequences are included in the tree. Finally, the method predicts a position for the root of the tree.

Using Example 2 from the above analysis:

Example: UPGMA Analysis

1. Sequences D and E are the most closely related. The branch distances d and e to the node below them are calculated as $d = e = n_{de}/2 = 5$ based on the assumption of an equal rate of change in each branch of the tree. The tree is often drawn in a form (Fig. 6.16a) where only the horizontal lines indicate branch lengths, but the branches are intended to be joined to a common node as in Figure 6.16B.

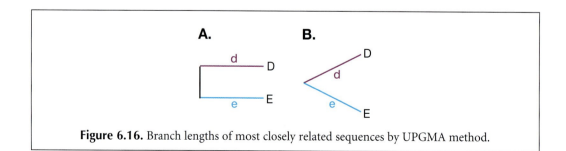

Figure 6.16. Branch lengths of most closely related sequences by UPGMA method.

2. Treating D and E as a composite sequence pair, find the next most related pair. The calculations will be similar to the FM method above and the distance between DE and C, $n_{DE,C} = 19$, will be the shortest one. Because we are assuming an equal rate of change in each branch of the tree, there will be two equal length branches, one including D and E and passing to a common node for C and DE, and a second from the common node to C. Some simple arithmetic gives $c = 19/2 = 9.5$ and $g = 9.5 - 5 = 4.5$ (Fig. 6.17).

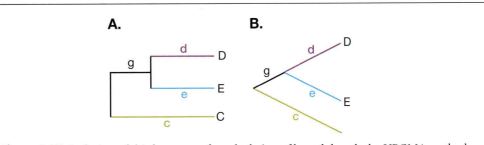

Figure 6.17. Inclusion of third sequence for calculation of branch lengths by UPGMA method.

3. With CDE now being treated as a composite trio of sequences, the next closest pair is A and B, giving an estimate of the distance between them and a common node in the tree of $a = b = n_{AB}/2 = 11$ (Fig. 6.18).

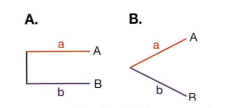

Figure 6.18. Inclusion of fourth and fifth sequences in UPGMA tree.

4. The final calculation is to take the average distance between the two composite sets of sequences CDE and AB. The average of n_{AC}, n_{AD}, n_{AE}, n_{BC}, n_{BD}, and $n_{BE} = 39 + 39 + 41 + 41 + 41 + 43 = 40.7$. One half of this distance $40.7/2 = 20.35$ is included in the part of the tree that goes from the root to CDE, and the other half goes from the root to AB. Note also that the presence of the root breaks the branch between AB and CDE, previously denoted f in this example, into two components $f1$ and $f2$. Hence, $f2 + g + d = 20.35$, $f2 = 20.35 - 4.5 - 5 = 10.85$, and $f1 + a = 20.35$, $f1 = 20.35 - 11 = 9.35$ (Fig. 6.19).

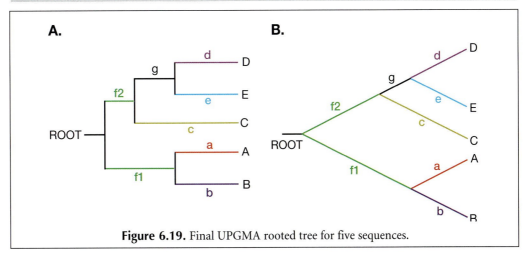

Figure 6.19. Final UPGMA rooted tree for five sequences.

The UPGMA method can lead to an erroneous tree if the rates of mutation in the branches of the tree are not uniform (Li and Graur 1991; Li 1997).

Choosing an Outgroup

If we have independently obtained information that certain sequences are more distantly related, a procedure may be followed which ensures that those sequences are added last to the tree and are closest to the root. This modification can improve the prediction of trees by the above methods by forcing the addition of the outgroup at a later stage in the procedure. One or more sequences of this type are referred to as an outgroup. Suppose, for example, that sequences A and B are from species that are known to have separated from the others at an early evolutionary time based on the fossil record. A and B may then be treated as an outgroup. Choosing one or more outgroups with the distance method can also assist with localization of the tree root (Swofford et al. 1996). The root will be placed between the outgroup and the node that connects the rest of the sequences. It is important that the sequence of the outgroup be closely related to the rest of the sequences, but also that there are significantly more differences between the outgroup and the other sequences than there are among the other sequences themselves. Choosing too distant a sequence as the outgroup may lead to incorrect tree predictions due to the more random nature of the differences between the distant outgroup and the other sequences (Li and Graur 1991; Li 1997). Multiple sequence changes at each site are more possible, and there has been more time for complex genetic rearrangements. For the same reason, using sequences that are too different in the distance method of phylogenetic prediction can lead to errors (Swofford et al. 1996). As the number of differences increases, the history of sequence changes at each site becomes more and more complex, and therefore much more difficult to predict. In choosing an outgroup, one is assuming that the evolutionary history of the gene under study is the same as that provided by the external information. If this assumption is incorrect, such as if horizontal gene transfer has occurred, an incorrect analysis could result.

Converting Sequence Similarity to Distance Scores

For determining phylogenetic relationships among a group of sequences, it is necessary to know the distances between the sequences. The majority of the available sequence alignments determine degree of similarity between sequences rather than distances. For simple scoring systems, similarity is a measure of the number of sequence positions that match in an alignment, whereas distance is the number of positions that are different and that must be changed to convert one sequence into the other. This difference reflects the number of changes that occurred since the sequences diverged from a common ancestor.

As outlined in Chapter 3, similarity methods provide an alignment score, and the significance of this score can be quite reliably calculated based on the probability that a score between unrelated sequences could achieve that score. What is needed is a way to convert such a score to a distance equivalent so that the appropriate phylogenetic analysis can be performed. A simple method, described and used above, is to count the number of different sequence pairs in an alignment. Another method is to convert the similarity score between two sequences to a normalized measure of similarity that varies from 0 for no similarity to 1 for full similarity. The distance can then be readily calculated.

Feng and Doolittle (1996) describe a method for calculating such a normalized score between a pair of aligned sequences. They calculate the similarity score between two sequences S_{real} for a given scoring matrix and gap penalty using a Needleman-Wunsch alignment algorithm (see Chapter 3). They then shuffle both sequences many times, align pairs of shuffled sequences using the same scoring system, and obtain a background average score S_{rand} for unrelated sequences. Finally, each sequence is aligned with itself to give a maximum score that could be obtained in an alignment of two identical sequences with the scoring system used, and the average of these two scores, S_{ident}, is calculated. The normalized similarity score S between the proteins is then given by

$$S = (S_{real} - S_{rand})/(S_{ident} - S_{rand}) \tag{1}$$

A different method for calculating S_{rand} from the scoring matrix, amino acid composition, and number of gaps in a multiple sequence alignment is also given (Feng and Doolittle 1996).

If, instead, a local alignment based on the Smith-Waterman algorithm is obtained (see Chapter 3), then the statistics of local similarity scores can be used. If λ and K have been calculated for a given scoring matrix and gap penalty combination, the standardized score of an alignment of score S_{rand} is given by

$$S' = \lambda S_{rand} - \log Kmn \tag{2}$$

where m and n are the sequence lengths. Recall that S' gives approximate probability of a higher score by $e^{-S'}$ (see Chapter 3, p. 109). A conservative value of 5 for S' corresponds to a probability of 7×10^{-3}. A value of S_{rand} is then given by

$$S_{rand}(p = 0.007) = 1/\lambda\,(5 + \log Kmn) \tag{3}$$

An expected value for S_{ident}, $S_{ident(calc)}$, is provided by the scoring matrix as the score for a match of identical amino acids (the scores along the diagonal of the log odds form of the amino acid substitution matrix) averaged over amino acid composition for the matrix. If s_{ii} is the score for a match and p_i is the proportion of each amino acid, the predicted score for an alignment of sequences of length m and n, $S_{ident(calc)}$, where n is the length of the shorter sequence, is given by

$$S_{ident(calc)} = n \sum_{i=1}^{20} p_i s_{ii} \tag{4}$$

where $\Sigma\, p_i = 1$. For the PAM250 matrix, the average expected score for a matched pair of identical amino acids is 4.95. Subtracting S_{rand} from this value is not appropriate because the score is not a local alignment score but a global one that grows proportional to sequence length. With the above changes, Equation 1 becomes

$$S = (S_{real} - S_{rand(p = 0.007)})/S_{ident(calc)} \tag{5}$$

Once the similarity score S has been obtained, it is tempting to calculate the distance between the sequences as $1 - S$. Recall that a simple model of amino acid substitutions is a constant probability of change per site per unit of evolutionary time. Accordingly, some of the observed substitutions in a sequence alignment represent a single amino acid change between the two sequences, but others represent two or more sequential changes. The model predicts that the expected number of 0, 1, 2, . . . substitutions is expected to follow the Poisson distribution, where D is the average number of substitutions. The calculated probability of zero changes is e^{-D}. The probability of one or more changes, which corresponds to S, is then given by $1 - e^{-D}$ such that

$$S = 1 - e^{-D} \tag{6}$$

Taking logarithms of both sides and rearranging then gives

$$D = -\log(S) \tag{7}$$

which is used to calculate D.

Example: Distance Calculation

Two sequences of length 250 have an alignment score of 700, using the PAM250 scoring matrix and gap penalties of -12, -2, which are small enough to give a long but local alignment score, then $\lambda = 0.145$ and $K = 0.012$ (Altschul and Gish 1996). Then $S_{\text{rand}(p = 0.007)} = 1 / 0.145 (5 + \log 0.012 \times 250 \times 250) = 80$ and $S_{\text{ident(calc)}} = 4.95 \times 250 = 1238$. Then, $S = (700 - 80) / 1238 = 0.50$, and $D = -\log 0.50 = \log 2 = 0.69$.

There are some additional points to make about the above procedure for calculating genetic distance from similarity scores:

1. Use of scoring matrices that are based on an evolutionary model are much preferred to matrices that are based on some other criterion. The Dayhoff PAM matrices meet this criterion but are based on a small data set. A more recent set of PAM matrices (Jones et al. 1992) discussed in Chapter 3 are based on a much larger data set and are based on the same evolutionary model as the Dayhoff matrices.

2. A scoring matrix that models the amino acid substitutions expected for a particular distance should be used. The PAM250 matrix models a separation giving only a remaining level 20% similarity. In the above example, the alignment should be rescored using the log odds PAM80 matrices, which model the expected substitution proteins that are 50% similar, and a better alignment score may be obtained. Suitable gap penalties will have to be found by trial and error, and statistical parameters will be calculated as described above. One must also be sure that the scoring system chosen provides a local alignment by demonstrating a logarithmic dependence of the growth of the alignment score on sequence length.

3. Note that Equation 7 provides an estimate of distance based on the observed similarity. The relationship only holds for sequences that are 50% or more similar. Beyond that point, so many multiple substitutions are possible that the distance essentially becomes 1.

4. When Feng and Doolittle perform distance calculations, they use multiple sequence alignments to assess the changes that occur in a family of related proteins. This method is a large improvement over aligning sequence pairs because

the presumed evolutionary changes can be seen in perspective of a whole related family of proteins. However, using multiple sequence alignment presents a brand new set of challenges that are discussed in Chapter 4.

The following sections describe two entirely different approaches for determining the evolutionary distance between related sequences.

Correction of Distances between Nucleic Acid Sequences for Multiple Changes and Reversions

In the above examples, the assumption is made that each observed sequence change represents a single mutational event. This assumption may be reasonable for sequences that are very much alike, but as the number of observed changes increases, the chance that two or more changes actually occurred at the same site and that the same site changed in both sequences increases. Some of the types of changes that may have occurred are illustrated in Figure 6.20. Note that of all the possible changes, only certain classes shown cause sequence variations.

In the PAM model of evolutionary change described in Chapter 3, such multiple evolutionary changes and reversions are taken into account for a fixed period of evolutionary time called 1 PAM, where 1 PAM roughly equals 10 million years (my). Such tables provide a way to score a sequence alignment by taking into account all possible changes that may have occurred. The PAM table is chosen that provides the highest log odds score between two sequences, and the PAM value of this table then provides a measure of the evolutionary distance between the sequences.

There are several models of evolutionary change of increasing complexity for correcting for the likelihood of multiple mutations and reversions in nucleic acid sequences. These models use a normalized distance measurement that is the average degree of change per length of aligned sequence. For example, in the 20-amino-acid-long sequence alignment given above, there are three changes between sequences A and B. Hence, $d_{AB} = n_{AB} / N = 3/20 = 0.15$.

The simplest model, called the Jukes-Cantor model, is that there is the same probability of change at each sequence position, and that once a mutation has occurred, that position is also just as likely to change again. The model also assumes that each base will eventually have the same frequency in DNA sequences (0.25) once equilibrium has been reached. It may be shown (Li and Graur 1991; Li 1997) that the average number of substitutions per site K_{AB} between two sequences A and B by this model is given by

$$K_{AB} = -3/4 \log_e [1 - 4/3 \, d_{AB}] \tag{8}$$

Thus, K_{AB} in the above example is $K_{AB} = -3/4 \log_e [1 - (4/3 \times 0.15)] = 0.17$, which is slightly greater than the observed number of changes (0.15) to compensate for some mutations that may have reverted. For more different sequences, such as A and D ($d_{AD} = 8/20 = 0.4$), the number of substitutions will be relatively higher than the observed number of changes. $K_{AD} = -3/4 \log_e [1 - (4/3 \times 0.4) = 0.57]$. Hence, the difference between the estimated and observed substitution rates will increase as the number of observed substitutions increases.

The Jukes-Cantor model has been modified to take into account unequal base frequencies (Swofford et al. 1996), which may be calculated from the multiple sequence alignment of the sequences.

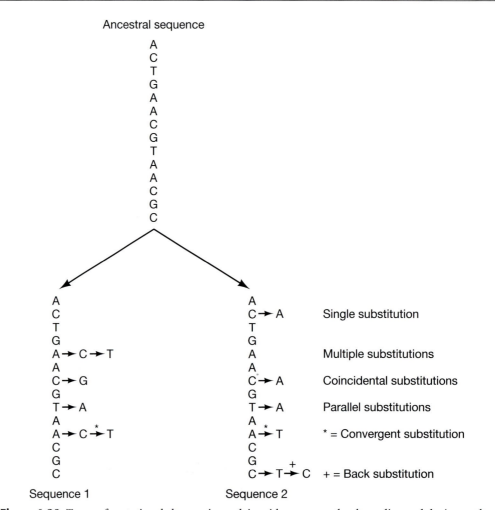

Figure 6.20. Types of mutational changes in nucleic acid sequences that have diverged during evolution. Note that the observed sequence changes between these homologous sequences represent only a fraction of the actual number of sequence variations that may have occurred during evolution and that multiple changes may have occurred at many sites. (Redrawn, with permission, from Li and Graur 1991 [copyright Sinauer Associates].)

$$K_{AB} = -B \log_e [1 - d_{AB}/B] \qquad (9)$$

where B is given by $B = 1 - (f_A^2 + f_G^2 + f_C^2 + f_T^2)$ and f_A is the frequency of A in the set of sequences, etc.

A slightly more complex model of change, the so-called Kimura two-parameter model (Kimura 1980), assumes that transition mutations should occur more often than transversions. However, there are four ways of obtaining a transition mutation A $\leftrightarrow$ G and C $\leftrightarrow$ T, but eight ways of making transversions, A $\leftrightarrow$ C, A $\leftrightarrow$ T, G $\leftrightarrow$ T, and G $\leftrightarrow$ C. Thus, in general, transversions can more readily be produced by multiple changes than transitions, and the frequency of each should be adjusted separately. This model also assumes that the eventual frequency of each base in the two sequences will be 1/4. In this case, it is necessary to calculate the proportion of transition and transversion mutations between two sequences. If the frequencies of transitions and transversions between two sequences A and B are $d_{ABtransition}$ and $d_{ABtransversion}$, respectively, if $a = 1 / (1 - 2d_{ABtransition} - d_{ABtransversion})$

and $b = 1 / (1 - 2d_{\text{ABtransversion}})$, and if the basic mutation rate to transitions and transversions is the same, the number of substitutions per site K_{AB} (Li and Graur 1991) is given by

$$K_{\text{AB}} = 1/2 \log_e (a) + 1/4 \log_e (b) \qquad (10)$$

For example, suppose that between two 20-nucleotide-long aligned sequences there are six transitions and two transversions, then $a = 1 / (1 - 2 \times 0.3 - 0.1) = 3.33$, $b = 1 / (1 - 2 \times 0.1) = 1.25$, and $K_{\text{AB}} = 1/2 \log_e (3.33) + 1/4 \log_e (1.25) = 0.66$. For comparison, by the Jukes-Cantor model, $K_{\text{AB}} = -3/4 \log_e [1 - 4/3 \times 8/20] = 0.57$. The larger predicted distance between A and B in the Kimura two-parameter model is due to the greater number of sequence changes in this model that could have given rise to the two observed transversion mutations.

The Jukes-Cantor and Kimura two-parameter models can be modified to take into account variations in the rates of mutation at different sites in the sequence alignment (see Swofford et al. 1996, p. 436), and there is also a Kimura three-parameter model that distinguishes between A $\leftrightarrow$ T / G $\leftrightarrow$ C transversions with A $\leftrightarrow$ C / G $\leftrightarrow$ T transversions. These various models are used in the distance methods for phylogenetic construction described above.

For distance calculations between sequences, these base-change models provide ways to improve estimates of the average mutation rate between sequences. They have less effect on phylogenetic predictions of closely related sequences and of the tree branch lengths, but a stronger effect on the more distantly related sequences.

Comparison of Protein Sequences and Protein-encoding Genes

One of the commonest types of phylogenetic comparisons made by biologists is to perform a multiple sequence alignment of a set of proteins using the BLOSUM50 or BLOSUM62 scoring matrix and then to design a phylogenetic tree using the neighbor-joining method. The fraction of sequence positions in an alignment that match provides a similarity score. Ambiguous matches and gaps may also be included in the scoring system for similarity. The distance, 1 minus the similarity score, is calculated and used to produce a tree. CLUSTALW and other programs described in Chapter 4 provide both an alignment and a tree.

Using amino acid variations for phylogenetic predictions offers several advantages. Amino acids confer structure and function to proteins. The order of variations in the tree may therefore provide information concerning the influence of the amino acids on function and of mutations associated with conservation of function and others with changes in function. The difficulty of using the above methods with protein sequences is that, in many cases, no evolutionary model of protein sequence variation is being used. Some amino acid substitutions are much more rare than others and should therefore reflect a longer evolutionary interval. Therefore, treating the substitutions equally may not provide the best phylogenetic prediction.

Another method for circumventing this problem is to use PAM scoring tables. Recall that as evolutionary distance between proteins increases, the expected pattern of amino acid changes varies. Rarer substitutions come into play, and the rate of increase of other changes with increasing time slows down. The Dayhoff PAM amino acid scoring matrices were designed to predict the expected substitutions for proteins separated by different evolutionary distances. The PAM score of the matrix that provides the best alignment score between two sequences reflects the evolutionary separation of the proteins, a distance of 1

PAM being approximately 10 my. Some phylogenetic programs use these original Dayhoff PAM tables. Another updated set of protein PAM tables based on changes in 40-fold more proteins (the PAM250 equivalent is called PET91) is also available (Jones et al. 1992). Some phylogenetic prediction methods use these PAM tables.

The PAM tables have been criticized for failure to take the mutational origin of amino acid changes into account. Although useful for analyzing amino acid variation, they do not allow for the multiple mutations required for some amino acid changes (see Chapter 3, p. 83). Amino acid variation arises through mutation and natural selection acting on DNA sequences. Some amino acid changes require several mutations in codons and should therefore be more rare than amino acid mutations, which require only one mutation in a codon.

Another method for comparing protein sequences is to assess the number of nucleic acid changes that are likely to generate the amino acid differences. In the original Fitch-Margoliash method, when only amino acid sequences were available, the distance between an amino acid pair was chosen to be the minimum number of base changes that would be required to change from a codon for the first amino acid into a codon for the second.

With the availability of the cDNA sequences that encode proteins, cDNA sequences may be compared instead of the amino acid sequences of the encoded proteins. Distance methods may be applied directly to the DNA sequence after the number of different positions in the sequences has been determined. If the protein sequences are very similar, most of the changes that will be observed are silent changes that do not change the amino acid and should provide an accurate representation of the phylogenetic history without the complications of evolutionary selection. However, as the amount of variation increases, the number of silent changes will increase and multiple mutations at some of these sites will occur, whereas at other sites, other more rare types of changes will appear. It is very difficult to make accurate predictions when faced with such variation in the rate of change at different sites. One method around this difficulty is to analyze changes in only the first and second base positions in each codon, ignoring the third position, which is the source of most silent mutations (Swofford et al. 1996). A comparison of nucleic acid sequences that encode proteins for mutations that either (1) change the amino acid or (2) do not change the amino acid may be made. Once these types of changes have been distinguished, phylogenetic predictions based on only one of them may be made.

A final type of correction that may be made to phylogenetic predictions is for the increase in multiple substitutions as the evolutionary distance between protein expected sequences increases. Although use of the PAM matrices provides this type of correction, another way is to adapt the Jukes-Cantor model for nucleic acid sequences to protein sequences. The correction to the distance is given by Equation 9, where $B = 19/20$ for the assumption of equal amino acid representation and $B = 1 - \Sigma f_{aai}$ for unequal representation of the amino acids, where f_{aai} is the frequency of amino acid i, and the sum is taken over all 20 amino acids. The second representation is, of course, much preferred, since amino acid frequencies in proteins vary.

Another correction that may be applied to protein distances is due to Kimura (1983). This correction is based on the Dayhoff PAM model of amino acid substitution. If K is the corrected distance and D the observed distance (number of exact matches between two sequences divided by total number of matched residues in alignment), then

$$K = -\ln(1 - D - 0.2\,D^2) \tag{11}$$

This formula may be used up to values of $D = 0.75$. Above this value, tables based on the Dayhoff PAM model at these distances are used. This correction is applied by

CLUSTALW, a commonly used program for multiple sequence alignment and phylogenetic analysis (Higgins et al. 1996).

Comparison of Open Reading Frames by Distance Methods

When nucleic acid sequences that encode proteins first became available, the appearance of synonymous substitutions that do not change the amino acid (silent changes) and nonsynonymous substitutions (replacement changes) that do change the amino acid was analyzed. Separate analyses of these two kinds of substitutions can help remove site-to-site variation in more closely related sequences and background noise of silent mutations in more distantly related sequences (Swofford et al. 1996).

One method of estimating the rates of synonymous and nonsynonymous mutations (Li et al. 1985; Li and Graur 1991; Li 1997) employs the following steps:

1. The fraction of substitutions at each codon position that can give rise to synonymous substitutions and the fraction that can give rise to nonsynonymous substitutions are counted. The first two positions of most codons count as two nonsynonymous sites because the amino acid will change regardless of the substitution. Similarly, many third-codon substitutions are synonymous. Other sites contribute synonymous and nonsynonymous substitutions. The total number of each of these two possible substitutions is determined for each sequence, and the average of these two values for the two sequences is then calculated. N_{syn} is the average number of synonymous sites and N_{nonsyn} is the average number of nonsynonymous sites in the two sequences.

2. Each pair of codons in the alignment is then compared to classify nucleotide differences into synonymous and nonsynonymous types. A single base difference can readily be designated as synonymous or nonsynonymous. When the codons differ by more than one substitution, all of the possible pathways of sequence change must be considered, and the number of synonymous and nonsynonymous changes in each pathway is identified. The average of each type of change in the two pathways is then calculated. Weights derived from the frequency of these pathways for known codon pairs may be used to derive this average, or else the pathways may be weighted equally. These calculations give the number of synonymous differences M_{syn} and the number of nonsynonymous differences M_{nonsyn} between the sequences.

3. The fraction of synonymous differences per synonymous site ($f_{syn} = N_{syn} / M_{syn}$) and the fraction of nonsynonymous differences per nonsynonymous site ($f_{nonsyn} = N_{nonsyn} / N_{nonsyn}$) are calculated. These fractions may then be corrected for the effect of multiple changes at the same site by the Jukes-Cantor formula (Eq. 8) or by some alternative method.

An alternative method for estimating synonymous and nonsynonymous substitutions (Li et al. 1985; Li and Graur 1991; Li 1993, 1997) is to classify each nucleotide position in the coding sequences as nondegenerate, twofold degenerate, or fourfold degenerate. The Genetics Computer Group program DIVERGE uses this method. A site is nondegenerate if all possible changes at this site are nonsynonymous, twofold degenerate if one of the three possible changes is synonymous, and fourfold degenerate if all possible changes are synonymous. For simplification, the third position of isoleucine codons (ATA, ATC, and ATT in the universal code) is treated as a twofold degenerate site even though in reality it is threefold degenerate. The number of each type of site in each of the two sequences is calculated and the average values for the two sequences are calculated. Each pair of codons in the sequence alignment is then compared to classify nucleotide differences as to type of site (nondegenerate, twofold degenerate, or fourfold degenerate) and as to whether the change is a transition or a transversion.

Calculation of Nonsynonymous and Synonymous Changes

To calculate these values, note that by definition all substitutions at nondegenerate sites are nonsynonymous, and all substitutions at fourfold degenerate sites are synonymous. At twofold degenerate sites, transitions nearly always produce synonymous changes, and transversions nearly always produce nonsynonymous changes. Hence, counting transitions and transversions at these sites provides a nearly exact count of the number of synonymous and nonsynonymous substitutions, respectively. One exception to this scoring scheme in the universal genetic code is that one type of transversion in the first position of the arginine codons produces a synonymous change, whereas the other transversion and the transition produce a synonymous change. Another exception is in the last position of the three isoleucine codons. When the codons differ by more than one substitution, a method similar to that described above is used to evaluate each possible pathway for changing one codon into the other, and the average of each type of change in the pathways is then calculated.

The scored codon differences are then used to calculate the proportions of each type of site that are transitions or transversions. The proportion of synonymous substitutions per synonymous site and the corresponding proportion for transversions may then be calculated. The two-parameter model of Kimura may be used to correct for multiple mutations and for differences between rates of transitions and transversions before these calculations are performed.

Example of Distance Analysis: Using the PHYLIP Programs DNADIST and FITCH (Fitch-Margoliash Distance Method)

A set of aligned DNA sequences was converted to the PHYLIP format and placed in a text file called infile in the same folder/directory as the programs (Fig. 6.21A). READSEQ may be used to produce a file with this format from a multiple sequence alignment. Note the required spacing of the sequences including spaces for a sequence name at the start of each row of sequence, and note that line 1 includes two numbers giving the number of sequences and the length of the alignment. Note also the presence of ambiguous sequence characters that are recognized appropriately by the program. Longer sequence alignments may be continued in additional blocks without the identifying names.

DNADIST was invoked, the program automatically read the infile, and after setting various options on a menu, an outfile was produced (Fig. 6.21B). This file was edited to remove all but the distance matrix shown. Note the required number on line 1 giving the number of taxa or sequences. Each distance is given twice as a mirror image about the upper-right to lower-left diagonal.

The predicted unrooted tree is given in the outfile and the treefile by the FITCH program. The average percent standard deviation of the predicted intersequence distance was 14, and 990 trees were analyzed to produce this result. The treefile was used as input to the program DRAWTREE, and shown in Figure 6.21C.

A. Sequences in Phylip format

```
  20    60
MACHIERH    AACNGGCCTT CTACTAGCCA TACACTACAC CGCAGACACC ACCCTAGCCT TTTCATCTGT
CIRCUS      AACTGGCCTN CTACTAGCAA CACACTATTC CGCAGACACT ACCCTGGCTT TCTCATCCGT
LOPHICTI    AACTGGCCTC CTACTGGCCA TGCACTACAC CGCAGACACA TCACTAGCCT TCTCGTCCGT
AQUILA      AACCGGCCTC CTATTAGCCA TACACTACAC GGCAGACACC ACCCTAGCCT TCTCATCCGT
ACCIPITE    AACCGGCCTC CTCCTAGCAA TACACTACAC CGAAGACACC ACCCTAGCCT TTTCATCAGT
BUTASTUS    AACCGGCCTC CTCCTAGCAA TACACTACAC CGCAGACACC ACCCTAGCCT TTTCATCAGT
HAERAETU    AACCGGCCTC CTACTAGCCA TGCACTACAC CGCAGACACC ACCCTAGCCT TCTCGTCCGT
```

B. DNA distances.

```
  20
MACHIERH    0.0000 0.1739 0.1705 0.0899 0.0899 0.0711 0.0899 0.1496 0.1292 0.1705 0.10
0.1292  0.1496
CIRCUS      0.1739 0.0000 0.2373 0.1921 0.2144 0.1921 0.1921 0.1292 0.1496 0.1496 0.21
0.2144  0.2853
LOPHICTI    0.1705 0.2373 0.0000 0.1674 0.2326 0.2102 0.0883 0.1885 0.1674 0.2557 0.18
0.1674  0.1468
AQUILA      0.0899 0.1921 0.1674 0.0000 0.1268 0.1073 0.0698 0.1268 0.1468 0.1885 0.08
0.0698  0.1674
ACCIPITE    0.0899 0.2144 0.2326 0.1268 0.0000 0.0169 0.1268 0.1468 0.1268 0.1674 0.14
0.1885  0.2326
BUTASTUS    0.0711 0.1921 0.2102 0.1073 0.0169 0.0000 0.1073 0.1268 0.1073 0.1468 0.12
0.1674  0.2102
HAERAETU    0.0899 0.1921 0.0883 0.0698 0.1268 0.1073 0.0000 0.1268 0.1073 0.1674 0.08
0.1268  0.1468
ELANUS      0.1496 0.2853 0.1468 0.1674 0.2326 0.2102 0.1468 0.2102 0.2326 0.2795 0.21
0.1268  0.0000
```

C. Fitch tree

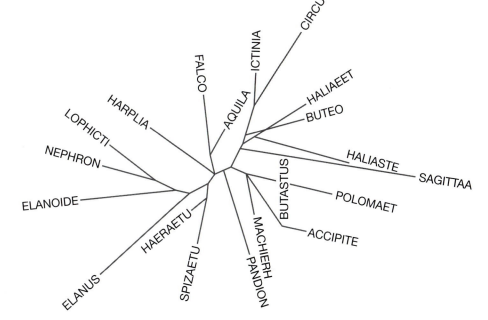

Figure 6.21. Tree predicted by FITCH (Fitch-Margoliash distance method) for the DNA sequences given in the example above.

THE MAXIMUM LIKELIHOOD APPROACH

This method uses probability calculations to find a tree that best accounts for the variation in a set of sequences. The method is similar to the maximum parsimony method in that the analysis is performed on each column of a multiple sequence alignment. All possible trees are considered. Hence, the method is only feasible for a small number of sequences. For each tree, the number of sequence changes or mutations that may have occurred to give the sequence variation is considered. Because the rate of appearance of new mutations is very small, the more mutations needed to fit a tree to the data, the less likely that tree (Felsenstein 1981). The maximum likelihood method resembles the maximum parsimony method in that trees with the least number of changes will be the most likely. However, the maximum likelihood method presents an additional opportunity to evaluate trees with variations in mutation rates in different lineages, and to use explicit evolutionary models such as the Jukes-Cantor and Kimura models described in the above section with allowances for variations in base composition. Thus, the method can be used to explore relationships among more diverse sequences, conditions that are not well handled by maximum parsimony methods. The main disadvantage of maximum likelihood methods is that they are computationally intense. However, with faster computers, the maximum likelihood method is seeing wider use and is being used for more complex models of evolution (Schadt et al. 1998). Maximum likelihood has also been used for an analysis of mutations in overlapping reading frames in viruses (Hein and Støvlbæk 1996). PAUP version 4 can be used to perform a maximum likelihood analysis on DNA sequences. The method has also been applied for changes from one amino acid to another in protein sequences.

PHYLIP includes two programs for this maximum likelihood analysis:

1. DNAML estimates phylogenies from nucleotide sequences by the maximum likelihood method, allowing for variable frequencies of the four nucleotides, for unequal rates of transitions and transversions, and for different rates of change in different categories of sites, as specified by the program.

2. DNAMLK estimates phylogenies from nucleotide sequences by the maximum likelihood method in the same manner as DNAML, but assumes a molecular clock.

One starts with an evolutionary model of sequence change that provides estimates of rates of substitution of one base for another (transitions and transversions) in a set of nucleic acid sequences, as illustrated in Table 6.5. The rates of all possible substitutions are chosen so that the base composition remains the same. The set of sequences is then aligned, and the substitutions in each column are examined for their fit to a set of trees that describe possible phylogenetic relationships among the sequences. Each tree has a certain likelihood based on the series of mutations that are required to give the sequence data. The probability of each tree is simply the product of the mutation rates in each branch of the tree, which itself is the product of the rate of substitution in each branch times the branch length. There are multiple sets of possible base changes within each tree to consider. For each column in the aligned sequences, the probability of each set of changes is found and the probabilities are then added to produce a combined probability that a given tree will produce that column in the alignment. A simple example of this approach is shown in Figure 6.22. Once all positions in the sequence alignment have been examined, the likelihoods given by each column in the alignment for each tree are multiplied to give the likelihood of the tree. Because these likelihoods are very small numbers, their logarithms are usually added to give the logarithm likelihood of each tree. The most likely tree given the data is then identified.

Table 6.5. *General model of sequence evolution*

Base	A	C	G	T
A	$-u(a\pi_C+b\pi_G+c\pi_T)$	$ua\pi_C$	$ub\pi_G$	$uc\pi_T$
C	$ug\pi_A$	$-u(g\pi_A+d\pi_G+e\pi_T)$	$ud\pi_G$	$ue\pi_T$
G	$uh\pi_A$	$uj\pi_G$	$-u(h\pi_A+j\pi_G+f\pi_T)$	$uf\pi_T$
T	$ui\pi_A$	$uk\pi_G$	$ul\pi_T$	$-u(i\pi_A+k\pi_G+l\pi_T)$

The table gives rates for any substitution in a nucleic acid sequence or for no substitution at all (the diagonal values). Base frequencies are given by π_A, π_C, π_G, and π_T, the mutation rate by u, and the frequency of change of any base to any other by a, b, $c..,l$. Rates of substitutions in one direction, i.e., A→G, are generally considered to be the same as that in the reverse direction so that $a = g$, $b = h$, etc. In the JC model these frequencies are all equal, and in the Kimura two-parameter there are only two frequencies, one for transitions (α) and the other for transversions (β), and the frequency for transitions is twice that for transversions. PAUP allows these numbers to be varied. This model assumes that changes in a sequence position constitute a Markov process, with each subsequent change depending only on the current base. Furthermore, the model assumes that each base position has the same probability of change in any branch of the tree (Swofford et al. 1996).

SEQUENCE ALIGNMENT BASED ON AN EVOLUTIONARY MODEL

A careful reading of these papers by those interested in evolutionary models of sequence changes is strongly recommended.

Thorne et al. (1991, 1992) have introduced a method of sequence alignment based on a model (Bishop and Thompson 1986) that predicts the manner in which DNA sequences change during evolution. Although this method has limitations and is only considered by these authors to be preliminary, it will be outlined here because of its relationship to the maximum likelihood method for phylogenetic analysis. The basis of this method is to devise a scheme for introducing substitutions, insertions, and gaps into sequences and to provide a probability that each of these changes occurs over certain periods of evolutionary time. Given each of these predicted changes, the method examines all the possible combinations of mutations to change one sequence into another. One of these combinations will be the most likely one over time. Once this combination has been determined, a sequence alignment and the distance between the sequences will be known. This method is different from the Smith-Waterman local alignment algorithm in identifying the most probable (maximum likelihood probability alignment) based on an evolutionary model of change in sequences, as opposed to a score based on observed substitutions in related proteins and a gap scoring system. The underlying mutational theory is, however, like those used to produce the PAM matrices for predicting changes in DNA and protein sequences.

Sequences are predicted to change by a Markov process (see Chapter 3 discussion of PAM matrices, p. 78) such that each mutation in the sequence is independent of previous mutations at that site or at other sites. For example, a given nucleotide at any sequence position can mutate into another at the same rate or may not change at all during a period of evolutionary time. This model is very similar to the PAM model of evolutionary change in proteins introduced by Dayhoff and discussed earlier. In the Thorne et al. (1991) model, single insertion–deletion events between any two nucleotides are modeled by a birth–death process that leaves the sequence length roughly the same. Longer insertion–deletion events were modeled in a similar way by considering the sequence to be composed of a set of fragments, and the rate of substitution of these fragments is allowed to vary (Thorne et al. 1992).

A set of transition probabilities for changing from one nucleotide to another or for introducing an insertion or deletion into a sequence is derived mathematically from the evolutionary model. The substitution probabilities are not unlike the substitution probabilities in the protein and DNA PAM matrices. An important difference between the PAM matrices and the transition probabilities is that the insertion/deletion probabilities have been derived from the evolutionary model rather than from the ad hoc gap penalty scoring system (penalty = gap opening penalty + gap extension penalty × length) that is commonly used to produce sequence alignments by dynamic programming. Two algorithms not unlike dynamic programming are then used, one to obtain a sequence alignment and the other to calculate the likelihood that the sequences are related (the likelihood of the

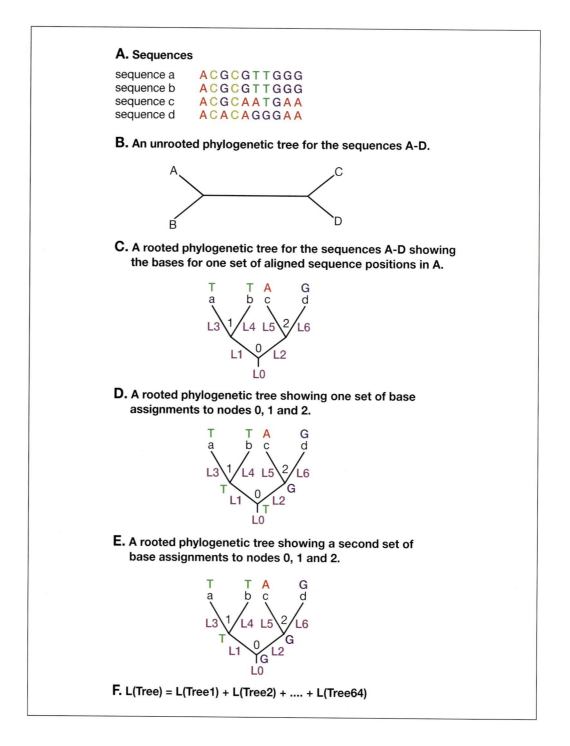

A. Sequences

sequence a ACGCGTTGGG
sequence b ACGCGTTGGG
sequence c ACGCAATGAA
sequence d ACACAGGGAA

B. An unrooted phylogenetic tree for the sequences A-D.

C. A rooted phylogenetic tree for the sequences A-D showing the bases for one set of aligned sequence positions in A.

D. A rooted phylogenetic tree showing one set of base assignments to nodes 0, 1 and 2.

E. A rooted phylogenetic tree showing a second set of base assignments to nodes 0, 1 and 2.

F. L(Tree) = L(Tree1) + L(Tree2) + ... + L(Tree64)

sequences) given the calculated set of parameters. The entries in the scoring matrices are likelihood scores (giving the highest probability of arriving at that position in the scoring matrix by a combination of mutations and gaps) and not a sum of weights for substitutions based on a scoring matrix. To estimate the likelihood of the sequences also requires that the number and types of substitutions, insertions, and deletions be optimized to find the most likely path for changing one sequence into another. This path then provides an indication of the evolutionary distance between the sequences.

Figure 6.22. Maximum likelihood estimation of phylogenetic tree. For the hypothetical sequences shown in A, one of three possible unrooted trees is shown in B. One column has been set aside for analysis. (C) One of five possible rooted derivatives of the unrooted tree is shown. The position of the root is not important since the likelihood of the tree is the same regardless of the root location. This property follows the assumption that the substitutions along each branch are considered to be a Markov chain with reversible steps (Felsenstein 1981). The bases from the marked alignment column are shown on the outer branches of this tree. Also shown are three interior nodes of the tree labeled 0, 1, and 2. The object is to consider every possible base assignment to these three nodes and then to calculate the likelihood of each choice. Since there are four possible bases for each of the three node positions of the tree, there are $4 \times 4 \times 4 = 64$ possible combinations. Also shown on the tree are six likelihood values L1–L5 for the probability of a base change per site along the respective branches of the tree, and a probability L0 for the base at node 0. These probabilities depend on the bases assigned to nodes 0, 1, and 2 and on the resulting types of base substitutions in the particular tree under consideration. The likelihood of a tree with a particular choice of bases at nodes 0, 1, and 2 is given by the product of the probability of the base at node 0 times the product of each of the substitution probabilities, or L(tree) = L0 $\times$ L1 $\times$ L2 $\times$ L3 $\times$ L4 $\times$ L5 $\times$ L6 (Felsenstein 1981). (D) A possible tree (tree1) with T assigned to nodes 0 and 1, and G assigned to node 2. L0 will be given by the frequency of T and will have an approximate value of 0.25. L2 will be the probability of a transversion of T to G, and L5 the probability of a transition of G to A in this tree. The remaining likelihoods will have an approximate value of unity with a small adjustment for the possibility that a mutation has occurred and then reverted to the original base so that no substitutions are observed. Assuming that the probabilities of the transition and transversion are 2×10^{-6} and 10^{-6}, respectively, the likelihood of tree1 is approximately $0.25 \times 2 \times 10^{-6} \times 10^{-6} = 5 \times 10^{-13}$. These numbers are usually very small and are therefore handled as logarithms in the computer. (E) Another possible arrangement of base assignments in tree2. The likelihood of this tree will take into account the probability of a G to T transversion (L1) and that of a G to A transition (L5). (F) The likelihood of the tree in B or the tree in C is given by the sum of the likelihoods of these two trees. To this sum is added the probability of the other 62 possible arrangements of bases. This calculation is repeated for all other columns in the multiple sequence alignment. The likelihood of the tree given the data in all of the aligned columns, that in the first column, or that in the second, etc., will be the sum of the likelihoods so calculated for each column. Each of the three possible trees for four sequences is then evaluated in this same manner and the one with the highest likelihood score is identified. These calculations can be computationally so intense for a large number of sequences that trees for a fraction of the sequences may first be found. The data for additional sequences will then be sequentially added to refine this initial tree. The procedure may then be repeated with a different starting group of sequences with the hope that the range of trees found will give an indication of the most likely tree (Felsenstein 1981). However, this procedure is not guaranteed to find the optimal tree. Additional calculations are made in the ML method. The probability of each branch in the tree is individually adjusted by a method similar to expectation maximization (see Chapter 3) to maximize the likelihood of the tree while holding the probability of the other branches at a constant value. The rate of evolution of each site or each column in the multiple sequence alignment is also allowed to vary. Otherwise, the method will be biased by sites that do not vary much and the information in variable sites may become lost, a problem shared with the maximum parsimony method. For an average number of mutations x over all branches, the number along an individual branch is assumed to vary according to the Poisson distribution $P(n) = e^{-x} x^n / n!$. A continuous variable giving the equivalent probability of observing a given number of changes along a particular branch for various average values of x (or a particular mutation rate along that branch) is given by the Γ distribution. These probabilities may then be used in calculations of tree likelihoods (Swofford et al. 1996).

RELIABILITY OF PHYLOGENETIC PREDICTIONS

As discussed earlier in this chapter, phylogenetic analysis of a set of sequences that aligns very well is straightforward because the positions that correspond in the sequences can be readily identified in a multiple sequence alignment of the sequences. The types of changes in the aligned positions or the numbers of changes in the alignments between pairs of sequences then provide a basis for a determination of phylogenetic relationships among the sequences by the above methods of phylogenetic analysis. For sequences that have diverged considerably, a phylogenetic analysis is more challenging. A determination of the sequence changes that have occurred is more difficult because the multiple sequence alignment may not be optimal and because multiple changes may have occurred in the aligned sequence positions. The choice of a suitable multiple sequence alignment method depends on the degree of variation among the sequences, as discussed in Chapter 4. Once a suitable alignment has been found, one may also ask how well the predicted phylogenetic relationships are supported by the data in the multiple sequence alignment.

In the bootstrap method, the data are resampled by randomly choosing vertical columns from the aligned sequences to produce, in effect, a new sequence alignment of the same length. Each column of data may be used more than once and some columns may not be used at all in the new alignment. Trees are then predicted from many of these alignments of resampled sequences (Felsenstein 1988). For branches in the predicted tree topology to be significant, the resampled data sets should frequently (for example, >70%) predict the same branches. Bootstrap analysis is supported by most of the commonly used phylogenetic inference software packages and is commonly used to test tree branch reliability. Another method of testing the reliability of one part of the tree is to collapse two branches into a common node (Maddison and Maddison 1992). The tree length is again evaluated and compared to the original length, and any increase is the decay value. The greater the decay value, the more significant the original branches. In addition to these methods, there are some additional recommendations that increase confidence in a phylogenetic prediction.

One further recommendation is to use at least two of the above methods (maximum parsimony, distance, or maximum likelihood) for the analysis. If two of these methods provide the same prediction, confidence in the prediction is much higher. Another recommendation is to pay careful attention to the evolutionary assumptions and models that are used for both sequence alignment and tree construction (Li and Graur 1991; Swofford et al. 1996; Li 1997).

COMPLICATIONS FROM PHYLOGENETIC ANALYSIS

The above methods provide a further level of sequence analysis by predicting possible evolutionary relationships among a group of related sequences. The methods predict a tree that shows possible ancestral relationships among the sequences. A phylogenetic analysis can be performed on proteins or nucleic acid sequences using any one of the three methods described above, each of which utilizes a different type of algorithm. The reliability of the prediction can also be evaluated.

The traditional use of phylogenetic analysis is to discover evolutionary relationships among species. In such cases, a suitable gene or DNA sequence that shows just enough, but not too much, variation among a group of organisms is selected for phylogenetic analysis. For example, analysis of mitochondrial sequences is used to discover evolutionary relationships among mammals. Two more recent uses of phylogenetic analysis are to analyze gene families and to trace the evolutionary history of specific genes. For example, database similarity searches discussed in Chapter 7 may identify several proteins in a plant genome that are similar to a yeast query protein. From a phylogenetic analysis of the protein family, the plant gene most closely related to the yeast gene and therefore most likely to have the same function can be determined. The prediction can then be evaluated in the laboratory. Tracking the evolutionary history of individual genes in a group of species can reveal which genes have remained in a genome for a long time and which genes have been horizontally transferred between species. Thus, phylogenetic analysis can also contribute to an understanding of genome evolution, as further explored in Chapter 10.

REFERENCES

Altschul S.F. and Gish G. 1996. Local alignment statistics. *Methods Enzymol.* **266:** 460–480.

Barns S.M., Delwiche C.F., Palmer J.D., and Pace N.R. 1996. Perspectives on archaeal diversity, thermophily and monophyly from environmental rRNA sequences. *Proc. Natl. Acad. Sci.* **93:** 9188–9193.

Bishop M.J. and Thompson E.A. 1986. Maximum likelihood alignment of DNA sequences. *J. Mol. Biol.* **190:** 159–165.

Brown J.R. and Doolittle W.F. 1997. *Archaea* and the procaryotic-to-eukaryote transition. *Microbiol. Mol. Biol. Rev.* **61:** 456–502.

Comeron J.M. and Kreitman M. 1998. The correlation between synonymous and nonsynonymous substitutions in *Drosophila:* Mutation, selection or relaxed constraints? *Genetics* **150:** 767–775.

Doolittle W.F. 1999. Phylogenetic classification and the universal tree. *Science* **284:** 2124–2128.

Felsenstein J. 1981. Evolutionary trees from DNA sequences: A maximum likelihood approach. *J. Mol. Evol.* **17:** 368–376.

———. 1988. Phylogenies from molecular sequences: Inferences and reliability. *Annu. Rev. Genet.* **22:** 521–565.

———. 1989. PHYLIP: Phylogeny inference package (version 3.2). *Cladistics* **5:** 164–166.

———. 1996. Inferring phylogeny from protein sequences by parsimony, distance and likelihood methods. *Methods Enzymol.* **266:** 368–382.

Feng D.F. and Doolittle R.F. 1996. Progressive alignment of amino acid sequences and construction of phylogenetic trees from them. *Methods Enzymol.* **266:** 368–382.

Fitch W.M. 1981. A non-sequential method for constructing trees and hierarchical classifications. *J. Mol. Evol.* **18:** 30–37.

Fitch W.M. and Margoliash E. 1987. Construction of phylogenetic trees. *Science* **155:** 279–284.

Hein J. and Støvlbæk J. 1996. Combined DNA and protein alignment. *Methods Enzymol.* **266:** 402–418.

Henikoff S., Greene E.A., Pietrokovski S., Bork P., Attwood T.K., and Hood L. 1997. Gene families: The taxonomy of protein paralogs and chimeras. *Science* **278:** 609–614.

Higgins D.G., Thompson J.D., and Gibson T.J. 1996. Using CLUSTAL for multiple sequence alignments. *Methods Enzymol.* **266:** 383–402.

Jin L. and Nei M. 1990. Limitations of the evolutionary parsimony method of phylogenetic analysis. *Mol. Biol. Evol.* **7:** 82–102.

Jones D.T., Taylor W.R., and Thornton J.M. 1992. The rapid generation of mutation data matrices from protein sequences. *Comput. Appl. Biosci.* **8:** 275–282.

Kimura M. 1980. A simple method for estimating evolutionary rates of base substitutions through comparative studies of nucleotide sequences. *J. Mol. Evol.* **16:** 111–120.

———. 1983. *The neutral theory of molecular evolution.* Cambridge University Press, Cambridge, United Kingdom.

Li W.-H. 1993. Unbiased estimation of the rates of synonymous and nonsynonymous substitution. *J. Mol. Evol.* **36:** 96–99.

———. 1997. *Molecular evolution.* Sinauer Associates, Sunderland, Massachusetts.

Li W.-H. and Graur D. 1991. *Fundamentals of molecular evolution*, pp. 106–111. Sinauer Associates, Sunderland, Massachusetts.

Li W.-H. and Gu X. 1996. Estimating evolutionary distances between DNA sequences. *Methods Enzymol.* **266:** 449–459.

Li W.-H., Wu C.I., and Luo C.C. 1985. A new method for estimating synonymous and nonsynonymous rates of nucleotide substitution considering the relative likelihood of nucleotide and codon changes. *Mol. Biol. Evol.* **2:** 150–174.

Maddison W.P. and Maddison D.R. 1992. MacClade: Analysis of phylogeny and character evolution (version 3). Sinauer Associates, Sunderland, Massachusetts.

Maidak B.L., Cole J.R., Parker C.T., Jr., Garrity G.M., Larsen N., Li B., Lilburn T.G., McCaughey M.J., Olsen G.J., Overbeek R., Pramanik S., Schmidt T.M., Tiedje J.M., and Woese C.R. 1999. A new version of the RDP (ribosomal database project). *Nucleic Acids Res.* **27:** 171–173.

Martin W. 1999. Mosaic bacterial chromosomes: A challenge en route to a tree of genomes. *Bioessays* **21:** 99–104.

Mayr E. 1998. Two empires or three? *Proc. Natl. Acad. Sci.* **95:** 9720–9723.

McDonald J.H. and Kreitman M. 1991. Adaptive protein evolution at the *Adh* locus in *Drosophila*. *Nature* **351:** 652–654.

Miyamoto M.M. and Cracraft J. 1991. *Phylogenetic analysis of DNA sequences.* Oxford University Press, New York.

Nielsen R. and Yang Z. 1998. Likelihood models for detecting positively selected amino acid sites and applications to the HIV-1 envelope gene. *Genetics* **148:** 929–936.

Pearson W.R., Robins G., and Zhang T. 1999. Generalized neighbor-joining: More reliable phylogenetic tree construction. *Mol. Biol. Evol.* **16:** 806–816.

Saitou N. 1996. Reconstruction of gene trees from sequence data. *Methods Enzymol.* **266:** 427–449.

Saitou N. and Nei M. 1987. The neighbor-joining method: A new method for reconstructing phylogenetic trees. *Mol. Biol. Evol.* **4:** 406–425.

Sankoff D. 1975. Minimal mutation trees of sequences. *SIAM J. Appl. Math.* **78:** 35–42.

Sattath S. and Tversky A. 1977. Additive similarity trees. *Psychometrika* **42:** 319–345.

Schadt E.E., Sinsheimer J.S., and Lange K. 1998. Computational advances in maximum likelihood methods for molecular phylogeny. *Genome Res.* **8:** 222–233.

Snel B., Bork P., and Huynen M.A. 1999. Genome phylogeny based on gene content. *Nat. Genet.* **21:** 108–110.

Swofford D.L., Olsen G.J., Waddell P.J., and Hillis D.M. 1996. Phylogenetic inference. In *Molecular systematics*, 2nd edition (ed. D.M. Hillis et al.), chap. 5, pp. 407–514. Sinauer Associates, Sunderland, Massachusetts.

Tatusov R.L., Koonin E.V., and Lipman D.J. 1997. A genomic perspective on protein families. *Science* **278:** 631–637.

Thorne J.L., Kishino H., and Felsenstein J. 1991. An evolutionary model for maximum likelihood alignment of DNA sequences. *J. Mol. Evol.* **33:** 114–134.

———. 1992. Inching toward reality: An improved likelihood model of sequence evolution. *J. Mol. Evol.* **34:** 3–16.

Woese C.R. 1987. Bacterial evolution. *Microbiol. Rev.* **51:** 221–271.

## CHAPTER 7

Database Searching for Similar Sequences

INTRODUCTION

DATABASE SIMILARITY SEARCHES have become a mainstay of bioinformatics. Large sequencing projects in which all the genomic DNA sequence of an organism is obtained have become quite commonplace. The genomes of a number of model organisms have been sequenced, including the budding yeast *Saccharomyces cerevisiae*, the bacterium *Escherichia coli*, the worm *Caenorhabditis elegans*, the fruit fly *Drosophila melanogaster*, and the human species *Homo sapiens*. These species have also been subjected to intense biological analysis to discover the functions of the genes and encoded proteins. Thus, there is a good deal of information available as to the biological function of particular sequences in model organisms that may be exploited to predict the function of similar genes in other organisms. In addition to genomic DNA sequences, complete cDNA copies of messenger RNAs that carry all the sequence information for the protein products have also been obtained for some of the expressed genes of various organisms. Translation of these cDNA copies provides a close-to-correct prediction of the sequence of the encoded proteins. Because obtaining intact cDNA sequences is laborious and time-consuming, a common practice is to make a library of partial cDNA sequences from the expressed genes, and then to perform high-throughput, low-accuracy sequencing of a large number of these partial sequences, known as expressed sequence tags (ESTs). The objective of an EST project is to find enough sequence of each cDNA and to have enough accuracy in the sequence that the amino acid sequence of a significant length of the encoded protein can be predicted. Overlapping ESTs can then be combined, and interesting ones can be found by database similarity searches. The full cDNA sequence of these genes of interest may then be obtained. Once all the sequence information is collected and placed in the sequence databases, the big task at hand is to search through the databases to locate similar sequences that are predicted to have a similar biological function through a close evolutionary relationship.

Sequence database searches can also be remarkably useful for finding the function of genes whose sequences have been determined in the laboratory. The sequence of the gene of interest is compared to every sequence in a sequence database, and the similar ones are identified. Alignments with the best-matching sequences are shown and scored. If a query sequence can be readily aligned to a database sequence of known function, structure, or biochemical activity, the query sequence is predicted to have the same function, structure, or biochemical activity. The strength of these predictions depends on the quality of the alignment between the sequences. As a rough rule, if more than one-half of the amino acid sequence of query and database proteins is identical in the sequence alignments, the prediction is very strong. As the degree of similarity decreases, confidence in the prediction also decreases. The programs used for these database searches provide statistical evaluations that serve as a guide for evaluation of the alignment scores.

Previous chapters have described methods for aligning sequences or for finding common patterns within sequences. The purpose of making alignments is to discover whether or not sequences are homologous or derived from a common ancestor gene. If a homology relationship can be established, the sequences are likely to have maintained the same function as they diverged from each other during evolution. If an alignment can be found that would rarely be observed between random sequences, the sequences are predicted to be related with a high degree of confidence. The presence of one or more conserved patterns in a group of sequence is also useful for establishing evolutionary and structure–function relationships among sequences.

The above methods of establishing sequence relationships have been utilized in database searches that are summarized in Table 7.1. In addition to standard searches of a sequence database with a query sequence (Table 7.1A), a matrix representation of a family of related protein sequences may be used to search a sequence database for additional proteins that are in the same family (Table 7.1B,C,D,), or a query protein sequence may be searched for the presence of sequence patterns that represent a protein family to determine whether the sequence belongs to that particular family (Table 7.1E). Genomic DNA sequences may also be searched for consensus regulatory patterns such as those representing transcription factor-binding sites, promoter recognition signals, or mRNA splicing sites; these types of searches are discussed in Chapter 8.

SEQUENCE SIMILARITY SEARCH WITH A SINGLE QUERY SEQUENCE

Searching a sequence database for sequences that are similar to a query sequence is the most common type of database similarity search. The search provides a list of database sequences with which the query sequence can be aligned. Once a list is available, additional searches may be performed using one of the initially found sequences as a query sequence. In this manner, the search may be expanded to find more distant relatives of the initial query sequence. Once a family of related sequences is found, the entire sequence may be aligned in a multiple sequence alignment, or the sequences may be analyzed for the occurrence of short regions of similarity, as described later in the chapter. Chapter 10 describes the use of those repetitive searches to identify families of paralogous proteins. Web sites and computational resources that support this type of database similarity searching are described in Table 7.2.

A common reason for performing a database search with a query sequence is to find a related gene in another organism. For a query sequence of unknown function, a matched gene may provide a clue as to function. Alternatively, a query sequence of known function (e.g., a yeast gene) may be used to search through sequences of a particular organism (e.g, a plant) to identify a gene that may have the same function. Sequences of an organism that are collected for such purposes include genomic sequences (sequences of BAC clones or the assembled sequence of an entire chromosome), EST sequences, and cDNA/protein sequences for particular genes. Database similarity searches may use one type of sequence (e.g., an EST sequence) to find matching EST sequences, genomic DNA sequences, or cDNA/protein sequences in the same organism. The Institute for Genomic Research (TIGR) has indexed a large number of EST sequences of model organisms in this manner (Table 7.2). These indexed databases may also be searched with a query sequence to identify related sequences.

ALLOWING FAST SEARCHES

When database searches were first attempted, machine size and speed were limiting factors that prevented use of a full alignment program, such as the dynamic programming algorithm, for each search. Although these considerations no longer apply due to the availability of more powerful machines, the sheer number of such searches that are presently performed on whole genomes creates a need for faster procedures. Hence, two methods that are at least 50 times faster than dynamic programming were developed. These methods follow a heuristic (tried-and-true) method that almost always works to find related sequences in a database search but does not have the underlying guarantee of an optimal solution like the dynamic programming algorithm. The first rapid search method was FASTA, which found short common patterns in the query and database sequences and

Table 7.1. *Types of database searches for proteins*

Type of search	Target database	Method	Type of query data	Examples of programs used, location (also see Tables 7.2, 7.4, 7.7, and 7.8)	Results of database search
A. Sequence similarity search with query sequence	protein sequence database (or genomic sequences[a])	search for database sequence that can be aligned with query sequence	single sequence, e.g., DAHQSNGA	FASTA (TFASTA[a]), SSEARCH http://fasta.bioch.virginia.edu/fasta/ BLASTP (TBLASTN[a]) http://www.ncbi.nlm.nih.gov/BLAST/ WU-BLAST http://blast.wustl.edu/	list of database sequences having the most significant similarity scores
B. Alignment search with profile (scoring matrix[b,d] with gap penalties)	protein sequence database	prepare profile from a multiple sequence alignment (Profilemake) and align profile with database sequence	profile representing gapped multiple sequence alignment, e.g., D-HQSNGA ESHQ-YTM EAHQSN-L EGVQSYSL	PROFILESEARCH ftp.sdsc.edu/pub/sdsc/biology	list of database sequences that can be aligned with the profile
C. Search with position-specific scoring matrix[c,d] (PSSM) representing ungapped sequence alignment (BLOCK)	protein sequence database	prepare PSSM from ungapped region of multiple sequence alignment or search for patterns of same length in unaligned sequences,[c] then use for database search	PSSM representing ungapped alignment, e.g., DAHQSN ESHQSY EAHQSN EGVQSY	MAST http://meme.sdsc.edu/meme/website/mast.html	list of database sequences with one or more patterns represented by PSSM but not necessarily in the same order
D. Iterative alignment search for similar sequences that starts with a query sequence, builds a gapped multiple alignment, and then uses the alignment to augment the search[d]	protein sequence database	uses initial matches to query sequence to build a type of scoring matrix and searches for additional matches to the matrix by an iterative search method[d]	builds matches to query sequence, e.g., DAHQSNGA iteration 1 H-SNGA EAHQSN-L → further iterations	PSI-BLAST http://www.ncbi.nlm.nih.gov/BLAST/	PSI-BLAST finds a set of sequences related to each other by the presence of common patterns (not every sequence may have same patterns).

| E. Search query sequence for patterns representative of protein families[e] | database of patterns found in protein families | search for patterns represented by scoring matrix or hidden Markov model (profile HMM)[e] | single sequence, e.g., DAHQSNGA | PROSITE http://www.expasy.ch/prosite INTERPRO http://www.ebi.ac.uk/interpro PFAM http://www.sanger.ac.uk/Pfam CDD/IMPALA http://www.ncbi.nlm.nih.gov/ Structure/cdd/cdd.shtml (also see Table 9.5) | list of sequence patterns found in query sequence |

[a] Searches of this type include the use of programs that search nucleic acid databases for matches to a query protein sequence by automatically translating the nucleic acid sequences in all six possible reading frames (TFASTA, TBLASTN). These searches may be useful when only genomic sequences or partial cDNA sequences (expressed sequence tag or EST sequences) of an organism are available. Genomic sequences that encode proteins may also have been found by gene prediction programs (Chapter 8). The predicted protein is then usually entered in the protein sequence databases. Matches to these predicted proteins may be found by searches of the protein sequence databases. These gene predictions are error-prone (see Chapter 8).

[b] A multiple sequence alignment that includes gaps may be represented by a profile, a type of scoring matrix discussed in Chapter 4, page 161. The consecutive rows of the matrix represent the multiple sequence alignment, and the column values represent the distribution of amino acids in each column of the alignment. The profile includes extra columns with gap opening and extension penalties. The profile is aligned to a sequence by sliding the profile along the sequence and finding the position with the best alignment score by means of a dynamic programming method. The alignment may include gaps in the database sequence. The best scoring alignments are with database sequences that have a pattern similar to that represented by the profile.

[c] The position-specific scoring matrix (PSSM), or weight matrix as it is sometimes called, is a representation of a multiple sequence alignment that has no gaps (a BLOCK). The matrix may be made from a multiple sequence alignment or by searching for patterns of the same length in a set of sequences using pattern-finding or statistical methods, e.g., expectation maximization, Gibbs sampling, ASSET, and by aligning these patterns, as discussed in Chapter 4. The consecutive columns of the matrix represent columns of the aligned patterns and the rows represent the distribution of amino acids in each column of the alignment. The PSSM columns include log odds scores for evaluating matches with a target sequence. The matrix is used to search a sequence for comparable patterns by sliding the matrix along the sequence and, at each position in the sequence, evaluating the match at each column position using the matrix values for that column. The log odds scores for each column are added to obtain a log odds score for the alignment to that sequence position. High log odds scores represent a significant match.

[d] Using a scoring matrix instead of a single query sequence can enhance a database search because the matrix represents the greater amount of sequence variation found in a multiple sequence alignment. Amino acid representation in each column of the alignment is also reflected in the matrix scores for that column; the more common an amino acid, the higher the score for a match to that amino acid. Note also that the matrix does not store any information about correlations between sequence positions. Thus, if two amino acids are commonly found together in the sequences at two positions of the alignment, these will each be independently scored by the matrix, but there will be no information as to their co-occurrence (or covariation) in the sequences. Since this type of information is missing, the matrix can give high scores to patterns that include new combinations of amino acids not found in the original set of sequences. Scoring covariation in sequence positions is discussed further in Chapters 5, 8, and 9.

[e] Pattern databases are described in Chapter 9.

Table 7.2. *Web resources for performing database searches with a simple query sequence*

Server/program	Web address or FTP site	Reference
BLAST—Basic Local Alignment Search Tool[a]	http://www.ncbi.nlm.nih.gov/BLAST FTP to ncbi.nlm.nih.gov/blast/executables	Altschul et al. (1990, 1997); Altschul and Gish (1996)
WU-BLAST[b]	sites that run WU-BLAST 2.0 are listed at http://blast.wustl.edu programs obtainable at http://blast.wustl.edu/ blast/executables with licensing agreement	Altschul et al. (1990, 1997); Altschul and Gish (1996)
FASTA[c]	http://fasta.bioch.virginia.edu/fasta FTP to ftp.virginia.edu/pub/fasta	Pearson (1995, 1996, 1998, 2000)
BCM Search Launcher (Baylor College of Medicine)	http://dot.imgen.bcm.tmc.edu:9331/	see Web site
TIGR gene indices search	http://www.tigr.org	see Web site

Additional resources for performing a database sequence search using a dynamic programming method are described in Table 7.7. There are also many other BLAST and FASTA servers on the Web, including ones for searches in specific organisms (see Chapter 10). The TIGR site is given as an example of such a site.

[a] A stand-alone BLAST server may also be established on a local machine running Windows, UNIX, or MacOS.

[b] Executable programs for UNIX platforms are available from the FTP site. Note the advice given to increase search speed in protein searches by an order of magnitude (http://blast.wustl.edu/blast/TO-FLY.html)

[c] Executable programs that run on PC, Macintosh, or UNIX platforms are available from the FTP site. The FASTA package also includes programs for performing pair-wise sequence alignments and for a statistical analysis of alignment scores (see Chapter 3). A number of Web sites offer FASTA database search, including the FASTA server and the BCM Search Launcher.

joined these into an alignment. BLAST, the next method, was similar to FASTA but gained a further increase in speed by searching only for rarer, more significant patterns in nucleic acid and protein sequences. BLAST is very popular due to availability of the program on the World Wide Web through a large server at the National Center for Biotechnology Information (NCBI) (http://ncbi.nlm.nih.gov) and at many other sites. The NCBI BLAST server site receives tens of thousands of requests a day. Both FASTA and BLAST have undergone evolution to recent versions that provide very powerful search tools for the molecular biologist and are freely available to run on many computer platforms. They are discussed further below.

With the more recent increased speed and size of computers and algorithmic improvements in the Smith-Waterman dynamic programming algorithm (described in Chapter 3), database similarity searches may also be performed by a search based on a full sequence alignment. The searches are 50-fold or more slower than FASTA and BLAST, but control experiments have revealed that more distantly related sequences will usually be found in a database search, provided that the appropriate statistical methods are used. A popular version of the Smith-Waterman program is SSEARCH (FTP to ftp.virginia.edu/pub/fasta), which is also available on Web sites but usually should be established on a local computer due to the length of time required for a search. Another recently introduced method for sequence alignment that has been used in database searches is the Bayes block aligner, described in Chapter 3 (p. 126). This program found more remotely similar sequences in protein families based on three-dimensional structure than did SSEARCH but is a much slower method (Zhu et al. 1998).

DNA VERSUS PROTEIN SEARCHES

One very important principle for database searches is to translate DNA sequences that encode proteins into protein sequences before performing a database search. DNA sequences comprise only four nucleotides, whereas protein sequences comprise 20 amino acids. Due to the fivefold larger variety of sequence characters in proteins, it is much easier to detect patterns of sequence similarity between protein sequences than between DNA sequences. Pearson (1995, 1996, 2000) has proven that searches with a DNA sequence encoding a protein against a DNA sequence database yield far fewer significant matches than searches using the corresponding protein sequence. To assist with an analysis based on translation of DNA sequences, both BLAST and FASTA provide programs that translate the query DNA sequence, the database DNA sequence, or both sequences in all six reading frames before making comparisons. An example of an exception to this rule would be a comparison of nucleic acid sequences in the same organism to locate other database entries of the same sequence. In such cases, a nucleic acid search would be needed.

When comparing methods of searching protein sequence databases, the sensitivity and selectivity of the methods should be considered. Sensitivity refers to the ability of the method to find most of the members of the protein family represented by the query sequence. Selectivity refers to the ability of the method not to find known members of other families as false positives. Ideally, both sensitivity and selectivity should be as high in quality as possible. A suitable method for describing both features is to describe the degree of coverage of families at a given level of false positives. Although similarity among many family members based on sequence similarity is readily identifiable, for some family members the similarity is weak and difficult to identify.

Identification of protein families is easier when the families are based on sequence similarity rather than on structural similarity, as discussed in detail in Chapter 9. Proteins that have the same structural features may have little, if any, sequence similarity. To facilitate a match of the query protein to a protein of known three-dimensional structure, protein sequences are grouped into families based on sequence similarity. All members of this family have sequence similarity with at least one of the remaining members, but not necessarily with all of the members, as illustrated in Figure 7.1. Families that include a protein of known three-dimensional structure are then identified. If a similarity search identifies a match of a query sequence with a member of such a protein family, the query sequence may be predicted to have a similar structure.

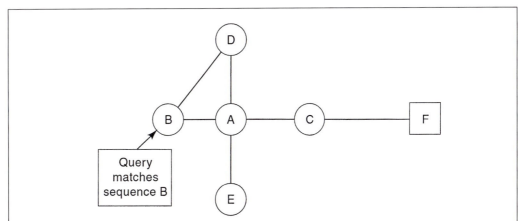

Figure 7.1. Structural prediction in database similarity searches. Sequences A–F refer to six members of a protein family defined by sequence similarity between some or all of the members. The sequences are represented as nodes on a graph, and similarity between sequences is represented by joining the nodes with a line (or edge). Note that not all nodes are joined. Thus, sequence A has detectable similarity to sequence B and to sequence C, but the relationship between sequence B and sequence C is not easily detectable. Suppose that sequence C can be aligned with sequence F, a protein of known three-dimensional structure. Hence, all members of this family may be predicted to have the same structure provided that the pair-wise alignments are significant and convincing. To gain further support for this prediction, improved alignments and identification of more family members to help bridge the similarity gaps are needed.

For protein sequence searches, two recent developments have greatly assisted with the finding of more distantly related sequences. First, combinations of amino acid substitution matrices and gap penalty scores that are most suitable for searches have been identified. Second, improved methods for establishing the statistical significance of a sequence alignment have been developed. Thus, whether a weak alignment between a query sequence and a database sequence is significant can be quite readily and confidently assessed. These topics are extensively discussed in Chapter 3 and on the book Web site and are reviewed below. Use of these new tools has also greatly improved the ability to balance sensitivity of a database search with selectivity.

SCORING MATRICES FOR SIMILARITY SEARCHES

There are a number of choices of amino acid substitution matrices for use in similarity searches of protein sequence databases (Henikoff and Henikoff 2000). The best performing matrices are now widely used, and they often are the default choice of the database search program. The most important consideration to be made is that the scoring matrix be in the log odds form so that statistical significance of the search results can be properly evaluated. In the log odds matrix, each matrix entry is the observed frequency of substitution of amino acids A and B for each other in proteins known to be related divided by the expected frequency of a chance substitution based on the frequency of A and B in proteins; the resulting ratio is then converted to a logarithm. The score is simply the logarithm of the odds that a pair of aligned amino acids is found because the sequences are related to a chance alignment of the pair in an alignment between unrelated sequences. The log odds form is useful because the probabilities that successive pairs in an alignment are related is the product of the odds of each pair. When log odds values are used, the probabilities may be found by addition in a much simpler calculation. Choice of the best scoring matrix for sequence alignments is discussed in detail in Chapter 3 and on the book Web site and is reviewed below.

PAM250 Scoring Matrix

For a long time, the Dayhoff PAM250 matrix was used for database searches. This scoring matrix is based on an evolutionary model that predicts the types of amino acid changes over long periods of time. The matrix is based on tallying the observed amino acid changes in a closely related group of proteins that were 85% identical. The proteins were organized into an evolutionary tree, and the predicted amino acid changes in the tree were used to estimate the frequency of substitution of each amino acid for another. These frequencies were then normalized to those expected if 1% of the sequence were to change, giving the PAM1 matrix. This level of change roughly corresponds to those amino acid changes expected over a period of 50 million years of evolutionary history. The substitution frequencies in the PAM1 matrix were then extrapolated to predict the changes occurring over longer periods of evolutionary time. For example, if D substitutes for E in the first PAM period, then in the second period, there is an additional chance that D might substitute for E. However, it is also possible that in a second PAM period the initial D substitution might revert to E or change to any other amino acid. As more time passes, the type and frequency of each substitution between the beginning and end of the time period will change. PAM250 represents a period of time at which only 20% of the amino acids will remain unchanged, but the expected frequencies are extrapolated many times from those observed in proteins that are 85% similar. Additional information concerning more recent substitution matrices that are based on an evolutionary model is discussed in Chapter 3 and on the book Web site. For many types of database searches, the PAM250 scoring matrix has been replaced by the BLOSUM matrices.

BLOSUM62 Scoring Matrix

The amino acid substitution matrix used by the BLAST programs is the BLOSUM62 scoring matrix. This matrix represents frequencies of amino acid substitutions observed in a large number of related proteins, including some quite similar and some quite different protein sequences. The observed substitutions are all lumped together to provide average frequencies of substitutions without regard to the degree of divergence between sequences. This approach appears to be more suitable for similarity searches in databases than using the Dayhoff PAM250 matrix, probably because sequences separated by any evolutionary distance may be more readily recognized. The Dayhoff matrices are also based on a much smaller data set than the BLOSUM62 matrix. The BLOSUM scoring matrices were generated by S. Henikoff and J.G. Henikoff (1992), who searched for common sequence patterns (blocks) of the same length among all of the related proteins in the Prosite catalog (see p. 428). They then added some additional related sequences in the current databases at the time and scored the columns in a multiple sequence alignment of these patterns for amino acid substitutions. In scoring the columns, some amino acid substitutions were much more common than others because many of the sequences had the same amino acid. The resulting BLOSUM matrices have a number to designate how much these repeated occurrences were weighted. The BLOSUM62 matrix uses only 62% of the repeats in one column and thereby reduces the relative weight given to those substitutions in the matrix. Another scoring matrix, BLOSUM50, which weights the repeated substitutions somewhat less, has been found to be more suitable for database searches by the FASTA and SSEARCH programs, which use different algorithms from BLAST. BLOSUM matrices give the best results when the appropriate gap opening and gap extension are used, as discussed in Chapter 3.

Other Scoring Matrices

In addition to the BLOSUM amino acid substitution matrices, a number of other scoring matrices have been devised. The usefulness of various combinations of search programs and substitution matrices for identifying that largest possible number of related sequences in a database search, including remotely related sequences, has been studied in considerable detail. These studies are extensively reviewed and referenced in Chapter 3 and on the book Web site.

LIMITING OUTPUT

Database similarity search programs tend to produce large volumes of output. It can become difficult to screen this volume of material and to assess whether or not the more remotely related sequences are really related to the query sequence. Thus, it is important to limit the sequence output; there are some relatively simple procedures that may be followed for each program, as described below. For searches of protein databases, avoid repetitive alignments with the same sequence by limiting searches to the protein sequence databases that are well curated, such as SwissProt and PIR, as opposed to translated GenBank sequences (the Genpept database).

METHODS

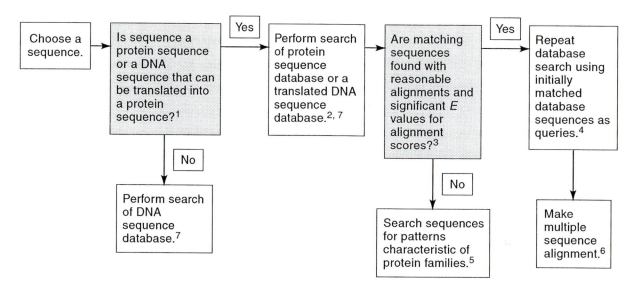

1. Translation of protein-encoding DNA sequences into protein sequences before performing sequence comparisons has been shown to be a more effective way to identify related genes than direct comparisons of untranslated DNA sequences. This method also corrects for different codon usage, base composition, and other DNA sequence variations by different organisms. However, to search for a matching DNA sequence in the same organism (e.g., a section of genomic DNA that is thought to encode a protein is used as a query against an EST database for the organism), a nucleic acid search is more appropriate. If the entire sequence does not encode a protein (e.g., the sequence is a genomic sequence that includes introns), the sequence can be translated in all six reading frames to locate open reading frames that may specify the amino acid sequence of a protein. The predicted translation product may then be compared to a protein sequence database or a DNA sequence database that is translated in all six reading frames. Alternatively, a gene annotation of the genomic DNA sequence—a predicted amino acid sequence for the protein encoded by the gene that has been entered into the protein sequence database—may be used, as described in Chapter 8. Masking low-complexity regions and sequence repeats in the query sequence is also necessary in many cases because such regions tend to give high-scoring alignments.

2. The carefully annotated protein sequence database (e.g., PIR, SwissProt) will provide a more manageable output list of matched sequences. However, investigators may also wish to expand the search to include predicted genes from gene annotations of genomic sequences (see note 1 and Chapter 8) that are frequently entered into the DNA sequence translation databases (e.g., DNA sequences in the GenBank DNA sequence databases automatically translated into protein sequences and placed in the Genpept protein sequence database). To compare a protein or predicted protein sequence to EST sequences of an organism, the ESTs should be translated into all six possible reading frames (Pearson 2000).

3. A matched database sequence that is listed should have a small E score and a reasonable alignment with the query sequence (or translations of protein-encoding DNA sequences should have these same features). The E (expect value) of the alignment score between the sequences gives the statistical chance that an unrelated sequence in the database or a random sequence could have achieved such a score with the query sequence, given as many sequences as there are in the database. The smaller the E score, the more significant the alignment. A cutoff value in the range of 0.01–0.05 is used (Pearson 1996). However, the alignment should also be examined for absence of repeats of the same residue or residue pattern because these patterns tend to give false high alignment scores. Filtering of low-complexity regions from the query sequence in a database search helps to reduce the number of false positives. The alignment should also be examined for reasonable amino acid substitutions and for the appearance of a believable alignment (see Chapter 3 flowchart for a summary). One of the sequences

may be shuffled many times, and each random sequence may be realigned with the other sequence to obtain a score distribution for a set of unrelated sequences. This distribution may then be used to evaluate the significance of the true alignment score (Chapter 3).

4. Including these extra steps may find additional members of a protein family that has too low a sequence similarity to the original query sequence to be detected in the first search.

5. These types of searches are discussed later in the chapter.

6. Methods and considerations that need to be made for producing a multiple sequence alignment are discussed in Chapter 4. Additional relationships among the matched sequences may be found by performing a phylogenetic analysis based on the multiple sequence alignments as described in Chapter 6. Such a phylogenetic analysis can reveal which sequence of several found in an organism is most closely related to a query sequence and therefore is the most likely of the group to have the same function as the query sequence.

7. For performing a large number of searches, there is a definite advantage to setting up the search programs on a local machine, especially since versions of the programs that run on most computer platforms are available. One can then set up batch commands or scripts (shell or Perl scripts) for processing the sequences and managing the returned data. The NCBI staff provides assistance in the form of SEALS (a system for analysis of lots of sequences) at http://ncbi.nlm.nih.gov/Walker/SEALS/index.html (Walker and Koonin 1997).

FASTA SEQUENCE DATABASE SIMILARITY SEARCH

FASTA is a program for rapid alignment of pairs of protein and DNA sequences. Rather than comparing individual residues in the two sequences, FASTA instead searches for matching sequence patterns or words, called *k*-tuples (Wilbur and Lipman 1983; Lipman and Pearson 1985; Pearson and Lipman 1988). These patterns comprise *k* consecutive matches in both sequences. The program then attempts to build a local alignment based on these word matches. Due to the ability of the algorithm to find matching sequences in a sequence database with high speed, FASTA is useful for routine searches of this type. Comparable methods are the BLAST algorithm, which is faster, and of comparable sensitivity for protein queries, and a Smith-Waterman dynamic programming algorithm, which is much slower but more sensitive when full-length protein sequences are used as queries. Detailed performance studies of these methods have been made, one showing that the Smith-Waterman dynamic programming algorithm and FASTA outperformed BLAST (Pearson 1995). The FASTA programs have all undergone recent enhancements that have improved detection of more remotely related sequences. For sequence fragments, FASTA is as good as Smith-Waterman methods. For DNA searches, FASTA is theoretically better able than BLAST to find matches because a *k*-tuple smaller than the minimum obligatory one of 7 (default size 11) for BLASTN (3 for TBLASTN, BLASTX, TBLASTX) may be used. For reviews on using FASTA, see Pearson (1995, 1996, 1998). The following information is largely based on these reviews and on information provided in the FASTA distribution package.

FASTA3

FASTA has gone through a series of updates and enhancements leading to version 3, denoted FASTA3. FASTA3 has improved methods of aligning sequences and of calculating the statistical significance of alignments. These changes result in a greatly increased ability of FASTA3 to detect distantly related sequences. The FASTA package is available by anonymous FTP from ftp.virginia.edu/pub/.

FASTA compares an input DNA or protein sequence to all of the sequences in a target sequence database and then reports the best-matched sequences and local alignments of these matched sequences with the input sequence. The input sequence is usually in the standard FASTA format, but it is also very easy to change sequence formats, as described in Chapter 2. FASTA finds sequence similarities between the query sequence and each database sequence in four steps illustrated in Figure 7.2.

In the initial stage of a search for regions of similarity, FASTA uses an algorithmic method known as hashing, illustrated in Table 7.3. In this method, a lookup table showing the positions of each word of length k, or k-tuple, is constructed for each sequence. The relative positions of each word in the two sequences are then calculated by subtracting the position in the first sequence from that in the second. Words that have the same offset position are in phase and reveal a region of alignment between the two sequences. Using hashing, the number of comparisons increases linearly in proportion to average sequence length. In contrast, the number of comparisons in dot matrix and dynamic programming methods increases between the square and the cube of the average sequence length. In FASTA, the k-tuple length is user-defined and is usually 1 or 2 for protein sequences (i.e., either the positions of each of the individual 20 amino acids or the positions of each of the 400 possible dipeptides are located). For nucleic acid sequences, the k-tuple is 4–6, and is much longer than for protein sequences because short k-tuples are much more common due to the four-letter alphabet of nucleic acids. The larger the k-tuple chosen, the more rapid, but less thorough, a database search.

Significance of FASTA Matches

The methods used by FASTA to report the significance of a database search were revised in later versions, and use of the latest version FASTA3 is strongly recommended. Similar methods are used by the database search program SSEARCH, which is based on a slower Smith-Waterman type of alignment. The statistical scores provide a reliable indication as to whether or not the alignment scores for sequences found in a database search are significant. This analysis provides the probability that scores between unrelated sequences could reach as high a value as those found for the higher-scoring alignments (Pearson 1998). The statistical distribution of scores found in a database search follows the extreme value distribution, described in detail in Chapter 3 (p. 96).

Recall that the parameters of the extreme value distribution, u and λ, vary with the length and composition of the sequences being compared, and also with the particular scoring system. In database searches, the expected score between the query sequence and an unrelated database sequence increases in proportion to the logarithm of the length of the database sequence. The parameters change when a different scoring system, e.g., a different scoring matrix or gap penalty, is used. FASTA calculates these parameters from the scores found with unrelated sequences during the database search. Some of the sequence scores in the database search arise from matches with related sequences and must be removed before the statistical calculations are performed. FASTA performs these tasks in the following manner:

1. The average score for database sequences in the same length range is determined.
2. The average score is plotted against the logarithm of average sequence length in each length range.
3. The points are then fitted to a straight line by linear regression.
4. A z score, the number of standard deviations from the fitted line, is calculated for each score.

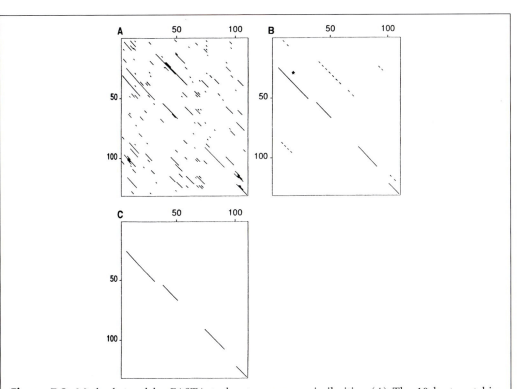

Figure 7.2. Methods used by FASTA to locate sequence similarities. (*A*) The 10 best-matching regions in each sequence pair are located by a rapid screen. First, all sets of *k* consecutive matches are found by a rapid method described below. For DNA sequences, *k* is usually 4–6 and for protein sequences, 1–2. Second, those matches within a certain distance of each other (for proteins, 32 for *k*=1 and 16 for *k*=2) are joined along with the region between them into a longer matching region without gaps. The regions with the highest density of matches are identified. The calculation is very much like a dot matrix analysis described in Chapter 3, but is calculated in fewer steps. The diagonals shown in *A* represent the locations of these common patterns initially found in the two sequences. (*B*) The highest-density regions of protein sequences identified in *A* are evaluated using an amino acid substitution matrix such as a PAM or BLOSUM scoring matrix. A corresponding matrix may also be used for DNA sequences. The highest-scoring regions, called the best initial regions (INIT1), are identified and used to rank the matches for further analysis. The best-scoring INIT1 region is shown marked by an asterisk. (*C*) Longer regions of identity of score INITN are generated by joining initial regions with scores greater than a certain threshold. The INITN score is the sum of the scores of the aligned individual regions less a constant gap penalty score for each gap introduced between the regions. Later versions of FASTA include an optimization step. When the INITN score reaches a certain threshold value, the score of the region is recalculated to produce an OPT score by performing a full local alignment of the region using the Smith-Waterman dynamic programming algorithm. By improving the score, this step increases the sensitivity but decreases the selectivity of a search (Pearson 1990). INITN and OPT scores are used to rank database matches. Finally, not shown, an optimal local alignment between the input query sequence and the best-scoring database sequences is performed based on the Smith-Waterman dynamic programming algorithm. (Reprinted, with permission, from Pearson and Lipman 1988.)

5. High-scoring, presumably related sequences, and also very low scoring alignments that do not fit the straight line are removed from consideration.

6. Steps 1–5 are repeated one or more times.

7. The known statistical distribution of alignment scores is used to calculate the probability that a *Z* score between unrelated or ran-

Table 7.3. *Lookup method for finding an alignment*

```
position    1  2  3  4  5  6  7  8  9  10  11
sequence  1 n  c  s  p  t  a  ·  ·  ·  ·   ·

position    1  2  3  4  5  6  7  8  9  10  11
sequence  2             a  c  s  p  r   k
```

amino acid	position in protein A	protein B	offset pos A - pos B
a	6	6	0
c	2	7	-5
k	–	11	
n	1	–	
p	4	9	-5
r	–	10	
s	3	8	-5
t	5	–	

```
Note the common offset for the 3 amino acids c, s, and p.
A possible alignment is thus quickly found

protein 1 n c s p t a
            | | |
protein 2 a c s p r k
```

Shown are fragments of two sequences that share a pattern c-s-p. All of the positions at which a given character is found are listed in a table. The positions of a given character in one of the sequences are then subtracted from the positions of the same character in the second sequence, giving an offset in location. When the offsets for more than one character are the same, a common word is present that includes those characters. Common words, or k-tuples, in two sequences are found by this method in a number of steps proportional to the sequence lengths.

dom sequences of the same lengths as the query and database sequence could be greater than z,

$$P\,(Z > z) = 1 - \exp(-e^{-1.2825z - 0.5772})_{ss} \qquad (1)$$

The derivation of this equation is given in Chapter 3, page 108.

The expectation E of observing, in a database of D sequences, no alignments with scores higher than z is given by e^{-DP} and that of observing at least one score z is $E \simeq 1 - e^{-DP}$. For $P<0.1$, this relationship is approximated by $E \simeq DP$ as indicated below.

$$E\,(Z > z) = D \times P\,(Z > z) \qquad (2)$$

8. Normalized similarity scores are calculated for each score by the formula $z' = 50 + 10z$. Thus, an alignment score with a standard deviation of 5 has a normalized score of 100. These normalized scores are reported in the program output.

9. The significance of an alignment score between a given sequence and a database sequence may be further analyzed by aligning a sequence with a shuffled library or a shuffled sequence with an unshuffled library (Pearson 1996) as described in Chapter 3, page 116.

An example of a database search with FASTA, vers. 3 is shown in Figure 7.3.

Versions of FASTA

There are several implementations of the FASTA algorithm (W. Pearson, release notes for FASTA vers. 3 and earlier releases; Pearson et al. 1997; Pearson 1998) using newly developed algorithms (Zhang et al. 1997):

1. FASTA compares a query protein sequence to a protein sequence library to find similar sequences. FASTA also compares a DNA sequence to a DNA sequence library.

2. TFASTA compares a query protein sequence to a DNA sequence library, after translating the DNA sequence library in all six reading frames.

3. FASTF/TFASTF and FASTS/TFASTS compare a set of short peptide fragments, as obtained from analysis of a protein, against a protein sequence database (FASTF/FASTS) or a DNA sequence database translated in all six reading frames (TFASTF/TFASTS). The FASTF programs analyze a set of fragments following cleavage and sequencing of protein bands resolved by electrophoreseis and the FASTS programs data from a mass spectrometry analysis of a protein. Note that a different sequence format is required to specify the separate peptides (see http://fasta.bioch.virginia.edu/fasta/).

Additional programs have been developed that are designed to align a DNA sequence with a protein sequence, allowing gaps and frameshifts. If a DNA sequence has a high possibility of errors, such as EST sequences, the translated sequence may be inaccurate due to amino acid changes or frameshifts. These programs are designed to go around such errors by allowing gaps and frameshifts in the alignments. FASTX and TFASTX allow only frameshifts between codons, whereas FASTY and TFASTY allow substitutions and frameshifts within a codon. These programs have been shown to be very useful for gene panning, the search for related sequences in EST databases (Retief et al. 1999).

1. FASTX and FASTY translate a query DNA sequence in all three reading forward frames and compare all three frames to a protein sequence database.

2. TFASTX and TFASTY compare a query protein sequence to a DNA sequence database, translating each DNA sequence in all six possible reading frames.

The above FASTA suite of programs is available as executable binary files for most computer systems including Windows, Macintosh, and UNIX platforms (ftp.virginia.edu/pub/FASTA).

The FASTA algorithm has also been adapted for searching through a pattern database instead of a sequence database (Ladunga et al. 1996). FASTA-pat and FASTA-swap are accessible at the Baylor College of Medicine Web site (http://dot.imgen.bcm.tmc.edu:9331/seq-search/Options/fastapat.html). Instead of comparing a query sequence to a sequence database, these programs compare the query sequence to a pattern database that contains patterns representative of specific protein families (see below, p. 326). A match between the query sequence and specific database patterns is an indication of a familial relationship between that sequence and the sequences from which those database patterns were generated.

Matching Regions of Low Sequence Complexity

FASTA and SSEARCH (described below) do not provide a method for avoiding low-complexity sequences or sequence repeats (Pearson 1998). Such regions can lead to higher scores between the query and database sequences than for other sequence pairs, thus giving the appearance that the sequences are related when they are actually not related. An

A. Score distribution

```
    z'    opt      E()
 < 20    176      0:==
   22      1      0:=                   one = represents 115 library sequences
   24      2      0:=
   26      2      2:*
   28     21     17:*
   30    109    102:*
   32    355    393:===*
   34   1053   1067:=========*
   36   2214   2191:===================*
   38   4004   3620:============================*===
   40   5285   5050:=====================================*==
   42   6350   6173:===============================================*==
   44   6716   6809:===============================================================*
   46   6847   6935:================================================================*
   48   6661   6640:===============================================================*
   50   5777   6059:======================================================= *
   52   5015   5327:==================================================  *
   54   4344   4550:===========================================  *
   56   3772   3801:====================================*
   58   3025   3120:=============================*
   60   2475   2528:======================*
   62   1968   2026:=================*
   64   1607   1612:==============*
   66   1362   1274:==========*
   68    983   1002:========*
   70    823    785:======*=
   72    597    614:=====*
   74    584    478:====*=
   76    456    372:===*
   78    306    289:==*
   80    238    225:=*=
   82    230    172:=*
   84    141    136:=*
   86    131    105:*=
   88     93     82:*            inset = represents 2 library sequences
   90     52     63:*
   92     55     49:*            :========================*===
   94     41     38:*            :==================*==
   96     37     29:*            :=============*====
   98     20     23:*            :=========  *
  100     20     17:*            :=======*=
  102     16     14:*            :======*=
  104      7     10:*            :====*
  106      9      8:*            :===*=
  108      7      6:*            :==*=
  110      3      5:*            :==*
  112      6      4:*            :=*=
  114      2      3:*            :=*
  116      4      2:*            :*=
  118      3      2:*            :*=
 >120     14      1:*            :*======
```

B. Fit of data to extreme value distribution.

```
26840295 residues in 74019 sequences
 statistics extrapolated from 50000 to 73831 sequences
 Expectation_n fit: rho(ln(x))= 5.9599+/-0.000515; mu= 7.4670+/- 0.029;
 mean_var=81.3676+/-15.767, Z-trim: 42   B-trim: 68 in 1/63
 Kolmogorov-Smirnov  statistic: 0.0106 (N=29) at   42
```

Figure 7.3. *Figure continues on next page.*

```
C.  Identification of database sequences which give high scoring alignment with probe sequence.

FASTA (3.14 April, 1998) function (optimized, BL50 matrix) ktup: 2
 join: 39, opt: 27, gap-pen: -12/ -2, width:  16 reg.-scaled

The best scores are:        initn  init1 opt    z-sc     E()
XPF_HUMAN   11/97  ( 905) 5893    5893 5893 6529.7   0
RA16_SCHPO  11/97  ( 892) 1569     519  749  827.2   2.1e-39
RAD1_YEAST  11/97  (1100)  975     362  619  681.7   2.7e-31
YIS2_YEAST  11/95  ( 993)   37      37  161  174.6   0.0047
YAXB_SCHPO  10/96  ( 578)   91      91  133  147.1   0.16
```

Figure 7.3. *Continued.* Example of a FASTA, Vers. 3 search. The SwissProt protein database was searched with the human XPF DNA repair protein on a local UNIX server with a locally written Web page interface. The recommended (default) BLOSUM50 amino acid scoring matrix and gap penalties of $-12/-2$ were used. Actual z scores are normalized to a mean of 50 and a standard deviation of 10 (normalized scores are indicated in this version of the program output in A as z', in B as z, and in C as Z). These values may be converted back to actual z scores for statistical calculations by subtracting 50 and dividing by 10. (*A*) Histogram of the normalized similarity scores and the expected score distribution. The first column gives the lower score in each range of scores, the second labeled "opt" is the number of optimized scores in that range, and the third labeled "E()" is the number of alignment scores expected to be in that range for unrelated sequences based on the extreme value distribution and the calculated values of u and λ. The "=" signs outline an approximate curve for the actual score distribution and the "*" gives the same information for the expected score distribution. Note the excellent agreement between the observed and expected numbers until a normalized score >120 is reached, at which point some high-scoring alignments are revealed. (*B*) An evaluation of the fit of the data to the expected curve is given by the Kolmogorov-Smirnov statistic, which compares the maximum deviation between the observed and expected values. In his FASTA distribution notes, W. Pearson indicates that statistic values <0.10 (for $N=30$) reveal excellent agreement. If this statistic is >0.2, he suggests repeating the analysis with higher gap penalties, e.g., -16, -4 rather than -12, -2. (*C*) Database sequences that have high normalized alignment scores are listed along with the raw init1, initn, opt, z' score, and E() for a z' score of that value. E() gives the probability that alignment of the query sequence with D database sequences unrelated to the query sequence could generate at least one such z' score. Note that the first row of scores is that for aligning the query sequence with a database copy of itself, followed by very high-scoring alignments to two yeast DNA repair genes on the next rows. (*D*) Smith-Waterman local alignments are shown along with additional information about the percent identity. A ":" in the alignment is an identity and "." is a conservative substitution. Included is a sketch indicating the extent to which the sequences can be locally aligned.
Figure continues on next page.

update of this feature can be anticipated in the near future. The BLAST2 programs described below filter regions of low complexity in both DNA and protein query sequences. Programs and Web sites for this purpose are described below in the description of BLAST. The program PRSS in the FASTA distribution package provides a straightforward way of establishing whether or not low complexity plays a role in the alignment score between two sequences. These programs shuffle the matching library sequences many times and realign each of the shuffled sequences with the query sequence. Two levels of shuffling are possible, one at the level of individual amino acids and a second at the level of sequence segments of a chosen length. The first method explores the possibility that restricted amino acid composition plays a role in the alignment, and the second that particular regions in the query sequence, such as sequence repeats, influence the score. If low complexity at either level is a problem, high scores will be produced when shuffled sequences are aligned with the query sequence. The distribution of scores from alignment between shuffled and query sequences is used to compute the statistical significance of the actual alignment score between the sequences. An example of using PRSS is presented in Chapter 3 (p. 116).

D. Local alignments of probe sequence and high-scoring database sequences.

```
>>XPF_HUMAN  11/97   ASCII   Len Q92889 homo sapi (905 aa)
 initn: 5893 init1: 5893 opt: 5893 Z-score: 6529.7 expect()     0
Smith-Waterman score: 5893;  100.000% identity in 905 aa overlap


>XPF_HU    1- 905:----------------------------------------:

               10        20        30        40        50
gi|284 MAPLLEYERQLVLELLDTDGLVVCARGLGADRLLYHFLQLHCHPACLVLV
       ::::::::::::::::::::::::::::::::::::::::::::::::::::
XPF_HU MAPLLEYERQLVLELLDTDGLVVCARGLGADRLLYHFLQLHCHPACLVLV
               10        20        30        40        50
......


>>RA16_SCHPO  11/97   ASCII   Len P36617 schizosa (892 aa)
 initn: 1569 init1: 519 opt: 749 Z-score: 827.2 expect() 2.1e-39
Smith-Waterman score: 1691;  34.056% identity in 922 aa overlap


>RA16_S    5- 896:----------------------------------------:

                 10        20        30        40
gi|284     MAPLLEYERQLVLELLDTDGLVVCARGLGADRLLYHFLQLHCHPAC
           :  :..:.  ::...  :::   :  :  ::.   ..   .  :.    :.
RA16_S METKVHLPLAYQQQVFNELIEEDGLCVIAPGLSLLQIAANVLSYFAVPGS
           10        20        30        40        50


           50        60        70        80        90
gi|284 LVLVLNTQPAEEEYFINQLKIEGVEHLPRRVTNEITSNSRYEVYTQGGVI
       :.:....   . : ....  ..:   :. .. ..:   : . .:..
RA16_S LLLLVGANVDDIELIQHEMESHLEKKLITVNTETMSVDKREKSYLEGGIF
              60        70        80        90       100


           100       110       120       130       140
gi|284 FATSRILVVDFLTDRIPSDLITGILVYRAHRIIESCQEAFILRLFRQKNK
       :::::::.:.::  ::... :::::..  .: :... .   :::.::.:.  ::
RA16_S AITSRILVMDLLTKIIPTEKITGIVLLHADRVVSTGTVAFIMRLYRETNK
```

Figure 7.3. *Continued.*

Recommended Steps for a FASTA Search

The following strategy is recommended for searches with FASTA for finding the most homologous sequences in a database search while avoiding false-negative matches (Pearson 1996, 2000):

1. Look for agreement between the real and theoretical distribution of scores. If the query sequence has a low-complexity, repeated domain or if the gap penalties are set too low, there may be an excess of unrelated sequences with E less than 0.1. If there is an excess of three- to fivefold more sequences than expected in the score range of 80–110, repeat the search after removing the low-complexity regions from the query sequence (see page 308 for a description of this method) or else increase the gap penalties from $-12/-2$ to $-14/-4$. Another test to apply is to examine the number of high-scoring unrelated sequences with E smaller than 1.0. If there are more than 5–10 such sequences, the analysis is suspect.

```
                       110        120        130        140        150
......

>>RAD1_YEAST  11/97  ASCII  Len P06777 saccharo (1100 aa)
 initn: 975 init1: 362 opt: 619 Z-score: 681.7 expect() 2.7e-31
Smith-Waterman score: 1366;  30.258% identity in 1008 aa overlap


>RAD1_Y     5- 892:----------------------------------------:

                                           10        20
gi|284                          MAPLLEYERQLVLE-LLDTDGLVVCARGL
                                  :  .....:  . :.  :.:..  ..::
RAD1_Y  EPDDIETSKPNINDIRPVDIQLTLPLPFQQKVVENSLITEDALIIMGKGL
           70        80        90        100       110

        30        40                  50                60
gi|284  GADRLLYHFLQLHCHPAC--------LVLVLNTQP--------AEEE--Y
        :    ..  ..:..   :.    ::::::..:.            : ::  .
RAD1_Y  GLLDIVANLLHVLATPTSINGQLKRALVLVLNAKPIDNVRIKEALEELSW
           120       130       140       150       160

                        70        80              90
gi|284  FINQLK------IEGVEHLPRRVTNEITSNS-----RYEVYTQGGVIFAT
        :  :   :     .:. ..: .:  :  .:..:    :  ..:  .:::..  :
RAD1_Y  FSNTGKDDDDTAVESDDELFERPFNVVTADSLSIEKRRKLYISGGILSIT
           170       180       190       200       210

          100       110       120       130       140
gi|284  SRILVVDFLTDRIPSDLITGILVYRAHRIIESCQEAFILRLFRQKNKRGF
        ::::::.:..:.   .    . .::.::  :   . .. .:.::::..:.::  ::
RAD1_Y  SRILIVDLLSGIVHPNRVTGMLVLNADSLRHNSNESFILEIYRSKNTWGF
           220       230       240       250       260
......
```

Figure 7.3. *Continued.*

2. Recall that the expect score *E* of a database match is the number of times that an unrelated database sequence would obtain a score higher than *z* just by chance. For a match to be significant, *E* should be < 0.01–0.05. If the search has correctly identified homologous sequences, the corresponding *E* values should be much less than 0.01, whereas scores between unrelated sequences should be much greater than this value; e.g, at least 0.5. If there are no *E* scores less than 0.1, the search has not found any sequences with significant similarity to the query sequence.

3. If there are no matches with *E* less than 0.1, repeat the search with FASTA with *k*-tuple = 1, or else use the Smith-Waterman dynamic programming method with a program such as SSEARCH. If the program now finds matches with *E* less than 0.02, the sequences may be homologous, if there is not a low-complexity region in the query sequence. Computer experiments with FASTA have revealed that sequences with scores of 0.2–10 may also be homologous but have marginal sequence similarity. For further study of this possibility, select some of these marginal sequences and use them as query sequences for additional database searches with FASTA. Additional family members with significant similarity may then be found.

4. Confirm homology of marginal matches by shuffling the query or database sequence many times to calculate the significance of the real alignment. The program PRSS described in Chapter 3, page 116, performs this task.

5. Protein sequence alignments with 50% identity in a short 20- to 40-amino-acid region are common in unrelated proteins. To be truly significant, the alignment should extend over a longer region.

BASIC LOCAL ALIGNMENT SEARCH TOOL (BLAST)

The BLAST algorithm was developed as a new way to perform a sequence similarity search by an algorithm that is faster than FASTA while being as sensitive. A powerful computer system dedicated to running BLAST has been established at NCBI, National Library of Medicine. Access to this BLAST system is possible through the Internet (http://www.ncbi.nlm.nih.gov/) as a Web site and through a BLAST E-mail server. There are also numerous other Web sites that provide a BLAST database search. In addition to the BLAST programs developed at the NCBI, an independent set of BLAST programs has been developed at Washington University (see Table 7.2). These programs perform similarity searches using the same methods as NCBI-BLAST and produce gapped local alignments. The statistical methods used to evaluate sequence similarity scores are different, and thus WU-BLAST and NCBI-BLAST can produce different results (see box below, point 11).

The BLAST Web server at http://www.ncbi.nlm.nih.gov/ is the most widely used one for sequence database searches and is backed up by a powerful computer system so that there is usually very little wait. Like FASTA, the BLAST algorithm increases the speed of sequence alignment by searching first for common words or *k*-tuples in the query sequence and each database sequence. Whereas FASTA searches for all possible words of the same length, BLAST confines the search to the words that are the most significant. For proteins, significance is determined by evaluating these word matches using log odds scores in the BLOSUM62 amino acid substitution matrix. For the BLAST algorithm, the word length is fixed at 3 (formerly 4) for proteins and 11 for nucleic acids (3 if the sequences are translated in all six reading frames). This length is the minimum needed to achieve a word score that is high enough to be significant but not so long as to miss short but significant patterns. FASTA theoretically provides a more sensitive search of DNA sequence databases because a shorter word length may be used. Like FASTA, the BLAST algorithm has gone through several developmental stages. The most recent gapped BLAST, or BLAST2, is recommended, as older versions of BLAST are reported to overestimate the significance of database matches (Brenner et al. 1998). The most important recent change is that BLAST reports the significance of a gapped alignment of the query and database sequences. Former versions reported several ungapped alignments, and it was more difficult to evaluate their overall significance. The statistical analysis of sequence alignments that made this change possible is discussed in detail in Chapter 3, page 97.

Steps Used by the BLAST Algorithm

Steps for searching a protein sequence database by a query protein sequence include the following (Altschul et al. 1990, 1994, 1997; BLAST Web server help pages):

1. The sequence is optionally filtered to remove low-complexity regions that are not useful for producing meaningful sequence alignments (see below).

2. A list of words of length 3 in the query protein sequence is made starting with positions 1, 2, and 3; then 2, 3, and 4, etc.; until the last 3 available positions in the sequence are reached (word length 11 for DNA sequences, 3 for programs that translate DNA sequences).

3. Using the BLOSUM62 substitution scores, the query sequence words in step 1 are evaluated for an exact match with a word in any database sequence. The words are also evaluated for matches with any other combination of three amino acids, the object being to find the scores for aligning the query word with any other three-letter word found in a database sequence. There are a total of 20 × 20 × 20 = 8000 possible match scores for this one sequence position. For example, suppose that the three-letter word PQG occurs in the query sequence. The likelihood of a match to itself is found in the BLOSUM62 matrix as the log odds score of a P-P match, plus that for a Q-Q match, plus that for a G-G match = 7 + 5 + 6 = 18. These scores are added because the BLOSUM62 matrix is made up of logarithms of odds of finding a match in sequences. To find three consecutive matches, the likelihoods of each pair are multiplied because we are asking that all characters match at the same time—the first pair and the second and the third. Adding logarithms of scores is the equivalent of multiplying the raw odds scores. Similarly, matches of PQG to PEG would score 15, to PRG 14, to PSG 13, and to PQA 12. For DNA words, a match score of +5 and a mismatch score of −4 are used, corresponding to the changes expected in sequences separated by a PAM distance of 40 (see p. 90).

4. A cutoff score called neighborhood word score threshold (T) is selected to reduce the number of possible matches to PQG to the most significant ones. For example, if this cutoff score T is 13, only the words that score above 13 are kept. In the above example, the list of possible matches to PQG will include PEG (15) but not PQA (12). The list of possible matching words is thereby shortened from 8000 of all possible to the highest scoring number of approximately 50.

5. The above procedure is repeated for each three-letter word in the query sequence. For a sequence of length 250 amino acids, the total number of words to search for is approximately 50 × 250 = 12,500.

6. The remaining high-scoring words that comprise possible matches to each three-letter position in the query sequence are organized into an efficient search tree for comparing them rapidly to the database sequences.

7. Each database sequence is scanned for an exact match to one of the 50 words corresponding to the first query sequence position, for the words to the second position, and so on. If a match is found, this match is used to seed a possible ungapped alignment between the query and database sequences.

8. (a) In the original BLAST method, an attempt was made to extend an alignment from the matching words in each direction along the sequences, continuing for as long as the score continued to increase, as illustrated below. The extension process in each direction was stopped when the accumulated score stopped increasing and had just begun to fall a small amount below the best score found for shorter extensions. At this point, a larger stretch of sequence (called the HSP or high-scoring segment pair), which has a larger score than the original word, may have been found.

```
        L   P       P   Q   G       L   L   QUERY SEQUENCE
        M   P       P   E   G       L   L   DATABASE SEQUENCE
                    <WORD>                  THREE LETTER WORD FOUND
                                            INITIALLY
                    7 2 6                   BLOSUM62 scores, word
                                            score = 15
        <------      ------>
EXTENSION TO LEFT    EXTENSION TO RIGHT
        2   7       7   2   6       4   4
        <           HSP         >   HSP SCORE = 9 + 15 + 8 = 32
```

(b) In the later version of BLAST, called BLAST2 or gapped BLAST, a different and much more time-efficient method is used. The method starts by making a list of high-scoring matching words, as in steps 1–4 above, with the exception that a lower value of T, the word cutoff score, such as 11 in our example, is used. This change results in a longer word list and matches to lower-scoring words in the database sequences. Matches between the query sequence and one database sequence are illustrated below. The x's mark positions of the words with scores at least as high as the new value of T. The object is to use these short matched regions lying on the same diagonal and within distance A of each other as the starting points for a longer ungapped alignment between the words. Once found, these joined regions are then extended using the method in part (a). Usually only a few such regions are extended. Because the new matches depend on finding two contiguous words, it is necessary to use a lower value of T to maintain the same level of sensitivity for detecting sequence similarity. The newly found diagonals are then scored by summing the scores of the individually matched sequence pairs (Fig. 7.4).

9. The next step is to determine whether each HSP score found by one of the above methods is greater in value than a cutoff score S. A suitable value for S is determined empirically by examining the range of scores found by comparing random sequences, and by choosing a value that is significantly greater. The high scoring pairs (HSPs) matched in the entire database are identified and listed.

10. BLAST next determines the statistical significance of each HSP score. A probability that two random sequences, one the length of the query sequence and the other the entire length of the database (which is approximately equal to the sum of the lengths of all of the database sequences), could achieve the HSP score is calculated. The topic of sequence statistics is discussed in detail in Chapter 3 and therefore the procedure is only reviewed briefly here. The main problem encountered is that scores between random sequences can reach extremely high values and become higher, the longer the random sequences. The probability p of observing a score S equal to or greater than x is given by the equation,

$$p\,(S \geq x) = 1 - \exp(-e^{-\lambda(x-u)}) \tag{3}$$

where $u = [\log\,(Km'n')]/\lambda$ and where K and λ are parameters that are calculated by BLAST for the amino acid substitution scoring matrix, n' is the effective length of the query sequence, and m' is the effective length of the database sequence. Methods for calculating the parameters K and λ are described in

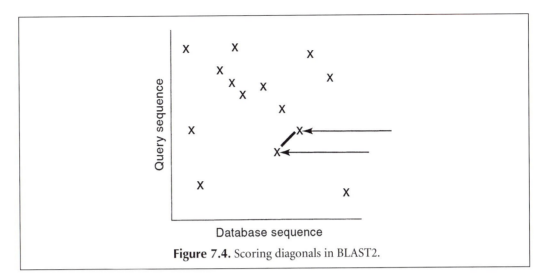

Figure 7.4. Scoring diagonals in BLAST2.

Chapter 3 and in greater detail on the book Web site.

The effective sequence lengths are the actual lengths of the query and database sequences less the average length of an alignment between two random sequences of the same length. m' and n' are calculated from the following relationship:

$$m' \approx m - (\ln Kmn)/H \qquad (4)$$

$$n' \approx n - (\ln Kmn)/H \qquad (5)$$

where H is the average expected score per aligned pair of residues in an alignment of two random sequences (Altschul and Gish 1996). H is calculated from the relationship $H = (\ln Kmn)/l$, where l is the average length of the alignment that can be achieved between random sequences of lengths m and n using the same scoring system as used in the database search. l is measured from actual alignments of random sequences. H is similar to the relative entropy of a scoring matrix described in Chapters 3 and 4, except that in this case, H is calculated from alignments of random sequences for a given scoring matrix, usually BLOSUM62. The basis for using these reduced lengths in statistical calculations is that an alignment starting near the end of one of the sequences is likely not to have enough sequence to build an optimal alignment. Using this correction also provides an improved match to statistical theory (Altschul and Gish 1996).

Note that the higher the value of H for a scoring matrix–gap penalty combination, the smaller the correction to the sequence length in Equations 4 and 5. Hence, to obtain alignments with shorter sequences, a scoring system with a higher H value is the most suitable combination. For example, for protein queries in the length range 50–85, the BLAST help pages recommend using BLOSUM80 with gap penalties $(-10,-1)$ instead of BLOSUM62 with gap penalties $(-11,-1)$ because the value of H is higher. To see these recommendations, click on the matrix link on the main BLAST page. For the BLOSUM62 scoring matrix and ungapped alignments, these values are $K = 0.14$ and $\lambda = 0.318$. The probability of the HSP score given by the above equation is adjusted to account for the multiple comparisons performed in the database search. The expectation E of observing a score $S \geq x$ in a database of D sequences is approximately given by the Poisson distribution,

$$\mathrm{E} \approx 1 - e^{-p(s > x)D} \tag{6}$$

and for $p < 0.1$, E is approximately pD. The expectation is the chance that a score as high as the one observed between two sequences will be found by chance in a search of a database of size D. Thus, $E = 1$ means that there is a chance that 1 unrelated sequence will be found in the database search. A similar expectation E is calculated by FASTA and SSEARCH.

11. Sometimes, two or more HSP regions that can be made into a longer alignment will be found, thereby providing additional evidence that the query and database sequences are related. In such cases, a combined assessment of the significance will be made. Suppose that two HSP scores are found; one set is 65 and 40, and the second 52 and 45. Which combination of scores is more significant, the one with the highest score (65 versus 52) or the one with the higher of the lower score of each set (45 versus 40)? Two methods have been used by BLAST for calculating this probability (Altschul and Gish 1996). One, the Poisson method, assumes that the probability of the multiple scores is higher when the lower score of each set is higher (45 is better than 40). The other, the sum-of-scores method, calculates the probability of the sum of the scores. In this example, $65 + 40 = 105$ is more significant than $52 + 45 = 97$. Earlier versions of NCBI-BLAST use the Poisson method; WU-BLAST (Washington University BLAST) and gapped BLAST use the sum-of-scores method. The most recent versions of NCBI-BLAST2 perform a local gapped alignment of the sequences and calculate the expect value of the alignment score. Such calculations became possible when it was realized that a statistical score could be calculated for gapped alignments (see Chapter 3, p. 112; Altschul and Gish 1996). To calculate the significance of the gapped alignment score, values of K and λ are determined on the basis of the alignment scores of random sequences using a combination of scoring matrix and gap penalties, and Equations 3 and 4 are then used.

12. Smith-Waterman local alignments are shown for the query sequence with each of the matched sequences in the database. Earlier versions of BLAST produced only ungapped alignments that included the initially found HSP. If two HSPs were found, two separate alignments were produced because the two regions could not be aligned without gaps. BLAST2 versions produce a single alignment with gaps that can include all of the initially found HSP regions. From the discussion of improvements in the dynamic programming alignment in Chapter 3 and on the book Web site, recall that the procedure of aligning of sequences may be divided into subalignments of the sequences, one starting at some point in sequence 1 and going to the beginning of the sequences, and another starting at the distal ends of the sequences and ending at the same position in sequence 1. A similar method is used to produce an alignment starting with the alignment between the central pair in the highest-scoring region of the HSP pattern as a seed for producing a gapped alignment of the sequences. The score of the alignment is obtained and the expect value for that score is calculated.

13. When the expect score for a given database sequence satisfies the user-selectable threshold parameter E, the match is reported. An example of a BLASTP v2 output file is shown in Figure 7.5.

A.

```
BLASTP 2.0.5 [May-5-1998]
Query= human XP-F repair gene            (905 letters)

Database: Non-redundant SwissProt sequences 74,596 sequences; 26,848,718 total letters
```

B.

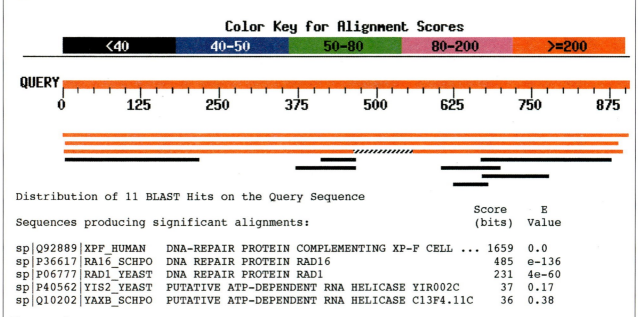

```
Distribution of 11 BLAST Hits on the Query Sequence
                                                    Score      E
Sequences producing significant alignments:        (bits)   Value

sp|Q92889|XPF_HUMAN   DNA-REPAIR PROTEIN COMPLEMENTING XP-F CELL ... 1659  0.0
sp|P36617|RA16_SCHPO  DNA REPAIR PROTEIN RAD16                        485   e-136
sp|P06777|RAD1_YEAST  DNA REPAIR PROTEIN RAD1                         231   4e-60
sp|P40562|YIS2_YEAST  PUTATIVE ATP-DEPENDENT RNA HELICASE YIR002C      37   0.17
sp|Q10202|YAXB_SCHPO  PUTATIVE ATP-DEPENDENT RNA HELICASE C13F4.11C    36   0.38
```

Figure continues on next page.

Figure 7.5. Example of BLASTP output. The BLAST server at http://www.ncbi.nlm.nih.gov/BLAST/, advanced version BLAST2 was given the human XP-F DNA repair sequence in FASTA format (providing the sequence accession number is another option). Program option BLASTP, database option SwissProt, and default program settings (gapped alignment, expectation value = 10, and low-complexity filtering), and 10 descriptions and alignments were chosen. Expectation value is the number of matches expected by chance between the query sequence and random or unrelated database sequences from a database of the size used. If the program is allowed to report all of the matches that it finds, the number found will include at least this many matches with unrelated sequences in the database. The BLAST Web site has excellent help pages that should be consulted, especially when new updates of BLAST have revised Web pages. For example, one revised page did not provide an option for changing the amino acid substitution matrix from the default BLOSUM62 scoring matrix. The comments given below summarize the results of the above BLAST2 search. (*A*) BLAST version number, query sequence, and sequence database are identified. (*B*) First, a graphical representation of the extent to which database sequences match the query sequence is shown. Note that three database sequences can be aligned with the entire length of the query sequence and are therefore likely to be highly significant alignments. Other alignments found are only with portions of the query sequence. The mouse may be used to go directly to the alignments represented in the graph. The scores of the requested 10 highest-scoring database sequences and the 1 identical database sequence are reported, one in each row. Each row includes the database sequence identifier where "sp" indicates a SwissProt match followed by SwissProt accession number and locus name, the score of the alignment in bits (see *C*), and the expectation value (*E*) of the alignment. *E* values of 0.0 and e^{-136} (which is 10^{-136}) in the first and second rows indicate that the match is highly significant. The first match is to the query sequence itself and the next two matches are closely related to the query sequence as indicated by their low *E* scores. If older versions of BLAST that give ungapped alignments or the ungapped option is used, or if the results are from BLASTX and TBLASTN searches, each row may have an additional column displaying *n*, the number of HSPs found and the probability of the sum of these HSP scores, as indicated in step 11 above. (*C*) Gapped alignments between the query sequence and the matched database sequences are shown. The query sequence is named as such and the database sequence is called the subject sequence. Note the filtering of a low-complexity region in the query sequence indicated by the replacement of sequence by X. Gaps are indicated by a dash. Shown in each alignment are the sequence ID and length, and the score of the alignment in bits ("score" is the sum of log odds scores of each matching amino acid pair in the alignment less gap penalties; the raw score in bits is the log odds score in units of logarithms to the base 2). The score shown in the program output is in units of normalized bits = ($\lambda \times$ raw score $-$ ln *K*) / ln 2. This number is independent of the scoring matrix used, but the raw score in bits is also shown in parentheses. The expectation value *E* of chance matches of unrelated sequences from a database of this size, percent identities in the alignment, percent positives in the alignment (identities plus positive scoring matches in the BLOSUM62 matrix), and percent of the alignment that is gaps are also shown. (*D*) Statistical information about the search is provided, including the numbers found in the steps outlined above. The statistical parameters *K* and λ, which are different for gapped and ungapped alignments (Chapter 3), and the gap penalty scores are also shown. This information will be useful as a basis for adjustment of the basic input parameters.

C.

```
sp|Q92889|XPF_HUMAN DNA-REPAIR PROTEIN COMPLEMENTING XP-F CELL (XERODERMA PIGMENTOSUM GROUP F COMPLEMENTING PROTI
(DNA EXCISION REPAIR PROTEIN ERCC-4) Length = 905  Score = 1659 bits (4249), Expect = 0.0  Identities = 838/905 (
Positives = 838/905 (92%)

Query:   1  MAPLLEYERQLVLELLDTDGLVVCARGLGADRLLYHFLQLHCHPACLVLVLNTQPAEEEY  60
            MAPLLEYERQLVLELLDTDGLVVCARGLGADRLLYHFLQLHCHPACLVLVLNTQPAEEEY
Sbjct:   1  MAPLLEYERQLVLELLDTDGLVVCARGLGADRLLYHFLQLHCHPACLVLVLNTQPAEEEY  60
.
.
Query: 301  SLRATEKAFGQNSGWLFLDSSTSMFINARARVYHLPDAXXXXXXXXXXXXXXXXXXXXXXX  360
            SLRATEKAFGQNSGWLFLDSSTSMFINARARVYHLPDA
Sbjct: 301  SLRATEKAFGQNSGWLFLDSSTSMFINARARVYHLPDAKMSKKKEKISEKMEIKEGEETKK  360
.
.
.

sp|P36617|RA16_SCHPO DNA REPAIR PROTEIN RAD16 Length = 892  Score =    485 bits (1236), Expect = e-136  Identities =
303/918 (33%), Positives = 497/918 (54%), Gaps = 76/918 (8%)

Query:   5  LEYERQLVLELLDTDGLVVCARGLGADRLLYHFLQLHCHPACLVLVLNTQPAEEEYFINQ  64
            L Y++Q+ EL++ DGL V A GL   ++  P  L+L++    + E     ++
Sbjct:   9  LAYQQQVFNELIEEDGLCVIAPGLSLLQIAANVLSYFAVPGSLLLVGANVDDIELIQHE   68
.
.
.

Query: 304  ------ATEKAFGQNSGWLFLDSSTSMFINARARVYHLPDAXXXXXXXXXXXXXXXXXXXXX  358
                  ++ + Q S WL LD++ M  AR RVY  +
Sbjct: 309  LSVNVSSYPSNAQPSPWLMLDAANKMIRVARDRVYKESEGPNMDAIP------------   355
.
.

sp|P06777|RAD1_YEAST DNA REPAIR PROTEIN RAD1 Length = 1100  Score =   231 bits (583), Expect = 4e-60  Identities =
136/369 (36%), Positives = 208/369 (55%), Gaps = 37/369 (10%)

Query: 559  LHEVEPRYVVLYDAELTFVRQLEIYRASRPGKPLRVYFLIYGGSTEEQRYLTALRKEKEA  618
            L E+ P Y+++++ +++F+RQ+E+Y+A      +VYF+ YG S EEQ +LTA+++EK+A
Sbjct: 704  LQEMMPSYIIMFEPDISFISFIRQIEVYKAIVKDLQPKVYFMYYGESIEEQSHLTAIKREKDA  763
.
```

Figure 7.5. *Continued.*

```
sp|P40340|TBP7_YEAST TAT-BINDING HOMOLOG 7  Length = 1379  Score = 31.7 bits (70), Expect = 5.6  Identities = 12.
(21%), Positives = 29/55 (51%)

Query:  625  EKASMVVPEEREGRDETNLDLVRGTASADVSTDTRKAGGQEQNGTQQSIVVDMRE  679
             +K  V+PE+  +E   +L++ T ++++++TD +   +E   + S+  + E
Sbjct: 1209  DKEKAVIPEDSGANEEYTTELIQATCTSEITTDDDERARKEPKENEDSLQTQVTE  1263

D.

Database: Non-redundant SwissProt sequences
Number of letters in database: 26,848,718
Number of sequences in database: 74,596
Lambda        K        H
0.320       0.136    0.394
Gapped
Lambda        K        H
0.270       0.0470   0.230
Matrix: BLOSUM62
Gap Penalties: Existence: 11, Extension: 1
Number of Hits to DB: 4277291
Number of Sequences: 74596
Number of extensions: 1706128
Number of successful extensions: 4638
Number of sequences better than 10.0: 12
Number of HSP's better than 10.0 without gapping: 4
Number of HSP's successfully gapped in prelim test: 8
Number of HSP's that attempted gapping in prelim test: 4616
Number of HSP's gapped (non-prelim): 16
length of query: 905
length of database: 26848718
effective HSP length: 55
effective length of query: 850
effective length of database: 22745938
effective search space: 19334047300
effective search space used: 19334047300
T: 11
A: 40
X1: 16 ( 7.4 bits)
X2: 38 (14.8 bits)
X3: 64 (24.9 bits)
S1: 41 (21.8 bits)
S2: 68 (30.9 bits)
```

Figure 7.5. *Continued.*

Sequence Filtering

The BLAST programs include a feature for filtering the query sequence through programs that search for low-complexity regions or for sequence repeats. Note that filtering is only applied to the query sequence and not to the database sequence. These regions are marked with an X (protein sequences) or N (nucleic acid sequences) and are then ignored by the BLAST program. Such regions tend to give high scores that do not reflect sequence similarity but rather the occurrence of low-complexity or repetitive sequences. Removing these types of sequences increases emphasis on the more significant database hits. The NCBI programs SEG and PSEG are used for amino acid sequences, and NSEG for nucleic acid sequences (Wootten and Federhen 1993, 1996). The SEG programs are available by anonymous FTP from ncbi.nlm.nih.gov/pub/seg, including documentation. The program DUST is also used for DNA sequences (see http://www.ncbi.nlm.nih.gov/BLAST/filtered.html).

Regions of low-complexity or repetitive sequences may be readily visualized in a dot matrix analysis of a sequence against itself (see Chapter 3, p. 63). Low-complexity regions with a repeat occurrence of the same residue can appear on the matrix as horizontal and vertical rows of dots representing repeated matches of one residue position in one copy of the sequence against a series of the same residue in the second copy. Repeats of a sequence pattern appear in the same matrix as short diagonals of identity that are offset from the main diagonal (see Fig. 3.6). Sequence complexity may also be analyzed by examining the fraction of all possible residues that are represented in a sequence window.

The compositional complexity in a window of sequence of length L is given by (Wootten and Federhen 1996)

$$K = 1/L \log_N (L!/\prod_{\text{all } i} n_i!) \tag{7}$$

where N is 4 for nucleic acid sequences and 20 for protein sequences, and n_i are the numbers of each residue in the window. K will vary from 0 for very low complexity to 1 for high complexity. Thus, complexity is given by:

For the sequence GGGG,
$L! = 4 \times 3 \times 2 \times 1 = 24$
$n_G = 4 \ n_C = 0 \ n_T = 0 \ n_A = 0$
$\prod_{\text{All } i} n_i = 4 \times 3 \times 2 \times 1 \times 0! \times 0! \times 0! = 24 \times 1 \times 1 \times 1 = 24$
$K = 1/4 \log_4 (24/24) = 0$
For the sequence CTGA,
$L! = 24$
$n_G = n_C = n_T = n_A = 1$
$\prod n_i = 1$
All $K = 1/4 \log_4 (24/1) = 0.573$

Compositional complexities are sometimes calculated to logarithms to the base 2 to produce scores in bit units. A sliding window (usually 12 residues) is moved along the sequence, and the complexity is calculated at each position. Regions of low complexity are identified, neighboring regions are then combined, and the resulting region is then reduced to a single optimal segment by a minimization procedure. SEG is used for analysis of either proteins or nucleic acids by the above methods. PSEG and NSEG are similar to SEG but are set up for analysis of protein and nucleic acid sequences, respectively. These versatile programs may also be used for locating specific sequence patterns that are characteristic of exons (Chapter 8) or protein structural domains (Chapter 9). In database searches involving comparisons of genomic DNA sequences with EST sequence libraries, use of repeat masking is important for filtering the output to the most significant matches (Claverie 1996).

In addition to low-complexity regions, BLAST will also filter out repeat elements (such as human SINE and LINE retroposons; see Chapter 10). Another filtering program for repeats of periodicity less than 10 residues (XNU; Claverie and States 1993) is used by the BLAST stand-alone programs, but is not available on the NCBI server.

Another Web server, RepeatMasker (http://ftp.genome.washington.edu/cgi-bin/) screens sequence for interdispersed repeats known to be present in mammalian genomes and also can filter out low-complexity regions (A.F.A. Smeet and P. Green, see Web site above). A dynamic programming search program (cross-match, P. Green, see Web site) performs a search of a repeat database with the query sequence (Claverie 1996). A database of repetitive elements (Repbase) maintained at http://www.girinst.org/!server/repbase.html) by the Genetics Information Research Institute (Jurka 1998) can also be used for this purpose.

Other BLAST Programs and Options

There are a number of different versions of the BLAST program for comparing either nucleic acid or protein query sequences with nucleic acid or protein sequence databases. If necessary, the programs translate nucleic acid sequences in all six possible reading frames to compare them to protein sequences. These BLAST programs are shown in Table 7.4 along with the types of alignment, gapped or ungapped, that they produce. Table 7.5 lists the databases that are available, and Table 7.6 lists the options and parameter settings that are available on the BLAST server. These various options may be chosen and are also described on the main BLAST Web page at http://www.ncbi.nlm.nih.gov/. The results produced by a sample BLASTP version 2 output are shown and described in Figure 7.5.

1. BLAST CLIENT (BLASTcl3) is a network-client BLAST that may be established on a local machine and used to access the BLAST2 server (FTP at ncbi.nlm.nih.gov/blast/network/netblast) rather than using a Web browser.

2. Stand-alone BLAST. Executable versions of all of the BLAST programs for Windows, Macintosh, and UNIX platforms are available (FTP at ncbi.nlm.nih.gov/blast).

3. BLAST E-mail server. When the BLAST server is busy so that the interactive Web page is slow and unresponsive, an alternative is to send the job by E-mail and to have the results returned by E-mail. A standard format is required in the E-mail message, as shown in Figure 7.6. The format changes periodically, therefore it is a good idea to send for the current format by sending the message help to the BLAST E-mail server, BLAST@ncbi.nlm.nih.gov. Note that there are obligatory and optional lines in the E-mail message.

Other BLAST-related Programs

1. BLAST-enhanced alignment utility (BEAUTY). BEAUTY adds additional information to BLAST search results, including figures summarizing the information on the locations of HSPs and any already known domains and sites present in the matching

Table 7.4. *BLAST programs provided by the National Center for Biotechnology Information*

Program	Query sequence	Database	Type of alignment[a]
BLASTP	protein	protein	gapped
BLASTN	nucleic acid	nucleic acid	gapped
BLASTX	translated nucleic acid[b]	protein	each frame gapped
TBLASTN	protein	translated nucleic acid[b]	each frame gapped
TBLASTX[c]	translated nucleic acid[b]	translated nucleic acid[b]	ungapped

[a] Type of alignment available between query and database sequences in BLAST2. A gapped alignment is usually preferred, if available. BLASTX and TBLASTN generate gapped alignment for each reading frame found and may use sum statistics. TBLASTX provides only ungapped alignments and sum statistics. Ungapped alignments available as option for BLASTP and BLASTN.

[b] Nucleic acid sequence is translated in all six possible reading frames and then compared to the protein sequence.

[c] TBLASTX is a heavy user of computer resources and therefore cannot be used with the nr nucleic acid database on the BLAST Web page.

Table 7.5. *Databases available on BLAST Web server*

Database	Description
A. Peptide sequence databases	
nr	translations of GenBank DNA sequences with redundancies removed, PDB, SwissProt, PIR, and PRF
month	new or revised entries or updates to nr in the previous 30 days
swissprot	latest release of the SwissProt protein sequence database[a]
Drosophila genome	provided by Celera and Berkeley Drosophila genome project
yeast	yeast (*Saccharomyces cerevisiae*) genomic sequences
E. coli	*E. coli* genomic sequences
pdb	sequences of proteins of known three-dimensional structure from the Brookhaven Protein Data Bank
yeast	yeast (*S. cerevisiae*) protein sequences
E. coli	*E. coli* genomic coding sequence translations
pdb	sequences of proteins of known three-dimensional structure from the Brookhaven Protein Data Bank
kabat [kabatpro]	Kabat's database of sequences of immunological interest
alu	translations of select *Alu* repeats from REPBASE, a database of sequence repeats
B. Nucleotide sequence databases	
nr	GenBank, EMBL, DDBJ, and PDB sequences with redundancies removed (EST, STS, GSS, and HTGS sequences excluded)
month	new or revised entries or updates to nr in the previous 30 days
dbest[b]	EST sequences from GenBank, EMBL, and DDBJ with redundancies removed
dbsts[b]	STS sequences from GenBank, EMBL, and DDBJ with redundancies removed
htgs[b]	high-throughput genomic sequences
kabat [kabatnuc]	Kabat's database of sequences of immunological interest
vector	vector subset of GenBank
mito	database of mitochondrial sequences
alu	select *Alu* repeats from REPBASE, a database of sequence repeats; suitable for masking *Alu* repeats from query sequences
epd	eukaryotic promoter database
gss[b]	genome survey sequences, includes single-pass genomic data, exon-trapped sequences, and *Alu* PCR sequences

[a] The SwissProt database is carefully curated but not always up to date because updates are released after longer intervals. SwissProt and PIR are the preferred protein databases for searches because the nr protein database is a composite of several databases and has duplicates of many sequences. Unfortunately, PIR is not provided as a separate choice on the database menu.

[b] Databases containing sequences that may have been less accurately determined.

Example of request to the BLAST email server. The choices are as listed in the above tables. Options are shown in Table 7.6.

A. Mandatory email format and commands

```
To:  BLAST@ncbi.nlm.nih.gov    email address of BLAST server
Subject:                       subject line ignored by server
Message:
PROGRAM BLASTn                 BLAST program to run
DATALIB nr                     database to be searched
BEGIN                          indicates start of sequence in FASTA format
>name of sequence              name of the query sequence after a '>'
tgcttggctgaggagccataggacgagagct  (the sequence itself)
caccaccatggacagcaaa
Blank line                     leave a blank line at end of sequence
```

B. Optional commands (see above tables for further explanation): these commands are inserted before the sequence.

```
NCBI_GI                NCBI ID to be displayed in output - useful for
                       sequence retrieval
HTML                   output suitable for viewing by web browser
DESCRIPTIONS 10        number of descriptions, limits output
ALIGNMENTS 10          number of alignments, limits output
EXPECT 0.5             number of matches to be expected by chance alone,
                       low number important for reducing output
MATRIX PAM120          symbol comparison table
FILTER SEG/NONE        remove low complexity regions from query sequence
                       type SEG, SEG+XNU or XNU for protein, DUST for DNA
                       sequences, or NONE for no filtering. XNU removes
                       segments consisting of short-periodicity internal
                       repeats
GCODE 1                alternate genetic code see Table 7.6.
PATH address           senders e-mail address
```

Figure 7.6. Example of request to the BLAST E-mail server.

Example: BEAUTY program output

A. Relative location of each HSP within query sequence with the sequence accession number linked to the individual reports listed below (also shown by BLAST v. 2).

Locally-aligned regions (HSPs) with respect to query sequence:

```
Locus_ID
gi|44804|lcl|2
sp|P13186|KIN2
sp|P27704|ERK3
gi|4229|lcl|13
gi|393281|lcl|
sp|P32361|IRE1
gi|450233|lcl|
pir||B40466|gi
sp|P08414|KCC4
gi|306479|lcl|
sp|P13185|KIN1
```

```
Query sequence:  |          |          |          |          |          | 224
                 0          50         100        150        200
```

B. Prosite patterns in query sequence are shown and a link is provided to the database entry for any matches.

Example of program output
Prosite Hits:

```
Query sequence:  |          |          |          |          |          | 224
                 0          50         100        150        200
```

PROTEIN_KINASE_TYR   Tyrosine protein kinases specific active 138..150

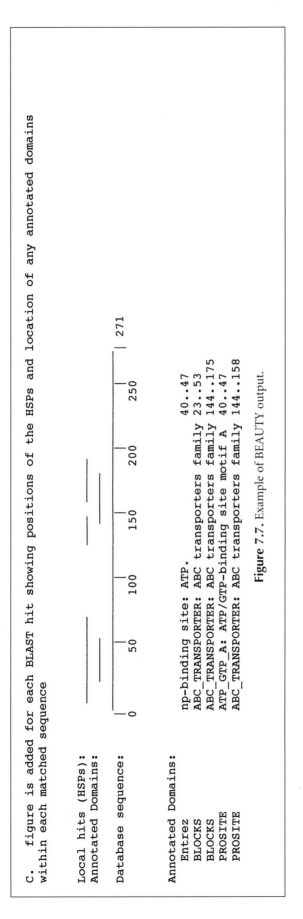

C. figure is added for each BLAST hit showing positions of the HSPs and location of any annotated domains within each matched sequence

Local hits (HSPs):
Annotated Domains:

Database sequence:

```
    |    |    |    |    |    |    | 271
    0    50  100  150  200  250
```

Annotated Domains:
```
Entrez     np-binding site: ATP.                         40..47
BLOCKS     ABC_TRANSPORTER: ABC transporters family      23..53
BLOCKS     ABC_TRANSPORTER: ABC transporters family      144..175
PROSITE    ATP_GTP_A: ATP/GTP-binding site motif A       40..47
PROSITE    ABC_TRANSPORTER: ABC transporters family      144..158
```

Figure 7.7. Example of BEAUTY output.

Table 7.6. *Options and parameter settings available on the BLAST server*

Parameter	Range of choices or values	Function
Descriptions	0–500	number of matching sequences to report
Alignments	0–500	number of alignments to show
Expect	0.001–1000	number of matches from unrelated sequences expected by chance from the selected database smaller values decrease chance of reporting of such matches
Filter	yes or no	removes regions of low sequence complexity from the query sequence because they can give misleading high scoring matches
NCBI-gi	yes or no	gi identifier shown in output
Genetic code	various codon use tables[a]	for translation of nucleic acid sequences
Graphical overview	yes or no	useful display of matches to the query sequence mouse may be used to show alignment
Advanced options	—	type into space provided[b]

[a] Codon tables include standard, vertebrate mitochondrial, yeast mitochondrial, mold mitochondrial, invertebrate mitochondrial ciliate nuclear, echinoderm mitochondrial, euplotid nuclear, bacterial, alternative yeast nuclear, ascidian mitochondrial, flatworm mitochondrial, and blepharisma macronuclear. These are numbered 1–15, respectively, for E-mail access.

[b] Options include (where n is an integer 0,1,2, . . .): $-G\ n$, penalty or cost to open a gap; $-E\ n$, penalty to extend a gap; $-q\ n$ penalty for a mismatch in BLASTN; $-r\ n$, match score in BLASTN; $-W\ n$, initial word size; $-v\ n$, number of descriptions; $b\ n$, number of alignments to show; and $-E\ r$, expect value where r is a real number such as 10.0. For example, to set the gap opening penalty to 10 and the gap extension penalty to 2, click the mouse on the advanced options form and then type $-G\ 10\ -E\ 2$. For more advanced searches of the entire proteome of an organism using stand-alone BLAST on a local machine, additional options must be used, for example, effective database size, to obtain reliable statistical results.

database sequences (Worley et al. 1995). To make this enhanced type of analysis possible, a database of domains and sites was created for use with the BEAUTY program. A new database of sequence domains and sites was made showing for each sequence in ENTREZ the possible location of patterns in the Prosite catalog, the BLOCKS database, and the PRINTS protein fingerprint database. This information is displayed in the following example of the program output (Fig. 7.7). The BEAUTY program is accessible on the BCM Search Launcher (http://dot.imgen.bcm.tmc.edu:9331/seq-search/protein-search.html).

2. BLAST searching with a Cobbler sequence. The BLOCKS server (http://www.blocks.fhcrc.org) offers a variety of BLAST searches that use as a query sequence a consensus sequence derived from multiple sequence alignment of a set of related proteins. This consensus sequence, called a Cobbler sequence (Henikoff and Henikoff 1997), is used to focus the search on residues that are in the majority in each column of the multiple sequence alignment, rather than on any one particular sequence. Hence, the search may detect additional database sequences with variation unlike that found in the original sequences, yet still representing the same protein family. An example of a Cobbler sequence is shown in the BLOCKS search example on page 325.

3. BLAST2. This program uses the BLASTP or BLASTN algorithms for aligning two sequences and may be reached at http://www.ncbi.nlm.nih.gov/gorf/bl2.html. This site should be useful for aligning very long sequences, but sequences >150 kb are not recommended.

DATABASE SEARCHES WITH THE SMITH-WATERMAN
DYNAMIC PROGRAMMING METHOD

The objective of similarity searching in a sequence database is to discover as many sequences as possible that are similar to the query sequence. For proteins, the resulting collection of sequences may represent a sequence family. Because there may be < 20% amino acid identity (an alignment has this many identical residues) between some family members, finding such distant relatives is a difficult task. The aforementioned programs, FASTA and BLAST, are designed to find database sequences related to a query sequence rapidly and with high reliability. They achieve their speed by searching first for short identical patterns in the query sequence and each database sequence and then by aligning the sequences starting at these patterns. Because patterns are very often found in related sequences, the methods work most of the time. FASTA and BLAST are not based on an algorithm that guarantees the best or optimal alignment, but instead on a heuristic method that works most of the time in practice; thus, they may fail to detect some distant sequence relationships.

The Smith-Waterman dynamic programming algorithm discussed in Chapter 3 is mathematically designed to provide the best or optimal local alignment between two sequences and is therefore expected to be the most reliable method for finding family members in a database search. Several studies discussed below have shown that such is the case. The disadvantage of using dynamic programming is that it is 50–100 times slower than FASTA and BLAST, and until recently a search could take up to several hours on a typical medium-sized machine. With the advent of faster and more powerful computers and improvements in the dynamic programming algorithm discussed in Chapter 3 (and on the book Web site), it is now possible to perform database searches in an hour or less. Some institutions have gone so far as to establish a powerful system of several computers linked together in a parallel architecture that allows a search to be performed within minutes. Several of these sites listed below offer public access through the Web (Table 7.5). It is important to examine the site for use of up-to-date databases and use of an appropriate statistical analysis. Detection of distant sequence relationships depends on use of the statistical methods that have been developed for BLAST and FASTA. For routine use of dynamic programming methods for database searches, establishing the program SSEARCH (FTP to ftp.virginia.edu/pub/fasta; Pearson 1991; Pearson and Miller 1992) and the appropriate sequence databases on a local UNIX server is recommended.

In several studies (Pearson 1995, 1996, 1998; Agarwal and States 1998; Brenner et al. 1998), it has been shown that using SSEARCH, which is based on the Smith-Waterman dynamic programming algorithm, is more suitable for identifying related proteins of limited sequence similarity than FASTA and BLAST in a database search. In several of these studies, known members of protein families are used as a query sequence searching for the remaining members in a protein sequence database. In another study, the performance of the sequence analysis methods was determined using protein sequences of known structural relationships (Brenner et al. 1998). The results are presented in terms of the sensitivity and selectivity of the algorithm, or the ability to identify correct family members, including some that are only weakly similar, without incorrectly identifying other unrelated proteins as members (Pearson 1995, 1998). The ability to discriminate true from false matches depends on the use of appropriate amino acid substitution matrices, gap opening and extension penalties that provide local alignments, and a careful statistical analysis of the search results using the extreme value distribution to predict scores from unrelated sequences (Brenner et al. 1998; Pearson 1998). The program SSEARCH has the necessary

```
SSEARCH version 3.1t02 March, 1998


xurtg.aa, 222 aa vs PIR NBRF  library
        opt        E()
< 20     16        0:=
  22      3        0:=                  one = represents 183 library sequences
  24     16        0:=
  26     29        2:*
  28     84       27:*
  30    290      163:*=
  32    703      629:===*
  34   1715     1705:=========*
  36   3245     3501:==================  *
  38   5497     5786:===============================*
  40   7563     8071:======================================    *
  42   9700     9866:=====================================================*
  44  10702    10883:============================================================*
  46  10965    11085:============================================================*
  48  10471    10612:==========================================================*
  50   9912     9684:====================================================*==
  52   8720     8514:==============================================*=
  54   7395     7272:========================================*=
  56   6194     6075:==================================*
  58   4897     4987:==========================*
  60   4252     4040:=====================*=
  62   3392     3239:=================*=
  64   2743     2576:==============*
  66   2077     2036:===========*
  68   1710     1601:========*=
  70   1309     1255:======*=
  72   1003      981:=====*
  74    767      765:====*
  76    586      595:===*
  78    454      463:==*
  80    338      359:=*
  82    277      275:=*
  84    207      218:=*
  86    173      168:*
  88    125      130:*                inset = represents 5 library sequences
  90     97      101:*
  92     70       78:*          :==============  *
  94     65       60:*          :==========*=
  96     57       47:*          :=========*==
  98     52       36:*          :======*===
 100     32       28:*          :=====*=
 102     26       22:*          :====*=
 104     16       17:*          :===*
 106      8       13:*          :==*
 108     14       10:*          :=*=
 110      7        8:*          :=*
 112     10        6:*          :=*
 114     14        5:*          :*==
 116      7        4:*          :*=
 118      4        3:*          :*
>120    216        2:*=         :*=========================================
40855328 residues in 118225 sequences
```

Figure 7.8. *Figure continues on next page.*

```
statistics extrapolated from 50000 to 118006 sequences
Expectation_n fit: rho(ln(x))= 7.5260+/-0.000579; mu= 1.3848+/- 0.033;
mean_var=57.7254+/-11.311, Z-trim: 110  B-trim: 0 in 0/58
Kolmogorov-Smirnov  statistic: 0.0114 (N=29) at  48

Smith-Waterman (3.1 March, 1997) function (BL50 matrix), gap-penalty: -12/-2 reg.-scaled
The best scores are:                                    s-w  z-sc E(118006)
P1;XURTG                                        ( 249) 1446 1896.7 9.1e-99
P1;A26653                                       ( 271) 1401 1836.7 2e-95
P1;C28946                                       ( 259) 1387 1818.7 2e-94
.
.
.
P1;1GSDB                                        ( 241) 1081 1416.6 5e-72

>>P1;A26653                                             (271 aa)
```

Figure 7.8. Example of SSEARCH. The PIR database was searched with the rat glutathione transferase sequence (EC 2.5.1.18). PIR was used to avoid multiple reports of the same sequence that is obtained with combined databases such as Genpept. SSEARCH was obtained from ftp.virginia.edu/FASTA and compiled on a local UNIX server. The PIR database was accessed by the program from the Genetics Computer Group sequence libraries, which are locally available. SSEARCH was run in the UNIX command line mode since a Web page interface was not available. The program output is very similar to that of FASTA which is described in detail in Fig. 7.3. Note that, if not specified otherwise, SSEARCH uses the BLOSUM50 scoring matrix with gap penalties $-12/-2$. Like FASTA, the program calculates the statistical parameters λ and K from the alignment scores calculated for 50,000 unrelated sequences, and then uses these parameters to calculate the E-value scores of the alignment scores with related sequences. The z values are calculated from a linear regression of the scores against the logarithm of the sequence length, and deviations from this line are converted to standard z scores, as described for FASTA. The glutathione transferases are a large and diverse group of sequences, some of which share very little sequence similarity with the others (Pearson 1996). The large number of normalized scores >120 indicates that a large number of related sequences were found in PIR. Only a few of the alignments are shown, and the alignment of the query sequence with itself is omitted. *Figure continues on next page.*

features and is available for database searches. The reliability of the statistical scores reported by FASTA, BLAST2, and SSEARCH has been determined using sequences of known structural relatedness as a guide. The E-value scores reported by FASTA and SSEARCH are reliable, with the number of false positives agreeing with the scores. BLAST2 E-value scores also appear to be reliable (see Brenner et al. 1998).

An example of an SSEARCH vers. 3 database search is given in Figure 7.8. Several guest Web sites for performing a database search with the Smith-Waterman dynamic programming algorithm are listed in Table 7.7.

DATABASE SEARCHES WITH THE BAYES BLOCK ALIGNER

From the discussion so far, it is apparent that the fastest and most convenient way to perform sequence database searches is with the FASTA and BLAST2 programs. The much slower Smith-Waterman dynamic programming programs, such as SSEARCH, may find more distantly related sequences. The significance of the alignment scores can be accurately evaluated by these programs. A even better method for detection of distant sequence relationships has been described; this is the Bayes block aligner (Zhu et al. 1998), which was previously discussed in Chapter 3 (p. 126). This program requires several series of computational steps roughly proportional to the product of the sequence lengths and is therefore considerably slower than SSEARCH. As an indication of length of time required, alignment of two standard-sized proteins scoring 7 blocks with all available BLOSUM or PAM matrices on the author's 500 MHz laptop with 500 megabytes of memory running the Linux operating system took less than 10 seconds.

```
     Z-score: 1836.7 expect() 2e-95
Smith-Waterman score: 1401;  96.396% identity in 222 aa overlap


                                                10        20        30
XURTG                                   MSGKPVLHYFNARGRMECIRWLLAAAGVEF
                                        ::::::::::::::::::::::::::::::
P1;A26 EECCLASSALPHACHAINYAHEPATICRATMSGKPVLHYFNARGRMECIRWLLAAAGVEF
         20        30        40        50        60        70


               40        50        60        70        80        90
XURTG   DEKFIQSPEDLEKLKKDGNLMFDQVPMVEIDGMKLAQTRAILNYIATKYDLYGKDMKERA
         .::.:::::::::::::::::::::::::::::::::::::::::::::::::::::::::
P1;A26  EEKLIQSPEDLEKLKKDGNLMFDQVPMVEIDGMKLAQTRAILNYIATKYDLYGKDMKERA
         80        90       100       110       120       130


               100       110       120       130       140       150
XURTG   LIDMYTEGILDLTEMIMQLVICPPDQKEAKTALAKDRTKNRYLPAFEKVLKSHGQDYLVG
        :::::.:::::::::::::::.::::::::::::::::::::::::::::::::::::::::
P1;A26  LIDMYSEGILDLTEMIIQLVICPPDQREAKTALAKDRTKNRYLPAFEKVLKSHGQDYLVG
         140       150       160       170       180       190


               160       170       180       190       200       210
XURTG   NRLTRVDIHLLELLLYVEEFDASLLTSFPLLKAFKSRISSLPNVKKFLQPGSQRKLPMDA
        ::::::::::::::::::::::::::::::::::::::::::::::::::::::::::  :::
P1;A26  NRLTRVDIHLLELLLYVEEFDASLLTSFPLLKAFKSRISSLPNVKKFLQPGSQRKPAMDA
         200       210       220       230       240       250


               220
XURTG   KQIEEARKIFKF
        ::::::::.:::
P1;A26  KQIEEARKVFKF
         260       270

.
.
.
>>P1;1GSDB                                              (241 aa)
 Z-score: 1416.6 expect() 5e-72
Smith-Waterman score: 1081;  77.670% identity in 206 aa overlap

                                                10        20        30
XURTG                                   MSGKPVLHYFNARGRMECIRWLLAAAGVEFDEK
                                        ::  :::::::::  ::::::::::::::.::
P1;1GS  HINETRANSFERASEAECCHAINBHMANAEKPKLHYFNARGRMESTRWLLAAAGVEFEEK
         10        20        30        40        50        60


               40        50        60        70        80        90
XURTG   FIQSPEDLEKLKKDGNLMFDQVPMVEIDGMKLAQTRAILNYIATKYDLYGKDMKERALID
        ::.: :::.::.:: :::::.:::::::::::::.::::::::.::.:::::.:::::::::
P1;1GS  FIKSAEDLDKLRNDGYLMFQQVPMVEIDGMKLVQTRAILNYIASKYNLYGKDIKERALID
         70        80        90       100       110       120


               100       110       120       130       140       150
XURTG   MYTEGILDLTEMIMQLVICPPDQKEAKTALAKDRTKNRYLPAFEKVLKSHGQDYLVGNRL
        ::  ::: :: :::. : .::::.:: :: :.. ::::::::::::::::::::::::::.:
P1;1GS  MYIEGIADLGEMILLLPVCPPEEKDAKLALIKEKIKNRYFPAFEKVLKSHGQDYLVGNKL
         130       140       150       160       170       180


               160       170       180       190       200       210
XURTG   TRVDIHLLELLLYVEEFDASLLTSFPLLKAFKSRISSLPNVKKFLQPGSQRKLPMDAKQI
        .:.::::.::: :::::::.::.::..::::::.:.:::.::.::::::::::: :: :::
P1;1GS  SRADIHLVELLYYVEELDSSLISSFPLLKALKTRISNLPTVKKFLQPGSPRKPPMD
         190       200       210       220       230       240


               220
XURTG   EEARKIFKF


222 residues in 1 query    sequences
40855328 residues in 118225 library sequences
```

Figure 7.8. *Continued.*

Table 7.7. *Examples of guest Web sites for performing a database search based on the Smith-Waterman dynamic programming algorithm*

Server/program	Reference	Web address
BCM Search Launcher (with programming links to several servers)	Baylor College of Medicine	http://dot.imgen.bcm.tmc.edu:9331/ seq-search/protein-search.html
bic-sw[a]	Bic server European Bioinformatics Institute	http://www.ebi.ac.uk/bic_sw/
Mpsearch[b]	National Institute of Agrobiological Resources, Tsukuba, Japan	http://www.dna.affrc.go.jp/htbin/mp_PP.pl
Scanps	G.Barton, European Bioinformatics Institute	http://barton.ebi.ac.uk; http://www.ebi.ac.uk/scanps
SSEARCH E-mail server	DNA Databank of Japan	http://www.ddbj.nig.ac.jp/E-mail/homology.html
Swat[c]	Phil Green, University of Washington	http://www.genome.washington.edu/UWGC/ analysistools/swat.htm

A comprehensive list of servers for these types of analyses may be found at http://www.sdsc.edu/ResTools/biotools/biotools1.html.

[a] Bic-sw provides a combination of amino acid scoring matrix and gap penalties and also length-normalized *z* scores (similar to FASTA and BLAST) which are most appropriate for resolving more distantly related sequences.

[b] MPSearch is an extremely fast implementation of the Smith-Waterman dynamic programming algorithm by J.F. Collins and S. Sturrock, Biocomputing Resource Unit, the University of Edinburgh, distribution rights by Oxford Molecular Ltd. An E-mail server is at http://www.gen-info.osaka-u.ac.jp/. Some versions of the Mpsearch algorithm at this site use the same penalty for all gaps, others use gap opening and extension penalties. The former is designed to find similar sequences in which gaps are less important in the alignment, the latter the more distant sequence alignments. An on-line manual is available at http://www.dna.affrc.go.jp/htdocs/ MPsrch/MPsrchMain.html. Current versions of these programs rank the sequences found by two kinds of scoring systems. A statistical analysis is performed but the scores do not appear to be length-normalized. Hence, the sensitivity of the program may not exceed that shown by FASTA (Pearson 1996).

[c] Includes Smith-Waterman and Needleman-Wunsch search algorithms. Calculates statistical significance using extreme value statistics (like FASTA and BLAST).

A Web page describing Bayesian bioinformatics and the source of the Bayes block aligner software is located at http://www.wadsworth.org/resnres/bioinfo/.

Evaluation of programs for finding related proteins is usually based on searches in databases for families using sequence similarity (Pearson 1998). A more difficult type of evaluation is based on the searches of structural databases (Brenner et al. 1998). In these databases, discussed in Chapter 9, the sequences have been organized into families having similar three-dimensional structures. Three of these databases representing groups of proteins that have less than 25%, 35%, or 45% identities (Hobohm et al. 1992) were searched using representatives of structural families in each. In each case, the block aligner slightly but significantly outperformed SSEARCH in finding structural relatives. For example, at the 1% false-positive level, the Bayes block aligner found an average of 14.4% of the proteins in the less-than-25% identity group, whereas SSEARCH with usual scoring matrix, gap penalties, and statistical score options found 12.9%, a difference of 1.5%. In addition, the Bayes block aligner can align sequences that have very little similarity but provide alignments that closely match those found by a careful structural analysis described in Chapter 9 using the VAST program (Madej et al. 1995). A similar study (Brenner et al. 1998) compared the ability of BLAST2, FASTA, and SSEARCH to identify proteins in the families of the SCOP structural database (Murzin et al. 1995, and see Chapter 9).

The Bayes block aligner uses a new method for producing sequence alignments. The method, discussed in detail in Chapter 3, starts by finding all possible blocks, which are patterns without gaps, that are located in two sequences. A large number of possible alignments between two sequences are generated by aligning combinations of blocks. Gaps will be present between the blocks, as illustrated in Figure 7.9. The sequence alignments are

```
            Sequence 1 xxxxxx---o--xxxxxxxxxxooo-oooxxxxxx
            Sequence 2 xxxxxxooooooxxxxxxxxxxxo-ooo-oxxxxxx
                       block1       block2          block3
```

Figure 7.9. Alignment found by the Bayes block aligner. The alignment between two sequences includes ungapped blocks (marked by x where aligned x's may be identical or substitutions; there will be at east one identity in each block used to identify the block) and intervening unaligned regions with gaps (marked by o for unaligned residue and − for a gap). These two regions are designed to represent conserved structural alignments in the protein core and variable surface loops, respectively. A large number of alignments of this type involving different combinations of blocks are found. These alignments are then evaluated by a set of scoring matrices. The best alignment is then derived by a Bayesian statistical analysis, described in Chapter 3.

scored only where the sequences are aligned in the blocks: There is no gap penalty as in the dynamic programming method of alignment.

Alignments are also scored differently by the Bayes block aligner than by the dynamic programming method. In the Bayes block aligner, a set of amino acid substitution matrices is used. Each scoring matrix models a different degree of substitution between the sequences, and the matrices that best represent this degree should give the highest alignment scores. When PAM-type matrices are used, the evolutionary distance between parts of sequences can be estimated knowing the best-scoring matrix. When the analysis has been completed, there are a large number of possibilities to sort out, including choices of block number, alignments, and scoring matrices.

By using a Bayesian statistical analysis of the results, it is possible to derive block alignments in which amino acids in each sequence are most often associated, regardless of the many possible choices. These alignments are represented as the posterior probability of aligning those amino acids given the initial preferences of block number and scoring matrices. It is these block alignments that are statistically the best representation of the alignment between two sequences. From the Bayesian analysis, the probability that the sequences are related may be calculated from the posterior probabilities of block number by examining the analysis for evidence that the block number is greater than zero. If this calculation, described in Chapter 3 (p. 130), yields a probability greater than 0.5, the Bayesian analysis supports a relationship between the sequences (Zhu et al. 1998)

DATABASE SEARCHES WITH A SCORING MATRIX OR PROFILE

The methods for database searching discussed so far in this chapter are based on using a single query sequence to search a sequence database. Another method of database searching is to use the variation found in a multiple sequence alignment of a set of related sequences to search for matching database sequences. This enhanced type of search will locate database sequences that match new combinations of sequence characters in the multiple sequence alignment. For example, if column 1 of a multiple sequence alignment includes the amino acids P and Q and column 2 the amino acids D and E, then database sequences that match all four combinations of these two amino acids can be found, whereas only the combinations found in the original sequences would be matched if single query sequences were to be used.

Multiple sequence alignments represent the occurrence of one or more patterns common to a set of sequences. These patterns may be relatively short or may include long conserved stretches of sequence. In Chapter 4, two methods for identifying a common set of patterns in sequences were described. The first extracts a set of patterns from a multiple sequence alignment, which can be produced by methods such as dynamic programming, genetic algorithms, or hidden Markov models. The second uses pattern finding and statistical methods, including expectation maximization and Gibbs sampling methods, to locate patterns in unaligned sequences. Hidden Markov models are also useful for representing a set of conserved patterns that includes gaps, in a protein family. The resulting PROFILE HMM (Durbin et al. 1998) may then be used to search a query sequence for matches to the set of patterns. Chapter 4 should be consulted for a discussion of these methods; the relevant programs and Web sites are described below.

To search a sequence database for matches to a set of patterns, the sequence information is stored as a matrix of 20 rows, one row for each amino acid, and n columns, one column for each column in the multiply aligned sequence patterns. In addition, there may be extra rows for ambiguous or unidentified symbols and, in the sequence profile matrix, there are rows for gap opening and extension penalties. Examples are shown in Figures 4.11 and 4.12 in Chapter 4.

The simplest scoring matrix, the position-specific scoring matrix (PSSM), represents an alignment of sequence patterns of the same length (no gaps). The production of a PSSM is also discussed in Chapter 4 (p. 192). To summarize, the sequence patterns are first aligned as a multiple sequence alignment so that corresponding residues are in the same column. Raw amino acid counts are first found by summing the numbers in each column of the alignment, and these numbers are placed in the corresponding columns of the scoring matrix, one for each amino acid in the designated row. These counts are then adjusted by a weighting method designed to prevent overrepresentation of the amino acids in the more closely related sequences. Otherwise, the matrix would be more tuned to those sequences than to the less-alike ones in the group. To these raw scores, additional counts are added based on previously observed general types of amino acid variations in alignments of related proteins. The idea behind this strategy is that the small number of sequences usually present in these alignments does not represent the full range of expected amino acid variations. Therefore, additional pseudocounts are added based on substitution patterns found in an amino acid substitution matrix or representative blocks in the BLOCKS database (Dirichlet mixtures). The statistical basis for adding counts is that including prior information in the form of pseudocounts should increase the sensitivity of the scoring matrix. The sum of the raw and additional counts in each column is then divided by the expected frequency of the amino acid from the data or from other sources. The resulting ratio represents the odds for finding a match of another related sequence to the column divided by the chance of a random match with an unrelated sequence. For ease in multiplying probabilities by adding their logarithms, each odds score is converted to a log odds score, usually to logarithms to the base 2. The log odds score for each column in the alignment is placed in the corresponding column of the matrix, and there is one row of scores for each amino acid that is the same width as the pattern window. The resulting PSSM is easy to align with a sequence, as discussed below.

SEARCHING SEQUENCE DATABASES WITH A POSITION-SPECIFIC SCORING MATRIX OR SEQUENCE PROFILE

Aligning a PSSM with a protein sequence is illustrated in Figure 7.10. Every possible sequence position is scored by sliding the matrix along the sequence one position at a time. The amino acid substitution scores in each column of the PSSM are used to evaluate each sequence position. Positions with the highest scores are the best matches of the corresponding set of sequence patterns with the sequence. In searches of a sequence database, those sequences with a region that is a close match to the pattern will produce the highest scores and may be readily identified.

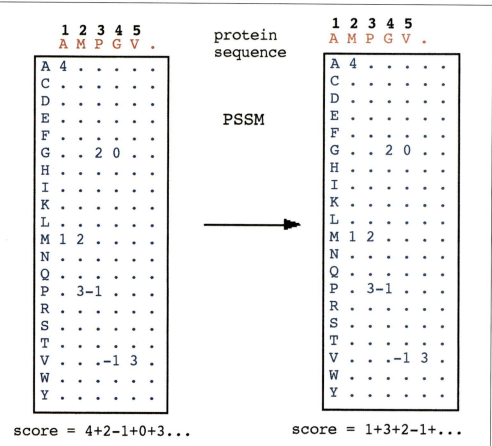

Figure 7.10. Scoring an alignment of a PSSM with a sequence. Only a few matrix values are shown for simplicity. The PSSM is first aligned with sequence position 1 so that columns 1–10 match positions 1–10 in the sequence. The column values that match each sequence position are then added to give a total log odds score for sequence position 1. The PSSM is then moved to sequence position 2 and a new score for matches with sequence positions 2–11 is calculated. This process if repeated until the last 10 sequence positions are matched. The highest scores represent the best matches of the motif represented by the PSSM to the sequence. This type of scoring system preferentially rewards matches to columns that have a conserved amino acid more than to matches with variable columns, and penalizes mismatches to columns that have a conserved amino acid more heavily than mismatches to variable columns.

Scoring matrices that correspond to a sequence profile also include two extra rows for gap penalty scores (sometimes these scores may be found in extra columns if the labeling of the rows and columns is reversed). When aligning this type of scoring matrix with a sequence, a similar procedure to the above is followed in that the score for matching the profile scoring matrix to each sequence character is calculated. In addition, a gap of any length may be inserted into the sequence or profile at that position, and the gap penalties are those given in the relevant column of the profile. The gap penalties are usually quite high with respect to the match scores, but are less when gaps were present in the original multiple sequence alignment. The problem of finding the best alignment between the profile and a given start position in the sequence is similar to the problem of aligning two sequences. As with alignment of sequences, the dynamic programming algorithm is used, except that the match scores and gap penalties are site-specific and are the values given in the profile columns.

Web sites and programs for finding common motifs and profiles in a set of related sequences or for searching a protein sequence database with these patterns are listed in Table 7.8. Also shown are sites that can be given an ambiguous pattern, called a regular expression, to use in a protein sequence database search. The first programs available for producing profiles and using these for sequence searches were Profilemake for making profiles from a multiple sequence alignment, Profilegap for aligning a profile with one or more sequences, and Profilesearch for searching a protein sequence database with a profile (Gribskov and Veretnik 1996). These programs are best known as components of the Genetics Computer Group suite of programs. Profiles produced by newer versions of these programs use evolutionary predictions of the amino acid changes in each column, which

Table 7.8. *Programs and Web sites for database similarity searches with a regular expression, motif, block, or profile*

Program	Database searched	Source or location of analysis
1. Regular expressions and motifs[a]		
EMOTIF Scan	SwissProt and Genpept	http://dna.stanford.edu/scan/
Prosite patterns	SwissProt and TrEMBL	http://www.expasy.ch/tools/scnpsit2.html

ISREC pattern-
    finding service
	SwissProt and non-
	    redundant EMBL database
		http://www.isrec.isb-sib.ch/software/PATFND_form.html

fpat	PDB SwissProt Genpept	http://www.ibc.wustl.edu/fpat/
PHI-BLAST	BLAST databases	http://www.ncbi.nlm.nih.gov/
MOTIF	SwissProt, PDB, PIR, PRF, Genes	http://www.motif.genome.ad.jp/MOTIF2.html
2. Blocks		
BLOCKS[b]	most databases	http://www.blocks.fhcrc.org/blockmkr/make_blocks.html
MAST[c]	most databases	http://meme.sdsc.edu/meme/website/
BLIMPS[d]	locally available databases	anonymous FTP ncbi.nlm.nih.gov/repository/blocks/unix/blimps
Probe[e]	BLAST databases	anonymous FTP ncbi.nlm.nih.gov/pub/neuwald/probe1.0
Genefind[f]	PIR	http://pir.georgetown.edu/gfserver
3. Profiles		
Profilesearch[g]	locally available databases	anonymous FTP ftp.sdsc.edu/pub/sdsc/biology/profile_programs
Profile-SS[h]	most databases	http://www.psc.edu/general/software/packages/profiless/profiless.html

These resources search for similarity to a sequence pattern. Resources for producing patterns from aligned or unaligned sequences are described in Chapter 4. An individual sequence may also be searched for matches to a motif database, and this procedure is discussed in Chapter 9. Additional resources for database searching are listed in Bork and Gibson (1996).

A statistical estimate of finding the site by random chance in a sequence is sometimes but not always given. Reading how these estimates are derived by the individual programs is strongly recommended. The statistical theory for sequence alignments described in Chapter 3 can be used in these types of analyses (Bailey and Gribskov 1998) but may not always be implemented.

[a] The Scan Web page shows how to compile a regular expression. Mismatches within the expression are allowed. The Prosite form of a regular expression is at http://www.expasy.ch/tools/scnpsit3.html. PHI-BLAST is a BLAST derivative that searches a given sequence for a regular expression and then searches iteratively for other sequences matching the pattern found, at each iteration including the newly found sequences to expand the search.

[b] The BLOCKS server will send a new block analysis to the MAST server.

[c] MAST is the Motif Alignment and Search Tool (Bailey and Gribskov 1998). Available protein databases are similar to those on the BLAST server. It is also possible to search translated nucleotide sequence databases.

[d] BLIMPS will prepare a PSSM from a motif and perform a database search with the PSSM (see README file on FTP site).

[e] PROBE (Neuwald et al. 1997) is described in the text.

[f] The GENEFIND site has the program MOTIFIND for Motif Identification by Neural Design (Wu et al. 1996). This motif finder uses a neural network design to generate motifs and a search strategy for those motifs. The method performed favorably in sensitivity and selectivity with others such as Blimps and Profilesearch and is in addition very fast. Neural networks are described in Chapters 8 and 9.

[g] Profilesearch is one of a set of programs in the GCG suite (see text). It is important to review the parameters of the program which if used inappropriately can lead to incomplete or low-efficiency searches (Bork and Gibson 1996).

[h] A version of Profilesearch running at the University of Pittsburgh Supercomputing Center.

improves the ability of the profile to find related proteins in a database search. Methods for making evolutionary profiles and for using them are discussed in Chapter 4 (p. 163). Profile searches may be performed at two supercomputer centers (Table 7.8). The standard Genetics Computer Group multiple sequence alignment format, the MSF file (described in Chapter 2), is used as input to these programs. READSEQ and other sequence reformatting programs can be used to change the sequence format to the MSF format (see Chapter 2), which can then be used with Profilemake.

There is a difference in the way PSSM and the profile matrix are generated that should influence the results of a database search. The PSSM treats all amino acids as being equal, so that matching an Ala with an Ala is as significant as matching a Cys with a Cys. Scores for amino acid substitutions are based on the distribution of amino acids in each column of the alignment on which the PSSM is based. Profile scores are also based on the distribution of amino acids in each column of the alignment, but the matrix values are also derived from an amino acid substitution matrix, such as the Dayhoff PAM matrices. Hence, the PSSM and profile methods should give different results.

To illustrate these methods, the results of finding blocks by the BLOCKS server, and of using motifs found by the MEME server, for a search of the SwissProt database by MAST are given below. The terms "blocks" and "motifs" are used interchangeably by these sites in that both mean a reasonably long ungapped pattern in a family of protein sequences. Once matching sequences have been found, other searches can be performed with sets of blocks or motifs of shorter or longer length and of more occurrences. Use of the program MACAW, which runs on many computer platforms with a Windows interface, is also very useful for exploring motif size and number (see Chapter 4, p. 177). This program can find motifs either by an alignment method using an amino acid scoring matrix or by the statistical Gibbs sampling method. The relative positions of the found motifs are shown on a graphical representation of the sequences.

Example: BLOCKS Server

This Web server takes a set of unaligned sequences and finds blocks of sequence (matching ungapped patterns) that are present in the sequences. A request to the BLOCKS Web site to find blocks in three sequences similar to the human *XPF* DNA repair gene was made, and an example of the program output is shown in the following example (Fig. 7.11). The sequences were input into a Web form in FASTA format. The server finds blocks by two methods, a pattern-searching method called MOTIF and a statistical method called the Gibbs sampler. These two methods are described in detail in Chapter 4 (pp. 171 and 177). The blocks found by each method may not be the same because each method uses a different algorithm. Examples of a representative block found by MOTIF, www.bloA is shown in Figure 7.11A. Also shown is a portion of a Cobbler sequence of one of the input sequences, xpf95.pro, an *Arabidopsis* gene. In the Cobbler sequence, the sequence in xpf95.pro corresponding to each block location has been replaced by a consensus sequence derived from bloA. These replaced regions are capitalized. In Figure 7.11B, an example of a Gibbs sampler block also called www.bloA is shown. A Cobbler sequence was also produced from these blocks (not shown). There are two options given after each list of blocks: (1) the Cobbler sequence may be used in a BLAST search and (2) the blocks may be sent to the MAST server to search a protein sequence database.

A.

BLOCKS from MOTIF

>www.blo xpf95.pro   957 a.a. family
3 sequences are included in 15 blocks

```
          www.bloA, width = 30
gi|131810|   113 GKGLGLLDIVANLLHVLATPTSINGQLKRA
gi|2842712    27 GLGADRLLYHFLQLHCHPACLVLVLNTQPA
 xpf95.pro    27 GLSLAKLIASLLILHSPSQGTLLLLLSPAA
```

COBBLER sequence from MOTIF
>www.blo gi|2842712 from 1 to 905 with embedded consensus blocks
maplleyerqlvlelldtdglvvcarGLGLGRLIYHFLLLHCHPTCTVLVLQTKPAeeeyfinqlkiegvehlprrvtne
itsnsRYKLYTSGGVLFITSRILIVDLLTDRIPPNRITGILVLNAHSIRENCNEAFILRIYRSKNSWGFIKAFSDRPQAF
VTGFchvervmrnlfvrklylwprfhvavnsfleqhkpEVVEIRVSMTNTMVGIQFAIMECLNACLKELKRHNPsleved
lslenaigkpfdktirhyldPLWHRLGYKTKQLVKDLKFLRHLLQYLVQYDCVDFlnlLEALKPTEKAKYQNSPWLFVDS
SYKVFDYAKKRVYhlpdakmskkekisekmeikegeetkkEYVLEENPKWEALTEILHEIeaenkesealggpGPVLVCC
SDDRTCMQLrdyitlgaeafllrlyrktfekdskaeevwmkfrkedsskrirkshkrpkdpqNKERHVDKARCTKKKkrk
ltltqmvgkpeeleeegdveegyrreissspescpeeikheefdvnlssdaafgilkepltiihpllgcsdpyaltrvlh
evePSYIIMYDPDLSFVRQLEVYKASNPGKPLKVYFLYYGESTEEQKYLTAIRREKEAFEKLIREKASMvvpeeregrde
tnldlvrgtasadvstdTRKAGGQQqngtqqsivvdmrefrselpslihrrgidiepvtlevgdyiltpemcverksiSD
LIGSLNNGRLYHQCEKMSRYYRYPVLLIEFDQDKSFSltsrgalfqeissndisskltlltlHFPRLRILWSPSPHATAE
IFTELKQNRDQPDaatalaitadsetlpesENYNPSPFEFLLKMPGVSKANYRSLMHKIKSFAELASLsqdeltsilgna
anakqlydfihtsfaevvskgkgkk

B.
          BLOCKS from GIBBS

>www.blo xpf95.pro   957 a.a. family
3 sequences are included in 10 blocks

```
          www.bloA, width = 45
```

```
gi|131810|   201 EKRRKLYISGGILSITSRILIVDLLSGIVHPNRVTGMLVLNADSL
gi|2842712    84 NSRYEVYTQGGVIFATSRILVVDFLTDRIPSDLITGILVYRAHRI
 xpf95.pro    85 NQRYSLYTSGSPFFITPRILIVDLLTQRIPVSSLAGIFILNAHSI
```

Figure 7.11. Example of BLOCK output.

Example: Motif Alignment and Search Tool (MAST) Server

The MAST server searches a protein database for best matches to a set of ungapped motifs or blocks (Bailey et al. 1997). The motifs may also be found by MEME by submitting unaligned sequences to the MEME server for analysis by a statistical method, the expectation maximization method, described in Chapter 4. A MEME output example is shown as Figure 4.15. The same three DNA repair sequences as used above were input into a Web form in FASTA format. To simplify output of many possible choices, MEME was asked for one motif per sequence and for up to six different motifs of short length. Once received by E-mail, the motif messages were saved to a local file, and then this file was submitted to the MAST server (http://www.sdsc.edu/MEME) using the Browse option of the Web page to read the newly made local file of the MEME output. Using a short file name with a short subdirectory listing was necessary because there is not much space on the Web form. It is also possible to submit an already-found motif in GCG, MEME, or PSSM format (see http://www.sdsc.edu/MEME/meme/website/motif-format.html). Another method for readily accessing MAST is through the BLOCKS server. As shown in Figure 7.11, unaligned sequences are searched for blocks by two methods, and from the BLOCKS Web page, the results may be immediately submitted to the MAST server. MAST uses the method shown in Figure 7.10 to align the blocks with each database sequence. If not specified otherwise, the output files are sent by E-mail as HTML files suitable for viewing by a Web browser. These files have some nice graphical features. These files are first saved to a file and then opened with a Web browser.

Alternatively, the files may be requested in text format, as was done below (Fig. 7.12). The initial list in the MAST output is of the motifs found by MEME. Note that motifs are given an ID number (1–6) that is used later in the MAST report. Section I then lists the scoring matches found in the SwissProt sequence database. The expect value is the number of unrelated sequences in a database of the size of SwissProt that would achieve a score as high as the one shown with the motifs used in the search and is based on the scores of individual motifs with the sequence using the extreme value distribution (Bailey and Gribskov 1998). The highest-scoring matches are with the two input DNA repair proteins but then there are several lower-scoring matches with other proteins that interact with DNA, suggesting a common structural motif; however, caution is necessary in interpreting these kinds of matches (Bork and Gibson 1996). One of the input sequences was that of an *Arabidopsis* DNA repair protein that is not reported because it is not in the database yet. Section II shows the locations of the motifs in each sequence. The motifs are shown in brackets and numbered as at the top of the file. Note that the order in the first three sequences is approximately the same, but that there are more and more variations going down the list, reflecting more divergence. Finally, in section III, the matched motifs are aligned with the matched sequence. At each aligned position, the motif number, the *P* value of each match, the motif sequence giving the best match between sequence and motif, and a plus sign to indicate sequence letters corresponding to a positive match score in the motif column are given. A diagram shows the order of the motifs found and a combined *P* (combined probability for matching all matrices to an unrelated sequence) and *E* score (expectation of finding these matches with an unrelated sequence in a database of the size searched). The combined probabilities are calculated using the extreme value distribution as used for determining the significance of FASTA and BLAST scores (Bailey and Gribskov 1998).

OTHER METHODS FOR COMPARING DATABASES OF SEQUENCES AND PATTERNS

One variation of the method for comparing sequences and patterns is to search a query sequence with a database of patterns (search type E, Table 7.1). If the sequence contains patterns representative of a protein family, the sequence is a candidate for membership in that same family. A large number of protein pattern databases are available (Table 9.5), most of them offering this type of search.

FASTA-pat and FASTA-swap are versions of FASTA that may be used for comparing a query sequence to a database of patterns characteristic of protein families. They are designed to search for remotely related protein sequences by a finely tuned system of

```
MAST - version 2.2
DATABASE swissprot contains 74596 sequences

      MOTIF WIDTH BEST POSSIBLE MATCH
      ----- ----- -------------------
        1    11   VGDYILTPDIC
        2     9   QCKMMSRYY
        3     8   YFMFYGES
        4     8   WPRFHVDV
        5     9   HFPRLRILW
        6     8   IVDMREFM

SECTION I: HIGH-SCORING SEQUENCES

SEQUENCE NAME             DESCRIPTION                       E-VALUE  LENGTH
-------------             -----------                       -------  ------
sp|Q92889|XPF_HUMAN       DNA-REPAIR PROTEIN COMPLEMENTING XP...  5.3e-35   905
sp|P06777|RAD1_YEAST      DNA REPAIR PROTEIN RAD1            1.1e-31  1100
sp|P36617|RA16_SCHPO      DNA REPAIR PROTEIN RAD16          8.4e-23   892
sp|Q07864|DPOE_HUMAN      DNA POLYMERASE EPSILON, CATALYTIC S...   0.62  2257

SECTION II: MOTIF DIAGRAMS

SEQUENCE NAME             E-VALUE    MOTIF DIAGRAM
-------------             -------    -------------
sp|Q92889|XPF_HUMAN       5.3e-35    181-[4]-405-[3]-71-[6]-20-[1]-20-[2]-41-
                                     [5]-114

sp|P06777|RAD1_YEAST      1.1e-31    298-[4]-433-[3]-75-[6]-20-[1]-20-[2]-55-
                                     [5]-146

sp|P36617|RA16_SCHPO      8.4e-23    185-[4]-241-[2]-134-[3]-83-[6]-20-[1]-20-
                                     [2]-41-[5]-106

sp|Q07864|DPOE_HUMAN         0.62    190-[6]-175-[2]-381-[2]-426-[5]-34-[4]-
                                     366-[2]-478-[6]-147

SECTION III: ANNOTATED SEQUENCES
```

Figure 7.12. Example of MAST output. *Figure continues on next page.*

amino acid matches. The FASTA algorithm normally identifies sequence similarity very rapidly by a method for finding common patterns, or *k*-tuples, in the same order in two sequences. In FASTA-pat and FASTA-swap, the same rapid method is used to find common patterns. FASTA-pat performs a faster method of comparing sequences to patterns by means of a lookup table, as described above (Table 7.3). FASTA-swap performs a more rigorous search for the most significant matches of sequence to patterns.

These programs use databases of patterns found in columns of multiple sequence alignments of related protein sequences. Multiple sequence alignments of a large number of protein families were prepared using the PIMA program (see Chapter 4, p. 160). From these alignments, a large number of conserved patterns were identified, and the pattern was placed in a new type of scoring matrices. Unlike PSSMs, the columns in these matrices only indicate whether or not a given amino acid is present; there is no score indicating frequency.

In addition to these pattern matrices, two log odds scoring matrices, weighted-match minimum average matrix (WMM) and empirical matrix (EMMA), were prepared from the scoring matrices. These scoring matrices are used by FASTA-pat and FASTA-swat for comparing a query sequence with a database of pattern matrices.

The scoring system takes into account the possibility that the substitution of amino acid *a* for amino acid *b* may not be as likely as the substitution of *b* for *a*. An example from

```
gi|548659|sp|P36617|RA16_SCHPO
  DNA REPAIR PROTEIN RAD16
  LENGTH = 892   COMBINED P-VALUE = 1.12e-27   E-VALUE =  8.4e-23
  DIAGRAM: 185-[4]-241-[2]-134-[3]-83-[6]-20-[1]-20-[2]-41-[5]-106

                                    [4]
                                    1.9e-07
                                    WPRFHVDV
                                    +++++ +
  151   TGFIKAFSDDPEQFLMGINALSHCLRCLFLRHVFIYPRFHVVVAESLEKSPANVVELNVNLSDSQKTIQSCLLTC

                                                        [2]
                                                        8.8e-05
                                                        QCKMMSRYY
                                                        +  ++++
  376   ETMLADTDAETSNNSIMIMCADERTCLQLRDYLSTVTYDNKDSLKNMNSKLVDYFQWREQYRKMSKSIKKPEPSK

                                                        [3]
                                                        6.8e-10
                                                        YFMFYGES
                                                        ++++++++
  526   NSIYIYSYNGERDELVLNNLRPRYVIMFDSDPNFIRRVEVYKATYPKRSLRVYFMYYGGSIEEQKYLFSVRREKD

                                                                    [6]
                                                                    3.6e-09
                                                                    IVDMREF
                                                                    +++ +++
  601   SFSRLIKERSNMAIVLTADSERFESQESKFLRNVNTRIAGGGQLSITNEKPRVRSLYLMFICIKTLKVIVDLREF

                        [1]                             [2]
                        6.0e-13                         8.5e-09
              M         VGDYILTPDIC                     QCKMMSRYY
              +         ++++++++ ++                     +++ ++ ++
  676   RSSLPSILHGNNFSVIPCQLLVGDYILSPKICVERKSIRDLIQSLSNGRLYSQCEAMTEYYEIPVLLIEFEQHQS

                        [5]
                        3.7e-08
                        HFPRLRILW
                        ++ +++ +
  751   FTSPPFSDLSSEIGKNDVQSKLVLLTLSFPNLRIVWSSSAYVTSIIFQDLKAMEQEPDPASAASIGLEAGQDSTN


CPU: ghidorah
Time 68.583141 secs.
```

Figure 7.12. *Continued.*

Ladunga et al. (1996) is informative. On the one hand, if an alignment column has 9 Cys and 1 Ala, the substitution of Ala for Cys in this column would be given a low substitution score because Cys is involved in disulfide bonds and this function cannot be replaced by Ala. Cys-to-Cys substitutions receive a high score for the same reason. On the other hand, if a column has 1 Cys and 9 Ala, then the Cys might readily substitute for the Ala, which has no comparable specific function. The substitution of Cys for Ala is considered to be a random insertion of no particular significance and is therefore given a corresponding likelihood score of zero. When aligning a query sequence to a pattern, a single amino acid in the sequence is matched to a series of possible substitutions in the pattern. WMM uses the minimum of the scores for aligning the amino acid in the query sequence with each of the amino acids in the pattern. WMM gives significantly better results than EMMA, probably because it is more finely tuned for detecting the types of variations in related sequences.

Program outputs of FASTA-pat and FASTA-swap are very similar to those of FASTA described above.

Another type of pattern database searching is to use a pattern query to search a database of patterns. The LAMA (Local Alignment of Multiple Alignments) server at the BLOCKS Web site, described below, performs such a search. A final variation is to use a query sequence, called a Cobbler sequence (see Fig. 7.11), modified by substituting a consensus sequence for the corresponding part of the sequence. The BLOCKS server automatically produces such sequences when generating new blocks from sequences, and they may be used for sequence database searches. Embedding consensus residues has been demonstrated to improve database searches by a query sequence (Henikoff et al. 1995).

LAMA is a type of analysis provided on the BLOCKS server (http://www.blocks. fhcrc.org/blockshelp/LAMA_help.html#LAMA) that compares a query PSSM representing a particular set of proteins with a database of such matrices to find related sets of proteins (Pietrokovski 1996). In this manner, new and larger related sets of proteins not identified previously might be discovered. Because the search is for matching sequence patterns instead of entire sequence alignments, there is an opportunity to analyze the evolution of function in different parts of a protein molecule (Henikoff et al. 1997). For example, a given group of proteins may be found to have two regions, one related to one particular group of proteins and a second related to another group. The LAMA program compares the scores found in each column of one PSSM to those in a second to discover whether there is any correlation. Examples of the procedure are given at http://www.blocks. fhcrc.org/blockshelp/LAMA_help.html#EXAMPLES.

PSI-BLAST, A Version of BLAST for Finding Protein Families

As described above, there are advantages to using a scoring matrix that represents conserved sequence patterns in a protein family instead of a single query sequence to search a sequence database. The search of sequence databases will thereby be expanded to identify additional related sequences that might otherwise be missed. The major difficulty with such an expanded search is that an alignment of related sequences must already be available to know the variations at each position in the query sequence. A new version of BLAST called position-specific-iterated BLAST, or PSI-BLAST, has been designed to provide information on this variation starting with a BLAST search by a single query sequence. A similar program, PHI-BLAST, performs a similar type of search starting with a specified pattern in a query sequence (see below).

The method used by PSI-BLAST involves a series of repeated steps or iterations. First, a database search of a protein sequence database is performed using a query sequence. Second, the results of the search are presented and can be assessed visually to see whether any database sequences that are significantly related to the query sequence are present. Third, if such is the case, the mouse is clicked on a decision box to go through another iteration of the search. The high-scoring sequence matches found in the first step are aligned, and, from the alignment, a type of scoring matrix that indicates the variations at each aligned position is produced. The database is then again searched with this scoring matrix. Thus, the search has been expanded to include sequences that match the variations found in the multiple sequence alignment at each sequence position. The results are again displayed, indicating any newly discovered sequences that are significantly related to the aligned sequences in addition to those found in the previous iteration. Again, an opportunity is given to go through another iteration of the program, but this time including any newly recruited sequences to refine the alignment. In this fashion, a new family of sequences that are significantly similar to the original query sequence can be found.

This new method was made possible by the development of the gapped BLAST program, which increased the speed of the BLAST algorithm by over one-half so that more sophisticated search routines of PSI-BLAST could be added without an overall loss of speed. PSI-BLAST may not be as sensitive as other pattern-generating and searching programs described in Chapter 4 and above, but the simplicity and ease of use of this program are very attractive features for exploring protein family relationships. In a comparison of the ability of PSI-BLAST with the Smith-Waterman dynamic programming program SSEARCH to identify members of 11 protein families defined by sequence similarity, PSI-BLAST found more sequences and, in some cases, many-fold more sequences, than SSEARCH and at a 40-fold greater speed.

A similar program, MAXHOM, has been described previously (Sander and Schneider 1991). The sequence alignment is built up in two steps. Matching sequences found in a database search are aligned by dynamic programming with a query sequence, and a profile is made from the alignment. A new round of sequences that match the updated profile are then picked from the SwissProt database (visit http://www.embl-heidelberg.de/predictprotein/predictprotein.html).

The main difficulty with searching for subtle sequence relationships based on similarity is determining the significance of the alignments that are found. Such similarities may be evidence of structural or evolutionary relationships, but they could also be due to matching of random variations that have no common origin or function (Bork and Gibson 1996). Protein structures are in general composed of a tightly packed core and outside loops. Amino acid substitutions within the core are common, but only certain substitutions will work at a given amino acid position in a given structure. Thus, sequence similarity is not usually a good indicator of structural similarity (see Chapter 9), and the alignments found need to be carefully evaluated before any firm conclusions can be drawn. Another difficulty with the PSI-BLAST approach is that the procedure follows a type of algorithm called a greedy algorithm. Put simply, once additional sequences that match the query are found, these newly found sequences influence the finding of more sequences like themselves, and so on. If a different but also related query sequence was used initially, a different group with possible overlaps with the first may be found. Thus, there is no guarantee that the alignments finally discovered represent the same set of related sequences. Nevertheless, PSI-BLAST potentially offers exciting opportunities to the curious but careful investigator. New types of relationships in the protein databases may be readily discovered and used to infer evolutionary origins of proteins (Tatusov et al. 1997).

The later steps of a PSI-BLAST search use a scoring matrix that represents the alignments found. PSI-BLAST has been engineered to find database matches to this matrix almost as rapidly as BLASTP finds matches to a query sequence. However, there are some differences between the matrix produced by PSI-BLAST and those produced by other matrix programs: (1) The matrix covers the entire length of the aligned sequences whereas other matrices cover only a short stretch of the alignment; (2) the same gap penalties are used throughout the procedure and there is no position-specific penalty as in other programs; (3) each subsequent alignment is based on using the query sequence as a master template for producing a multiple sequence alignment of the same length as the query sequence. Columns in the alignment involve varying numbers of sequences depending on the extent of the local alignment of each sequence with the query, and columns with gaps in the query sequence are ignored. Sequences >98% similar to the query are not included to avoid biasing the matrix. Thus, the multiple sequence alignment is a compilation of the pairwise alignments of each matching database sequence with the query sequence and is not a true multiple sequence alignment, as illustrated below. The resulting alignment provides the columns for the scoring matrix

```
xxxxxxxxxxxxxxxx       query sequence with no gaps
  xx-xxxx              alignment of sequence 1
        xxx-x              alignment of sequence 2
   xxxx-xx              alignment of sequence 3
----------------       columns of the PSI-BLAST
                       alignment
```

Once the alignment has been found, the frequencies of amino acids in each column are adjusted by weighting the sequences to reduce the influence of the more-alike sequences, and by adding more counts (pseudocounts) representing other amino acid substitutions found among the observed types in order to increase the statistical power of the matrix. These procedures are discussed in Chapter 4 (p. 192). The resulting scores in each column of the scoring matrix are scaled using the same scaling factor λ as the BLOSUM62 scoring matrix so that a threshhold value T for HSPs and other statistical parameters used by BLASTP may also be used by PSI-BLAST. At each iteration, previously matched sequences with an E value less than 0.001 are used to produce the next alignment, but this value may also be changed. PSI-BLAST is in a state of evolution, and the Web page should be consulted for recent improvements. An example of a PSI-BLAST result is shown in Figure 7.13.

Pattern-Hit Initiated BLAST (PHI-BLAST)

This program functions much like PSI-BLAST except that the query sequence is first searched for a complex pattern provided by the investigator (Zhang et al. 1998). The subsequent search for similarity in the protein sequence database is then focused on regions containing the pattern. Thus, the method provides an opportunity to explore variations of a known pattern in the sequence database. This program is accessible from the BLAST server at http://www.ncbi.nlm.nih.gov/.

The chosen query sequence is first searched for a particular pattern or class of patterns called a regular expression, which allows for a wide range of pattern-matching options. The Prosite catalog also uses regular expressions to describe variability in the amino acid patterns for the active sites of proteins. For example, the expression [LIVMF]-G-E-x-[GAS]-[LIVM]-x(5,11)-R-[STAQ]-A-x-[LIVMA]-x-[STACV] means: one of LIVMF in the first position, followed by G and then E, followed by any single character (indicated by x), followed by one of GAS and then by one of LIVM, followed by any 5–11 characters indicated by x(5,11), then by R, one of STAQ, then A, then any single character, then one of LIVMA, then any single character, and finally by one of STACV. More information about these patterns may be provided by the investigator in a standard file, as described on the Prosite Web site (http://www.expasy.ch/prosite/.

PROBE

PROBE is a database search tool that is similar to PSI-BLAST but performs a more complex and rigorous type of data analysis (Neuwald et al. 1997). Like PSI-BLAST, the program PROBE starts with a single query sequence and searches for family members by a BLASTP search. After removing the most-alike sequences, PROBE constructs an alignment model by means of a Bayesian statistical approach that uses both a Gibbs sampling procedure and the genetic algorithm (both methods are described in Chapter 4) to sort the pat-

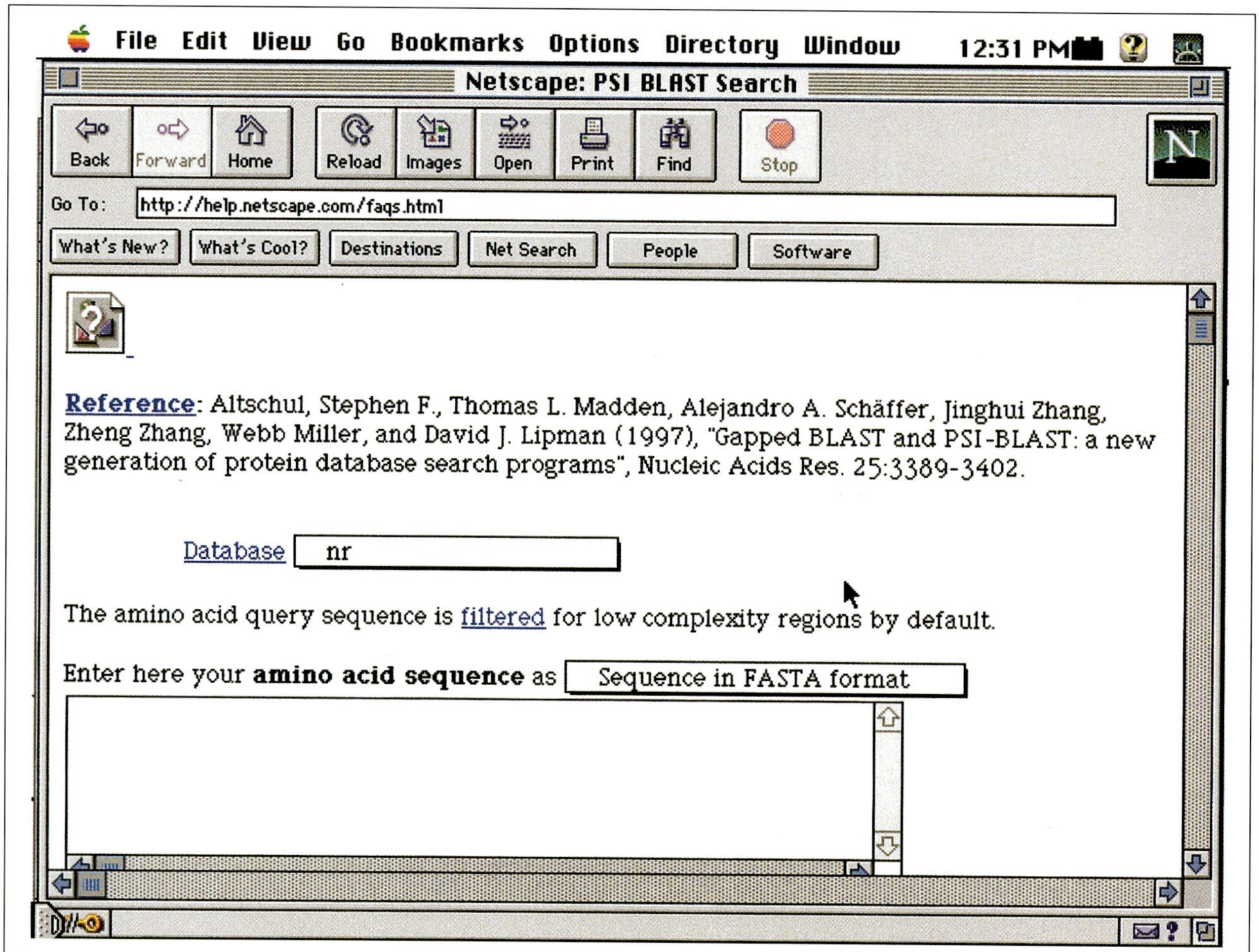

Figure 7.13. Example of PSI-BLAST search. The sequence of the *Arabidopsis XPF* DNA repair gene was used to query the SwissProt database, with an *E* setting of 0.01, requesting 10 descriptions and alignments with otherwise the recommended default program settings. The initial iteration found three matching sequences, and these were used to enter iteration 1. Iteration 1 did not produce any additional matches at the chosen level of significance, and the program indicated that the search had converged with no more sequences at the chosen level of significance. Therefore, for iteration 2 the sequences scoring worse than the threshhold were used. Since only those lower-scoring sequences that have an alignment with the query could influence the result, this option could potentially find additional sequences. A yeast transport protein was then reported. With another iteration using the four sequences above threshhold, another set of sequences was now pulled into the high-scoring group. This search therefore revealed that the SwissProt database has three other sequences strongly related to the query sequence but that other sequences of less-significant similarity were also present. *Figure continues on next page.*

terns in all possible combinations in order to find the most significant set. As in PSI-BLAST, the alignment model is then used as a query for additional database sequences. PROBE provides a new and powerful approach toward finding a sequence family and is available by anonymous FTP from ncbi.nlm.nih.gov/pub/neuwald/.

SUMMARY

As the sequence databases continue to increase in size, for the most part with genomic DNA sequences of unknown function, it is important to have a set of computational tools for predicting the functions of these sequences. The first choice is usually to go to the

```
<Psi-BLAST output example>
Psi-BLAST initial iteration
sp|Q92889|XPF_HUMAN  DNA-REPAIR PROTEIN COMPLEMENTING XP-F CELL ...   504  e-142
sp|P06777|RAD1_YEAST  DNA REPAIR PROTEIN RAD1                         300  6e-81
sp|P36617|RA16_SCHPO  DNA REPAIR PROTEIN RAD16                        231  3e-60


Psi-BLAST iteration 1
with sequences scoring better than E threshhold


Converged
sp|Q92889|XPF_HUMAN  DNA-REPAIR PROTEIN COMPLEMENTING XP-F CELL ...  1020  0.0
sp|P06777|RAD1_YEAST  DNA REPAIR PROTEIN RAD1                         953  0.0
sp|P36617|RA16_SCHPO  DNA REPAIR PROTEIN RAD16                        897  0.0


Psi-BLAST iteration 2
with sequences scoring worse than E threshhold

sp|Q92889|XPF_HUMAN   DNA-REPAIR PROTEIN COMPLEMENTING XP-F CELL ...  1020  0.0
sp|P06777|RAD1_YEAST  DNA REPAIR PROTEIN RAD1                         967  0.0
sp|P36617|RA16_SCHPO  DNA REPAIR PROTEIN RAD16                        939  0.0
sp|P25386|USO1_YEAST   INTRACELLULAR PROTEIN TRANSPORT PROTEIN USO1    53  3e-06


Psi-BLAST iteration 3
with sequences scoring better than E threshhold

sp|Q92889|XPF_HUMAN    DNA-REPAIR PROTEIN COMPLEMENTING XP-F CELL ...  1007  0.0
sp|P06777|RAD1_YEAST   DNA REPAIR PROTEIN RAD1                          950  0.0
sp|P36617|RA16_SCHPO   DNA REPAIR PROTEIN RAD16                         884  0.0
sp|P25386|USO1_YEAST   INTRACELLULAR PROTEIN TRANSPORT PROTEIN USO1     294  5e-79
sp|Q08696|MST2_DROHY   AXONEME-ASSOCIATED PROTEIN MST101(2)             52  4e-06
sp|Q62209|SCP1_MOUSE   SYNAPTONEMAL COMPLEX PROTEIN 1 (SCP-1 PROT...    49  5e-05
sp|Q03410|SCP1_RAT   SYNAPTONEMAL COMPLEX PROTEIN 1 (SCP-1 PROTEIN)     49  5e-05
sp|Q02224|CENE_HUMAN   CENTROMERIC PROTEIN E (CENP-E PROTEIN)           45  5e-04
```

Figure 7.13. *Continued.*

BLAST Web site because a variety of database searches are possible against regularly updated databases and can be performed with rapid turnaround time. This chapter has discussed a variety of additional resources for such searches, most available on Web sites or available for setup on a local computer system. For extensive searching, establishment of the databases and programs on a local system is a reasonable and achievable option. It is then possible to set up batch files or scripts that automate the searches. These searches generate large amounts of information that needs to be organized into a database.

Some of the most interesting matches are those to more distantly related sequences. A short alignment region between a query and a database sequence is usually not biologically significant, even though there may be a number of identities in the alignment. If additional sequences can be found that share the same alignment, however, it is possible that the pattern represents a common structure in a family of related proteins. There are, in addition, databases of conserved patterns in protein families, and it has been estimated that about one-half of these patterns can be linked to a protein structural fold. Thus, it is very worthwhile to follow the distant relationships further with the eventual goal of trying to discover a relationship to a protein of known structure. There are some excellent computer tools available to the molecular biologist for finding conserved patterns in protein families and for searching new sequences with these patterns, and it can be anticipated that the number will continue to grow. There are a large number of Web servers for this purpose, and these are described in Chapter 9.

As methods are used to search for related sequences, it is important to keep an eye on the statistical significance of the matches and the plausibility of the observed amino acid substitutions from a structural perspective. It is quite easy to end up with a group of sequences that are related to each other but not to the query sequence. There are presently no guides as to which of the above methods is most likely to work. The best advice is to go further than the basic methods and Web sites by becoming familiar with the range of available methods.

REFERENCES

Agarwal P. and States D.J. 1998. Comparative accuracy of methods for protein sequence similarity search. *Bioinformatics* **14:** 40–47.

Altschul S.F. and Gish W. 1996. Local alignment statistics. *Methods Enzymol.* **266:** 460–480.

Altschul S.F., Boguski M.S., Gish W., and Wootton J.C. 1994. Issues in searching molecular sequence databases. *Nat. Genet.* **6:** 119–129.

Altschul S.F., Gish W., Miller W., Myers E.W., and Lipman D.J. 1990. Basic local alignment search tool. *J. Mol. Biol.* **215:** 403–410.

Altschul S.F., Madden T.L., Schaffer A.A., Zhang J., Zhang Z., Miller W., and Lipman D.J. 1997. Gapped BLAST and PSI-BLAST: A new generation of protein database search programs. *Nucleic Acids Res.* **25:** 3389–3402.

Bailey T.L. and Gribskov M. 1998. Combining evidence using p-values: Application to sequence homology searches. *Bioinformatics* **14:** 48–54.

Bailey T.L., Baker M.E., and Elkan C.P. 1997. An artificial intelligence approach to motif discovery in protein sequences: Application to steriod dehydrogenases. *J. Steroid Biochem. Mol. Biol.* **62:** 29–44.

Bork P. and Gibson T. 1996. Applying motif and profile searches. *Methods Enzymol.* **266:** 162–184.

Brenner S.E., Chothia C., and Hubbard T.J. 1998. Assessing sequence comparison methods with reliable structurally identified distant evolutionary relationships. *Proc. Natl. Acad. Sci.* **95:** 6073–6078.

Claverie J.-M. 1996. Effective large-scale sequence similarity searches. *Methods Enzymol.* **266:** 212–227.

Claverie J.-M. and States D.J. 1993. Information enhancement methods for large scale sequence analysis. *Comput. Chem.* **17:** 191–201.

Durbin R., Eddy S., Krogh A., and Mitchison G. 1998. Profile HMMs for sequence families. In *Biological sequence analysis. Probabilistic models of proteins and nucleic acids,* chap. 5. Cambridge University Press, United Kingdom.

Gribskov M. and Veretnik S. 1996. Identification of sequence pattern with profile analysis. *Methods Enzymol.* **266:** 198–212.

Henikoff S. and Henikoff J.G. 1992. Amino acid substitution matrices from protein blocks. *Proc. Natl. Acad. Sci.* **89:** 10915–10919.

———. 1997. Embedding strategies for effective use of information from multiple sequence alignments. *Protein Sci.* **6:** 698–705.

———. 2000. Amino acid substitution matrices. *Adv. Protein Chem.* **54:** 73–97.

Henikoff S., Henikoff J.G., Alford W.J., and Pietrokovski S. 1995. Automated construction and graphical presentation of protein blocks from unaligned sequences. *Gene* **163:** GC17–26.

Henikoff S., Greene E.A., Pietrokovski S., Bork P., Attwood T.K., and Hood L. 1997. Gene families: The taxonomy of protein paralogs and chimeras. *Science* **278:** 609–614.

Hobohm U., Scharf M., Schneider R., and Sander C. 1992. Selection of representative protein data sets. *Protein Sci.* **1:** 409–417.

Jurka J. 1998. Repeats in genomic DNA, mining and meaning. *Curr. Opin. Struct. Biol.* **8:** 333–337.

Ladunga I., Wiese B.A., and Smith R.F. 1996. FASTA-SWAP and FASTA-PAT: Pattern database searches using combinations of aligned amino acids, and a novel scoring theory. *J. Mol. Biol.* **259:** 840–854.

Lipman D.J. and Pearson W.R. 1985. Rapid and sensitive protein similarity searches. *Science* **227:** 1435–1441.

Madej T., Gibrat J.F., and Bryant S.H. 1995. Threading a database of protein cores. *Proteins* **23**: 356–369.

Murzin A.G., Brenner S.E., Hubbard T., and Chothia C. 1995. SCOP: A structural classification of proteins database for the investigation of sequences and structures. *J. Mol. Biol.* **247**: 536–540.

Neuwald A.F., Liu J.S., Lipman D.J., and Lawrence C.E. 1997. Extracting protein alignment models from the sequence database. *Nucleic Acids Res.* **25**: 1665–1677.

Pearson W.R. 1990. Rapid and sensitive sequence comparison with FASTP and FASTA. *Methods Enzymol.* **183**: 63–98.

———. 1991. Searching protein sequence libraries: Comparison of the sensitivity and selectivity of the Smith-Waterman and FASTA algorithms. *Genomics* **11**: 635–650.

———. 1995. Comparison of methods for searching protein sequence databases. *Protein Sci.* **4**: 1150–1160.

———. 1996. Effective protein sequence comparison. *Methods Enzymol.* **266**: 227–258.

———. 1998. Empirical statistical estimates for sequence similarity searches. *J. Mol. Biol.* **276**: 71–84.

———. 2000. Flexible sequence similarity searching with the FASTA3 program package. *Methods Mol. Biol.* **132**: 185–219.

Pearson W.R. and Lipman D.J. 1988. Improved tools for biological sequence comparison. *Proc. Natl. Acad. Sci.* **85**: 2444–2448.

Pearson W.R. and Miller W. 1992. Dynamic programming algorithms for biological sequence comparison. *Methods Enzymol.* **210**: 575–601.

Pearson W.R., Wood T., Zhang Z., and Miller W. 1997. Comparison of DNA sequences with protein sequences. *Genomics* **46**: 24–36.

Pietrokovski S. 1996. Searching databases of conserved sequence regions by aligning protein multiple-alignments. *Nucleic Acids Res.* **24**: 3836–3845.

Retief J.D., Lynch K.R., and Pearson W.R. 1999. Panning for genes — A visual strategy for identifying novel gene orthologs and paralogs. *Genome Res.* **9**: 373–382.

Sander C. and Schneider R. 1991. Database of homology derived protein structures and the structural meaning of sequence alignment. *Proteins* **9**: 56–68.

Schäffer A.A., Wolf Y.I., Ponting C.P., Koonin E.V., Aravind L., and Altschul S.F. 1999. IMPALA: Matching a protein sequence against a collection of PSI-BLAST constructed position-specific score matrices. *Bioinformatics* **15**: 1000–1011.

Tatusov R.L., Koonin E.V., and Lipman D.J. 1997. A genomic perspective on protein families. *Science* **278**: 631–637.

Walker D.R. and Koonin E.V. 1997. SEALS: A system for easy analysis of lots of sequences. *Ismb* **5**: 333–339.

Wilbur W.J. and Lipman D.J. 1983. Rapid similarity searches of nucleic acid and protein data banks. *Proc. Natl. Acad. Sci.* **80**: 726–730.

Wootton J.C. and Federhen S. 1993. Statistics of local complexity in amino acid sequences and sequence databases. *Comput. Chem.* **17**: 149–163.

———. 1996. Analysis of compositionally biased regions in sequences. *Methods Enzymol.* **266**: 554–571.

Worley K.C., Wiese B.A., and Smith R.F. 1995. BEAUTY: An enhanced BLAST-based search tool that integrates multiple biological information resources into sequence similarity search results. *Genome Res.* **5**: 173–184.

Wu C.H., Zhao S., Chen H.L., Lo C.J., and McLarty J. 1996. Motif identification neural design for rapid and sensitive protein family search. *Comput. Appl. Biosci.* **12**: 109–118.

Zhang Z., Pearson W., and Miller W. 1997. Aligning a DNA sequence with a protein sequence. *J. Comput. Biol.* **4**: 333–443.

Zhang Z., Schäffer A.A., Miller W., Madden T.L., Lipman D.J., Koonin E.V., and Altschul S.F. 1998. Protein sequence similarity searches using patterns as seeds. *Nucleic Acids Res.* **26**: 3986–3990.

Zhu J., Liu J.S., and Lawrence C.E. 1998. Bayesian adaptive sequence alignment algorithms. *Bioinformatics* **14**: 25–39.

Gene Prediction

W<small>ITH THE ADVENT OF</small> whole-genome sequencing projects, there is considerable use for computer programs that scan genomic DNA sequences to find genes, particularly those that encode proteins. Once a new genomic sequence has been obtained, the most likely protein-encoding regions are identified and the predicted proteins are then subjected to a database similarity search, as described in the previous chapter. The genomic DNA sequence is then annotated with information on the exon–intron structure and location of each predicted gene along with any functional information based on the database searches. This procedure is summarized in the gene prediction flowchart (p. 346).

In this chapter, I first discuss methods of predicting the genes that encode proteins and then the identification of sequences, such as promoters, that regulate the activity of protein-encoding genes. The prediction of genes that specify classes of RNA molecules is discussed in Chapter 5. The organization of genomes is discussed in Chapter 10. There are many computer programs and Web sites for gene prediction, and representative examples are shown in Table 8.1.

INTRODUCTION

The simplest method of finding DNA sequences that encode proteins is to search for open reading frames, or ORFs. An ORF is a length of DNA sequence that contains a contiguous set of codons, each of which specifies an amino acid. There are six possible reading frames in every sequence, three starting at positions 1, 2, and 3 and going in the 5' to 3' direction of a given sequence, and another three starting at positions 1, 2, and 3 and going in the 5' to 3' direction of the complementary sequence. In prokaryotic genomes, DNA sequences that encode proteins are transcribed into mRNA, and the mRNA is usually translated directly into proteins without significant modification. The longest ORFs running from the first available Met codon on the mRNA to the next stop codon in the same reading frame generally provide a good, but not assured, prediction of the protein-encoding regions (see Table 8.1 for Web sites that provide a more detailed analysis). A reading frame of a genomic sequence that does not encode a protein will have short ORFs due to the presence of many in-frame stop codons. An example of a search of the *Escherichia coli lac* operon for ORFs is shown in Figure 8.1. These predictions have to take into account the observation in *E. coli* and its phages of the presence of multiple genes on mRNA and sometimes of overlapping genes in which two different proteins may be encoded in different reading frames of the same mRNA, either on the same or complementary DNA strands. In eukaryotes, prediction of protein-encoding genes is a more difficult task.

In eukaryotic organisms, transcription of protein-encoding regions initiated at specific promoter sequences is followed by removal of noncoding sequence (introns) from pre-mRNA by a splicing mechanism, leaving the protein-encoding exons. Once the introns have been removed and certain other modifications to the mature RNA have been made, the resulting mature mRNA can be translated in the 5' to 3' direction, usually from the first start codon to the first stop codon. As a result of the presence of intron sequences in the genomic DNA sequences of eukaryotes, the ORF corresponding to an encoded gene will be interrupted by the presence of introns that usually generate stop codons.

Three types of posttranscriptional events influence the translation of mRNA into protein and the accuracy of gene prediction. First, the genetic code of a given genome may vary from the universal code (see Table 8.1 for reference Web sites). For the most part, the universal genetic code, shown in Table 8.2, is used.

Table 8.1. *Programs and Web pages for sequence translation and related information*

Name of translation site	Web address	Reference
Arabidopis intron splice site table	http:// www.Arabidopsis.org/splice_site.html	see Web site
Codon usage database	http://www.kazusa.or.jp/codon/	see Web site
Ecoparse for finding *E. coli* genes based on HMM model	mail server described at http://www.cbs.dtu.dk/krogh/EcoParse.info	Krogh et al. (1994)
EST-GENOME for alignment of EST/cDNA and genomic sequences	http://www.hgmp.mrc.ac.uk/Registered/Option/ est_genome.html	see Web site; also see Florea et al. (1998)
Exon recognizer, including GeneScope	http://gf.genome.ad.jp/	see Web page
FGENES and related programs that use linear discriminant analysis or hidden Markov models[a]	http://genomic.sanger.ac.uk/gf/gf.shtml	Solovyev et al. (1995); see Web site
FINEX–exon intron boundary analysis	http://www.icnet.uk/LRITu/projects/finex/	Brown et al. (1995)
GeneFinder access site at Baylor College of Medicine	http://dot.imgen.bcm.tmc.edu:9331/ gene-finder/gf.html	collection of methods
Genehacker for microbial genomes based on HMMs	http://www-scc.jst.go.jp/sankichi/GeneHacker/	Hirosawa et al. (1997)
GeneID-3 Web server using rule-based models, and GeneID+[b]	http://www1.imim.es/geneid.html Mail server at geneid@darwin.bu.edu	Guigó et al. (1992); Guigó (1998)
GeneMark and GeneMark.hmm[c] uses hidden Markov models	http://genemark.biology.gatech.edu/GeneMark/; http://www2.ebi.ac.uk/genemark/	Lukashin and Borodovsky (1998)
GeneMark home page (see webgenemark)	http://genemark.biology.gatech.edu/GeneMark/	Borodovsky and McIninch (1993)
GeneParser[a,b] Web page, uses combination of neural network and dynamic programming methods	http://beagle.colorado.edu/~eesnyder/GeneParser.html	Snyder and Stormo (1993, 1995)
Genescan using Fourier transform of DNA sequences to find characteristic patterns	http://202.41.10.146/GS.html	Tiwari et al. (1997)
GeneScope	http://gf.genome.ad.jp/genescope/; see Exon recognizer	Murakàmi and Takagi (1998)
Genetic code variations	http://www.ncbi.nlm.nih.gov/htbin-post/Taxonomy/ wprintgc?mode=c	
Genie for finding human genes in 10-kb DNAs and in *Drosophila* by hidden Markov models and neural networks	http://www.cse.ucsc.edu/~dkulp/cgi-bin/genie http://www.fruitfly.org/seq_tools/genie.html http://www.tigem.it/TIGEM/HTML/Genie.html	Kulp et al. (1996); Reese et al. (1997, 2000)
GenLang using linguistic methods	http://www.cbil.upenn.edu/	Dong and Searls (1994)
GenScan based on probabilistic model of gene structure for vertebrate, *Drosophila*, and plant genes	http://genes.mit.edu/GENSCAN.html	Burge and Karlin (1998)
Genseqer for aligning genomic and EST sequences	http://gremlin1.zool.iastate.edu/cgi-bin/gs.cgi	see Web site and Splicepredictor
Glimmer uses interpolated Markov models for prokaryotic translation	http://www.tigr.org/softlab/ and http://www.cs.jhu. edu/labs/compbio/glimmer.html	Salzberg et al. (1998)

Continued.

Table 8.1. *Continued.*

Name of translation site	Web address	Reference
GrailII[a,b] prediction by neural networks based on scores of characteristic sequence patterns and composition	http://compbio.ornl.gov/	Uberbacher and Mural (1991); Uberbacher et al. (1996)
Hexon for exon prediction by linear discriminant analysis	see GeneFinder access site	Solovyev et al. (1994)
Human splice sites with decision tree analysis[d]	http://sol2.ebi.ac.uk/projects/Events/gene/ genepred-thanaraj.html	Thanaraj (1999)
INFO for finding splice junctions by database similarity search	http://elcapitan.ucsd.edu/~info/	Laub and Smith (1998)
INFOGENE: a database of known gene structures and predicted genes	http://genomic.sanger.ac.uk/inf/infodb.shtml	Solovyev and Salamov (1999)
Initiation codon analysis	http://www.ncbi.nlm.nih.gov/htbin-post/ Taxonomy/wprintgc?mode=c	see Web site
Microbial genome coding region identification based on Markov chains of order 5	http://igs-server.cnrs-mrs.fr/~audic/selfid.html	Audic and Claverie (1998)
Morgan for finding vertebrate genes by decision tree classification[d]	http://www.cs.jhu.edu/labs/compbio/morgan.html	see http://www.cs.jhu.edu/labs/ compbio/morgan.html#refs; Salzberg (1998); Searls (1998)
MZEF uses quadratic discriminant analysis for human, mouse, *Arabidopsis,* and *S. pombe* exons	http://argon.cshl.org/genefinder/	Zhang (1997)
NetGene uses neural networks for analysis of splice sites in human, *C. elegans*, and *Arabidopsis* genes	http://www.cbs.dtu.dk/services/NetGene2/	Brunak et al. (1991); Hebsgaard et al. (1996)
NetPlantGene	http://www.cbs.dtu.dk/services/NetPGene/	see NetGene
NetStart uses neural networks for gene prediction in vertebrate and *Arabidopsis* genes	http://www.cbs.dtu.dk/services/NetStart/	Pedersen and Nielsen (1997)
Procrustes based on comparison of related genomic sequences	http://www-hto.usc.edu/software/procrustes/	Gelfand et al. (1996)
Push-button Gene Finder for gene identification using Markov and hidden Markov models	http://www.cse.ucsc.edu/research/compbio/pgf/	see Web site
Splice Predictor for plants uses trained logitlinear models	http://gremlin1.zool.iastate.edu/cgi-bin/sp.cgi	Brendel and Kleffe (1998); Brendel et al. (1998)
Splicing Sites by neural network at LBNL	http://www.fruitfly.org/seq_tools/splice.html	see Genie
Translate tool at ExPASy	http://www.expasy.ch/tools/dna.html	see Web site
Translation machine on the Web at EBI	http://www2.ebi.ac.uk/translate/	see Web site

Table 8.1. *Continued.*

Name of translation site	Web address	Reference
Translation of large genome sequences on the Web	http://alces.med.umn.edu/rawtrans.html	see Web site
Veil (Viterbi exon-intron locator) uses hidden Markov models for vertebrate DNA	http://www.cs.jhu.edu/labs/compbio/veil.html	Henderson et al. (1997)
Webgene, a set of gene prediction tools and concurrent database similarity searches	http://www.itba.mi.cnr.it/webgene/	see Web site
Webgenemark and Webgenemark.hmm[c]	http://genemark.biology.gatech.edu/GeneMark/	see GeneMark; Lukashin and Borodovsky (1998)
Yeast splice sites by M. Ares Jr. laboratory	http://www.cse.ucsc.edu/research/compbio/yeast_introns.html	Spingola et al. (1999)

Abbreviations: (LBNL) Lawrence Berkeley National Laboratory.

Lists of Web sites for gene recognition and splice site prediction with references and program availability are also available at http://linkage.rockefeller.edu/wli/gene/programs.html, http://www.bork.embl-heidelberg.de/genepredict.html, http://www.hgc.ims.u-tokyo.ac.jp/~katsu/genefinding/programs.html and http://www-hto.usc.edu/software/procrustes/links.html. A more detailed list of programs for gene recognition has been prepared (Burset and Guigó 1996).

Performance comparisons are given at http://igs-server.cnrs-mrs.fr/igs/banbury/, http://www.cs.jhu.edu/labs/compbio/veil.html#perf, http://www1.imim.es/courses/SeqAnalysis/GeneIdentification/Evaluation.html, and are also described in many of the references (see, e.g., Snyder and Stormo 1993; Zhang 1997).

[a] Programs that assemble exons into predicted genes.

[b] Prediction can be enhanced through database similarity searches. GeneParser 3 has this option.

[c] The GeneMark.hmm program is designed to use additional information at the 5′ end of bacterial sequences.

[d] A decision tree analysis has features in common with the phlyogenetic analysis described in Chapter 5 and also with the discriminant analysis described in the text. Scorable features of sequences in coding versus noncoding regions are used as a basis for optimally classifying the sequences into sets. Cutoff values for these features are then used as a basis for scoring unknown sequences as coding or noncoding. These criteria are applied in a sequential order much like starting at the root of a tree and passing through a series of nodes. At each node a further criterion is applied that is the basis for moving along one branch from that node and moving to the next node. Eventually, a terminal branch is reached that is labeled with a decision. In this case, the label is a YES if the sequence is coding, a splice site, or whatever test is being applied because it meets the criteria applied in passing through the decision nodes on the tree, or NO, the sequence is not coding, etc., and because it does not meet the applied criteria.

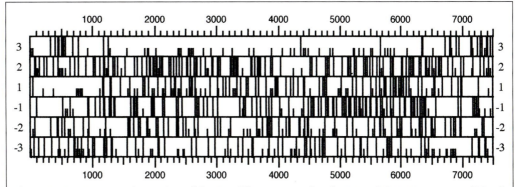
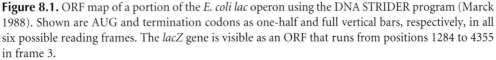

Figure 8.1. ORF map of a portion of the *E. coli lac* operon using the DNA STRIDER program (Marck 1988). Shown are AUG and termination codons as one-half and full vertical bars, respectively, in all six possible reading frames. The *lacZ* gene is visible as an ORF that runs from positions 1284 to 4355 in frame 3.

Table 8.2. *The universal or standard genetic code*

UUU-Phe	F	UCU-Ser	S	UAU-Tyr	Y	UGU-Cys	C
UUC-Phe	F	UCU-Ser	S	UAU-Tyr	Y	UGU-Cys	C
UUA-Leu	L	UCA-Ser	S	UAA-	TER	UGA-	TER
UUG-Leu	L	UCG-Ser	S	UAG-	TER	UGG--Trp	W
CUU-Leu	L	CCU-Pro	P	CAU-His	H	CGU-Arg	R
CUC-Leu	L	CCU-Pro	P	CAU-His	H	CGC-Arg	R
CUA-Leu	L	CCA-Pro	P	CAA-Gln	Q	CGA-Arg	R
CUG-Leu	L	CCG-Pro	P	CAG-Gln	Q	CGG-Arg	R
AUU-Ile	I	ACU-Thr	T	AAU-Asn	N	AGU-Ser	S
AUC-Ile	I	ACC-Thr	T	AAC-Asn	N	AGC-Ser	S
AUA-Ile	I	ACA-Thr	T	AAA-Lys	K	AGA-Arg	R
AUG-MET	M	ACG-Thr	T	AAG-Lys	K	AGG-Arg	R
GUU-Val	V	GCU-Ala	A	GAU-Asp	D	GGU-Gly	G
GUC-Val	V	GCC-Ala	A	GAC-Asp	D	GGC-Gly	G
GUA-Val	V	GCA-Ala	A	GAA-Glu	E	GGA-Gly	G
GUG-Val	V	GCG-Ala	A	GAG-Glu	E	GGG-Gly	G

Shown are each codon and the three-letter and one-letter codes for each encoded amino acid. ATG is the usual START codon and the three TER codons cause translational termination.

It is important to be aware of cellular organelles and organisms in which the genetic code varies so that the correct translation may be made.

Second, one tissue may splice a given mRNA differently from another, thus creating two similar but also partially different mRNAs encoding two related but partially different proteins (Lopez 1998). Understanding the molecular interactions between RNA and the RNA-binding proteins that perform these modifications is an area of active investigation. Availability of this information will assist in the prediction of such variations. Third, mRNAs may be edited, changing the sequence of the mRNA and, as a result, of the encoded protein (see, e.g., Gray and Covello 1993; Paul and Bass 1998; Morse and Bass 1999). Such changes also depend on interaction of RNA with RNA-binding proteins.

TESTING THE RELIABILITY OF AN ORF PREDICTION

DNA sequences that encode protein are not a random chain of available codons for an amino acid, but rather an ordered list of specific codons that reflect the evolutionary origin of the gene and constraints associated with gene expression. This nonrandom property of coding sequences can be used to advantage for finding regions in DNA sequences that encode proteins (see Fickett and Tung 1992). Each species also has a characteristic pattern of use of synonymous codons; i.e., codons that stand for the same amino acid (Table 8.3) (Wada et al. 1992). There are also different patterns of use of codons in strongly versus weakly expressed genes, as, for example, in *E. coli*. Also in *E. coli*, there is a strong preference for certain codon pairs within a coding region and for certain codons to be next to the termination codon. Some of this preference is due to constraints in amino acid sequences in proteins and some to the influence of a given codon on the translation of neighboring codons (Gutman and Hatfield 1989). There is also a strong preference for codon pairs in eukaryotic exons that has been very useful for distinguishing exons and introns in eukaryotic genomic DNAs, as described later in this chapter. Organisms with a high genome content of GC have a strong bias of G and C in the third codon position (for review, see Von Heijne 1987; Rice et al. 1991).

Table 8.3. *Codon usage table*

UUU-Phe	16.6	26.0	UCU-Ser	14.5	23.6	UAU-Tyr	12.1	18.8	UGU-Cys	9.7	8.0
UUC-Leu	20.7	18.2	UCC-Ser	17.7	14.2	UAC-Tyr	16.3	14.7	UGC-Cys	12.4	4.7
UUA-Leu	7.0	26.3	UCA-Ser	11.4	18.8	UAA-TER	0.7	1.0	UGA-TER	1.3	0.6
UUG-Leu	12.0	27.1	UCG-Ser	4.5	8.6	UAG-TER	0.5	0.5	UGG-Trp	13.0	10.3
CUU-Leu	12.4	12.2	CCU-Pro	17.2	13.6	CAU-His	10.1	13.7	CGU-Arg	4.7	6.5
CUC-Leu	19.3	5.4	CCC-Pro	20.3	6.8	CAC-His	14.9	7.8	CGC-Arg	11.0	2.6
CUA-Leu	6.8	13.4	CCA-Pro	16.5	18.2	CAA-Gln	11.8	27.5	CGA-Arg	6.2	3.0
CUG-Leu	40.0	10.4	CCG-Pro	7.1	5.3	CAG-Gln	34.4	12.2	CGG-Arg	11.6	1.7
AUU-Ile	15.7	30.2	ACU-Thr	12.7	20.2	AAU-Asn	16.8	36.0	AGU-Ser	11.7	14.2
AUC-Ile	22.3	17.1	ACC-Thr	19.9	12.6	AAC-Asn	20.2	24.9	AGC-Ser	19.3	9.7
AUA-Ile	7.0	17.8	ACA-Thr	14.7	17.7	AAA-Lys	23.6	42.1	AGA-Arg	11.2	21.3
AUG-MET	22.2	20.9	ACG-Thr	6.4	8.0	AAG-Lys	33.2	30.8	AGG-Arg	11.1	9.3
GUU-Val	10.7	22.0	GCU-Ala	18.4	21.1	GAU-Asp	22.2	37.8	GGU-Gly	10.9	23.9
GUC-Val	14.8	11.6	GCC-Ala	28.6	12.6	GAC-Asp	26.5	20.4	GGC-Gly	23.1	9.7
GUA-Val	6.8	11.7	GCA-Ala	15.6	16.2	GAA-Glu	28.6	45.9	GGA-Gly	16.4	10.9
GUG-Val	29.3	10.7	GCG-Ala	7.7	6.1	GAG-Glu	40.6	19.1	GGG-Gly	16.5	6.0

Shown are frequency of each codon per 100,000 codons obtained from http://www.kazusa.or.jp/codon/ for *Homo sapiens*; columns 2, 5, 8, and 11, and for *Saccharomyces cerevisiae*, columns 3, 6, 9, and 12.

On the basis of these characteristics of protein-encoding sequences, three tests of ORFs have been devised to verify that a predicted ORF is in fact likely to encode a protein (Staden and McLachlan 1982; Staden 1990). The first test is based on an unusual type of sequence variation that is found in ORFs; namely, that every third base tends to be the same one much more often than by chance alone (Fickett 1982). This property is due to nonrandom use of codons in ORFs and is true for any ORF, regardless of the species. No information about nucleotide or codon preference is needed for this analysis. The program TEST-CODE, which is available in the Genetics Computer Group suite of programs (http://www.gcg.com), provides a plot of the nonrandomness of every third base in the sequence. An example of TESTCODE output is shown in Figure 8.2. The second test is an analysis to determine whether the codons in the ORF correspond to those used in other genes of the same organism (Staden and McLachlan1982). For this test, information on codon use for an organism is necessary, such as shown in Table 8.3 for human and yeast genes, averaged over all genes. In addition, there may be variations in codon use by different genes of an organism providing a type of gene regulation. An example of the analysis of an *E. coli* gene for the presence of more and less frequently used *E. coli* codons is shown in Figure 8.3. A parameter that reflects the frequency of codon use may also be calculated, as in the Genetics Computer Group CODONFREQUENCY program. Third, the ORF may be translated into an amino acid sequence and the resulting sequence then compared to the databases of existing sequences. If one or more sequences of significant similarity are found, there will be much more confidence in the predicted ORF (Gish and States 1993).

EUKARYOTIC GENES HAVE REPEATED SEQUENCE ELEMENTS THAT PROBABLY REFLECT NUCLEOSOME STRUCTURE

Eukaryotic DNA is wrapped around histone–protein complexes called nucleosomes. As a result, some of the base pairs in the major or minor grooves of the DNA molecule face the nucleosome surface and others face the outside of the structure. Binding sites for some

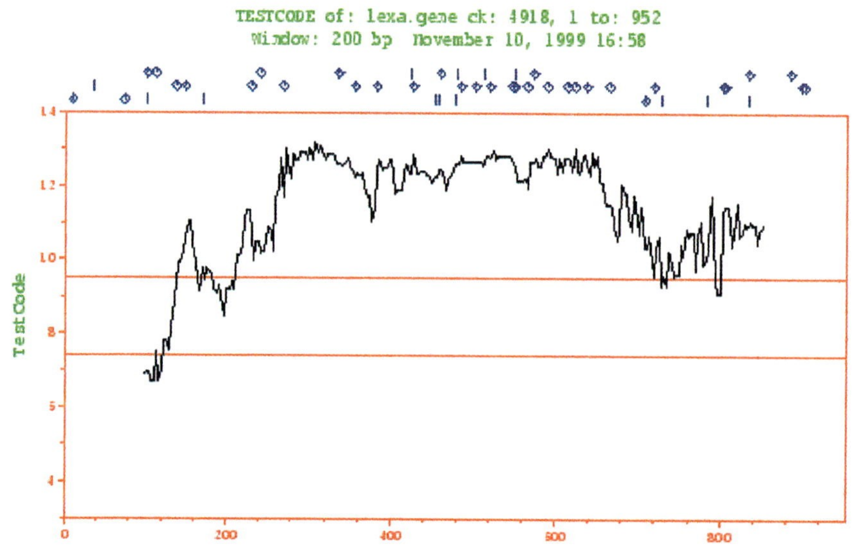

Figure 8.2. TESTCODE analysis of the *E. coli lexA* gene, which is known to extend from positions 102 to 707 in the sequence shown. The TESTCODE statistic (Fickett 1982; for comparison, see Staden 1990) was plotted for each base position in a sliding window of 200 nucleotides. The TESTCODE statistic is found in the following way: (1) The number of each base is counted at every third position starting at positions 1, 2, and 3, and going to the end of the sequence window; (2) the asymmetry statistic for each base is calculated as the ratio of the maximum count of the three possible reading frames divided by the minimum count for the same base plus 1; (3) the frequency of each base in the window is also calculated; (4) the resulting asymmetry and frequency scores are then converted to probabilities of being found in a codon region (found from an analysis of known coding and noncoding regions); and (5) the probabilities are multiplied by weighting factors that are summed. Weighting factors are chosen so that the resulting sum best discriminates coding from noncoding sequences. A value of >0.95 classifies the sequence as coding, and <0.74 classifies the sequence as noncoding. These cutoff values are indicated by red horizontal lines. TESTCODE was run and displayed using TESTCODE in the Genetics Computer Group suite of programs. Above the plot, short vertical lines indicate possible start codons, and diamonds indicate possible stop codons.

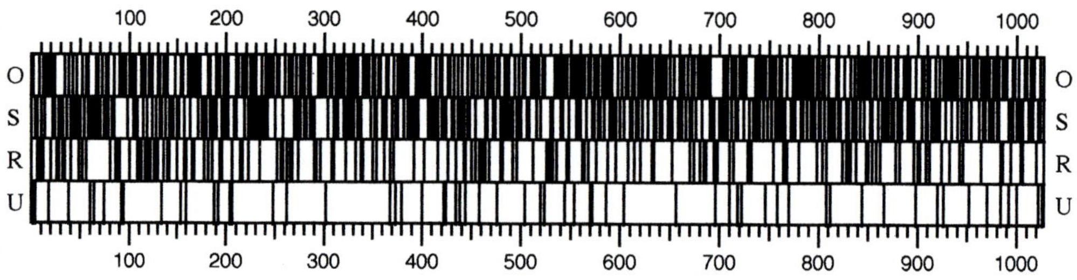

Figure 8.3. Analysis of *E. coli lacZ* gene for occurrence of frequent and infrequent codons using the codon adaptation analysis feature of DNA STRIDER. The positions of common (O for optimum), less common (S for suboptimal), rare (R), and unique (U, which includes the three stop codons, the AUG Met codon and the UGG Trp codon) codons along the sequence are shown, starting at the first nucleotide in the sequence and analyzing three at a time. These first three classes correspond, respectively, to codon adaptation values (Sharp and Li 1987) of >0.9, 0.1–0.9, and <0.1. The gene is obviously represented by commonly used codons.

proteins that regulate transcription may therefore be hidden on the inside of the structure. Nucleosomes located in the promoter region are remodeled in a manner that can influence the availability of binding sites for regulatory proteins, making them more or less available (Carey and Smale 2000).

The computational background of this model is that repeated patterns of sequence have been found in the introns and exons and near the start site of transcription of eukaryotic genes by hidden Markov model (HMM) analysis (Baldi et al. 1996; for a detailed analysis, see Baldi and Brunak 1998; see also Chapter 4, p. 185) and other types of pattern-searching methods (Ioshikhes et al. 1996). These sequences appear to be correlated with the position of nucleosomes and are not found in prokaryotic DNA (Stein and Bina 1999). An example of the HMM is shown in Figure 8.4. These patterns appear with a periodicity of 10; that is,

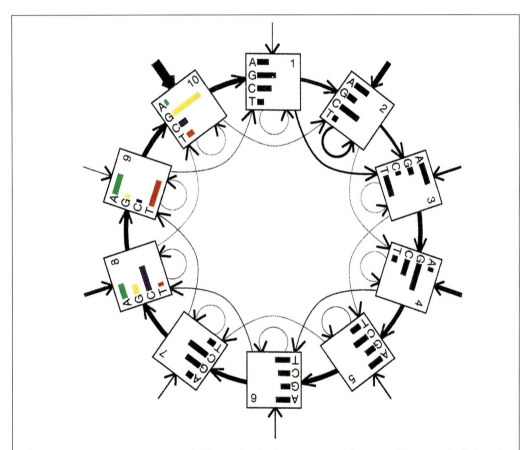

Figure 8.4. A hidden Markov model (HMM) of eukaryotic internal exons. This HMM is designed to detect a statistically significant frequency of the same base at intervals of 10 bp in sequences. Imagine feeding an exon sequence into the part of the sequence shown by the heaviest arrow at 11 o'clock on the circle and then threading the sequence clockwise around the circle, noting the base at each subsequent position in the sequence, and recording that information in the corresponding box (the state of the HMM). If there is a small repeated pattern of a few bases at every tenth position in the sequence starting at the same position from the start of the exon sequence, the distribution of bases in some of the boxes will begin to reflect that pattern. Hence, there is a repeated pattern of not-T (i.e., A, C, or G), A or T, then a G. By a slightly more sophisticated analysis similar to that discussed in Chapter 4 (p. 187), the model can be used to show that the same pattern may start at other positions with respect to the start of the sequence (other arrows feeding into the circle) and also that some sequence positions in the circle may be skipped (arrows going around some of the states) or extra sequence may be found (loop arrow returning to same state). A similar pattern is found in introns and also around the start site of transcription. This structure is modulated by histone-modifying systems as one means of gene regulation in eukaryotes. (Redrawn, with permission, from Baldi et al. 1996 [copyright Academic Press, London].)

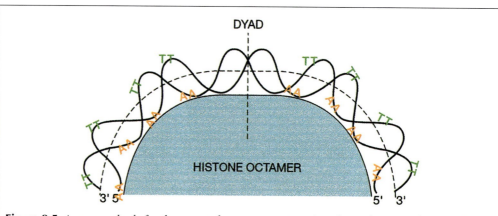

Figure 8.5. A proposed role for the repeated sequence patterns in eukaryotic genes. Shown is the portion of a DNA molecule wrapped around a nucleosome. The actual length of DNA around the nucleosome will be approximately 145 bp. The repeated patterns found in eukaryotic genes (including not-T, A or G, G) and AA/TT dinucleotides influence the bendability of the DNA strand in which they are located, and hence will facilitate the folding of DNA around a nucleosome. (Redrawn, with permission, from Ioshikhes et al. 1996 [copyright Academic Press, London].)

the number of base pairs expected in a single turn of the DNA double-stranded helix around a nucleosome. The patterns found in promoter sequences are those that bend more easily when located in the major groove of DNA and are thought to be located on the inside of the bent molecule (Ioshikhes et al. 1996; Pederson et al. 1998), as shown in Figure 8.5. Using these observations, a model has been proposed that sequence patterns located downstream from the transcription start site are suitable for positioning of nucleosomes, whereas upstream regions do not show the necessary patterns (Pederson et al. 1998).

Loops of chromatin are attached to the nuclear matrix by relatively short (100–1000 bp long) sequences called matrix attachment regions (MARS) or scaffold-associated regions (SARS). These regions are considered to be an indicator of the presence of expressed genes. Although the sequence of only a small number of such regions has been determined, several characteristic sequence patterns have been identified. The program MARS-FINDER (see Table 8.6 for Web site) searches for sequences that have a high representation of such sites in genomic DNA (Singh et al. 1997).

METHODS

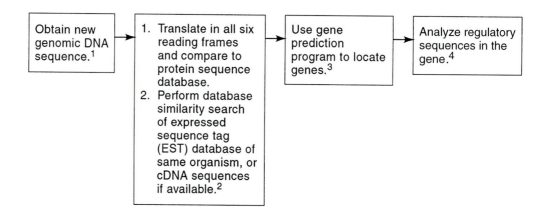

1. The purpose of gene prediction is to identify regions of genomic DNA that encode proteins, although searches for RNA-encoding genes are also performed (see Chapter 5). The genomic DNA sequence may be that of an insert of genomic DNA in a bacterial artificial chromosome (BAC) or similar vector or that of an assembled chromosome or chromosomal fragment. Genome sequencing centers often search through newly acquired sequences with gene prediction programs and then annotate the sequence database entry with this information. This annotation includes gene location, gene structure (positions of predicted exons/introns and regulatory sites), and any matches of the translated exons with the protein sequence databases. The amino acid sequence of the predicted gene may also be entered in the protein sequence databases. Because the standards for identification are not uniform, and because gene predictions can be incorrect, it is a good idea to reconfirm any gene prediction of interest, perform alignments of the predicted sequence with matching database sequences to confirm statistical and biological significance (as described in Chapters 3 and 7), and confirm the predicted gene sequence by cDNA sequencing. If EST sequences are available in a sufficient coverage of the genome, these are also useful for confirmation of predicted gene sequences. For an example of the gene annotation procedure that was followed for the *Drosophila melanogaster* genome sequence, see Adams et al. (2000). The final goal of the gene annotation procedure for an organism is to produce a genome database that includes a rich supply of biological information on the function of each gene, as discussed in Chapter 10. This information will come from laboratory experimentation and manual entry of relevant published data into the genome database.

2. Database similarity searches of this type are described in the flowchart of Chapter 7. For genes of prokaryotic organisms, step 1 identifies open reading frames (ORFs, a series of amino acid-specifying codons) that encode a protein similar to one found in another organism. ORFs without a similar gene in another organism may also be found, as described in the text. Genes of eukaryotic organisms often have intron and exon sequences in the genomic DNA sequence. Step 1 provides the approximate locations of exons that encode a protein similar to one in another organism. Eukaryotic genomes may also have ORFs that do not match a database sequence, and these ORFs may or may not encode a protein. In the Genome Annotation Assessment Project (GASP) of the *Drosophila* genome, one study showed that combining gene prediction methods with homology searches generally provides a reliable annotation method (Birney and Durbin 2000). Step 2 is an additional type of database similarity search that identifies protein-encoding ORFs. Because cDNA sequences and partial cDNA sequences correspond to exons, genomic ORFs that can be aligned to these expressed gene sequences include exon sequences. This analysis can be enhanced by using databases of indexed genes in which overlapping ESTs have been identified (see flowchart, Chapter 7). EST_GENOME is a program for aligning EST and cDNA sequences to genome sequences (Table 8.1). Collections of EST sequences for an organism are often only partial collections; thus, failure to find a matching EST is not a sufficient criterion for rejecting an ORF by this test. Searching the EST collections of related organisms, e.g., another mammal or plant, may be helpful in identifying such missing EST sequences. An additional type of gene analysis is to use an already-identified ORF as a query sequence in a database search against the entire proteome (all of the predicted proteins) of an organism to find families of paralogous genes, as described in Chaper 10.

3. There are a large number of gene prediction programs available (Table 8.1). They all have in common to varying degrees the ability to differentiate between gene sequences characteristic of exons, introns, splicing sites, and other regulatory sites in expressed genes from other non-gene sequences that lack these patterns. Because these gene sequences as well as gene structure (the number and sizes of exons and introns) vary from one organism to another (see Fig. 10.3), a program trained on one organism, e.g., the bacterium *E. coli* or the worm *Caenorhabditis elegans*, is not generally useful for another organism, e.g., another bacterial species or the fruit fly *D. melanogaster*. Reliability tests of gene prediction programs have shown that the available methods for predicting known gene structure are, in general, error-prone. Referring to Web sites with this information (Table 8.1) or performing one's own reliability check is recommended. Some "reliability checks" should be eyed with suspicion because they are based on a comparison of new predictions with previous gene annotations. When gene predictions are made using gene-sized rather than large-sized, multigene sequence genomic DNA fragments, the predictions are generally more reliable (see text).

4. In prokaryotes, the predicted genes may have conserved sequence patterns such as those for promoter recognition by RNA polymerases and transcription factors, for ribosomal binding to mRNA, or for termination of transcription, as found in the model prokaryote *E. coli*. Similarly, in eukaryotes, the

region at the 5′ end of the gene may also have characteristic sequence patterns such as a high density and periodicity of putative transcription-factor-binding sites and sequence patterns characteristic of RNA polymerase II promoters. These types of analyses are enhanced by searching for similar sequence patterns in genes that are regulated by the same set of environmental conditions or that are expressed in the same tissue. Regulatory predictions are enhanced when information about conserved oligomers found in the promoters of co-regulated genes is available, as described in the text.

GENE PREDICTION IN MICROBIAL GENOMES

Predicting protein-encoding genes is generally easier in prokaryotic than eukaryotic organisms because prokaryotes generally lack introns and because several quite highly conserved sequence patterns are found in the promoter region and around the start sites of transcription and translation, at least in the *E. coli* model of prokaryotes. When a set of different patterns characteristic of a gene are found in the same order and with the same spacing in an unknown sequence, the prediction is more reliable than if only one pattern is found, and this type of information can be obtained in *E. coli*.

An example of the regulatory sequences for an *E. coli* gene, the *lexA* gene, is shown in Figure 8.6. Note the presence of the −10 and −35 regions (yellow) that mark the site of interaction with RNA polymerase, and the ribosomal binding site on the mRNA product (green) that is complementary to the ribosomal RNA. The ORF that encodes the LexA product is also indicated (blue). Also shown are three potential binding sites for LexA product to the promoter region, recognizable by searching for a consensus of known LexA-binding sites. Note that these sites are inverted repeats; i.e., the sequence on the forward and reverse sequence is approximately the same. This feature with minor variations is not uncommon in the binding sites of proteins that regulate transcription and is a reflection of the binding of a dimer of LexA protein to the two sites, which produces a stronger interaction than binding of a single monomer to a single site. The sites in the *lexA* promoter region represent a form of self-regulation. The two downstream sites have been shown to bind the protein and to act as a repressor that prevents further transcription. The binding at two sites may be cooperative in that two dimer molecules are more effective at preventing transcription than one, possibly because the bound proteins interact, thus making the overall binding to the promoter region stronger.

In the case of a number of other genes, binding of a regulatory protein such as LexA to a recognizable target sequence activates transcription by stimulating the binding of RNA polymerase. The consensus patterns for these various regulatory sites may be found by sequence alignment and statistical and neural network methods. These methods are discussed in Chapters 3 and 4, and also later in this chapter. Ribosomal binding sites were the first to be modeled by a neural network with no hidden layer (or perceptron), which is also discussed below (Stormo et al. 1982; Bisant and Maizel 1995).

HMMs are also used for modeling a multiple sequence alignment of many proteins and for use in identification of more members of the same family of proteins (see Chapter 3 for details.)

The highly conserved features of *E. coli* genes have made gene identification methods an attractive possibility. One such method is that of HMMs. Here a model of an *E. coli* gene is made and then expanded to include multiple genes and the sequences between the genes. The model shown in Figure 8.7 is an example of a simple HMM of a bacterial genome as a DNA molecule that is densely packed with genes with relatively short intergenic sequences and no introns. This model will read through a sequence of unknown gene composition and find the genes, i.e., a series of codons that specify amino acids flanked by start and stop codons, that are most like a set of known gene sequences and flanking regions that have been used to train or calibrate the model. Because codon usage and flanking sequence will probably vary from one genome to the next, the model trained with *E. coli* genes may not work for finding genes in other organisms. The reliability of the model depends on the

DNA PATTERNS IN THE *E. coli lexA* GENE

GENE SEQUENCE	PATTERN

```
  1 GAATTCGATAAATCTCTGGTTTATTGTGCAGTTTATGGTT          CTGNNNNNNNNNNNCAG
                                        TT           TTGACA
 41 CCAAAATCGCCTTTTGCTGTATATACTCACAGCATAACTG          CTGNNNNNNNNNNNCAG
    CCAA -35             -10  TATACT        >         TATAAT, > mRNA start
 81 TATATACACCCAGGGGGCGGAATGAAAGCGTTAACGGCCA          CTGNNNNNNNNNNNCAG
            +10  GGGGG Ribosomal binding site        GGAGG
121 GGCAACAAGAGGTGTTTGATCTCATCCGTGATCACATCAG
161 CCAGACAGGTATGCCGCCGACGCGTGCGGAAATCGCGCAG          ATG
201 CGTTTGGGGTTCCGTTCCCAAACGCGGCTGAAGAACATC
241 TGAAGGCGCTGGCACGCGCAAAGGCGTTATTGAAATTGTTTC
281 CGGCGCATCACGCGGGATTCGTCTGTTGCAGGAAGAGGAA
321 GAAGGGTTGCCGCTGGTAGGTCGTGTGGCTGCCGGTGAAC
361 CACTTCTGGCGCAACAGCATATTGAAGGTCATTATCAGGT          OPEN READING FRAME
401 CGATCCTTCCTTATTCAAGCCGAATGCTGATTTCCTGCTG
441 CGCGTCAGCGGGGATGTCGATGAAAGATATCGGCATTATGG
481 ATGGTGACTTGCTGGCAGTGCATAAAAACTCAGGATGTACG
521 TAACGGTCAGGTCGTTGTCGCACGTATTGATGACGAAGTT
561 ACCGTTAAGCGCCTGAAAAAACAGGGCAATAAAGTCGAAC
601 TGTTGCCAGAAAATAGCGAGTTTAAACCAATTGTCGTTGA
641 CCTTCGTCAGCAGAGCTTCACCATTGAAGGGCTGGCGGTT
681 GGGGTTATTCGCAACGGCGACTGGCTGTAACATATCTCTG          TAA
721 AGACCGCGATGCCGCCTGGCGTCGCGGTTTGTTTTTCATC
761 TCTCTTCATCAGGCTTGTCTGCATGGCATTCCTCACTTCA
801 TCTGATAAAGCACTCTGGCATCTCGCCTTACCCATGATTT
841 TCTCCAATATCACCGTTCCGTTGCTGGGACTGGTCGATAC
881 GGCGGTAATTGGTCATCTTGATAGCCCGGTTTATTTGGGC
921 GGCGTGGCGGTTGGCGCAACGGCGGACCAGCT
```

Shown are matches to approximate consensus binding sites for LexA repressor (CTGNNNNNNNNNNNCAG), the -10 and -35 promoter regions relative to the start of the mRNA (TTGACA and TATAAT), the ribosomal binding site on the mRNA (GGAGG), and the open reading frame (ATG...TAA). Only the second two of the predicted LexA binding sites actually bind the repressor.

Figure 8.6. The promoter and open reading frame of the *E. coli lexA* gene.

accuracy of the gene start and stop information that is used for the training or calibration step and on the number of such genes used for training. For *E. coli*, the positions of many genes have been accurately determined. For other microbial genomes, this information is not as available, and genes predicted by alignment of the predicted proteins with *E. coli* proteins have to be used. Similar models of gene structure have been developed for other microbial genomes.

The HMM model shown in Figure 8.7 assumes that there is no relationship between each codon and later codons in the sequence; i.e., that the choice of each codon is independent of the rest. This model of genes as a Markov chain may not be fully correct because there may be long-distance correlations between some positions due to requirements for mRNA structure or translation. However, using this simplifying assumption, useful gene models can be produced. Analyses of sequential codons in genes have shown that some pairs are found at a greater frequency and others at a lesser frequency than expected by chance alone (Gutman and Hatfield 1989; Farber et al. 1992). Hence, a more appropriate choice is to design a model that uses sequence information from the previous five instead of the previous two bases to make what is called a fifth-order Markov model. In such a model, the frequency of hexamers is used to differentiate between coding and noncoding sequences. A version of GeneMark (Borodovsky and McIninch 1993) called GeneMark.HMM uses a HMM of this type to search for *E. coli* genes (Lukashin and Borodovsky 1998).

From an information perspective, as the number of consecutive sequence positions being compared in two sequences is increased, the chance of being able to find similarities above background noise increases. For example, when using the dot matrix method for comparing sequences, a sliding window in which words of length *n* are compared is used to locate the most significant matches. In comparing codon and noncoding sequences, a comparison of three consecutive positions at a time can be used to find ORFs as uninterrupted runs of amino-acid-specifying codons. Extending the number of positions to a number greater than three, such as four to six, increases the chances of discovering higher-order sequence correlations in coding sequences that may be used to distinguish them from noncoding sequences.

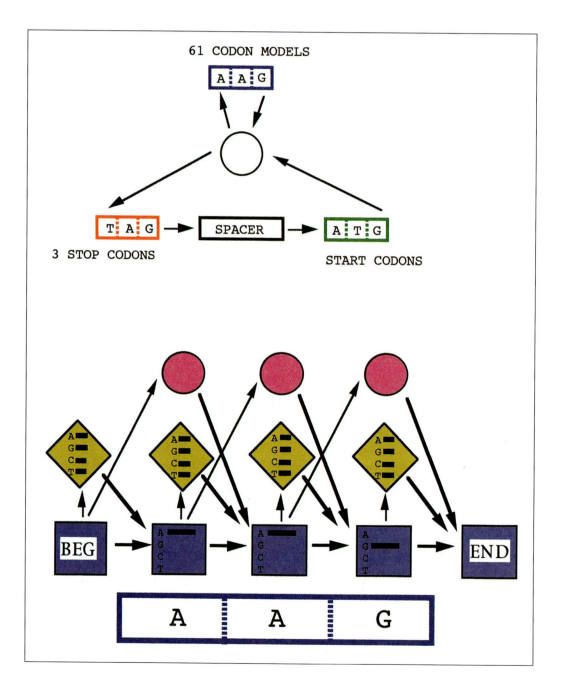

For fifth-order Markov models to give accurate gene predictions, there must be many representatives of each hexameric sequence in genes, and if there is not, the method will be statistically limited. A new type of model, the interpolated Markov model (IMM; e.g., Glimmer; see Table 8.1), overcomes this difficulty of finding a sufficient number of patterns by searching for the longest possible patterns that are represented in the known gene sequences up to a length of eight bases. Thus, if there are not enough hexameric sequences, then pentamers or smaller may be more highly represented, and in other cases many representative patterns even longer than six bases may be found. In general, the longer the patterns, the more accurate the prediction. The IMM combines probability estimates from the different-sized patterns, giving emphasis to longer patterns and weighting more heavily the patterns that are well represented in the training sequences (Salzberg et al. 1998).

Both GeneMark.HMM and IMM find genes in microbial genomes with an apparent high degree of accuracy, assuming that gene predictions made by other methods such as sequence similarity of the translated proteins to known *E. coli* proteins are accurate. Therefore, these methods can be expected to produce reliable predictions of genes that do not match previously identified protein sequences. A further improvement of the prediction of the bacterial start codon position has been found (Hannenhalli et al. 1999). This method sorts through a set of predictions for the start codon in a set of sequences, where the actual signal is known. These predictions depend on weighting each of a set of input sequence information. The weights are adjusted so that the predictions are made more accurate by a method called mixed integer programming.

Compare this gene model with the model for protein sequence alignments shown on page 186, Figure 4.16.

Figure 8.7. HMM of an *E. coli* gene (Krogh et al. 1994). This model is designed to generate a sequence of amino-acid-encoding codons of the approximate length of an *E. coli* gene starting with an ATG codon and ending with a stop codon. A set of predicted genes are separated by intergenic spacer regions of the range of lengths actually found between *E. coli* genes. Variations in this basic model are described in the text. The model is first trained on a set of known *E. coli* gene sequences with flanking sequences. The training step is performed in very much the same manner as that described in Chapter 4 for multiple sequence alignment. The trained model may then be used to find the most probable set of genes in *E. coli* genomic sequences of unknown gene composition. The model for each codon (lower part of diagram) is represented by a set of round, diagonal, or square boxes representing match, insert, and delete states, respectively. The model shown is that for the AAG codon. Each of the 61 codons has a similar structure. If a sequence were extremely accurate, only match states would be needed in the model. The insert and delete states allow an ORF with an extra or missing base to be recognized. Similarly, the inclusion of alternative bases in each match state allows for errors in base identification. Stop codons and initiation codons are assumed to be correctly represented in each sequence and no allowance for errors is made. Hence, errors in these codons would lead to an incorrect prediction. Each match and insert state has a certain probability of producing an A, another probability for producing a G, and so on. The delete state does not produce a letter but instead acts to skip a sequence position. Directional arrows (transitions) give the probability of passing from one state to another in the model. Thus, if one state generates an A with probability of 1.0, the transition probability to the next state is 0.9, and the next state generates a G with probability 0.98, then the probability of AG is $1.0 \times 0.9 \times 0.98 = 0.88$. The model is entered at any position (upper part of diagram) and the arrows designate possible paths through the model between successive states. The central state represented by a circle does not generate a sequence position but acts as a junction between adjacent codons. For the model to generate a sequence, the probability of a codon following another codon must be quite high. Hence, the transition probability of going from the junction to a codon is much higher than for going to a stop codon. Once a stop codon has been reached, a sequence representing an intergenic spacer region is generated. Within this region is a model for sequences that are found upstream from the ATG codon for the next gene, such as the Shine-Dalgarno ribosomal binding site and other sequence information (see Hayes and Borodovsky 1998). The presence of this sequence increases the probability for a downstream gene.

GENE PREDICTION IN EUKARYOTES

A simple method for discovering protein-encoding genes within a eukaryotic genomic sequence is to perform a sequence database search by translating the sequence in all possible reading frames and comparing the sequence to a protein sequence database using the BLASTX or FASTX programs described in Chapter 7. Alternatively, if a genomic sequence is to be scanned for a gene encoding a particular protein, the protein can be compared to a nucleic acid sequence database that includes genomic sequences and is translated in all six possible reading frames by the TBLASTN or TFASTX/TFASTY programs. For proteins that are highly conserved, these methods can give a very good, albeit approximate, indication of the gene structure. If the proteins are not highly conserved, or if the exon structure of a gene is unusual, these methods may not work.

Additional information as to the locations of genes in genomic DNA sequences may be found by using cDNA sequences of expressed genes (see flowchart). An enhanced method (Pachter et al. 1999) for finding eukaryotic genes rapidly is to prepare a dictionary of sequence words (4-letter words in a protein sequence database, 11-letter words in an EST database) and to use these dictionaries to compare a genomic DNA sequence to the expressed gene and protein sequence databases.

The commonly used methods for eukaryotic gene prediction depend on training a computer program to recognize sequences that are characteristic of known exons in genomic DNA sequences. The program is then used to predict the positions of exons in unknown genomic sequences and to join these exons into a predicted gene structure. Predictions depend on analysis of a variety of sequence patterns that are characteristic of known genes in a particular organism. These include patterns characteristic of exons, intron–exon boundaries, and upstream promoter sequences. As more sequences are collected for specific organisms and the actual structures of additional genes become known, these prediction methods should become more reliable. Patterns that specify RNA splice sites are poorly conserved with only a few identical positions. Therefore, the positions of intron–exon boundaries cannot be defined precisely by simple pattern-searching methods. Neural networks (described below and in Chapter 9) provide a method of sequence analysis that has the capability of finding complex patterns and relationships among sequence positions that may not be obvious. The available methods also depend on the analysis of windows of sequence in genomic DNA to determine whether these regions are likely to be coding or noncoding. Regions that encode proteins are found to have characteristic patterns reflecting preferential codon usage and codon neighbors. These observations have led to the widely used analysis of 6-mers in DNA sequences as a basis for gene prediction.

For RNA PolII genes, gene prediction programs give possible locations of exons that can then be joined to predict the sequence of the mRNA of the gene. This sequence will include an upstream 5′ region (5′ untranslated region, UTR) extending from the start site of transcription to the initiation codon, the ORF for the protein ending in a translational termination codon, and the downstream region (3′UTR) extending to the termination of transcription in the region where the signal for polyadenylation of the mRNA may be found. The initiation site for translation in eukaryotic mRNAs is usually the AUG codon nearest the 5′ end of the mRNA, but sometimes downstream AUG codons still close to the 5′ end of the mRNA may also be used (Kozak 1999).

As examples of the types of analyses that are available, two types of gene prediction methods, neural networks and pattern discrimination methods, are described below. Other methods and Web sites for finding genes in eukaryotic DNA are described in Table 8.1.

Neural Networks

Grail II

Grail II provides analyses of protein-coding regions, poly(A) sites, and promoters; constructs gene models; predicts encoded protein sequences; and provides database searching capabilities. A list of most likely exon candidates is first established, and these are evaluated further using the neural network described in Figure 8.8. The algorithm makes its final prediction by picking the best candidates. A dynamic programming approach is then used to define the most probable gene models (Uberbacher et al. 1996).

Input for Grail II includes several indicators of sequence patterns. These inputs include several from different types of analyses, including a Markov model for gene recognition that, in principle, resembles the one shown in Figure 8.7, and inputs from two additional neural networks that evaluate the region for potential splice sites. One important indicator is the in-frame 6-mer preference score. Recall that the occurrence of codon pairs in coding regions is not random, whereas in noncoding regions their occurrence is more random.

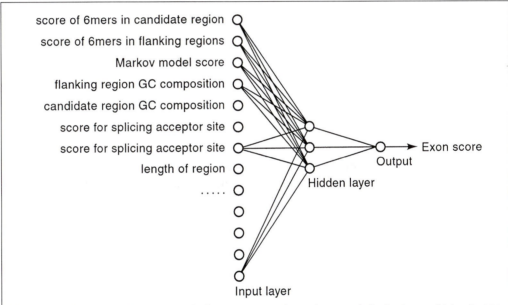

Compare the use of neural networks for gene prediction with that for protein secondary structure prediction shown in Figure 9.29 (p. 453).

Figure 8.8. The Grail II system for finding exons in eukaryotic genes (Uberbacher and Mural 1991; Uberbacher et al. 1996). The method uses a neural network to identify patterns characteristic of coding sequences. The method has similarities to and differences from that used for predicting secondary structure of proteins and described in Chapter 9. Similarities include the use of three layers, an input layer for the data with the data coming from a candidate exon sequence, and a hidden layer for discerning relationships among the input data. An output layer comprising one neuron indicates whether or not the region is likely to be an exon. Each neuron receives information from a set in the layer above, some with a positive value and others with a negative value, sums these values; and then converts them to an output of approximately 0 or 1. The system is trained using a set of known coding sequences, and as each sequence is utilized, the strengths and types of connections (positive or negative) between the neurons are adjusted, decreasing or increasing the signal to the next neuron in a manner that produces the correct output. The major difference between neural networks for exon and secondary structure prediction is that the exon prediction uses sequence pattern information as input whereas secondary structure prediction uses a window of amino acid sequence in the protein. In Grail II, a candidate sequence is evaluated by calculating pattern frequencies in the sequence and applying these values to the neural network. If the output is close to a value of 1, then the region is predicted to be an exon.

Consequently, higher frequencies of 6-mers in genomic DNA that are more commonly found in coding regions can be an indicator of the presence of an exon. For various organisms, tables have been constructed giving the frequency of each 6-mer (base 1 of first codon to base 3 of second, base 2 of first codon to base 1 of the third codon, and so on) of known cDNAs divided by the frequency of the 6-mer in noncoding DNA. The logarithm of this ratio gives what is called an in-frame preference value for the 6-mer. These 6-mer preference scores increase as GC composition rises, thus increasing the preference scores of a 6-mer with GC richness. Grail II automatically corrects for this increase to put predictions from GC-rich regions on an even footing with other regions.

The log ratios for a potential ORF starting at base 1 in the test sequence, another for an ORF starting at base 2, and a third starting at base 3 are calculated by adding the logarithms of these individual 6-mers. These sums provide a log likelihood score for an exon starting at the first, second, or third positions in the given genomic sequence. These likelihoods are further modified by including conditional information on the likelihood of the next 5 bases on coding and noncoding regions, given the current 6-mer. The probability of an exon starting at base 1 is then given by a Bayesian formulation

$$P = a_1 / a + C n_1 \tag{1}$$

where a_1 is the score for an exon starting at base 1; a is the sum of scores for base 1, base 2, and base 3; n_1 is the score for a noncoding region starting at base 1; and C is the ratio of coding to noncoding bases in the organism. This value is used as the score of 6-mers in the candidate region (Uberbacher et al. 1996). A similar score is calculated for the regions 60 bases on each side of the candidate region. If these regions also appear to be encoding exons, the examined region will be enlarged and the prediction repeated. In this manner, a given exon candidate sequence will be enlarged until the coding signals from flanking sequences are no longer to be found.

GeneParser

This program predicts the most likely combination of exons and introns in a genomic sequence by a dynamic programming approach. Dynamic programming was introduced in Chapter 3 as a way for aligning sequences to obtain a most likely alignment for a given scoring system with scores for matches, mismatches, and gaps. The alignment up to a given set of sequence positions is stored in a scoring matrix, and the dynamic programming algorithm provides a method for finding the best score at that position. GeneParser uses a likelihood score for each sequence position being in an intron or exon. The intron and exon positions are then aligned with the constraint that they must alternate within a gene structure. In this manner, a combination of the most likely intron and exon regions that comprise a gene structure is found. GeneParser includes one other novel feature, a scheme for adjusting the weights used for several types of sequence patterns that make up the intron and exon scores.

A neural network is used to adjust the weights given to the sequence indicators of known exon and intron regions, including codon usage, information content (see Chapter 4, p. 195), length distribution, hexamer frequencies, and scoring matrices (see Chapter 4, p. 192) for splicing signals. The integration of the dynamic programming and neural network methods works as follows:

1. The characteristics described above of a set of intron sequences and a second set of exon sequences are determined. For example, a table of hexamer frequencies is prepared.

2. For a training gene sequence, a series of indicator matrices is prepared. The sequence is listed both down the side of the matrix and across the top. Each position in one of the matrices representing positions a and b in the sequence gives the likelihood for an exon or intron that starts at position a and ends at position b. One such matrix would be the likelihood of an exon based on hexamer frequency in the a–b interval. Another matrix (or the other half of the same matrix, since only one half is needed for exon values) gives the likelihood of an intron based on the same criterion. Other sets of matrices for the sequence based on compositional complexity, length distribution or exons, or splice signals on weight matrices are also prepared.

3. The a,b values in the above indicator matrices for exons are each transformed by a weight and bias, and the sum of the weighted values is obtained. An initial arbitrary set of weights is chosen for each type of sequence information. These weights are later adjusted until they provide the correct gene structure of the training sequence. This sum (s) is then further transformed to a number (L) that is either close to 0 or 1 by using the neural network gating function $L = 1/[1 - e^{-s}]$. The transformed a,b values are then placed in another matrix L_E that gives the weighted score for exons going from position a to position b in the sequence. A similar set of transformed values for an intron at position a,b, but not necessarily weighted the same way, is placed in another matrix L_I at position a,b (which can be the other half of L_E since only half of the L_E matrix is needed). The reason for this transformation is to use the information at a later stage as input to a neural network, in the same manner as used in neural networks for prediction of protein secondary structure and discussed in Chapter 9.

4. Dynamic programming is used to predict by the most compatible number and lengths of introns in the training gene up to any position j in the sequence.

5. Steps 2–4 are repeated for each training sequence.

6. The accuracy of the predictions is then determined.

7. If a certain required level of accuracy is not achieved, a neural network similar to that described above for Grail II is used to adjust the weights used for the input exon and intron features.

8. If the required level of accuracy is reached, the method is ready to be used for determining the structure of an unknown genomic DNA sequence.

Pattern Discrimination Methods

Discrimination methods applied to DNA sequences are statistical methods used for classifying the sequence based on one or more observed sequence patterns. For gene prediction, features of patterns found in genomic sequences are examined statistically to determine whether they are like those found in coding sequences. One such feature that is characteristic of coding sequences is the 6-mer exon preference score (EPS) described above. Another is a score for a 3′-flanking splice site (3′SS) calculated in a similar manner. In effect, the distribution of these two scores and a number of others is obtained for a large set of known exons and also for a set of noncoding sequences. Using the EPS and 3′SS as examples, the pair of scores for each sequence is plotted on a graph and each point is labeled as coding or noncoding, as illustrated in Figure 8.9. A line is then positioned between the two groups of sequences. A sequence of unknown coding capability is similarly analyzed to determine whether the features of the sequence place it on one. HEXON and FGENEH (combines exon prediction into a gene structure) use linear discriminant

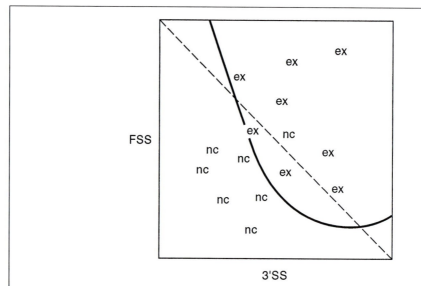

Similar types of discriminant analyses are used to classify microarray data (Fig. 10.11, p. 522).

Figure 8.9. Analysis of candidate sequences for exon status by a discriminant function. Up to nine different pattern features of sequences are analyzed in coding and noncoding sequence. Shown is a plot of two of these features for several exon (ex) and noncoding (nc) sequences. The object of the discriminant analysis is to define a boundary between these two groups of sequences such that they are maximally separated, or that the sum of distances from a boundary line to each point is a minimum. A linear discriminant analysis (Solovyev et al. 1994) assumes that the covariations among the data are the same for the exon and noncoding sequences and provides a straight line boundary (*dotted straight line*) between the two sets of data. Such a boundary may miss some of the data points. A quadratic discriminant analysis (Zhang 1997) is more flexible, does not assume a similar covariation in the exon and noncoding sequences, and provides a curved boundary formed by a quadratic equation that can, in principle, provide a better separation of the groups (*solid line*). Once these boundary lines have been calculated, the EPS and 3′SS values of a query sequence will indicate whether the sequence belongs to the exon group or noncoding group of sequences. For an actual analysis, multiple analyses are performed on a candidate sequence leading to a more complex, multidimensional type of analysis.

analysis (Solovyev et al. 1994) and MZEF uses quadratic discriminant analysis (see Table 8.1) (Zhang 1997).

EVALUATION OF GENE PREDICTION METHODS

A comparison of the above methods for accuracy and reliability must take into account the type of analysis, whether neural network, linear discriminant, or other; the number and types of sequences used for training and evaluation; and the method used for evaluation. In addition, choice of program variables by the user will affect the predictions that are made. As more gene sequences become known, more are becoming available for training and evaluation. The ideal method for evaluation uses a known set of gene structures for training the method and a second set that is not used in the training or similar to those used in the training for evaluation (Burset and Guigó 1996). The evaluation is usually more stringent if the evaluation set includes a gene and neighboring sequence rather than just the sequence between the first and last exons. A current evaluation of most methods is available at the Web sites for these methods listed in the footnotes to Table 8.1. These evaluations are most useful when different prediction methods are used in combination.

The method for evaluation is similar to that used for testing the reliability of protein secondary structure prediction as described in Chapter 9 (Mathews 1975; Burset and Guigó 1996). The program, which is trained on a set of sequences from a given organism, is used to predict the exons, or set of exons, comprising a gene of a set of genomic evaluation sequences from the same organism. An evaluation is then made of the number of true positives (TP) where the length and end sequence positions are correctly predicted, the number of over-predicted positive predictions or false positives (FP), true negative (TN), and number of underpredicted residues as misses or false negative (FN) predictions. The following calculations are made: (1) Number of actual positives is AP = TP + FN; (2) the number of actual negatives is AN = FP + TN; (3) the predicted number of positives is PP = TP + FP; and (4) the predicted number of negatives is PN = TN + FN. The sensitivity of a method SN is given by SN true positives/actual positives = TP / (TP + FN), the specificity by SP = true positives/predicted positives = TP / (TP + FP), and a correlation coefficient CC by

$$CC = [(TP)(TN) - (FP)(FN)] / \sqrt{[(AN)(PP)(AP)(PN)]}$$

By this coefficient, a method given all correct gene predictions would score 1, and the worst possible prediction would be −1. In tests of this kind on three sets of human sequences, GeneParser, GenID, and Grail gave (1) sensitivities of 0.68–0.75, 0.65–0.67, and 0.48–0.65; (2) specificities of 0.68–0.78, 0.74–0.78, and 0.86–0.87; and (3) correlation coefficients of 0.66–0.69, 0.66–0.67, and 0.61–0.72, respectively, for the accuracy of finding the correct nucleotide ends of exons. GeneParser was also shown to be more reliable for genes with short exons and least reliable for genes with long exons (Snyder and Stormo 1993).

A detailed evaluation of the available gene prediction programs has been performed, and the correlation coefficent was found to lie between 0.6 and 0.7, and the fraction of correctly found exons was generally less than 50%. The performance decreased when longer test sequences were used and when a 1% level of artificial frameshift mutations was introduced. Programs including protein sequence database searches (GeneID+ and GeneParser3) showed substantially greater accuracy (Burset and Guigó 1996). These studies therefore indicate that gene prediction programs reliably locate genomic regions that encode genes, but they provide an only approximate indication of the gene structure. In a later similar study using the same data set as the above study, and comparing Grail II, FGENEH, and MZEF, these numbers were: (1) sensitivities 0.79, 0.93, 0.95; (2) specificities 0.92, 0.93, 0.95; and (3) 0.83, 0.85, 0.89, respectively (Zhang 1997).

To illustrate the results obtained by the gene prediction programs, an *Arabidopsis* genomic sequence was submitted to several Web servers, as shown in Table 8.4. Because the cDNA sequence was also available, the accuracy of the programs could be determined. There is a computer program designed for aligning the cDNA and genomic DNA sequences of a gene (Florea et al. 1998; and see Table 8.1). As shown, the results of the analyses vary considerably and the program variables must sometimes be optimized to find the correct translation. In this case, GeneMark gave a fully accurate translation of the sequence. Other programs, such as NetPlantGene, gave a large number of possible exon–intron boundaries including some of the actual ones.

PROMOTER PREDICTION IN *E. COLI*

The method that has most often been used to analyze *E. coli* promoters is to align a set of promoter sequences by the position that marks the known transcription start site (TSS)

Table 8.4. *Example of exons predicted in an* Arabidopsis *genomic sequence by gene prediction programs*

cDNA	Netgene[b]	GeneMark[c]	FgeneP[d]	GeneScan	Mzeff[e]
345[a]–1210	x 1210	345–1210	345–1210	530–1210	276–1210
1290–1513	1290 1513	1290–1513	x 1513	1242–1513	1290–1513
1611–1696	1611* 1696	1611–1696	x x	1611–1696	1611–1696
1880–2029	1880* 2034	1880–2029	x x	1880–2029	1880–2029
2143–2880	2143 2880	2143–2880	x 2880	2143–2880	x 2880
3143–3253	x 3253	3143–3253	x x	x x	3143–3253
3339–3599	3339* 3599	3339–3599	3339-3599	3339–3599	3339–3599
3698–3921	3698 3921	3698–3921	3698-3921	3698–3921	3698–3921
4010–4217	4010 x	4010–4220[f]	x x	4010–4220[f]	x x

This test is given as an example and should not be taken as a measure of the reliability of these programs. The Web sites were provided with the genomic sequences of the *Arabidopis UVH1* gene with approximately 250 bp upstream from the first exon and 200 bp downstream beyond the last exon. As indicated in the text, these programs are more reliable when they are presented with short genomic sequences, as was done in this example. The consensus splice sites for *Arabidopsis* may be found at http://genome-www.stanford.edu/Arabidopsis/splice_site.html. A more detailed assessment of the reliability of gene prediction programs on Arabidopsis genomic sequences has been published (Pavy et al. 1999).

[a] Predicted.

[b] NetPlantGene was used. This program predicts intron–exon and exon–intron junctions and not most probable combinations of the two. In this case many false-positive intron–exon junctions were predicted with low probability. The highest scoring junctions are marked by *. x are actual sites not predicted. The intron–exon junctions are predicted much more reliably, and three false positives were reported.

[c] GeneMark shows a remarkably good frequency of prediction for these exons and usually joins the exons in the correct reading frame, but not always. Therefore, some parts of the predicted protein sequence are not correct.

[d] x are actual sites not predicted. Exon start sites of 1370–1513 and 2779–2880 were found illustrating a difficulty with finding exon start sites.

[e] The prior probability was set at 0.6–0.8 to obtain these results. The higher this value, the lower the level of discrimination used, the more sensitive the test, and the greater the number of exons that is predicted. x was not predicted; instead a start site of 2709 was predicted. This program predicts internal exons only.

[f] The 4220 end includes the termination codon.

and then to search for conserved regions in the sequences. Following such an alignment, *E. coli* promoters are found to contain three conserved sequence features: a region approximately 6 bp long with consensus TATAAT at position −10 (the Pribnow box), a second region approximately 6 bp long with consensus TTGACA at position −35, and a distance between these regions of approximately 17 bp that is relatively constant (see Fig. 8.6 for an example). A weaker region exists around +1, the designation given to the start of transcription, and an AT-rich region is found before the −35 region (Hawley and McClure1983; Mulligan and McClure 1986). The sequences changed to some extent as the number of sequences and the types of promoters analyzed were varied. For example, promoters that are activated by transcription factors have more variable sequences (Hertz and Stormo 1996). The RegulonDB (http://www.cifn.unam. mx/Computational_Biology/regulondb/; Salgado et al. 1999), Dpinteract (http://arep.med.harvard.edu/dpinteract-database; Robison et al. 1998), and regulatory site database (Thieffry et al. 1998; http://www.cifn.unam.mx/Computational_Biology/E.coli-predictions) have been developed with information on the *E. coli* genome. With the availability of a large number of prokaryotic genomes (see Chapter 10 and http://www.tigr.org/tdb/mdb/mdb.html), a similar analysis of the genes and regulatory sites in these other genomes has become possible.

The aligned promoter regions provide a consensus sequence that may be used to search for matching regions as potential promoters in *E. coli* sequences. Each column in the alignment gives the variation found in that position of the promoter. Programs such as the

Genetics Computer Group program FINDPATTERNS and PatScan (http://www-unix.mcs.anl.gov/compbio/PatScan/HTML/patscan.html) may be used to search for matches to the consensus sequence or the variation found in each column in a target DNA sequence. The difficulty with using the consensus sequence to search for new promoters is that most sequence positions in the aligned regions vary to some extent, and some regions are much less variable than others; e.g., the first, second, and sixth positions in the −10 region.

An alternative is to use the search features of FINDPATTERNS and PatScan that allow alternative symbols at one sequence position, repeats of a symbol, inverted repeats, gaps, and so on. For example, providing the pattern GAT (TG, T, G) {1,4} to FINDPATTERNS means to search for GAT followed by a TG, or a T, or a G repeated up to four times. These types of pattern expressions are similar to regular expressions that are used to specify PROSITE patterns in protein sequences and to inititate PHI-BLAST searches of protein sequence databases (see Chapter 7, p. 331). Although these expressions are extremely useful for locating complex regulatory patterns in DNA sequence, they do not take into account the frequency of each residue at each pattern position. What is needed is a more quantitative way to use these known sequence variations to search a target sequence. The scoring matrix method provides such an analysis.

The Scoring Matrix Method Used with Aligned Promoter Sequences

A more complex type of promoter analysis used for both prokaryotic and eukaryotic sequences is a scoring or weight matrix. This kind of matrix was previously described in Chapter 4 (p. 192) as a method for representing the variation in a set of sequence patterns in a multiple sequence alignment, and in Chapter 7 (p. 320) as a tool for finding additional sequences with the same pattern in a database search. The scoring matrix has also been used to analyze promoters, ribosomal binding sites, and eukaryotic splice junctions (Staden 1984).

An example using a scoring matrix for representing the −10 region of *E. coli* promoters is illustrated in Table 8.5. In this example, *N* sequences have been aligned by their –10 regions and a count of each base in each column of the alignment has been made. These counts are converted to frequencies. For example, if 79 of 100 sequences have a T in column 1, the frequency of T in column 1 of the matrix is 0.79. Similarly, a T occurs in column 2 with a frequency of 0.94. These frequencies are converted into log odds scores, as described in Table 8.5. An example of using the scoring matrix in Table 8.5 to locate the most likely −10 sites in a query sequence is shown in Figure 8.10. The matrix is moved along the query sequence one position at a time. At each position, the base in the sequence is noted and the corresponding score of that base in the matrix is then used. This procedure is repeated for the remaining positions. The log odds scores are then added to obtain a combined log odds score for the particular position in the sequence that is a −10 region in a promoter. The sum of the log odds scores in bits may be converted to odds scores by the formula odds score $= 2^{(\text{log odds score})}$ or if the log odds score is in nats, by the formula odds score $= e^{(\text{log odds score})}$. These numbers vary from small fractions to large numbers reflecting variations in the likelihood of a −10 region at each sequence position.

The odds scores at every possible matching location along the sequence may be used to find the probability of each sequence location. The odds scores are first summed to give sum *S*. The odds score at a particular location of six bases in the sequence divided by *S* then provides a probability that the location is a −10 region. To give a simple example, of the three matches in Figure 8.10, the probability of the match at the third location shown is $391/[(1/786)+(1/630)+(391)] = 1.000$.

Table 8.5. *A scoring matrix representing the frequency of DNA bases found in the −10 position in* E. coli *promoters*

A. Fraction of each base at each column of the aligned promoters in the −10 region

Position	A	C	G	T
1	0.02	0.09	0.10	0.79
2	0.94	0.02	0.01	0.03
3..6	. . . .	. . . .	. . . .	. . . .

B. Log odds score

Position	A	C	G	T
1	−3.80	−1.49	−1.34	1.67
2	1.92	−3.81	−4.81	−3.22
3	−0.06	−0.81	−0.66	0.81
4	1.24	−1.00	−0.72	−0.89
5	1.02	−0.35	−1.00	−0.56
6	−4.81	−3.22	−4.81	1.95

(*A*) Frequency of each base found, showing two positions as examples. (*B*) Conversion of frequencies to log odds scores. The first step is to convert the frequency of each base at each sequence position into an odds score. The odds score is simply the frequency observed in the column divided by the frequency expected, or the background frequency of the base, usually averaged over the genome. Thus, if the position frequency is 0.79 and the background 0.25, the odds score is 0.79/0.25 = 3.16. This number means that if a sequence is being examined for the presence of a promoter, and a T is present in the sequence at predicted position 1, the odds of the sequence representing a promoter (a win) to the sequence not representing a promoter (a loss) is 3.16/1. Finally, the odds score is converted to a log odds score by taking the logarithm of the odds score, usually to the base 2 (units of bits) and sometimes to the natural logarithm (units of nats). As described in Chapter 4, bit units have a special meaning in information theory. They represent the number of questions that must be asked to decide whether or not the base in the column of the scoring matrix is a match to the aligned sequence position. This number is called the information content of the matrix position. On the one hand, if all four bases are equally represented in the matrix position, the number of questions that must be asked is two. The first question might be is the sequence position one of A or T, or one of G and C. The second question will then find the correct base. On the other hand, if only one base is found in the matrix position, then no question need be asked of the sequence position. The fewer questions that have to be asked, the more information in the matrix, and the more discriminatory it is for distinguishing real matches from random matches (Schneider et al. 1986). A set of log odds scores for the major six positions in the −10 region of E. coli promoters is shown (Hertz and Stormo 1996). In the actual matrices that are used, an additional 6–12 base positions that flank these major positions are also used. There is a zero occurrence of one particular base in the matrix, thus creating a problem because the logarithm of zero is infinity. In this case, a single count is substituted for the zeros and the resulting small fraction will calculate to a large negative log odds score. Alternatively, a large negative log odds score may be used at such positions in a scoring matrix.

Another formula for calculating the scoring matrix value of base i in column j, $w_{i,j}$, is given by

$$w_{i,j} = \log\,[(n_{i,j} + P_i)/\{(N + 1)P_i\}] \approx \ln\,(f_{i,j} / P_i)$$

where $n_{i,j}$ is the count of base i in column j, P_i is the background frequency of base i, N is the total number of sequences, and $f_{i,j} = n_{i,j}/N$ (Hertz and Stormo 1999). Bucher (1990) uses the formula

$$w_{i,j} = \log\,[(n_{i,j}/P_i) + (s/100)] + C_j$$

where s is a smoothing percentage for the column values and C_j is a column-specific constant. Bucher sometimes also uses dinucleotide composition for calculating the background base frequency to accommodate local sequence complexity (Bucher 1990). These formulas both accommodate zero occurrences of a base by adding a small value in a scoring matrix to zero positions. Another method is to add pseudocounts to these positions, as described in Chapter 4 (p. 193).

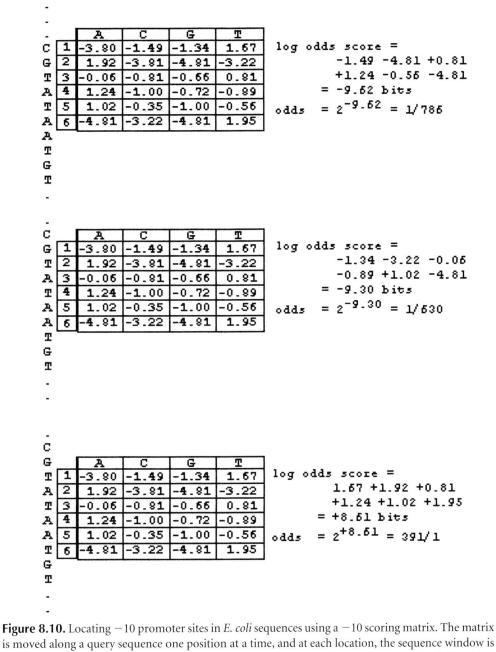

Figure 8.10. Locating −10 promoter sites in *E. coli* sequences using a −10 scoring matrix. The matrix is moved along a query sequence one position at a time, and at each location, the sequence window is scored for a match to the matrix by summing the log odds scores. A cutoff score may be defined that permits recognition of known promoter sequences while minimizing the prediction of false-positive sites.

For scoring *E. coli* sequences for the presence of promoters, scoring matrices for a 35-bp region encompassing the −35 region, a 19-bp region encompassing the −10 region, and a 12-bp region encompassing the +1 region are each applied to both strands of a query DNA sequence. Each matrix will provide a distribution of odds scores that predict possible locations for matches to itself in the query sequence. These matches are then examined for spacings that are characteristic of the known promoter sequences. The region between the −10 and −35 regions varies from 15 to 21 but is usually 17, and the region between −10 and +1 is 4–8 bp. When a suitably oriented combination of high-scoring matches is found, the log odds scores of each sequence region are added. From this sum, a penalty may be subtracted if the distance between the −10 and −35 regions is not an optimal 17 in length or if the distance between the −10 and +1 regions is not optimal (Hertz and Stormo 1996). The resulting log odds score represents an overall likelihood that a test sequence includes regions characteristic of *E. coli* promoters in the correct spacing. A similar application of weight matrices for identifying the start of prokaryotic genes (program ORPHEUS) has been described previously (Frishman et al. 1998).

Reliability of the Matrix Method

The reliability of a combination of scoring matrices for promoter prediction can be assessed by comparing the range of scores found in a set of known promoters versus scores in a set of random sequences. A threshold score that is achieved by most of the known promoters, but only by a small number of false positives in random sequences, may then be chosen (Bucher 1990; Hertz and Stormo 1996). For example, 0.048 or 0.27 of the positions in random sequences may achieve such a score, when compared to promoters that are not activated by additional transcription factors (are more alike) versus all promoters (are more variable), respectively. When a lower threshold is chosen that gives a lower false prediction rate of 0.0005, 0.26 of all promoters and 0.60 of activated promoters achieve such a score. To try to improve the predictive values, the lengths of the scoring matrices and the gap penalty values have been varied, but the predictive value of the matrices is not much improved above these values.

There are several reasons that matrix methods do not achieve a better prediction of *E. coli* promoters. The first is that the matrix method adds the scores for each sequence position, whereas in reality, one position in the −10 region, for example, may play a role in one stage of transcription such as promoter recognition by RNA polymerase, whereas another may play a role in a subsequent stage of transcription, such as initiation of transcription or elongation of the mRNA. Matching positions with these types of functional separations are not expected to be additive, as assumed by the matrix method. A second difficulty that the matrix method shares with most other methods of promoter prediction is that all promoters are treated as being in the same class, whereas different forms of RNA polymerase that are complexed with a set of transcriptional activators (σ factors) may have preference for different sequence positions in the promoter region. With the whole genome of *E. coli* now available for analysis (see http://www.genetics.wisc.edu), such additional classification may become a possibility (Hertz and Stormo 1996). A third difficulty is that the promoter sequence is treated as a Markov chain, meaning that each sequence position acts independently of the others so that a match at each position may be individually scored without reference to the other positions. According to a statistical mechanical theory discussed below, the most conserved positions are thought to act independently. However, as evidenced by the fact that some weight matrices are not efficient in locating matching sites, there may be correlations between the sequence positions so that covariation of the bases at these positions occurs at frequencies greater than expected by chance. Such correlations are not easily found in a small number of training sequences. Methods include using decision trees and locating specific oligonucleotides, discussed later in the chapter. A number of ways to improve matrix methods, including corrections for base composition, utilizing a different number of matrix positions, have been tried, but none of these is significantly better than the basic scoring matrix described above. In addition to the matrix methods, a number of additional methods for predicting *E. coli* promoters and other regulatory sites have been developed, but without much improvement over the scoring matrix method (Hertz and Stormo 1996).

A second method that has been used for promoter prediction is the use of neural networks, which are described in Chapter 9, page 450. In this case, a neural network is trained to distinguish *E. coli* sequences from nonpromoter sequences (Horton and Kanehisa 1992; Pedersen et al. 1996). The network is like that used for prediction of protein secondary structure and is trained by similar methods. Horton and Kanehisa used a network lacking a hidden layer, called a perceptron (see Fig. 8.11). This type of network scans the sequence to be analyzed using a sliding window and at each location reads each of the sequence positions within the window. Some positions within the window may not be counted corresponding to the spaces between the conserved regions. The sequence characters are given a simple identification scheme to avoid any bias (e.g., A is 1000, G 0100, etc.) and the sum of these sequence values after weighting is used as input for a single output neuron, which produces a number close to 1 if the region is within a promoter or 0 if the region is not in

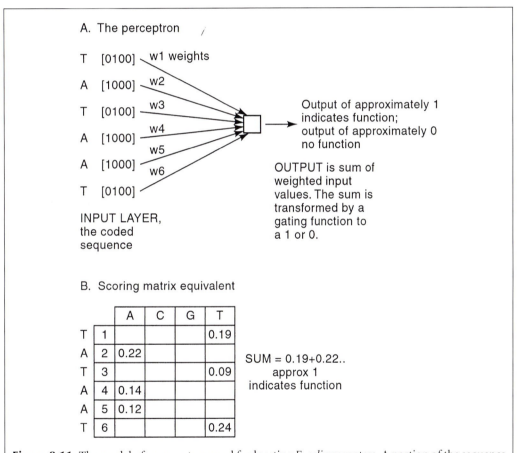

Figure 8.11. The model of a perceptron used for locating *E. coli* promoters. A portion of the sequence in the −10 region that is to be scanned is shown for illustrative purposes. (*A*) A known promoter sequence is encoded and used as input into a single output neuron. The input signals are weighted, the weighted values summed, and the sum transformed into a number that is approximately 0 or 1. The network is trained by starting with an initial set of weights, then adjusting each weight by a small amount until the correct output is found for as many of the training sequences as possible. (*B*) The trained perceptron used for scanning unknown sequences for promoter-like patterns is a special type of scoring matrix. In this case, the matrix is aligned with the sequence, and the matrix values that match the sequence are added. If the sum of numbers is approximately 1, then a promoter is predicted, and if zero, then a promoter is not predicted. The difference between this matrix and the scoring matrix described above is that in this matrix each position within the sequence window is given a different weight.

a promoter. The network is trained on known promoter sequences by adjusting the weights of the input sequence positions so that the output produces the correct response. However, the perceptron method was not found to be any more effective than scoring matrices for finding *E. coli* promoters.

Finding Less-conserved Binding Sites for Regulatory Proteins in Sequences That Do Not Readily Align

In the above example of finding consensus binding sites for RNA polymerase in *E. coli* promoters, the sequences could be quite readily aligned by the transcriptional start site and the −10 and −35 regions. The binding sites for other regulatory proteins, such as the LexA protein described above, are also quite readily found because the sequence of the binding sites is conserved. However, in many other cases, particularly those for eukaryotic transcription factor binding sites described later in this chapter, the sites vary considerably and the surrounding regions are also variable so that it is impossible to find conserved positions in the binding site by aligning the sequences. Thus, methods are needed to find a common but degenerate pattern in sequence fragments that are expected to carry a binding site but that cannot be aligned.

The problem is similar to that described in Chapter 4 for finding patterns that are common to a set of related protein sequences that cannot be readily aligned. However, there is one important difference. In proteins, there are a possible 20 amino acids in each matching position of the sequence pattern, but in DNA-binding sites there are only four possible bases in the pattern—the alphabet is much smaller in DNA sequences. Hence, it is more difficult to detect DNA sequence patterns above background noise. Some of the statistical methods used for finding protein patterns, e.g., expectation maximization and hidden Markov models, are also used for identifying DNA patterns in unaligned DNA sequences.

The expectation maximization method is described in Chapter 4. Briefly, an initial scoring matrix of estimated length is made by a guessed alignment of the known promoter sequences (the expectation step). The scoring matrix is then used to scan each sequence in turn, and the probability of a match to each position in each sequence is calculated as discussed above. The scoring matrix is then updated by the sequence pattern found at each scanned position times the probability of a match to that position (the maximization step). The two steps are repeated until there is no improvement. The method has been adapted to find multiple patterns separated by a variable spacer region, to take into account the −10 and −35 regions of *E. coli* promoters (Cardon and Stormo 1992). These studies have provided useful information as to which positions in the promoter sequences provide information that enhances specificity. Hidden Markov models such as those described in Chapter 4 (p. 185) and earlier in this chapter have also been used for prokaryotic promoter prediction (Pedersen et al. 1996). In principle, because HMM methods are based on the expectation maximization method, they should be comparable in effectiveness to the EM method.

Another statistical method of finding patterns in unaligned sequences has also been used for DNA sequences. In one case, this method was used with a dinucleotide analysis to reduce background noise (Ioshikhes et al. 1999). A Gibbs sampling method that takes into account additional features of DNA sequences such as inverted repeats has been described (Zhang 1999b). Align Ace is a program designed for promoter analysis that uses a Gibbs sampling strategy (see Table 10.1E). The inverted repeat feature is designed to identify binding sites of regulatory proteins that are inverted repeats, like LexA-binding sites in Figure 8.6.

A different method has been developed for searching through a set of unaligned sequences for a common but degenerate sequence pattern (Stormo and Hartzell 1989;

Hertz et al. 1990). The program developed for this purpose, consensus, was used to produce a set of scoring matrices for eukaryotic transcription-factor-binding sites (Chen et al. 1995). Recently, a theory was developed that allows a statistical evaluation of the results (Hertz and Stormo 1999). In its simplest form, illustrated in Figure 8.11, a sliding window of sequence in each of the sequences is matched against similar windows in the remaining sequences, searching for the best scoring matrix, as judged by the information content of the matrix (p. 195). There is no allowance made for gaps, and the choice of a base at each matrix position is assumed to be independent of the other positions, although the development of methods for including such features has been described previously (Hertz and Stormo 1995). In consensus, parameters such as window width, whether or not each sequence can contribute at most one word, whether or not there are additional words after an initial one, whether or not words overlap, whether or not the complementary sequence is used, and the maximum number of alignments to be saved are set by the user. In a related program, wconsensus, the optimum window size is not set by the user. Instead, biases are used and subtracted from the information content of each column in the scoring matrix to make the amount of information a smaller number, called the crude information content. The object is to reduce the average alignment score to a negative value so that an interesting alignment appears as a positive score, much like the procedure used in the Smith-Waterman algorithm for sequence alignment by dynamic programming. wconsensus finds the scoring matrix that maximizes this crude information content. At the same time, wconsensus also saves the flanking sequence regions from each sequence included in the matrix. As more sequences are added, these regions may also become incorporated into the alignment and help to locate additional matching regions.

The time required for computing these patterns is extensive and increases as a linear function of the number of sequences and as the square of the sequence lengths. The programs accept user input to reduce the computational time. These programs are not guaranteed to provide the best possible matrix, but by trying out several reasonable values for user-provided variables, there is a strong possibility of finding the best matrix. Associated with these programs is a statistical evaluation of each matrix. If I is the information content of the matrix calculated and N the number of sequences used to create the matrix, the probability of obtaining a greater product $I \times N$ from random sequences of the same length and base composition is determined. This procedure is similar in principle to the methods used to evaluate scores found in sequence alignments and database searches, except that the statistical models are quite complex (Hertz and Stormo 1999). Similar numerical methods for calculating the significance of scoring matrices and matches to scoring matrices have also been developed (Staden 1989). Thus, different matrices found by using different matrix widths, base compositions, and other variables may be evaluated for significance, and the best ones chosen. The consensus programs run under the UNIX operating system and are available by anonymous FTP from beagle.colorado.edu in the directory /pub/consensus.

Binding sites for repressors and activators of *E. coli* and other bacteria have been analyzed for conserved patterns by the above methods. An example is the set of bacterial and bacteriophage genes that is repressed by the *E. coli lexA* gene product (Lewis et al. 1994). As illustrated in Figure 8.6, these genes carry the binding site for LexA repressor, which is located in the vicinity of the promoter and transcription start site and has the consensus sequence CTGTNNNNNNNNNNCAG. The more conserved positions in the binding site contribute the most to the binding of the LexA protein to these sites and, in general, the closer the binding site to consensus, the more tightly bound the protein to that site. Similar observations of several transcriptional regulators and promoters of *E. coli* have led to a statistical mechanical theory that the most conserved positions each independently contribute the most binding energy to the interaction (Berg and von Hippel 1987; Fields et al. 1997; Stormo and Fields 1998).

PROMOTER PREDICTION IN EUKARYOTES

Transcriptional Regulation in Eukaryotes

The regulation of transcription of protein-encoding genes by RNA polymerase II (RNA PolII) involves the interaction of a large number of protein complexes, called transcription factors (TFs), with each other and with DNA-binding sites in the promoter region. The regions upstream from the start point of transcription, but also just downstream, influence the regulation and degree of expression of the gene. The region immediately upstream, the core promoter, has DNA-binding sites to which a preinitiation complex comprising RNA PolII and TFIIA, B, D, E, F, and H binds (Tjian 1996).

The position of binding sites is given with reference to the start site of transcription (TSS). A box defined as TATA is present in about 75% of vertebrate RNA PolII promoters. A TATA box HMM trained on vertebrate promoter sequences has the consensus sequence TATAWDR (W = A/T, D is not C, R is G or A) starting at approximately −17 bp from TSS (Bucher 1990; http://www.epd.isb-sib.ch/promoter_elements/). This sequence is thought to position the initiation complex around TSS. A component of TFIID, TATA-binding protein (TBP), recognizes and binds to this sequence. INR is a loosely defined sequence around TSS that also influences the start position of transcription and may be recognized by other protein subunits of TFIID (Chalkley and Verrijzer 1999).

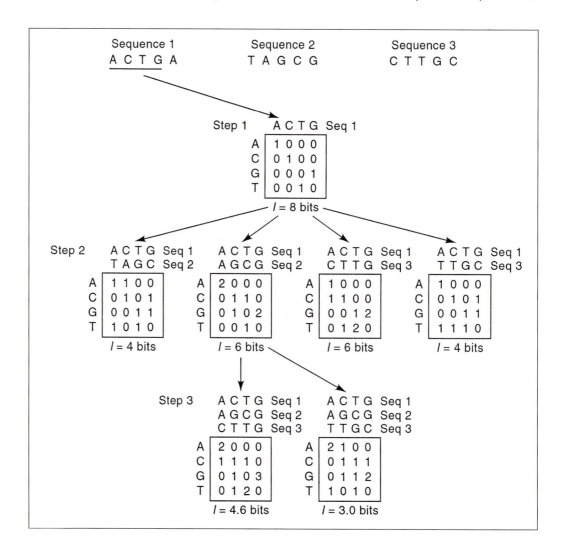

Another conserved sequence lying upstream of TATA and present in about one-half of vertebrate promoters is the CCAAT box, which is thought to be the site of binding of additional proteins that influence preinitiation and later stages of transcription. Another conserved regulatory site is the GC box. These boxes lie at variable distances from TSS and function in either orientation. Weight matrices that describe them have been produced (Bucher 1990).

The region upstream of the core promoter and other enhancer sites in the neighborhood of a gene also influences gene expression. A variety of transcription factors, some affected by environmental influences such as hormone levels, bind to DNA-binding sites in these regions. These factors can also form large multiprotein complexes that interact with a preinitiation complex to induce or repress transcription. These interactions can cause remodeling of the local nucleosome structure by histone acetylation or deacetylation, conformational changes in the transcription complex, and possibly phosphorylation of

Figure 8.12. The Hertz, Stormo, and Hartzell method for locating common DNA-binding sites for regulatory proteins in unaligned sequences (Hertz and Stormo 1999). This example illustrates how the algorithm compares a fixed window of sequence (length 4 in this example) in a set of sequences assumed to carry one site for a DNA-binding protein that cannot be readily found by aligning the sequences. The object is to find the 4-mer in each sequence that constitutes as nearly identical a pattern as can be found in all of the sequences. The user specifies the number of matrices that can be saved by the program for further analysis. Redundant matrices are eliminated. *Step 1.* The sequence of the first four bases from sequence 1 is first chosen. An analysis of only this one window is shown in this example. Normally the program would start with all possible 4-long words in each of the sequences, thus producing a total number of 6 possible step 1 matrices in this example. *Step 2.* The sequence window chosen in step 1 is moved across sequence 2, then sequence 3, and so on until all possible windows in all sequences have been selected. If a sufficient number of saved matrices is specified, this procedure would be repeated for all of the six saved matrices in step 1. Only one matrix is shown for illustration purposes. At each selected position, the number of matches with sequence 1 is recorded in a scoring matrix. The amount of sequence conservation in each column is calculated as the information content (*Ic*) of the column, and the *Ic* values for each column are then added to give *I* of the matrix. The best-scoring matrix is chosen. Calculation of the information content of a scoring matrix is discussed in detail in Chapter 4 (p. 195). Given a position in a test sequence that is being examined for a match to a matrix column, the maximum uncertainty of a matrix column is the number of questions that must be asked to find a match to the position in a test sequence. Uncertainty is zero if only one base is represented and 2 if all four bases are represented equally. Information content of a column is 2 minus the uncertainty of the column. For example, as each column in the first matrix in step 2 requires a single question to identify a match to a sequence (for column 1, one question must be asked: "Does the matching sequence position have an A or a T?"), then *I* of the matrix is $1+1+1+1=4$. The first column of the second matrix in step 2 has two As, and no other base is represented. Because no question need be asked, *I* is 2. A general method for calculating the amount of information in a column *c* is given by $Ic = \Sigma_i \{f_{ic} \log (f_{ic}/p_i)\}$ where f_{ic} is the fraction of each base in the column and p_i is the background frequency of base *i* in the sequences. If logarithms to the base 2 are used, then *I* units are in bits, and if natural logarithms are used, *I* units are in nats. *Step 3.* The sequence windows found in the highest-scoring matrix in step 2 are now compared to all other possible windows in the remaining sequences. In this case, only one sequence remains and the next high-scoring matrix is identified. Only one matrix is shown as an example; the maximum number that can be used for further analysis will be determined by the specified number of matrices that can be saved by the program. Additional steps (not shown) are then used to compare this best matrix with any remaining sequences until all have been included. The final matrices provide a consensus sequence by using the base in each column that has the highest score. The algorithm is greedy because the development of the highest-scoring matrix depends on matches found in ancestor matrices based on a smaller number of alignments. On the basis of this limitation and constraints provided by the user such as window size or matrix bias (see text), and on the number of matrices saved, the algorithm is not guaranteed to provide the optimal matrix for a large number of sequences.

RNA PolII. The independent binding of proteins to separate DNA sites in the initiation and upstream control regions is cooperative in that the binding of one protein to one site enhances the binding of another protein molecule to a second site. In this manner, a series of weak interactions between individual components is amplified by protein–protein interactions to give an overall strong binding of the complex to the promoter.

An example of a mammalian gene with multiple regulatory elements that have been defined by experiment is shown in Figure 8.13. The gene is the rat *pepCK* gene, which encodes phosphoenol pyruvate kinase, a major enzyme for metabolism of glucose in mammals. This gene is regulated by four different hormones—glucocorticoids, glucagon, retinoic acid, and insulin—through a system of binding sites for particular transcription factors in the promoter region. The response of the cell to these agents involves binding of the hormone to a specific receptor protein and the subsequent binding of the hormone–receptor complex to specific sequences called response elements (REs) in the promoter region of responsive genes. The *pepCK* gene also responds to the level of cyclic AMP (cAMP) through a similar interaction. In addition, the gene has other characteristic and essential sequence features for RNA PolII recognition, such as the TATA box, the initiation region (INR) that includes the transcription start site (TSS) at +1. The REs are flanked by binding sites for other transcription factors that influence the effect of the bound receptor through protein–protein interactions.

Thus, many different transcription factors may be involved in the regulation of a particular eukaryotic gene. The sequence of the DNA-binding site recognized by many of these TFs is not known, or only a few sites are known, thus limiting the ability to predict promoter-binding sites for these TFs. In some cases, enough DNA-binding sites are known to produce a weight matrix, described earlier in this chapter (Table 8.5). However, such scoring matrices tend to be much more variable than prokaryotic matrices, so that the matrix is less useful for discriminating true binding sites from random sequence variation.

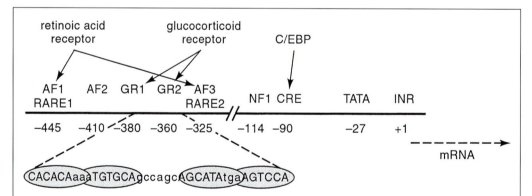

Figure 8.13. Regulatory elements in the promoter of the rat *pepCK* gene. This gene has been analyzed experimentally for the presence of transcription-factor-binding sites. The relative positions of these elements in a fusion of the *pepCK* promoter to a reporter gene are illustrated (Yamada et al. 1999). The glucocorticoid response unit (GRU) includes three accessory factor binding sites (AF1, AF2, and AF3), two glucocorticoid response elements (GR1 and GR2), and a cAMP response element (CRE). A dimer of glucocorticoid receptor bound to each GR element is depicted. The retinoic response unit (RAU) includes two retinoic acid response elements (RARE1 and RARE2) that coincide with the AF1 and AF3, respectively (Sugiyama et al. 1998). The sequences of the two GR sites and the binding of the receptor to these sites are shown. These sites deviate from the consensus sites and depend on their activity on accessory proteins bound to other sites in the GRU. This dependence on accessory proteins is reduced if a more consensus-like (canonical) GR element comprising the sequence TGTTCT is present. The CRE that binds factor C/EBP is also shown.

Such a matrix can be used to predict putative binding sites for a TF in a particular promoter. Because TF binding sites may be detectable on either forward and complementary strands or present on both forward and complementary strands in a repeated configuration, both strands of the test sequence are generally searched for binding sites. Interpolated HMMs described previously for identifying prokaryotic genes have also been used for eukaryotic promoter identification (Ohler et al. 1999). This method identifies the most informative lengths of sequence in promoters and uses them for promoter prediction in test sequences.

As shown in Figure 8.13, binding sites for TFs cluster in the promoter region. This clustering is the basis of one method for promoter prediction discussed below. A search for binding sites in the EPD database (Table 8.6) and a human first exon database showed that tandem binding sites for the same TF that are approximately 10 bp apart and expressing with a periodicity of 145 bp can be detected. Such studies confirm that searching for multiple TF binding sites can provide a more reasonable prediction of promoter function (Ioshikhes et al. 1999).

Complexes of TFs bound to DNA can either activate or repress transcription through their interaction with RNA PolII. Some quite remarkable variations of this theme can occur (Yamamoto et al. 1998). First, changes in the RE or in the binding of nearby accessory proteins can determine whether the binding of glucocorticoid response elements (GR) activates or represses transcription. Second, the GR can influence transcription simply by forming a complex with other factors and without binding to DNA itself. Thus, predicting the regulatory behavior solely on the basis of finding REs in a promoter region is probably not feasible without additional consideration of interactions among the regulatory elements and proteins themselves (Bucher et al. 1996).

RNA PolII Promoter Classification

Eukaryotic promoter sequences show variation not only between species, but also among genes within a species. A gene that is regulated by a certain set of signals during development will have a significantly different promoter than a second gene that responds to a different set of signals. For this reason, a set of promoters in an organism that share a regulatory response have been analyzed, as these promoters are expected to share common regulatory elements. Such an analysis has been performed on the genes expressed in skeletal muscle. Binding sites for TFs in skeletal muscle promoters are used to make scoring matrices, which are then used to find other muscle-regulated genes in genomic sequences. The ability of individual scoring matrices to locate signals in known muscle promoters, while at the same time not finding signals in control promoters, is determined. The alignment scores for each matrix are then weighted in favor of the most informative matrices. The sum of these weighted scores gives a value between 0 (no promoter) and 1 (has promoter function), called the logit value of the promoter. Similar promoters from closely related species are also used to enhance the ability of the method to discriminate muscle promoters from other promoters in a method described as phylogenetic footprinting (Wasserman and Fickett 1998).

Because the usefulness of different scoring matrices for TF binding sites is variable, other methods have been devised for weighting the scores obtained for an individual weight matrix on test sequences. An additional development includes a new algorithm for determining the cutoff value using the background rate estimated on non-promoters (see TFBIND in Table 8.6). Scores of matches of weight matrices to test sequences follow the extreme value distribution (p. 326), and have also been used to evaluate matches (Claverie 1994; Claverie and Audic 1996). The application of neural networks for devising a

Table 8.6. *Promoter prediction programs, Web pages, and related information*

Name	Web address	Reference
BDNA video analysis of transcription factor binding sites using conformational and physicochemical DNA features	see GeneExpress	Ponomarenko et al. (1999)
ConsInspector–see Transfac database[a]	http://www.gsf.de/biodv/consinspector.html	
Core-Promoter–for finding RNAPII promoters of human genes by quadratic discriminant analysis	http://argon.cshl.org/genefinder/ CPROMOTER/index.htm	Zhang (1998a, b)
EPD Eukaryotic promoter database	http://www.epd.isb-sib.ch/; http://www.epd.isb-sib.ch/ promoter_elements/	Bucher (1990); Périer et al. (1999, 2000)
EpoDB genes expressed during vertebrate erythropoiesis	http://www.cbil.upenn.edu/	Stoeckert et al. (1999)
FastM for transcription factor binding sites	http://genomatix.gsf.de/cgi-bin/fastm2/fastm.pl	Klingenhoff et al. (1999)
GeneExpress analysis of transcriptional regulations with TRRD database	http://wwwmgs.bionet.nsc.ru/systems/ GeneExpress/	Kolchanov et al. (1999a, b)
Genome inspector for combined analysis of multiple signals in genomes	http://www.gsf.de/biodv/genomeinspector.html	Quandt et al. (1997)
GrailII[b] prediction of TSS by neural networks based on scores of characteristic sequence patterns and composition	http://compbio.ornl.gov/ see also book Web site	Uberbacher and Mural (1991); Uberbacher et al. (1996)
MAR-FINDER for finding matrix attachment regions	http://www.ncgr.org/MarFinder/	Kramer et al. (1997); Singh et al. (1997)
MatInd—see Transfac database		
MatInspector[a]–see Transfac database	http://www.gsf.de/biodv/matinspector.html (for downloading) http://www.gsf.de/cgi-bin/matsearch.pl (for interactive web page)	
Nuclear (including glucocorticoid) receptor resource[c]	http://nrr.georgetown.edu/GRR/GRR.html	Martinez et al. (1997)
Mirage (Molecular Informatics Resource for the Analysis of Gene Expression)[d]	http://www.ifti.org/	see Web page
NNPP Promoter Prediction by Neural Network for prokaryotes or eukaryotes	http://www.fruitfly.org/seq_tools/promoter.html	Reese et al. (1996)
NSITE–search for TF binding sites or other consensus regulatory sequences	http://genomic.sanger.ac.uk/gf/gf.shtml	see Web site
OOTFD Object-Oriented Transcription Factor Database	http://www.ifti.org/cgi-bin/ifti/ootfd.pl	Ghosh (1998)
PLACE plant *cis*-acting regulatory elements	http://www.dna.affrc.go.jp/htdocs/PLACE/	Higo et al. (1999)
PlantCARE plants *cis*-acting regulatory elements	http://sphinx.rug.ac.be:8080/ PlantCARE/index.htm	Rombauts et al. (1999)
Pol3scan for RNAP III/tRNA promoter sequences using pattern scoring matrices	http://irisbioc.bio.unipr.it/genomics.html	Pavesi et al. (1994)
Polyadq for locating polyadenylation sites	http://argon.cshl.org/tabaska/ polyadq_form.html	Tabaska and Zhang (1999)

Continued.

Table 8.6.   *Continued.*

Name	Web address	Reference
Promoter element weight matrices and HMMs	http://www.epd.isb-sib.ch/promoter_elements/	Bucher (1990)
Promoter II for recognition of PolII sequences by neural networks	http://www.cbs.dtu.dk/services/promoter/	Knudsen (1999)
PromoterScan[e]	http://cbs.umn.edu/software/proscan/promoterscan.htm	Prestridge (1995) and see Web site
RegScan for promoter classification	http://wwwmgs.bionet.nsc.ru/mgs/programs/classprom/	Babenko et al. (1999)
Sequence walkers for graphical viewing of the interaction of regulatory protein with DNA binding site	http://www-lecb.ncifcrf.gov/~toms/walker/narcoverlogowalker.html	Schneider (1997)
Signal scan for transcriptional elements	http://bimas.dcrt.nih.gov:80/molbio/signal/	Prestridge (1991, 1996)
TargetFinder for promoter searching in selected annotated sequences	http://hercules.tigem.it/TargetFinder.html	Lavorgna et al. (1999)
TESS for searching for transcription factor binding sites	http://www.cbil.upenn.edu/tess/	Schug and Overton (1997a, b)
Tfbind for transcription factor binding sites	http://tfbind.ims.u-tokyo.ac.jp	Tsunoda and Takagi (1999)
Thyroid receptor resource[c]	http://xanadu.mgh.harvard.edu/receptor/trrfront.html	see Web page
Transfac programs providing search for TF binding sites. MatInd for making scoring matrices and MatInspector for searching for matches to matrices	http://www.gsf.de/cgi-bin/matsearch.pl	see http://www.gsf.de/biodv/staff_pub.html; Knüppel et al. (1994); Quandt et al. (1995); Heinemeyer et al. (1999); Klingenhoff et al. (1999)
TRRD transcriptional regulatory region database; see GeneExpress		Kolchanov et al. (1999a)
TSSG, like TSSW but based on sequences from a different promoter database	http://genomic.sanger.ac.uk/gf/gf.shtml; http://dot.imgen.bcm.tmc.edu:9331/seq-search/gene-search.html	see Web site
TSSW; recognition of human PolII promoter region and start of transcription by linear discriminant function analysis	http://genomic.sanger.ac.uk/gf/gf.shtml; http://dot.imgen.bcm.tmc.edu:9331/seq-search/gene-search.html	see Web site
Yeast cell cycle gene retrieval and promoter analysis	http://www.ncbi.nlm.nih.gov/CBBresearch/Landsman/Cell_cycle_data/upstream_seq.html; http://www.ncbi.nlm.nih.gov/CBBresearch/Landsman/Cell_cycle_data/	Wolfsberg et al. (1999)
Yeast cell cycle analysis project	http://genome-www.stanford.edu/cellcycle/info	Spellman et al. (1998)

Multiple methods of analysis are offered at sites http://dot.imgen.bcm.tmc.edu:9331/seq-search/gene-search.html and on http://genomic.sanger.ac.uk/gf/gf.shtml. Lists of Web sites are given at: http://linkage.rockefeller.edu/wli/gene/programs.html. A comparison of many of the promoter prediction programs included in this table and several additional ones on a small number of promoter-containing sequences not used in program training is available (Fickett and Hatzigeorgiou 1997).

[a] MatInspector DOS, Windows 95 and NT, and Mac versions and ConsInspector DOS and Mac versions available by FTP from ariane.gsf.de/pub/.

[b] GrailII must be given both gene and promoter sequences.

[c] Includes links to other receptor databases.

[d] The transcriptional informatics site MIRAGE includes links to regulatory data sites and programs.

[e] Accepts one person at a time; DOS version also available.

weighting scheme as used in the gene prediction program GeneParser would be another method for weighting a group of scoring matrices to give maximum discrimination between promoter and non-promoter sequences.

Gene microarrays discussed in Chapter 10 (p. 519) can assist with discovering which genes are regulated in the same manner and therefore should have binding sites for the same TFs (Cho et al. 1998; Eisen et al. 1998; Claverie 1999; Golub et al. 1999; Zhang 1999a,b). The promoter regions of these genes can be compared. The 5-mer content promoter sequences of yeast genes that are co-regulated during the cell cycle have been analyzed by the program wconsensus (described above, p. 365) and a Gibbs DNA sampler (similar to the Gibbs motif sampler described in Chapter 4 but adapted for DNA sequences) (Spellman et al. 1998; Zhang 1999b) and the results are available on a Web site (see Table 8.6).

In another study, pentamers and hexamers that are overrepresented among the upstream regions of cell-cycle-regulated genes were identified using a simple statistical sampling procedure. The sequences are divided into two sets; one set comprises cell cycle genes and the second set is the rest of the genome. A hexamer is then counted in both sets. The background number in the control set is used to identify overrepresented oligonucleotides in the cell cycle set. The actual number counted in the cell cycle genes is then compared to this expected value using a Chi-square test. For example, the hexamer ACGCGT is found with a variable location and orientation in the promoters of many cell cycle genes that are expressed during the late phase of the G_1 phase of the yeast cell cycle, whereas the pentamer CCCTT is located at positions -104 to -202 in one orientation in early G_1 (Wolfsberg et al. 1999). These types of analyses, which are available on Web sites (Table 8.6), demonstrate that computational analysis of the promoters of co-related genes reveals the presence of highly representative sequence patterns. Although some of these patterns correspond to the binding sites of transcription factors, others play a role that has yet to be determined. A similar method of oligomer counting has been used to identify overrepresented oligonucleotides with intron-containing genes in yeast and also to identify signals for localization of RNAs to mitochondria (Jacobs Anderson and Parker 2000). Hence, the oligonucleotide scoring method shows considerable promise for the identification of regulatory sites in co-regulated genes.

Prediction Methods for RNA PolII Promoters

A number of methods for predicting the location of RNA PolII promoters in genomic DNA have been derived. Several Web sites that offer an analysis are listed in Table 8.6. Also shown in this table are a number of Web sites that provide databases and information on TFs and their DNA-binding sites and other information related to transcriptional regulation in eukaryotes. A test analysis of these and several additional programs not listed in the table on a small number of new promoter sequences has been described previously (Fickett and Hatzigeorgiou 1997). The programs predicted 13–54% of the TSSs correctly, but each program also predicted a number of false-positive TSSs.

Samples of methods of analysis and programs included in Table 8.6 are listed below (for additional information on program availability, see Fickett and Hatzigeorgiou 1997; Frech et al 1997).

1. Use of a neural network trained on the TATA and Inr sites, allowing for a variable spacing between the sites (NNPP) or a neural network–genetic algorithm approach to identify conserved patterns in RNA PolII promoters and conserved spacing among the patterns (PROMOTER2.0).

2. Recognition of a TATA box using a weight matrix and an analysis of the density of TF sites. The density of TF sites at least 50 bp apart in known promoter sequences of the

eukaryotic promoter database (EPD) and on a set of non-promoter primate sequences from GenBank is compared and used to produce a promoter recognition profile (PromoterScan).

3. Use of a linear discriminant function as described above for gene prediction, but in this case, used for distinguishing features of promoter sequences from non-promoter sequences. The function is based on a TATA box score, triplet base-pair preferences around TSS, hexamer frequencies in consecutive 100-bp upstream regions, and TF binding-site prediction (TSSD and TSSW).

4. A quadratic discriminant analysis similar to that described above for gene prediction, but in this case, applied to variable lengths of sequence in the promoter region. The frequency of pentamers in a contiguous set of thirteen 30-bp windows and also in a second set of five 45-bp windows in the same 240-bp region was compared. This double-overlapping window appeared to reduce the background noise and to enhance the transcriptional signal from the promoter region (CorePromoter).

5. Searches of weight matrices for different organisms against a test sequence (TFSearch/TESS). Use of user-provided limits on type of weight matrix, key set of matches (core similarity) to individual matrices, and range of match scores (matrix similarity), and also generation of new matrices (MatInspector and ConsInspector).

6. Evaluation of test sequences for the presence of clustered groups or modules of TF binding sites that are characteristic of a given pattern of gene regulation (FastM).

REFERENCES

Adams M.D., Celniker E., Holt R.A., Evans C.A., Gocayne J.D., Amanatides P.G., Scherer S.E., Li P.W., Hoskins R.A., Galle R.F., George R.A., et al. 2000. The genome sequence of *Drosophila melanogaster*. *Science* **287:** 2185–2196.

Audic S. and Claverie J.M. 1998. Self-identification of protein-coding regions in microbial genomes. *Proc. Natl. Acad. Sci.* **95:** 10026–10031.

Babenko V.N., Kosarev P.S., Vishnevsky O.V., Levitsky V.G., Basin V.V., and Frolov A.S. 1999. Investigating extended regulatory regions of genomic DNA sequences. *Bioinformatics* **15:** 644–653.

Baldi P. and Brunak S. 1998. *Bioinformatics: The machine learning approach.* MIT Press, Cambridge, Massachusetts.

Baldi P., Brunak S., Chauvin Y., and Krogh A. 1996. Naturally occurring nucleosome positioning signals in human exons and introns. *J. Mol. Biol.* **263:** 503–510.

Berg O.G. and von Hippel P.H. 1987. Selection of DNA binding sites by regulatory proteins. Statistical-mechanical theory and application to operators and promoters. *J. Mol. Biol.* **193:** 723–750.

Birney E. and Durbin R. 2000. Using GeneWise in the *Drosophila* annotation experiment. *Genome Res.* **10:** 547–548.

Bisant D. and Maizel J. 1995. Identification of ribosome binding sites in *Escherichia coli* using neural network models. *Nucleic Acids Res.* **23:** 1632–1639.

Borodovsky M. and McIninch J. 1993. GeneMark: Parallel gene recognition for both DNA strands. *Comput. Chem.* **17:** 123–133.

Brendel V. and Kleffe J. 1998. Prediction of locally optimal splice sites in plant pre-mRNA with applications to gene identification in *Arabidopsis thaliana* genomic DNA. *Nucleic Acids Res.* **26:** 4748–4757.

Brendel V., Kleffe J., Carle-Urioste J.C., and Walbot V. 1998. Prediction of splice sites in plant pre-mRNA from sequence properties. *J. Mol. Biol.* **276:** 85–104.

Brown N.P., Whittaker A.J., Newell W.R., Rawlings C.J., and Beck S. 1995. Identification and analysis of multigene families by comparison of exon fingerprints. *J. Mol. Biol.* **249:** 342–359.

Brunak S., Engelbrecht J., and Knudsen S. 1991. Prediction of human mRNA donor and acceptor sites from the DNA sequence. *J. Mol. Biol.* **220:** 49–65.

Bucher P. 1990. Weight matrix descriptions of four eukaryotic RNA polymerase II promoter elements derived from 502 unrelated promoter sequences. *J. Mol. Biol.* **212:** 563–578.

Bucher P., Fickett J.W., and Hatzigeorgiou A. 1996. Computational analysis of transcriptional regulatory elements: A field in flux. *Comput. Appl. Biosci.* **12:** 361–362.

Burge C.B. and Karlin S. 1998. Finding the genes in genomic DNA. *Curr. Opin. Struct. Biol.* **8:** 346–354.

Burset M. and Guigó R. 1996. Evaluation of gene structure prediction programs. *Genomics* **34:** 353–367.

Cardon L.R. and Stormo G.D. 1992. Expectation maximization algorithm for identifying protein-binding sites with variable lengths from unaligned DNA fragments. *J. Mol. Biol.* **223:** 159–170.

Carey M. and Smale S.T. 2000. *Transcriptional regulation in eukaryotes: Concepts, strategies, and techniques.* Cold Spring Harbor Laboratory Press, Cold Spring Harbor, New York.

Chalkley G.E. and Verrijzer C.P. 1999. DNA binding site selection by RNA polymerase II TAFs: A TAF(II)250-TAF(II)150 complex recognizes the initiator. *EMBO J.* **18:** 4835–4845.

Chen Q.K., Hertz G.Z., and Stormo G.D. 1995. MATRIX SEARCH 1.0: A computer program that scans DNA sequences for transcriptional elements using a database of weight matrices. *Comput. Appl. Biosci.* **11:** 563–566.

Cho R.J., Campbell M.J., Winzeler E.A., Steinmetz L., Conway A., Wodicka L., Wolfsberg T.G., Gabrielian A.E., Landsman D., Lockhart D.J., and Davis R.W. 1998. A genome-wide transcriptional analysis of the mitotic cell cycle. *Mol. Cell* **2:** 65–73.

Claverie J.-M. 1994. Some useful statistical properties of position-weight matrices. *Comput. Chem.* **18:** 287–294.

———. 1999. Computational methods for the identification of differential and coordinated gene expression. *Hum. Mol. Genet.* **8:** 1821–1832.

Claverie J.-M. and Audic S. 1996. The statistical significance of nucleotide position-weight matrix matches. *Comput. Appl. Biosci.* **12:** 431–439.

Dong S. and Searls D.B. 1994. Gene structure prediction by linguistic methods. *Genomics* **23:** 540–551.

Eisen M.B., Spellman P.T., Brown P.O., and Botstein D. 1998. Cluster analysis and display of genome-wide expression patterns. *Proc. Natl. Acad. Sci.* **95:** 14863–14868.

Farber R., Lapedes A., and Sirotkin K. 1992. Determination of eukaryotic protein coding regions using neural networks and information theory. *J. Mol. Biol.* **226:** 471–479.

Fickett J.W. 1982. Recognition of protein coding regions in DNA sequences. *Nucleic Acids Res.* **10:** 5303–5318.

Fickett J.W. and Hatzigeorgiou A.G. 1997. Eukaryotic promoter recognition. *Genome Res.* **7:** 861–878.

Fickett J.W. and Tung C.S. 1992. Assessment of protein coding measures. *Nucleic Acids Res.* **20:** 6441–6450.

Fields D.S., He Y., Al-Uzri A.Y., and Stormo G.D. 1997. Quantitative specificity of the Mnt repressor. *J. Mol. Biol.* **271:** 178–194.

Florea L., Hartzell G., Zhang Z., Rubin G.M., and Miller W. 1998. A computer program for aligning a cDNA sequence with a genomic DNA sequence. *Genome Res.* **8:** 967–974.

Frech K., Quandt K., and Werner T. 1997. Finding protein-binding sites in DNA sequences: The next generation. *Trends Biochem. Sci.* **22:** 103–104.

Frishman D., Mironov A., Mewes H.W., and Gelfand M. 1998. Combining diverse evidence for gene recognition in completely sequenced bacterial genomes. *Nucleic Acids Res.* **26:** 2941–2947.

Gelfand M.S., Mironov A.A., and Pevzner P.A. 1996. Gene recognition via spliced sequence alignment. *Proc. Natl. Acad. Sci.* **93:** 9061–9066.

Ghosh D. 1998. OOTFD (object-oriented transcription factors database): An object-oriented successor to TFD. *Nucleic Acids Res.* **26:** 360–362.

Gish W. and States D.J. 1993. Identification of protein coding regions by database similarity search. *Nat. Genet.* **3:** 266–272.

Golub T.R., Slonim D.K., Tamayo P., Huard C., Gaasenbeek M., Mesirov J.P., Coller H., Loh M.L., Downing J.R., Caligiuri M.A., Bloomfield C.D., and Lander E.S. 1999. Molecular classification of cancer: Class discovery and class prediction by gene expression monitoring. *Science* **286:** 531–537.

Gray M.W. and Covello P.S. 1993. RNA editing in plant mitochondria and chloroplasts. *FASEB J.* **7:** 64–71.

Guigó R. 1998. Assembling genes from predicted exons in linear time with dynamic programming. *J. Comput. Biol.* **5:** 681–702.

Guigó R., Knudsen S., Drake N., and Smith T. 1992. Prediction of gene structure. *J. Mol. Biol.* **226:** 141–157.

Gutman G.A. and Hatfield G.W. 1989. Nonrandom utilization of codon pairs in *Escherichia coli*. *Proc. Natl. Acad. Sci.* **86:** 3699–3703.

Hannenhalli S.S., Hayes W.S., Hatzigeorgiou A.G., and Fickett J.W. 1999. Bacterial start site prediction. *Nucleic Acids Res.* **27:** 3577–3582.

Hawley D.K. and McClure W.R. 1983. Compilation and analysis of *Escherichia coli* promoter DNA sequences. *Nucleic Acids Res.* **11:** 2237–2255.

Hayes W.S. and Borodovsky M. 1998. Deriving ribosome binding site (RBS) statistical models from unannotated DNA sequences and the use of the RBS model for N-terminal prediction. *Pac. Symp. Biocomput.*, 1998: 279–290.

Hebsgaard S.M., Korning P.G., Tolstrup N., Engelbrecht J., Rouze P., and Brunak S. 1996. Splice site prediction in *Arabidopsis thaliana* DNA by combining local and global sequence information. *Nucleic Acids Res.* **24:** 3439–3452.

Heinemeyer T., Chen X., Karas H., Kel A.E., Kel O.V., Liebich I., Meinhardt T., Reuter I., Schacherer F., and Wingender E. 1999. Expanding the TRANSFAC database towards an expert system of regulatory molecular mechanisms. *Nucleic Acids Res.* **27:** 318–322.

Henderson J., Salzberg S., and Fasman K.H. 1997. Finding genes in DNA with a hidden Markov model. *J. Comput. Biol.* **4:** 127–141.

Hertz G.Z. and Stormo G.D. 1995. Identification of consensus patterns in unaligned DNA and protein sequences: A large deviation statistical basis for penalizing gaps. In *Proceedings of the 3rd International Conference on Bioinformatics and Genome Research* (ed. H.A. Lim and C.R. Cantor), pp. 201–216. World Scientific, Singapore.

———. 1996. *Escherichia coli* promoter sequences: Analysis and prediction. *Methods Enzymol.* **273:** 30–42.

———. 1999. Identifying DNA and protein patterns with statistically significant alignments of multiple sequences. *Bioinformatics* **15:** 563–577.

Hertz G.Z., Hartzell G.W., III, and Stormo G.D. 1990. Identification of consensus patterns in unaligned DNA sequences known to be functionally related. *Comput. Appl. Biosci.* **6:** 81–92.

Higo K., Ugawa Y., Iwamoto M., and Korenaga T. 1999. Plant *cis*-acting regulatory DNA elements (PLACE) database: 1999. *Nucleic Acids Res.* **27:** 297–300.

Hirosawa M., Sazuka T., and Yada T. 1997. Prediction of translation initiation sites on the genome of *Synechocystis* sp. strain PCC6803 by hidden Markov model. *DNA Res.* **4:** 179–184.

Horton P.B. and Kanehisa M. 1992. An assessment of neural network and statistical approaches for prediction of *E. coli* promoter sites. *Nucleic Acids Res.* **20:** 4331–4338.

Ioshikhes I., Trifonov E.N., and Zhang M.Q. 1999. Periodical distribution of transcription factor sites in promoter regions and connection with chromatin structure. *Proc. Natl. Acad. Sci.* **96:** 2891-2895.

Ioshikhes I., Bolshoy A., Derenshteyn K., Borodovsky M., and Trifonov E.N. 1996. Nucleosome DNA sequence pattern revealed by multiple alignment of experimentally mapped sequences. *J. Mol. Biol.* **262:** 129–139.

Jacobs Anderson J.S. and Parker R. 2000. Computational identification of *cis*-acting elements affecting post-transcriptional control of gene expression in *Saccharomyces cerevisiae*. *Nucleic Acids Res.* **28:** 1604–1617.

Klingenhoff A., Frech K., Quandt K., and Werner T. 1999. Functional promoter modules can be detected by formal models independent of overall nucleotide sequence similarity. *Bioinformatics* **15:** 180–186.

Knudsen S. 1999. Promoter2.0: For the recognition of PolII promoter sequences. *Bioinformatics* **15:** 356–361.

Knüppel R., Dietze P., Lehnberg W., Frech K., and Wingender E. 1994. TRANSFAC retrieval program: A network model database of eukaryotic transcription regulating sequences and proteins. *J. Comput. Biol.* **1:** 191–198.

Kolchanov N.A., Ananko E.A., Podkolodnaya O.A., Ignatieva E.V., Stepanenko I.L., Kel-Margoulis O.V., Kel A.E., Merkulova T.I., Goryachkovskaya T.N., Busygina T.V., Kolpakov F.A., Podkolodny N.L., Naumochkin A.N., and Romashchenko A.G. 1999a. Transcription regulatory regions database (TRRD): Its status in 1999. *Nucleic Acids Res.* **27:** 303–306.

Kolchanov N.A., Ponomarenko M.P., Frolov A.S., Ananko E.A., Kolpakov F.A., Ignatieva E.V., Podkolodnaya O.A., Goryachkovskaya T.N., Stepanenko I.L., Merkulova T.I., Babenko V.V., Ponomarenko Y.V., Kochetov A.V., Podkolodny N.L., Vorobiev D.V., Lavryushev S.V., Grigorovich D.A., Kondrakhin Y.V., Milanesi L., Wingender E., Solovyev V., and Overton G.C. 1999b. Integrated databases and computer systems for studying eukaryotic gene expression. *Bioinformatics* **15:** 669–686.

Kozak M. 1999. Initiation of translation in prokaryotes and eukaryotes. *Gene* **234:** 187–208.

Kramer J.A., Singh G.B., and Krawetz S.A. 1997. Computer assisted search for sites of nuclear matrix attachment. *Genomics* **33:** 302–308.

Krogh A., Mian I.S., and Haussler D. 1994. A hidden Markov model that finds genes in *E. coli* DNA. *Nucleic Acids Res.* **22:** 4768–4778.

Kulp D., Haussler D., Reese M.G., and Eeckman F.H. 1996. A generalized hidden Markov model for the recognition of human genes in DNA. *Ismb* **4:** 134–142.

Laub M.T. and Smith D.W. 1998. Finding intron/exon splice junctions using INFO, INterruption Finder and Organizer. *J. Comput. Biol.* **5:** 307–321.

Lavorgna G., Guffanti A., Borsani G., Ballabio A., and Boncinelli E. 1999. TargetFinder: Searching annotated sequence databases for target genes of transcription factors. *Bioinformatics* **5:** 172–173.

Lewis L.K., Harlow G.R., Gregg-Jolly L.A., and Mount D.W. 1994. Identification of high affinity binding sites for LexA which define new DNA damage-inducible genes in *Escherichia coli*. *J. Mol. Biol.* **241:** 507–523.

Lopez A.J. 1998. Alternative splicing of pre-mRNA: Developmental consequences and mechanisms of regulation. *Annu. Rev. Genet.* **32:** 279–305.

Lukashin A.V. and Borodovsky M. 1998. GeneMark.hmm: New solutions for gene finding. *Nucleic Acids Res.* **26:** 1107–1115.

Marck C. 1988. DNA Strider: A 'C' program for the fast analysis of DNA and protein sequences on the Apple Macintosh family of computers. *Nucleic Acids Res.* **16:** 1829–1836.

Martinez E., Moore D.D., Keller E., Pearce D., Robinson V., MacDonald P.N., Simons S.S., Jr., Sanchez E., and Danielsen M. 1997. The nuclear receptor resource project. *Nucleic Acids Res.* **25:** 163–165.

Mathews B. 1975. Comparison of the predicted and observed secondary structure of T4 phage lysozyme. *Biochim. Biophys. Acta* **405:** 442–451.

Morse D.P. and Bass B.L. 1999. Long RNA hairpins that contain inosine are present in *Caenorhabditis elegans* poly(A)$^+$ RNA. *Proc. Natl. Acad. Sci.* **96:** 6048– 6053.

Mulligan M.E. and McClure W.R. 1986. Analysis of the occurrence of promoter-sites in DNA. *Nucleic Acids Res.* **14:** 109–126.

Murakami K. and Takagi T. 1998. Gene recognition by combination of several gene-finding programs. *Bioinformatics* **14:** 665–675.

Ohler U., Harbeck S., Niemann H., Noth E., and Reese M.G. 1999. Interpolated Markov chains for eukaryotic promoter recognition. *Bioinformatics* **15:** 362–369.

Pachter L.K., Batzoglou S., Spitkovsky B.I., Banks E., Lander E.S., Leitman D.J., and Berger B. 1999. A dictionary-based approach for gene annotation. *J. Comput. Biol.* **6:** 419–430.

Paul M.S. and Bass B.L. 1998. Inosine exists in mRNA at tissue-specific levels and is most abundant in brain mRNA. *EMBO J.* **17:** 1120–1127.

Pavesi A., Conterio F., Bolchi A., Dieci G., and Ottonello S. 1994. Identification of new eukaryotic tRNA genes in genomic databases by a multistep weight matrix analysis of transcriptional control regions. *Nucleic Acids Res.* **22:** 1247–1256.

Pavy N., Rombauts S., Dehais P., Mathe C., Ramana D.V., Leroy P., and Rouze P. 1999. Evaluation of gene prediction software using a genomic data set: Application to *Arabidopsis thaliana* sequences. *Bioinformatics* **15:** 887–899.

Pedersen A.G. and Nielsen H. 1997. Neural network prediction of translation initiation sites in eukaryotes: Perspectives for EST and genome analysis. *Ismb* **5:** 226–233.

Pedersen A.G., Baldi P., Brunak S., and Chauvin Y. 1996. Characterization of prokaryotic and eukaryotic promoters using hidden Markov models. *Ismb* **4:** 182–191.

Pedersen A.G., Baldi P., Chauvin Y., and Brunak S. 1998. DNA structure in human RNA polymerase II promoters. *J. Mol. Biol.* **281:** 663–673.

Périer R.C., Junier T., Bonnard C., and Bucher P. 1999. The eukaryotic promoter database (EPD): Recent developments. *Nucleic Acids Res.* **27:** 307–309.

Périer R.C., Praz V., Junier T., Bonnard C., and Bucher P. 2000. The eukaryotic promoter database (EPD). *Nucleic Acids Res.* **28:** 302–303.

Ponomarenko J.V., Ponomarenko M.P., Frolov A.S., Vorobyev D.G., Overton G.C., and Kolchanov N.A. 1999. Conformational and physicochemical DNA features specific for transcription factor binding sites. *Bioinformatics* **15:** 654–668.

Prestridge D.S. 1991. SIGNAL SCAN: A computer program that scans DNA sequences for eukaryotic transcriptional elements. *Comput. Appl. Biosci.* **7:** 203–206.

————. 1995. Prediction of Pol II promoter sequences using transcription factor binding sites. *J. Mol. Biol.* **249:** 923–932.

————. 1996. SIGNAL SCAN 4.0: Additional databases and sequence formats. *Comput. Appl. Biosci.* **12:** 157–160.

Quandt K., Grote K., and Werner T. 1997. GenomeInspector: A new approach to detect correlation patterns of elements on genomic sequences. *Comput. Appl. Biosci.* **12:** 405–413.

Quandt K., Frech K., Karas H., Wingender E., and Werner T. 1995. MatInd and MatInspector: New, fast, and versatile tools for detection of consensus matches in nucleotide sequence data. *Nucleic Acids Res.* **23:** 4878–4884.

Reese M.G., Harris N.L., and Eeckman F.H. 1996. Large scale sequencing specific neural networks for promoter and splice site recognition. In *Biocomputing: Proceedings of the 1996 Pacific Symposium* (ed. L. Hunter and T.E. Klein). World Scientific, Singapore.

Reese M.G., Eeckman F.H., Kulp D., and Haussler D. 1997. Improved splice site detection in Genie. *J. Comput. Biol.* **4:** 311–323.

Reese M.G., Kulp D., Tammana H., and Haussler D. 2000. Genie — Gene finding in *Drosophila melanogaster*. *Genome Res.* **10:** 529–538.

Rice P.M., Elliston K.E., and Gribskov M. 1991. DNA. In *Sequence analysis primer* (ed. M. Gribskov and J. Devereux), pp. 43–49. Stockton Press, New York.

Robison K., McGuire A.M., and Church G.M. 1998. A comprehensive library of DNA-binding site matrices for 55 proteins applied to the complete *Escherichia coli* K-12 genome. *J. Mol. Biol.* **284:** 241–254.

Rombauts S., Dehais P., Van Montagu M., and Rouze P. 1999. PlantCARE, a plant *cis*-acting regulatory element database. *Nucleic Acids Res.* **27:** 295–296.

Salgado H., Santos A., Garza-Ramos U., van Helden J., Diaz E., and Collado-Vides J. 1999. RegulonDB (version 2.0): A database on transcriptional regulation in *Escherichia coli*. *Nucleic Acids Res.* **27:** 59–60.

Salzberg S. 1998. Decision trees and Markov chains for gene finding. In *Computational methods in molecular biology* (ed. S.L. Salzberg et al.), chap. 10, pp. 187–203. Elsevier, Amsterdam, The Netherlands.

Salzberg S., Delcher A., Kasif S., and White O. 1998. Microbial gene identification using interpolated Markov models. *Nucleic Acids Res.* **26:** 544–548.

Schneider T.D., Stormo G.D., Gold L., and Ehrenfeucht A. 1986. Information content of binding sites on nucleotide sequences. *J. Mol. Biol.* **188:** 415–431.

Schug J. and Overton G.C. 1997a. Modeling transcription factor binding sites with Gibbs sampling and minimum description length encoding. *Ismb* **5:** 268–271.

————. 1997b. TESS: Transcription element search software on the WWW. Computational Biology and Informatics Laboratory, University of Pennsylvania School of Medicine (Technical Report CBIL-TR-1997-1001-v0.0).

Schneider T.D. 1997. Sequence walkers: A graphical method to display how binding proteins interact with DNA or RNA sequences (erratum appears in *Nucleic Acids Res.* [1998] **26:** following 1134). *Nucleic Acids Res.* **25:** 4408–4415.

Searls D. 1998. Grand challenges in computational biology. In *Computational methods in molecular biology* (ed. S.L. Salzberg et al.), chap. 1, pp. 1–10. Elsevier, Amsterdam, The Netherlands.

Sharp P.M. and Li W.H. 1987. The codon adaptation index — A measure of directional synonymous codon usage bias, and its potential applications. *Nucleic Acids Res.* **15:** 1281–1295.

Singh G.B., Kramer J.A., and Krawetz S.A. 1997. Mathematical model to predict regions of chromatin attachment to the nuclear matrix. *Nucleic Acids Res.* **25:** 1419–1425.

Snyder E.E. and Stormo G.D. 1993. Identification of coding regions in genomic DNA sequences: An application of dynamic programming and neural networks. *Nucleic Acids Res.* **21:** 607–613.

————. 1995. Identification of protein coding regions in genomic DNA. *J. Mol. Biol.* **248:** 1–18.

Solovyev V.V. and Salamov A.A. 1999. INFOGENE: A database of known gene structures and predicted genes and proteins in sequences of genome sequencing projects. *Nucleic Acids Res.* **27:** 248–250.

Solovyev V.V., Salamov A.A., and Lawrence C.B. 1994. Predicting internal exons by oligonucleotide composition and discriminant analysis of spliceable open reading frames. *Nucleic Acids Res.* **22:** 5156–5163.

————. 1995. Identification of human gene structure using linear discriminant functions and dynamic programming. *Ismb* **3:** 367–375.

Spellman P.T., Sherlock G., Zhang M.Q., Iyer V.R., Anders K., Eisen M.B., Brown P.O., Botstein D., and Futcher B. 1998. Comprehensive identification of cell cycle-regulated genes of the yeast *Saccharomyces cerevisiae* by microarray hybridization. *Mol. Biol. Cell* **9:** 3273–3297.

Spingola M., Grate L., Haussler D., and Ares M., Jr. 1999. Genome-wide bioinformatic and molecular analysis of introns in *Saccharomyces cerevisiae*. *RNA* **5:** 221–234.

Staden R. 1984. Computer methods to locate signals in nucleic acid sequences. *Nucleic Acids Res.* **12:** 505–519.

————. 1989. Methods for calculating the probabilities of finding patterns in sequences. *Comput. Appl. Biosci.* **5:** 89–96.

————. 1990. Finding protein coding regions in genomic sequences. *Methods Enzymol.* **183:** 163–180.

Staden R. and McLachlan A.D. 1982. Codon preference and its use in identifying protein coding regions in long DNA sequences. *Nucleic Acids Res.* **10:** 141–156.

Stein A. and Bina M. 1999. A signal encoded in vertebrate DNA that influences nucleosome positioning and alignment. *Nucleic Acids Res.* **27:** 848–853.

Stoeckert C.J., Jr., Salas F., Brunk B., and Overton G.C. 1999. EpoDB: A prototype database for the analysis of genes expressed during vertebrate erythropoiesis. *Nucleic Acids Res.* **27:** 200–203.

Stormo G.D. and Fields D.S. 1998. Specificity, free energy and information content in protein-DNA interactions. *Trends Biochem. Sci.* **23:** 109–113.

Stormo G.D. and Hartzell G.W., III. 1989. Identifying protein-binding sites from unaligned DNA fragments. *Proc. Natl. Acad. Sci.* **86:** 1183–1187.

Stormo G.D., Schneider T.D., Gold L., and Ehrenfeucht A. 1982. Use of the 'Perceptron' algorithm to distinguish translational initiation sites in *E. coli*. *Nucleic Acids Res.* **10:** 2997–3011.

Sugiyama T., Scott D.K., Wang J.C., and Granner D.K. 1998. Structural requirements of the glucocorticoid and retinoic acid response units in the phosphoenolpyruvate carboxykinase gene promoter. *Mol. Endocrinol.* **12:** 1487–1498.

Tabaska J.E. and Zhang M.Q. 1999. Detection of polyadenylation signals in human DNA sequences. *Gene* **231:** 77–86.

Thanaraj T.A. 1999. A clean data set of EST-confirmed splice sites from *Homo sapiens* and standards for clean-up procedures. *Nucleic Acids Res.* **27:** 2627–2637.

Thieffry D., Salgado H., Huerta A.M., and Collado-Vides J. 1998. Prediction of transcriptional regulatory sites in the complete genome sequence of *Escherichia coli* K-12. *Bioinformatics* **14:** 391–400.

Tiwari S., Ramachandran S., Bhattacharya A., Bhattacharya S., and Ramaswamy R. 1997. Prediction of probable genes by Fourier analysis of genomic sequences. *Comput. Appl. Biosci.* **13:** 263–270.

Tjian R. 1996. The biochemistry of transcription in eukaryotes: A paradigm for multisubunit regulatory complexes. *Philos. Trans. R. Soc. Lond. B Biol. Sci.* **351:** 491–499.

Tsunoda T. and Takagi T. 1999. Estimating transcription factor bindability on DNA. *Bioinformatics* **15:** 622–630.

Uberbacher E.C. and Mural R.J. 1991. Locating protein-coding regions in human DNA sequences by a multiple sensor-neural network approach. *Proc. Natl. Acad. Sci.* **88:** 11261–11265.

Uberbacher E.C., Xu Y., and Mural R.J. 1996. Discovering and understanding genes in human DNA sequence using GRAIL. *Methods Enzymol.* **266:** 259–281.

von Heijne G. 1987. *Sequence analysis in molecular biology — Treasure trove or trivial pursuit*, pp. 50–54. Academic Press, San Diego, California.

Wada K., Wada Y., Ishibashi F., Gojobori T., and Ikemura T. 1992. Codon usage tabulated from the GenBank genetic sequence data. *Nucleic Acids Res.* (suppl.) **20:** 2111–2118.

Wasserman W.W. and Fickett J.W. 1998. Identification of regulatory regions which confer muscle-specific gene expression. *J. Mol. Biol.* **278:** 167–181.

Wolfsberg T.G., Gabrielian A.E., Campbell M.J., Cho R.J., Spouge J.L., and Landsman D. 1999. Candidate regulatory sequence elements for cell cycle- dependent transcription in *Saccharomyces cerevisiae*. *Genome Res.* **9:** 775–792.

Yamada K., Duong D.T., Scott D.K., Wang J.C., and Granner D.K. 1999. CCAAT/enhancer-binding protein beta is an accessory factor for the glucocorticoid response from the cAMP response element in the rat phosphoenolpyruvate carboxykinase gene promoter. *J. Biol. Chem.* **274:** 5880–5887.

Yamamoto K.R., Darimont B.D., Wagner R.L., and Iniguez-Lluhi J.A. 1998. Building transcriptional regulatory complexes: Signals and surfaces. *Cold Spring Harbor Symp. Quant. Biol.* **63:** 587–598.

Zhang M.Q. 1997. Identification of protein coding regions in the human genome based on quadratic discriminant analysis (erratum appears in *Proc. Natl. Acad. Sci.* [1997] **94:** 5495). *Proc. Natl. Acad. Sci.* **94:** 565–568.

———. 1998a. Identification of human gene core promoters in silico. *Genome Res.* **8:** 319–326.

———. 1998b. A discrimination study of human core-promoters. *Pac. Symp. Biocomput.,* 1998: 240–251.

———. 1999a. Large-scale gene expression data analysis: A new challenge to computational biologists. *Genome Res.* **9:** 681–688.

———. 1999b. Promoter analysis of co-regulated genes in the yeast genome. *Comput. Chem.* **23:** 233–250.

## CHAPTER 9

Protein Classification and Structure Prediction

INTRODUCTION

O NE OF THE MAJOR GOALS OF BIOINFORMATICS is to understand the relationship between amino acid sequence and three-dimensional structure in proteins. If this relationship were known, then the structure of a protein could be reliably predicted from the amino acid sequence. Unfortunately, the relationship between sequence and structure is not that simple. Much progress has been made in categorizing proteins on the basis of structure or sequence, and this type of information is very useful for protein modeling. A review of protein synthesis and structure is therefore in order.

PROTEIN STRUCTURE PREDICTION

The polypeptide chain is first assembled on the ribosome using the codon sequence on mRNA as a template, as illustrated in Figure 9.1. The resulting linear chain forms secondary structures through the formation of hydrogen bonds between amino acids in the chain. Through further interactions among amino acid side groups, these secondary structures then fold into a three-dimensional structure. Chaperone proteins and membranes may assist with this process. For the protein to have biological activity, processing of the protein by cleavage or chemical modification may also be necessary. Therefore, protein structure is largely specified by amino acid sequence, but how one set of interactions of the many possible occurs is not yet fully understood (Branden and Tooze 1991).

Some protein sequences have distinct amino acid motifs that always form a characteristic structure. Prediction of these structures from sequence is quite achievable using presently available methods and information. For most proteins, however, the accuracy of secondary structure prediction is approximately 70–75%. Methods for matching sequence to three-dimensional structure have been formulated, but they are not yet very reliable. However, great forward strides have been made, and there is a very active community of structural biochemists and bioinformaticists working on improvements. The need for such an effort is revealed by the rapid increases in the number of protein sequences and structures.

As of June 2000, more than 12,500 protein structures had been deposited in the Brookhaven Protein Data Bank (PDB), and 86,500 protein sequence entries were in the SwissProt protein sequence database, a ratio of approximately 1 structure to 7 sequences. The number of protein sequences can be expected to increase dramatically as more sequences are produced by research laboratories and the genome sequencing projects. As more and more sequences and structures have been found, there have been some quite

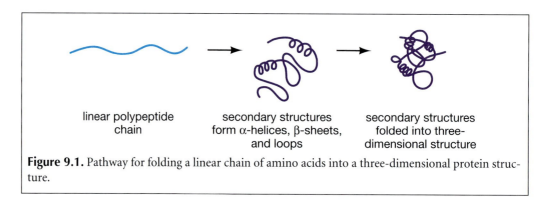

linear polypeptide chain

secondary structures form α-helices, β-sheets, and loops

secondary structures folded into three-dimensional structure

Figure 9.1. Pathway for folding a linear chain of amino acids into a three-dimensional protein structure.

remarkable revelations that make the goals of reliable structure prediction more within reach.

It has first been estimated that there are approximately 1,000 protein families composed of members that share detectable sequence similarity (Dayhoff et al. 1978; Chothia 1992). Thus, as new protein sequences are obtained, they will be found to be similar to other sequences already in the databases and can be expected to share structural features with these proteins. Whether this low number represents physical restraints in folding the polypeptide chain into a three-dimensional structure or merely the selection of certain classes of three-dimensional structure by evolution has yet to be discovered (Gibrat et al. 1996). The sequence alignment, motif-finding, block-finding, and database similarity search methods described in Chapters 3, 4, and 7 may be used to discover these familial relationships. Understanding these relationships is fundamentally important because this information can greatly assist with structural predictions. As discussed below, information from amino acid substitutions at a particular sequence position as obtained from a multiple sequence alignment has been found to increase significantly the prediction of secondary structures from protein sequences. A second major advance in protein structure analysis has been the revelation that proteins adopt a limited number of three-dimensional configurations.

Protein structures include a core region comprising secondary structural elements packed in close proximity in a hydrophobic environment. Specific interactions between the amino acid side chains occur within this core structure. At a given amino acid position in a given core, the amino acids that can substitute are limited by space and available contacts with other nearby amino acids. Outside of the core are loops and structural elements in contact with water, other proteins, and other structures. Amino acid substitutions in these regions are not as restricted as in the core. Through a close comparison of a newly generated three-dimensional structure with previously found structures, the new structure has often been found to fold into α-helical and β-sheet structural elements in the same order and spatial configuration as one or more structures already in the structural database. Proteins that show such structural similarities often do not share any detectable sequence similarity in these same regions. Hence, entirely different sequences can fold into similar three-dimensional configurations. Databases of these common structural features have been prepared and are available on Web sites described later in this chapter.

The finding that only certain amino acids can be substituted at each position in a particular protein core underscores two difficulties in using sequence alignments to make structural predictions. First, because a different set of substitutions apply to each position in each protein core, standard amino acid substitution matrices such as the Dayhoff PAM matrices and the BLOSUM matrices, described in Chapter 3, may not provide an alignment that has structural significance. The substitutions used to produce these tables are averaged over many sequence alignments, representing observed substitutions in both core regions and loops of sequence families.

Scoring matrices that represent a conserved region in the multiple sequence alignment of a set of similar proteins may also be produced, as described in Chapters 4 and 7. These matrices store information on the amino acid variation found in each column of the multiple sequence alignment. They are powerful tools for searching a new protein sequence for the presence of a sequence pattern that is similar to those in the original set of proteins. These scoring matrices include the profile, which represents gapped alignments, and the position-specific scoring matrix (PSSM), which represents ungapped alignments. A conserved region with gaps in a multiple sequence alignment may also be represented by a profile hidden Markov model (profile HMM), which provides a probability-based model of the multiple sequence alignment. Like the scoring matrices, the profile HMM representation of a sequence alignment can be used to identify related sequences. These methods are discussed in detail in Chapters 4 and 7.

Scoring matrices and profile HMMs can provide a direct link between sequence and structure. If one of the sequences that is represented by the matrix or profile model has a known three-dimensional structure, then any other sequences that match the model are also predicted to have the same structure. Conversely, if the model can be shown to match a protein of known structure, a sequence–structure link may be made. A related method is to produce a HMM (also called a discrete state-space model in the protein structure literature) for a set of proteins that belong to a structural family. These models include information on amino acid preference for positions in secondary structures. A query sequence can then be searched by a set of such models to determine whether the sequence has sequence patterns that represent the structure. A range of Web sites provide a variety of these types of analyses (see Fig. 9.30).

A second difficulty in making sequence alignments reflect structural similarity is that gaps in the alignment should be confined to regions not in the core. Alignments that reflect structures in core regions should have few if any gaps. Some multiple sequence alignment programs such as CLUSTALW (see Chapter 4, p. 153) and the Bayes block aligner (Chapter 3, p. 124) do provide for such variation in gap placement. These programs place alignment gaps where the alignment scores are low and, from a structural viewpoint, represent variable loops. The profile models described above also accommodate such variations of placement.

In addition to sequence-by-sequence alignment and sequence-by-structure alignment, it is also possible to perform a structure-by-structure alignment. In this type of alignment, sequential positions of the backbone carbon atoms for each amino acid in the two sequences are compared to determine whether the chain of atoms is tracing the same path in space. If two or more similar paths are found in the same relative positions and orientations, the structures corresponding to those paths are similar. From these methods, discussed below, databases of structural elements have been made and are available to the laboratory.

What is a reasonable goal for protein structure prediction from the perspective of a molecular biologist? The most satisfying result is to find sequence and structural alignments of a newly identified protein with a protein of known three-dimensional structure. Even if such a prediction can be made, the positions of individual amino acids will probably not be accurately known. If the sequence identity is 50% or better, one sequence can be superimposed on the structure of the other sequence and the predicted structure will be quite accurate. If the sequence identity is greater than 30%, it may be possible to identify common structural features, but it will become more difficult to identify the precise positions of the amino acids in the structure as sequence identity decreases.

A special case is the use of the term "structural homology," meaning as it did with sequence homology that the sequences were derived from a common ancestor, as evidenced by their having significant sequence similarity. As described above, two proteins may have significant structural similarity but no detectable sequence similarity. Therefore, it may be incorrect to refer to these proteins as homologous in the absence of evidence that they are derived from a common ancestor.

The prediction of protein structure is an active and promising area of research. As more three-dimensional structures are found and the computational tools for predicting structure are improved, structural predictions will undoubtedly improve. The existence of new groups of proteins for structural analysis is suggested by the existence of genome "ORFans" that may represent new sets of families (superfamilies) with a unique structure and function (Fischer and Eisenberg 1999). One group of investigators that works on protein classification has developed a protein structure initiative to identify new protein targets for structural analysis (http://www.structuralgenomics.org/main.html). A method for estimating the probability for a protein to have a new fold has been described previously (Portugaly and Linial 2000). The Human Proteome/Structural Genomics Pilot Project (http://proteome.bnl.gov), a consortium of Brookhaven National Laboratory, the Rockefeller University, and Albert Einstein College of Medicine, is examining the feasibility of high-throughput determination of three-dimensional structures of proteins starting with genomic sequences.

With a larger set of protein models, the usefulness of structure prediction is increased even further (Pennisi 1998). Many additional methods for structural classification of proteins and for displaying the structures have meanwhile been devised, and the Web has provided these resources to the research community. Formerly, special software and hardware were required to view structures. Now, there are a variety of visualization tools that work with a Web browser and allow one to view a molecule in three dimensions, to compare structures, and to perform other useful procedures. A representation of several useful Web sites for protein structure analysis is given in Table 9.1.

In this chapter, basic features of protein structure and structural terminology and the terms describing them are first reviewed. Some terms refer to sequence similarity, some to structural similarity, and some to both sequence and structure, and it is important not to confuse them.

Table 9.1. *Main Web sites for protein structural analysis*

Name of resource	Resources available	Internet address
Protein data bank (PDB) at the State University of New Jersey (Rutgers)[a]	atomic coordinates of structures as PDB files, models, viewers, links to many other Web sites for structural analysis and classification	http://www.rcsb.org/pdb; also at mirror Web sites (Berman et al. 2000)
National Center for Biotechnology Information Structure Group	Molecular Modelling Database (MMDB), Vector Alignment Search Tool (VAST) for structural comparisons, viewers, threader software	http://www.ncbi.nlm.nih.gov/Structure/
Structural Classification of Proteins at Cambridge University	SCOP database of structural relationships among known protein structures classified by superfamily, family, and fold	http://scop.mrc-lmb.cam.ac.uk/scop; also at Web mirror sites
Biomolecular Structure and Modelling group at the University College, London	CATH database, a hierarchical domain classification of protein structures by class, architecture, fold family and superfamily, other databases and structural analyses, threader software	http://www.biochem.ucl.ac.uk/bsm; also at Web mirror sites
European Bioinformatics Institute, Hinxton, Cambridge	databases, TOPS protein structural topology cartoons, Dali domain server, and FSSP database[b]	http://www2.ebi.ac.uk/
The PredictProtein server at the European Molecular Biology Laboratory at Heidelberg, Germany	important site for secondary structure prediction by PHD, predator, TOPITS, threader	http://cubic.bioc.columbia.edu/predictprotein; also at Web mirror sites[c]
Swiss Institute of Bioinformatics, Geneva	basic types of protein analysis[d] databases, the Swiss-Model resource for prediction of protein models, Swiss-PdbViewer	http://www.expasy.ch/

Additional sites are listed in the text. In addition to these sites, there are a number of Web sites and courses that discuss protein structure. The Swiss Institute for Bioinformatics (ISREC server) provides a tutorial on Principles of Protein Structure, Comparative Protein Modelling, and Visualisation at http://www.expasy.ch/swissmod/course/course-index.htm. There is also a Web course in protein structure at Birkbeck College http://www.cryst.bbk.ac.uk/teaching/.

[a] A summary of the PDB entries is provided at http://www.biochem.ucl.ac.uk/bsm/pdbsum/ (Laskowski et al. 1997).

[b] 3Dee database of protein domains at http://barton.ebi.ac.uk/servers/3Dee.html. Dali domain server is at http://www2.embl-ebi.ac.uk/dali/domain/ and FSSP database at http://www2.embl-ebi.ac.uk/dali/fssp/fssp.html.

[c] Also performed at the structure prediction server at http://www.doe-mbi.ucla.edu/people/frsvr/frsvr.html.

[d] This site offers a series of basic types of protein analysis to assist with protein identification, including identification by amino acid composition, charge, size, and sequence fingerprint. Predictions of posttranslational modifications and oligosaccharide structures are also available.

Other terms that are used to describe protein structure and the methods for displaying and comparing protein structures are described. Some of the more easily found structures and then the methods used to predict secondary and three-dimensional structures are discussed. A flowchart showing the steps to be followed to analyze a new protein sequence is included at the beginning of the Methods section (p. 399). The chapter concludes with a discussion of methods used to evaluate the success of these predictions.

REVIEW OF PROTEIN STRUCTURE AND TERMINOLOGY

Proteins are chains of amino acids joined by peptide bonds, as illustrated in Figure 9.2. Many conformations of the chain are possible due to the rotation of the chain about each C_α atom. It is these conformational variations that are responsible for differences in the three-dimensional structures of proteins. Each amino acid in the chain is polar, i.e., it has separated positive and negatively charged regions with a chemically free C=O group, which can act as a hydrogen bond acceptor, and an NH group, which can act as a hydrogen bond donor. These groups interact in protein structures. The 20 amino acids found in proteins can be grouped according to the chemistry of their R groups, as depicted in Table 9.2. The R side chains also play an important structural role. Special roles are played by glycine, which does not have a side chain and can therefore increase local flexibility in structures, and cysteine, which can react with another cysteine to form a cross-link that can stabilize the protein structure.

Much of the protein core comprises regular secondary structures, α helices and β sheets, folded into a three-dimensional configuration. In these secondary structures, regular patterns of H bonds are formed between neighboring amino acids, and the amino acids

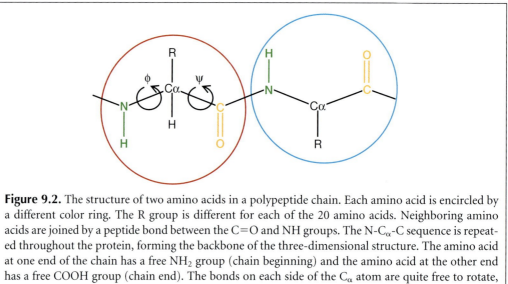

Figure 9.2. The structure of two amino acids in a polypeptide chain. Each amino acid is encircled by a different color ring. The R group is different for each of the 20 amino acids. Neighboring amino acids are joined by a peptide bond between the C=O and NH groups. The N-C_α-C sequence is repeated throughout the protein, forming the backbone of the three-dimensional structure. The amino acid at one end of the chain has a free NH_2 group (chain beginning) and the amino acid at the other end has a free COOH group (chain end). The bonds on each side of the C_α atom are quite free to rotate, but many combinations of angles are not possible for most amino acids due to spatial constraints from the R group and neighboring positions in the chain. The conformation of the protein backbone in space is determined by the angles of these bonds, Φ of the bond between the N and C_α atoms and Ψ of the bond between the C_α and C of the C=O group, also named C_β. The distribution of these two angles for the amino acids in a particular protein is often plotted on a graph called a Ramachandran plot. The angle Ω of the peptide bond joining the C=O and NH groups (not shown) is nearly always 180°.

Table 9.2. *Chemical properties of the 20 amino acids*

Chemical group	Amino acid (one-letter code)	Name
Hydrophobic		
	A	alanine
	V	valine
	Y	phenylalanine
	P	proline
	M	methionine
	I	isoleucine
	L	leucine
Charged		
	D	aspartic acid
	E	glutamic acid
	K	lysine
	R	arginine
Polar		
	S	serine
	T	threonine
	Y	tyrosine
	H	histidine
	C	cysteine
	N	asparagine
	Q	glutamine
	W	tryptophan
Glycine		
	G	glycine
Cross-linking		
	−	cysteine + cysteine

have similar Φ and Ψ angles, as depicted in Figure 9.3. The formation of these structures neutralizes the polar groups on each amino acid. The secondary structures are tightly packed in the protein core in a hydrophobic environment. Each amino acid side group has a limited volume to occupy and a limited number of possible interactions with other near-by side chains, a situation that must be taken into account in molecular modeling and alignments.

α Helix

The α helix depicted in Figure 9.3A is the most abundant type of secondary structure in proteins. The helix has 3.6 amino acids per turn with an H bond formed between every fourth residue; the average length is 10 amino acids (3 turns) or 10 Å but varies from 5 to 40 (1.5 to 11 turns). The alignment of the H bonds creates a dipole moment for the helix with a resulting partial positive charge at the amino end of the helix. Because this region has free NH_2 groups, it will interact with negatively charged groups such as phosphates. The commonest location of α helices is at the surface of protein cores, where they provide an interface with the aqueous environment. The inner-facing side of the helix tends to have hydrophobic amino acids and the outer-facing side hydrophilic amino acids. Thus, every third of four amino acids along the chain will tend to be hydrophobic, a pattern that can be quite readily detected. In the leucine zipper motif, a repeating pattern of leucines on the facing sides of two adjacent helices is highly predictive of the motif. A helical-wheel plot can be used to show this repeated pattern (see below). Other α helices buried in the protein core or in cellular membranes have a higher and more regular distribution of

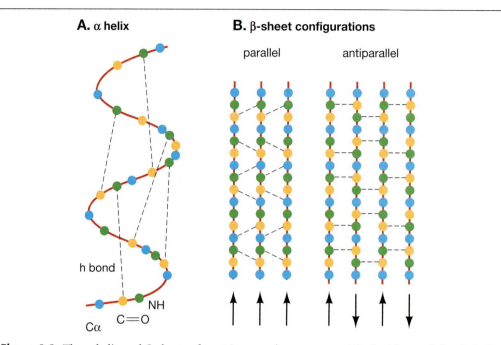

Figure 9.3. The α helix and β sheets of protein secondary structure. The backbone of the chain is shown in red, the C$_\alpha$ atoms and the C=O and NH groups are shown in blue, yellow, and green, respectively. (*A*) In the α helix, note that each C=O group at amino acid position *n* in the sequence is hydrogen-bonded with the NH group at position *n* + 4. There are 3.6 residues per turn. The helix is usually right-handed, but short sections of 3–5 amino acids of left-handed helices occur occasionally. The average Φ and Ψ angles of the amino acids in the right-handed helix are approximately 60° and 40°, respectively. The R side chains of the amino acids are on the outside of the helix. (*B*) The β sheet is made up of β strands that are portions of the protein chain. The strands may run in the same (parallel) or opposite (antiparallel) chemical directions (or a mixture of the two), and the pattern of hydrogen bonds is different in each case and also varies in antiparallel strands.

hydrophobic amino acids, and are highly predictive of such structures. Helices exposed on the surface have a lower proportion of hydrophobic amino acids. Amino acid content can be predictive of an α-helical region. Regions richer in alanine (A), glutamic acid (E), leucine (L), and methionine (M) and poorer in proline (P), glycine (G), tyrosine (Y), and serine (S) tend to form an α helix. Proline destabilizes or breaks an α helix but can be present in longer helices, forming a bend. There are computer programs for predicting quite reliably the general location of α helices in a new protein sequence.

β Sheet

β Sheets are formed by H bonds between an average of 5–10 consecutive amino acids in one portion of the chain with another 5–10 farther down the chain, as shown in Figure 9.3B. The interacting regions may be adjacent, with a short loop in between, or far apart, with other structures in between. Every chain may run in the same direction to form a parallel sheet, every other chain may run in the reverse chemical direction to form an antiparallel sheet, or the chains may be parallel and antiparallel to form a mixed sheet. As illustrated in Figure 9.3, the pattern of H bonding is different in the parallel and antiparallel configurations. Each amino acid in the interior strands of the sheet forms two H bonds with neighboring amino acids, whereas each amino acid on the outside strands forms only one bond with an interior strand. Looking across the sheet at right angles to the strands, more distant strands are rotated slightly counterclockwise to form a left-handed twist, which is apparent in some of the structures shown below. The C_α atoms alternate above and below the sheet in a pleated structure, and the R side groups of the amino acids alternate above and below the pleats. The Φ and Ψ angles of the amino acids in β sheets vary considerably in one region of the Ramachandran plot (see Fig. 9.2 legend). It is more difficult to predict the location of β sheets than of α helices. The situation improves somewhat when the amino acid variation in multiple sequence alignments is taken into account.

Loop

Loops are regions of a protein chain that are (1) between α helices and β sheets, (2) of various lengths and three-dimensional configurations, and (3) on the surface of the structure. Hairpin loops that represent a complete turn in the polypeptide chain joining two antiparallel β strands may be as short as two amino acids in length. Loops interact with the surrounding aqueous environment and other proteins. Because amino acids in loops are not constrained by space and environment as are amino acids in the core region, and do not have an effect on the arrangement of secondary structures in the core, more substitutions, insertions, and deletions may occur. Thus, in a sequence alignment, the presence of these features may be an indication of a loop. The positions of introns in genomic DNA sometimes correspond to the locations of loops in the encoded protein. Loops also tend to have charged and polar amino acids and are frequently a component of active sites. A detailed examination of loop structures has shown that they fall into distinct families.

Coil

A region of secondary structure that is not a helix, a sheet, or a recognizable turn is commonly referred to as a coil.

PROTEIN CLASSIFICATION

Proteins may be classified according to both structural and sequence similarity. For structural classification, the sizes and spatial arrangements of secondary structures described in the above section are compared in known three-dimensional structures. For classification by sequence similarity, alignments of protein sequences are made using the methods described in Chapters 3 and 4. Classification based on sequence similarity was historically the first to be used. Initially, similarity based on alignments of whole sequences was performed. Later, proteins were classified on the basis of the occurrence of conserved amino acid patterns. Databases that classify proteins by one or more of these schemes are available.

In considering protein classification schemes, it is important to keep several observations in mind. First, two entirely different protein sequences from different evolutionary origins may fold into a similar structure. Conversely, the sequence of an ancient gene for a given structure may have diverged considerably in different species while at the same time maintaining the same basic structural features. Recognizing any remaining sequence similarity in such cases may be a very difficult task. Second, two proteins that share a significant degree of sequence similarity either with each other or with a third sequence also share an evolutionary origin and should share some structural features also. However, gene duplication and genetic rearrangements during evolution may give rise to new gene copies, which can then evolve into proteins with new function and structure. Examples of these events are discussed at the beginning of Chapter 2 and in Chapters 6 and 10. To make assessments of protein structure, a number of terms that describe protein similarity and structural relationships are used.

Terms Used for Classifying Protein Structures and Sequences

The more commonly used terms for describing evolutionary and structural relationships among proteins are listed below. Many additional terms are used to describe various kinds of structural features found in proteins. Descriptions of such terms may be found at the CATH Web site (http://www.biochem.ucl.ac.uk/bsm/cath/lex/glossary.html), the Structural Classification of Proteins (SCOP) Web site (http://pdb.wehi.edu.au/scop/gloss.html and Web mirror sites), and a Glaxo-Wellcome tutorial on the Swiss bioinformatics Expasy Web site (http://www.expasy.ch/swissmod/course/course-index.htm).

Active site is a localized combination of amino acid side groups within the tertiary (three-dimensional) or quaternary (protein subunit) structure that can interact with a chemically specific substrate and that provides the protein with biological activity. Proteins of very different amino acid sequences may fold into a structure that produces the same active site.

Architecture describes the relative orientations of secondary structures in a three-dimensional structure without regard to whether or not they share a similar loop structure. In contrast, a fold is a type of architecture that also has a conserved loop structure. Architecture is a classification term used by the CATH database (http://www.biochem.ucl.ac.uk/bsm/cath/).

Blocks is a term used to describe a conserved amino acid sequence pattern in a family of proteins. The pattern includes a series of possible matches at each position in the represented sequences, but there are not any inserted or deleted positions in the pattern or in the sequences. By way of contrast, sequence profiles are a type of scoring matrix that represents a similar set of patterns that includes insertions and deletions. Profile HMMs are hidden Markov models of such gapped patterns (see Chapters 4 and 7). There are 2,290 HMM profile models in Pfam release 5.4 described below.

Class is a term used to classify protein domains according to their secondary structural content and organization. Four classes were originally recognized by Levitt and Chothia (1976), and several others have been added in the SCOP database described below. Three classes are given in the CATH database: mainly-α, mainly-β, and α–β, with the α–β class including both alternating α/β and $\alpha+\beta$ structures. Thus, class 4 of the SCOP database is included in class 3 of the CATH database.

Core is the portion of a folded protein molecule that comprises the hydrophobic interior of α helices and β sheets. The compact structure brings together side groups of amino acids into close enough proximity so that they can interact. When comparing protein structures, as in the SCOP database, core refers to the region common to most of the structures that share a common fold or that are in the same superfamily. In structure prediction, core is sometimes defined as the arrangement of secondary structures that is likely to be conserved during evolutionary change (Madej et al. 1995). A library of protein cores designated LPFC is maintained at Stanford University at http://www-camis.stanford.edu/projects/helix/LPFC/ and is based on multiple sequence alignments using amino acid scoring matrices based on structural substitutions.

Domain (sequence context). See Homologous domain.

Domain (structural context; also see Homologous domain entry) refers to a segment of a polypeptide chain that can fold into a three-dimensional structure irrespective of the presence of other segments of the chain. The separate domains of a given protein may interact extensively or may be joined only by a length of polypeptide chain. A protein with several domains may use these domains for functional interactions with different molecules. 3Dee, a database of protein domain definitions, is provided at http://barton.ebi.ac.uk/servers/3Dee.html/. A structural classification of protein domains is maintained at http://www2.embl-ebi.ac.uk/dali/domain/ (Holm and Sander 1998), and ddbase, a database of protein domains, may be found at http://www-cryst.bioc.cam.ac.uk/~ddbase. Another domain database may be found at http://www3.icgeb.trieste.it/ (Pongor et al. 1993).

Family (sequence context), as defined originally by Dayhoff et al. (1978), is a group of proteins of similar biochemical function that are more than 50% identical when aligned. This same cutoff is still used by the Protein Information Resource (PIR). A protein family comprises proteins with the same function in different organisms (orthologous sequences) but may also include proteins in the same organism (paralogous sequences) derived from gene duplication and rearrangements (Henikoff et al. 1997). If a multiple sequence alignment of a protein family reveals a common level of similarity throughout the lengths of the proteins, PIR refers to the family as a homeomorphic family. The aligned region is referred to as a homeomorphic domain, and this region may comprise several smaller homology domains that are shared with other families. Families may be further subdivided into subfamilies or grouped into superfamilies based on respective higher or lower levels of sequence similarity (Barker et al. 1995; http://www-nbrf.georgetown.edu/). The SCOP database described below (release 1.50) reports 1296 families and the CATH database (version 1.7 beta), also described below, reports 1846 families.

When the sequences of proteins with the same function are examined in greater detail, some are found to share high sequence similarity. They are obviously members of the same family by the above criteria. However, others are found that have very little, or even insignificant, sequence similarity with other family members. In such cases, the family relationship between two distant family members A and C can often be demonstrated by finding an additional family member B that shares significant similarity with both A and C (Pearson 1996; Park et al. 1997). Thus, B provides a connecting link between A and C. Another approach is to examine distant alignments for highly conserved matches (Patthy 1987, 1996).

At a level of identity of >50%, proteins are likely to have the same three-dimensional structure, and the identical atoms in the sequence alignment will also superimpose within approximately 1 Å in the structural model (Holm and Sander 1994). Thus, if the structure of one member of a family is known, a reliable prediction may be made for a second member of the family, and the higher the identity level, the more reliable the prediction. Protein structural modeling can be performed by examining how well the amino acid substitutions fit into the core of the three-dimensional structure.

Family (structural context), as used in the FSSP database (Holm and Sander 1998) and the DALI/FSSP Web site (see below), refers to two structures that have a significant level of structural similarity but not necessarily significant sequence similarity.

Fold is a term with similar meaning to structural motif, but in general refers to a somewhat larger combination of secondary structural units in the same configuration. Thus, proteins sharing the same fold have the same combination of secondary structures that are connected by similar loops. An example is the Rossman fold comprising several alternating α helices and parallel β strands. In the SCOP, CATH, and FSSP databases described below, the known protein structures have been classified into hierarchical levels of structural complexity with the fold as a basic level of classification. From a survey of the currently known protein structures in the Brookhaven Protein Data Bank (Holm and Sander 1998), approximately 500 independent folds have been identified. The number of distinct folds in the SCOP database is 548 (release 1.50) and the number of the equivalent topological families in the CATH database is 580 (version 1.70 beta release). These databases are described below. **Foldon** is a related term that has been used to describe an independently folding unit (Panchenko et al. 1996, 1997).

Homologous domain (sequence context, also see Domain, structural context) refers to an extended sequence pattern, generally found by sequence alignment methods, that indicates a common evolutionary origin among the aligned sequences. A homology domain is generally longer than motifs. The domain may include all of a given protein sequence or only a portion of the sequence. Some domains are complex and made up of several smaller homology domains that became joined to form a larger one during evolution. A domain that covers an entire sequence is called the homeomorphic domain by PIR (Barker et al. 1996; see http://www-nbrf.georgetown.edu/).

Module is a region of conserved amino acid patterns comprising one or more motifs and considered to be a fundamental unit of structure or function. The presence of a module has also been used to classify proteins into families.

Motif (sequence context) refers to a conserved pattern of amino acids that is found in two or more proteins. In the Prosite catalog, a motif is an amino acid pattern that is found in a group of proteins that have a similar biochemical activity, and that often is near the active site of the protein. Examples of sequence motif databases are the Prosite catalog (http://www.expasy.ch/prosite) and the Stanford Motifs Database (http://dna.stanford.edu/emotif/).

Motif (structural context) refers to a combination of several secondary structural elements produced by the folding of adjacent sections of the polypeptide chain into a specific three-dimensional configuration. An example is the helix-loop-helix motif. Structural motifs are also referred to as supersecondary structures and folds.

Position-specific scoring matrix (sequence context, also known as weight or scoring matrix) represents a conserved region in a multiple sequence alignment with no gaps. Each matrix column represents the variation found in one column of the multiple sequence alignment.

Position-specific scoring matrix—3D (structural context) represents the amino acid variation found in an alignment of proteins that fall into the same structural class. Matrix columns represent the amino acid variation found at one amino acid position in the aligned structures (Kelley et al. 2000).

Primary structure refers to the linear amino acid sequence of a protein, which chemically is a polypeptide chain composed of amino acids joined by peptide bonds.

Profile (sequence context) is a scoring matrix that represents a multiple sequence alignment of a protein family. The profile is usually obtained from a well-conserved region in a multiple sequence alignment. The profile is in the form of a matrix with each column representing a position in the alignment and each row one of the amino acids. Matrix values give the likelihood of each amino acid at the corresponding position in the alignment. The profile is moved along the target sequence to locate the best scoring regions by a dynamic programming algorithm. Gaps are allowed during matching and a gap penalty is included in this case as a negative score when no amino acid is matched. A sequence profile may also be represented by a hidden Markov model, referred to as a profile HMM.

Profile (structural context) is a scoring matrix that represents which amino acids should fit well and which should fit poorly at sequential positions in a known protein structure. Profile columns represent sequential positions in the structure, and profile rows represent the 20 amino acids. As with a sequence profile, the structural profile is moved along a target sequence to find the highest possible alignment score by a dynamic programming algorithm. Gaps may be included and receive a penalty. The resulting score provides an indication as to whether or not the target protein might adopt such a structure.

Quaternary structure is the three-dimensional configuration of a protein molecule comprising several independent polypeptide chains. A Web site for predicting quaternary structure is described at http://msd.ebi.ac.uk/Services/Quaternary/quaternary.html. A database of experimentally identified interacting domains of protein subunits (DIP) is maintained at http://dip.doe-mbi.ucla.edu; Xenarios et al. 2000; also see Table 9.5).

Secondary structure refers to the interactions that occur between the C=O and NH groups on amino acids in a polypeptide chain to form α helices, β sheets, turns, loops, and other forms, and that facilitate the folding into a three-dimensional structure.

Superfamily is a group of protein families of the same or different lengths that are related by distant yet detectable sequence similarity. Members of a given superfamily thus have a common evolutionary origin. Originally, Dayhoff defined the cutoff for superfamily status as being the chance that the sequences are not related of $<10^{-6}$, on the basis of an alignment score (Dayhoff et al. 1978). Proteins with few identities in an alignment of the sequences but with a convincingly common number of structural and functional features are placed in the same superfamily. At the level of three-dimensional structure, superfamily proteins will share common structural features such as a common fold, but there may also be differences in the number and arrangement of secondary structures. The PIR resource uses the term homeomorphic superfamilies to refer to superfamilies that are composed of sequences that can be aligned from end to end, representing a sharing of single sequence homology domain, a region of similarity that extends throughout the alignment. This domain may also comprise smaller homology domains that are shared with other protein families and superfamilies. Although a given protein sequence may contain domains found in several superfamilies, thus indicating a complex evolutionary history, sequences will be assigned to only one homeomorphic superfamily based on the presence of similarity throughout a multiple sequence alignment. The superfamily alignment may also include regions that do not align either within or at the ends of the alignment (Barker et al. 1995, 1996; http://www-nbrf.georgetown.edu/). In contrast, sequences in the same family align well throughout the alignment. The SCOP Web site reports 820 superfamilies (release 1.50), and the CATH Web site (version 1.7 beta) reports 900 superfamilies (sites described below).

Supersecondary structure is a term with similar meaning to a structural motif.

Tertiary structure is the three-dimensional or globular structure formed by the packing together or folding of secondary structures of a polypeptide chain.

Classes of Protein Structure

From the work of Levitt and Chothia (1976), four principal classes of protein structure were recognized based on the types and arrangements of secondary structural elements. These classes are described and illustrated below. In addition, several other classes recognized in the SCOP database discussed below (p. 402) (Murzin et al. 1995) are also included. Examples of this classification are taken from Branden and Tooze (1991).

1. Class α comprises a bundle of α helices connected by loops on the surface of the proteins (see Fig. 9.4).

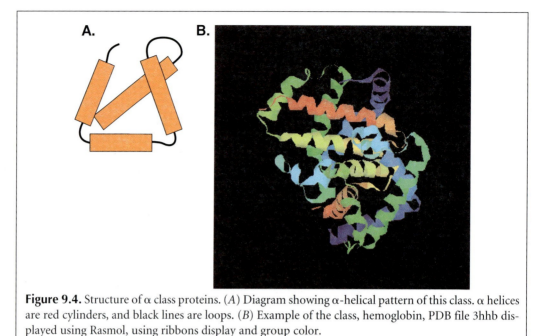

Figure 9.4. Structure of α class proteins. (*A*) Diagram showing α-helical pattern of this class. α helices are red cylinders, and black lines are loops. (*B*) Example of the class, hemoglobin, PDB file 3hhb displayed using Rasmol, using ribbons display and group color.

2. Class β comprises antiparallel β sheets, usually two sheets in close contact forming a sandwich (see Fig. 9.5). Alternatively, a sheet can twist into a barrel with the first and last strands touching. Examples are enzymes, transport proteins, antibodies, and virus coat proteins such as neuraminidase.

3. Class α/β comprises mainly parallel β sheets with intervening α helices, but may also have mixed β sheets (see Fig. 9.6). In addition to forming a sheet in some proteins in this class, as illustrated below, in others parallel β strands may form into a barrel structure that is surrounded by α helices (not shown). This class of proteins includes many metabolic enzymes.

4. Class α + β comprises mainly segregated α helices and antiparallel β sheets (Fig. 9.7).

5. Multidomain (α and β) proteins comprise domains representing more than one of the above four classes.

6. Membrane and cell-surface proteins and peptides excluding proteins of the immune system comprise this class (see Fig. 9.8).

Protein Databases

A protein can be analyzed in the laboratory at the levels of sequence and structure. The amino acid sequence and the atomic coordinates of each atom in the structure are unique to each protein. The sequence is obtained in the molecular biology laboratory as a DNA sequence and translated into the amino acid sequence of the encoded protein (see Chapter 8). DNA sequences are deposited in the DNA sequence databases such as GenBank and EMBL, where they are automatically translated to produce the Genpept and TrEMBL protein databases, respectively. Sometimes protein fragments are also sequenced, and matches with DNA sequence databases are used to identify the encoding gene (Chapter 8). The encoded proteins are additionally annotated in databases such as SwissProt and PIR as described in Chapter 2.

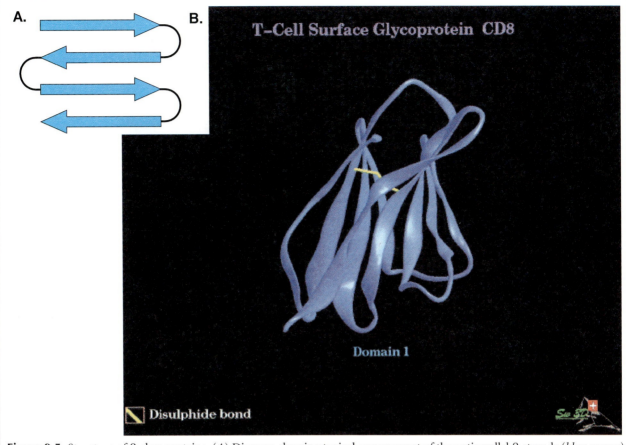

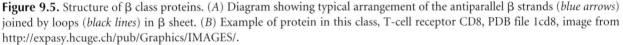

Figure 9.5. Structure of β class proteins. (*A*) Diagram showing typical arrangement of the antiparallel β strands (*blue arrows*) joined by loops (*black lines*) in β sheet. (*B*) Example of protein in this class, T-cell receptor CD8, PDB file 1cd8, image from http://expasy.hcuge.ch/pub/Graphics/IMAGES/.

The three-dimensional structure of a protein is usually obtained by making crystals of the protein and using X-ray diffraction to determine the positions of molecules that are fixed within the crystal. The technique of nuclear magnetic resonance (NMR) is also used to obtain protein structures. Once the three-dimensional coordinates of each atom in the protein molecule have been found, a table of these coordinates is deposited with the Brookhaven Data Bank as a PDB entry. PDB entries such as shown in Table 9.3 give the atomic coordinates of the amino acids in proteins, protein fragments, or proteins bound to substrates or inhibitors. PDB files may be easily retrieved from the PDB Web site (http://www.rcsb.org/pdb/) and displayed with a molecular viewer such as Rasmol. Structural information may also be stored in forms other than PDB, but PDB is the most accessible for the molecular biologist. There are three different kinds of databases that provide an analysis of proteins, one kind for sequences, a second for structures, and a third for comparing sequences and structures.

As more and more protein structures have been solved by X-ray crystallographic and NMR methods, these structures have been classified by various means into structural databases. This classification is based on comparison and alignment of the protein structures. The types, order, connections, and relative positions of secondary structures are compared using the known atomic coordinates of atoms in each structure and methods described below. This type of information can then be combined with sequence information to identify other proteins that might have similar structural features.

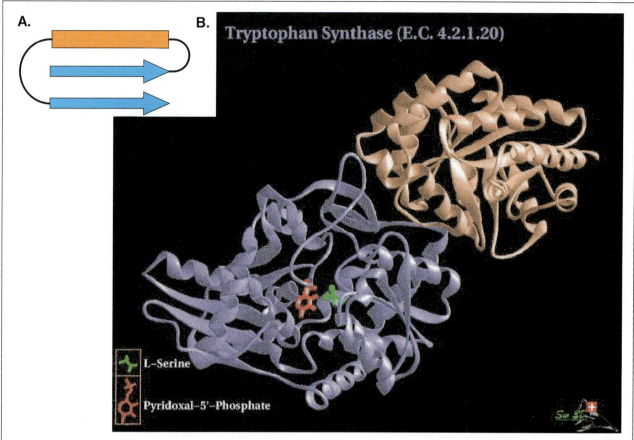

Figure 9.6. Structure of α/β class proteins. (*A*) Diagram showing one possible configuration of parallel β strands (*blue arrows*) in a β sheet and an intervening α helix (*red cylinder*), joined by loops (*black lines*). (*B*) Example of protein in this class, tryptophan synthase β subunit obtained from http://expasy.hcuge.ch/pub/Graphics/IMAGES/.

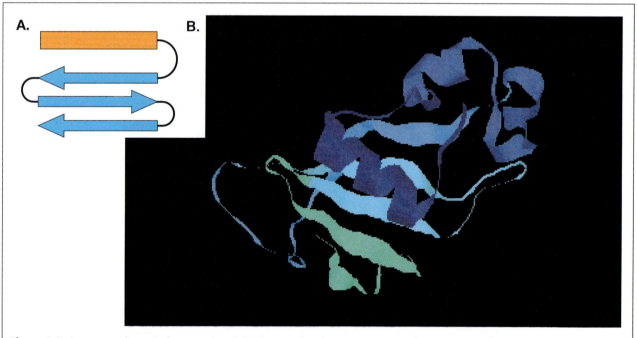

Figure 9.7. Structure of α + β class proteins. (*A*) Diagram showing arrangement of typical motif of antiparallel β strands (*blue arrows*) in β sheet and segregated from α helix (*red cylinder*) and showing loops (*black lines*). (*B*) Example of protein in this class, G-specific endonuclease complex with deoxy-dinucleotide inhibitor, PDB file 1rnb viewed with Rasmol.

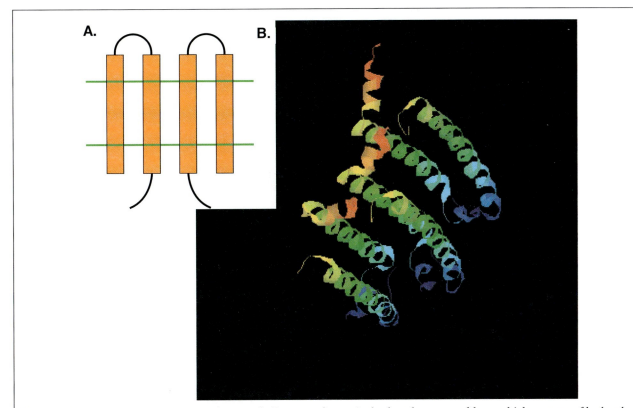

A.

B.

Figure 9.8. Structure of membrane proteins. α helices are of a particular length range and have a high content of hydrophobic amino acids traversing a membrane, features that make this class readily identifiable by scanning a sequence for these hydrophobic regions (see below). (*A*) Diagram showing typical arrangement of membrane-traversing, hydrophobic α helices (*red*). Membrane bilayer shown as green lines. (*B*) Example of protein in this class, integral membrane light-harvesting complex, PDB file 1kzu viewed with Rasmol.

Another type of protein sequence analysis is a sequence alignment of protein sequences discussed in Chapter 3 or a search for similar sequences in the sequence databases, as described in Chapter 7. The alignment will reveal any significant similarity and the degree of amino acid identity between two sequences. Similarity may be present throughout the sequences or localized to certain regions. Localization of sequence similarity can best be performed by global and local sequence alignment methods, as discussed in Chapter 3. The stronger the similarity and identity, the more similar are the three-dimensional folds and other structural features of the proteins. Another level of sequence analysis is examining a group of sequences for common amino acid patterns. Methods for finding different types of patterns, including motifs (short gapped or ungapped patterns), blocks (ungapped patterns), and patterns with gaps (represented by profile scoring matrices and profile HMMs) are discussed in Chapter 4. These patterns may be obtained from sequences of proteins that are already known to have the same function, or they may be obtained by statistical or pattern-finding methods of any set of sequences of biological interest. Depending on the extent and significance of these patterns and additional information about the function of the proteins, their presence may or may not represent structural similarity or an evolutionary relationship among the proteins. A combined form of sequence and structural alignments provides an additional level of analysis.

When proteins of unknown structure are similar to a protein of known structure at the sequence level, multiple sequence alignment and pattern analysis can be used to predict the

Table 9.3. *Brookhaven Protein Data Bank (PDB) entry 3hhb for deoxy hemoglobin*

Header compnd	Oxygen transport hemoglobin (deoxy)							13-Jul-93		2hhb
ATOM	1	N	VAL A	1		5.428	17.064	5.060	1.00	41.29
ATOM	2	CA	VAL A	1		6.168	18.292	4.856	1.00	41.33
ATOM	3	C	VAL A	1		7.676	18.056	5.068	1.00	31.64
ATOM	4	O	VAL A	1		8.120	17.488	6.076	1.00	38.31
ATOM	5	CB	VAL A	1		5.644	19.268	5.884	1.00	52.26
ATOM	6	CG1	VAL A	1		6.044	20.696	5.512	1.00	52.75
ATOM	7	CG2	VAL A	1		4.124	19.120	6.000	1.00	58.75
ATOM	8	N	LEU A	2		8.444	18.512	4.116	1.00	27.63
ATOM	9	CA	LEU A	2		9.896	18.420	4.308	1.00	33.62
ATOM	10	C	LEU A	2		10.360	19.592	5.216	1.00	32.51
ATOM	11	O	LEU A	2		10.128	20.760	4.900	1.00	31.03
ATOM	12	CB	LEU A	2		10.568	18.584	2.932	1.00	34.38
ATOM	13	CG	LEU A	2		10.284	17.488	1.924	1.00	32.23
ATOM	14	CD1	LEU A	2		11.032	17.676	0.580	1.00	36.30
ATOM	15	CD2	LEU A	2		10.576	16.136	2.560	1.00	38.42

Shown is the initial part of the entry showing ATOM records that provide cartesian coordinates of all atoms in the first two amino acids Val and Leu. The last columns give the occupancy and temperature factor for each atom. The occupancy gives the frequency with which the atom is present in the crystal and is usually 1. The temperature gives a measurement of the uncertainty of the position of the atom due to the motion of the atom in the crystal. The units of temperature are Angstroms squared. A typical value of a crystal at room temperature at 2 Å resolution is 20 Å; the higher this value for an atom, the more uncertain the position of that atom. Structural entries sometimes provide the author's assignment of a secondary structure to each amino acid.

structures of these proteins. Databases of such related proteins are available. In another type of analysis, called threading, the sequence of amino acids in a protein of unknown structure is tested for ability to fit into a known three-dimensional structure. The size and chemistry of each amino acid R group and proximity to other amino acids are taken into account. This analysis provides a method for aligning a sequence with a structure.

METHODS

1. Amino acid sequences of proteins are derived from translation of cDNA sequences or predicted gene structures in genomic DNA sequences. Partial sequences are also derived by translation of expressed sequence tag (EST) sequences or genomic DNA sequences in all six reading frames. These predictions can be improved when genomic and EST sequences can be aligned and when overlapping EST sequences are identified by gene indexing, as described in Chapters 7 and 8.

2. The sequence is used as a query in a database similarity search against the proteins in the Protein Data Bank (PDB), all of which have a known three-dimensional structure. A significant alignment of the query sequence with a PDB sequence is evidence that the query sequence has a similar three-dimensional structure. If a relationship with a PDB protein is not found, then a second database similarity search against a protein sequence database such as SwissProt can be performed. Matching sequences including both closely related and more distantly related ones can then be used in a search against PDB sequences. The PSI-BLAST tool described in Chapter 7 automates and enhances the process of finding related sequences in the protein database. The goal is to discover one or more database sequences that are related both to the query and to a PDB sequence, as illustrated in Figure 7.1.

3. If the database similarity search reveals a significant alignment between the query sequence and a PDB sequence, the alignment between the sequences can be used to position the amino acids of the query sequence in the same approximate three-dimensional structure. Testing the significance of alignment scores is discussed in Chapter 3.

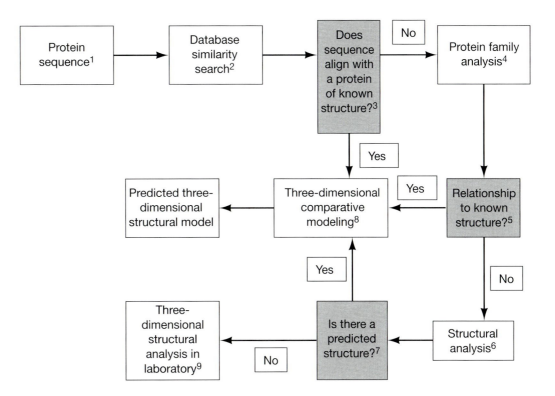

4. Proteins have been classified into families on the basis of sequence similarity. The relationships are depicted in a multiple sequence alignment of the proteins, as described in Chapter 4. Proteins of known three-dimensional structure have also been classified into fold families on the basis of a common arrangement of secondary structures. Sequences of proteins in the same fold family are often not similar, so they cannot be aligned. However, the individual proteins in a particular fold family are often members of families based on sequence similarity. Hence, these similar sequences are also predicted to have the same structural fold as the fold family. The goal of this step in the flowchart is to exploit these structure–sequence relationships. Two questions are addressed: (1) Is the new protein a member of a protein family based on sequence similarity? (2) Does the matched family have a predicted structural fold? The first question is usually addressed by analyzing the test sequence for patterns that represent each family using PSSMs, profile HMMs, and other tools, as described in Chapter 7. Web sites such as Interpro (Table 9.5) include a large, composite collection of patterns and will search a new sequence for matches. 3D-PSSM (Table 9.5) includes a powerful set of scoring matrices based on structural alignments for use in three-dimensional structure prediction. These Web sites usually provide links to related fold families, thus identifying a predicted structural fold for the new protein. Other Web sites employ a cluster analysis of proteins based on pair-wise alignment scores of all of the proteins in the SwissProt database. These sites offer an alternative method for finding relationships between a new sequence and all of the other sequences in SwissProt, and thus for discovering a link to a known protein structure.

5. If the family analysis reveals that the new protein is a member of a family that is predicted to have a structural fold, multiple sequence alignments of these proteins can be used for structural modeling.

6. This step in the flowchart includes several different types of analyses that are described below in the chapter. First, the presence of small amino acid motifs in a protein can be an indicator of a biochemical function. The Prosite catalog can be used to search a new protein sequence for motifs. Second, spacing and arrangement of specific amino acids, e.g., hydrophobic amino acids, provides important structural clues that can be used for modeling. Third, the tendency of certain amino acid combinations to occur in a given type of secondary structure provides methods for predicting where these structures are likely to occur in a new sequence. Fourth, the structural fold families described in note 4 above have been represented by PSSMs and by HMMs that capture the tendency to find each amino acid at a particular position in a structural fold and variations in the fold itself. Other models of three-

dimensional structure represent the size and chemistry of amino acids or the energetic stability associated with amino acid interactions. A new protein sequence can be aligned with these models to determine whether the sequence matches one of them, a procedure known as threading a sequence into a structure.

7. The structural analysis in step 6 provides clues as to the presence of active sites, regions of secondary and three-dimensional structure, and the order of predicted secondary structures. If these predictions are convincing enough, it may be possible to identify a new protein as a member of a known structural class.

8. Sequence or structural alignments of the new protein with a protein of known structure provide a starting three-dimensional model of the protein. By using computer graphics and protein modeling software, the amino acids can then be positioned to accommodate available space and interactions with neighboring amino acids.

9. Proteins that fail to show any relationship to proteins of known structure are candidates for structural analysis. There are approximately 500–600 known fold families, and new structures are frequently found to have an already known structural fold. Accordingly, protein families with no relatives of known structure may represent a novel structural fold.

VIEWING PROTEIN STRUCTURES

The first major step in displaying a structure is to identify the correct PDB identification code for the structural file. Most sites provide a browser program for searching the structural database for the name of the protein, organism, or other identifying features (see below). There may be a number of choices from which to choose, including domains, folds, or protein fragments, or structures of the protein bound to a substrate or inhibitor. Some databases also include the predicted structure of mutant proteins. The available choices need to be screened carefully for the correct one.

A number of molecular viewers are freely available and run on most computer platforms and operating systems, including Microsoft Windows, Macintosh, and UNIX X-Windows. These programs convert the atomic coordinates into a view of the molecule. They may also recompute information to remove inconsistencies in the database or to supply missing information (Hogue and Bryant 1998a,b). Viewers also provide ways to manipulate the molecule, including rotation, zooming, and creating two images that provide a stereo view. Rotating a molecule by dragging the mouse across the image can illustrate the three-dimensional structure. Viewers can also be used to show a structural alignment of two or more structures or a predicted structure. Unless a very high-resolution view is needed, the simplest way to use a viewer is through a network browser. The browser may be readily configured to run a viewer program automatically when the particular file format used by the viewer is being downloaded from the remote computer. Most sites that provide protein structural files provide several formats allowing a choice of viewers, and they also provide Web links to other sites from which the viewer program may be downloaded. The viewer option usually appears once a particular structural file has been chosen. Shown in Table 9.4 are some representative viewers that are commonly used and their features.

The correct processing of files with molecular structural information through the Web or through E-mail attachments is made possible by the chemical MIME (multipurpose internet mail extension) project (http://www.ch.ic.ac.uk/chemime/iupac.html). This project acts as a repository for standard types of MIME files. As an example, if the start of the file includes the label chemical/x-pdb (MIME type chemical and subtype x-pdb), the file is a text file in the Brookhaven Protein Data Bank file format, and a viewer for a pdb file such as Rasmol or Chime is needed. Files intended for viewing by Rasmol may also be indicated by MIME type application/x-rasmol and the pdb file may also be identified by the file-

Table 9.4. *Programs for viewing protein molecules*

Viewer	Web location	Features
Chime	http://www.umass.edu/microbio/chime/	A Web browser plug-in that can be used to display and manipulate structures inside a Web page. There are many mouse-driven controls. Excellent for lecture presentations.
Cn3d[a]	http://www.ncbi.nlm.nih.gov/Structure/ (Hogue 1997)	Provides viewing of three-dimensional structures from Entrez and MMDB.[a] Cn3D runs on Windows, MacOS, and Unix; simultaneously displays structural and sequence alignments; can show multiple superimposed images from NMR studies.
Mage	http://kinemage.biochem.duke.edu/website/kinhome.html (see Richardson and Richardson 1994)	Standard molecular viewing features with animation and kaleidoscope effects.
Rasmol[b]	http://www.umass.edu/microbio/rasmol/ (Sayle and Milner-White 1995)	Most commonly used viewer for Windows, MacOS, UNIX, and VMS operating systems. Performs many functions.
Swiss 3D viewer, Spdbv	http://www.expasy.ch/spdbv/mainpage.html (Guex and Peitsch 1997)	Protein models can be built by structural alignments; calculates atomic angles and distances, threading, energy minimization, and interacts with the Swiss Model server.

Additional viewers are accessible from the referenced Web sites. Viewer functions usually include wireframe of C_α backbone, ribbon of secondary structures, space-filling displays, color schemes to illustrate features such as residues, structures, temperature, mouse-drag rotation, several views including stereo, zooming, and exporting to graphic file formats. Assistance with these viewers is provided at the following Web sites for obtaining molecular coordinates: Molecules R Us at NIH, http://molbio.info.nih.gov/cgi-bin/pdb, and NCBI, http://www.ncbi.nlm.nih.gov/Structure/. A large list of available graphics viewers may be found at http://www.csb.yale.edu/user-guides/graphics/csb_hm_graph.html.

[a] The NCBI structure group has established a new format for databases called ASN.1 (see Chapter 2). The PDB files have been converted into this format to create another database MMDB (Molecular Modelling DataBase) that is highly suitable for structural alignments by vector methods described below. Ambiguities in PDB entries have been made explicit in the MMDB database (Hogue and Bryant 1998a,b; http://www.ncbi.nlm.nih.gov/Structure/).

[b] Rasmol and other viewers as well have many features in the molecular viewing window in addition to those described above. These additional features are accessible through a command line window that appears when the program is running.

name extension pdb. There are also additional chemical MIME formats. For Cn3D, chemical/ncbi-asn1-binary and val are the MIME type and filename extension, respectively. Cn3D files are sent as a binary file rather than a text file, meaning that some bytes include characters other than the standard ASCII characters. For MAGE, chemical/x-kinemage and kin are used. Molecules may also be viewed by means of programs called applets written in the JAVA programming language. These programs are sent at the same time as the molecular coordinates and are run by the browser.

In addition to retrieving the three-dimensional coordinates of a molecule, already prepared graphic views of molecules may be obtained from many of the Web sites that provide pdb files. The following FTP site contains a database of stored image files: http://www.expasy.ch/databases/swiss-3dimage/IMAGES/. These views include two file formats commonly used on the Web, the JPEG (Joint Photographic Experts Group) format and GIF (graphics interchange format). These formats produce images of a reasonably high quality but have varying levels of detail and resolution. A higher resolution and more detailed rendition of the molecule will have a larger file size and take longer to retrieve over the Internet. These files may be compressed to a smaller size by graphic format conversion programs. Programs such as Raster3D (http://www.bmsc.washington.edu/raster3d/) and Molscript (http://www.avatar.se/molscript/) produce very high-quality images in a number of different formats. These programs require graphics work stations and a more sophisticated level of programming experience.

PROTEIN STRUCTURE CLASSIFICATION DATABASES

The following databases are accessible on the Web and provide up-to-date structural comparisons for the proteins currently in the Brookhaven PDB and access to the sequences of these proteins. The methods used to classify the protein structures in these databases vary from manual examination of structures to fully automatic computer algorithms. Hence, although one can expect to find roughly the same groupings in each database, there will be some structural relationships that are only identified by one of these methods. Each database has useful information that may be lacking in the others. The MMDB and SARF databases (4 and 5 below) are based on a rapid structural alignment method that is designed to find the most significant alignments in the structural databank. The SCOP, CATH, and FSSP databases (1, 2, and 3) are based on different comparison methods and are likely to provide additional complementary information on relationships among protein structures. These classification schemes have been reviewed previously (Swindells et al. 1998).

1. *The SCOP database.* The SCOP (*s*tructural *c*lassification *o*f *p*roteins) database (Murzin et al. 1995; Brenner et al. 1996), based on expert definition of structural similarities, is located at http://scop.mrc-lmb.cam.ac.uk/scop/. Following classification by class, SCOP additionally classifies protein structures by a number of hierarchical levels to reflect both evolutionary and structural relationships; namely family, superfamily, and fold. Shown in Figure 9.9 is an example of the lineage for the all α class, globin-like fold, globin-like superfamily, globin, and phycocyanin families, and finally protein domains such as hemoglobin 1 which can be viewed by individual entry in PDB using a molecular viewer.

2. *The CATH database.* The CATH (*c*lassification by *c*lass, *a*rchitecture, *t*opology, and *h*omology) protein structure database resides at University College, London (Orengo et al. 1997; http://www.biochem.ucl.ac.uk/bsm/cath/). Proteins are classified first into hierarchical levels by class, similar to the SCOP classification except that α/β and α+β proteins are considered to be in one class. Instead of a fourth class for α+β proteins, the fourth class of CATH comprises proteins with few secondary structures. Following class, proteins are classified by architecture, fold, superfamily, and family. Similar structures are found by the program SSAP, described on page 419. An example of a CATH entry is shown in Figure 9.10.

3. *The FSSP database.* The FSSP (*f*old classification based on *s*tructure-*s*tructure alignment of *p*roteins) is based on a structural alignment of all pair-wise combinations of the proteins in the Brookhaven structural database by the structural alignment program DALI (Holm and Sander 1996; http://www2.embl-ebi.ac.uk/dali/fssp/fssp.html). PDB has a number of redundant structures of proteins whose sequences and structures are 25% or more identical. A subset of representative structures in PDB without these redundant entries was first produced by aligning all of the PDB structures with DALI. Each protein in the subset was then subdivided into individual domains. These domains were then aligned structurally with DALI to identify the common folds. Redundant folds were again eliminated, and a set of representative folds was chosen. From 8320 PDB entries, 947 representative structures, 1484 domains, and 540 structurally distinct fold types were identified in 1997 (Holm and Sander 1998). These fold types represent a unique configuration of secondary structural elements in the domains. For example, one fold might be composed of helix-strand-helix-6 strands joined by loops in a particular configuration.

   Corresponding to each representative fold type, there is a cluster of folds that are of the same approximate structure. The domains that have a given cluster of folds are structurally related, and the cluster is represented by structural alignments of these domains. The higher the statistical score for a given domain alignment and corresponding fold (higher Z value), the greater the degree to which the atoms occupy similar structural positions. Z values >16 indicate a very good structural alignment, 8–16 a less good alignment, until a level of 2, which indicates the lowest level of alignment detection, is reached. Thus, fold clusters may be organized in a hierarchical fashion with folds represented by the most low-scoring alignments at the top of the hierarchy, as illustrated in Figure 9.11, FSSP, part D.

In addition, the sequences of the 1000 representative structures were used as probes for a sequence similarity search of the SwissProt protein sequence database. The database search program MAXHOM, which begins with a sequence similarity search and then with an expanded profile search, was used, as discussed in Chapter 7. The resulting homology-derived structures of proteins (HSSP) database (Sander and Schneider 1991; Dodge et al. 1998; http://www.sander.ebi.ac.uk/hssp/) contains lists of similar proteins, one list for each representative structure. Given the PDB database number of a known structure, the program will show the closest representative structures, and one or more may be chosen. The program will then show any significant structural alignments between the chosen representative and other representative structures in FSSP. A structural alignment between the chosen representative and each of the matching proteins in the HSSP database entry for that representative may be selected. An example of searching for a structural and sequence similarity using the FSSP and HSSP databases is shown in Figure 9.11.

4. *MMDB (molecular modelling database).* Proteins of known structure in the Brookhaven PDB have been categorized into structurally related groups in MMDB by the VAST (*Vector Alignment Search Tool*) structural alignment program (Madej et al. 1995). VAST aligns three-dimensional structures based on a search for similar arrangements of secondary structural elements (see Fig. 9.12). This method provides a method for rapidly identifying PDB structures that are statistically out of the ordinary. MMDB has been further incorporated into the ENTREZ sequence and reference database at http://www.ncbi.nlm.nih.gov/Entrez (Hogue et al. 1996). Accordingly, it is possible to perform a simultaneous search for similar sequences and structures, designated neighbors, at the ENTREZ Web site. Structural neighbors within MMDB are based on detailed residue-by-residue alignments.

5. *The SARF database.* The SARF (*spatial arrangement of backbone fragments*) database at http://www-lmmb.ncifcrf.gov/~nicka/sarf2.html/ (Alexandrov and Fischer 1996) also provides a protein database categorized on the basis of structural similarity. Like VAST, SARF can find structural similarity rapidly based on a search for secondary structural elements. These structural hierarchies found by this method are in good agreement with those found in the SCOP, CATH, and FSSP databases with several interesting differences. The method also found several new groupings of structural similarity. The SARF Web site provides a similarity-based tree of structures at http://www-lmmb.ncifcrf.gov/ ~nicka/ tree.html/ and some excellent representations of overlaid structures.

ALIGNMENT OF PROTEIN STRUCTURES

As more and more protein structures, as well as access to recently developed and rapid methods for comparing protein structures, have become available on the Web, alignment of protein structures has become a task achievable by laboratories not trained in the techniques of structural biology. To perform a sequence alignment, the amino acid sequence of one pro-

Structural Classification of Proteins

home	mail	help	root	up	expand	collapse

Fold: Globin-like

core: 6 helices; folded leaf, partly opened;

Lineage:

1. Root: scop
2. Class: All alpha
3. Fold: Globin-like
   core: 6 helices; folded leaf, partly opened;

Superfamilies:

1. Globin-like (2)
   1. Globins (37)
      Heme-binding protein
      1. Hemoglobin I
         1. ark clam (*Scapharca inaequivalvis*) (6) □
         2. clam (*Lucina pectinata*) (2) □
      2. Glycera globin
         1. marine bloodworm (*Glycera dibranchiata*) (2) □
      3. Myoglobin
         1. sperm whale (*Physeter catodon*) (76) □
         2. sea hare (*Aplysia limacina*) (6) □
         3. common seal (*Phoca vitulina*) (1) □
         4. pig (*Sus scrofa*) (8) □
         5. horse (*Equus caballus*) (6) □
         6. human (*Homo sapiens*) (5) □
         7. asian elephant (*Elephas maximus*) (1) □
         8. Loggerhead sea turtle (*Caretta caretta*) (2) □
         9. yellowfin tuna (*Thunnus albacares*) (1) □
      4. Erythrocruorin
         1. Midge (*Chironomous thummi thummi*), Fraction III (4) □
      5. Leghemoglobin
         1. yellow lupin (*Lupinus luteus I*) (17) □
         2. Soybean (*Glycine max*), isoform A (2) □
      6. Hemoglobin, alpha-chain
         1. human (*Homo sapiens*) (39) □ □
         2. horse (*Equus caballus*) (3) □
         3. deer (*Odocoileus virginianus*) (1) □
         4. bovine (*Bos taurus*) (1) □
            □

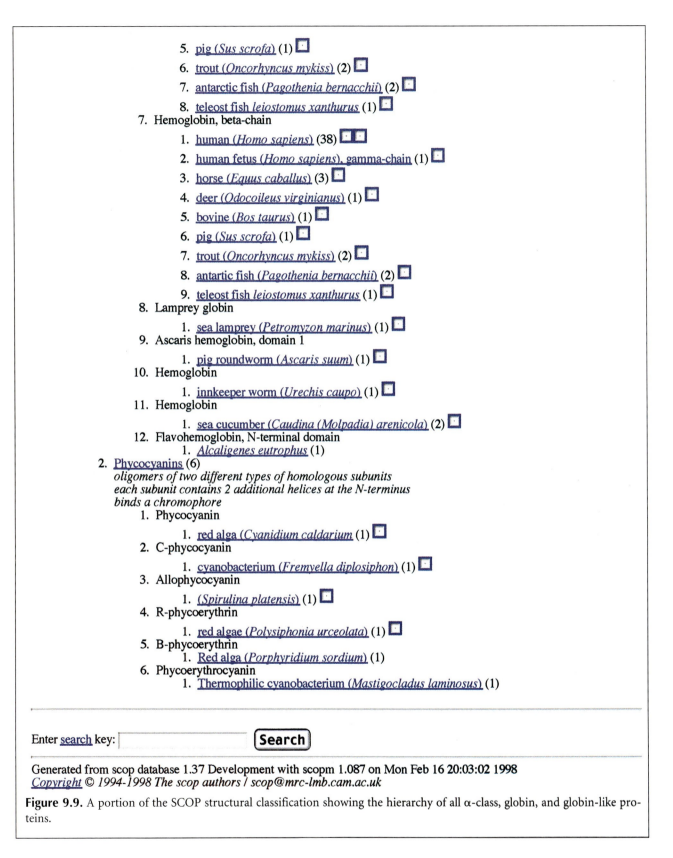

    5. <u>pig (*Sus scrofa*)</u> (1)

    6. <u>trout (*Oncorhyncus mykiss*)</u> (2)

    7. <u>antarctic fish (*Pagothenia bernacchii*)</u> (2)

    8. <u>teleost fish *leiostomus xanthurus*</u> (1)

7. Hemoglobin, beta-chain

    1. <u>human (*Homo sapiens*)</u> (38)

    2. <u>human fetus (*Homo sapiens*), gamma-chain</u> (1)

    3. <u>horse (*Equus caballus*)</u> (3)

    4. <u>deer (*Odocoileus virginianus*)</u> (1)

    5. <u>bovine (*Bos taurus*)</u> (1)

    6. <u>pig (*Sus scrofa*)</u> (1)

    7. <u>trout (*Oncorhyncus mykiss*)</u> (2)

    8. <u>antartic fish (*Pagothenia bernacchii*)</u> (2)

    9. <u>teleost fish *leiostomus xanthurus*</u> (1)

8. Lamprey globin

    1. <u>sea lamprey (*Petromyzon marinus*)</u> (1)

9. Ascaris hemoglobin, domain 1

    1. <u>pig roundworm (*Ascaris suum*)</u> (1)

10. Hemoglobin

    1. <u>innkeeper worm (*Urechis caupo*)</u> (1)

11. Hemoglobin

    1. <u>sea cucumber (*Caudina (Molpadia) arenicola*)</u> (2)

12. Flavohemoglobin, N-terminal domain

    1. *Alcaligenes eutrophus* (1)

2. <u>Phycocyanins</u> (6)

   oligomers of two different types of homologous subunits
   each subunit contains 2 additional helices at the N-terminus
   binds a chromophore

   1. Phycocyanin

    1. <u>red alga (*Cyanidium caldarium*</u> (1)

   2. C-phycocyanin

    1. <u>cyanobacterium (*Fremyella diplosiphon*)</u> (1)

   3. Allophycocyanin

    1. <u>(*Spirulina platensis*)</u> (1)

   4. R-phycoerythrin

    1. <u>red algae (*Polysiphonia urceolata*)</u> (1)

   5. B-phycoerythrin

    1. <u>Red alga (*Porphyridium sordium*)</u> (1)

   6. Phycoerythrocyanin

    1. <u>Thermophilic cyanobacterium (*Mastigocladus laminosus*)</u> (1)

Enter <u>search</u> key: [   ] **Search**

Generated from scop database 1.37 Development with scopm 1.087 on Mon Feb 16 20:03:02 1998
<u>*Copyright*</u> *© 1994-1998 The scop authors / scop@mrc-lmb.cam.ac.uk*

Figure 9.9. A portion of the SCOP structural classification showing the hierarchy of all α-class, globin, and globin-like proteins.

A. **CATH Search**

Search results for 2reb

Name: Rec a protein (e.c.3.4.99.37)
Source: *(escherichia coli)*

Summary:

for multi-chain proteins, click on any chain to see a more detailed description...

PDB Code	Chain	Status
2reb	-	In CATH

---

2 assigned domains

Domain 1: residues 27 to 269
Goto CATH entry

Class	3	Alpha Beta
Architecture	40	3-Layer(aba) Sandwich
Topology	50	Rossmann fold (Nitrogenase Molybdenum-Iron Protein, subunit A, domain 3)
Homologous superfamily	1200	2reb domain 1

Domain 2: residues 270 to 328
Goto CATH entry

Class	3	Alpha Beta
Architecture	30	2-Layer Sandwich
Topology	250	Rec A Protein, domain 2
Homologous superfamily	10	2reb domain 2

---

cath@biochem.ucl.ac.uk

Figure 9.10. CATH entry for *E. coli* RecA protein (PDB 2reb). (*A*) CATH classification of the protein. (*B*) Ancillary information provided by CATH database including structure, sequence-secondary structure alignment, a structural image, and links to other databases. *Figure continues on next pages.*

---

tein is written above the amino acid sequence of a second protein. Similar or identical amino acids are placed in the same columns and gaps are placed at positions where there is no matching character. In performing structural alignments, the three-dimensional structure of one protein domain is superimposed upon the three-dimensional structure of a second protein domain, fitting together the atoms as closely as possible so that the average deviation between them is minimum. Sequence alignments are performed to discover sequence similarity, and structural alignments are done to discover structural similarity (evidence that the structures share a common fold). New structural relationships are being constantly discovered. Just as a laboratory may discover a remote sequence similarity between two protein domains reflecting a family or superfamily relationship, so may the same laboratory discover a previously unknown structural relationship between two proteins.

There is one important difference between sequence and structural similarity, however. Statistically significant sequence similarity is an indicator of an evolutionary relationship between sequences. In contrast, significant structural similarity is common, even among

B.

☒ View 1   # PDB code: 2reb

☒ READ ME   ☒ RasMol script   ☒ RasMol   ☒   ☒ VRML v.1.0

Self-cleavage stimulation

Structure: *Rec a protein*
Source: *(Escherichia coli)*

Resolution: 2.30Å. **R-factor:** 0.210.

Authors: R.M.Story, T.A.Steitz - **Date:** 06-Mar-92

Further information: PDB header (including references), 3DB Browser and coords, complete MacroMolecule, MMDB entry, CATH and SCOP classifications, FSSP structural alignments, PROCHECK summary, PDBREPORT, PROMOTIF analyses.

Enzyme Classification number: *(from PDB file:* E.C.3.4.99.37 *)*

SWISS-PROT entry: RECA_ECOLI.

Molecule(s) in PDB file 2reb:

☒ . *Protein:* **303** residues

- **CATH classification** :
  Domain 1: **3.40.560.10** -> **Class:** *Alpha Beta.* **Architecture:** *3-Layer(aba) Sandwich.*
  Domain 2: **3.30.250.10** -> **Class:** *Alpha Beta.* **Architecture:** *2-Layer Sandwich.*

  ☒ RasMol domains   Protein coloured by domain.

  ☒ Secondary structure plot

- **PROMOTIF** summary :
  3 sheets, 14 strands, 14 helices, 16 beta turns, 5 gamma turns, 8 beta bulges, 6 beta hairpins, 2 beta alpha beta units.
- **TOPS** protein topology cartoon

  ☒ SAS - annotated FASTA alignment of related sequences in the **PDB**

- **PROSITE** pattern present in this chain:-

  ☒ PROSITE pattern PS00321   **PS00321** - **RECA.**
      Ala214->Arg222: ALKFYASVR

  Help!

- **MolScript** picture (PostScript file)

☒ . **136** water molecules.

Enter new **PDB code**

Figure 9.10. *Continued.*

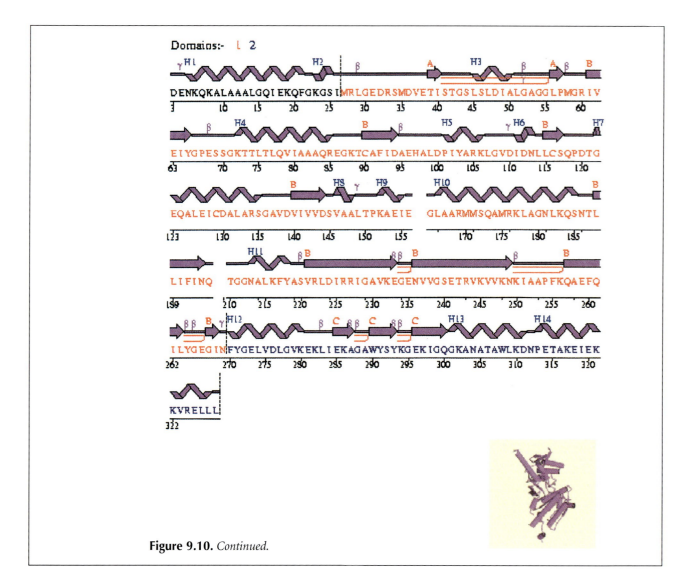

Domains:- 1 2

H1
DENKQKALAAALGQI EKQFGKGS I MRLGEDRSMDVETI STGSLSLDI ALGAGGLPMGR IV
3   10   15   20   25   30   35   40   45   50   55   60

H4   B   β   H5   γ H6   B   H7
EI YGPESSGKTTLTLQVI AAAQREGKTCAFI DAEHALDP I YARKLGVDI DNLLCSQPDTG
63   70   75   80   85   90   95   100   105   110   115   130

B   H8 γ H9   H10   B
EQALEI CDAL ARSGAVDVI VVDSVAALTPKAEI E   GLAARMMSQAMRKLAGNLKQSNTL
123   130   135   140   145   150   155   170   175   180   185

H11   β B   ββ B   β   B
LI FINQ TGGNALKFYASVRLDI RRI GAVKEGENVVGSETRVKVVKNKI AAPFKQAEFQ
189   210   215   220   225   230   235   240   245   250   255   260

ββ B γ H12   β C ββ C ββ C   H13   H14
I LYGEG IN FYGELVDLGVKEKLI EKAGAWYSYKGEKIGQGKANATAWLKDNP ETAKEI EK
262   270   275   280   285   290   295   300   305   310   315   320

KVRELLL
322

Figure 9.10. *Continued.*

Figure 9.11. FSSP report for structures in PDB that are similar to PDB entry 1tupA. 1tupA is the structure of a mutant p53 protein from a cancer cell line complexed with DNA. (A) PDB representative entries that are structurally similar to 1tupA. Shown is a table of other PDB representatives that show structural homology with 1tupA. Ticking boxes in the first column chooses entries for viewing alignments or structures. The columns are the PDB code of the matched representative structure (STRID), a statistical measure of the strength of the structural similarity (Z score), the root-mean-square-deviation between the aligned C_α atoms in Angstroms (RMSD), and the number of aligned C_α atoms in the superimposed structures (LALI), the length of the second protein in the second, matched structure (LSEQ2), percent sequence identity over the aligned positions (%IDE), and the name of the aligned protein. Structures that align with 1tupA with a Z score of at least 2 and representing atomic separation distances that are at least two standard deviations above background values for unrelated sequences are shown. Clicking the choice 3D superimposition with the mouse provides a PDB-like table of atomic coordinates of each separate structure. The overlaid structures can then be viewed as separate chains by a molecular viewing program (not illustrated). Choosing multiple alignment provides a structural alignment of the entries indicating which amino acids occupy similar relative positions in the structures, as shown in B. Choosing multiple families provides a multiple sequence alignment of the protein with similar sequences >25% identical in the protein sequence database, as shown in D.

A. # FSSP: select structural neighbours of 1tupA

Please cite: L. Holm and C. Sander (1996) Science 273(5275):595-60.

Select (check) structural neighbours to display

| 3D superimposition | Multiple alignment | Multiple families | Reset selection |

	STRID2	Z	RMSD	LALI	LSEQ2	%IDE	PROTEIN
☐	1tupA	38.8	0.0	196	196	100	tumor suppressor p53 DNA (5'-d(t
☐	1tsrA	38.5	0.0	196	196	100	p53 tumor suppressor DNA
☐	1ycsA	34.6	0.4	191	191	100	p53 fragment 53bp2 fragment (p53
☐	1tupB	33.9	0.8	194	194	100	tumor suppressor p53 DNA (5'-d(t
☐	1tsrB	33.9	0.8	194	194	100	p53 tumor suppressor DNA
☐	1tsrC	33.6	0.8	194	195	100	p53 tumor suppressor DNA
☐	1tupC	33.6	0.8	194	195	100	tumor suppressor p53 DNA (5'-d(t
☐	1a02N	7.4	3.8	139	280	8	nfat fragment (nf-at) biological
☐	1bglA	6.3	4.7	93	997	9	beta-galactosidase
☐	1a3gA	6.1	3.7	119	285	5	nuclear factor-kappa-b p52 fragm
☐	1rhoA	6.1	3.0	102	145	4	rho gdp-dissociation inhibitor 1
☐	3dpa	5.4	3.3	103	218	6	PapD
☐	1ctn	5.0	2.9	86	538	11	Chitinase a (ph 5.5, 4 degrees c
☐	1mspA	5.0	3.0	93	124	5	major sperm protein (msp)
☐	1mfa	4.8	2.7	88	229	7	Fv fragment (murine se155-4) com
☐	1f13A	4.8	3.9	116	721	8	cellular coagulation factor xiii
☐	1ddt	4.7	3.7	114	523	7	Diphtheria toxin (dimeric)
☐	1eut	4.7	3.6	97	601	15	sialidase (neuraminidase)
☐	1hcz	4.7	3.5	105	250	6	cytochrome f
☐	1xbrA	4.6	3.5	106	184	6	t protein fragment DNA
☐	1cdy	4.6	2.6	78	178	10	t-cell surface glycoprotein cd4
☐	1clc	4.4	3.4	91	541	3	endoglucanase celd (1,4-beta-d-g
☐	1ten	4.4	2.7	83	89	1	Tenascin (third fibronectin type
☐	1fnf	4.4	2.8	83	368	10	fibronectin
☐	1tcrA	4.4	3.5	92	202	7	alpha, beta t-cell receptor (vb8
☐	1neu	4.3	3.0	88	115	5	myelin p0 protein fragment
☐	1vcaA	4.3	3.1	81	199	6	human vascular cell adhesion mol
☐	1lla	4.3	2.8	98	600	7	Hemocyanin (subunit type ii)
☐	1nkr	4.3	2.8	86	195	9	p58-cl42 kir fragment (killer ce
☐	1tf4A	4.2	3.6	106	605	8	t. fusca endoEXO-CELLULASE E4 CA
☐	1tvdA	4.0	3.1	88	116	7	t cell receptor fragment (es204
☐	1ah1	4.0	3.2	92	129	12	ctla-4 fragment (cd152) biologic
☐	1bec	3.9	3.4	99	238	5	14.3.D t cell antigen receptor M
☐	1aohA	3.9	3.3	100	143	11	cellulosome-integrating protein

Continues on next page

☐	2mcm	3.9	3.2	83	112	9	Macromomycin
☐	1oakL	3.8	3.1	90	212	3	nmc-4 igg1 fragment von willebra
☐	1rsy	3.8	3.0	86	135	7	Synaptotagmin i (first c2 domain
☐	1bf5A	3.8	3.4	119	545	7	stat-1 biological_unit DNA
☐	1igtB	3.8	3.8	96	444	5	igg2a intact antibody - mab231
☐	7ahlA	3.8	3.3	110	293	5	alpha-hemolysin (alphatoxin) bio
☐	1cd8	3.8	3.4	89	114	9	Cd8 (t cell c0-receptor, n-termi
☐	1nbcA	3.8	3.5	90	155	8	cellulosomal scaffolding protein
☐	1cto	3.7	3.4	88	109	6	granulocyte colony-stimulating f
☐	1exg	3.7	3.1	88	110	11	Exo-1,4-beta-d-glycanase (cellul
☐	1tit	3.6	2.7	77	89	5	titin, i27 (connectin i27, titin
☐	1gof	3.6	3.0	77	639	8	Galactose oxidase (ph 4.5)
☐	1hnf	3.6	3.3	75	179	7	Cd2 (human)
☐	2ncm	3.6	3.2	84	99	10	neural cell adhesion molecule fr
☐	1ebaA	3.6	3.2	77	212	8	epo receptor fragment (ebp) epo
☐	1edhA	3.6	2.9	82	211	6	e-cadherin (epithelial cadherin
☐	1kb5B	3.6	3.2	86	117	4	kb5-c20 t-cell antigen receptor
☐	1cfb	3.6	3.0	82	205	5	Drosophila neuroglian (chymotryp
☐	1bihA	3.5	3.4	83	391	4	hemolin
☐	1cid	3.5	2.5	67	177	9	Cd4 (domains 3 and 4)
☐	1axiB	3.4	3.4	81	191	4	growth hormone (hgh) Mutant grow
☐	1amx	3.4	3.8	91	150	7	collagen adhesin fragment (cbd19
☐	1xsoA	3.4	3.1	86	150	8	Cu, zn superoxide dismutase
☐	1zxq	3.4	3.2	82	192	7	intercellular adhesion molecule-
☐	1bquA	3.3	4.3	75	208	4	gp130 fragment
☐	1jrhI	3.2	3.0	74	95	7	antibody a6 fragment interferon-
☐	1tlk	3.2	3.1	77	103	11	Telokin
☐	1qba	3.2	3.6	98	858	9	chitobiase (beta-n-acetylhexosam
☐	1pamA	3.1	3.0	69	686	13	cyclodextrin glucanotransferase
☐	1rlw	3.1	3.3	79	126	15	phospholipase a2 fragment (calb
☐	1bdyA	3.1	3.8	81	123	4	protein kinase c fragment (pkc)
☐	1who	3.0	3.1	69	94	6	allergen phl p 2 (phl p ii)
☐	1itbB	3.0	3.6	85	310	5	interleukin-1 beta biological_un
☐	1kcw	3.0	3.7	98	1017	6	ceruloplasmin biological_unit
☐	1bd9A	2.9	3.4	101	180	6	phosphatidylethanolamine binding
☐	1agdB	2.8	3.4	73	99	4	b*0801 fragment (b8) beta-2 micr
☐	1iakA	2.8	3.3	74	182	7	mhc class ii i-ak hen eggwhite l
☐	1plc	2.8	3.7	82	99	5	Plastocyanin (cu2+, ph 6.0)
☐	1acc	2.8	3.1	83	665	5	anthrax protective antigen (pa)
☐	1ksr	2.7	3.3	80	100	8	gelation factor fragment (abp-12
☐	1bhgA	2.7	3.6	92	611	7	beta-glucuronidase (gus gene pro
☐	1nwpA	2.7	3.4	88	128	6	azurin
☐	1djxB	2.7	3.3	82	561	10	phosphoinositide-specific phosph
☐	1aac	2.7	2.7	72	104	7	amicyanin
☐	1kum	2.6	3.6	76	108	9	glucoamylase fragment (1,4-alpha

Figure 9.11. *Continued.*

☐	1aozA	2.6	3.6	65	552	11	Ascorbate oxidase
☐	1aol	2.5	4.0	93	227	10	gp70 fragment (su)
☐	1ar1B	2.5	5.2	89	252	10	cytochrome c oxidase (cytochrome
☐	1ahsA	2.4	4.4	84	126	6	african horse sickness virus (se
☐	1iakB	2.4	3.6	70	185	10	mhc class ii i-ak hen eggwhite l
☐	4kbpA	2.3	3.2	93	424	5	purple acid phosphatase
☐	1preA	2.3	4.4	71	449	11	proaerolysin
☐	1cwpA	2.3	2.9	72	149	14	cowpea chlorotic mottle virus (c
☐	7paz	2.3	3.8	74	123	6	pseudoazurin Mutant biological_u
☐	1dupA	2.2	3.7	86	136	2	deoxyuridine 5'-triphosphate nuc
☐	1cyx	2.2	3.7	85	158	4	cyoa fragment Mutant biological_
☐	1etb1	2.2	3.5	71	118	9	Transthyretin (prealbumin) mutan
☐	8atcB	2.2	3.0	61	146	5	Aspartate carbamoyltransferase (
☐	1lktA	2.1	3.8	73	104	2	tailspike protein fragment (late
☐	1svb	2.1	2.5	62	395	15	tick-borne encephalitis virus gl
☐	1bvp1	2.1	4.2	86	349	7	Bluetongue virus 10 (usa) vp7 (b
☐	1nls	2.1	4.3	91	237	10	concanavalin a biological_unit
☐	1aly	2.1	3.8	86	146	7	cd40 ligand fragment
☐	1rcy	2.1	3.7	81	151	8	rusticyanin biological_unit
☐	1ciy	2.0	4.4	103	577	6	cryia(a)
☐	1bv8	2.0	3.8	93	137	9	alpha-2-macroglobulin fragment
☐	2cbp	2.0	3.4	65	96	6	cucumber basic protein
☐	2tbvA	2.0	3.5	76	283	4	Tomato bushy stunt virus

return to FSSP home page / Dali Domain Dictionary

(C) L. Holm, EMBL-EBI, Hinxton, May 1996

Figure 9.11. *Continued.*   *Continues on next page*

proteins that do not share any sequence similarity or evolutionary relationship. Thus, structural similarity may or may not be an indicator of an evolutionary relationship. Further light may be shed on this question by a close examination of the similarity. The similarity may be quite simple, such as a common arrangement and spacing of several secondary structural elements. Alternatively, there may be a highly significant alignment of many of the proteins through the same sequence of secondary structures and loops, and many of the atoms in the two proteins may be quite superimposable. Such structural closeness may be an indication of a possible evolutionary relationship. The results of a search for remote sequence similarity by sensitive statistical methods (Gibbs sampling, expectation maximization methods, and Bayesian alignment methods discussed in Chapter 4) may be found to provide further support for such a possibility. The ability to make such comparisons has depended on the development and availability of fast and efficient methods for performing structural comparisons.

Structural comparison methods share some of the features of methods for comparing sequences, but with additional considerations. For comparing two sequences, one searches for a row of amino acids in one sequence that matches a row in the second, allowing for substitutions and the insertion of gaps in one sequence to make up for extra characters in the other. For comparing structures, positions of atoms in two three-dimensional structures are compared. These methods initially examine the positions of secondary structural elements, α helices and β strands, within a protein domain to determine whether or not

B. FSSP: structural neighbours of 1tupA

Please cite: L. Holm and C. Sander (1996) Science 273(5275):595-60.

Structural alignment by Dali

Notation: Uppercase: structurally equivalent with 1tupA; lowercase: structurally non-equivalent with 1tupA

```
Identities computed with respect to sequence: (1) 1tupA
Colored by: identity+property


               1 [         .         .         .         .         :         . 60
1 1tupA 100.0%   SSSVPS------------------------------------------------------
2 2tbvA   2.9%   gvtvtshreyltqvnnssgfvvnggivgnslqlnpsngtlfswlpalasnfdqysfnsvv

              61           .         .         .         1         .         . 120
1 1tupA 100.0%   ------------------------------------------------------------
2 2tbvA   2.9%   ldyvplcgttevgrvalyfdkdsqdpepadrvelanfgvlketapwaeamlriptdkvkr

             121           .         .         :         .         .         . 180
1 1tupA 100.0%   ------------------------------------------------------------
2 2tbvA   2.9%   ycndsatvdqklidlgqlgiatyggagadavgelflarsvtlyfpqptntlkrldltgsl

             181           .         2         .         .         .         . 240
1 1tupA 100.0%   QKTYQGSYGFRLGFLHSGTAKSVTCTYSPALNKMFCQLAKTCPVQLWVDSTPPPGTRVRA
2 2tbvA   2.9%   ADATGP---GYLV-----------------ltRTPT---vLTHTFRA-----tgTFNLS

             241           :         .         .         .         .         3 300
1 1tupA 100.0%   MAIYKQSQHMTEVVRRCPHHERCSDSDGLAPPQHLIRVEGNL----RVEYLDDRNTFR-H
2 2tbvA   2.9%   GGL-------------------------rcltSLTLGATgavviNDILAIdnvgtasD

             301           .         .         .         .         .         . 360
1 1tupA 100.0%   SVVVPYEPPEVGSDCTTIHYNYMCNSSCMGGMNRRPILTIITLEDSSGNLLGRNSFEVRV
2 2tbvA   2.9%   YFLNCTVSS----LPATVTFTVSG----------vAAGILLVGRARANvvnll-------

             361           .         ] 375
1 1tupA 100.0%   CACPGRDRRTEEENL
2 2tbvA   2.9%   ---------------
```

mview 1.16 Copyright (c) Nigel P. Brown, EMBL-EBI 1997.

return to FSSP home page / Dali Domain Dictionary

(C) L. Holm, EMBL-EBI, Hinxton, May 1996

Figure 9.11. *Continued.*

(B) Two representative structures that can be aligned with 1tupA with a high level of significance. Amino acid colors reflect side-chain chemistry and use the multiple alignment display program of N.P. Brown (Brown et al. 1998), which can be obtained from the author (see FSSP Web site). The aligned amino acids represent a structural alignment obtained with program DALI, not a sequence alignment. The capitalized amino acids match 1tupA structurally; lowercase amino acids do not match. Note that the percent sequence identity between the p53 sequence of 1tupA and the other two proteins is quite low at 11% for chitinase A (structure 1ctn) and 15% for sialidase-neuraminidase (structure 1eut).

c. FSSP Fold Tree

The FSSP database

The FSSP database includes all protein chains from the Protein Data Bank which are longer than 30 residues. The chains are divided into a **representative set** and **sequence homologs** of structures in the representative set. Sequence homologs have more than 25 % sequence identity, and the representative set contains no pair of such sequence homologs. An all-against-all structure comparison is performed on the representative set. The resulting alignments are reported in the FSSP entries for individual chains. In addition, FSSP entries include the structure alignments of the search structure with its sequence homologs.

Reference

L. Holm and C. Sander (1998) Touring protein fold space with Dali/FSSP. Nucl. Acids Res. 26, 316-319.

Availability

Free academic use. No commercial use. No incorporation into other databases.

This table

is a fold classification of the representative set. A hierarchical clustering method is used to construct a tree based on the structural similarities from the all-against-all comparison. Family indices are constructed by cutting the tree at levels of 2, 3, 4, 5, 10 and 15 standard deviations above database average.

Related tables

PROTEIN INDEX is sorted according to PDB codes. See the accompanying README file for additional information.

Hyperlinks

Click on Family index for a summary of aligned pairs. Click on PDB-code for the complete FSSP entry. Click on alignment to view the structural alignments.

Family index	PDB-code	Alignments	compound
1.1.1.1.1.1	1af5	alignment	"i-crei (DNA endonuclease i-crei) Mutant"
2.1.1.1.1.1	1iba	alignment	"glucose permease fragment"
3.1.1.1.1.1	1aie	alignment	"p53 fragment"
4.1.1.1.1.1	1bba	alignment	"Bovine pancreatic polypeptide (bpp) (NMR, mean struct
4.1.2.1.1.1	__1ppt	alignment	"Avian pancreatic polypeptide"
5.1.1.1.1.1	1emn	alignment	"fibrillin fragment"
6.1.1.1.1.1	1hcgB	alignment	"Blood coagulation factor xa"
7.1.1.1.1.1	1pft	alignment	"tfiib fragment (pftfiibn)"
8.1.1.1.1.1	1ayp	alignment	"RNA polymerase ii fragment"
8.1.1.2.1.1	___1tfi	alignment	"Transcriptional elongation factor sii (tfiis, nucleic
9.1.1.1.1.1	1baf	alignment	"stat-4 fragment"

Figure 9.11. *Continued.*   *Continues on next page*

115.1.1.1.1.1	2plc	alignment	"phosphatidylinositol-specific phospholipase c (pi-plc
115.1.1.2.1.1	___1uroA	alignment	"uroporphyrinogen decarboxylase (uro-d, urod) biologic
115.1.1.2.1.2	_____1a0cA	alignment	"xylose isomerase (glucose isomerase) biological_unit'
115.1.1.2.1.2	_____4xis	alignment	"Xylose isomerase complex with xylose and mnCl2"
115.1.1.2.2.1	____1aq0A	alignment	"1,3-1,4-beta-glucanase (1,3-1,4-beta-d-glucan 4-gluco
115.1.1.2.2.1	_____1balA	alignment	"beta-galactosidase"
115.1.1.2.2.1	_____1bhgA	alignment	"beta-glucuronidase (gus gene product) biological_unit
115.1.1.2.2.1	_____1ceo	alignment	"cellulase celc (1,4-beta-d-glucan-glucanohydrolase, ε
115.1.1.2.2.1	_____1eceA	alignment	"endocellulase e1 fragment (endo-1,4-beta-d-glucanase)
115.1.1.2.2.1	_____1ed9	alignment	"endoglucanase a fragment (endo-(1,4)-beta-glucanase,
115.1.1.2.2.1	_____1qowA	alignment	"beta-glycosidase biological_unit"
115.1.1.2.2.1	_____2myr	alignment	"myrosinase (thioglucoside glucohydrolase) biological_
115.1.1.2.2.2	_____1byb	alignment	"Beta-amylase reacted with 200 mm maltose and complexe
115.1.1.2.2.2	_____1xyzA	alignment	"1,4-beta-d-xylan-xylanohydrolase (endo-1\,4-beta-xylc
115.1.1.2.2.3	_____1qba	alignment	"chitobiase (beta-n-acetylhexosaminidase, n-acetyl-bet
115.1.1.2.2.4	_____1cnv	alignment	"concanavalin b"
115.1.1.2.2.4	_____1ctn	alignment	"Chitinase a (ph 5.5, 4 degrees c)"
115.1.1.2.2.4	_____1nar	alignment	"Narbonin"
115.1.1.2.2.4	_____2ebn	alignment	"Endo-beta-n-acetylglucosaminidase f1 (endoglycosidase
115.1.1.2.3.1	___1onrA	alignment	"transaldolase b"
115.1.1.2.4.1	____1nsj	alignment	"phosphoribosyl anthranilate isomerase (prai)"
115.1.1.2.4.2	____1igs	alignment	"indole-3-glycerolphosphate synthase (igps)"
115.1.1.2.4.2	____1pii	alignment	"N-(5'phosphoribosyl)anthranilate isomerase complex wi
115.1.1.2.4.2	____2tysA	alignment	"tryptophan synthase Mutant biological_unit"
115.1.1.2.4.3	____1aj2	alignment	"dihydropteroate synthase (dhps) biological_unit"
115.1.1.2.4.4	____1aw5	alignment	"5-aminolevulinate dehydratase (porphobilinogen syntho
-			
-			
407.1.1.1.1.1	1hev	alignment	"Hevein (NMR, 6 structures)"
407.1.1.1.2.1	____9waaA	alignment	"Wheat germ agglutinin (isolectin 2)"

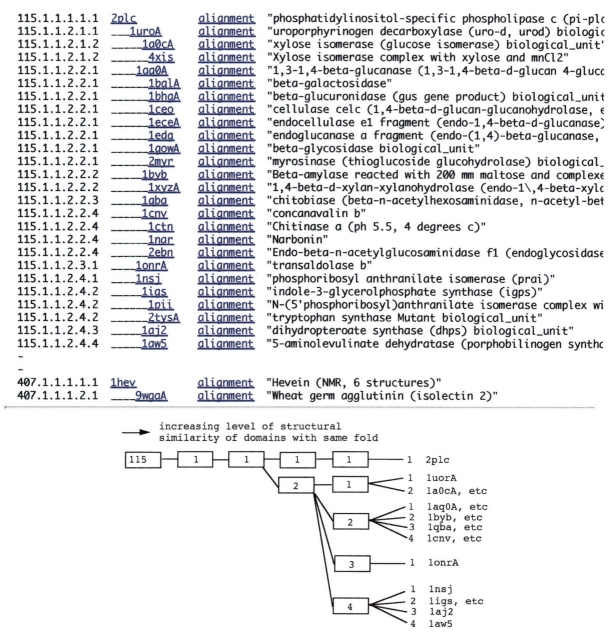

Figure 9.11. *Continued.*

(*C*) Hierarchical clustering of folds and domains. All of the current folds represented by domain alignments in FSSP have been organized into a dendogram that indicates the relationships among them. The dendogram for fold 115 is first illustrated, and a tabular representation is then shown. Domains are identified by the PDB file from which they were derived. If only one of several domains is represented by the fold, the domain is identified by the PDB file name plus a letter code; e.g, 1uorA is a domain of the structure 1uor. Domains that are grouped on the right are the most structurally alike and give a high statistical score for an alignment of the representative fold (a certain combination of secondary structures in space and their connections) when they are aligned with the DALI program. Although these domains have very little sequence similarity, their very close structural similarity suggests that they could possibly be homologous and represent a superfamily. Domains that are joined in deeper branches of the dendogram, e.g., 1uorA and 1aq0A, are less structurally alike, and the score for their alignment is lower. Although domain 1plc has the same fold as the rest of the domains, its atoms align the least well with the other domains. The structures and alignments represented can be viewed by the links on the Web page. The page is accessible from the main page of the FSSP database.

D. FSSP: family alignment around 1tupA

Please cite: L. Holm and C. Sander (1996) Science 273(5275):595-60.

Structures aligned by Dali with sequence neighbours from HSSP

Notation: parent structure[:domain-identifier]|Swissprot-identifier; uppercase: structurally equivalent to 1tupA; lowercase: bounds sequence insertion; - deletion from HSSP; ~ structurally nonequivalent to 1tupA

Identities computed with respect to sequence: (1) 1tupA
Colored by: identity+property

```
                     1 [      .         .         .         .         :         .  60
 1 1tupA        100.0%  SSSVPSQKTYQGSYGFRLGFLHSGTAKSVTCTYSPALNKMFCQLAKTCPVQLWVDSTPPP
 2 1tupA|p53_human 100.0%  SSSVPSQKTYQGSYGFRLGFLHSGTAKSVTCTYSPALNKMFCQLAKTCPVQLWVDSTPPP
 3 1tupA|p53_macmu  97.4%  SSSVPSQKTYHGSYGFRLGFLHSGTAKSVTCTYSPDLNKMFCQLAKTCPVQLWVDSTPPP
 4 1tupA|p53_cerae  97.4%  SSSVPSQKTYHGSYGFRLGFLHSGTAKSVTCTYSPDLNKMFCQLAKTCPVQLWVDSTPPP
 .

44 1tupA|p53_oryla  59.4%  ~TTVPVTTDYPGSYELELRFQKSGTAKSVTSTYSETLNKLYCQLAKTSPIEVRVSKEPPK
45 1tupA|p53_plafe  58.0%  SSTVPVVTDYPGEYGFQLRFQKSGTAKSVTSTFSELLKKLYCQLAKTSPVEVLLSKEPPQ

                    61      .         .         .         .         1         .  120
 1 1tupA        100.0%  GTRVRAMAIYKQSQHMTEVVRRCPHHERCSDSDGLAPPQHLIRVEGNLRVEYLDDRNTFR
 2 1tupA|p53_human 100.0%  GTRVRAMAIYKQSQHMTEVVRRCPHHERCSDSDGLAPPQHLIRVEGNLRVEYLDDRNTFR
 3 1tupA|p53_macmu  97.4%  GSRVRAMAIYKQSQHMTEVVRRCPHHERCSDSDGLAPPQHLIRVEGNLRVEYSDDRNTFR
 4 1tupA|p53_cerae  97.4%  GSRVRAMAIYKQSQHMTEVVRRCPHHERCSDSDGLAPPQHLIRVEGNLRVEYSDDRNTFR
 .

44 1tupA|p53_oryla  59.4%  GAILRATAVYKKTEHVADVVRRCPHHQN---EDSVEHRSHLIRVEGSQLAQYFEDPYTKR
45 1tupA|p53_plafe  58.0%  GAVLRATAVYKKTEHVADVVRRCPHHQT---EDTAEHRSHLIRLEGSQRALYFEDPHTKR

                   121      .         .         .         :         .         .  180
 1 1tupA        100.0%  HSVVVPYEPPEVGSDCTTIHYNYMCNSSCMGGMNRRPILTIITLEDSSGNLLGRNSFEVR
 2 1tupA|p53_human 100.0%  HSVVVPYEPPEVGSDCTTIHYNYMCNSSCMGGMNRRPILTIITLEDSSGNLLGRNSFEVR
 3 1tupA|p53_macmu  97.4%  HSVVVPYEPPEVGSDCTTIHYNYMCNSSCMGGMNRRPILTIITLEDSSGNLLGRNSFEVR
 4 1tupA|p53_cerae  97.4%  HSVVVPYEPPEVGSDCTTIHYNYMCNSSCMGGMNRRPILTIITLEDSSGNLLGRNSFEVR
 .

44 1tupA|p53_oryla  59.4%  QSVTVPYEPPQPGSEMTTILLSYMCNSSCMGGMNRRPILTILTLET-EGLVLGRRCFEVR
45 1tupA|p53_plafe  58.0%  QSVTVPYEPPQLGSETTAILLSFMCNSSCMGGMNRRQILTILTLETPDGLVLGRRCFEVR

                   181      .         ]  196
 1 1tupA        100.0%  VCACPGRDRRTEEENL
 2 1tupA|p53_human 100.0%  VCACPGRDRRTEEENL
 3 1tupA|p53_macmu  97.4%  VCACPGRDRRTEEENF
 4 1tupA|p53_cerae  97.4%  VCACPGRDRRTEEENF
 .

44 1tupA|p53_oryla  59.4%  ICACPGRDRKTEEES~
45 1tupA|p53_plafe  58.0%  VCACPGRDRKTDEES~
```

mview 1.16 Copyright (c) Nigel P. Brown, EMBL-EBI 1997.

return to FSSP home page / Dali Domain Dictionary

(C) L. Holm, EMBL-EBI, Hinxton, May 1996

Figure 9.11. *Continued.*

(*D*) Structural alignment of 1tupA with other protein sequences in SwissProt that are similar in sequence to p53. The information on matching sequences is stored in the HSSP database described in the text. Shown on each row of the alignment are the PDB structure identification, matched SwissProt sequence, percent sequence identity between the sequence of the structural entry and the SwissProt sequence, and the multiple sequence alignment of the sequence with the other matching SwissProt sequences based on a structural alignment. This alignment reveals which amino acid residues in these proteins are predicted to occupy the same structural position. Sequence notations are indicated on the page. Only a portion of the alignment is shown.

⊗ VAST Structure Neighbors

Structures similar to MMDB 2890, 2REB domain 1

Rec A Protein (E.C.3.4.99.37)

View / Save Alignments   ⊗ New   *Get Cn3D 2.0 Now!*

Options: **Viewer:**   **Complexity:**

◉ Launch Viewer   ◉ Cn3D v2.0 (asn.1)   ◉ Aligned Chains only   ◉ Alpha Carbons only
○ See File   ○ Mage (Kinemage)   ○ All Chains   ○ All Atoms
○ Save File   ○ (PDB)

	PDB C D	RMSD	NRES	%Id	Description
☐	1REA 1	0.3	275	100.0	Rec A Protein (E.C.3.4.99.37) Complex With Adenosine Diphosphate (Rec A-Adp)
☐	1SKY E 5	3.1	160	14.4	Crystal Structure Of The Nucleotide Free Alpha3beta3 Sub-Complex Of F1-Atpase From The Thermophilic Bacillus Ps3
☐	1THM	4.1	99	9.1	Thermitase (E.C.3.4.21.66)
☐	1UAA B 5	2.5	93	17.2	Structure Of The Rep Helicase-Single Stranded Dna Complex At 3.0 Angstroms Resolution
☐	1UAG 2	4.5	81	8.6	Udp-N-Acetylmuramoyl-L-Alanine:d-Glutamate Ligase
☐	1POX A 3	3.1	106	4.7	Pyruvate Oxidase (E.C.1.2.3.3) Mutant With Pro 178 Replaced By Ser, Ser 188 Replaced By Asn, And Ala 458 Replaced By Val (P178s,S188n,A458v)
☐	1CL1 A 1	4.2	106	8.5	Cystathionine Beta-Lyase (Cbl) From Escherichia Coli
☐	1A5T 1	2.7	114	14.9	Crystal Structure Of The Delta Prime Subunit Of The Clamp-Loader Complex Of Escherichia Coli Dna Polymerase Iii
☐	1FTS 2	4.3	106	16.0	Signal Recognition Particle Receptor From E. Coli
☐	1GPL 1	3.4	79	8.9	Rp2 Lipase
☐	1AAT 1	3.1	88	9.1	Cytosolic Aspartate Aminotransferase (E.C.2.6.1.1) Complex With 2-Oxo-Glutaric Acid
☐	1RLA B	3.8	97	6.2	Three-Dimensional Structure Of Rat Liver Arginase, The Binuclear Manganese Metalloenzyme Of The Urea Cycle
☐	2TPL A 3	3.8	86	8.1	Tyrosine Phenol-Lyase From Citrobacter Intermedius Complex With 3-(4'-Hydroxyphenyl)propionic Acid, Pyridoxal-5'-Phosphate And Cs+ Ion
☐	2DRI 1	3.2	72	12.5	D-Ribose-Binding Protein Complexed With Beta-D-Ribose
☐	8ABP 1	3.1	72	11.1	L-Arabinose-Binding Protein (Mutant With Met 108 Replaced By Leu) (M108L) Complex With D-Galactose
☐	1CYD D	4.4	97	8.3	Carbonyl Reductase Complexed With Nadph And 2-Propanol
☐	1N2C E	2.8	80	16.2	Nitrogenase Complex From Azotobacter Vinelandii Stabilized By

Figure 9.12. Example of searching for structural neighbors identified by the VAST algorithm. Shown is the result of a search for neighbors to chain 1 of the *E. coli* RecA protein structure (PDB identifier 2reb). If the rightmost column box next to any of the listed structural neighbors and the view/save structures box are sequentially checked, then an overlay view of the structures is provided by ENTREZ for viewing by Cn3d or Mage. In the output table of structural neighbors, PDB is a four-character PDB-identifier of the structural neighbor, C is the PDB chain name, D is the MMDB domain identifier, RMSD is the root mean square deviation in Angstroms between the superimposed atoms, NRES is the number of equivalent pairs of C_α atoms super-

					Adp-Tetrafluoroaluminate
☐	1NGS A 2	2.7	87	12.6	Complex Of Transketolase With Thiamin Diphosphate, Ca2+ And Acceptor Substrate Erythrose-4-Phosphate
☐	1GDH A 1	3.0	65	7.7	D-Glycerate Dehydrogenase (Apo Form) (E.C.1.1.1.29)
☐	1A4S A 2	3.0	70	11.4	Betaine Aldehyde Dehydrogenase From Cod Liver
☐	1AK1 2	3.2	70	5.7	Ferrochelatase From Bacillus Subtilis
☐	1ZIN 1	2.9	80	13.8	Adenylate Kinase With Bound Ap5a
☐	1MIO D 13	3.0	59	5.1	Nitrogenase Molybdenum-Iron Protein
☐	1RRF	3.5	52	9.6	Non-Myristoylated Rat Adp-Ribosylation Factor-1 Complexed With Gdp, Monomeric Crystal Form
☐	1RVV 1	2.1	63	14.3	SynthaseRIBOFLAVIN SYNTHASE COMPLEX OF BACILLUS SUBTILIS
☐	1CEY	2.5	58	10.3	Chey Complexed With Magnesium (Nmr, 46 Structures)
☐	1DEK A 1	2.7	77	13.0	Deoxynucleoside Monophosphate Kinase Complexed With Deoxy-Gmp
☐	1MIO D 11	3.0	70	4.3	Nitrogenase Molybdenum-Iron Protein
☐	1RAA C 5	2.1	45	17.8	Aspartate Transcarbamoylase (E.C.2.1.3.2) (Aspartate Carbamoyltransferase) (T State) Complexed With Ctp (Fast Cooling Sa Refinement After 250 Steps Of Equilibration Of Preliminary Refined Model)
☐	1BNC A 1	2.8	54	11.1	Mol_id: 1; Molecule: Biotin Carboxylase; Chain: A, B; Ec: 6.3.4.14
☐	1UAG 3	3.0	62	14.5	Udp-N-Acetylmuramoyl-L-Alanine:d-Glutamate Ligase
☐	1MIO C 9	2.3	52	0.0	Nitrogenase Molybdenum-Iron Protein
☐	1IIB A	2.2	41	14.6	Crystal Structure Of Iibcellobiose From Escherichia Coli
☐	1IOW 1	2.0	36	5.6	Complex Of Y216f D-Ala:d-Ala Ligase With Adp And A Phosphoryl Phosphinate
☐	1BPL A 1	2.5	60	6.7	Glycosyltransferase
☐	1XAN 3	1.7	30	13.3	Human Glutathione Reductase In Complex With A Xanthene Inhibitor

Display / Sort Hits

Display Subset: **Sorted by:**   **Column Format:**

◉ Non-redundant; BLAST p-value 10e-7   ◉ VAST Score   ◉ RMSD, NRES, %Id

◯ Non-redundant; BLAST p-value 10e-40   ◯ VAST P-value   ◯ All values

◯ Non-redundant; BLAST p-value 10e-80   ◯ Rmsd

◯ Non-identical sequences   ◯ Aligned residues

◯ All of MMDB   ◯ Identities

imposed between the two structures, %Id is the percent identical residues in the aligned sequence region, and description of the neighbor is taken from the PDB database entry. The options in the lower part of the page influence the number of matches reported in the table, and this number may be varied by mouse-clicking the "display subset" box. The MMDB database is organized into groups based on sequence similarity, and only a representative member of each group is included. Groups are based on extensive BLAST searches for sequence similarity followed by clustering by a neighbor-joining procedure (see Chapter 6). Several different levels of clustering based on different ranges of BLAST scores are shown. The lower the score chosen, the more group members are reported. Note that the format of a structural entry in the MMDB database is different from that in the PDB database and requires visualization by the Cn3d viewer.

the number, type, and relative positions of these elements are similar or if the proteins have a similar architecture. Distances between the C_α or C_β atoms within these structures are then examined in detail to determine the degree to which the structures may be superimposed. If a few elements can be aligned and are joined by a similar arrangement of loops, the proteins share a common fold. As the arrangement, joining, and alignment of secondary structural elements within the proteins increase, the degree of structural similarity between the proteins becomes more and more convincing and significant.

To specify a three-dimensional structure, positions of molecules are expressed as x, y, and z cartesian coordinates within a fixed frame of reference, as shown in Figure 9.13. The direction of the bond angles and the interatomic distances between amino acids along the polypeptide chain may also be represented as a vector. Secondary structures can also be represented by a vector that starts at the beginning of the secondary structural element, extends for the length of the element, and has a direction that reveals the orientation of the element in the overall structure. Comparison of these structural representations in two proteins provides a framework for comparing the structures of the proteins. In many structural comparison methods, distances between C_α or between C_β atoms in two protein structures are used for comparison purposes. A more detailed comparison of the structures can be made by adding information on side chains such as the amount of outside area of the side chain

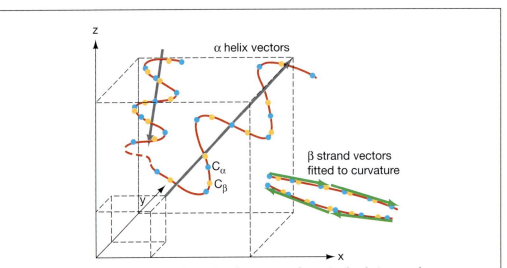

Figure 9.13. Alignment of the three-dimensional structure of proteins by their secondary structures. Representation of arrangement of secondary structures in three-dimensional space is shown on a two-dimensional projection. In the structural alignment programs VAST and SARF, the atoms of each secondary structural element in each protein are replaced by a vector of position, length, and direction determined by the positions of the C_α or C_β atoms along the element. Shown are projections of two α helices and two β strands and their vector representations as gray and green arrows, respectively, from a common x, y, z cartesian coordinate system. The three-dimensional cartesian coordinates of the start and end of one α-helical vector are diagrammed as wide dashed lines. Only these two sets of coordinates are needed to specify the location of the vector, whereas many such sets are required to locate the C_α or C_β atoms in the corresponding α helix. An element that is curved is approximated by two or more sequential vectors, as depicted for the two β strands, which are bent due to the twist of their composite β sheet. The joining of the helices by a short loop is also recognized by the algorithm. The vector representations of two proteins are then compared. If the type and arrangement of the elements are similar in two proteins within a reasonable margin of error and level of significance, the three-dimensional structures of the proteins are predicted to be similar.

that is buried under other molecules so that the chain is not accessible to water molecules. Distances and bond angles to other atoms in the structure may also be compared. Several of the parameters used for structural comparisons may also be used to classify the environment of a particular amino acid, e.g., a buried, hydrophobic amino acid in a β strand.

There are two reasons that it is more difficult to align structures than sequences. First, a similar structure may form by many different foldings of the amino acid C_α backbone. As a result, matched regions may not necessarily be in the same order in the two proteins so that two matching segments are often separated by unmatched segments. Second, although the local environments of many molecules in two proteins may be similar, there may also be some local differences. For example, central positions, but not the ends, of secondary structures in two proteins may match closely. For this reason, structural alignment methods often smooth out the comparisons by comparing several molecules at the same time and choosing an average result.

Structural biologists have been working on the problem of finding similar structural features in proteins for a long time, and a variety of methods have been devised for performing comparisons of protein structures (for review, see Blundell and Johnson 1993; Holm and Sander 1994, 1996; Alexandrov and Fischer 1996; Gibrat et al. 1996; Orengo and Taylor 1996). A complete discussion of this subject is beyond the scope of this text. Programs publicly accessible at Web sites, SSAP and DALI, and two programs that utilize a fast search for common arrangements of secondary structures, VAST and SARF, are described below.

Dynamic Programming

Algorithms like those used for sequence alignment have also been used for aligning structures. For aligning sequences, the object is to bring as many identical or similar sequence characters into vertical register in the alignment with a minimum cost of insertions and deletions. For aligning structures, the local environment of each amino acid expressed in interatomic distances, bond angles, or R group is given a coded value or vector representation that reflects the environment of that amino acid. Alternatively, a scoring matrix much like the amino acid scoring used for sequence alignments may be made. For protein structures, each sequential column in the scoring matrix gives a score for the fit of any of the 20 amino acids to a single position in the structure (more on matrices below). An optimal alignment between these sets of values by dynamic programming is then found.

The alignment program SSAP (*s*econdary *s*tructure *a*lignment *p*rogram) uses a method called double dynamic programming to produce a structural alignment between two proteins (Taylor and Orengo 1989; Orengo et al. 1993; Orengo and Taylor 1996). A local structural environment is independently defined for each residue in each sequence, and the method then matches residues by comparing these structural environments. The environment assigned to each amino acid takes into account the degree of burial in the hydrophobic core and type of secondary structure. As in sequence alignment by dynamic programming, a scoring matrix is derived and the highest-scoring regions in this matrix define the optimal structural alignment of the two proteins. One of the environmental variables that is used is a representation of the geometry of the protein by drawing a series of vectors from the C_β atoms of an amino acid to the C_β atoms of all of the other amino acids in the protein. If the resulting geometric views in two protein structures are similar, the structures must also be similar. The double dynamic programming method of aligning structures using C_β vectors is illustrated in Figure 9.14.

Because each sequential pair of amino acids is compared, an alignment will be possible only if the two protein chains follow the same approximate conformational changes throughout their lengths. If the proteins follow the same changes along some of their lengths, then diverge, then return again, it is difficult to align them through the divergent region by the above method, as described. The problem is similar to trying to choose a gap penalty for sequence alignments, but in the structural case, many kinds of rearrangements are possible.

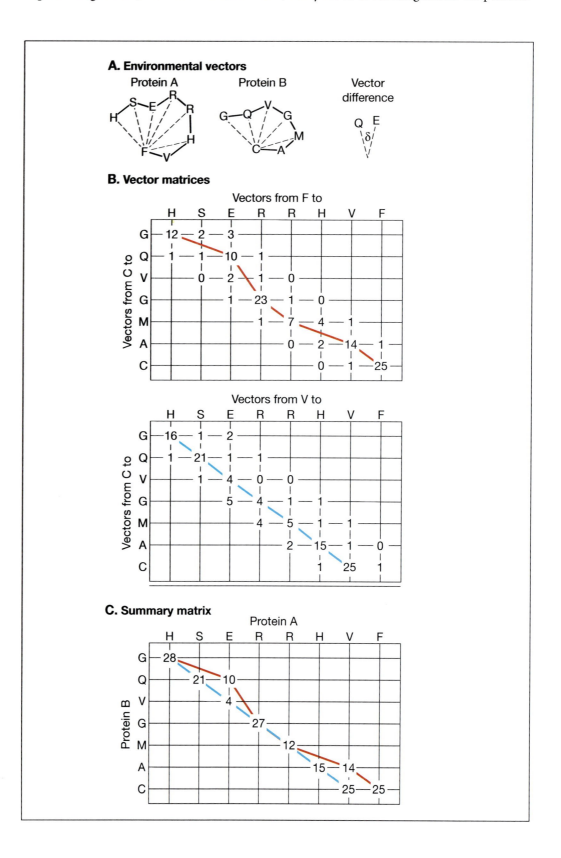

Another version of SSAP (SSAP1) has been developed for identifying conserved folds/motifs, and this method circumvents the above alignment problem. This program uses all of the vector matrix values in the summary matrix and then uses a local alignment version of the dynamic programming algorithm to locate the most alike regions in the structures. The algorithm has been greatly speeded up by comparing only pairs of amino acids with similar torsional angles (Φ and Ψ) and extent of residue burial/lack of water accessibility. SSAP is used to cluster proteins in the CATH database in a fully automated manner (Orengo et al. 1997).

Distance Matrix

The distance method uses a graphic procedure very similar to a dot matrix to identify the atoms that lie most closely together in the three-dimensional structure. If two proteins have a similar structure, the graphs of these stuctures will be superimposable. Distances between C_α atoms along the polypeptide chain and between C_α atoms within the protein structure can be compared by a two-dimensional matrix representation of the structure, as shown in Figure 9.15. Instead of aligning environmental variables of each successive amino acid in

◄──

Figure 9.14. The double dynamic programming method for structural alignment. (*A*) Vectors from the C_β atom of one amino acid to a set of other nearby amino acids in each of two protein segments are shown as two-dimensional projections. These vectors are given the same coordinate axes. Hence, one vector may be subtracted from the other to compare the relative positions of the C_β atoms in the two protein segments, shown in *A* as a vector difference. The smaller the differences, the more alike the structures. In SSAP, the vectors are subtracted (the resulting difference is δ) and the difference added to an empirically derived number, 10. The resulting value is then divided into a second empirically derived number, 500, to give a score S for the vector difference. For example, if the vector difference is 10°, then $S = 500 / (10 + 10) = 25$. (*B*) Two vector matrices that represent differences between the geometric view from one amino acid position in one protein and a view for one amino acid in the second protein. The set of vectors of one protein are listed across the top of the matrix and the set for the other are listed down the right side. The matrix is then filled with scores of vector differences. For example, if the vector from F to H in protein 1 less the vector from C to G in protein 2 is 31°, then the score placed in the upper right corner is $S = 500/(31 + 10) = 12$. The remaining difference scores are calculated in a similar manner. Although vectors to neighboring amino acids are shown in this example, vectors to immediate neighbor positions are actually not used to reduce effect of local secondary structure. An optimal alignment, shown as a red path through the matrix, is then found through the vector matrix by a global form of the dynamic programming algorithm, using a constant deletion penalty of 50. For performing a structural alignment by this method, a similar set of vector differences are determined between the next amino acid V in protein A and the amino acid in protein B, as shown in the lower matrix in *B*, and an optimal path (*blue*) is obtained. This procedure is repeated until vector views between all amino acid positions have been compared. Two vector matrices are shown, comparing one position in protein A to each of two positions in protein B. (*C*) The resulting alignments (shown as red and blue paths) and the scores on the alignment path are transferred to a summary matrix. If two optimal alignment paths cross the same matrix position, the scores of those positions in the two alignments are summed. One part of the alignment path (*black*) is found in both comparisons, thereby providing corroborative evidence of vector similarity in these regions. In the example shown, the sum of the upper right positions in the two vector matrices is 12 + 16 = 28. When all of the alignments have been placed into the summary matrix, a second dynamic programming alignment is performed through this matrix. The final alignment found represents the optimal alignment between the protein structures. The logarithm of the final score is scaled such that a maximum value of 100 is possible. An adjusted score of 80 indicates a close structural relationship; one of 60–70 indicates a probable common fold. Other types of environmental variables other than the position of the C_β atoms in this example may also be aligned with this double dynamic programming method, as described in the text. (Adapted from an example in Orengo and Taylor 1996.)

two protein structures, the distance matrix method compares geometric relationships between the structures without regard to alignment. The sequence of the protein is listed both across the top and down the side of the matrix. Each matrix position represents the distance between the corresponding C_α atoms in the three-dimensional structure. The smallest distances represent the more closely packed atoms within secondary structures and regions of tertiary structure. Positions of closest packing are marked with a dot to highlight them, much as in a dot matrix. Distance matrices are produced for each three-dimensional structure of interest. Similar groups of secondary structural elements are superimposed as closely as possible into a common core structure by minimizing the sum of the atomic distances between the aligned C_α atoms. The method is outlined in Figure 9.15.

The program DALI (*distance alignment* tool) uses this method to align protein structures (Vriend and Sander 1991). The existing structures have been exhaustively compared to each other by DALI and the results organized into a database, the FSSP database, which may be accessed at http://www2.embl-ebi.ac.uk/dali/fssp/fssp.html. A newly found structure may be compared to the existing database of protein structures using DALI at http://www2.embl-ebi.ac.uk/dali/. The network version of DALI uses fast comparison methods to determine whether a new structure is similar to one already present in the FSSP database.

Alignment in DALI

The assembly step of the original DALI algorithm uses a Monte Carlo simulation that performs a random search strategy for submatrices that can be aligned using the similarity score defined below as a guide. The algorithm is similar to the genetic and simulated annealing algorithms (Chapter 4) in using a probabilistic method to improve previously found alignments. There is no existing algorithm for direct alignment of two structures; such an algorithm would have to find the closest alignment of two sets of points in three-dimensional space, a very difficult problem computationally. Hence, the need for an approximate solution. Other methods for aligning structures that are described below also use simulations to find alignments. The Internet version of the DALI program utilizes more rapid search methods than those described above to compare new structures to existing structures in the FSSP database, but the overall analysis is very similar.

The similarity score for a structural alignment of two proteins by the distance method is based on the degree to which all of the matched elements can be superimposed. In the example shown in Figure 9.16, the score for a matching set of helices is the sum of the similarity scores of all of the atom pairs using a particular scheme for scoring each pair. Suppose that two helices a and b have been found to interact in protein A, and that a pair of helices a' and b' in protein B are superimposable on a and b. A certain pair of C_α atoms that are very close in the model, one in helix a (i^A) and a second in helix b (j^A), is identified. This set will correspond to a matched pair i^B in helix a' and j^B in helix b' of protein B. If the distance between i^A and j^A is d_{ijA} and the distance between i^B and j^B is d_{ijB}, then the similarity score for this pair of atoms is derived from the fractional deviation $|d_{ijA} - d_{ijB}| / d_{ij^*}$, where d_{ij^*} is the average of d_{ijA} and d_{ijB}. If two atom pairs can be superimposed, they are given a threshold similarity score of 0.20; otherwise they are given a similarity score of the threshold less the above fractional deviation. A deviation of 0.20 will correspond to adjacent β strands matching to within 1 Å and to α helices and helix strands matching to within 2–3 Å. As these scores are summed over all of the atoms in the matching helices, the contributions of more distant atoms are down-weighted by an exponential factor to allow for bending and other distortions. The result of using this scoring system is that the similarity score for matching the two helix pairs in proteins A and B will increase in proportion to the number of superimposable atoms in the two helices. As additional matching elements are added to the structural alignment of the two proteins, the similarity scores for matching each individual pair of secondary structures are added to give a higher similarity score that reflects the full alignment of the structures.

The DALI method provides one convenient method, in addition to the others described herein, to compare a new structure to existing structures in the Brookhaven structural database, and is accessible from a Web site.

Fast Structural Similarity Search Based on Secondary Structure Analysis

One class of structural alignment methods performs a comparison of the types and arrangements of α helices and β strands in one protein structure with the α helices and β strands in a second structure, as well as the ways in which these elements are connected (for review, see Gibrat et al. 1996). If the elements in two structures are similarly arranged, the corresponding three-dimensional structures are also similar. Because there are relatively few secondary structural elements in proteins and the relative positions of these elements may be quite adequately described by vectors giving their position, direction, and length, vector methods provide a fast and reliable way to align structures. It is a much simpler computational problem to compare vector representations of secondary structures than to compare the positions of all of the C_α or C_β atoms in those structures. If an element of a given type and orientation within a given tolerance level is found in the same relative position in both structures, they possess a basic level of structural similarity. Elements that do not match within the tolerance level are not considered to be structurally similar. VAST and SARF are examples of programs that are available on the Web that use this methodology (Hogue et al. 1996; Alexandrov and Fischer 1996; see http://www.ncbi.nlm.nih.gov/ Entrez and http://www-lmmb.mcifcrf.gov/~nicka/sarf2.html/). Vector methods do not use the structure authors' assigned secondary structures in the PDB entry, but rather use automatic methods to assign secondary structure based on the molecular coordinates of atoms on the structure. Different methods are used for defining the number and extent of secondary structural elements and for the thresholds that make up an acceptable match (Bryant and Lawrence 1993; Madej et al. 1995; Gibrat et al. 1996; Alexandrov and Fischer 1996). Until one of these methods is shown to be superior, it is advisable to try all to increase the chance of a finding a biologically important match.

Once individually aligning sets of secondary structural elements have been identified, they are clustered into larger alignment groups. For example, if three matching sets of α helices have been found in two structures, a similarly oriented group of three α helices must be present in the structures. The same arrangements of a small number of secondary structural elements are commonly found in protein structures, thus this method often finds new occurrences of a previously found arrangement. An arrangement with a large number of secondary elements is less common and therefore more significant. This clustering step generates a large number of possible groups of secondary structural elements from which the most likely ones must be selected. Some methods use the clusters with the largest number of secondary structures as the most significant. Other methods perform a more detailed analysis of the aligned secondary structures. For example, the atomic coordinates of an α helix in one protein structure will be aligned with those of the matched α helix in the second structure, and the root mean square deviation (rmsd) will be calculated. The quality of this new alignment provides an indication of which secondary structure

A.

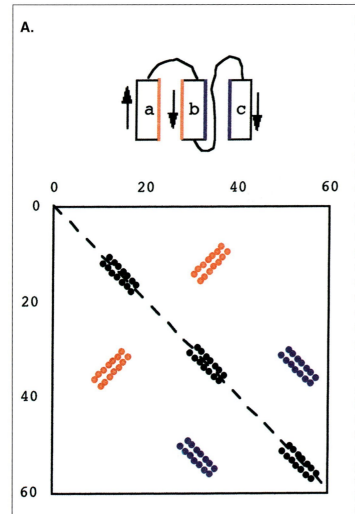

Figure 9.15. Distance matrix of hypothetical three-helix structure. (*A*) Matrix positions that represent closest distances of approximately <12 Å between the C_α atoms in the known three-dimensional structures of the protein are marked by filling them with dots. Positions marked with black dots drawn just above the main downward-pointing diagonal (*dashed line*) from upper left to lower right represent amino acid sequential positions aa1-aa2, aa2-aa3, etc., that are close to each other because they are in the α helix. Marked regions of shortest C_α–C_α distances along this diagonal thus indicate positions of the α helices. Other marked diagonal regions (*red and blue dots*) indicate tertiary structural interactions, including those between adjacent secondary structural elements. Helices a and b are close to each other and have opposite chemical polarities so that aa10-aa11-aa12 . . . are close to aa40-aa39-aa38 . . . . on the red surface of the helices. An upward-running diagonal (*red dots*) from lower left to upper right reveals this spatial relationship. Helices b and c are also close to each other but have the same polarity so that aa30-aa31-aa32 . . . are close to aa50-aa51-aa52 . . . , producing a downward-directed diagonal (*blue dots*). If another protein has a matrix pattern similar to that of the above example, then the two protein structures have the same three-helical arrangement and the loops joining the helices are of approximately the same length and conformation. The distance alignment method will find such three-helix patterns, even when the loop patterns are not similar. (*B*) Search for a common structural pattern in proteins A and B by DALI. A hypothetical example of a three-helix architecture is again used. In the top row, DALI first searches the entire distance matrix of protein A for a set of matching helices, a and b, indicated by an upward-directed diagonal whose position is the intersection of the locations of the helices in the sequence of protein A (*left column*). A similar search is performed for a corresponding pair of helices a′ and b′ in the distance matrix of protein B. In practice, the algorithm breaks down each full-sized matrix into a set of overlapping submatrices of size 6 × 6 amino acids. Distance patterns within the submatrices from each protein are then compared to locate similar structural configurations. Some matches will be longer than 6 amino acids and will therefore be found in several neighboring submatrices. A computationally sophisticated assembly step in the algorithm (see below) combines these overlaps into a complete structural alignment. Once found, individual matches are assembled. If a pair of helices is found in each structure, a beginning structural alignment of the sequences may be made (*right column*). A search for a third pair of helices c and c′ that interact with helices b and b′ in proteins A and B, respectively, is then made, as illustrated in the second row. A hypothetical pair common to A and B is shown. In this case, the order of regions b′ and c′ on the sequence of protein B is reversed from that of b and c. The composite matrices and alignment of all helices a, b, c and a′, b′, c′ are shown in the third row. Only the top one-half of the matrix is shown, leaving out the mirror image. Finally, DALI removes the insertions and deletions in the matrices and rearranges the sequence of the protein B to produce a parallel alignment of the elements in the two sequences (*bottom row*). By following these steps, an alignment of helices a, b, and c and a′, b′, and c′ in structures A and B is found by DALI, but the arrangements of sequences that produce this common architecture are different. Structural features that include β strands in proteins are found in the same manner. (Diagram derived from Holm and Sander 1993, 1996.)

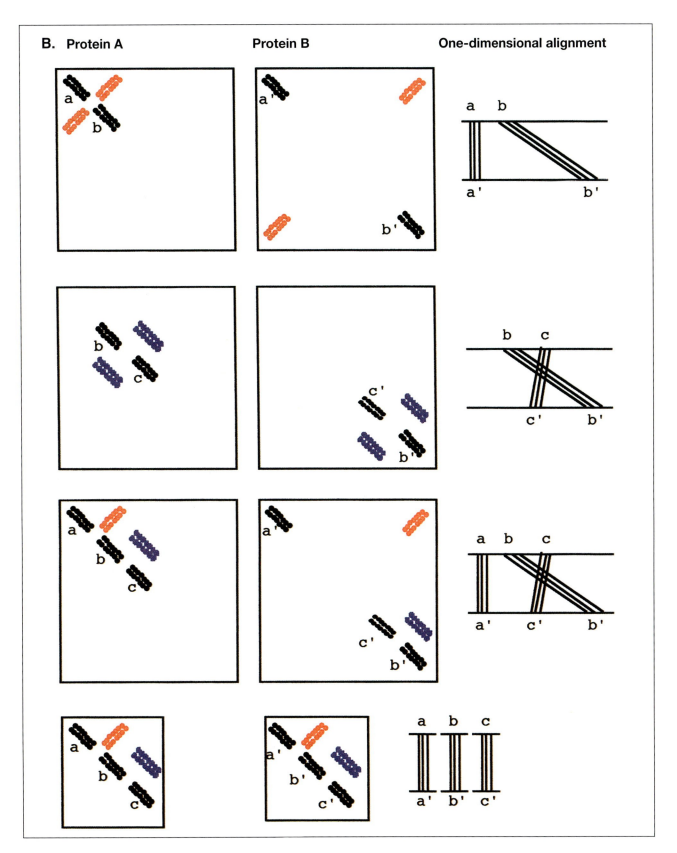

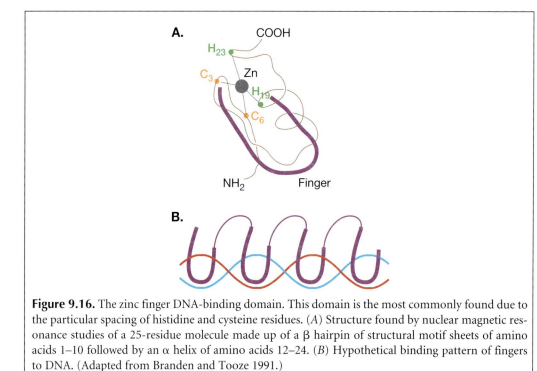

Figure 9.16. The zinc finger DNA-binding domain. This domain is the most commonly found due to the particular spacing of histidine and cysteine residues. (*A*) Structure found by nuclear magnetic resonance studies of a 25-residue molecule made up of a β hairpin of structural motif sheets of amino acids 1–10 followed by an α helix of amino acids 12–24. (*B*) Hypothetical binding pattern of fingers to DNA. (Adapted from Branden and Tooze 1991.)

clusters are the most feasible. After starting with an alignment that includes the highest matching number of elements, the VAST algorithm examines alternative alignments that might increase the alignment score using the Gibbs sampling algorithm described in Chapter 4.

Like other structural alignment methods, VAST and SARF are available on Web pages and may be used for comparing new structures to the existing databases or for viewing structural similarities within the existing databases. An important aspect of searches for structural similarity by the vector method and other methods is the extent of the alignment found, or as Gibrat et al. (1996) state, is the alignment "surprising"?

Significance of Alignments of Secondary Structure

As in sequence alignment, it is important to estimate the reliability or statistical significance of a structural alignment. The problem is to determine the probability with which a given cluster of secondary structural elements would be expected between unrelated structures. The analogous problem with sequences is to determine whether or not an alignment score between two test sequences would also be found between random or unrelated sequences.

When comparing the arrangement of secondary elements in protein structures, a very large number of possible alignments are commonly found (Gibrat et al. 1996). The probability of a chance alignment of a few elements in two large but structurally unrelated proteins that have many such elements is quite high. Therefore, alignment of only a few elements in an actual comparison of two test sequences is not particularly significant. The probability of an alignment between most of the elements in large, unrelated proteins, however, is extremely low. Hence, such an alignment between structures is highly significant. The problem of significance thus boils down to assessing the number of possible ways of aligning elements in two unrelated proteins.

For calculating the probability of an alignment, the VAST algorithm uses a statistical theory very similar to that of the BLAST algorithm to calculate this probability. Recall that BLAST calculates a probability (or expect value) that a sequence alignment score at least as high as that found between a test sequence and a database sequence would also be found by alignment of random sequences. Sequence alignment scores are derived by using amino acid substitution matrices and suitable alignment gap penalties, and the probabilities that alignments of random sequences could score as high as actual scores are calculated using the extreme value distribution. The equivalent VAST score is the number of superimposed secondary structural elements found in comparing two structures. The greater the number of elements that can be aligned, the more believable and significant the alignment. The statistical significance of a score is the likelihood that such a score would be seen by chance alignment of unrelated structures. This likelihood is calculated from the product of two numbers—the probability that such a score would be found by picking elements randomly from each protein domain and the number of alternative element pair combinations. Thus, if the chance of picking the number of matching elements found is 10^{-8} and the number of combinations is 10^4, the likelihood of an alignment of the same number of elements between unrelated structures is $10^{-8} \times 10^4 = 10^{-4}$.

Displaying Protein Structural Alignments

The programs and Web sites that perform a structural alignment or that provide access to databases of similar structures will transmit coordinates of the matched regions. The aligned regions may then be viewed with a number of molecular viewing programs, including Rasmol, Cn3d, and Spdbv. Cn3d also shows a second window with the matching sequence alignment, and aligned structures may be highlighted starting from this window. The program JOY provides a method for annotating sequence alignments with three-dimensional structural information (http://www-cryst.bioc.cam.ac.uk/~joy; Mizuguchi et al. 1998b).

STRUCTURAL PREDICTION

Use of Sequence Patterns for Protein Structure Prediction

Although the sequences of 86,000 proteins are available, the structures of only 12,500 of these proteins are known. The increasing rate of genome sequencing can also be expected to outpace the rate of solving protein structures. Protein structural comparisons described above have shown that newly found protein structures often have a similar structural fold or architecture to an already-known structure. Thus, many of the ways that proteins fold into a three-dimensional structure may already be known. Structural comparisons have also revealed that many different amino acid sequences in proteins can adopt the same structural fold, and these sequences have been organized into databases described above. Further examination of sequences in structures has also revealed that the same short amino acid patterns may be found in different structural contexts. Amino acid sequences present in secondary structures have been entered into databases that are useful for structure prediction. Many proteins in the sequence databases also have conserved sequence patterns upon which they may be further categorized.

If two proteins share significant sequence similarity, they should also have similar three-dimensional structures. The similarity may be present throughout the sequence lengths or in one or more localized regions having relatively short patterns that may or may not be interrupted with gaps. When a global sequence alignment is performed, if more than 45% of the amino acid positions are identical, the amino acids should be quite superimposable in the three-dimensional structure of the proteins. Thus, if the structure of one of the aligned proteins is known, the structure of the second protein and the positions of the identical amino acids in this structure may be reliably predicted. If less than 45% but more than 25% of the amino acids are identical, the structures are likely to be similar, but with more variation at the lower identity levels at the corresponding three-dimensional positions.

Protein Classification Schemes

Proteins have been classified on the basis of sequence similarity or the presence of common amino acid patterns. First, they have been organized into families and superfamilies on the basis of the level of sequence similarity in sequence alignments. The current method of organizing proteins by this method at the Protein Information Resource (PIR) (http://www-nbrf.georgetown.edu) is that each entry in the PIR protein sequence database is searched against the remaining entries using the FASTA algorithm. Similar sequences are then aligned with the Genetics Computer Group multiple sequence alignment program PILEUP. This level of comparison based on sequence alignment was originally made by the PIR founded by M. Dayhoff. Using present-day classification schemes (Barker et al. 1996), families are composed of proteins that align along their entire lengths with a level of sequence identity of usually 50% or better.

More recent analyses of amino acid patterns in protein sequences have revealed that many proteins are made up of modules, short regions of similar amino acid sequence that correspond to a particular function or structure. Furthermore, sets of proteins from widely divergent biological sources may share several such modules and the modules may not be in the same order. Hence, it has become necessary to redefine the concepts of family and superfamily. Proteins that comprise the same set of similar homology domains (extended regions of sequence similarity) in the same order are referred to as homeomorphic protein families. Protein families, members of which have the same domains in the same order, but also have dissimilar regions, are designated as a homeomorphic superfamily (Barker et al. 1996). The superfamily classification of a newly identified protein sequence may be analyzed at several Web sites (Table 9.5).

The second method of classifying proteins is based on the presence of amino acid patterns. Proteins with the same biochemical function have been examined for the presence of strongly conserved amino acid patterns that represent an active site or other important feature. The resulting database is known as the Prosite catalog (A. Bairoch and colleagues; Hofmann et al. 1999) (Table 9.5). Proteins have also been categorized on the basis of the occurrence of common amino acid patterns—motifs and conserved gapped and ungapped regions in multiple sequence alignments. These patterns are found by extracting them from multiple sequence alignments, by pattern-finding algorithms that search unaligned sequences for common patterns, and by several statistical methods that search through unaligned sequences. The patterns vary in length, presence of gaps, and degree of substitution. The algorithms that are used include pattern-finding methods, hidden Markov models, the expectation maximization method, and the Gibbs sampling method. These methods and the computer programs and Web sites that provide them are described in Chapter 4. Listed in Table 9.5 are several databases that categorize proteins based on the occurrence of common patterns. Also shown are databases of amino acid patterns that determine cellular localization of proteins or sites of protein modification (signal or transit peptides). FSSP, a structural family database, is listed in this table because it includes links to information on sequence families and superfamilies.

A given protein sequence may be classified by using one of the resources in Table 9.5 for sequence patterns that are characteristic of a group or family of proteins. Because most of these databases are derived by quite different methods of pattern analysis, statistics, and database similarity searching, they can be expected to provide complementary information. Thus, a given database may include a sequence pattern that is not identified in others, and this pattern may provide an important link to structure or function for one group of proteins. Another database may provide patterns more suitable for classifying a different group of proteins. Therefore, a wise choice would be to use as many of these resources as possible for classifying a new sequence. However, note the availability of Web sites that have combined the resources of separate protein classification databases into a single database (e.g., INTERPRO; Table 9.5). In one new field of endeavor, protein taxonomy, genomic databases that list the entire set of proteins produced by a particular organism are searched for matches. Such searches can provide a wealth of information on protein evolution (Pellegrini et al. 1999).

Clusters

Another, more recently introduced, method for classifying proteins is to use clustering methods. In these methods, every protein in a sequence database such as SwissProt is compared to every other sequence using a database search method including the BLAST, FASTA, and Smith-Waterman dynamic programming methods described in Chapter 7. Thus, each protein in the database receives a sequence similarity score with every other sequence. A similar method is used to identify families of paralogous proteins encoded by a single genome (p. 501). Matching sequences are further aligned by a pair-wise alignment program like LALIGN to recalculate the significance of the alignment score (see Chapter 3 flowchart, p. 58). In a cluster analysis, sequences are represented as vertices on a graph, and those vertices representing each pair of related sequences are joined by an edge that is weighted by the degree of similarity between the pair (see Fig. 10.4). In a first step, the clustering algorithm detects the sets of proteins that are joined in the graph by strongly weighted edges. In subsequent steps, relationships between the initial clusters found in the first step are identified on the basis of weaker, but still significant, connections between them. These related clusters are then merged in a manner that maximizes the strongest global relationships (see Web sites for ProtoMap and SYSTERS; Table 9.5). Clustering has been used to identify groups of proteins that lack a relative with a known structure and hence are suitable for structural analysis (Portugaly and Linial 2000). Additional information on clustering methods is provided in Chapter 10.

Proteins Comprise Motifs, Modules, and Other Sequence Elements of Structural Significance

The above analysis describes the types and distribution of motifs in proteins from the same or different organisms. A motif can represent an individual folded structure or active-site residues. Several different motifs widely separated in the same protein sequence are often found. These motifs represent conserved regions that lie in the core of the protein structure. Hence, their presence in two sequences predicts a common structural core (for review, see Henikoff et al. 1997).

Table 9.5. *Databases of patterns and sequences of protein families*

Name	Web address	Description	Reference
3D-Ali	http://www.embl-heidelberg.de/argos/ali/ali_info.html	aligned protein structures and related sequences using only secondary structures assigned by author of the structures	Pascarella and Argos (1992)
3D-PSSM	http://www.bmm.icnet.uk/3dpssm	uses a library of scoring matrices based on structural similarity given in the SCOP classification scheme (p. 402) for alignment with matrices based on sequence similarity	Kelley et al. (2000)
BLOCKS	http://blocks.fhcrc.org/	ungapped blocks in families defined by the Prosite catalog	Henikoff and Henikoff (1996); Henikoff et al. (1998)
COGS (Clusters of Orthologous Groups database and search site)	http://www.ncbi.nlm.nih.gov/COG	clusters of similar proteins in at least three species collected from available genomic sequences	Tatusov et al. (1997)
DIP (Database of Interacting Proteins)	http://dip.doe-mbi.ucla.edu	database of interacting proteins	Xenarios et al. (2000)
eMOTIF	http://dna.Stanford.EDU/emotif/	common and rare amino acid motifs in the BLOCKS and HSSP databases	Nevill-Manning et al. (1998)
HOMSTRAD	http://www-cryst.bioc.cam.ac.uk/~homstrad/	structure-based alignments organized at the level of homologous families[a]	Mizuguchi et al. (1998a)
HSSP	http://swift.embl-heidelberg.de/hssp/ http://www.sander.ebi.ac.uk/hssp/	sequences similar to proteins of known structure	Dodge et al. (1998)
INTERPRO integrated resource of protein domains and functional sites[b]	http://www.ebi.ac.uk/interpro	combination of Pfam, PRINTS, Prosite, and current SwissProt/TrEMBL sequence	see Web site
LPFC	http://www-camis.stanford.edu/projects/helix/LPFC/	a library of protein family cores based on multiple sequence alignment of protein cores using amino acid substitution matrices based on structure (see Chapter 3)	see Web page
NetOGly 2.0 prediction server	http://www.cbs.dtu.dk/services/NetOGlyc/	predicts glycosylation sites in mammalian proteins by neural network analysis	Hansen et al. (1997)
NNPSL	http://predict.sanger.ac.uk/nnpsl/	predicts subcellular location of proteins by neural network	see Web site
Pfam	http://www.sanger.ac.uk/Pfam	profiles derived from alignment of protein families, each one composed of similar sequence and analyzed by hidden Markov models	Sonnhammer et al. (1998)
PIR	http://www-nbrf.georgetown.edu/pirwww/pirhome.shtml	family and superfamily classification based on sequence alignment	Barker et al. (1996)
PRINTS	http://www.biochem.ucl.ac.uk/bsm/dbbrowser/PRINTS/PRINTS.html	protein fingerprints or sets of unweighted sequence motifs from aligned sequence families	Attwood et al. (1999)

Table 9.5. *Continued.*

Name	Web address	Description	Reference
PROCLASS	http://www-nbrf.georgetown.edu/ gfserver/proclass.html	database organized by Prosite patterns and PIR superfamilies; neural network system for protein classification into superfamily	Wu (1996); Wu et al. (1996)
PRODOM	http://protein.toulouse.inra.fr/ prodom.html	groups of sequence segments or domains from similar sequences found in SwissProt database by BLASTP algorithm; aligned by multiple sequence alignment	Corpet et al. (1998)
Prosite	http://www.expasy.ch/prosite	groups of proteins of similar biochemical function on basis of amino acid patterns	Bairoch (1991); Hofmann Bairoch et al. (1999)
ProtoMap	http://protomap.cornell.edu	classification of SwissProt and TrEMBL proteins into clusters	Yona et al. (1999)
PSORT	http://psort.nibb.ac.jp	predicts presence of protein localization signals in proteins	see Web site
SignalP Web server	http://www.cbs.dtu.dk/services/ SignalP/	predicts presence and location of signal peptide cleavage sites in proteins of different organisms by neural network analysis	Nielsen et al. (1997)
SMART	http://smart.embl-heidelberg.de	database of signaling domain sequences with accurate alignments	Schultz et al. (1998)
SYSTERS	http://www.dkfz-heidelberg.de/tbi/ services/cluster/systersform	classification of all sequences in the SwissProt database into clusters based on sequence similarity	Krause et al. (2000)
TargetDB	http://molbio.nmsu.edu:81/	database of peptides that target proteins to cellular locations	see Web site

A list of Web sites with protein sequence/structure databases is maintained at http://www.imb-jena.de/ImgLibDoc/help/db/. Many protein family databases are accessible through the European Bioinformatics Institute (http://srs.ebi.ac.uk/). Information on the available protein family databases is also found on the MetaFam site at http://metafam.ahc.umn.edu/.

[a] Sequence alignments of each family shown with residues labeled by solvent accessibility, secondary structure, H bonds to main-chain amide or carbonyl group, disulfide bond, and positive Φ angle.

[b] A combination of Pfam 5.0, PRINTS 25.0, Prosite 16, and current SwissProt and TrEMBL data. Additional merges with other protein pattern databases are planned.

A more detailed analysis of motifs has revealed that they are components of a more fundamental unit of structure and function, the protein module. Proteins may have several modules corresponding to different units of function, and these modules may be present in a different order (Henikoff et al. 1997). These diverse arrangements suggest that a biologically important module has been repeatedly employed in protein evolution by gene duplication and rearrangement mechanisms that are discussed in Chapter 6 and Chapter 10. The presence of modules also provides a further system of protein classification into module-based families.

An example of an important motif is the C_2H_2 (2 cysteines and 2 histidines) zinc finger DNA-binding motif Xfin of *Xenopus laevis* illustrated in Figure 9.16. The zinc finger is one of the most commonly identified motifs, in part due to the characteristic spacing of C and H residues in the motif sequence. As indicated in Figure 9.17, the zinc atom forms bonds with these residues to create the finger-like projection. When present in tandem copies, the finger is thought to lie in an alternating pattern in the major groove of DNA. A simple plot

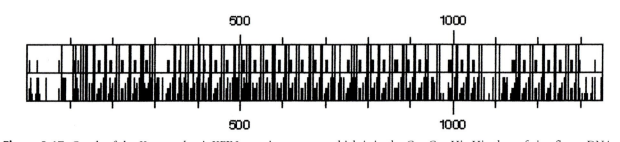

Figure 9.17. Graph of the *Xenopus laevis* XFIN protein sequence which is in the Cys-Cys-His-His class of zinc finger DNA-binding proteins (Branden and Tooze 1991). The graph was produced using the AA Window, Cys + His map option of DNA STRIDER vers. 1.2 on a Macintosh computer. The bottom panel shows amino acids Y, C, F, L, and H, respectively, as bars of increasing length. The top panel shows H and C as half- and full bars, respectively. The fingers appear in the top panel as double half-bars (two Cys residues separated by 2 amino acids) followed by double full bars (two His residues separated by 2 amino acids). This type of graphic representation is extremely useful for visualizing amino acid patterns in proteins.

of the positions of C and H residues on the protein sequence as shown in Figure 9.17 provides a very simple way to locate zinc fingers in a protein sequence.

Pfam is a Web site that provides a listing of proteins that carry the zinc finger sequence motif. As shown in Figure 9.18, the zinc finger is one of the most commonly recognized motifs, and proteins that carry the motif have been classified into a family. Two other families of zinc finger proteins with 4 cysteine or 3 cysteine and 1 histidine residues interacting with the Zn atom, and additional variations in the basic structure of zinc fingers, have also been identified. Descriptions and alignments of these proteins are provided at the Pfam Web site, as illustrated in Figure 9.19. Other families in the Pfam classification are given a description that best reflects the extent and complexity of the conserved sequence patterns, be it a domain, module, repeat, or motif. In general, all of these patterns represent a conserved unit of structure or function.

Structural Features of Some Proteins Are Readily Identified by Sequence Analysis

The above section indicates that a newly identified protein may be classified on the basis of the presence of sequence motifs, modules, or other sequence elements that represent structure or function. The zinc finger motif is one structural motif that may be readily identified on the basis of the order and spacing of a conserved pattern of cysteine and histidine residues in the sequence. Other classes of proteins have characteristic amino acid composition and patterns such that the structure can often be reliably predicted from the amino acid sequence. Some other examples of structure recognition on the basis of sequence are given below.

Leucine zippers and coiled coils. The leucine zipper motif is typically made up of two antiparallel α helices held together by interactions between hydrophobic leucine residues located at every seventh position in each helix, as illustrated in Figure 9.20A. The zipper holds protein subunits together. The leucines are located at approximately every two turns of the α helix. It is this repeated occurrence of leucines that makes the motif readily identifiable. In the transcription factors Gcn4, Fos, Myc, and Jun, the binding of the subunits forms a scissor-like structure with ends that lie on the major groove of DNA, as shown in Figure 9.20B. If the amino acids in each helical region are plotted as a spiral of 3.6 amino acid residues per turn, representing a view looking down the helix from the end starting at residue 1 on the inside of the spiral, then the result shown in Figure 9.20C is found. The leucine residues are found on approximately the same side of the helix, slightly out of phase

☒ [WashU]

| Pfam (St. Louis) | Pfam (Cambridge) | Pfam (Stockholm) | HMMER software | WashU Dept. of Genetics |
| Home | Analyze a sequence | Browse alignments | Text search | Swisspfam | Help & more information |

Pfam 3.4: available alignments and models

The families are grouped under the first letter of their name, regardless of case. All families starting with a number are found in 'Number'

Available sections: Numbers A B C D E F G H I J K L M N O P Q R S T U V W X Y Z Top twenty families

Top twenty families							
Name	**acc number**	**#seed**	**#full**	**av. len**	**av. %id**	**structure**	**Description**
GP120	PF00516	24	13408	131 aa	53%	1gc1	Envelope glycoprotein GP120
zf-C2H2	PF00096	200	4991	23 aa	35%	1zaa	Zinc finger, C2H2 type
ig	PF00047	65	3495	65 aa	20%		Immunoglobulin domain
RuBisCO_large	PF00016	17	3007	401 aa	77%	3rub	Ribulose bisphosphate carboxylase, large chain
pkinase	PF00069	67	2942	212 aa	24%		Eukaryotic protein kinase domain
cytochrome_b_N	PF00033	9	2866	152 aa	69%		Cytochrome b(N-terminal)/b6/petB
EGF	PF00008	73	2388	34 aa	35%	1apo	EGF-like domain
Collagen	PF01391	15	2125	59 aa	42%		Collagen triple helix repeat (20 copies)
fn3	PF00041	109	2103	85 aa	20%		Fibronectin type III domain
efhand	PF00036	86	1773	28 aa	27%	1osa	EF hand
LRR	PF00560	300	1753	47 aa	23%	1bnh	Leucine Rich Repeat (2 copies)
MHC_II_beta	PF00969	165	1688	44 aa	66%	1seb	Class II histocompatibility antigen, beta domain
zf-CCHC	PF00098	122	1678	17 aa	57%	1ncp	Zinc finger, CCHC class
ank	PF00023	95	1663	33 aa	27%		Ank repeat
rvp	PF00077	34	1508	95 aa	79%	1ida	Retroviral aspartyl proteases

Figure 9.18. The Pfam Web mirror site at Washington University (http://pfam.wustl.edu/browse.shtml). Shown are the 20 most common protein families classified according to the motifs that are present. Note the presence of the Pfam entry for zf-C2H2, the name assigned to the C_2H_2 (2 cysteines and 2 histidines) Zn finger DNA-binding motif, accession no. PF00096. Any family may be examined by clicking the mouse on the first letter of the family name. Fig. 9.19 is an example of the entry for the PF00096.

Continues on next page

WD40	PF00400	37	1482	39 aa	24%		WD domain, G-beta repeat
homeobox	PF00046	45	1431	49 aa	41%	1ahd	Homeobox domain
7tm_1	PF00001	64	1423	230 aa	19%		7 transmembrane receptor (rhodopsin family)
gag_p17	PF00540	4	1319	104 aa	77%	2hmx	gag gene protein p17 (matrix protein).
oxidored_q1	PF00361	33	1315	220 aa	32%	1min	NADH-Ubiquinone/plastoquinone (complex I), various chains

1407 families.

Figure 9.18. *Continued.*

with the rotational symmetry of the helix. The predicted structure is that of a coiled coil, as shown in Figure 9.20D (Branden and Tooze 1991).

Coiled-coil structures typically comprise two to three α helices coiled around each other in a left-handed supercoil in a manner that slightly distorts the helical repeat so that it is 3.5 residues per turn instead of the usual 3.6, or an integral number of 7 residues every second turn (Lupas 1996). They occur in fibrous proteins such as keratin and fibrinogen, and are also thought to occur in leucine zippers, as there is a repeat of leucine at every seventh residue (Branden and Tooze 1991). If the spiral wheel in Figure 9.20C is plotted so that there are 7 residues every second turn instead of 7.2, then the residues align more uniformly on one face of the helix. Consequently, the leucine zipper has been hypothesized to adopt a coiled-coil structure.

Coiled-coil regions may be predicted by searching for the 7-residue (heptad) periodicity observed in the sequence of these proteins. Naming these respective positions a, b, c, d, e, f, and g, then a and d are usually hydrophobic amino acids and the remaining amino acids are hydrophilic because coiled coils are generally fibrous, solvent-exposed structures. As more and more of these sequential patterns are observed along a sequence, one can be more convinced that the prediction is reliable. If there are at least 5–10 of these heptads and the hydrophobicity pattern is strongly conserved, the prediction is a good one. Poorer quality patterns come into doubt.

A program COILS2 has been developed for predicting coiled-coil regions with greater reliability than simple pattern searching for heptad repeats (Lupas et al. 1991; Lupas 1996; program description at http://www.embl-heidelberg.de/predictprotein/). There are two Web sites for predicting the occurrence of coiled-coil regions in protein sequences using the COILS program—http://www.isrec.isb-sib.ch/software/software.html and http://www.embl-heidelberg.de/predictprotein/predictprotein.html. The program may also be obtained from these sites for running on a local server. Central to the method is the generation of a profile scoring matrix, with each column showing the distribution of amino acids in each of the seven positions, a–g, found in all of the known coiled-coil proteins.

Pfam 3.4 (St. Louis) : <u>Home</u> I <u>Analyze a sequence</u> I <u>Browse alignments</u> I <u>Text search</u> I <u>Swisspfam</u> I <u>Help</u> I

Pfam entry: zf-C2H2

```
Accession number:        PF00096
Definition:              Zinc finger, C2H2 type
Author:                  Bateman A, Boehm S, Sonnhammer ELL
Source of seed members:  Boehm S
Alignment method of seed: Manual
HMM build command line:  hmmbuild HMM SEED
HMM build command line:  hmmcalibrate --seed 0 HMM
Gathering method:        hmmsearch -T 15 --domT 5
Trusted cutoffs:         15.00 5.00
Noise cutoffs:           14.80 17.50
Reference Number:        [1]
Reference Medline:       97315340
Reference Title:         Variations of the C2H2 zinc finger motif in the yeast genome
Reference Title:         and classification of yeast zinc finger proteins.
Reference Author:        Boehm S, Frishman D, Mewes HW;
Reference Location:      Nucleic Acids Res 1997;25:2464-2469.
Database Reference:      PROSITE; PDOC00028;
Database Reference:      PRINTS; PR00048;
Database Reference:      SCOP; 1zaa; fa; [SCOP-USA][CATH-PDBSUM]
Comment:                 The C2H2 zinc finger is the classical zinc finger domain.
Comment:                 The two conserved cysteines and histidines co-ordinate a
Comment:                 zinc ion. The following pattern describes the zinc finger.
Comment:                 #-X-C-X(1-5)-C-X3-#-X5-#-X2-H-X(3-6)-[H/C]
Comment:                 Where X can be any amino acid, and numbers in brackets
Comment:                 indicate the number of residues. The positions marked # are
Comment:                 those that are important for the stable fold of the zinc
Comment:                 finger. The final position can be either his or cys.
Comment:                 The C2H2 zinc finger is composed of two short beta strands
Comment:                 followed by an alpha helix. The amino terminal part of the
Comment:                 helix binds the major groove in DNA binding zinc fingers.
Number of members:       4991
```

Retrieve a Pfam alignment for zf-C2H2

Which alignment: **Full alignment** ⬍

What format: **Plain text** ⬍

☐ Output straight text. (Default is HTML-ized text.)

Retrieve alignment **Reset**

Pfam 3.4 (St. Louis) : <u>Home</u> I <u>Analyze a sequence</u> I <u>Browse alignments</u> I <u>Text search</u> I <u>Swisspfam</u> I <u>Help</u> I

Comments, questions, flames? Email <u><pfam@genetics.wustl.edu></u>.

Figure 9.19. The Pfam entry for family zf-C2H2 (accession no. PF00096). The mouse was clicked on the entry for zf-C2H2 shown in the above figure. The Pfam database is based on a statistical analysis of sequences with the same motif using hidden Markov models. The result is a profile of the sequences with matches, mismatches, and gaps. The entry describes how this profile was produced by the HMMER program, and also provides references and a link to a multiple sequence alignment of the sequences. As discussed in Chapter 3, this hidden Markov model of the sequences can be used to produce the multiple sequence alignment by choosing the most probable path through the model.

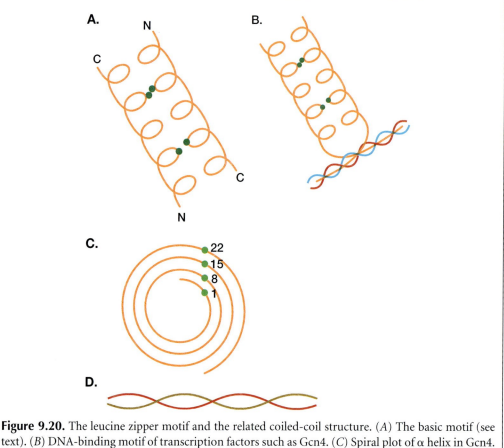

Figure 9.20. The leucine zipper motif and the related coiled-coil structure. (*A*) The basic motif (see text). (*B*) DNA-binding motif of transcription factors such as Gcn4. (*C*) Spiral plot of α helix in Gcn4. (*D*) A left-handed coiled-coil structure. (*Orange*) Helical backbones; (*green*) leucine residues; (*red and blue*) dsDNA strands.

Thus, column 1 representing residue a and column 4 representing residue 4 will show a high score for hydrophobic residues, but the other positions will show greater variability. These matrices may be 7, 14, 21, or 28 residues long for scoring shorter or longer sequence regions. Separate profile matrices of this kind have been prepared for two-stranded coiled coils, three-stranded coiled coils, bundles, and coiled-coils with parallel or antiparallel strands. There are also two scoring matrices in use, MTK derived from the sequences of myosins, tropomyosins, and keratins (intermediate filaments type I and II), and MTIDK, a new matrix derived from a much larger number of sequences. These matrices appear to give better predictions of different sets of structures for an unknown reason. The profile scoring matrix is moved along the candidate sequence one position at a time, calculating a score at each position. This score for each window is then given a probability P based on the distribution of scores found in coiled-coil and globular (not coiled-coil) proteins using the same matrix and other comparative conditions, where

$$P = G_{cc} / (R\, G_g + G_{cc}) \tag{1}$$

and G_{cc} and G_g are probabilities derived from the statistical distribution of scores of all coiled-coil and globular proteins, respectively, and R is the predicted ratio of coiled-coil to globular residues in GenBank. These scores will vary with each window size and option chosen, and the scores may be normalized to give a better impression of their range. A false positive can occur with sequences that have a biased amino acid distribution; these false positives can be identified by the program option of weighting the two hydrophobic positions a and d the same as the five hydrophilic positions b, c, e, f, and g. Normally, these positions are weighted 2.5 times more heavily during the scoring procedure. False positives will continue to have a high score whereas true positives will not.

For candidate protein sequences, Lupas recommends using both types of weighting and both MTK and MTIDK matrices. The program reliably predicts known coiled-coil regions (Lupas 1996). An example of the program output from the ISREC Web site is shown in Figure 9.21, using as input the sequence of Gcn4 (identified as GCN4_YEAST in the SwissProt database), which has a leucine zipper region. The protein is scanned for the number of occurrences of coiled coils in a sliding window of 7, 14, 21, or 28 residues.

Another method for predicting coiled coils is based on an analysis of correlations between pairs of amino acids (Berger et al. 1995), and the program is accessible at http://dot.imgen.bcm.tmc.edu:9331/seq-search/struc-predict.html.

Transmembrane-spanning Proteins

The all-α superfamily of membrane proteins (see classification of membrane proteins at the SCOP structural database at http://scop.mrc-lmb.cam.ac.uk/scop/) is composed of proteins that traverse membranes back and forth through a series of α helices comprising amino acids with hydrophobic side chains. The typical length, 20–30 residues, and strong hydrophobicity of these helices provide a simple method for scanning a candidate sequence for such features. An example of such a structure is illustrated in Figure 9.22.

Membrane-spanning hydrophobic α helices can be quite accurately located by scanning for hydrophobic regions about 19 residues in length in the amino acid sequence (Kyte and Doolittle 1982). The occurrence of such regions in a candidate protein of unknown structure is a good indicator that the region spans a membrane. In Figure 9.23, such an analysis is shown for subunit M of the above molecule. Membrane-spanning helices are different from α helices that are located on the surface of a protein structure. The surface helices tend to have hydrophobic residues located on the core-facing side (inside) and the hydrophilic residues on the solvent-facing side (outside) of the helix. These surface-exposed helices can be recognized by this separation of hydrophobic residues through a helical moment analysis described below. Membrane α helices are more like α helices that are buried in the structural core of a protein, which also have a high proportion of amino acids with hydrophobic side groups located throughout their lengths. In an effort to distinguish different classes of α helices, several methods for improving the prediction of transmembrane regions have been devised and are available on Web sites.

One such method is one of the program choices of the PHD (profile-fed neural network system from *Hei*delberg) server for protein structure prediction at http://www.embl-heidelberg.de/predictprotein/predictprotein.html. The membrane-spanning helix predict program is named PHDhtm (PHD for *h*elical *t*rans*m*embrane proteins). Briefly, a machine learning method called a neural network (see below) is trained to recognize the sequence patterns and sequence variations of a set of α-helical transmembrane proteins of known three-dimensional structure. A candidate sequence is then scanned for the presence of similar sequence variations and a prediction is made as to the occurrence and location of α-helical domains in the candidate protein. The specific steps were as follows. First, each of the small number of structurally identified α-helical transmembrane proteins was used to

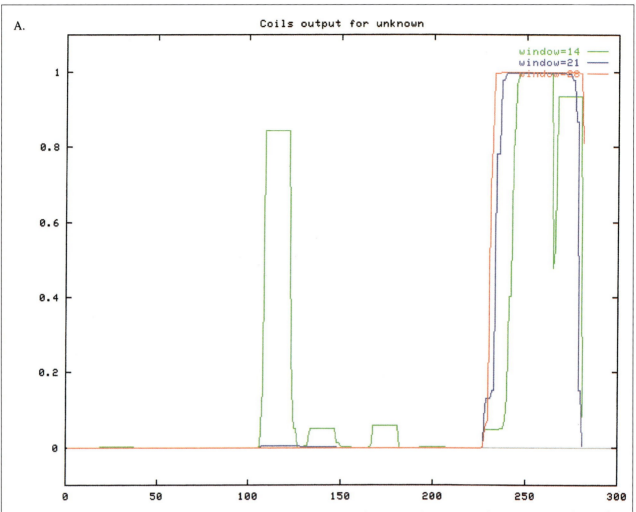

Figure 9.21. Prediction of coiled-coil regions by the COILS2 program. The Gcn4 protein was used as input to a Web page listed in the text. (*A*) Plot of probability of residue in coiled-coil structure versus residue number obtained from the ISREC server. (*B*) Partial list of scores by residue number. Analysis obtained from the Predict Protein server. Note that highest probabilities are obtained with a window of 28 amino acids. The expected order of amino acids in the coiled coil is a, b, c, d, e, f, and g. This order does not start until residue 244. Amino acids found at position 243 and at lower numbered positions are characteristically found at other places in the coiled-coil heptad.

search the SwissProt protein sequence database for additional sequences in this superfamily using the BLAST or FASTA algorithms. Second, the sequences found were assembled first into a multiple sequence alignment and then into a motif by the program MAXHOM. Sequences less than 30% identical, and therefore least likely to be in the superfamily, were not included. The most-alike sequences in the alignment were also removed to provide a representative and statistically reasonable range of amino acid substitutions in each column of the motif. The neural network was then trained to differentiate between columns in the motif representing the α-helical domains and the flanking nonhelical domains. The training method is described in greater detail below. The orientation of the predicted α-helical domains with respect to the inside (cytoplasmic) or outside of the membrane is also predicted based on the observed preponderance of positively charged amino acids on the cytoplasmic side of solved structures (Rost et al. 1995). An illustrative example of a PHD-htm analysis on protein 1prc_M is shown in Figure 9.24. As shown, the program correctly predicts five transmembrane helices, but positions of the ends of these helices are not

```
B.
COILS version 2.1
using MTK matrix.
weights: a,d=2.5 and b,c,e,f,g=1.0
Input file is /home/phd/server/work/predict_h24138-21300.fasta
>prot (#) ppOld, gcn4 /home/phd/server/work/predict_h24138
  Residue        Window=14              Window=21              Window=28
              Score  Probability     Score  Probability     Score  Probability
  ·
  ·

  239 A      g 1.240   0.011       d 1.529   0.419       d 1.546   0.852
  240 R      a 1.240   0.011       e 1.581   0.585       e 1.546   0.852
  241 R      e 1.446   0.051       f 1.581   0.585       f 1.546   0.852
  242 S      f 1.446   0.051       g 1.581   0.585       g 1.546   0.852
  243 R      g 1.450   0.052       a 1.581   0.585       a 1.551   0.862
  244 A      b 1.529   0.093       b 1.607   0.664       b 1.643   0.968
  245 R      c 1.592   0.145       c 1.607   0.664       c 1.643   0.968
  246 K      d 1.669   0.238       d 1.607   0.664       d 1.643   0.968
  247 L      e 2.433   0.994       e 1.843   0.978       e 1.984   1.000
  248 Q      f 2.433   0.994       f 1.988   0.997       f 2.041   1.000
  249 R      g 2.433   0.994       g 2.018   0.998       g 2.052   1.000
  250 M      a 2.433   0.994       a 2.054   0.999       a 2.052   1.000
  251 K      b 2.433   0.994       b 2.054   0.999       b 2.052   1.000
  252 Q      c 2.433   0.994       c 2.054   0.999       c 2.052   1.000
  253 L      d 2.433   0.994       d 2.054   0.999       d 2.052   1.000
  254 E      e 2.433   0.994       e 2.054   0.999       e 2.052   1.000
  255 D      f 2.433   0.994       f 2.054   0.999       f 2.052   1.000
  256 K      g 2.433   0.994       g 2.054   0.999       g 2.052   1.000
  257 V      a 2.433   0.994       a 2.054   0.999       a 2.052   1.000
  258 E      b 2.433   0.994       b 2.054   0.999       b 2.052   1.000
  259 E      c 2.433   0.994       c 2.054   0.999       c 2.052   1.000
  260 L      d 2.433   0.994       d 2.054   0.999       d 2.052   1.000
  261 L      e 2.433   0.994       e 2.054   0.999       e 2.052   1.000
  262 S      f 2.421   0.993       f 2.054   0.999       f 2.052   1.000
  263 K      g 2.421   0.993       g 2.054   0.999       g 2.052   1.000
  ·
  ·
  ·
  271 V      a 2.026   0.848       a 2.004   0.998       a 2.052   1.000
  272 A      b 2.026   0.848       b 1.968   0.996       b 2.052   1.000
  273 R      c 2.026   0.848       c 1.943   0.994       c 2.052   1.000
  274 L      d 2.026   0.848       d 1.943   0.994       d 2.052   1.000
  275 K      e 2.026   0.848       e 1.883   0.987       e 2.052   1.000
  276 K      f 2.026   0.848       f 1.883   0.987       f 2.052   1.000
  277 L      g 2.026   0.848       g 1.776   0.948       g 1.986   1.000
  278 V      a 2.026   0.848       a 1.776   0.948       a 1.949   1.000
  279 G      b 2.026   0.848       b 1.631   0.732       b 1.868   0.999
  280 E      c 2.026   0.848       c 1.631   0.732       c 1.868   0.999
  281 R      a 1.378   0.030       d 1.090   0.003       d 1.381   0.263
```

Figure 9.21. *Continued.*

always correctly predicted, as revealed by a lack of correlation between the predicted regions (H) and the known regions (*).

A second method for prediction of transmembrane α helices is by the TMpred server. This method scans a candidate sequence for matches to a sequence scoring matrix obtained by aligning the sequences of all of the transmembrane α-helical regions that are known from structures. These sequences have been collected into a database (TMbase) of such sequences. An example of a transmembrane analysis of 1prc_M by this method is shown in Figure 9.25. As shown, the program correctly predicted five α-helical transmembrane segments. Two alternative models were predicted, the first more highly favored, but neither one matched the known ends of these regions. These examples serve to illustrate that these methods can be expected to identify membrane-spanning α-helical proteins quite reliably

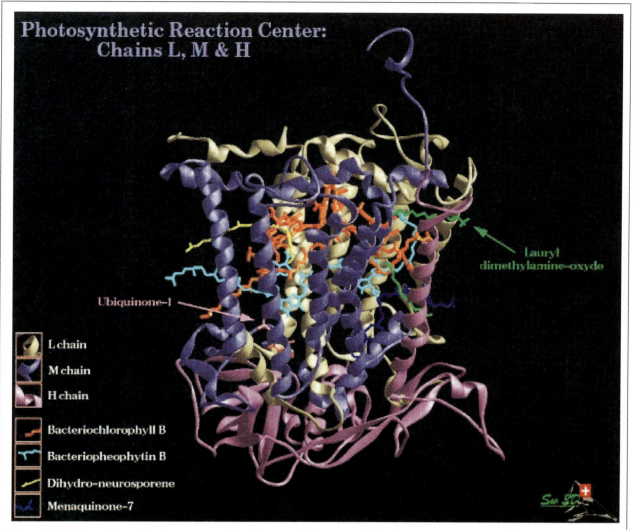

Figure 9.22. Three-dimensional structure of the photosynthetic center of *Rhodopseudomonas viridis*. The three subunits each cross the membranes of hollow vesicles found in these bacteria at approximately right angles, two of them back and forth multiple times. The light-harvesting pigments chlorophyll and pheophytin are bound between these helices. These membrane-spanning regions are 25–29 amino acids long and are composed of α helices. There is an abundance of hydrophobic amino acids in these helices. Hence, a hydrophobicity plot of the protein chain will show peaks centered on the position of the helices, as shown in Fig. 9.23. (Image from http://expasy.hcuge.ch/pub/Graphics/IMAGES/)

but not the ends of such regions. A simple hydrophobicity plot may also be used as shown in Table 9.3. The number and extent of these regions can also be predicted from the peaks in this plot. This method is unsuitable for scanning genomic sequences for possible membrane-spanning proteins; the automatic methods are much more suitable for this purpose.

Prediction of Protein Secondary Structure from the Amino Acid Sequence

Accurate prediction as to where α helices, β strands, and other secondary structures will form along the amino acid chain of proteins is one of the greatest challenges in sequence analysis. At present, it is not possible to predict these events with very high reliability. As methods have improved, prediction has reached an average accuracy of 64–75% with a higher accuracy for α helices, depending on the method used. These predictive methods

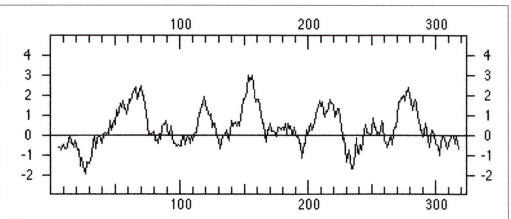

Figure 9.23. Hydrophobicity plot of subunit M of the photosynthetic center of *Rhodopseudomonas viridis* illustrated in Fig. 9.22. The hydrophobicity plotting program of DNA Strider 1.2 on a Macintosh computer was used with Kyte-Doolittle hydrophobicity values (Table 9.6) and a sliding window of 19, chosen to detect the approximate length of membrane-spanning α helices of 20–30 residues. The plot reliably predicts the five hydrophobic membrane-spanning helices of this protein, which are located in the three-dimensional structure at amino acid positions 52–78, 110–139, 142–167, 197–225, and 259–285. The SwissProt entry for this protein is numbered P06010 and the protein ID is RCEM_RHOVI. The program TGREASE, which is available in the FASTA suite of programs, will provide a similar plot on Macintosh or PC computers.

can be made especially useful when combined with other types of analyses discussed in this chapter. For example, a search of a sequence database or a protein motif database for matches to a candidate sequence may discover a family or superfamily relationship with a protein of known structure. If significant matches are found in regions of known secondary or three-dimensional structure, the candidate protein may share the three-dimensional structural features of the matched protein. Several Web sites provide such an enhanced analysis of secondary structure. These sites and others that provide secondary structure analysis of a query protein are given in Table 9.7. The main methods of analyses used at these sites are described below.

Methods of structure prediction from amino acid sequence begin with an analysis of a database of known structures. These databases are examined for possible relationships between sequence and structure. When secondary structure predictions were first being made in the 1970s and 1980s, only a few dozen structures were available. This situation has now changed with present databases including approximately 500 independent structural folds. The combination of more structural and sequence information presents a new challenge to investigators who wish to develop more powerful predictive methods.

The ability to predict secondary structure also depends on identifying types of secondary structural elements in known structures and determining the location and extent of these elements. The main types of secondary structures that are examined for sequence variation are α helices and β strands. Early efforts focused on more types of structures, including other types of helices, turns, and coils. To simplify secondary structure prediction, these additional structures that are not an α helix or β strand were subsequently classified as coils. Assignment of secondary structure to particular amino acids is sometimes included in the PDB file by the investigator who has solved the three-dimensional structure. In other cases, secondary structure must be assigned to amino acids by examination of the structural coordinates of the atoms in the PDB file. Methods for comparing three-dimensional structures, described above, frequently assign these features automatically, but not always

```
          ...,....1....,....2....,....3....,....4....,....5....,....6
AA       |ADYQTIYTQIQARGPHITVSGEWGDNDRVGKPFYSYWLGKIGDAQIGPIYLGASGIAAFA|
PHD htm  |                                       HHHHHHHHHHHHHHHHHHHH|
                                                 ********

          ...,....7....,....8....,....9....,...10...,...11...,...12
AA       |FGSTAILIILFNMAAEVHFDPLQFFRQFFWLGLYPPKAQYGMGIPPLHDGGWWLMAGLFM|
PHD htm  |HHHHHHHHHHHHHHHHHHHH                            HHHHHHHHHH|
          ****************                                ***********

          ...,...13...,...14...,...15...,...16...,...17...,...18
AA       |TLSLGSWWIRVYSRARALGLGTHIAWNFAAAIFFVLCIGCIHPTLVGSWSEGVPFGIWPH|
PHD htm  |HHHHHHHHHHH        HHHHHHHHHHHHHHHHHHHHHHHHHHH    HHHH |
          ****************  ************************

          ...,...19...,...20...,...21...,...22...,...23...,...24
AA       |IDWLTAFSIRYGNFYYCPWHGFSIGFAYGCGLLFAAHGATILAVARFGGDREIEQITDRG|
PHD htm  |                HHHHHHHHHHHHHHHHHHHHHHHHHHHHH|
                           ***************************

          ...,...25...,...26...,...27...,...28...,...29...,...30
AA       |TAVERAALFWRWTIGFNATIESVHRWGWFFSLMVMVSASVGILLTGTFVDNWYLWCVKHG|
PHD htm  |                HHHHHHHHHHHHHHHHHHHHHHHHHHHHHHHHH|
                           *************************

          ...,...31...,...32...,...33...,...34...,...35...,...36
AA       |AAPDYPAYLPATPDPASLPGAPK|
```

Figure 9.24. Analysis of a known transmembrane protein by PHDhtm program at the predict protein server at Heidelberg. The same protein used for the above hydrophobicity analysis in Fig. 9.25 hydro and also the blue protein in Fig. 9.23 (structural name 1prc_M, SwissProt P06010) was submitted to the server at http://www.embl-heidelberg.de/predictprotein/ppDoPred.html, choosing the transmembrane prediction option on the expert page and minimizing program output. Additional program output, including probabilities, assignment of inside and outside domain (topology), and neural network details are not shown. Note that the protein 1prc_M is listed on the server as one of the proteins that was used to train the neural network. Hence, using this protein is a biased test of program accuracy, which is claimed in more objective tests to identify residues in the transmembrane helices with 95% reliability and helical transmembrane proteins at 86% accuracy (Rost et al. 1995, 1996; Rost 1996). The predicted helical regions are shown by an H and the known regions in the three-dimensional structure, obtained from the SwissProt entry for the protein, are shown by an asterisk.

in the same manner. Hence, some variation is possible, and deciding which is the best method can be difficult. The DSSP database of secondary structures and solvent accessibilities is a useful and widely used resource for this purpose (Kabsch and Sander 1983; http://www.sander.ebi.ac.uk/dssp/). This database, which is based on recognition of hydrogen-bonding patterns in known structures, distinguishes eight secondary structural classes that can be grouped into α helices, β strands, and coils (Rost and Sander 1993). A more recently described automatic method makes predictions in accord with published assignments (Frishman and Argos 1995).

The assumption on which all the secondary structure prediction methods are based is that there should be a correlation between amino acid sequence and secondary structure. The usual assumption is that a given short stretch of sequence may be more likely to form one kind of secondary structure than another. Thus, many methods examine a sequence window of 13–17 residues and assume that the central amino acid in the window will adopt a conformation that is determined by the side groups of all the amino acids in the window. This window size is within the range of lengths of α helices (5–40 residues) and β strands (5–10 residues).

There is evidence that more distant interactions within the primary amino acid chain may influence local secondary structure. The same amino acid sequence up to 5 (Kabsch

```
2 possible models considered, only significant TM-segments used

STRONGLY preferred model: Amino-terminus inside
Five  strong transmembrane helices
 # from   to length    actual
 1   52   71 (20)      52- 78
 2  110  132 (23)     110-139
 3  146  166 (21)     142-167
 4  199  219 (21)     197-225
 5  268  289 (22)     259-285

alternative model
 five strong transmembrane helices
 # from   to length    actual
 1   53   71 (19)      52- 78
 2  113  129 (17)     110-139
 3  144  161 (18)     142-167
 4  201  225 (25)     197-225
 5  268  289 (22)     259-285
```

Figure 9.25. Analysis of known transmembrane protein by TMPRED. The same protein used above for the PHDhtm analysis (structural name 1prc_M, SwissProt P06010) was submitted to the server at http://www.ch.embnet.org /software/TMPRED_form.html. Shown are two predicted structural models; score and topology information are not included. Known locations of the α helices are shown in last column for comparison.

Table 9.6. *Hydrophobicity scales for the amino acids*

Residue		Value
Ala	A	1.8
Arg	R	−4.5
Asn	N	−3.5
Asp	D	−3.5
Cys	C	2.5
Gln	Q	−3.5
Glu	E	−3.5
Gly	G	−0.4
His	H	−3.2
Ile	I	4.5
Leu	L	3.8
Lys	K	−3.9
Met	M	1.9
Phe	F	2.8
Pro	P	−1.6
Ser	S	−0.8
Thr	T	−0.7
Trp	W	−0.9
Tyr	Y	−1.3
Val	V	4.2

These values are based on adjusted values derived from several sets of experimental measurements (Kyte and Doolittle 1982). The most hydrophobic amino acids are printed in green, the least hydrophobic amino acids in red. A number of additional scales are also available (von Heijne 1987).

Table 9.7. *Selected programs for performing protein secondary structure prediction*

Program	Web address	Method	Reference
Baylor College of Medicine (BCM)	http://dot.imgen.bcm.tmc.edu:9331/ seq-search/struc-predict.html	collection of methods and linked to other servers	see Web site and text
DSC	http://www.bmm.icnet.uk/dsc/	linear discrimination	King et al. (1997)
J-Pred structure prediction server	http://jura.ebi.ac.uk:8888/	NNSSP, DSC, Predator, Mulpred,[b] Zpred,[c] Jnet,[e] and PHD	Cuff et al. (1998); and see text
NNPRED	http://www.cmpharm.ucsf.edu/ ~nomi/nnpredict.html	neural networks enhanced to detect sequence periodicity	Kneller et al. (1990)
NPS@ server, MLR combination for secondary structure prediction[a]	http://pbil.ibcp.fr/NPSA/	combination of prediction methods using multivariate linear regression to optimize the predictions	Guermeur et al. (1999)
Protein Sequence Analysis (PSA) System[d]	http://bmerc-www.bu.edu/psa/ index.html	discrete space models (hidden Markov models) for patterns of α helices, β strands, tight turns, and loops in specific structural classes	Stultz et al. (1993, 1997); White et al. (1994)
PREDATOR	http://www.embl-heidelberg.de/ argos/predator/predator_info. html	based on analysis of long- and short-range amino acid interactions and alignments of sequence pairs	Frishman and Argos (1995, 1996, 1997)
Predict Protein server	http://www.embl-heidelberg.de/ predictprotein/predictprotein. html; see also mirror sites	neural networks of multiple sequence alignment	Rost and Sander (1994); Rost (1996)
PSSP	http://dot.imgen.bcm.tmc.edu:9331/ seq-search/struc-predict.html	nearest neighbor enhanced by non-intersecting local and multiple sequence alignments	Salamov and Solovyev (1995, 1997)
Simpa96	http://pbil.ibcp.fr/NPSA/	nearest-neighbor method	Levin (1997)
SOPM, SOPMA	http://pbil.ibcp.fr/NPSA/	nearest-neighbor method based on sequence alignments	Geourjon and Deleage (1994, 1995)
SSP	http://dot.imgen.bcm.tmc.edu:9331/ seq-search/struc-predict.html	linear discriminant analysis based on amino acid composition of local and adjacent regions	see H option for this program on Web page
UCLA-DOE structure prediction server	http://www.doe-mbi.ucla.edu/ people/frsvr/frsvr.html	collection of methods and linked to other servers	Fischer and Eisenberg (1996)

[a]Consensus option provides a user-defined combination of methods.
[b]See Cuff et al. (1998).
[c]Zpred server is also available at http://kestrel.ludwig.ucl.ac.uk/zpred.html. The program predicts secondary structure based on physicochemical information and GOR prediction scores.
[d] This server will also predict 3D structural class.
[e] Jnet uses multiple sequence alignments and a trained neural network to make secondary structure predictions (Cuff and Barton 2000).

and Sander 1984) and 8 (Sudarsanam 1998) residues in length can be found in different secondary structures. An 11-residue-long amino acid "chameleon" sequence has been found to form an α helix when inserted into one part of a primary protein sequence and a β sheet when inserted into another part of the sequence (Minor and Kim 1996). More distant interactions may account for the observation that β strands are predicted more poorly by analysis of local regions (Garnier et al. 1996). However, the methods that have been used to predict the secondary structure of an amino acid residue all perform less well when amino acids more distant than in the small window of sequence are used.

The number of possible amino acid combinations in a sequence window of 17 amino acids is very large ($17^{20} = 14 \times 10^{24}$). If many combinations influence one type of secondary structure, examination of a large number of protein structures is required to discover the significant patterns and correlations within this window. Earlier methods for predicting secondary structure assumed that each amino acid within the sequence window of 13–17 residues influences the local secondary structure independently of other nearby amino acids; i.e., there is no interaction between amino acids in influencing local secondary structure. Later methods assumed that interactions between amino acids within the window could play a role.

Neural network models described below have the ability to detect interactions between amino acids in a sequence window, including conditional interactions. A hypothetical example of the interactions that might be discovered illustrates the possibilities. If the central amino acid in the sequence window is Leu and if the second upstream amino acid toward the amino terminus is Asn, the Leu is in an α helix; however, if the neighboring amino acid is not Asn, the Leu is in a β strand. In another method of secondary structure prediction, the nearest-neighbor method, sequence windows in known structures that are most like the query sequence are identified. This method bypasses the need to discover complex amino acid patterns associated with secondary structure. Protein secondary structure has also been modeled by hidden Markov models, also described as discrete state-space models, which are described below (Stultz et al. 1993; White et al. 1994).

Accuracy of Secondary Structure Prediction

One method of assessing accuracy of secondary structure prediction is to give the percentage of correctly predicted residues in sequences of known structure, called Q_3. This measure, however, is not very effective by itself, because even a random assignment of structure can achieve a high score by this test (Holley and Karplus 1991). Another measure is to report the fraction of each type of predicted structure that is correct. A third method is to calculate a correlation coefficient for each type of predicted secondary structure (Mathews 1975). The coefficient indicating success of predicting residues in the α-helical configuration, C_α, is given by

$$C_\alpha = (p_\alpha n_\alpha - u_\alpha o_\alpha) / \sqrt{([n_\alpha + u_\alpha] [n_\alpha + o_\alpha] [p_\alpha + u_\alpha] [p_\alpha + o_\alpha])} \qquad (2)$$

where p_α is the number of correct positive predictions, n_α is the number of correct negative predictions, o_α is the number of overpredicted positive predictions (false positives), and u_α is the number of underpredicted residues (misses). The closer this coefficient is to a value of 1, the more successful the method for predicting a helical residue. An overall level of prediction accuracy does not provide information on the accuracy of the number of predicted secondary structures, and their lengths and location in the sequence. One simple index of success is to compare the average of the predicted lengths with the known average (Rost and Sander 1993).

Another factor to consider in prediction accuracy is that some protein structures are more readily predictable than others, such that the spectrum of test proteins chosen will influence the frequency of success. A representative set of proteins that have limited similarity will provide the most objective test. Rost and Sander (1993) have chosen a set of 126 globular and 4 membrane proteins that have less than 25% pair-wise similarity and have used this set for training and testing neural network models. A newer set of 540 structurally distinct fold types in the FSSP database provides an even larger set of training and test structures of unique structure and sequence (Holm and Sander 1998). In the often-used jackknife test, one protein in a set of known structure is left out of a calibration or training step of the program being tested. The rest of the proteins are used to predict the structure of the left-out one, and the procedure is cycled through all of the sequences. The overall frequency of success of predicting the secondary structural features of the left-out sequence is used as an indicator of success. An even more comprehensive approach to the problem of accuracy is to examine the predictions for different structural classes of proteins. Because some classes are much more difficult to predict, the overall success rate with respect to protein class is an important index of success. Prediction accuracy is discussed further below.

A valuable addition to secondary structure prediction is giving the degree of reliability of the prediction at each position. Some prediction methods produce a score for each of the three types of structures (helix, strand, coil or loop) at each residue position. If one of these scores is much higher than the other two, the score is considered to be more reliable, and a high reliability index may be assigned that reflects high confidence in the prediction. If the scores are more similar, the index is lower. By examining predictions for known structures, as in a jackknife experiment, the accuracy of these reliability indices may be determined. What has been found is that a prediction with a high index score is much more accurate (Yi and Lander 1993; and see PHD server below), thus increasing confidence in the prediction of these residues.

Methods for Secondary Structure Prediction

Three widely used methods of protein secondary structure prediction, (1) the Chou-Fasman and GOR methods, (2) neural network models, and (3) nearest-neighbor methods, are discussed below. An additional method that models structural families by hidden Markov models is then described. These methods can be further enhanced by examining the distribution of hydrophobic, charged, and polar amino acids in protein sequences.

Chou-Fasman/GOR Method

The Chou-Fasman method (Chou and Fasman 1978) was based on analyzing the frequency of each of the 20 amino acids in α helices, β sheets, and turns of the then-known relatively small number of protein structures. It was found, for example, that amino acids Ala (A), Glu (E), Leu (L), and Met (M) are strong predictors of α helices, but that Pro (P) and Gly (G) are predictors of a break in a helix. A table of predictive values for each type of secondary structure was made for each of the α helices, β strands, and turns. To produce these values, the frequency of amino acid i in structure s is divided by the frequency of all residues in structure s. The resulting three structural parameters ($P\alpha$, $P\beta$, and Pt) vary roughly from 0.5 to 1.5 for the 20 amino acids.

To predict a secondary structure, the following set of rules is used. The sequence is first scanned to find a short sequence of amino acids that has a high probability for starting a nucleation event that could form one type of structure. For α helices, a prediction is made when four of six amino acids have a high probability >1.03 of being in an α helix. For β strands, the presence in a sequence of three of five amino acids with a probability of >1.00 of being in a β strand predicts a nucleation event for a β strand. These nucleated regions are extended along the sequence in each direction until the prediction values for four amino acids drops below 1. If both α-helical and β-strand regions are predicted, the higher probability prediction is used.

Turns are predicted somewhat differently. Turns are modeled as a tetrapeptide, and two probabilities are calculated. First, the average of the probabilities for each of the four amino acids being in a turn is calculated as for α helix and β strand predictions. Second, the probabilities of amino acid combinations being present at each position in the turn tetrapeptide (i.e., the probability that a particular amino acid such as Pro is at position 1, 2, 3, or 4 in the tetrapeptide) are determined. These probabilities for the four amino acids in the candidate sequence are multiplied to calculate the probability that the particular tetrapeptide is a turn. A turn is predicted when the first probability value is greater than the probabilities for an α helix and a β strand in the region and when the second probability value is greater than 7.5×10^{-5}. In practice, the Chou-Fasman method is only about 50–60% accurate in predicting secondary structural domains.

Garnier et al. (1978) developed a somewhat more involved method for protein secondary structure prediction that is based on a more sophisticated analysis. The method is called the GOR (Garnier, Osguthorpe, and Robson) method. Whereas the Chou-Fasman method is based on the assumption that each amino acid individually influences secondary structure within a window of sequence, the GOR method is based on the assumption that amino acids flanking the central amino acid residue influence the secondary structure that the central residue is likely to adopt. In addition, the GOR method uses principles of information theory to derive predictions (Garnier et al. 1996).

As in the Chou-Fasman method, known secondary structures are scanned for the occurrence of amino acids in each type of structure. However, the frequency of each type of amino acid at the next 8 amino-terminal and carboxy-terminal positions is also determined, making the total number of positions examined equal to 17, including the central one. In the original GOR method, three scoring matrices, containing in each column the probability of finding each amino acid at one of the 17 positions, are prepared. One matrix corresponds to the central (eighth) amino acid being found in an α helix, the second for the amino acid being in a β strand, the third a coil, and the fourth, a turn. Later versions omitted the turn calculation because these were the most variable features and were consequently the most difficult to predict. A candidate sequence is analyzed by each of the three to four matrices by a sliding window of 17 residues. Each matrix is positioned along a candidate sequence and the matrix giving the highest score predicts the structural state of the central amino acid. At least 4 residues in a row have to be predicted as an α helix and 2 in a row for a β strand for a prediction to be validated.

Matrix values are calculated in somewhat the same manner as amino acid substitution matrices (described in Chapter 3), in that matrix values are calculated as log odds units representing units of information. The information available as to the joint occurrence of secondary structural conformation S and amino acid a is given by (Garnier et al. 1996)

$$I\,(S;\,a) = \log\,[\,P(S\mid a)\,/\,P(S)\,] \tag{3}$$

where $P(S \mid a)$ is the conditional probability of conformation S given residue a, and $P(S)$ is the probability of conformation S. By Bayes' rule (see Chapter 3, p. 120), the probability of conformation S given amino acid a, $P(S \mid a)$ is given by

$$P(S \mid a) = P(S,\,a)\,/\,P(a) \tag{4}$$

where $P(S, a)$ is the joint probability of S and a and $P(a)$ is the probability of a. These probabilities can be estimated from the frequency of each amino acid found in each structure and the frequency of each amino acid in the structural database. Given these frequencies,

$$I(S; a) = \log(f_{S,a} / f_S) \tag{5}$$

where $f_{S,a}$ is the frequency of amino acid a in conformation S and f_S is the frequency of all amino acid residues found to be in conformation S.

The GOR method maximizes the information available in the values of $f_{S,a}$ and avoids data size and sampling variations by calculating the information difference between the competing hypotheses that residue a is in structure S, $I(S;a)$, or that a is in a different conformation (not S), $I(\text{not } S;a)$. This difference $I(\Delta S;a)$ is calculated from Equation 5 with simple substitutions by

$$\begin{aligned} I(\Delta S; a) &= I(S; a) - I(\text{not } S; a) \\ &= \log\{P(S,a)/[1 - P(S,a)]\} + \log\{[1 - P(S)/P(S)]\} \end{aligned} \tag{6}$$

which is derived from the observed amino acid data as

$$I(\Delta S; a) = \log[f_{S,a} / (1 - f_{S,a})] + \log[(1 - f_S)/f_S] \tag{7}$$

where the frequency of finding amino acid a not in conformation S is $1 - f_{S,a}$ and of not finding any amino acid in conformation S is $1 - f_S$. Equation 6 is used to calculate the information difference for a series of x consecutive positions flanking sequence position m,

$$I(\Delta S_m; a_1,..a_X) = \log[P(S_m,a_1,..a_X)/(1 - P(S_m,a_1,..a_X))] + \log[1 - P(S)/P(S)] \tag{8}$$

from which the following ratio of the joint probability of conformation S_m given $a_1,..a_X$ to the joint probability of any other conformation may be calculated

$$P(S_m,a_1,..a_X)/[1-P(S_m,a_1,..a_X)] = \{P(S)/[1-P(S)]\}\, e^{-I(\Delta Sm; a1,..aX)} \tag{9}$$

Searching for all possible patterns in the structural database would require an enormous number of proteins. Hence, three simplifying approaches have been taken. First, it was assumed in earlier versions of GOR that there is no correlation between amino acids in any of the 17 positions (both the flanking 8 positions and the central amino acid position), or that each amino acid position had a separate and independent influence on the structural conformation of the central amino acid. The steps are then: (1) Values for $I(\Delta S; a)$ in Equation 7 are calculated for each of the 17 positions; (2) these values are summed to approximate the value of $I(\Delta S_m; a_1,..a_X)$ in Equation 8; (3) the probability ratios in Equation 9 are calculated.

The second assumption used in later versions of GOR was that certain pair-wise combinations of an amino acid in the flanking region and central amino acid influence the conformation of the central amino acid. This model requires a determination of the frequency of amino acid pairs between each of the 16 flanking positions and the central one, both for when the central residue is in conformation S and when the central residue is not in conformation S. Finally, in the most recent version of GOR, the assumption is made that certain pair-wise combinations of amino acids in the flanking region, or of a flanking amino acid and the central one, influence the conformation of the central one. Thus, there are $17 \times 16/2 = 136$ possible pairs to use for frequency measurements and to examine for correlation with the conformation of the central residue. With the advent of a large number of protein structures, it has become possible to assess the frequencies of amino acid combinations and to use this information for secondary structural predictions. The GOR method predicts 64% of the residue conformations in known structures and quite drastically (36.5%) underpredicts the number of residues in β strands.

Use of the Chou-Fasman and GOR methods for predicting the secondary structure of the α subunit of *Salmonella typhimurium* tryptophan synthase is illustrated in Figure 9.26. In this particular case, the positions of the secondary structures predicted by either of these methods are very similar to those in the solved crystal structure (Branden and Tooze 1991). However, tests of the accuracy of these methods using sequences of other proteins whose structures are known have shown that the Chou-Fasman method is only about 50–60% accurate in predicting the structural domains. The methods are most useful in the hands of a knowledgeable structural biologist, and have been used most successfully in polypeptide design and in analysis of motifs for organelle transport (Branden and Tooze 1991). A useful approach is to analyze each of a series of aligned amino acid sequences and then to derive a consensus structural prediction.

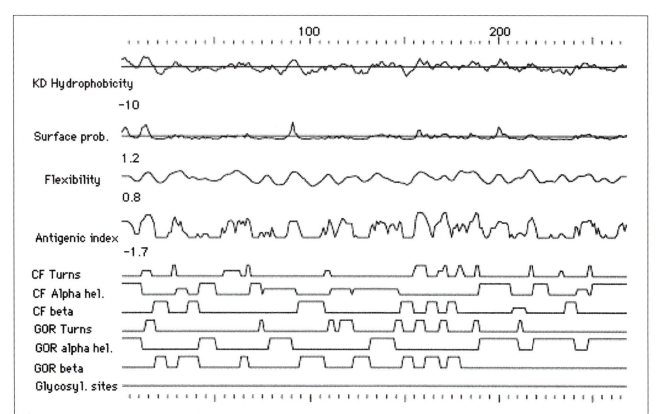

Figure 9.26. Example of the secondary structure predictions for the α subunit of *S. typhimurium* tryptophan synthase by the Chou-Fasman and GOR methods included in the Genetics Computer Group suite of programs. The predictions are shown on the lower panels, labeled as CF for the Chou-Fasman method (Chou and Fasman 1978) and GOR (referred to as GOR I) for the Garnier, Osguthorpe, and Robson method (Garnier et al. 1978). This protein is in the α-β class with an α/β barrel type of structure comprising eight parallel β strands and eight α helices in an alternating pattern and three additional α helices, and is shown in Fig. 9.6. The predicted structure is quite acccurate and represents the correct pattern of secondary structure.

Patterns of Hydrophobic Amino Acids Can Aid Structure Prediction

Prediction of secondary structure can be aided by examining the periodicity of amino acids with hydrophobic side chains in the protein chain. This type of analysis was discussed above in the prediction of transmembrane α-helical domains in proteins. Hydrophobicity tables that give hydrophobicity values for each amino acid are used to locate the most hydrophobic regions of the protein (Table 9.6) (see Lüthy and Eisenberg 1991). As for secondary structure prediction, a sliding window is moved across the sequence and the average hydrophobicity value of amino acids within the window is plotted. A hydrophobicity plot of the α subunit of *S. typhimurium* tryptophan synthase is included in the first panel of Figure 9.26.

Similar methods for predicting surface peptides including antigenic sites, chain flexibility, or glycosylation sites are also illustrated in Figure 9.26. These methods use the chemical properties of amino acid side chains to predict the location of these amino acids on the surface or buried within the core structure.

The location of hydrophobic amino acids within a predicted secondary structure can also be used to predict the location of the structure. One type of display of this distribution is the helical wheel or spiral display of the amino acids in an α helix, as shown in Figure 9.27. This use of this display was described above as a way to visualize the location of leucine residues on one face of the helix in a leucine zipper structure. There is also a tendency of hydrophobic residues located in α helices on the surface of protein structures to face the core of the protein and for polar and charged amino acids to face the aqueous environment on the outside of the α helix. This arrangement is also revealed by the helical wheel display shown in Figure 9.27. Another type of display, the hydrophobic moment display, is shown in Figure 9.28. The contours in this plot show positions in the amino acid sequence where hydrophobic amino acids tend to segregate to opposite sides of a structure plotted against various angles of rotation from one residue to the next along the protein chain. For α helices, the angle of rotation is 100 degrees and for β strands, 160 degrees. The analysis in the figure predicts, for example, an α helix at approximate sequence position 165 that has segregated hydrophobic amino acids on one helix face. Helix α5 runs from positions 160 to168 in the crystal structure of this protein.

Secondary Structure Prediction by Neural Network Models

The most sophisticated methods that have been devised to make secondary structural predictions for proteins use artificial intelligence, or so-called neural net algorithms. An earlier method of this type examined patterns that represent secondary structural features like the Chou-Fasman method. However, this method went farther and tried to locate these patterns in a particular order that coincides with a known domain structure. Patterns typical of α/β proteins (Cohen et al. 1983), turns in globular proteins (Cohen et al. 1986), or helices in helical proteins (Presnell et al. 1992) may be located and used to predict secondary structure with increased confidence. The program MACMATCH, which combines these methods with a neural network approach to predict the secondary structure of globular proteins on a Macintosh computer, has been described (Presnell et al. 1993).

In the neural network approach, computer programs are trained to be able to recognize amino acid patterns that are located in known secondary structures and to distinguish these patterns from other patterns not located in these structures. There are many examples of the use of this method to predict protein structures (see, e.g., Qian and Sejnowski 1988; Muggleton et al. 1992; Stolorz et al.1992; Rost and Sander 1993), which have been reviewed (Holley and Karplus 1991; Hirst and Sternberg 1992). The early methods are reported to be up to 63–64% accurate. These methods have been improved to a level of over 70% for globular proteins by the use of information from multiple sequence alignments (Rost and Sander 1993, 1994). Two Web sites that perform a neural network analysis for protein secondary structure prediction are PHD (Rost and Sander 1993; Rost 1996; http://www.embl-

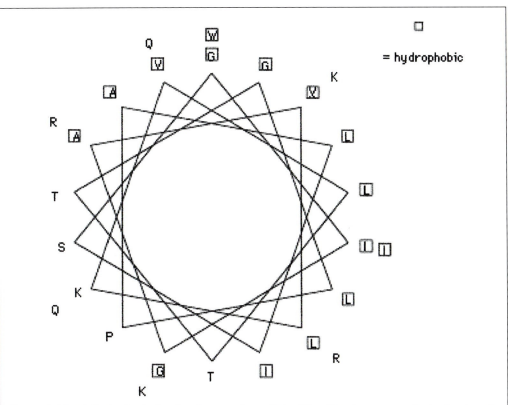

Figure 9.27. Helical wheel plot for the protein melittin. The plot shown was obtained using the Genetics Computer Group HELICALWHEEL program. The diagram shows the relative positions of amino acids in an end-on view of an α helix with the angle of rotation of 100° between adjacent amino acids in α helices (the angle would be 160° for β strands). The hydrophobic amino acids Leu (L), Ile (I), and Val (V) are primarily located on one side of the helix, thereby illustrating the amphiphobic nature of the helix.

heidelberg.de/predictprotein/predictprotein.html) and NNPREDICT (Kneller et al. 1990; http://www.cmpharm.ucsf.edu/~nomi/nnpredict.html). These neural network models are theoretically able to extract more information from sequences than the information theory method described above (Qian and Sejnowski 1988). Neural networks have also been used to model translational initiation sites and promoter sites in *E. coli*, splice junctions, and specific structural features in proteins, such as α-helical transmembrane domains. These applications are discussed elsewhere in this chapter and in Chapter 8.

Neural network models are meant to simulate the operation of the brain. The complex patterns of synaptic connections among a large number of neurons are presumed to underlie the functions of the brain. Some groups of neurons are involved in collecting data as environmental signals, others in processing data, and yet others in providing a response to the signals. Neural networks are an attempt to build a similar kind of learning machine where the input is a 13–17-amino-acid length of sequence and the output is the predicted secondary structure of the central amino acid residue. The object is to train the neural network to respond correctly to a set of such flanking sequence fragments when the secondary structural features of the centrally located amino acid are known. The training is designed to achieve recognition of amino acid patterns associated with secondary structure. If the neural network has sufficient capacity for learning, these patterns may potentially include complex interactions among the flanking amino acids in determining secondary structures. However, two studies with neural networks described below have so far not found evidence for such interactions.

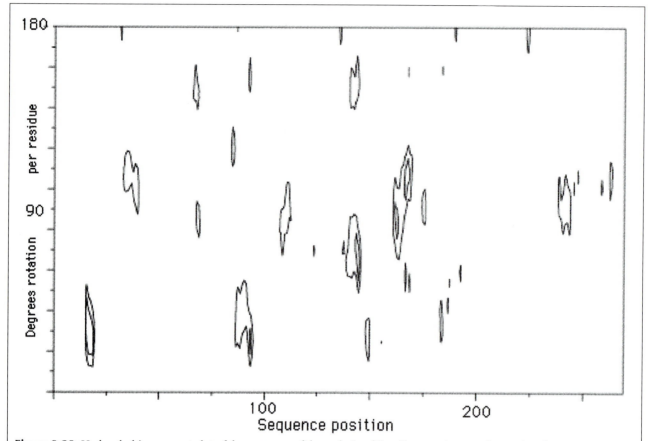

Figure 9.28. Hydrophobic moment plot of the sequence of the α chain of *E. coli* tryptophan synthase using the Genetics Computer Group MOMENT program. The moment uses the hydrophobicity values shown in Table 9.6 to measure the tendency of hydrophobic residues to be located on one face of a secondary structural element (Lüthy and Eisenberg 1991). The values are normalized so that the mean value is 0 and the standard deviation is 1. The moment is calculated for a window of 10 residues, 5 on each side of every amino acid position and for every possible rotational angle between adjacent residues. The angle is 100° for α helices and 160° for β strands. When one contour is shown, the moment values are 0.35; when two contours are shown, values are 0.35 (outer) and 0.45 (inner).

A typical neural network model used for protein secondary structure prediction is illustrated in Figure 9.29. A sliding window of 13–17 amino acid residues is moved along a sequence. The sequence within each window is read and used as input to a neural network model previously trained to recognize the secondary structure most likely to be associated with that pattern. The model then predicts the secondary structural configuration of the central amino acid as α helix, β strand, or other. Rules or another trained network are then applied that make the prediction of a series of residues reasonable. For example, at least 4 amino acids in a row should be predicted as being in an α helix if the prediction is to make structural sense.

The model comprises three layers of processing units—the input layer, the output layer, and the so-called hidden layer between these layers. Signals are sent from the input layer to the hidden layer and from the hidden layer to the output layer through junctions between the units. This configuration is referred to as a feed-forward multilayer network. The input layer of units reads the sequence, one unit per amino acid residue, and transmits information on the amino acid at that location. A small window of sequence is read at a time and information is sent as signals through junctions to a number of sequential units in the hidden layer by all of the input units within the window, as shown by the lines joining units in Figure 9.29. These signals are each individually modified by a weighting factor and then

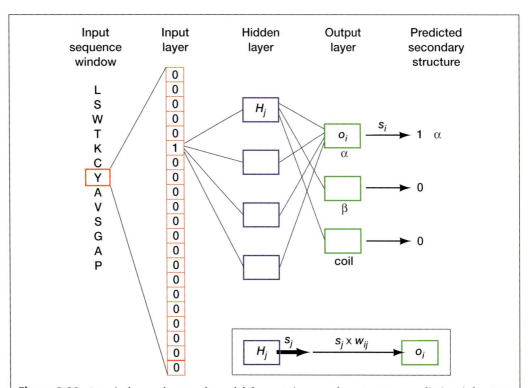

Figure 9.29. A typical neural network model for protein secondary structure prediction (after Rost and Sander 1993). Functions of the input (*red boxes*), hidden (*blue boxes*), and output (*green boxes*) layers are described in the text. There is one input unit for each amino acid in the sequence window of 13 (*first column*). Each input amino acid unit is made up of 21 input positions, one for each amino acid and one for a padding space when the window overlaps the end of the sequence. Other positions may be added to provide additional information. The positions each send information to the hidden unit layer. In a simple input coding system, only one of the 20 components in a given input unit has a value of 1. Shown is an example where the component for Y is turned on while the rest of the components are 0 (*second column*). When padding for the end of sequence is required, only the padding space is set to 1. When a sequence profile is used as input (not shown), each position is filled with the frequency of the amino acid in the corresponding column of the sequence profile or with a coded form of this frequency, and the numbers of insertions and deletions are added in two extra positions. Another position is used to indicate the amount of information due to the presence of conserved amino acids in the column. Signals from each position in each input unit are weighted as they proceed to units of the hidden layer. A signal from a component of one input unit will receive a different weight for each connection to a hidden unit. Each hidden unit sums the signals (s_{in}) received from the input layer and then transforms the sum using the trigger function $s_{out} = 1/(1 + e^{-ksin})$ to produce an output signal that is between and close to either 0 and 1, simulating the firing of a neuron. Strong signals are transformed by this function to a number approximately equal to 1 and weak or negative values to 0. As the constant k increases, discrimination between strong and weak signals is increased. The hidden layer output signals are weighted and sent to three output units, representing prediction of an α helix, β strand, or coil (loop) for the secondary structural configuration of the central amino acid in the window. The sum of these signals is transformed to values between 0 and 1. An output signal close to 1 is a prediction for the amino acid to have the corresponding structural configuration; a weak signal close to zero is no prediction. The example shown predicts an α-helical configuration for Y. Predictions for a series of adjacent windows are sorted out by applying rules or by additional neural networks. The insert illustrates the operation of the back-propagation algorithm that is used to train the network and is described by an example in the text (p. 455).

Information content of an alignment is discussed in Chapter 4, page 195.

added together to give a total input signal into each hidden unit. Sometimes a bias is added to this sum to influence the response of the unit. The resulting signal is then transformed by the hidden unit into a number that is very close either to a 1 or to a zero (or sometimes to a −1). A mathematical function known as a sigmoid trigger function, simulating the firing or nonfiring states of a neuron, is used for this transformation. Signals from the hidden units are then sent to three individual output units, each output unit representing one type of secondary structure (helix, strand, or other). Each signal is again weighted, the input signals are summed, and each of the three output units then converts the combined signal into a number that is approximately a 1 or a 0. An output signal that is close to 1 represents a prediction of the secondary structural feature represented by that output unit and a signal near to the value 0 means that the structure is not predicted.

When hidden layers are included, a neural network model is capable of detecting higher levels of interaction among amino acids that influence secondary structure. For example, particular combinations of amino acids may produce a particular type of secondary structure. To resolve these patterns, a sufficient number of hidden units is needed (Holley and Karplus 1991); the number varies from 2 to a range of 10–40. An interesting side effect of adding more hidden units is that the neural network memorizes the training set but at the same time is less accurate with test sequences. This effect is revealed by using the trained network to predict the same structures used for training. The number correct increases by over 20% as the number of hidden units increases from 0 to 10. In contrast, accuracy of prediction of test sequences not used for training decreased 3% (Holley and Karplus 1991).

Without hidden layers, the neural network model is known as a perceptron, and has a more limited capacity to detect such combinations. In two studies, networks with no hidden units were as successful in predicting secondary structure as those with hidden units. In addition, the number of hidden units was increased to as many as 60 in one study (Qian and Sejnowski 1988) and 20 in another (Holley and Karplus 1991) without significantly changing the level of success. These observations imply that the influence of local sequence on secondary structure is the additive influence of individual residues and that there is no higher level of interaction among these residues. To detect such interactions, however, requires a large enough training set to provide a significant number of examples, and these conditions may not have been met. These same studies examined the effect of input window size and found that a maximum information for secondary structure prediction seems to be located within a window of 13–17 amino acids, as larger windows do not increase accuracy. However, small windows were less effective, suggesting that they have insufficient information, and below a window size of 5, success at predicting β strands was decreased.

Training the neural network model is the process of adjusting the values of the weights used to modify the signals from the input layer to the hidden layer and from the hidden layer to the output layer. The object is to have these weights balance the input signals so that the model output correctly identifies the known secondary structure of the central amino acid in a sequence window of a protein of known structure. Because there may be thousands of connections between the various units in the network, a systematic method is needed to adjust these values. Initially, the weights are assigned a constant or random value (typical range −0.1 to +0.1). The sliding window is then positioned along one of the training sequences. The predicted output for a given sequence window is then compared to the known structure of the central amino acid residue. The model is adjusted to increase the chance of predicting the correct residue. The adjustment involves changing the weighting of propagated signals by a method called the back-propagation algorithm. This procedure is repeated for all windows in all of the training sequences. The better the model, the more predicted structures that will be correct. Conversely, the worse the model, the more predictions that will be incorrect. The object then becomes to minimize this incorrect number. The error E is expressed as the square of the total number of incorrect predictions by the output units.

Use of a perceptron for analyzing regulatory DNA sequences is illustrated in Figure 8.11 (p. 363).

When the back-propagation algorithm is applied, the weights are adjusted by a small amount to decrease errors. A window of a training sequence is used as input to the network, and the predicted and expected (known) structures of the central residue are compared. A set of small corrections is then made to the weights to improve an incorrect prediction, or the weights are left relatively unchanged for a correct prediction. This procedure is repeated using another training sequence until the number of errors cannot be reduced further. A large number of training cycles representing a slow training rate is an important factor for training the network to produce the smallest number of incorrect predictions. Not all of the training sequences may be used—a random input of training patterns may be used and sometimes these may be chosen from subsets of sequences that represent one type of secondary structure to balance the training for each type of structure. The back-propagation algorithm examines the contribution of each connection in the network on the subsequent levels and adjusts the weight of this connection, if needed to improve the predictions. The following example illustrates the operation of the algorithm.

**Example: Back-propagation Algorithm Used to
Train the Neural Network (Rost and Sander 1993)**

Consider an output unit O_i as shown in Figure 9.29. Let us assume that this unit predicts whether or not the central residue in the scanned sequence window is an α-helical secondary structure. The output signal from this unit is s_i which, if close to 1, predicts an α-helical structure or, if close to 0, does not predict an α helix. The network has been provided with a training sequence and it is known whether or not the central amino acid actually is found in an α helix. If the structure is an α helix, then the output of O_i should be close to 1, and if not, then close to zero. d_i is the expected or desired output of O_i and $d_i = 1$ if a helix is expected and 0 if not. The output of O_i is determined by the sum of the inputs received from each of the hidden units with which O_i is connected. The hidden units each emit a signal close to 0 or 1, and each signal is separately weighted as it passes from the hidden unit to O_i. Focus on one of the hidden units H_j that is connected to O_i and emits a signal s_j that is modified by weight w_{ij}. The signal arriving at O_i is thus $s_j \times w_{ij}$, as illustrated in the insert in Figure 9.29. The problem at hand is to adjust or not to adjust w_{ij} so that the output of O_i (s_i) is close to the desired value, d_i. The value of w_{ij} is adjusted according to a procedure known as gradient descent that is given by the formula

$$\Delta w_{ij} = w_{ij} - n\, \partial E/\partial w_{ij} + m \tag{10}$$

where the partial derivative of the error E with respect to w_{ij}, $\partial E/\partial w_{ij}$, is calculated by

$$\partial E/\partial w_{ij} = (s_i - d_i)\, s_i\, (1 - s_i)\, s_j \tag{11}$$

and where n is the rate of training (typical value 0.03) and m is a smoothing factor that allows a carryover of a fraction of previous values of w_{ij} (typical value 0.2). Suppose, for example, that s_j was sent from H_j to O_i as 0.2 and that d_i is 1, so that s_j is not contributing correct information. Then $\partial E/\partial w_{ij} = (0.2 - 1) \times 0.2 \times 0.8 \times 0.2 = -0.0256$. w_{ij} will then be increased in Equation 10 by the rate of training times this value adjusted by m for contributions from any previous value of w_{ij}. Adjusting the weights of connections between the input and hidden layers uses a more detailed formula that takes into account the effects of both the signal sent from the input unit to the hidden units and that of the hidden unit on each of the output units.

The PHDsec program in the PHD system described above in the section on prediction of transmembrane-spanning proteins (p. 437) is an example of a neural network program for protein secondary structure prediction (Rost and Sander 1993; Rost 1996). The Web address of this resource is http://www.emblheidelberg.de/predictprotein/predictprotein. html. PhDsec uses a procedure similar to that used by PHDhtm. A BLAST search of the input sequence is conducted to identify similar but not closely identical sequences, and a multiple alignment of the sequences is transformed into a sequence profile. This profile is then used as input to a neural network trained to recognize correlations between a window of 13 amino acids and the secondary structure of the central amino acid in the window. The neural network model is as the one shown in Figure 9.29. Program output includes a reliability index of each estimate on a scale of 1 (low reliability) to 9 (high reliability). These reliabilities (not shown) are obtained as normalized scores derived from the output values of the three units in the output layer of the network. The highest output value is compared to the next lowest value and the difference is normalized to give the reliability index. These indices are a useful way to examine the predictions in closer detail.

Example: Program Output from the PHD Server at
http://www.embl-heidelberg.de/ /predictprotein/predictprotein.html/

The input sequence was the α subunit of *S. typhimurium* tryptophan synthase (SwissProt ID TRPA_SALTY, accession P00929), which was originally included in the training sequences. (*A*) Secondary structure prediction. (H) α helix; (E) β strand; (L) loop; (.) no prediction. Rel is the reliability of the prediction on a scale of 0–9 (highest). Rel is based on the difference between the maximal and the second largest output signals from the network. The accuracies (Q_3) to be expected for each reliability value have been calculated and are shown in *B*. For example, when a Rel score of 9 is given a prediction for a residue, that prediction is known to be 94.2% accurate. Note that for a Rel of 9 the prediction of helix residues is almost 100% accurate but that prediction of strand residues is about 70% accurate. prH, prE, and prL are probabilities of helix, strand, and loop (no prediction), respectively, based on accuracy of the predictions shown in *B*. Subset is a listing of the more reliable predictions and includes all residues for which the expected accuracy is > 82%.

A. Predicted secondary structure and reliability of prediction.

```
                ....,....1....,....2....,....3....,....4....,....5....,....6
        AA   |MERYENLFAQLNDRREGAFVPFVTLGDPGIEQSLKIIDTLIDAGADALELGVPFSDPLAD|
        PHD  |  HHHHHHHHHH    EEEEEE     HHHHHHHHHH      EEEE          |
        Rel  |934899999996348872799842489984587999999997399668944767784689|
detail:
        prH-|036899999987531100000000000016788999999998300000000001113210
        prE-|000000000000000005798865300000000000000000000178863111000000
        prL-|963100000012368883100123689983211000000001699720036877886789
subset: SUB  |L..HHHHHHHH..LLL.EEEE...LLLL.HHHHHHHHHHH.LLLEEE..LLLLL.LLL|

                ....,....7....,....8....,....9....,....10...,....11...,....1
        AA   |GPTIQNANLRAFAAGVTPAQCFEMLALIREKHPTIPIGLLMYANLVFNNGIDAFYARCEQ|
        PHD  |  HHHHHHHHHHH    HHHHHHHHHHHHH      EEEEEE  HHHH  HHHHHHHHHH|
        Rel  |737899999999828962999999999999728999389988342244125599999999|
detail:
        prH-|258899999999841115899999999999984100000000023566432699999998
        prE-|000000000000000000000000000000000038998511000000000000000000
        prL-|741100000000158873000000000000158998610001355422457200000000
subset: SUB  |L.HHHHHHHHHH.LLL.HHHHHHHHHHHH.LLLL.EEEEE.......LHHHHHHHHHH|
```

```
              2...,....13...,....14...,....15...,....16...,....17...,....1
          AA  |VGVDSVLVADVPVEESAPFRQAALRHNIAPIFICPPNADDDLLRQVASYGRGYTYLLSRS
          PHD |H   EEEEE   HHHHHHHHHHHHH    EEEEE    HHHHHHHHHH    EEEEEE
          Rel |2695189946993225999999987299279997699987999999999829918999834
detail:
          prH-|5200000000000655799999999884000000000000189999999984000000000001
          prE-|0002589996200000000000000048999720000000000000000048999862
          prL-|4797410027993442000000001159951000179998100000000159951000135
subset: SUB |.LLL.EEE.LLL...HHHHHHHHHH.LL.EEEEELLLLLHHHHHHHHHH.LL.EEEEE..|


              8...,....19...,....20...,....21...,....22...,....23...,....2
          AA  |GVTGAENRGALPLHHLIEKLKEYHAAPALQGFGISSPEQVSAAVRAGAAGAISGSAIVKI|
          PHD |       HHHHHHHHHHHHH    EEEE    HHHHHHHHHH    EEE HHHHH|
          Rel |567544421256999999999981899727776366682899999999953997266149999|
detail:
          prH-|111222234577899999999985100000000000389999999973000011169999
          prE-|1111001000000000000000000001577732100000000000000015773100000
          prL-|677666655422000000000014899831126778510000000269974114200000
subset: SUB |LLLL......HHHHHHHHHHHH.LLLL.EEE.LLL.HHHHHHHHHH.LLL.EE..HHHHH|


              4...,....25...,....26...,....27
          AA  |IEKNLASPKQMLAELRSFVSAMKAASRA|
          PHD |HHH    HHHHHHHHHHHHHHHHH    |
          Rel |99725991889999999999998315899|
detail:
          prH-|99842005889999999999998642100
          prE-|00000000000000000000000000000
          prL-|00156994110000000000001356899
subset: SUB |HHH.LLL.HHHHHHHHHHHHHHH..LLLL|
```

B. The accuracy of secondary structure predictions for a given reliability index. % res is the % predicted residues that receive a given reliability index, Qtot is the overall prediction accuracy, H%obs and E%obs are the observed percentage of helix and strand (known structure), H%prd and E%prd are the predicted percentage of helix and strand predictions that are associated with such an index (provided by E-mail from server).

index	0	1	2	3	4	5	6	7	8	9
%res	100.0	99.2	90.4	80.9	71.6	62.5	52.8	42.3	29.8	14.1
Qtot	72.1	72.3	74.8	77.7	80.3	82.9	85.7	88.5	91.1	94.2
H%obs	70.4	70.6	73.7	77.1	80.1	83.1	86.0	89.3	92.5	96.4
E%obs	61.5	61.7	63.7	66.6	69.1	71.7	74.6	77.0	77.8	68.1
H%prd	77.8	78.0	80.0	82.6	84.7	86.9	89.2	91.3	93.1	95.4
E%prd	64.5	64.7	67.8	71.0	74.2	77.6	81.4	85.1	89.8	93.5

Nearest-neighbor Methods of Secondary Structure Prediction

Like neural networks, nearest-neighbor methods are also a type of machine learning method. They predict the secondary structural conformation of an amino acid in the query sequence by identifying sequences of known structures that are similar to the query sequence (Levin et al. 1986; Salzberg and Cost 1992; Zhang et al. 1992; Yi and Lander 1993; Salamov and Solovyev 1995, 1997; Frishman and Argos 1996). A large list of short sequence fragments is made by sliding a window of length n (e.g., $n = 16$) along a set of approximately 100–400 training sequences of known structure but minimal sequence similarity to each other, and the secondary structure of the central amino acid in each window is recorded. A window of the same size is then selected from the query sequence and compared to each of the above sequence fragments, and the 50 best-matching fragments are identified. The frequencies of the known secondary structure of the middle amino acid in each of these matching fragments (f_α, f_β, and f_{coils}) are then used to predict the secondary structure of the middle amino acid in the query window. As with other secondary structure prediction programs, the predicted secondary structure of a series of residues in the query sequence is subjected to a set of rules or used as input to a neural network to make a final prediction for each amino acid position.

Although not implemented in the most available programs, a true estimate of probability of the above set of frequencies may be obtained by identifying sets of training sequences that give the same value of $(f_\alpha + f_\beta + f_{coils})^{1/2}$. The frequencies of the secondary structures predicted by this group then give true estimates for p_α, p_β, and p_{coils} for the targeted amino acid in the query sequence (Yi and Lander 1993). Predictions based on the highest probabilities have been shown to be the most accurate, with the top 28% of the predictions being 86% accurate and the top 43% being 81% accurate. In addition, this method of calculating probability possesses more information than single-state predictions. Using this method, therefore, a substantial proportion of protein secondary structures can be predicted with high accuracy (Yi and Lander 1993, 1996).

The several nearest-neighbor programs that have been developed for secondary structure prediction (see Table 9.7) differ largely in the method used to identify related sequences in the training set. Originally, an amino acid scoring matrix such as a BLOSUM scoring matrix was used (Zhang et al. 1992). Distances between sequences based on a statistical analysis of the training sequences have also been proposed (Salzberg and Cost 1992). Use of a scoring matrix (Bowie et al. 1991, 1996) based on a categorization of amino acids into local structural environments, discussed below, in conjunction with a standard amino acid scoring matrix increased the success of the predictions (Yi and Lander 1993; Salamov and Solovyev 1995, 1997). Yet further increases in success have been achieved by aligning the query sequence with the training sequences to obtain a set of nonintersecting alignments with windows of the query sequence (as described in Chapter 3, p. 75), and of using a multiple sequence alignment as input with amino-terminal and carboxy-terminal positions of α helices and β strands and β turns treated as distinctive types of secondary structure (Salamov and Solovyev 1997).

The program PREDATOR (Table 9.7) is based on an analysis of amino acid patterns in structures that form H-bond interactions between adjacent β strands (β bridges) and between amino acid n and $n + 4$ on α helices (Frishman and Argos 1995, 1996). The H-bond pattern between parallel and antiparallel β strands is different (Fig. 9.3) and two types of antiparallel patterns have been recognized. By utilizing such information combined with substitutions found in sequence alignments, the prediction success of PREDATOR has been increased to 75% (Frishman and Argos 1997). Examples of the NNSSP (Salamov and Solovyev 1997) and PREDATOR (Frishman and Argos 1997) program outputs are given on page 459.

Example: NNSSP and PREDATOR Output

Two of the most accurate nearest-neighbor prediction programs are (1) NNSSP (accuracy to 73.5%) shown is the program output from http://dot.imgen.bcm.tmc.edu:9331/seq-search/struc-predict.html, choosing the PSSP/NNSSP option. PredSS is the predicted secondary structure by NNSSP (a = α; b = β; c = COILS). The output probabilities Prob a and Prob b give a normalized score by converting the values of f_α, f_β, and f_{coils} to a scale of 0–9. (2) PREDATOR (accuracy 75%) applies the FSSP assignments of secondary structure to the training sequences. PREDATOR does not provide a normalized score. PREDATOR predictions from http://www.embl-heidelberg.de/argos/predator/predator_info.html are shown below NNSP prediction on each line (H = α; E = β). The input sequence was the α subunit of *S. typhimurium* tryptophan synthase (SwissProt ID TRPA_SALTY, accession P00929), which is in the training sequences because the three-dimensional structure is known.

```
nnssp  Sat Mar 13 15:49:19 CST 1999
TS_subunit_alpha
 L=  268 SS content: a-  0.56 b=  0.08 c=  0.36
               10        20        30        40        50
PredSS     aaaaaaaaaaaaa    bbbbbb   aaaaaaaaaaaaaaaaaaaaa
AA seq     MERYESLFAQLKERKEGAFVPFVTLGDPGIEQSLKIIDTLIEAGADALEL
Prob a     99999999999974211100000010001688899999999974578863
Prob b     00000000000000001277788741000100000000000000001122
Predator   ___HHHHHHHHHHHHHH_EEEEEE_____HHHHHHHHHH_____

               60        70        80        90       100
PredSS              aaaaaaaaaa    aaaaaaaaaaaaa      bbba
AA seq     GIPFSDPLADGPTIQNATLRAFAAGVTPAQCFEMLALIRQKHPTIPIGLL
Prob a     11111111100124568899887311058899999999852000111133
Prob b     23221101100012110000001111000000000000002335544
Predator   _____HHHHHHHH____HHHHHHHHHHH____HHHH
             110       120       130       140       150
PredSS     aaaaaaa   aaaaaaaaaa   bbbbb       aaaaaaa
AA seq     MYANLVFNKGIDEFYAQCEKVGVDSVLVADVPVEESAPFRQAALRHNVAP
Prob a     54555453447899999988400100000111222234788998731111
Prob b     32112211000000000000001116898632211010000000000123
Predator   HHHHH_____HHHHHHHHH___EEEEEE_____HHHHHHHH___E

             160       170       180       190       200
PredSS     bbb     aaaaaaaaa    bbbb       aaaaaaaaaaaaaaaaa
AA seq     IFICPPNADDDLLRQIASYGRGYTYLLSRAGVTGAENRAALPLNHLVAKL
Prob a     00000000158999999731111212235211125556654388899999
Prob b     89852000000000000110113677531112211100112200000000
Predator   EEE_____HHHHHHHH___EEEEE_____HHHHH_____HHHHHH

             210       220       230       240       250
PredSS     aaa         aaaaaaaaaa      aaaaaaaaaaa   aaa
AA seq     KEYNAAPPLQGFGISAPDQVKAAIDAGAAGAISGSAIVKIIEQHINEPEK
Prob a     88632100111110111478999998745312222678888997542588
Prob b     00000001334332000000000000012211101000000000000000
Predator   HHH_____HHHHHHH_____HHHHHHHHH__HHH

             260
PredSS     aaaaaaaaaaaaaaaaaa
AA seq     MLAALKVFVQPMKAATRS
Prob a     989999998878898663
Prob b     000000000000000011
Predator   HHHHHHHH_____
```

Hidden Markov Model (Discrete-Space Model)

HMMs have been used to model alignments of three-dimensional structure in proteins (Stultz et al. 1993; Hubbard and Park 1995; Di Francesco et al. 1997, 1999; FORREST Web server at http://absalpha.dcrt.nih.gov:8008/). In one example of this approach, the models are trained on patterns of α helices, β strands, tight turns, and loops in specific structural classes (Stultz et al. 1993, 1997; White et al. 1994), which then may be used to provide the most probable secondary structure and structural class of a protein. The manner by which protein three-dimensional domains can be modeled is illustrated in Figure 9.30. An example of the class prediction by the Protein Sequence Analysis (PSA) server at Boston University is shown in Figure 9.31.

Prediction of Three-dimensional Protein Structure

Because the number of ways that proteins can fold appears to be limited, there is considerable optimism that ways will be found to predict the fold of any protein, just given its amino acid sequence. Structural alignment studies have revealed that there are more than 500 common structural folds found in the domains of the more than 12,500 three-dimensional structures that are in the Brookhaven Protein Data Bank. These studies have also revealed that many different sequences will adopt the same fold. Thus, there are many combinations of amino acids that can fit together into the same three-dimensional conformation, filling the available space and making suitable contacts with neighboring amino acids to adopt a common three-dimensional structure. There is also a reasonable probability that a new sequence will possess an already identified fold. The object of fold recognition is to discover which fold is best matched. Considerable headway toward this goal has been made.

Sequence alignment can be used to identify a family of homologous proteins that have the same sequence, and presumably a similar three-dimensional structure. As discussed above, there are many databases that link sequence families to the known three-dimensional structure of a family member. The structure of even a remote family or superfamily member can be predicted through such sequence alignment methods. When the sequence of a protein of unknown structure has no detectable similarity to other proteins, other methods of three-dimensional structure prediction may be employed. One such method is sequence threading.

In threading, the amino acid sequence of a query protein is examined for compatibility with the structural core of a known protein structure. Recall that the protein core is made up of α helices, β strands, and other structural elements folded into a compact structure. The environment of the core is strongly hydrophobic with little room for water molecules, extra amino acids, or amino acid side chains that are not able to fit into the available space. Side chains must also make contact with neighboring amino acid side chains in the structure, and these contacts are needed for folding and stability. Threading methods examine the sequence of a protein for compatibility of the side groups with a known protein core. The sequence is "threaded" into a database of protein cores to look for matches. If a reasonable degree of compatibility is found with a given structural core, the protein is predicted to fold into a similar three-dimensional configuration. Threading methods are undergoing a considerable degree of evolution at the present time. An excellent description of algorithms for threading is found in Lathrop et al. (1998). Presently available methods require considerable expertise with protein structure and with programming. However, there are some sites where the analysis may be performed on a Web server, as shown in Table 9.8.

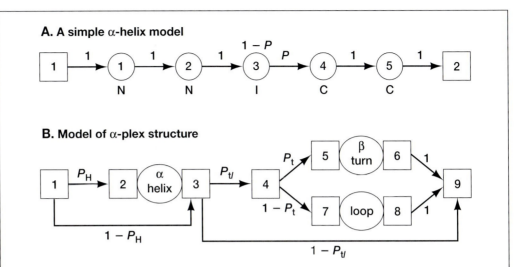

A. A simple α-helix model

B. Model of α-plex structure

Structural domains have also been modeled by position-specific scoring matrices (3D-PSSM, see Table 9.5). Hidden Markov models have also been used to model multi-domain proteins in which domains are represented by PSSM (Meta-MEME, p. 190).

Figure 9.30. Hidden Markov model (discrete state-space model) of protein three-dimensional structure (from Stultz et al 1993; White et al. 1994). (*A*) Example of model for one of the basic structural elements, an α helix. Squares indicate junctions in the model used to connect this model to models for other elements, including other α helices, β strands, β turns, and loops. Circles represent hidden states in the model. Those circles labeled N (1 and 2) are amino-terminal amino acids, I (3) is an internal amino acid, and C (4 and 5) are carboxy-terminal amino acids in the α helix. As in hidden Markov models that represent multiple sequence alignments, each of these states has a distribution of amino acids as found in aligned structures. For example, the frequency of H in an amino-terminal position in an α helix is 0.015, whereas in a carboxy-terminal amino acid, it is 0.028 in the proteins modeled by Stultz et al. (1993). At portions of internal amino acids, additional amino acids may be added to the α model. Arrows indicate a path to be followed through the states of the model, starting and ending with a numbered junction. Usually, the transition probability from one state to the next is 1, but at internal positions two alternative transition probabilities are possible, one (P) adds only one internal amino acid, whereas a second adds an additional internal amino acid with probability $1 - P$. (*B*) Model for a particular structural complex includes elements such as the α-helix model described in *A*. This particular model is of an α-helix complex named an α plex. These plexes are then combined to produce a model of a three-dimensional structural domain. The oval α helix is a condensed representation of the finite-state α-helix model shown in *A* with functions at each end. Similar oval representations for models of a β turn and a loop are also shown in this example. The arrows between junctions indicate transitions from one type of plex to another, and in several places there are two to three alternative possibilities in this model. Note that the plex can start with an α helix with transition probability P_H, but can also skip the α helix with probability $1 - P_H$. The remaining paths in the model include the possibilities of an α turn, a loop, or of no further elements. To turn domain models into a predictive tool, they are trained on a set of known proteins of that type. In this procedure, transitional probabilities between junctions, states, and plexes are adjusted using the sequences and structures of the training sequences until the model is optimal for distinguishing those domains from other types of domains. A structural prediction for a new protein sequence is then made by finding the most probable path of the sequence through a set of domain models and choosing the model that gives the best alignment with the new sequence. The procedure of calculating the probability of paths through each structural model and of finding the most probable path is similar to that used for sequence alignment models discussed in Chapter 4 (see Fig. 4.16 on p. 186). Only a limited number of domains have been modeled by this approach.

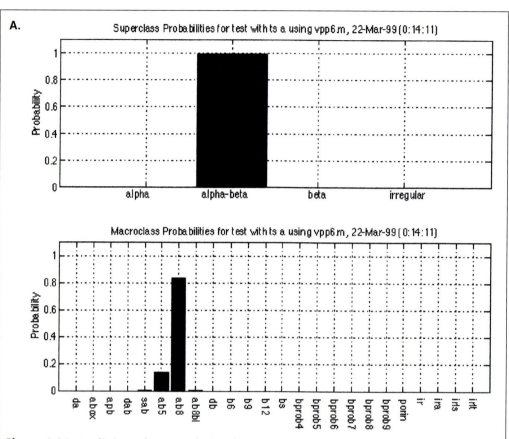

Figure 9.31. Prediction of structural class by the Protein Sequence Analysis (PSA) System at http://bmerc-www.bu.edu/psa/index.html. The analysis is based on the training of hidden Markov models for each structure by the sequences that specify those structures as described in Fig. 9.30 and in the text. The input sequence was the α subunit of *S. typhimurium* tryptophan synthase (SwissProt ID TRPA_SALTY, accession P00929), which was correctly identified as an α-β class protein. Proteins that are homologous to the input sequence were excluded from the analysis. Hence, although the structure was available, it was not used to produce the correct model. (*A*) The posterior probability of the protein being in a particular structural class as defined by the server is given, class α-β and macroclass ab8. (*B*) Probability plot of sequence being in a helix, strand, or turn. (*C*) Contour probability plot of secondary structural features combined with information about amphipathicity (the segregation of hydrophobic and nonhydrophobic residues to opposite faces of secondary structures). Areas of high probability appear as dark regions with closely spaced contour lines at probability increments of 0.1. Identification of structural class greatly facilitates the identification of secondary structural features. These figures were returned from the server by E-mail as postscript (ps) files.

Continues on next page

There are two methods in common use for deciding whether or not a given protein sequence is compatible with a known structural core, the environmental template (or structural profile) method and the contact potential method. In the environmental template method (Bowie et al. 1991, 1996; see also Ouzounis et al. 1993; Johnson et al. 1996), the environment of each amino acid in each known structural core is determined, including the secondary structure, the area of the side chain that is buried by closeness to other atoms, types of nearby side chains, and other factors. On the basis of these descriptions at each site, the position is classified into one of 18 types, 6 representing increasing levels of residue burial and fraction of surface covered by polar atoms combined with three classes of secondary structure. Each amino acid is then assessed for its ability to fit into that type

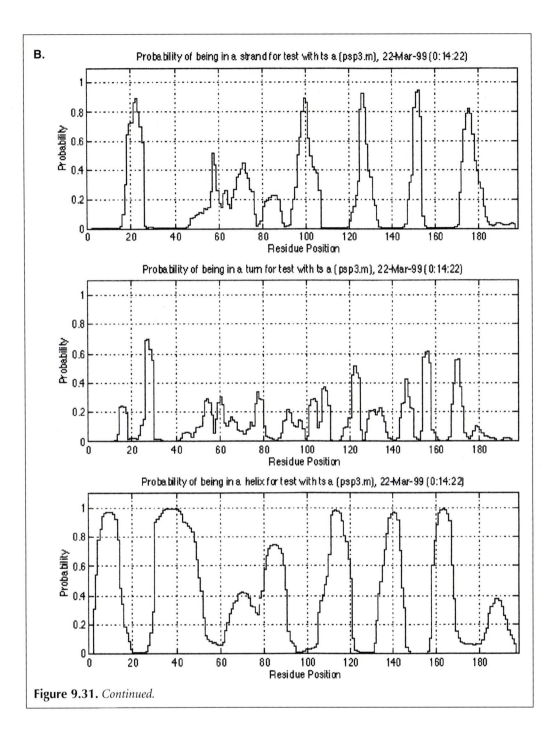

Figure 9.31. *Continued.*

of site in the structure. For example, if the side group is buried, another amino acid with a hydrophobic side chain may fit best into the structure at that position. The sequence of the protein is then aligned with a series of such environmentally defined positions in the structure to see whether a series of amino acids in the sequence can be aligned with the assigned structural environments of a given protein core. The procedure is then repeated for each core in the structural database, and the best matches of the query sequence to the core are identified. In the residue–residue contact potential method, the number and closeness of contacts between amino acids in the core are analyzed (Sippl 1990; Jones et al. 1992; Sippl and Weitckus 1992; Bryant and Lawrence 1993). The query sequence is evaluated for

C.

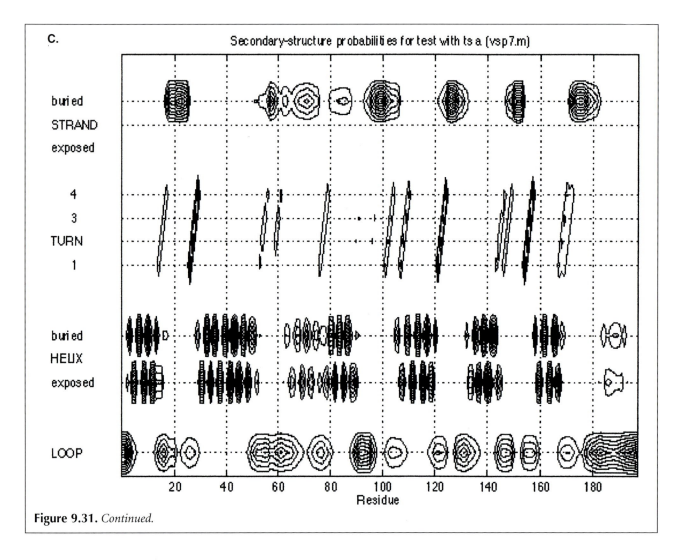

Figure 9.31. *Continued.*

amino acid interactions that will correspond to those in the core and that will contribute to the stability of the protein. The most energetically stable conformations of the query sequence thereby provide predictions of the most likely three-dimensional structure.

Structural Profile Method

In the structural profile method, predictions as to which amino acids might be able to fit into a given structural position are in the form of a sequence profile. This method assumes that if the query protein folds the same way as a target structure, the environments of the amino acids will be in the same linear order as they are in the target. In the normal scoring matrix, it is assumed that a given amino acid substitution always has the same likelihood of every occurrence of the substitution. However, in protein three-dimensional structures, a given substitution may have quite different effects depending on where in the structure and in which structure the substitution occurs. In a loop, where there are not many chemical and physical constraints, the substitution may usually not have any deleterious effects on the overall structure of the protein. In contrast, the same substitution in protein cores, where there are many restraints, may sometimes be possible without deleterious effects, but in other cases may be extremely deleterious. Thus, a sequence profile giving values for substitutions at each amino acid position is made for each core in the PDB.

Table 9.8. *Threading servers and program sources*

Program	Web address	Method	Reference
123D	http://www-lmmb.ncifcrf.gov/~nicka/123D.html	contact potentials between amino acid side groups	Alexandrov et al. (1996)
3D-PSSM	http://www.bmm.icnet.uk/~3dpssm	sequence-structure using position-specific scoring matrices	Russell et al. (1997)
Honig lab	http://honiglab.cpmc.columbia.edu/	threading methods using biophysical properties	see Web site
Libra I	http://www.ddbj.nig.ac.jp/htmls/E-mail/libra/LIBRA_I.html	target sequence and 3D profile are aligned by dynamic programming	Ota and Nishikawa (1997)
NCBI structure site	http://www.ncbi.nlm.nih.gov/Structure/RESEARCH/threading.html	Gibbs sampling algorithm used to align sequence and structure[a]	Bryant (1996)
Profit	http://lore.came.sbg.ac.at/home.html	fold recognition by the contact potential method	M. Sippl (see Web site)
Threader 2	http://insulin.brunel.ac.uk/threader/threader.html	prediction by recognition of the correct fold from a library of alternatives	Jones et al. (1995)
TOPITS	http://www.embl-heidelberg.de/predictprotein/doc/help05.html #P5 adv prd topits	detects similar motifs of secondary structure and accessibility between a sequence of unknown structure and a known fold	Rost (1995a,b)
UCLA-DOE structure prediction server	http://www.doe-mbi.ucla.edu/people/frsvr/frsvr.html	fold-recognition using 3D profiles and secondary structure prediction methods	Fischer and Eisenberg (1996)

Information on the research groups that work on structure prediction may be found at the CASP2 Web sites accessible at http://predictioncenter.llnl.gov/.

[a]Program has to be set up on a UNIX server.

These profiles, one for each core in the database, are then used to score the query sequence to be modeled for compatibility with that core.

The structural three-dimensional profile is a table of scores with one row for each amino acid position in the core and a column for each possible amino acid substitution at that position plus two columns for deletion penalties at that site, as shown in Figure 9.32. Each position in the core is assigned to one of 18 classes of structural environment. The scores in each row reflect the suitability of a given amino acid for that particular environment. The penalty at each core position reflects the acceptability of an insertion or deletion of one or more amino acids at that position in the structure. If the position is within the core, these penalties are generally high to reflect incompatibility with the structure, but lower for positions on the surface of the core and within loop regions. The dynamic programming algorithm is used to identify an optimal, best-scoring alignment, much as in aligning sequences by dynamic programming (discussed in Chapter 3). If a target structure is found to have a significantly high score, the new sequence is predicted to have a fold similar to that of the target core.

An entire database of sequences may be matched to a given structural profile to find the most compatible, a procedure called inverse folding. The alignment score for each protein is determined and then converted to a Z score, the number of standard deviations from the mean score for all of the sequences. The highest scoring sequences are the most compatible with a given structure (Bowie et al. 1996).

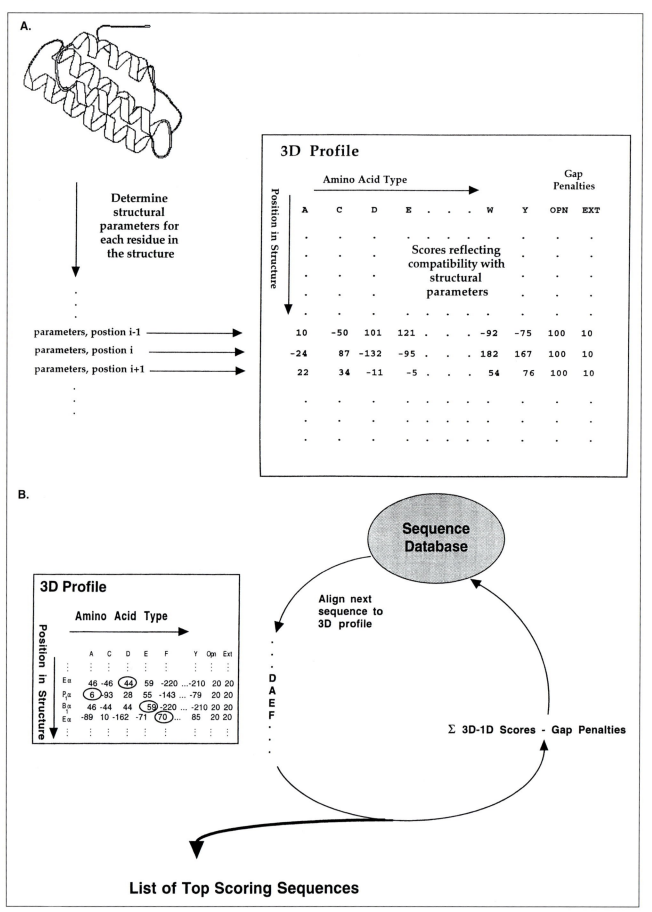

The above three-dimensional profile provides a discrete list of scores for matching one-dimensional sequence to a three-dimensional structure. This profile undergoes sharp transitions in values as the structural environment changes. Improved performance has been achieved by smoothing the values in these transitional regions to give a more gradual change using a Fourier analysis. Another improvement in the profile representation of protein three-dimensional structures, known as the *residue pair preference profile* (R3P) method, has been introduced (Wilmanns and Eisenberg 1993, 1995; Bowie et al. 1996).

R3P takes into account the amino acid neighbors, main-chain conformations, and secondary structure of each residue in the structure. Recall that to make amino acid scoring matrices for sequence–sequence comparisons, the frequency of amino acid substitutions in alignments is counted in sequence alignments. These frequencies are then divided by the expected frequency of finding the amino acids together in an alignment by chance. The ratio of the observed to expected counts is an odds score, and this score is usually converted to a log odds score for convenience in combining likelihood scores by adding their logarithms. Similarly, in the R3P method of making a three-dimensional scoring profile, the frequency of finding a particular pair of interacting amino acids, each with a particular structural feature, is calculated from the number of occurrences in known structures. For example, how often does amino acid a in an α helix interact with amino acid b in a β strand? This observed frequency of interaction in a specific structural configuration is then divided by the frequency of finding a and b interacting in any configuration, and the result is converted to a log odds score.

The pair preference log odds score S (aa_i,s_i,aa_j,s_j) for the amino acids aa_i and aa_j having properties c_i and c_j, respectively, is given by

$$S\ (aa_i,c_i,aa_j,c_j) = \ln\ [\ P\ (aa_i,c_i,aa_j,c_j)\ /\ P\ (aa_i,aa_j)] \tag{12}$$

where P (aa_i,c_i,aa_j,c_j) is the frequency of amino acids aa_i and aa_j having properties c_i and c_j, respectively, and P (aa_i,aa_j) is the frequency of finding an amino acid pair aa_i and aa_j. The score for position aa_i is then given by a weighted sum of all scores for the interacting pairs with aa_i.

$$S\ (aa_i) = \Sigma\ w_j\ S\ (aa_i,c_i,aa_j,c_j)/\ \Sigma\ w_j \tag{13}$$

where w_j is a weight representing the compatibility of the environment residue with its own local environment.

Amino acid interactions of a given amino acid residue in a particular core are then analyzed. To determine the neighbors of a given amino acid in the structure, a sphere of radius 12 Å is drawn centered on the C_β atom (see Fig. 9.2). If the C_β atom of another residue in the structure falls within this sphere, they may be interacting. A cylinder of radius 1.6 Å is then drawn between the C_β atoms and, if no H bonds or any other

---

Figure 9.32. The structural three-dimensional profile. (*A*) Generation of a three-dimensional profile of a structural core. (*B*) Screening sequences for compatibility with 3D profile. The methods of analysis are described in the text. (*A* and *B*: Redrawn, with permission, from Bowie et al. 1996 [copyright Academic Press].)

residue falls within this cylinder, the amino acid pair is considered to be interacting. This procedure is repeated for each amino acid that falls within the sphere, resulting in a defined list of approximately 8 amino acid pairs that are close enough without barriers to prevent interaction with the given residue. The amino acid type and one of several structural properties for the residue in question and for each interacting residue are then obtained. For example, the secondary structure of the two residues (α helix, β strand, or other) may be taken into account, giving 20 $\times$ 3 possible combinations of amino acid and secondary structure. Structural properties of the interacting residue may instead include the backbone dihedral angles Φ and Ψ (see Fig. 9.2) and the number of neighboring residues.

The structural configurations of the given residue and each interacting neighbor are determined. From this information, a score for this interaction can be found from the above analysis. Scores for all of the remaining interacting residues in their particular configurations can be found and then added to give a log odds score for the given amino acid site in the core. This score represents the likelihood of finding such a set of amino acid neighbors in their respective configurations in known protein structures. A value is then determined for various amino acid substitutions or for placing an insertion or deletion at that site. A similar set of scores is then obtained for each position in the protein core to generate a three-dimensional profile matrix based on the neighboring interactions. Three such profiles have been generated, one for each type of structural property in the amino acid pairs—backbone angles, secondary structure, and the number of neighboring residues for the interacting amino acid. A combined three-dimensional profile using elements of these residue pair preference profiles and those of the neighborhood three-dimensional profiles has also been used.

Sequence–structure alignments produced by the R3P method can be improved by an iterative procedure. In the initial alignment between a sequence and the three-dimensional profile of a core, predictions are made as to which residues will interact in the modeled three-dimensional structure. This feature provides information for improving the alignment. Likelihood scores for the predicted interactions can be calculated in the same way as described above for the amino acid interactions in the core. The scores for these interactions may then be summed, as before. In this case, these scores are weighted before summing to reduce the influence of those neighboring amino acids that are not in a compatible environment (Bowie et al. 1996). In evaluations of the R3P method with known three-dimensional structures, alignments are 50% or more correct on average for sequences whose three-dimensional structure pairs superimpose with a root mean square (rms) deviation of 1.97 Å or less (Wilmanns and Eisenberg 1995). Sequence–structure alignments may be further improved by including in the analysis the predicted secondary structure of the input sequence, with further improvements in fold assignment of 25% (Fischer and Eisenberg 1996).

One disadvantage of the structural profile method and the use of environmental variables is that these properties are statistically associated with the original sequence. Hence, the method retains a preference for matching the original sequence of the core protein. On the other hand, the success of present methods of three-dimensional structure prediction depends on a certain minimal level of similarity. The sequence of environmental patterns in the query sequence and the structure must also be in the same order throughout the sequence for the method to work. However, as discussed above for the SSAP alignment program, this problem may be circumvented by using local alignments.

Contact Potential Method

In this method, each structural core is represented as a two-dimensional contact matrix. The method is very similar to that used by the distance matrix method of the program DALI and illustrated in Figure 9.15. A simple matrix is produced with the amino acids in the structure listed across the rows and down the columns. In each matrix position, the distance between the corresponding pair of amino acids in the structure is placed. The amino acids in closest contact are immediately recognizable, and a group produces recognizable patterns. The object is to superimpose sets of amino acid pairs in the query sequence on to the distance matrix of the core. As shown in Figure 9.15, part B, sequences that fold into a similar structure should show similar contacts, although the amino acids that make up each structural feature do not have to be in the same linear order in both sequences. However, a large number of contacts must be analyzed to find the correct alignment.

To find the best combinations, the approximate conformational energies of each predicted pair are summed to predict the conformational stability of the predicted structure. Contacts have been extensively analyzed, and lookup tables with energies associated with these contacts have been produced. Hence, the energetic contributions of many possible combinations of pairs can be tested in a relatively short period of time. Computer experiments have revealed that contact energies can be used to choose the correct core in a structural database. Supporters of this method claim that the method can detect structural similarity in proteins that do not share any detectable sequence similarity. However, as shown in the next section, in truly blind experiments, the reliability of predictions drops when there is less than 25% sequence identity. A possible limitation to this analysis is that the energy associated with an isolated amino acid pair is assumed to be similar to that found in known protein structures. Recent experiments have suggested that the conformational energy of groups of amino acids larger than two may provide a more reliable prediction.

Example: Structure Prediction by Web Servers That Provide a Threading Service

These results were sent by E-mail. (A) Structure prediction by Libra (Table 9.8) and (B) UCLA-DOE structure prediction server. This server also provides a Web page for each match giving the results of other types of sequence database searches, secondary structure analyses, and the TOPITS server results. This analysis is of the α subunit of S. typhimurium tryptophan synthase (SwissProt ID TRPA_SALTY, accession P00929). The Web addresses and methods used by these servers are given in Table 9.8.

A.

```
----------------------------------------------------
         Forward Folding Search by LIBRA I
----------------------------------------------------
       LIBRA I was written by M. Ota in 1994-97

gap1= 2.400 gap9= 4.800 gapE= 0.400 npdb= 1389
gap1: Gap opening penalty for exposed sites
gap9: Gap opening penalty for buried sites
gapE: Gap extension penalty
npdb: Number of the structural templates

[Input sequence was the a subunit of S. typhimurium tryptophan synthase, Swiss-prot ID TRPA_SALTY,
accession P00929]

Compatible structures are:

  Rk StrC  Protein                         Lsr Lal    Rsc     SD    Rs/N    ID%
   1 2tsyA TRYPTOPHAN SYNTHASE;            262 268  -154.2  -5.57  -0.576   83.2
   2 1tlfA TRYPTIC CORE FRAGMENT OF THE L  296 284   -99.4  -2.94  -0.350   12.0
   3 2liv- LEUCINE(SLASH)*ISOLEUCINE(SLAS  344 275   -95.4  -2.75  -0.347    9.1
.
.
Key:
Rk  : Rank position
StrC: Structural code
Lsr : Length of the structural template
Lal : Length of the aligned region
Rsc : Raw score of the structural template
SD  : Standardized score
Rs/N: Raw score (Rsc) normalized by the alignment length (Lal)
ID% : Sequence identity

3D-1D alignments are:

  1 2tsyA structure vs your tsa sequence

3177238623 7322149997 9899888615 1775499469 6466899979
1AAAAAAAAA AAAgleeBBB BBBBelegeA AAAAAAAAAA AAAlegeBBB
MERYENLFAQ LNDRREGAFV PFVTLGDPGI EQSLKIIDTL IDAGADALEL
:::::: :::: :   :  :   : ::::: :::::::::::: ::::::::::: :  ::::::::
MERYESLFAQ LKERKEGAFV PFVTLGDPGI EQSLKIIDTL IEAGADALEL
.
.

Key:
1st line: Accessibility of the aligned site;
          1(exposed to water)-9(buried in protein)
2nd line: Local conformation of the aligned site;
          A(alpha), B(beta), gel(coils classified by the dihedral angles)
3rd line: Sequence of the template structure
4th line: Match site of the alignment
5th line: Query sequence
```

B.

```
Most similar fold: 1wsya
TRYPTOPHAN SYNTHASE (E.C.4.2.1.20)

RANK Z-SCORE        FOLD LENGTHALI %ID
   1   77.23        1wsya   248     85
   2    5.21        1aj0    232     24
   3    4.64        1ad1a   222     18.
.
.

LEGEND:
COL. 1: RANK. The ranks are obtained by sorting the fold library,
            by Z-SCORES, in decreasing order. Only the 15
            structures that are most compatible to your sequence
            are shown.
COL. 2: Z-SCORE. The z-scores are  computed using
            the distribution of raw scores (not shown) of all folds.
```

```
COL. 3: FOLD. Protein Data Bank codes for the coordinates of the 3D
                structures.
COL. 4: LENGTHALI. The number of residues from your sequence that were
                aligned to the fold.
COL. 5: % ID. Percentage of identical residues in the alignment.
[Description of each match not shown]

RELIABILITY OF THIS PREDICTION:
            With this method the confidence threshold
            is a z-score of  4.8 +- 1.0.
    YOUR HIGHEST SCORING FOLD IS ABOVE THIS THRESHOLD


Below is the alignment of your sequence with the top hit structure.
In the near future, all the alignments in a more readable format will be made available.


hhhhhhhhhhhhh    bbbbbbbb    hhhhhhhhhhhhh    bbbb
MERYESLFAQLKERKEGAFVPFVTLGDPGIEQSLKIIDTLIEAGADALEL
||||||  ||||||   |  |||||||||||||||||||||||||  ||||||||||
MERYENLFAQLNDRREGAFVPFVTLGDPGIEQSLKIIDTLIDAGADALEL
  hhhhhhhhhh      bbbbbb     hhhhhhhhhhhhh      bbb
·
·
```

EVALUATING THE SUCCESS OF STRUCTURE PREDICTIONS

As the above methods were developed, they were tested for ability to predict a structure that was already known. The structure to be predicted may be left out of the learning step so that the method has not been trained in any recognizable way to identify the correct structure. However, when the result is already known, there is always a possibility that the method was helped in some unintended way to identify the correct structure. A totally blind test of prediction accuracy provides a more objective test. A series of contests called CASP (critical assessment of structure prediction) was conceived in which structural biologists who were about to publish a structure were asked to submit the corresponding sequence for structure prediction by the contestants. The predictions were then compared with the newly determined structures. The newest CASP3 competition is given on a Web site (http://predictioncenter.llnl.gov/casp3/results/access.cgi). The results of earlier projects are given at http://predictioncenter.llnl.gov/casp1/ and http://predictioncenter.llnl.gov/casp2/. The contest involved a large number of research groups using a variety of methods including threading techniques. In one report of CASP3, the authors suggest that although there was overall progress from CASP1 to CASP2, there was little additional progress from CASP2 to CASP3. However, some improvement can be argued in CASP3 since the targets were more difficult (Sippl et al. 1999).

In the CASP2 conference, 32 groups made a total of 369 predictions on 15 different targets. There were two goals for each group: (1) to predict the correct three-dimensional fold of the target protein as the most similar known structures and (2) to predict the alignment of the sequence to the fold accurately. Once the structures of the prediction targets became available, the structure was aligned by DALI, SSAP, and VAST with all entries in the structural database to determine the closest matching structures that should have been found and also the sequence–structure alignment. The predictions were then compared to these alignments and evaluated for accuracy by specific criteria (Levitt 1997; Marchler-Bauer et al. 1997). This task was a most difficult one because different groups of investigators made predictions for different groups of proteins, some proteins much more difficult to predict than others. The range of sequence identity of target sequences to a known structure varied from 20% to 85%. The most difficult to predict and also the least successfully predicted were those that have less than 25% identity to any other protein of known structure. The easiest and most successfully predicted were those with sequence similarity above 25% (Martin et al. 1997).

The results of the CASP2 contest have been published by the participants in a special issue of *PROTEINS: Structure, Function and Genetics*, Suppl. 1, 1997, which provides details of the threading methods used. A similar volume discusses progress of CASP3 (*PROTEINS*, Suppl. 3, 1999). Threading methods improved considerably in performance in the 2-year period between the CASP1 and CASP2 meetings. A large number of groups using threading methods recognized the easier targets and performed much better than using simple sequence alignments (Levitt 1997). The advantages of using distant sequence homology and human knowledge of protein structure to predict three-dimensional structure was demonstrated by Murzin and Bateman (1997), who made the largest number of correct predictions. Their method uses the SCOP database, which organizes all known protein folds according to their structural and evolutionary relationships, for manual predictions. Their approach correctly assigned into an existing SCOP superfamily all six targets that were attempted, and found a homologous protein with a very similar structure. Local alignments between the target sequence and the corresponding protein superfamily were also among the most accurate. Several threading groups that were among the best performers are given in Table 9.8. At the present time, these methods are most suitable for modeling sequences that are recognizably similar to a known structure. These results confirm an earlier analysis that threading algorithms are quite disappointing in performance (Lemer et al. 1995). Improvements have been achieved by using a set of multiply aligned sequences instead of a single sequence (Defay and Cohen 1996; Ortiz et al. 1998).

STRUCTURAL MODELING

In the above section, detecting sequence similarity between a query sequence and a sequence of known structure plays an important role in successful structure prediction. Database searches as described in Chapter 6 provide alignments of a query sequence with a database of sequences, and can be used to search a database of protein sequences restricted to those of known structure. Hence, any alignment provides an indication as to which amino acids in the query may occupy a particular position in a structure. A search of this kind may be enhanced by superimposing the query sequence onto the molecular backbone of the matched sequence to produce a PDB file suitable for analysis by a three-dimensional viewer. An example of this type of analysis is provided by the Swiss-model Web site (Table 9.9). Molecular distances, angle, and energies of the superimposed sequence may then be analyzed and manipulated by the SPDBV viewer (Table 9.4). Additional Web sites for molecular modeling are listed in Table 9.9.

Table 9.9. *Web sites for predicting structural features of a query sequence*

Site	Web address	Description	Reference
Modeller	http://guitar.rockefeller.edu/ modeller/modeller.html	dynamic programming alignment of sequences and structures and molecular dynamics methods	Sali et al. (1995)
Swiss-model	http://www.expasy.ch/swissmod/ SWISS-MODEL.html	sequence alignment of query with sequences of known structure	Peitsch (1996)
Whatif	http://www.cmbi.kun.nl/whatif/	flexible molecular graphics rendering of models	Rodriguez et al. (1998)

SUMMARY AND FUTURE PROSPECTS

This chapter has described a number of methods for predicting protein structure from amino acid sequence. The best approach is to locate a link by sequence analysis between a new protein and a protein of known structure. Even a marginal sequence alignment with a protein of known structure can provide a feasible structural model. Databases that organize proteins into clusters and families with links to known protein structure are also a valuable resource for structure prediction. Proteins that represent new structural folds and domains can be readily identified in these databases, and these proteins can then be targeted for structural analysis by laboratory methods. Meanwhile, the methods for secondary structure and threading analysis (fitting a sequence to a structure) can provide useful predictions, although with variable levels of reliability. Increased confidence should come when several methods give a similar prediction.

The analysis of genomes described in Chapter 10 offers an additional opportunity for protein analysis. Functions of proteins can be discovered through conserved patterns of gene regulation and organization on the chromosomes of related organisms. The function and structure of a protein in one organism can then be predicted based on the function and structure of a functionally similar protein in a second organism.

REFERENCES

Alexandrov N.N. and Fischer D. 1996. Analysis of topological and nontopological structural similarities in the PDB: New examples with old structures. *Proteins* **25:** 354–365.

Alexandrov N.N., Nussinov R., and Zimmer R.M. 1996. Fast protein fold recognition via sequence to structure alignment and contact capacity potentials. *Pac. Symp. Biocomput.* 1996: 53–72.

Attwood T.K., Flower D.R., Lewis A.P., Mabey J.E., Morgan S.R., Scordis P., Selley J., and Wright W. 1999. PRINTS prepares for the new millennium. *Nucleic Acids Res.* **27:** 220–225.

Bairoch A. 1991. PROSITE: A dictionary of sites and patterns in proteins. *Nucleic Acids Res.* (suppl.) **19:** 2241–2245.

Barker W.C., Pfeiffer F., and George D.G. 1995. Superfamily and domain. In *Methods in protein structure analysis* (ed. M.Z. Atassi and E. Appella), pp. 473–481. Plenum Press, New York.

———. 1996. Superfamily classification in the PIR-international protein sequence database. *Methods Enzymol.* **266:** 59–71.

Berger B., Wilson D.B., Wolf E., Tonchev T., Milla M., and Kim P.S. 1995. Predicting coiled coils by use of pairwise residue correlations. *Proc. Natl. Acad. Sci.* **92:** 8259–8263.

Berman H.M., Westbrook J., Feng Z., Gilliland G., Bhat T.N., Weissig H., Shindyalov I.N., and Bourne P.E. 2000. The protein data bank. *Nucleic Acids Res.* **28:** 235–242.

Blundell T.L. and Johnson M.S. 1993. Catching a common fold. *Protein Sci.* **2:** 877–883.

Bowie J.U., Lüthy R., and Eisenberg D. 1991. A method to identify protein sequences that fold into a known three-dimensional structure. *Science* **253:** 164–170.

Bowie J.U., Zhang K., Wilmanns M., and Eisenberg D. 1996. Three-dimensional profiles for measuring compatibilty of amino acid sequence with three-dimensional structure. *Methods Enzymol.* **266:** 598–616.

Branden C. and Tooze J. 1991. *Introduction to protein structure.* Garland Publishing, New York.

Brenner S.E., Chothia C., Hubbard T.J., and Murzin A.G. 1996. Understanding protein structure: Using Scop for fold interpretation. *Methods Enzymol.* **266:** 635–643.

Brown N.P., Leroy C., and Sander C. 1998. MView: A web compatible database search or multiple alignment viewer. *Bioinformatics* **14:** 380–381.

Bryant S.H. 1996. Evaluation of threading specificity and accuracy. *Proteins* **26:** 172–185.

Bryant S.H. and Lawrence C.E. 1993. An empirical energy function for threading protein sequence through the folding motif. *Proteins Struct. Funct. Genet.* **16:** 92–112.

Chothia C. 1992. Proteins. One thousand families for the molecular biologist. *Nature* **357:** 543–544.

Chou P.Y. and Fasman G.D. 1978. Prediction of the secondary structure of proteins from their amino acid sequence. *Adv. Enzymol. Relat. Areas Mol. Biol.* **47:** 45–147.

Cohen F.E., Abarbanel R.M., Kuntz I.D., and Fletterick R.J. 1983. Secondary structure assignment for a/β proteins by a combinatorial approach. *Biochemistry* **22:** 4894–4904.

———. 1986. Turn prediction in proteins using a pattern-matching approach. *Biochemistry* **25:** 266–275.

Corpet F., Gouzy J., and Kahn D. 1998. The ProDom database of protein domain families. *Nucleic Acids Res.* **26:** 323–326.

Cuff J.A. and Barton G.J. 2000. Application of multiple sequence alignment profiles to improve protein secondary structure prediction. *Proteins* **40:** 502–511.

Cuff J.A., Clamp M.E., Siddiqui A.S., Finlay M., and Barton G.J. 1998. JPred: A consensus secondary structure prediction server. *Bioinformatics* **14:** 892–893.

Dayhoff M.O., Barker W.C., Hunt L.T., and Schwartz R.M. 1978. Protein superfamilies. In *Atlas of protein sequence and structure* (ed. M.O. Dayhoff), vol. 5, suppl. 3, pp. 9–24. National Biomedical Research Foundation, Georgetown University, Washington, D.C.

Defay T.R. and Cohen F.E. 1996. Multiple sequence information for threading algorithms. *J. Mol. Biol.* **262:** 314–323.

Di Francesco V., Munson P.J., and Garnier J. 1999. FORESST: Fold recognition from secondary structure predictions of proteins. *Bioinformatics* **15:** 131–140.

Di Francesco V., Geetha V., Garnier J., and Munson P.J. 1997. Fold recognition using predicted secondary structure sequences and hidden Markov models of protein folds. *Proteins* (suppl. 1): 123–128.

Dodge C., Schneider R., and Sander C. 1998. The HSSP database of protein structure-sequence alignments and family profiles. *Nucleic Acids Res.* **26:** 313–315.

Fischer D. and Eisenberg D. 1996. Fold recognition using sequence-derived predictions. *Protein Sci.* **5:** 947–955.

———. 1999. Finding families for genomic ORFans. *Bioinformatics* **15:** 759–762.

Frishman D. and Argos P. 1995. Knowledge-based protein secondary structure assignment. *Proteins* **23:** 566–579.

———. 1996. Incorporation of non-local interactions in protein secondary structure prediction from the amino acid sequence. *Protein Eng.* **9:** 133–142.

———. 1997. Seventy-five percent accuracy in protein secondary structure prediction. *Proteins* **27:** 329–335.

Garnier J., Gibrat J.-F, and Robson B. 1996. GOR method for predicting protein secondary structure from amino acid sequence. *Methods Enzymol.* **266:** 540–553.

Garnier J., Osguthorpe D.J., and Robson B. 1978. Analysis of the accuracy and implications of simple methods for predicting the secondary structure of globular proteins. *J. Mol. Biol.* **120:** 97–120.

Geourjon C. and Deleage G. 1994. SOPM: A self-optimized method for protein secondary structure prediction. *Protein Eng.* **7:** 157–164.

———. 1995. SOPMA: Significant improvements in protein secondary structure prediction by consensus prediction from multiple alignments. *Comput. Appl. Biosci.* **11:** 681–684.

Gibrat J.-F., Madej T., and Bryant S.H. 1996. Surprising similarities in structure comparison. *Curr. Opin. Struct. Biol.* **6:** 377-385.

Guermeur Y., Geourjon C., Gallinari P., and Deleage G. 1999. Improved performance in protein secondary structure prediction by inhomogeneous score combination. *Bioinformatics* **15:** 413–421.

Guex N. and Peitsch M.C. 1997. SWISS-MODEL and the Swiss-PdbViewer: An environment for comparative protein modeling. *Electrophoresis* **18:** 2714–2723.

Hansen J.E., Lund O., Rapacki K., and Brunak S. 1997. O-GLYCBASE version 2.0: A revised database of O-glycosylated proteins. *Nucleic Acids Res.* **25:** 278–282.

Henikoff J.G. and Henikoff S. 1996. Blocks database and its applications. *Methods Enzymol.* **266:** 88–105.

Henikoff S., Pietrokovski S., and Henikoff J.G. 1998. Superior performance in protein homology detection with the Blocks database servers. *Nucleic Acids Res.* **26:** 309–312.

Henikoff S., Greene E.A., Pietrokovski S., Bork P., Attwood T.K., and Hood L. 1997. Gene families: The taxonomy of protein paralogs and chimeras. *Science* **278:** 609–614.

Hirst J.D. and Sternberg M.J. 1992. Prediction of structural and functional features of protein and nucleic acid sequences by artificial neural networks. *Biochemistry* **31:** 7211–7218.

Hofmann K., Bucher P., Falquet L., and Bairoch A. 1999. The PROSITE database, its status in 1999. *Nucleic Acids Res.* **27:** 215–219.

Hogue C.W. 1997. Cn3D: A new generation of three-dimensional molecular structure viewer. *Trends Biochem. Sci.* **22:** 314–316.

Hogue C.W. and Bryant S.H. 1998a. Structure databases. *Methods Biochem. Anal.* **39:** 46–73.

———. 1998b. Structure databases. In *Bioinformatics: A practical guide to the analysis of genes and proteins* (ed. A.D. Baxevanis and B.F. Ouellette), pp. 46–73. Wiley-Liss, New York.

Hogue C.W., Ohkawa H., and Bryant S.H. 1996. A dynamic look at structures: WWW-Entrez and the molecular modeling database. *Trends Biochem. Sci.* **21:** 226–229.

Holley L.H. and Karplus M. 1991. Neural networks for protein structure prediction. *Methods Enzymol.* **202:** 204–224.

Holm L. and Sander C. 1993. Protein structure comparison by alignment of distance matrices. *J. Mol. Biol.* **233:** 123–138.

———. 1994. Searching protein structure databases has come of age. *Proteins* **19:** 165–173.

———. 1996. Mapping the protein universe. *Science* **273:** 595–603.

———. 1998. Touring protein fold space with Dali/FSSP. *Nucleic Acids Res.* **26:** 316–319.

Hubbard T.J. and Park J. 1995. Fold recognition and ab initio structure predictions using hidden Markov models and β-strand pair potentials. *Proteins* **23:** 398–402.

Johnson M.S., May A.C., Ridionov M.A., and Overington J.P. 1996. Discrimination of common protein folds: Application of protein structure to sequence/structure comparisons. *Methods Enzymol.* **266:** 575–598.

Jones D.T., Miller R.T., and Thornton J.M. 1995. Successful protein fold recognition by optimal sequence threading validated by rigorous blind testing. *Proteins* **23:** 387–397.

Jones D.T., Taylor W.R., and Thornton J.M. 1992. A new approach to protein fold recognition. *Nature* **358:** 86–89.

Kabsch W. and Sander C. 1983. Dictionary of protein secondary structure: Pattern recognition of hydrogen-bonded and geometrical features. *Biopolymers* **22:** 2577–2637.

———. 1984. On the use of sequence homologies to predict protein structure: Identical pentapeptides can have completely different conformations. *Proc. Natl. Acad. Sci.* **81:** 1075–1078.

Kelley L.A., MacCallum R.M., and Sternberg M.J. 2000. Enhanced genome annotation using structural profiles in the program 3D-PSSM. *J. Mol. Biol.* **299:** 499–520.

King R.D., Saqi M., Sayle R., and Sternberg M.J. 1997. DSC: Public domain protein secondary structure prediction. *Comput. Appl. Biosci.* **13:** 473–474.

Kneller D.G., Cohen F.E., and Langridge R. 1990. Improvements in protein secondary structure prediction by an enhanced neural network. *J. Mol. Biol.* **214:** 171–182.

Krause A., Stoye J., and Vingron M. 2000. The SYSTERS protein sequence cluster set. *Nucleic Acids Res.* **28:** 270–272.

Kyte J. and Doolittle R.F. 1982. A simple method for displaying the hydropathic character of a protein. *J. Mol. Biol.* **157:** 105–132.

Laskowski R.A., Hutchinson E.G., Michie A.D., Wallace A.C., Jones M.L., and Thornton J.M. 1997. PDBsum: A web-based database of summaries and analyses of all PDB structures. *Trends Biochem. Sci.* **22:** 488–490.

Lathrop R.H., Rogers R.G., Jr., Bienkowska J., Bryant B.K.M., Buturović L.J., Gaitatzes C., Nambudripad R., White J.V., and Smith T.F. 1998. Analysis and algorithms for protein sequence-structure alignment. *New Compr. Biochem.* **32:** 237–283.

Lemer C.M., Rooman M.J., and Wodak S.J. 1995. Protein structure prediction by threading methods: Evaluation of current techniques. *Proteins* **23:** 337–355.

Levin J.M. 1997. Exploring the limits of nearest neighbour secondary structure prediction. *Protein Eng.* **10:** 771–776.

Levin J.M., Robson B., and Garnier J. 1986. An algorithm for secondary structure determination in proteins based on sequence similarity. *FEBS Lett.* **205:** 303–308.

Levitt M. 1997. Competitive assessment of protein fold recognition and alignment accuracy. *Proteins Struct. Funct. Genet.* (suppl. 1): 92–104.

Levitt M. and Chothia C. 1976. Structural patterns in globular proteins. *Nature* **261:** 552–558.

Lupas A. 1996. Prediction and analysis of coiled-coil structures. *Methods Enzymol.* **266:** 513–525.

Lupas A., Van Dyke M., and Stock J. 1991. Predicting coiled coils from protein sequences. *Science* **252:** 1162–1164.

Lüthy R. and Eisenberg D. 1991. Protein. In *Sequence analysis primer* (ed. M. Gribskov and J. Devereux), pp. 61–87. Stockton Press, New York.

Madej T., Gibrat J.-F., and Bryant S.H. 1995. Threading a database of protein cores. *Protein Struct. Funct. Genet.* **23:** 356–369.

Marchler-Bauer A., Levitt M., and Bryant S.H. 1997. A retrospective analysis of CASP2 threading predictions. *Proteins* (suppl. 1): 83–91.

Martin A.C., MacArthur M.W., and Thornton J.M. 1997. Assessment of comparative modeling in CASP2. *Proteins* (suppl. 1):14–28.

Mathews B. 1975. Comparison of the predicted and observed secondary structure of T4 phage lysozyme. *Biochim. Biophys. Acta* **405:** 442–451.

Minor D.L., Jr. and Kim P.S. 1996. Context-dependent secondary structure formation of a designed protein sequence. *Nature* **380:** 730–734.

Mizuguchi K., Deane C.M., Blundell T.L., and Overington J.P. 1998a. HOMSTRAD: A database of protein structure alignments for homologous families. *Protein Sci.* **7:** 2469–2471.

Mizuguchi K., Deane C.M., Blundale T.M., Johnson M.S., and Overington J.P. 1998b. JOY: Protein sequence-structure representation. *Bioinformatics* **14:** 617–623.

Muggleton S., King R.D., and Sternberg M.J. 1992. Protein secondary structure prediction using logic-based machine learning. *Protein Eng.* **5:** 647–657.

Murzin A.G. and Bateman A. 1997. Distant homology recognition using structural classification of proteins. *Proteins* (suppl. 1): 105–112.

Murzin A.G., Brenner S.E., Hubbard T., and Chothia C. 1995. SCOP: A structural classification of proteins database for the investigation of sequences and structures. *J. Mol. Biol.* **247:** 536–540.

Nevill-Manning C.G., Wu T.D., and Brutlag D.L. 1998. Highly specific protein sequence motifs for genome analysis. *Proc. Natl. Acad. Sci.* **95:** 5865–5871.

Nielson H., Engelbrecht J., Brunak S., and von Heijne G. 1997. Identification of prokaryotic and eukaryotic signal peptides and prediction of their cleavage sites. *Protein Eng.* **10:** 1–6.

Orengo C.A. and Taylor W.R. 1996. SSAP: Sequential structure alignment program for protein structure comparison. *Methods Enzymol.* **266:** 617–635.

Orengo C., Flores T.P., Taylor W.R., and Thornton J.M. 1993. Identification and classification of protein fold families. *Protein Eng.* **6:** 485–500.

Orengo C.A., Michie A.D., Jones S., Jones D.T., Swindells M.B., and Thornton J.M. 1997. CATH — A hierarchic classification of protein domain structures. *Structure* **5:** 1093–1108.

Ortiz A.R., Kolinski A., and Skolnick J. 1998. Fold assembly of small proteins using Monte Carlo simulations driven by restraints derived from multiple sequence alignments. *J. Mol. Biol.* **277:** 419–448.

Ota M. and Nishikawa K. 1997. Assessment of pseudo-energy potentials by the best-five test: A new use of the three-dimensional profiles of proteins. *Protein Eng.* **10:** 339–351.

Ouzounis C., Sander C., Scharf M., and Schneider R. 1993. Prediction of protein structure by evaluation of sequence-structure fitness. Aligning sequences to contact profiles derived from three-dimensional structures. *J. Mol. Biol.* **232:** 805–825.

Panchenko A.R., Luthey-Schulten Z., and Wolynes P.G. 1996. Foldons, protein structural modules, and exons. *Proc. Natl. Acad. Sci.* **93:** 2008–2013.

Panchenko A.R., Luthey-Schulten Z., Cole R., and Wolynes P. 1997. The foldon universe: A survey of structural similarity and self-recognition of independently folding units. *J. Mol. Biol.* **272:** 95–105.

Park J., Teichmann S.A., Hubbard T., and Chothia C. 1997. Intermediate sequences increase the detection of homology between sequences. *J. Mol. Biol.* **273:** 349–354.

Pascarella S. and Argos P. 1992. A data bank merging related protein structures and sequences. *Protein Eng.* **5:** 121–137.

Patthy L. 1987. Detecting homology of distantly related proteins with consensus sequences. *J. Mol. Biol.* **198:** 567–577.

———. 1996. Consensus approaches in detection of distant homologies. *Methods Enzymol.* **266:** 184–198.

Pearson W.R. 1996. Effective protein sequence comparison. *Methods Enzymol.* **266:** 227–258.

Pellegrini M., Marcotte E.M., Thompson M.J., Eisenberg D., and Yeatts T.O. 1999. Assigning protein functioning by comparative genome analysis: Protein phylogenetic profiles. *Proc. Natl. Acad. Sci.* **96:** 4255–4288.

Pennisi E. 1998. Taking a structured approach to understanding proteins. *Science* **279:** 978–979.

Peitsch M.C. 1996. ProMod and Swiss-Model: Internet-based tools for automated comparative protein modelling. *Biochem. Soc. Trans.* **24:** 274–279.

Pongor S., Skerl V., Cserzo M., Hatsagi Z., Simon G., and Bevilacqua V. 1993. The SBASE domain library: A collection of annotated protein segments. *Protein Eng.* **6:** 391–395.

Portugaly E. and Linial M. 2000. Estimating the probability for a protein to have a new fold: A statistical computational model. *Proc. Natl. Acad. Sci.* **97:** 5161–5166.

Presnell S.R., Cohen B.I., and Cohen F.E. 1992. A segment-based approach to protein secondary structure prediction. *Biochemistry* **31:** 983–993.

———. 1993. MacMatch: A tool for pattern-based protein secondary structure prediction. *Comput. Appl. Biosci.* **9:** 373–374.

Qian N. and Sejnowski T.J. 1988. Predicting the secondary structure of globular proteins using neural network models. *J. Mol. Biol.* **202:** 865–884.

Richardson D.C. and Richardson J.S. 1994. Kinemages — Simple macromolecular graphics for interactive teaching and publication. *Trends Biochem. Sci.* **19:** 135–138.

Rodriguez R., Chinea G., Lopez N., Pons T., and Vriend G. 1998. Homology modeling, model and software evaluation: Three related resources. *Bioinformatics* **14:** 523–528.

Rost B. 1995a. In *Protein folds. A distance-based approach* (ed. H. Bohr and S. Brunak), pp. 132–151. CRC Press, Boca Raton, Florida.

———. 1995b. TOPITS: Threading one-dimensional predictions into three-dimensional structures. *Ismb* **3:** 314–321.

———. 1996. PHD: Predicting one-dimensional protein structure by profile-based neural networks. *Methods Enzymol.* **266:** 525–539.

Rost B. and Sander C. 1993. Improved prediction of protein secondary structure by use of sequence profiles and neural networks. *Proc. Natl. Acad. Sci.* **90:** 7558–7562.

———. 1994. Combining evolutionary information and neural networks to predict protein secondary structure. *Proteins* **19:** 55–72.

Rost B., Casadio R., and Fariselli P. 1996. Refining neural network predictions for helical transmembrane proteins by dynamic programming. *Ismb* **4:** 192–200.

Rost B., Casadio R., Fariselli P., and Sander C. 1995. Transmembrane helices predicted at 95% accuracy. *Protein Sci.* **4:** 521–533.

Russell R.B., Saqi M.A., Sayle R.A., Bates P.A., and Sternberg M.J. 1997. Recognition of analogous and homologous protein folds: Analysis of sequence and structure conservation. *J. Mol. Biol.* **269:** 423–439.

Salamov A.A. and Solovyev V.V. 1995. Prediction of protein secondary structure by combining nearest-neighbor algorithms and multiple sequence alignments. *J. Mol. Biol.* **247:** 11–15.

———. 1997. Protein secondary structure prediction using local alignments. *J. Mol. Biol.* **268:** 31–36.

Sali A., Potterton L., Yuan F., van Vlijmen H., and Karplus M. 1995. Evaluation of comparative protein modeling by MODELLER. *Proteins* **23:** 318–326.

Salzberg S. and Cost S. 1992. Predicting protein secondary structure with a nearest-neighbor algorithm. *J. Mol. Biol.* **227:** 371–374.

Sander C. and Schneider R. 1991. Database of homology-derived protein structures. *Proteins Struct. Funct. Genet.* **9:** 56–68.

Sayle R.A. and Milner-White E.J. 1995. RASMOL: Biomolecular graphics for all. *Trends Biochem. Sci.* **20:** 374.

Schultz J., Milpetz F., Bork P., and Ponting C.P. 1998. SMART, a simple modular architecture tool: Identification of signaling domains. *Proc. Natl. Acad. Sci.* **95:** 5857–5864.

Sippl M.J. 1990. Calculation of conformational ensembles from potentials of mean force. An approach to the knowledge-based prediction of local structures in globular proteins. *J. Mol. Biol.* **213:** 859–883.

Sippl M.J. and Weitckus S. 1992. Detection of native-like models for amino acid sequences of unknown three-dimensional structure in a data base of known protein conformations. *Proteins* **13:** 258–271.

Sippl M.J., Lackner P., Domingues F.X., and Koppensteiner W.A. 1999. An attempt to analyse progress in recognition from CASP1 to CASP3. *Proteins* (suppl. 3) **37:** 226–230.

Sonnhammer E.L., Eddy S.R., Birney E., Bateman A., and Durbin R. 1998. Pfam: Multiple sequence alignments and HMM-profiles of protein domains. *Nucleic Acids Res.* **26:** 320–322.

Stolorz P., Lapedes A., and Xia Y. 1992. Predicting protein secondary structure using neural net and statistical methods. *J. Mol. Biol.* **225:** 363–377.

Stultz C.M., White J.V., and Smith T.F. 1993. Structural analysis based on state-space modeling. *Protein Sci.* **2:** 305–314.

Stultz C.M., Nambudripad R., Lathrop R.H., and White J.V. 1997. Predicting protein structure with probabilistic models. *Adv. Mol. Cell Biol.* **22B:** 447–506.

Sudarsanam S. 1998. Structural diversity of sequentially identical subsequences of proteins: Identical octapeptides can have different conformations. *Proteins* **30:** 228–231.

Swindells M.B., Orengo C.A., Jones D.T., Hutchinson E.G., and Thornton J.M. 1998. Contemporary approaches to protein structure classification. *BioEssays* **20:** 884–891.

Tatusov R.L., Koonin E.V., and Lipman D.J. 1997. A genomic perspective on protein families. *Science* **278:** 631–637.

Taylor W.R. and Orengo C.A. 1989. Protein structure alignment. *J. Mol. Biol.* **208:** 1–22.

von Heijne G. 1987. *Sequence analysis in molecular biology — Treasure trove or trivial pursuit,* pp. 81–121. Academic Press, San Diego, California.

Vriend G. and Sander C. 1991. Detection of common three-dimensional substructures in proteins. *Proteins* **11:** 52–68.

White J.V., Stultz C.M., and Smith T.F. 1994. Protein classification by stochastic modeling and optimal filtering of amino-acid sequences. *Math. Biosci.* **119:** 35–75.

Wilmanns M. and Eisenberg D. 1993. Three-dimensional profiles from residue-pair preferences: Identification of sequences with β/α-barrel fold. *Proc. Natl. Acad. Sci.* **90:** 1379–1383.

———. 1995. Inverse protein folding by the residue pair preference profile method: Estimating the correctness of alignments of structurally compatible sequences. *Protein Eng.* **8:** 627–639.

Wu C. 1996. Gene classification artificial neural system. *Methods Enzymol.* **266:** 71–88.

Wu C., Zhao S., and Chen H.L. 1996. A protein class database organized with ProSite protein groups and PIR superfamilies. *J. Comput. Biol.* **3:** 547–561.

Xenarios I., Rice D.W., Salwinski L., Baron M.K., Marcotte E.M., and Eisenberg D. 2000. DIP: The database of interacting proteins. *Nucleic Acids Res.* **28:** 289–291.

Yi T.M. and Lander E.S. 1993. Protein secondary structure prediction using nearest-neighbor methods. *J. Mol. Biol.* **232:** 1117–1129.

———. 1996. Iterative template refinement: Protein-fold prediction using iterative search and hybrid sequence/structure templates. *Methods Enzymol.* **266:** 322–339.

Yona G., Linial N., and Linial M. 1999. ProtoMap: Automatic classification of protein sequences, a hierarchy of protein families, and local maps of the protein space. *Proteins* **37:** 360–378.

Zhang X., Mesirov J.P., and Waltz D.L. 1992. Hybrid system for protein secondary structure prediction. *J. Mol. Biol.* **225:** 1049–1063.

## CHAPTER 10

Genome Analysis

INTRODUCTION

A MAJOR APPLICATION OF BIOINFORMATICS IS analysis of the full genomes of organisms that have been sequenced starting in the late 1990s, including microbial genomes, the budding yeast *Saccharomyces cerevisiae*, the nematode worm *Caenorhabditis elegans*, the plant *Arabidopsis thaliana*, the fruit fly *Drosophila*, and the human genome. Many additional genome sequencing projects are either being planned or are already under way.

The genome is defined as the sum of the genes and intergenic sequences of the haploid cell (Bernardi 1995).

Traditional genetics and molecular biology have been directed toward understanding the role of a particular gene or protein in an important biological process. A gene is sequenced to predict its function or to manipulate its activity or expression. In contrast, the availability of genome sequences provides the sequences of all the genes of an organism so that important genes influencing metabolism, cellular differentiation and development, and disease processes in animals and plants, can be identified and the relevant genes manipulated.

The challenge is to identify those genes that are predicted to have a particular biological function and then to design experiments to test that prediction. This analysis depends on gene prediction using gene models for each organism followed by sequence comparisons between the predicted proteins with other proteins whose function is known from biological studies. To facilitate such comparisons, the genomes of a number of model organisms about which a great deal of biological information is available have been sequenced. Many years of genetic and biochemical research of these model organisms—the bacterium *Escherichia coli*, *S. cerevisiae*, *C. elegans*, *A. thaliana*, and *D. melanogaster*—have led to the accumulation of a large amount of information on gene organization and function. The mouse *Mus musculus* is a genetic model for humans because the two species are so closely related through evolution. A newly identified gene in another organism can be compared to the existing database of information to find whether it has a similar function. Genes involved in human disease, for example, are sometimes found to be similar to a fruit fly gene at the protein sequence level (for an example of how significant this kind of analysis can be, see Rubin et al. 2000). The genetic effects of mutations in the fruit fly's gene will then provide a biochemical, cellular, or developmental model for the human disease. Interestingly, it has not been possible to identify the function of all the genes in model organisms. As a result, a similar gene or family of genes may be found in several organisms, including a model organism, but the function is not known because the gene functions have not yet been analyzed. Hence, continued biological analysis of model organisms in those areas that are not tractable by the tools of bioinformatics has many important applications.

Tracing the phylogenetic history of such uncharacterized genes, characterized genes, and gene domains and gene linkages in diverse organisms is one of the most interesting and challenging aspects of genome analysis. In addition, even though a gene that specifies an important biological function has not been identified, the gene can be traced in individuals using sequence variations that occur among individuals in a population, called sequence polymorphisms. In humans, for example, single nucleotide polymorphisms (SNPs) can be found throughout the genome, including some that are positioned adjacent to an important disease gene. If a particular G $\rightarrow$ A polymorphism is right next to a defective tumor suppressor gene, for example, that polymorphism serves as a genetic marker for the presence of the defective gene. The applicable genetic principle, genetic linkage, is that closely linked genes seldom become separated by genetic recombination from one generation to the next. Another example of such linked polymorphisms is in crop plants. Features such as plant height and amount of seed produced are influenced by variations in sets of genes, called quantitative trait loci (QTL). Inheritance of QTLs can be traced from one generation to the next using sequence polymorphisms that are linked to the favored genetic variation without having to wait to observe the effects on plant growth.

The availability of genome sequences greatly facilitates the discovery and utilization of these sequence polymorphisms. It is recognized that some types of genetic variation, including specific human diseases, are best understood at the genome-wide level. The duplication of genes, gene segments, and gene clusters provides opportunities for recombination events that can cause changes in gene copy number or loss of gene function (Lupski 1998).

In summary, the availability of genome sequences provides an unprecedented opportunity to explore genetic variability both between organisms and within the individual organism. We now turn to a comparison of the main features of the genomes that have been sequenced. One major task is to identify the genes that encode proteins and to identify the function of as many of these proteins as possible by database similarity searches.

The entire set of proteins of an organism, including those known from biological studies and those predicted by bioinformatics, is the proteome of the organism.

The proteome may be compared to itself to identify paralogs, families of proteins that have arisen by gene duplication. One proteome may also be compared to another proteome to discover orthologous genes that have kept the same function, genes that have become fused to make a larger protein (or split into two to make two separate proteins), new arrangements of protein domains, and amplification of protein families to perform a new type of biological function (e.g., cell-to-cell communication during development of a multicellular organism). A representative collection of the large number of Web resource pages and references is shown in Table 10.1. This table is divided into six parts, A–F, dealing with resources for prokaryotic genomes (A) which have been the subject of intense sequence analysis, all model organisms (B), human genome and the related mouse genome (C), genome relationships (D), proteome and gene expression analysis (E), and functional characterization of genes (F). Since these sites are constantly being revised, this table will be periodically updated on the book Web site.

GENOME ANATOMY

Early biologists examining a particular plant, animal, or yeast cell using a microscope observed a nucleus (in a eukaryotic cell) with a specific number of chromosomes of variable length and morphology that could be seen at certain stages of cell division. The chromosomes comprised linear DNA molecules in a tightly compact form that was wrapped around protein complexes, called the nucleosome. Nuclei and chromosomes were not observed in bacteria (a prokaryotic cell), but when bacterial DNA was eventually detected, the molecule was usually circular and was also in a compacted form. The following sections outline the structure and composition of prokaryotic and eukaryotic genomes.

Prokaryotic Genomes

The first bacterial genome to be sequenced was that of *Hemophilus influenzae*, a mild human pathogen (Fleischmann et al. 1995). This project was carried out at the Institute of Genomics Research (TIGR, http://www.tigr.org) in part to prove a new genome sequencing method—the shotgun method. A large number of random overlapping fragments were sequenced and then a consensus sequence of the entire 1.8×16^6-bp chromosome of *Hemophilus* was assembled by computer, excepting several regions that had to be assembled manually. Once available, open reading frames were identified, and these were compared to the existing proteins by a database similarity search (see Chapter 7). Approximately 58% of the 1743 predicted genes matched genes of another species, the bacterial

Table 10.1. *Web resources and references for genome information and analysis*

A. Prokaryotic genomes[a]

MAGPIE: Multipurpose Automated Genome Project Investigation Environment (Gaasterland and Sensen 1998)	http://genomes.rockefeller.edu/magpie
Microbial genome databases	http://www.ncbi.nlm.nih.gov:80/PMGifs/Genomes/micr.html
	http://www.techfak.uni-bielefeld.de/techfak/persons/chrisb/ResTools/biotools/biotools10.html
	http://www-nbrf.georgetown.edu/pir/genome.html#PROK
Comparative genome analysis in P. Bork laboratory (see Web site)	http://www.bork.embl-heidelberg.de/Genome/
TIGR: The Comprehensive Microbial Resource Home Page—the omniome	http://www.tigr.org/tigr-scripts/CMR2/CMRHomePage.spl
U.S. Dept. of Energy Joint Genome Initiative	http://www.jgi.doe.gov/

[a] Also see the COG and PEDANT sites in part D.

B. Genomic databases of model organisms and other genome databases

Arabidopsis thaliana genome displayer	http://www.kazusa.or.jp/kaos
A. thaliana information resource TAIR	http://www.arabidopsis.org/
Caenorhabditis elegans (worm) database	http://www.wormbase.org/
C. elegans chromosomes	ftp://ftp.sanger.ac.uk/pub/databases/C.elegans_sequences/CHROMOSOMES/
C. elegans genome project	http://www.sanger.ac.uk/Projects/C_elegans/
C. elegans proteome database	http://www.sanger.ac.uk/Projects/C_elegans/wormpep/
	http://www.proteome.com/YPDhome.html
Dictyostelium discoideum genome information	http://www.biology.ucsd.edu/others/dsmith/dictydb.html
Drosophila melanogaster Berkeley Drosophila genome project	http://www.fruitfly.org/
D. melanogaster chromosomes	http://flybase.bio.indiana.edu/maps/fbgrmap.html
D. melanogaster: Flybase, a genomic database	http://flybase.bio.indiana.edu/
E. coli genome project	http://www.genetics.wisc.edu/
E. coli genome and proteome database GenProtEC	http://genprotec.mbl.edu/
E. coli index	http://web.bham.ac.uk/bcm4ght6/res.html
Genome databases at NCBI[a]	http://www.ncbi.nlm.nih.gov/Genomes/index.html
	http://www.ncbi.nlm.nih.gov/Entrez/Genome/main_genomes.html
	http://www.ncbi.nlm.nih.gov:80/PMGifs/Genomes/org.html
Genome databases other than NCBI[a]	http://www.techfak.uni-bielefeld.de/techfak/persons/chrisb/ResTools/biotools/biotools10.html
	http://www-nbrf.georgetown.edu/pir/genome.html
Genome list at NIH	http://molbio.info.nih.gov/molbio/db.html
Mitochondrial DNA Database MitBASE	http://www3.ebi.ac.uk/Research/Mitbase/mitbase.pl
Mouse (*Mus musculus*) genome informatics	http://www.informatics.jax.org/
Plant genome projects supported by the plant genome initiative of the U.S. National Science Foundation	http://www.nsf.gov/bio/dbi/pgrsites.htm
Organelle genome sequences	http://www.ncbi.nlm.nih.gov/PMGifs/Genomes/organelles.html
	http://www-nbrf.georgetown.edu/pir/genome.html
Parasite genome databases and genome research resources	http://www.ebi.ac.uk/parasites/parasite-genome.html
Retroviral genotyping and analysis site	http://www.ncbi.nlm.nih.gov/retroviruses/
Rice (*Oryza sativa*) genome project	http://rgp.dna.affrc.go.jp/
Saccharomyces cerevisiae: View of 16 chromosomes	http://genome-www.stanford.edu/Saccharomyces/MAP/GENOMICVIEW/GenomicView.html
S. cerevisiae, YPD Yeast Proteome database, a commercial database	http://www.proteome.com/YPDhome.html
S. cerevisiae (budding yeast) database SGD	http://genome-www.stanford.edu/Saccharomyces/

[a] The National Center for Biotechnology Information, National Library of Medicine, Bethesda, Maryland

Continued.

Table 10.1. *Continued*

C. Human and mouse genome comparisons

Celera Genomics: The company that assembles genome sequences by automated fragment assembly	http://www.celera.com/
Comparison of human (*Homo sapiens*) and mouse (*M. musculus*) chromosomes	http://www.bioscience.org/urllists/chromos.htm, http://www.ncbi.nlm.nih.gov/ Homology/
	http://infosrv1.ctd.ornl.gov/TechResources/Human_Genome/publicat/ 97pr/05g_mous.html
	http://srs.ebi.ac.uk/, databanks link, MOUSE2HUMAN
Cooperative Human Linkage Center: mouse-clickable map of chromosomes	http://lpg.nci.nih.gov/html-chlc/ChlcIntegratedMaps.html
Draft Human Genome Browser	http://genome.ucsc.edu/goldenPath/hgTracks.html
Human sequence polymorphisms, mutations, and mapping	http://srs.ebi.ac.uk/, databanks link
Human EST project	http://genome.wustl.edu/est/esthmpg.html
Human genome resources at NCBI	http://www.ncbi.nlm.nih.gov/genome/guide/
Human genome research sites provided by Oak Ridge National Labs	http://www.ornl.gov/hgmis/centers.html
Mouse (*M. musculus*) chromosomes: mouse-clickable map	http://brise.ujf-grenoble.fr/~mongelar/clickclientsideV2bis.html
On-line inheritance in man: Johns Hopkins University and NCBI	http://www3.ncbi.nlm.nih.gov/Omim/
Whitehead Institute for Biomedical Research	http://www.ornl.gov/hgmis/research/centers.html

D. Gene and genome relationships and proteome[a] analysis

Alfresco: Visualization tool for genome comparison	http://www.sanger.ac.uk/Software/Alfresco/
allgenes.org: A comprehensive gene index (catalog) derived from ESTs and predicted genes	http://www.allgenes.org/
CGAP: Cancer genome anatomy project	http://www.ncbi.nlm.nih.gov/CGAP
COG (cluster of orthologous groups): A gene classification system (Tatusov et al. 1997, 2000)	http://www.ncbi.nlm.nih.gov/COG/
Comparative DNA analysis across genomes (genome signatures by nucleotide compositional analysis)[b]	Karlin et al. (1998)
DOGS: Database of genome sizes	http://www.cbs.dtu.dk/databases/DOGS/index.html
E-CELL: A modeling and simulation environment for biochemical and genetic processes (Tomita et al. 1999)	http://www.e-cell.org
FAST_PAN for automatic searches of online EST databases to identify new family members (paralogs) (Retief et al. 1999)	http://www.uvasoftware.org/
GeneCensus Genome Comparisons by encoded protein structures	http://bioinfo.mbb.yale.edu/genome/
GeneQuiz: An integrated system for large-scale biological sequence analysis and data management (Andrade et al. 1999; Hoersch et al. 2000)	http://jura.ebi.ac.uk:8765/ext-genequiz/
Genes and disease: Map location on human chromosomes	http://www.ncbi.nlm.nih.gov/disease/
Genome channel at Oak Ridge National Laboratories	http://compbio.ornl.gov/channel/
GOLD™: Genomes OnLine Database (Kyrpides 1999)	http://wit.integratedgenomics.com/GOLD/

Continued.

Table 10.1. *Continued*

D. Gene and genome relationships and proteome[a] analysis *(continued)*

IMGT ImMunoGeneTics Database specializing in Immunoglobulins, T-cell receptors, and Major Histocompatibility Complex (MHC) of all vertebrate species (Ruiz et al. 2000)	http://www.ebi.ac.uk/imgt/index.html
KEGG: Kyoto Encyclopedia of Genes and Genomes (Kanehisa and Goto 2000)	http://www.genome.ad.jp/kegg/
MIA Molecular Information Agent: A Web server that searches biological databases for information on a macromolecule	http://mia.sdsc.edu/
Orthologous gene alignments at TIGR	http://www.tigr.org/tdb/toga/orth_tables.html
PEDANT: A protein extraction, description, and analysis tool	http://pedant.mips.biochem.mpg.de/
SEQUEST for identification of proteins following mass spectrometry (Link et al. 1999)	http://thompson.mbt.washington.edu/sequest/
STRING Search Tool for Recurring Instances of Neighboring Genes (see Web page) (Snel et al. 2000b)	http://www.Bork.EMBL-Heidelberg.DE/STRING/
Taxonomy browser at the NCBI arranges genomes taxonomically for sequence retrieval	http://www.ncbi.nlm.nih.gov/Taxonomy/taxonomyhome.html/
UniGene System gene-oriented clusters of GenBank sequences useful for gene identification	http://www.ncbi.nlm.nih.gov/UniGene/
U.S. Dept. of Agriculture, Agricultural Research Service reference site for plant and animal genomes (also see TAIR in model genomes); includes international links	http://genome.cornell.edu/

[a] The full complement of proteins produced by an organism, many following gene prediction.

[b] Whole genomes may be compared at the level of dinucleotide composition, codon usage, strand asymmetry for transcription, and rare oligonucleotides. For example, the dinucleotide TA is underrepresented in most prokaryotic and eukaryotic genomes but not in the genomes of several archaea.

E. Metabolism and regulation,[a] functional genomics

2D gel analysis of proteins: List of organisms	http://www.expasy.ch/ch2d/2d-index.html
AlignAce for promoter analysis of coordinately regulated genes, e.g., microarrays by Gibbs sampling (Roth et al. 1998; Hughes et al. 2000; McGuire et al. 2000)	http://atlas.med.harvard.edu/download/
ArrayExpress database at European Bioinformatics Institute for microarray analysis	http://www.ebi.ac.uk/arrayexpress/
BRITE: Database of protein-protein interactions and cross-reference links (see KEGG)	http://www.genome.ad.jp/brite/brite.html
Ecocyc electronic encyclopedia of genes and metabolism of *E. coli* (Karp et al. 2000)	http://ecocyc.PangeaSystems.com/ecocyc/
EpoDBis: A database of genes that relate to vertebrate red blood cells (Erythropoiesis) (Stoeckert et al. 1999)	http://www.cbil.upenn.edu/EpoDB/index.html
Expression Profiler tools for analysis and clustering of gene expression and sequence data	http://ep.ebi.ac.uk/
Functional genomics sites	http://www.ornl.gov/hgmis/publicat/hgn/hgnarch.html#fg
GeneCensus Genome Comparisons by encoded protein structures	http://bioinfo.mbb.yale.edu/genome/

Continued.

Table 10.1. *Continued*

E. Metabolism and regulation,[a] functional genomics *(continued)*

GENECLUSTER; Tamayo et al. (1999)	http://www.genome.wi.mit.edu/MPR/software.html
GeneRAGE for sequence clustering and domain detection; Enright and Ouzounis (2000)	available from authors
GeneX: A Collaborative Internet Database and Toolset for Gene Expression Data	http://www.ncgr.org/research/genex/
MetaCyc metabolic encyclopedia (see EcoCyc)	http://ecocyc.PangeaSystems.com/ecocyc/
Microarray guide: P. Brown lab	http://cmgm.stanford.edu/pbrown/
Microarray project at NIH	http://www.nhgri.nih.gov/DIR/LCG/15K/HTML/
Microarray software	http://rana.lbl.gov/
microarrays.org: A new public source for microarraying information, tools, and protocols	http://www.microarrays.org/
SMART: For the study of genetically mobile protein domains (Schultz et al. 2000)	http://smart.embl-heidelberg.de/
SWISS-2DPAGE: Two-dimensional polyacrylamide gel electrophoresis database (Hoogland et al. 2000)	http://www.expasy.ch/ch2d/
TIGR: Annotation and gene indexing resources, including analysis of the transcribed sequences represented in the public EST databases.	http://www.tigr.org/tdb/tgi.shtml
WIT (What is there?): Interactive metabolic reconstruction on the Web (Overbeek et al. 2000)	http://wit.mcs.anl.gov/WIT2/
Yeast (*S. cerevisiae*) transcriptome	http://bioinfo.mbb.yale.edu/genome/
Yeast genome (*S. cerevisiae*) on a chip	http://cmgm.stanford.edu/pbrown/yeastchip.html

[a] Identification of regulatory sequences is discussed in Chapter 8, and programs for analysis of eukaryotic promoters are listed in Table 8.6 and on page 371.

F. Gene nomenclature, functional characterization, and genome database development

A. thaliana nomenclature	http://www.arabidopis.org/links/nomenclature.html
Genome Annotation and Information Analysis GAIA (Bailey et al. 1998)	http://www.cbil.upenn.edu/gaia2/gaia
GeneQuiz: An integrated system for large-scale biological sequence analysis and data management (Andrade et al. 1999; Hoersch et al. 2000)	http://jura.ebi.ac.uk:8765/ext-genequiz//genequiz.html
GFF (Gene-Finding Features): Specification for describing genes and other features of genomics	http://www.sanger.ac.uk/Software/GFF/
GO (gene ontology) controlled vocabulary	http://genome-www.stanford.edu/GO/
K2 system for support of distributed heterogeneous database and information resource integration	http://www.cbil.upenn.edu/
Kleisli Project: A tool for broad-scale integration of databanks across the Internet (see Chung and Wong 1999)	http://sdmc.krdl.org.sg/kleisli/
MAGPIE: Multipurpose Automated Genome Project Investigation Environment (Gaasterland and Sensen 1998)	http://www.rockefeller.edu/labheads/gaasterland/gaasterland.html, http://genomes.rockefeller.edu/magpie/index.html, see http://magpie.genome.wisc.edu/tools.html
Mendel Plant Gene Nomenclature Database	http://genome-www.stanford.edu/Mendel/
RefSeq and LocusLink: A curated set of reference sequences with map locations, a foundation for functional annotation of the human genome (Pruitt et al. 2000)	http://www.ncbi.nlm.nih.gov/LocusLink/refseq.html
TAMBIS: A conceptual model of molecular biology and bioinformatics and methods for querying the model (Baker et al. 1999)	http://img.cs.man.ac.uk/tambis/

Prior to the sequencing of H. influenzae, the first free-living organism to be sequenced, a large number of viruses had been sequenced. Many of these organisms also serve as model systems for studying replication and gene expression. As an example, the nucleotide sequence of bacteriophage lambda was completed by Sanger et al. (1982). A simple way to retrieve sequences of viral and other extrachromosomal genetic elements such as organelles is through the National Center for Biotechnology Information (NCBI) taxonomy browser at http://www.ncbi.nlm.nih.gov/Taxonomy/taxonomyhome.html.

species *E. coli* K-12 that had been the subject of many years of genetic and biochemical research. The identification of these genes allowed the investigators to construct some of the biochemical pathways of the *Hemophilus* cell. The function of the other 42% of the *Hemophilus* genes could not be identified, although some of them were similar to the 38% of *E. coli* genes that were also of unknown function. Other unique sequences that appeared to be associated with the ability of the organism to behave as a human pathogen were also found.

The success of sequencing the *Hemophilus* genome in a relatively short time and with a modest budget heralded the sequencing of a large number of additional prokaryotic organisms (see Table 10.1A; de Bruijn et al. 1998). To date, the genomes of 31 of these species have been sequenced. Organisms were selected for sequencing based on at least three criteria: (1) They had been subjected to a good deal of biological analysis, e.g., *E. coli* and *Bacillus subtilis*, and thus were model prokaryotic organisms; (2) they were an important human pathogen, e.g., *Mycobacterium tuberculosis* (tuberculosis) and *Mycoplasma pneumoniae* (pneumonia); or (3) they were of phylogenetic interest. Analysis of the ribosomal RNA molecules of prokaryotes and eukaryotes had led to the prediction of three main branches in the tree of life represented by Archaea, the Bacteria, and the Eukarya.

For genome sequencing projects, organisms have been sampled from throughout the tree (see Fig. 6.3, p. 243), including some that are in deeper branches of the tree and that have growth properties reminiscent of an ancient environment. A summary of the genome size and composition of a representative list of prokaryotes is given in Table 10.2.

As these genome sequences were collected, they were annotated. Annotation involves identifying open reading frames in the genome sequence using the predicted protein as query sequences in a database similarity search and then adding any significant matches to the genome sequence entry in the sequence database. More sophisticated methods of

Table 10.2. *Features of representative prokaryotic genomes*

Organism (reference)	Phylogenetic group	Genome size (Mbp) (no. protein-encoding genes)	Novel functions
Escherichia coli (Blattner et al. 1997)	Bacteria	4.6 (4288)	model organism
Methanococcus jannaschii (Bult et al. 1996)	Archaea	1.66 (1682)[a]	grows at high temperature and pressure and produces methane
Hemophilus influenzae (Fleischmann et al. 1995)	Bacteria	1.83 (1743)	human pathogen
Mycoplasma pneumoniae (Himmelreich et al. 1996)	Bacteria	0.82 (676)	human pathogen that grows inside cells; metabolically weak
Bacillus subtilis (Kunst et al. 1997)	Bacteria	4.2 (4098)	model organism
Aquifex aeolicus (Deckert et al. 1998)	Bacteria	1.55 (1512)[b]	ancient species, grows at high temperature and can grow in a hydrogen, oxygen, carbon dioxide atmosphere in the presence of only mineral salts
Synechocystis sp. (Kaneko et al. 1996a,b)	Bacteria	3.57 (3168)	ancient organism that produces oxygen by light-harvesting; may have oxygenated atmosphere

The genome in each case is contained on a single circular DNA molecule except where noted. Another bacterial species, *Deinococcus radiodurans*, has two chromosomes of sizes 2.6 and 0.4 Mbp and two additional elements of size 0.17 Mb and 46 Kbp (http://www.tigr.org). Other bacterial species have linear chromosomes (for review, see Volff and Altenbuchner 2000).

[a] *M. jannaschii* has a small and a large extrachromosomal element.
[b] *A. aeolicus* has a single extrachromosomal element.

Prokaryotic organisms are included in the Archaea and Bacteria phylogenetic groups.

searching for protein families described in Chapters 7 and 9 are also used for annotation. In examining the results of such analysis, it is important to look for the method used, the statistical significance of the result, and the overall degree of confidence in the alignments. The analysis should be repeated if necessary. Annotation errors occur when the above criteria are not followed (Kyrpides and Ouzounis 1999). Computational resources listed in Table 10.1 can facilitate the analysis of bacterial genomes. GeneQuiz is an example of such a resource. Also shown in Table 10.1A are Web sites that provide a complete annotation of the prokaryotic genomes that have been sequenced.

Eukaryotic Genomes

In addition to having linear chromosomes within a nucleus, and differing from prokaryotic genomes in this respect, eukaryotic genomes commonly have tandem repeats of sequences and include introns in protein-coding genes.

Sequence Repeats

Centromeres hold newly replicated daughter chromosomes together and serve as a point of attachment for pulling the chromosomes apart during cell division.

Because of the skewed base composition of regions that have repeats, they may be purified by virtue of having different buoyant densities and are known as satellite DNA. The sequences fall into different types, each with a different repeat unit of length 5–200 bp. Most of this repetitive DNA is found near the centromere. Also found in eukaryotic genomes are minisatellites made up of repeat units of up to 25 bp and microsatellites composed of repeat units of 4 bp or less. Microsatellite repeats are found at the ends of eukaryotic chromosomes at the telomeres, which in humans comprise hundreds of copies of a 6-bp repeat TTAGGG.

Telomeres are necessary for chromosomal replication.

In nondividing cells, a mixture of lightly and darkly stained chromosomal regions called heterochromatin and euchromatin, respectively, are observed. The centromeric and telomeric regions are located in the heterochromatin, which is in a compact configuration and is thought not to be transcribed. Genes that are transcribed are located in the less compact euchromatin, to which regulatory proteins have access (for review, see Brown 1999).

Transposable Elements

These elements can comprise a large proportion of the eukaryotic genome as repetitive sequences. Transposable elements (TEs) are thought to play an important role in the evolution of these genomes (Kidwell and Lisch 1997, 2000). TEs are DNA sequences that can move from one chromosomal location to another faster than the chromosome can replicate. Hence, TEs have the potential to increase in number until they comprise a large proportion of the genome sequence, a feature already observed in many plants and animals. They remain detectable in the genome until they blend into the background sequence by mutation. The presence of these elements may be demonstrated using programs for detection of low-complexity regions in sequences (see Chapter 6, p. 308). The percentage of genomes that are composed of TEs is depicted in Figure 10.1. For example, more than one-third of the human genome consists of interspersed repetitive sequences derived from TEs.

Eukaryotic TEs fall into two main classes according to sequence similarity and the mechanism of transposition. Class I elements encode a reverse transcriptase and use RNA-mediated mechanisms of transcription. There are three main subclasses of these TEs—the long terminal repeat (LTR) retrotransposons, retroposons, and retrovirus-like elements with LTRs. The LTR retrotransposons are related by genetic structure to retroviruses. The retroposons include short (80–300 bp long) interspersed nuclear elements (SINES) and

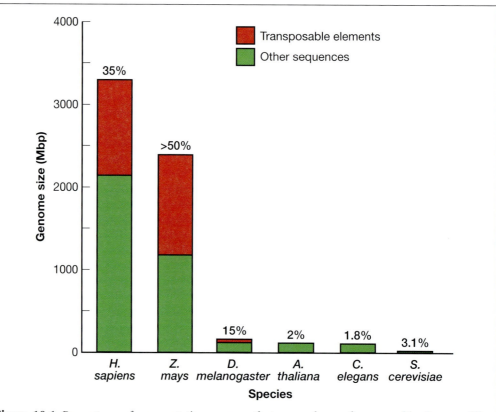

Figure 10.1. Percentages of representative genomes that are made up of transposable elements. The genomes include those of humans, maize, the fruit fly *Drosophila*, the model plant *Arabidopsis*, the nematode *C. elegans*, and budding yeast *S. cerevisiae*, respectively. (Adapted, with permission, from Kidwell and Lisch 2000 [copyright Elsevier Science].)

long (6–8 kbp long) interspersed nuclear elements (LINES). The types of transposable elements that are present in high copy numbers in mammalian genomes are illustrated in Figure 10.2. Ten percent of the human genome comprises one particular family of the SINE element, designated Alu (1.2 million copies) and 14.6% of one particular LINE designated LINE1 (593,000 copies)(Smit 1996).

Vertebrate chromosomes have long (>300 kb) regions of distinct GC richness, repeat content, and gene density, designated isochores in a model of genome organization proposing that genomes are made up of distinct segments of unique composition (Bernardi 1995). Human and mouse chromosomal regions that have a low density of genes are AT-rich and have more Alu or B1/B2 (SINES) than LINE1 elements, whereas the reverse is true for regions that have a high gene density, and those regions are more GC-rich (Henikoff et al. 1997).

The other class of TEs, class II, is made up of elements that employ a DNA-based mechanism of transposition. The human genome contains about 200,000 copies of this class of elements that probably predate human evolution (Smit 1996). Class II elements also include the Activation-Dissociation (Ac-Ds) family in maize and the P element in *Drosophila*.

A third category of TEs has features of both class I and class II TEs. These miniature, inverted repeat TEs (MITES) are 400 bp in length and were discovered in diverse flowering plants where they are frequently associated with regulatory regions of genes. Hence, they could be exerting an influence on regulation of gene expression (Kidwell and Lisch 1997).

The abundance of TEs in the genomes of humans, yeast, maize, and *E. coli* is illustrated in Figure 10.3. The following features are apparent: (1) TEs are present in all of the chromosomes, ranging from bacteria to humans, but their abundance varies; (2) TEs can com-

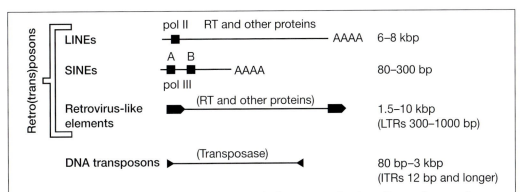

Figure 10.2. Transposable elements that produce high-copy-number interdispersed repeats in mammalian genomes. Shown are class of element, a representation of the structure, size of element plus, in some cases, size of terminal repeats. ■ RNA polymerase II or III promoter; ▶ long terminal repeat (LTR); ▶, ◀ inverted terminal repeats; RT reverse transcriptase. Parentheses above elements indicate protein found in autonomous elements. (Redrawn, with permission, from Smit 1996 [copyright Elsevier Science].)

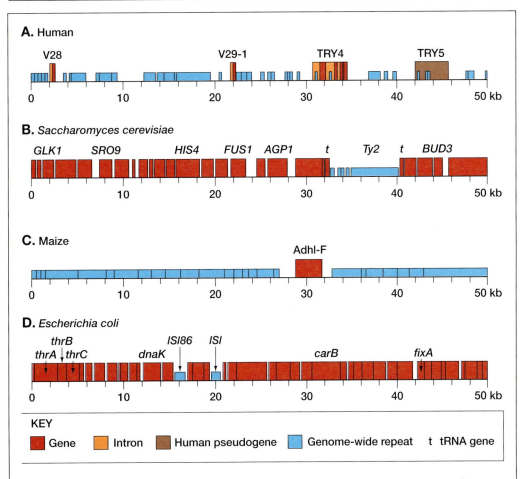

Figure 10.3. Comparison of genome composition in four genomes. (*A*) Human β T-cell receptor locus on chromosome 7. V28 and V29.1 encode parts of the β T-cell receptor proteins that are joined during development of the immune system (Rowen et al. 1996). TRY4, the gene for trypsinogen, and TRY5, a pseudogene related to the trypsinogen family, are not related to the receptor sequence. Why they are located here is not known. (*B*) Segment of yeast chromosome III (Oliver et al. 1992). (*C, D*) 50-kb fragments of the maize and *E. coli* chromosomes, respectively (SanMiguel et al. 1996; Blattner et al. 1997). The maize repeats are LTR retrovirus-like elements (Fig. 10.2) that have inserted within the last 3 million years (SanMiguel et al. 1998). (Redrawn, with permission, from Brown 1999 [copyright Wiley-Liss].)

prise a large portion of the genomes of higher eukaryotes, both plants and animals. Thus, only a small fraction of the genome of these organisms carries gene sequences.

Gene Structure Varies in Eukaryotes

Complex intron arrangements are often found. RNA of organelles can have introns with introns (Copertino and Hallick 1993), and nuclear genomes can encode genes in which one gene, including introns, is encoded within the introns of a second gene (see, e.g., Cawthon et al. 1990).

Eukaryotic genes that encode proteins are interrupted by introns of varying length and number. In *S. cerevisiae* (budding yeast), only a small fraction of the genes contain introns, and there are a total of 239 introns in the entire genome. In contrast, in individual human genes, introns may be present in numbers exceeding 100 and comprise more than 95% of the gene. Introns can remain at a corresponding position in a eukaryotic gene for long periods of evolutionary time. The origin of introns in eukaryotic genes is not understood but has been accounted for by two models. The "introns-early" view proposes that introns were used to assemble the first genes from sets of ancient conserved exons, whereas the "introns-late" view proposes that introns broke up previously continuous genes by inserting into them (Gilbert et al. 1997).

The intron structure of genes in a particular eukaryote is used for predicting the location of genes of genome sequences. Other features of eukaryotic genes in a particular organism that are useful for gene prediction include the consensus sequences at exon–intron and intron–exon splice junctions, base composition, codon usage, and preference for neighboring codons. Computational methods described in Chapter 8 incorporate this information into a gene model that may be used to predict the presence of genes in a genome sequence. Although not always correct, these methods provide a useful annotation of a new genome sequence, and in combination with database similarity searches

Table 10.3. *Number of genes predicted to encode proteins in model organisms and humans*

Organism	Biological features	Haploid genome size (Mb)	Predicted number of genes
Arabidopsis thaliana	plant with small genome; genes for metabolism, development by hormones and cell-cell interactions and environmental responses	130	~25,000[a]
Caenorhabditis elegans	worm (nematode) genes for development by a unique cell lineage, nervous system, and reproduction	100	18,424
Drosophila melanogaster	fruit fly; model for developmental processes by hormones and cell-cell interactions	180	13,601
Escherichia coli	bacterium; genes for growth on external sources of energy, transport of molecules through cell membrane, metabolic pathways, and replication as a single cell	4.7	4,288
Homo sapiens (human)	duplicates many gene functions in other model organisms and in addition includes control of higher brain functions	3×10^3	120,000[b]
Saccharomyces cerevisiae	budding yeast; genes for existence as a single-celled organism with the basic structure and organization of the eukaryotic cell	13.5	6,241

Examples of other model organisms that are to be sequenced include the mouse (*Mus musculus*), 3,300 Mb, and rice (*Oryza sativa*), 565 Mb. The mouse genome is a model for the human genome with which it shares a large amount of sequence homology and local gene order. The rice genome is a model for the cereal crops such as wheat (*Triticum aestivum,* genome size 1,700 Mb). The cultivated grasses all share similar genes, and cultivation has resulted in changes in the same genes (Paterson et al. 1995). Plant genomes in general vary in genome size due to the presence of repetitive elements including the number of copies of haploid chromosomes. Wheat, for example, has a hexaploid constitution (for review, see Devos and Gale 2000). The largest plant genomes are members of the Liliaceae family (>87,000 Mb) (see Bennetzen 2000).

[a] Based on the annotation of chromosomes 2 and 4 (Kaneko et al. 1999; Lin et al. 1999).

[b] Based on analysis of 2,000,000 carefully indexed ESTs (Liang et al. 2000). This is higher than previous estimates based on annotation of chromosome 22 (45,000).

described below, provide an indication of the genetic potential of an organism. Numbers of predicted genes estimated from the complete genome sequence of four model eukaryotic organisms are given in Table 10.3. The number of predicted genes in *E. coli* is also given for comparison. Due to the compact gene density in *E. coli* (see Fig. 10.3), there is about one gene per kb of genome sequence. Yeast is about twofold less compact than *E. coli*. Of the remaining genomes, *C. elegans* and *A. thaliana* have approximately the same density of genes (one gene per 6 kb), *Drosophila* being the least dense (one gene per 14 kb). One-sixth of the *Drosophila* sequence is composed of TEs and one-third is heterochromatic regions that do not include genes. Hence, in the euchromatic regions, the gene density in the *Drosophila* genome is one gene per 9 kb. Despite the fact that the lower number of predicted genes in *Drosophila* is smaller than that of the other genomes, the amount of functional diversity, as evidenced by protein family representation, is similar (Adams et al. 2000). Assessment of genome functional diversity is discussed in the following sections.

Pseudogenes

New gene functions are thought to be gained by duplication of an existing gene creating two tandem copies. Functional differentiation then occurs between the copies by mutation and selection. However, because most mutations are deleterious, and because only one gene copy may be needed for function, there is a strong tendency of one copy to accumulate mutations that render the gene nonfunctional. Accordingly, pseudogenes are DNA sequences that were derived from a functional copy of a gene but which have acquired mutations that are deleterious to function (Li 1997). In Figure 10.3A, the pseudogene *TRY5* is similar to the nearby functional gene *TRY4*.

There is also a second type of pseudogene found in eukaryotic genomes called a processed pseudogene. Processed pseudogenes are also derived from a functional gene, but they do not contain introns and lack a promoter; hence, they are not expressed. The origin of these pseudogenes is probably due to reverse transcription of the mRNA of the functional gene and insertion of the cDNA copy into a new chromosomal location by a LINE1 (Fig. 10.2) reverse transcriptase (Weiner 2000).

SEQUENCE ASSEMBLY AND GENE IDENTIFICATION

As discussed in Chapter 2, sequencing of genomes depends on the assembly of a large number of DNA reads into a linear, contiguous DNA sequence. The cost and efficiency of this process has been greatly improved by automatic methods of sequence assembly, first used for the sequencing of the bacterium *H. influenzae* (see Prokaryotic genome, p. 481). This same method of assembly was also used, in part, to complete the sequencing of the *Drosophila* (Myers et al. 2000) and human genomes in a timely manner.

As illustrated and explained in the Chapter 10 flowchart (p. 492), each genome sequence is scanned for protein-encoding genes using gene models trained on known gene sequences from the same organism. Methods of gene prediction in eukaryotic genomic DNA are discussed in Chapter 5 (for RNA-encoding genes) and Chapter 8 (for protein-encoding genes).

Identification of the function of protein-encoding genes is discussed in the Chapter 10 flowchart and in Chapter 7. For a new genome, each predicted gene is translated into a protein sequence; the collection of protein sequences encoded by the genome is the proteome of the organism. As illustrated in Figure 10.4, left panel, every protein in the proteome is then used as a query sequence in a database similarity search. Matching database sequences are realigned with the query sequence to evaluate the extent and significance of the alignment, as described in Chapter 2.

Screening the predicted protein sequences against an expressed sequence tag (EST) library confirms the prediction and expression of the gene (see Adams et al. 2000). The collective information on proteome function can then be further analyzed by self-comparison to find duplicated genes (paralogs) and by a proteome-by-proteome comparison to identify orthologs, genes that have maintained the same function through speciation, and other sequence and evolutionary relationships that are important for metabolic, regulatory, and cellular functions. These proteome comparisons are described in the next section.

METHODS

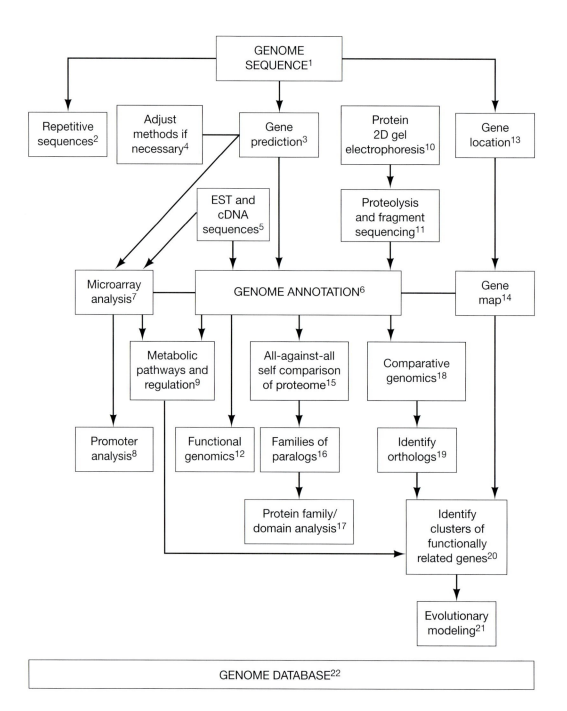

1. Genome sequences are assembled from DNA sequence fragments of approximate length 500 bp obtained using DNA sequencing machines as described in Chapter 2. Chromosomes of a target organism are purified, fragmented, and subcloned in fragments of size hundreds of kbp in bacterial artifical chromosomes (BACs). The BAC fragments are then further subcloned as smaller fragments into plasmid vectors for DNA sequencing (although the ends of BACs may also be sequenced as a way to circumvent problems with sequence repeats; see Myers et al. 2000). Full chromosomal sequences are then assembled from the overlaps in a highly redundant set of fragments by an automatic computational method (Myers et al. 2000) or from the fragment order on a physical map.

2. Eukaryotic genomes comprise classes of repeated elements, including tandem repeats present in centromeres and telomeres, dispersed tandem repeats (minisatellites and macrosatellites), and interdispersed TEs. TEs can comprise one-half or more of the genome sequence. Analysis of sequence repeats is discussed in Chapters 3 and 7. Identification of classes of repeated elements is aided by searchable databases discussed in Chapter 7 (p. 309).

3. Gene identification in prokaryotic organisms is simplified by their lacking introns. Once the sequence patterns that are characteristic of the genes in a particular prokaryotic organism (e.g, codon usage, codon neighbor preference) have been found, gene locations in the genome sequence can be predicted quite accurately. The presence of introns in eukaryotic genomes makes gene prediction more involved because, in addition to the above features, locations of intron–exon and exon–intron splice junctions must also be predicted. Methods of gene prediction in prokaryotes and eukaryotes are discussed in Chapter 8.

4. Gene prediction methods involve training a gene model (e.g., a hidden Markov model or neural network, see Chapter 8) to recognize genes in a particular organism. Due to variations in gene codon preferences and splice junctions (see note 3, Fig. 10.3), a model must usually be trained for each new genome.

5. Since gene prediction methods are only partially accurate (for review, see Bork 1999; see Chapter 8), gene identification is facilitated by high-throughput sequencing of partial cDNA copies of expressed genes (called expressed sequence tags or EST sequences). Presence of ESTs confirms that a predicted gene is transcribed. A more thorough sequencing of full-length cDNA clones may be necessary to confirm the structure of genes chosen for a more detailed analysis.

6. The amino acid sequence of proteins encoded by the predicted genes is used as a query of the protein sequence databases in a database similarity search. A match of a predicted protein sequence to one or more database sequences not only serves to identify the gene function, but also validates the gene prediction. Pseudogenes, gene copies that have lost function, may also be found in this analysis. Only matches with highly significant alignment scores and alignments (see Chapter 3, page 58) should be included. The genome sequence is annotated with the information on gene content and predicted structure, gene location, and functional predictions. The predicted set of proteins for the genome is referred to as the proteome. Accurate annotation is extremely important so that others users of the information are not misinformed. Procedures for searches starting with genome, EST, and cDNA sequence are described in Chapter 8. Usually, not all query proteins will match a database sequence. Hence, it is important to extend the analysis by searching the predicted protein sequence for characteristic domains (conserved amino acid patterns that can be aligned) that serve as a signature of a protein family or of a biochemical or structural feature (see note 17). A further extension is to identify members of protein families or domains that represent a structural fold using the computational tools described in Chapter 9. This additional information also needs to be accurately described and the significance established.

7. Microarray analysis provides a global picture of gene expression for the genome by revealing which genes are expressed at a particular stage of the cell cycle or developmental cycle of an organism, or genes that respond to a given environmental signal to the same extent. This type of information provides an indication as to which genes share a related biological function or may act in the same biochemical pathway and may thereby give clues that will assist in gene identification.

8. Genes that are found to be coregulated either by a microarray analysis or by a protein two-dimensional analysis should share sequence patterns in the promoter region that direct the activity of transcription factors. The types of analyses that are performed are discussed in Chapter 8 (pp. 357–373), and additional tools for analyzing coregulated genes are listed in Table 10.1E.

9. As genes are identified in a new genome sequence, some will be found that are known to act sequentially in a metabolic pathway or to have a known role in gene regulation in other organisms. From this information, the metabolic pathways and metabolic activities of the organism will become apparent. In some cases, the apparent absence of a gene in a well-represented pathway may lead to a more detailed search for the gene. Clustering of genes in the pathway on the genome of a related organism can provide a further hint as to where the gene may be located (see note 20).

10. Individual proteins produced by the genome can be separated to a large extent by this method and specific ones identified by various biochemical and immunological tests. Moreover, changes in levels of proteins in response to an environment signal can be monitored in much the same way as a microarray analysis is performed. Microarrays only detect untranslated mRNAs, whereas a two-dimensional gel protein analysis detects translation products, thus revealing an additional level of regulation. Resources for analysis of regulation by this method are given in Table 10.1D.

11. Protein spots may be excised from a two-dimensional protein gel (see note 10) and subjected to a combination of amino acid sequencing and cleavage analyses using the techniques of mass spectrometry and high-pressure liquid chromatography. Genome regions that encode these sequences can then be identified and the corresponding gene located. A similar method may be used to identify the gene that encodes a particular protein that has been purified and characterized in the laboratory. The computational methods are described in Chapter 7 (p. 295, FASTA tools) and Table 10.1D.

12. Functional genomics involves the preparation of mutant or transgenic organisms with a mutant form of a particular gene usually designed to prevent expression of the gene. The gene function is revealed by any abnormal properties of the mutant organism. This methodology provides a way to test a gene function that is predicted by sequence similarity to be the same as that of a gene of known function in another organism. If the other organism is very different biologically (comparing a predicted plant or animal gene to a known yeast gene), then functional genomics can also shed light on any newly acquired biological role. When two or more members of a gene family are found (see notes 16 and 17), rather than a single match to a known gene, the biological activity of these members may be analyzed by functional genomics to look for diversification of function in the family.

13. Since the entire genome sequence is available, as each gene is identified, the relative position of the gene will be known.

A more detailed analysis of the relative amount of sequence variability in a chromosomal region within populations of closely related species can reveal the presence of genes that are under selection. These regions will not have the expected amount of variability given their linkage: They are in a state of linkage disequilibrium. An example is the BRCA1 (breast cancer 1) gene of humans and chimpanzees (Huttley et al. 2000).

14. A map showing the location of each identified gene is made. These relative positions of genes can be compared to similar maps of other organisms to identify rearrangements that have occurred in the genome. Gene order in two related organisms reflects the order that was present in a common ancestor genome. Chromosomal breaks followed by a reassembly of fragments in a different order can produce new gene maps. These types of evolutionary changes in genomes have been modeled by computational methods (p. 512). Gene order is revealed not only by the physical order of genes on the chromosome, but also by genetic analysis. Populations of an organism show sequence variations that are readily detected by DNA sequencing and other analysis methods. The inheritance of genetic diseases in humans and animals (e.g., cancer and heart disease), and of desirable traits in plants, can be traced genetically by pedigree analysis or genetic crosses. Sequence variations (polymorphisms) that are close to (tightly linked) a trait may be used to trace the trait by virtue of the fact that the polymorphism and the trait are seldom separated from one generation to the next. These linked polymorphisms may then be used for mapping and identifying important genes.

15. A comparison is made in which every protein is used as a query in a similarity search against a database composed of the rest of the proteome, and the significant matches are identified by a low expect value ($E < 10^{-6}$ was used in a recent analysis by Rubin et al. [2000]). Since many proteins comprise different combinations of a common set of domains, proteins that align along most of their lengths (80% identity is a conservative choice) are chosen to select those that have a conserved domain structure.

16. A set of related proteins identified in step 15 is subjected to a cluster analysis in order to identify the most closely related groups of proteins and to avoid domain-matching. This group of proteins is derived from a gene family of paralogs that have arisen by gene duplication.

17. Each protein in the predicted proteome is again used as a query of a curated protein sequence database such as SwissProt in order to locate similar domains and sequences. The domain composition of each protein is also determined by searching for matches in domain databases such as Interpro, described in Table 9.5. The analysis reveals how many domains and domain combinations are present in the proteome, and reveals any unusual representation that might have biological significance. The number of expressed genes in each family can also be compared to the number in other organisms to determine whether or not there has been an expansion of the family in the genome.

18. Comparative genomics is a comparison of all the proteins in two or more proteomes, the relative locations of related genes in separate genomes, and any local groupings of genes that may be of functional or regulatory significance.

19. Orthologs are genes that are so highly conserved by sequence in different genomes that the proteins they encode are strongly predicted to have the same structure and function and to have arisen from a common ancestor through speciation. To identify orthologs, each protein in the proteome of an organism is used as a query in a similarity search of a database comprising the proteomes of one or more different organisms. The best hit in each proteome is likely to be with an ortholog of the query gene. In comparing two proteomes, a common standard is to require that for each pair of orthologs, the first of the pair is the best hit when the second is used to query the proteome of the first. To find orthologs, very low E value scores ($E < 10^{-20}$) for the alignment score and an alignment that includes 60–80% of the query sequence are generally required in order to avoid matches to paralogs. Although these requirements for classification of orthologs are very stringent, a more relaxed set of conditions will lead to many more false-positive predictions. In bacteria, the possibility of horizontal transfer of genes between species also has to be considered (p. 508).

20. In related organisms, both gene content of the genome and gene order on the chromosome are likely to be conserved. As the relationship between the organisms decreases, local groups of genes remain clustered together, but chromosomal rearrangements move the clusters to other locations. In microbial genomes, genes specifying a metabolic pathway may be contiguous on the genome where they are coregulated transcriptionally in an operon by a common promoter. In other organisms, genes that have a related function can also be clustered. Hence, the function of a particular gene can sometimes be predicted, given the known function of a neighboring, closely linked gene. Genomes are also compared at the level of gene content, predicted metabolic functions, regulation as revealed by microarray analysis, and others. These comparisons provide a basis for additional predictions as to which genes are functionally related. Gene fusion events that combine domains found in two proteins in one organism into a composite protein with both domains in a second organism are also found and provide evidence that the proteins physically interact or have a related function.

21. Evolutionary modeling can include a number of types of analyses including (1) the prediction of chromosomal rearrangements that preceded the present arrangement (e.g., a comparison of mouse and human chromosomes), (2) analysis of duplications at the protein domain, gene, chromosomal, and full genome level, and (3) search for horizontal transfer events between separate organisms.

22. Due to the magnitude of the task, the earlier stages of genome analysis including gene prediction and database similarity searches are performed automatically with little human intervention. The genome sequence is then annotated with any information found without involving human judgment. The types of genome analyses in the flowchart also provide many predictions and give rise to many preliminary hypotheses regarding gene function and regulation. As more detailed information is collected by laboratory experiment and by a closer examination of the sequence data, this information needs to be linked to the genome sequence. In addition, the literature, past and present, needs to be scanned for information relevant to the genome. A carefully crafted database that takes into account the entire body of information should then be established. In addition to information on the specific genome of interest, the database should include cross-references to other genomes. To facilitate such intergenome comparisons, common gene vocabularies have been proposed. This slow, expensive, and time-consuming phase of genome analysis is of prime importance if the genome information is to be available in an accurate form for public use.

A. Types of proteome analysis

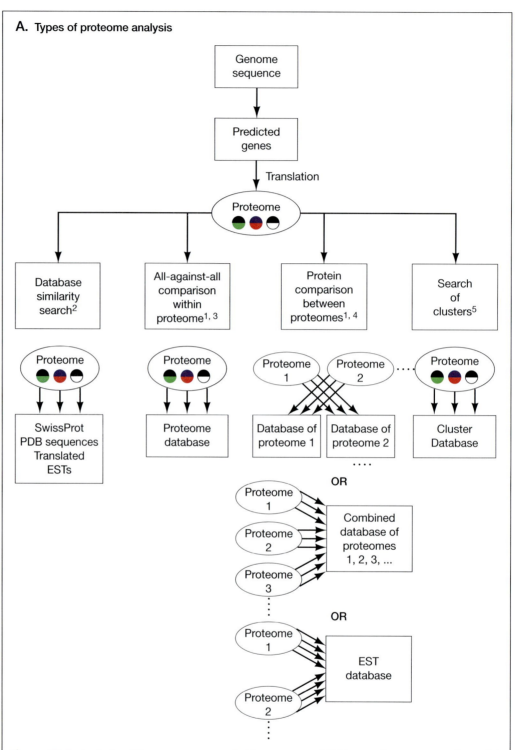

Figure 10.4. Analysis of the proteome encoded by genomes. (*A*) Types of proteome analyses. (*B*) Examples of database hits resulting from domain structure of proteins. (*C*) Cluster analysis of similar sequences. (*D*) Domain identification.

Notes:

1. Due to the large number of comparisons that must be made in these types of analyses (as many as 20,000 by 20,000 sequences) and due to the volume of program output, the procedure must be automated on a local machine using Perl scripts or a similar method and a database system. For BLAST, setting an effective database size appropriate for each search and program is important for obtaining a correct statistical evaluation of alignment scores. The bioperl project provides valuable resources for this purpose (http://www.bioperl.org).

B. Examples of database hits resulting from domain structure of proteins[6]

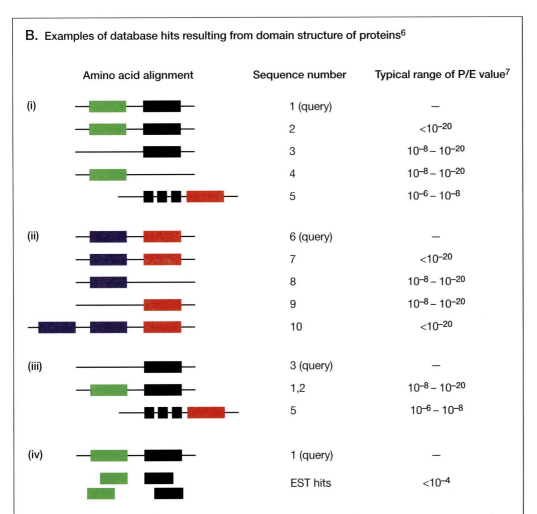

Amino acid alignment	Sequence number	Typical range of P/E value[7]
(i)	1 (query)	—
	2	$<10^{-20}$
	3	$10^{-8} - 10^{-20}$
	4	$10^{-8} - 10^{-20}$
	5	$10^{-6} - 10^{-8}$
(ii)	6 (query)	—
	7	$<10^{-20}$
	8	$10^{-8} - 10^{-20}$
	9	$10^{-8} - 10^{-20}$
	10	$<10^{-20}$
(iii)	3 (query)	—
	1,2	$10^{-8} - 10^{-20}$
	5	$10^{-6} - 10^{-8}$
(iv)	1 (query)	—
	EST hits	$<10^{-4}$

2. Each protein encoded by the genome is used as a query in database similarity searches to identify similar database proteins, some having a known structure or function. Additional searches of EST databases can be used to identify additional relatives of the query sequence. These searches and evaluation of the alignment scores of matching sequences are described in Chapter 7.

3. An all-against-all analysis requires first making a database of the proteome. This database is then sequentially searched by each individual protein sequence of the proteome using a rapid database similarity search tool such as BLAST, WU-BLAST, or FASTA. The scoring systems of these programs vary and are described in Chapter 7. Note also that P values of WU-BLAST (Chervitz et al. 1998) are similar to E values of NCBI BLAST (Rubin et al. 2000) for values of P and $E < 0.05$. This analysis generates a matrix of alignment scores, each with an E value and corresponding alignment for each pair of proteins. Recall that the E value of an alignment score is the probability that an alignment score as good as the one found would be observed between two random or unrelated sequences in a search of a database of the same size. The lower the E value, the more significant the alignment between a pair of matching sequences. In an all-against-all comparison within one proteome, significantly matched pairs of sequences may be paralogs that originated from a gene duplication event in this genome or the genome of an ancestor organism. Unique proteins can be identified through their not matching any other protein. A conservative cutoff E value (e.g., 10^{-6}; Rubin et al. 2000) limits the matches to the most significant ones, which are then clustered into families as described below and in the text.

4. To perform a between-proteome analysis, proteome databases are made for the known and predicted genes of two or more genomes. Both single (Chervitz et al. 1998) and combined proteome databases may be made (Rubin et al. 2000). Each protein of one proteome is then selected in turn as a query of the proteome of another organism or the combined proteome of a group of organ-

isms. As in an all-by-all protein comparison within a proteome, a matrix of alignment scores with *E* values is made, and the most closely related sequences in the two organisms are identified. This analysis can predict orthologs, i.e., proteins that have an identical function attributable to descent of the respective genes from a common ancestor. The types of criteria used in bioinformatics to define orthologs include (1) reciprocal database searches with one sequence as query give a best hit of the other sequence (Tatusov et al. 1997); (2) the alignment of the sequences includes at least 80% of each sequence (Chervitz et al. 1998; Rubin et al. 2000); and (3) the sequences are clustered when all matching sequences are subjected to a cluster analysis. The likelihood of orthology is also increased if a set of orthologous pairs are linked together on the respective genomes. The types of analyses are discussed further in the text.

5. The cluster search option is most useful for prokaryotic organisms. Each protein in the proteome is used as a query of a database of protein clusters using the program COGNITOR (Table 10.1, COGs entry). These clusters are composed of orthologous pairs of sequence defined by criterion 1, described in note 4. The database was made by performing an all-by-all genome comparison across a spectrum of prokaryotic organisms and a portion of the yeast proteome (Tatusov et al. 2000). Orthologous pairs of sequence were then merged with clusters or orthologous pairs (COGs) for multiple proteomes as described in the text. COGs have been linked to classes of biochemical function (Tatusov et al. 1997). Hence, matching a query sequence to the COG can potentially identify unique orthologs in another proteome that may have the same function. The COGs database is designed to provide a preliminary indication of orthologous relationships that can be tested by more detailed similarity searches, sequence alignments, and phylogenetic analysis of the matching sequences.

6. Due to the modular nature of proteins, several types of matches may be identified in the all-against-all and between-proteome comparisons. Each colored box represents a hypothetical conserved domain that is matched in the search. The dotted box (sequence 5) represents a less similar domain that will not align as well. Highest-scoring matches corresponding to matching of multiple domains present in the query and in the matched sequence ([i] and [ii], sequence pairs 1 and 2, 6 and 7, etc.). The alignment scores of these pairs should have extremely low *E* values. A multidomain query protein will also match database proteins that have a single domain (as in sequences 1 and 3, 6 and 8). Because only one domain is represented by the alignment, the alignment will in general be shorter and have a poorer (higher) *E* value score than a multidomain alignment. The analysis will also identify matches of a query with a database protein that has two or more copies of query sequence domain (sequence 10). Query sequences with a minimal domain representation (ii) will not score particularly well with any sequence (sequence 3). Duplicate comparisons generated by the method are eliminated. When only an EST library of an organism is available, the proteome may be compared to this library. However, since these databases are generally not complete and any alignments are shorter, it is diffcult to compare these results with the full proteome comparisons. From a biological standpoint, ESTs define expressed genes, whereas proteomes are predicted genes.

7. WU-BLAST produces *P* scores and BLAST (NCBI) *E* scores where $E = -\ln(1 - P)$. For values less than 0.05, $E = P$. The score ranges depicted in this column are hypothetical examples. The choice of a $<10^{-20}$ score is a conservative one for identification of orthologs that should have a similar domain structure, as do the sequences in this example (see Chervitz et al. 1998; Rubin et al. 2000). To define these groups, the distribution of hits below different thresholds should be examined, as in the above references. The higher cutoff score for EST matches is used because the search of an EST database may only produce short alignments.

8. Shown are two representations of the sequence relationships found in part *B*. In (i) the sequences, color coded to represent domain structure, are represented by vertices on graph. In comparing the graphic (i) and single linkage (ii) clusters, note that in (i) each sequence has multiple edges representing links to related sequences, whereas in (ii) the sequences are only connected to one branch on the outermost part of the tree.

9. The sequence alignments found above represent the presence of one or more conserved domains in each cluster or group of clusters. These clusters are next analyzed for the presence of known domains by searches of domain databases as described in Chapter 9. This analysis identifies the number and types of domains that are shared between organisms, or that have been duplicated in proteomes to produce paralogs.

C. Cluster analysis of similar sequences[8]

(i) Graphic representation

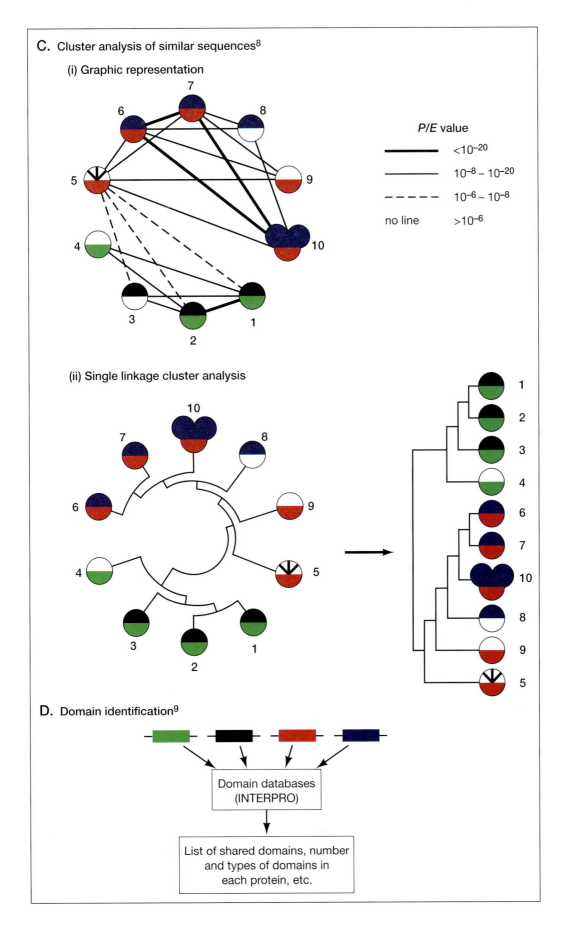

P/E value

———	$<10^{-20}$
——	$10^{-8} - 10^{-20}$
- - -	$10^{-6} - 10^{-8}$
no line	$>10^{-6}$

(ii) Single linkage cluster analysis

D. Domain identification[9]

Domain databases
(INTERPRO)

List of shared domains, number
and types of domains in
each protein, etc.

COMPARATIVE GENOMICS

Comparative genomics includes a comparison of gene number, gene content, and gene location in both prokaryotic and eukaryotic groups of organisms. The availability of complete genome sequences makes possible a comparison of all of the proteins encoded by one genome, the proteome of that organism, with those of another. Because the genome sequence provides both the sequence and the map location of each gene, both the sequence and location can be compared. Sequence comparisons provide information on gene relationships—the number of genes in two organisms that are so similar that they must have the same function and evolutionary history—these genes are orthologs (Fitch 1970). Map locations of orthologous genes may also be compared. If a set of genes is grouped together at a particular chromosomal location, and if a set of similar genes is also grouped together in the genome of another organism, these groups share an evolutionary history.

Proteins may also be clustered into families on the basis of either sequence or structural similarity, as discussed in Chapter 9. Proteins are modular and often comprise separate domains. The number of protein sequences that are available is sufficient to determine that domain shuffling occurs in evolution—domains appear or disappear in particular families, become combined to make new families, or else become separated into two different proteins that are predicted to interact (Snel et al. 2000a). The comparisons of proteomes of different organisms can identify the type of domain changes and also provide an indication as to what biological role they may have in a particular organism.

The assortment and reassortment of protein domains takes place in individual genomes. Proteins with new functions are produced by a gene duplication event in which two tandem copies of a gene are produced (see Fig. 3.3, p. 55). Through mutation and natural selection, one of the copies can develop a new function, leaving the other copy to cover for the original function. However, because most mutations are deleterious to function, often one of the copies becomes a pseudogene. Not all gene duplications are thought to have the above effects. Another scenario is that two duplicated genes both undergo change, but interactions between the proteins stabilize the original function and support the evolution of new ones (Force et al. 1999).

The processes of domain assortment and gene duplication produce families of proteins in organisms. Following speciation, a newly derived genome will inherit the families of ancestor organisms, but will also develop new ones to meet evolutionary challenges. Comparison of each of the proteins encoded by an organism with every protein, an all-against-all comparison, reveals which protein families have been amplified and what rearrangements have occurred as steps in the evolutionary process. When two or more proteins in the proteome share a high degree of similarity because they share the same set of domains (illustrated in Fig. 10.4B), they are likely to be paralogs (Fitch 1970), genes that arose by gene duplication events. Proteins that align over shorter regions share some domains, but also may not share others. Although gene duplication events could have created such variation, other rearrangements may have also occurred, blurring the evolutionary history.

The following sections describe methods to compare prokaryotic and eukaryotic genomes for orthologs and paralogs. It is important to keep in mind the predictive nature of these types of analyses. Decisions about gene relationships depend on careful manual inspection of sequence alignments (Huynen et al. 2000).

Proteome Analysis

All-against-all Self-comparison Reveals Numbers of Gene Families and Duplicated Genes

A comparison of each protein in the proteome with all other proteins distinguishes unique proteins from proteins that have arisen from gene duplication, and also reveals the number of protein families. The domain content of these proteins may also be analyzed. One type of all-against-all proteome comparison is described in Figure 10.4A, second panel. In this analysis, each protein is used as a query in a similarity search against the remaining proteome, and the similar sequences are ranked by the quality and length of the alignments found. The search is conducted in the manner described in Chapter 7, with each alignment score receiving a statistical evaluation (P or E value). As shown in Figure 10.4B, a match between a query sequence and another proteome sequence with the same domain structure will produce a high-scoring, highly significant alignment. These proteins are designated paralogs because they have almost certainly originated from a gene duplication event. Lower-scoring, less significant alignments may have identified proteins that share domains but not the high degree of sequence similarity that is apparent in the best-scoring alignments. These may also be paralogs, but they may also have a complicated history of domain shuffling that is difficult to reconstruct.

Cluster analysis. To sort out relationships among all of the proteins that are found to be related in a series of searches of the types shown in Figure 10.4B, they are subjected to a clustering analysis shown in Figure 10.4C. Only the relationships revealed by the hypothetical set of searches illustrated in part B are shown. Some of the proteins may have other relationships, which are not depicted in order to simplify the example.

Clustering organizes the proteins into groups by some objective criterion. One criterion for a matching protein pair is the statistical significance of their alignment score (the P or E value from BLAST searches). The lower this value, the better the alignment. There will be a cutoff P or E value at which the matches in the BLAST search are no longer considered significant. A value of P or $E > 0.01$–0.05 is usually the point at which the alignment score is no longer considered to be significant in order to focus on a more closely related group of proteins. A second criterion for clustering proteins is the distance between each pair of sequences in a multiple sequence alignment. The distance is the number of amino acid changes between the aligned sequences.

Clustering by making subgraphs. Figure 10.4 indicates two ways of clustering related sequences based on the above criteria. Part (i) is a graph in which each sequence is a vertex and each pair of sequences that is matched with a significant alignment score is joined by an edge that is weighted according to the statistical significance of the alignment score. One way to identify the most strongly supported clusters is simply to remove the most weakly supported edges in the graph, in this case the alignments with the highest P/E scores (dotted edges). As weaker and weaker links are removed, the remaining combinations of vertices and edges represent most strongly linked sequences. This type of analysis was performed on an initial collection of *E. coli* genes by Labedan and Riley (1995). Their analyses revealed that *E. coli* genes clustered in this manner encode proteins already known to belong to the same broad functional category, EC number, or to have a similar physiological function. For another approach to identify orthologs in microbial genes, see Bansal (1999).

Another method for clustering similar sequences that are likely to be paralogs is described in Rubin et al. (2000). In this method, edges of E value $> 1 \times 10^{-6}$ are removed. The remaining graph is then broken down into subgraphs comprising sequences that share a significant relationship to each other but not to other sequences. The criterion chosen is that the group should mutually share at least two-thirds of all of the edges from this group to all proteins in the proteome. If two proteins A and B share a domain but do not share another domain in A, and if A shares this other domain with a number of other sequences, the algorithm would tend not to cluster A with B (Rubin et al. 2000). Thus, the algorithm favors the selection of proteins with the same domain structure reflecting that these proteins are the most likely ones to be paralogs.

Clustering by single linkage. A second method for clustering related sequences is shown in Figure 10.4C, part (ii). This method is based on the distance criterion for sequence relationships described above. First, a group of related sequences found in the all-against-all proteome comparison is subjected to a multiple sequence alignment usually by CLUSTALW (Chapter 4, p. 154). A distance matrix that shows the number of amino acid changes between each pair of sequences is then made. This matrix is then used to cluster the sequences by a neighbor-joining algorithm. This procedure and the algorithms are the same as those used to make a phylogenetic tree by the distance methods, described in Chapter 8. These methods produce a tree (Fig. 10.4C, part ii, left) or a different representation of the tree called a dendrogram (Fig. 10.4C, part ii, right), that minimizes the number of amino acid changes that would generate the group of sequences. The tree is also defined as a minimum spanning tree (Duran and Odell 1974). The tree and dendrogram cluster the sequences into the most closely related groups. Branches joining the least related sequences may be removed, thus leaving two sub-trees with a small group of sequences. As smaller groups are chosen, the most strongly supported clusters are likely to be made up of paralogs. However, it is not easy to distinguish sequences that are paralogs, i.e., share several domains, from those that share domains but that also share other domains with more distantly related sequences without inspection of the alignments. GeneRage (Table 10.1E) provides an automatic system for classifying protein data sets by means of an iterative refinement approach using local alignments, matrix methods, and single-linkage clustering.

Core proteome. The above types of all-against-all analyses provide an indication as to the number of protein/gene families in an organism. This number represents the core proteome of the organism from which all biological functions have diversified. A representative sample is shown in Table 10.4.

In *Hemophilus*, 1247 of the total number of 1709 proteins do not have paralogs (Rubin et al. 2000). The core proteomes of the worm and fly are similar in size but with a greater number of duplicated genes in the worm. It is quite remarkable that the core proteome of the multicellular organisms (worm and fly) is only twice that of yeast.

Table 10.4. Numbers of gene families and duplicated genes in model organisms (Rubin et al. 2000)

Organism	Total number of genes	Number of gene families[a]	Number of duplicated genes[b]
Hemophilus influenzae (bacteria)	1709	1425[c]	284
Saccharomyces cerevisiae (yeast)	6241	4383	1858
Caenorhabditis elegans (worm)	18,424	9453	8971
Drosophila melanogaster (fly)	13, 600	8065	5536

[a] The number of clustered groups in the all-against-all analysis using the algorithm described in the text. This number represents the core proteome of the organism.

[b] Count of number of duplicated genes within the protein family clusters.

[c] 178 families have paralogs.

Grouping Sequences

The problem of deciding which sequences to include in the same group or cluster and which to separate into different groups or clusters is a recurring one. The conservative approach is to group only very similar sequences together. However, in making a conservative multiple sequence alignment with only very alike sequences, it is not possible to analyze the evolutionary divergence that may have occurred in a family of proteins. Furthermore, if a matrix or profile model is made from this alignment, that model will not be useful for identifying more divergent members of a family. The adventurous approach is to choose a set of marginally alignable sequences to pursue the difficult task of making a multiple sequence alignment and then to make profile models that may recognize divergence but will also give false predictions. The best method to choose is somewhere between the conservative and adventurous methods. Divergence is necessary, but the sequences chosen should be clearly related based on inspection of each pair-wise alignment and a statistical analysis. Clustering analyses of the sequences can also be useful. Questionable sequences can be left out of the analysis at one stage and added in a second to determine what effect they have on the model.

Between-proteome Comparisons Identify Orthologs, Gene Families, and Domains

Comparisons between proteomes of organisms are illustrated by the third panel in Figure 10.4A. In this analysis, each protein in the proteome is used as a query in a database similarity search against another proteome or combined set of proteomes. When the proteome of an organism is not available, an EST database may be searched for matches, but the type of search is less informative than a full-genome comparison (see below). As in the all-against-all search for paralogs, the search should identify highly conserved proteins of similar domain structure and other similar proteins that show variation in the domain structure as illustrated in Figure 10.4B. A pair of proteins in two organisms that align along most of their lengths with a highly significant alignment score are likely to be orthologs, proteins that share a common ancestry and that have kept the same function following speciation. These proteins perform the core biological functions shared by all organisms, including DNA replication, transcription, translation, and intermediary metabolism. They do not include the proteins unique to the biology of a particular organism.

Other matching sequences in this class could also be orthologs, but could also represent a match between a sequence in proteome A to a paralog of a true ortholog of the sequence in proteome B. In one method designed to identify true orthologs, the most closely related pairs of sequences in proteomes A and B are identified. Two proteins, X in proteome A and Y in proteome B, are predicted to be an orthologous pair if reciprocal searches of proteome A with Y and proteome B with X each produce the highest-scoring match with the other protein. Furthermore, the E value for each alignment should be < 0.01 and the alignment should extend over 60% of each protein (Huynen and Bork 1998).

In another method to identify the mostly closely related sequences in different proteomes, Chervitz et al. (1998) kept only matched sequences with a very conservative P value for the alignment score. The steps for identifying a group of related sequences between the yeast and worm proteomes were as follows:

1. Choose a yeast protein and perform a database similarity search of the worm proteome using WU-BLAST, a yeast-versus-worm search.

2. Group the worm sequences that match the yeast query sequence with a high P value (10^{-10} to 10^{-100}) and include the yeast query sequence in the group.

3. From the group in proteome B, choose a worm sequence and make a search of the yeast proteome, using the same P value limit as in step 2.

4. Add any matching yeast sequence to the grouping made in step 2.

5. Repeat steps 3 and 4 for all initially matched worm sequences.

6. Repeat steps 1–5 for every yeast protein.

7. Perform a comparable worm-versus-yeast analysis as outlined in steps 1–6.

8. Coalesce the groups of related sequences and remove any redundancies so that every sequence is represented only once.

9. Eliminate any matched pairs in which less than 80% of each sequence is in the alignment.

The above steps locate groups of highly related sequences in two proteomes based on high-scoring alignments among the group. These groups are then subjected to the single linkage cluster analysis described above and illustrated in Figure 10.4C. The analysis creates a multiple sequence alignment and a tree/dendrogram representation of sequence relationships very similar to that produced in a phylogenetic analysis. Orthologs appear as nearest neighbors on the tips of this tree.

The results of the above analysis with the yeast and worm proteomes are shown in Table 10.5. The numbers of sequence groups decrease about fivefold as the stringency of the E value of the required scores decreases from 10^{-10} to 10^{-100}, and a similar effect is observed for the subcategories shown in the table. Given that these sequences also align to the extent of 80%, they represent highly conserved sets of genes.

Clusters of orthologous groups. As described above, a pair of orthologous genes in two organisms share so much sequence similarity that they may be assumed to have arisen from a common ancestor gene. When entire proteomes of the two organisms are available, orthologs may be identified. Using the protein from one of the organisms to search the proteome of the other for high-scoring matches should identify the ortholog as the highest-scoring match, or best hit. However, in many cases, each of the orthologs belongs to a family composed of paralogous sequences related to each other by gene duplication events. Hence, in the above database search, the ortholog will not only match the orthologous sequence in the second proteome but also these other paralogous sequences. The objective of the clusters of orthologous groups (COG) approach is to identify all matching proteins in the organisms, defined as an orthologous group related by both speciation and gene duplication events. Related orthologous groups in different organisms are then clustered together to form a COG that will include both orthologs and paralogs. These clusters cor-

Table 10.5. *Numbers of closely related yeast and worm sequences*

Cut-off P value	$< 10^{-10}$	$< 10^{-20}$	$< 10^{-50}$	$< 10^{-100}$
Total number of sequence groups	1171	984	552	236
Number of groups with more than two members	560	442	230	79
Number and percent of all yeast proteins (6217) represented in groups	2697 (40)	1848 (30)	888 (14)	330 (5)
Number and percent of all worm proteins represented in groups	3653 (19)	2497 (13)	1094 (6)	370 (2)

Adapted, with permission, from Chervitz et al. 1998 (copyright AAAS).

respond to classes of metabolic function. A database produced by analysis of the available microbial genomes and part of the yeast genome has been made, and a newly identified microbial protein may be used as a query to search this database (see Table 10.1D). Any significant matches found will provide an indication as to the metabolic function of the query protein (Tatusov et al. 1997).

To produce COGs, similarity searches were performed among the proteomes of phylogenetically distinct clades of prokaryotes (see Fig. 6.3, p. 243 for a tree). Orthologous pairs were first defined by the best hits in reciprocal searches. A cluster of three orthologs in three different species was then represented as a triangle on a diagram. Some triangles included a common side, representing the presence of the same orthologous pair in a comparison of four or more organisms. Triangles with this feature were merged into a cluster similar in appearance to Figure 10.4C(i). Paralogs defined by sets of three matching sequences in the selected organisms were also added to these clusters. Paralogs may include a best hit or a high-scoring match of one of the sequences by another, but the reciprocal match can have low similarity that does not have to be significant (Koonin et al. 1998). Sixty percent of the original set of 720 COGs does not include paralogs, or includes paralogs from one lineage only, suggesting that there has not been extensive duplication of this group.

Some of the clusters defined in this manner include proteins having a different domain structure, as illustrated in Figure 10.4B. In other cases, examination of sequence similarity between some pairs of paralogs reveals that a particular paralog has disappeared in a particular lineage. The affected COGs have been modified to reflect more accurately the domain organization of proteins and loss of paralogs. Finally, some additional COGs not represented in the data set were produced by single linkage cluster analysis as described in Figure 10.4C and in the above sections (Tatusov et al. 1997). The proteins encoded by 13 prokaryotic organisms have been analyzed for COG relationships (Koonin et al. 1998). A COG analysis provides an initial assessment of the genome composition of prokaryotic organisms and should be followed by a more detailed analysis as described above for the worm and yeast genomes.

Comparison of proteomes to EST databases of an organism. For many eukaryotic organisms, the complete genome sequence is not available. What is available for some of these organisms is a large collection of EST sequences obtained by random sequencing of cDNA copies of cell mRNA sequences. These sequences are single DNA sequence reads that contain a small fraction of incorrect base assessments, insertions, and deletions. Many sequences arise from near the 3' end of the mRNA, although every effort is usually made to read as far 5' as possible into the upstream portion of the cDNA. Because not all of the genes may be expressed in the tissues chosen for analysis, the library will often not be complete. EST libraries are useful for preliminary identification of genes by database similarity searches as described in Chapter 7. A more detailed analysis may then be made by cloning and sequencing the intact cDNA.

An EST database of an organism can be analyzed for the presence of gene families, orthologs, and paralogs. A protein from the yeast or fly proteome, for example, can be used as a query of a human EST database by translating each EST sequence in all six possible reading frames. The program TBLASTN is frequently used for this purpose. The TFASTX and TFASTY programs are designed to accommodate the errors inherent in EST sequences (p. 295). The limitations to whole-proteome searches against EST libraries are that the short length of the translated EST sequence (the equivalent of 100–150 amino acids) will only match a portion of the query protein; for example, a domain or part of domain as illustrated in Figure 10.4B. Hence, it is not possible to impose the requirement of alignment with 60–80% of the query sequence that greatly improved the prediction of orthologs. Predictions of EST relationships can be improved by identifying overlapping EST sequences so that a longer alignment can be produced, as discussed in Chapter 7. Another method is to perform an exhaustive search for a protein family, described next.

Searching for orthologs to a protein family in an EST database. Searches of EST databases for matches to a query sequence routinely produce large amounts of output that must be searched manually for significant hits. Retief et al. (1999) have described an automatic method utilizing a computer script, FAST-PAN, that scans EST databases with multiple queries from a protein family, sorts the alignment scores, and produces charts and alignments of the matches found. An example of using this method is shown in Figure 10.5. A chart showing the *E* value, percent identity, fraction of query sequence matched, and type of query matched (color coded) is shown in Figure 10.5A.

In an example by Retief et al. (1999), the large family of known glutathione transferase proteins was first subjected to multiple sequence alignment, and a phylogenetic tree was made by distance methods to identify classes of proteins within the family. These proteins represented a broad range of phylogenetic context and included classes with sometimes less than 20% identity. The object was to choose class representatives for a similarity search of mammalian EST databases for paralogs and to decide which of these sequences were orthologs.

A computer script is a set of computer commands that are placed in a disk file. When the script is run, the commands are executed in the order given by the script. For example, the script may include collecting EST sequence by FTP, analyzing them by TBLASTN or TFASTY, collecting the alignment scores, ordering them, and making charts. The Perl programming language is used for producing such scripts.

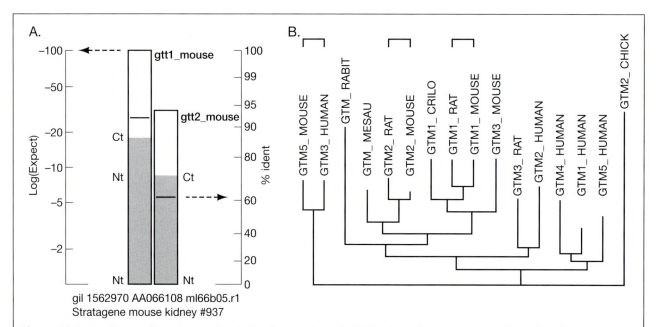

Figure 10.5. Prediction of paralogs and orthologs by searches of EST databases by gene panning (Retief et al. 1999). In this analysis, one class of glutathione transferase family members was used as queries to search mammalian EST databases for highly significant matches using TFASTY3 (Chapter 7). FAST_PAN is a Perl-script program (see Table 10.1D) that automatically searches EST databases as they are updated and compiles the results of the search. (*A*) Display of protein class matched (color), log Expect value (height of bar), length of query sequence matched (height of color bar), and percent identity (position of horizontal line in bar) on one graph as produced by FAST_PAN. Note that the log scales clearly reveal the lowest *E* value and highest identity matches. Shown are matches of two mouse ESTs to a query sequence. (*B*) Example of phylogenetic analysis to predict orthologs between species (bracketed). Amino acid sequences of ESTs in the matched regions were aligned, and this alignment was then used to direct an alignment of the EST codons. A phylogenetic tree was produced by the aligned EST sequences by the maximum likelihood method using the program DNAML in the PHYLIP package. As discussed by the authors, this method allows researchers to search rapidly and easily through EST databases to identify matching sequences and to examine the quality of the alignments found. In this example, a large number of glutathione transferase members were used as queries, allowing an exhaustive search of the EST database for representative family members. (Redrawn, with permission, from Retief et al. 1999.)

A novel feature of these searches was to use a lower-scoring PAM matrix to search for paralogs of a recently evolved group of sequences. Use of an appropriate PAM matrix that matches the expected evolutionary separation of a group of sequences provides an improved higher-scoring alignment, as described in detail in Chapter 3 (p. 82). ESTs with a high percent identity with the query sequence, a long alignment with the query sequence, and a very low *E* value of the alignment score represent groups of paralogous and orthologous genes. To identify orthologs as the most closely related sequence, ESTs were aligned using the amino acid alignment as a guide, and a phylogenetic tree was produced by the maximum likelihood method. This method, described in Chapter 6 (see flowchart for Chapter 6), is suitable for a divergent but recently evolved group of sequences. The predicted tree shown in Figure 10.5B predicts those pairs of sequences that are likely to be orthologous.

Family and Domain Analysis

As shown in the flowchart (p. 492), gene identification of predicted proteins in the genome is designed to discover the metabolic features of an organism. An important feature of proteins discussed in Chapter 9 is their organization into domains that represent modules of structure and function. Different proteins are mosaics of domains that occur in different combinations in a given protein. In a particular organism or group of organisms, one particular domain can be expanded to perform a particular function. Comparison of the domain content of an entire proteome with that of another proteome can reveal the biological roles of diverse domains in different organisms. Extensive comparisons for both prokaryotic and eukaryotic genomes have been performed (Chervitz et al. 1998; Huynen and Bork 1998; Rubin et al. 2000). A descriptive list of protein domain databases that may be used for such an analysis is given in Table 9.5. In a detailed analysis of the fly, worm, and yeast proteomes, 744 families and domains were common to all three organisms. More than 2000 fly and worm proteins are multidomain proteins, compared to about one-third this number in yeast (Rubin et al. 2000). Tekaia et al. (1999) have introduced the concept of a genome tree. A tree or dendrogram based on the proportion of proteins in one organism that is shared by another organism is produced by the single linkage clustering method described in Figure 10.4C.

Ancient Conserved Regions

Phylogenetically diverse groups of organisms have been analyzed for the presence of conserved proteins and protein domains that have been conserved over long periods of evolutionary time, called ancient conserved regions or ACRs (Green et al. 1993). The method involves database similarity searches of the SwissProt database with human, worm, yeast, or *E. coli* genes and identification of matches with sequences from a different phylum than the query sequence. An analysis of ACRs that predate the radiation of the major animal phyla some 580–540 million years ago suggested that 20–40% of coding sequences are ACRs. For example, a search with 1916 *E. coli* proteins detected 266 ACRs found in 439 sequences, roughly one-quarter of the SwissProt database. These ACRs may represent proteins present at the time of the prokaryotic–eukaryotic divergence.

With the later addition of complete genome sequences of phylogenetically diverse prokaryotic organisms, the number of ACRs may be estimated by the proportion of genes that match database sequence of known function. For the hyperthermophilic archaea *Pyro-*

coccus hirokoshii (Kawarabayasi et al. 1998), this proportion was 20%, perhaps representing an ancient set of prokaryotic ACRs. COGs described above represent sets of proteins that are conserved across distant phylogenetic lineages. For 11 prokaryotic genomes, the proportion of genes represented in COGs is approximately 50% (Koonin et al. 1998), and other studies suggest that as many as 70% of prokaryotic genomes contain ACRs (Koonin and Galperin 1997). However, one needs to take into account that horizontal transfer of genetic material discussed below increases the sharing of genes by different lineages of prokaryotes.

Horizontal Gene Transfer

The genomes of most organisms are derived by vertical transmission, the inheritance of chromosomes from parents to offspring from one generation to the next. However, in rare instances, genomes may also be modified by horizontal (sometimes called lateral) gene transfer (HT), the acquisition of genetic material from a different organism. The transferred material then becomes a permanent addition to the recipient genome. Although these exchanges do not occur very often on a generation-to-generation basis, a significant number can occur over a period of hundreds of millions of years. An extreme example is the proposed endosymbiont origin of mitochondria in eukaryotic cells and chloroplasts in plants. The endosymbiont theory proposes that these organelles were transferred from free-living bacteria to another organism with which they shared a symbiotic relationship (see Chapter 6 in Brown 1999).

Horizontal gene transfer is a significant source of genome variation in bacteria (for review, see Ochmann et al. 2000), allowing them to exploit new environments. Such transfer is rendered possible by a variety of natural mechanisms in bacteria for transferring DNA from one species to another. Detection of HT is made possible by the fact that each genome of each bacterial species has a unique base composition. Hence, transfer of a portion of a genome from one organism to another can generally be detected as an island of sequence of different composition in the recipient. If the amino acid composition of transferred genes is typical, these islands may be detected by a codon usage analysis as described in Chapter 8. Very ancient transfers may not be detectable because the base composition and codon usage of the transferred DNA will eventually blend into those of the recipient organism. The time of transfer of DNA may be estimated by the degree to which the composition of the HT DNA has blended into that of the recipient genome. Comparisons of completely sequenced bacterial genomes have revealed that they are mosaics of ancestral and horizontally transferred sequences. The proportion of the genome due to HT sequences also varies considerably roughly in proportion to genome size. A total of 12.8% of the genome of *E. coli* is due to HT DNA (the highest level found), whereas it is 0.0% in *Mycoplasma genitalium*, whose genome is less than one-quarter the size of that of *E. coli*. Mycoplasma have lost many of the genes needed to be a free-living organism and instead depend on nutrients provided by the interior of the host cell. Hence, these organisms would not be expected to carry any extra unnecessary genetic baggage. HT DNA contributes in a major way to the disease-producing ability of pathogenic bacteria, and this DNA frequently has flanking direct repeats characteristic of transposable elements. Note that when genes are clustered on the chromosome of the donor organism (described below), the recipient organism may gain an entire metabolic pathway from another by means of horizontal transfer. Hence, clustering in combination with horizontal transfers provides an evolutionary mechanism for altering metabolic pathways in diverse organisms.

Gene Annotation

Accurate annotation of genome sequences is an important first step in genome analysis. As described earlier, annotation is based on finding significant alignment to sequences of known function in database similarity searches. Matches of lesser significance provide only a tentative or hypothetical prediction and should be used as a working hypothesis of function (see Kyrpides and Ouzonis 1999). Computational tools such as MAGPIE and GENEQUIZ described below are designed to assist with accurate genome annotations.

FUNCTIONAL CLASSIFICATION OF GENES

Once sequences have been annotated, a useful next step is to classify the annotated genes by function. Genes that are significantly similar in an organism, i.e., paralogous sequences, frequently are found to have a related biological function. This discovery follows the expected origin of paralogs by gene duplication events, leaving one copy to perform the original function and producing a second copy to develop a new function not too distant from the original one under evolutionary selection. An early classification scheme for eight related groups of *E. coli* genes included categories for enzymes, transport elements, regulators, membranes, structural elements, protein factors, leader peptides, and carriers. Ninety percent of *E. coli* genes related by significant sequence similarity fell into these same broad categories (Labedan and Riley 1995).

The Enzyme Commission numbers formulated by the Enzyme Commission of the International Union of Biochemistry and Molecular Biology provide a detailed way to classify enzymes based on the biochemical reactions they catalyze (Webb 1992; Tipton and Boyce 2000). The designation ECa.b.c.d gives the following information: (a) one of six main classes of biochemical reactions, (b) the group of substrate molecule or the nature of chemical bond that is involved in the reaction, (c) designation for acceptor molecules (cofactors), and (d) specific details of the biochemical reaction. Using this system to compare sequence-related pairs of *E. coli* genes, Labedan and Riley (1995) found that 70% of them shared the first two EC designators (a and b) in the annotation of the corresponding genes, thereby indicating that they catalyze biochemically similar reactions. A third measure of functional similarity is based on a physiological characterization of *E. coli* proteins into 118 possible categories (e.g., DNA synthesis, TCA cycle, etc.) (Riley 1993). Approximately one-quarter of *E. coli* genes fall into the same category by this scheme.

An alternative approach to classification of genes that encode enzymes is to examine relationships among multiple enzymes that perform the same biochemical function in the same organism. Although catalyzing the same reaction, these enzymes showed variations in metabolic regulation of their activity. More than one-half of multiple enzymes in *E. coli* share significant sequence similarity; i.e., they are paralogs. However, the remainder do not share any sequence similarity. Either they were acquired by horizontal transfer from another bacterial species or the two enzymes were formed by convergent evolution from two different genetic starting points (Riley 1998). Accordingly, sequence similarity is frequently a good indicator of related biochemical function, but two enzymes that perform the same biochemical task may not share sequence similarity of evolutionary history.

Other functional classification schemes for genes include a broader category for genes involved in the same biological process, e.g., a three-group scheme for energy-related, information-related, and communication-related genes has also been used. By this scheme, plants devote more than one-half of their genome to energy metabolism, whereas animals devote one-half of their genome to communication-related functions (Ouzounis et al. 1996). Another scheme, described below, is to identify proteins that physically interact in a structure or biochemical pathway.

A system for functional annotation of the yeast genome has also been produced (Cherry et al. 1997) and used in a comparison of the yeast and worm proteomes (see SGD, Table 10.4B) (Chervitz et al. 1998). *D. melanogaster* genes were classified using the Gene Ontology (GO) classification scheme (Adams et al. 2000), a collaboration among yeast, fly, and mouse informatics groups to develop a general classification scheme useful for several genomes (see GO site, Table 10.1F). This classification scheme provides a description of gene products based on function, biological role, and cellular location.

GENE ORDER (SYNTENY) IS CONSERVED ON CHROMOSOMES OF RELATED ORGANISMS

Two species that have recently diverged from a common ancestor might be expected to share a similar set of genes and also similar chromosomes with these genes positioned along the chromosomes in the same order. Over evolutionary time, the sequence of each pair of genes will slowly diverge, as the species diverge and other changes such as gene duplication and gene loss change the gene content. In addition, the order of genes also changes over evolutionary time as a result of chromosomal rearrangements. These rearrangements may be modeled by occasional chromosomal breaks, random with respect to chromosomal location, and by random rejoining of the fragments by a DNA repair mechanism. Rearrangements may be analyzed by comparing the location of orthologs, genes of highly conserved sequence and function in prokaryotic and eukaryotic proteomes from different phylogenetic lineages.

Colinearity of gene order is referred to as synteny, and a conserved group of genes in the same order in two genomes as a syntenic group or cluster.

Two important observations have been made with regard to gene order: First, order is highly conserved in closely related species but becomes changed by rearrangements over evolutionary time. As more and more rearrangements occur, there will no longer be any correspondence in the order of orthologous genes on the chromosome of one organism with that of a second organism. Second, groups of genes that have a similar biological function tend to remain localized in a group or cluster. Examples of these observations and their significance are described below.

Chromosomal Rearrangements

In Figure 10.6, a genome plot of the positions of orthologs and paralogs on the genomes of two related bacteria, *Mycoplasma pneumoniae* and *Mycoplasma genitalium*, both human pathogens, is shown (Himmelriech et al. 1997). This plot is very similar to the dot matrix plot used for sequence alignment (see Chapter 3), except that in this case a dot or symbol is shown at the intersection of the position of one member of an orthologous pair of sequences on genome 1 and the position of the other member of the pair on genome 2. The plot clearly shows that large sections of chromosome are conserved but also that a number of rearrangements have occurred, making the gene order different from that of the other genome and from the common ancestor of these two organisms. In contrast, a similar plot of orthologous genes in the genomes of the bacterial species *E. coli* and *H. influenzae* appears quite random (Tatusov et al. 1996), even though the organisms are only slightly more distant in evolution than the two *Mycoplasma* species. However, on close inspection of gene function and order, similarities can be found. By classifying genes using a nine-

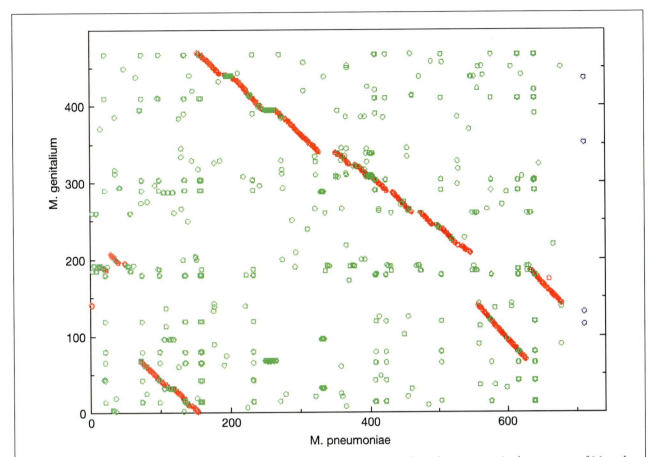

Figure 10.6. Genome plot of orthologous genes. Alignment of orthologous and paralogous genes in the genomes of *Mycoplasma genitalium* and *Mycoplasma pneumoniae* (Table 10.1A, comparative genome analysis in P. Bork laboratory). Horizontal axis is genome position in *M. pneumoniae*, vertical axis is genome position of *M. genitalium*. Positions of orthologs are shown in red, paralogs in green. Orthologous genes are in the same order in both genomes except for several chromosomal rearrangements. These genes are defined by high *E* values in database searches in which one of an orthologous pair is used as query of the proteome of the other species. Proteins should also align along 60% of the length of each (Huynen and Bork 1998). Paralogs are proteins that have striking, high-scoring similarity but are not the highest scoring in reciprocal proteome searches. Note also the occurrence of paralogs within the conserved stretches of orthologs, presumably representing gene duplication in these regions. In contrast to this conserved order of gene position in the *Mycoplasma* species, the orthologous genes in two other equally related species, *E. coli* and *Hemophilus influenzae*, show no detectable conservation of order on a similar genome plot.

class functional classification scheme (see above), several genes falling into the same functional category are clustered together on the chromosomes of both of these organisms, and the clusters are in a similar order (Ouzounis et al. 1996). Comparison of the number of rearrangements in a given period of evolutionary history may vary significantly from one organism to the next. In one analysis of prokaryotic organisms of diverse phylogenetic origin, it has been shown that if gene A has a neighboring gene B, then if an ortholog of A occurs in another genome, there is an increased probability of an ortholog of B also occurring in the other organism. However, the B ortholog is less likely to be a neighbor of the A ortholog of the genome of the second species if the two species are more divergent (Huynen and Bork 1998).

The TIGR Web site (Table 10.1D) includes a resource for comparing any two prokaryotic genomes of the 30+ available by means of a genome plot, as shown in Figure 10.6. In general, the order of orthologs is not well conserved in prokaryotes when the genomes have diverged sufficiently that the orthologs have <50% identity (Huynen and Bork 1998).

A similar conservation of gene order also appears to be present in closely related eukaryotic genomes. The evidence is based on chromosome painting experiments in which DNA from a section of a chromosome of one organism is labeled and then hybridized to chromosomes of a second organism. Regions of the second chromosome that are labeled reveal the presence of a homologous region. Although this method does not have the precision and sensitivity of sequence analysis methods, these experiments reveal that eukaryotic chromosomes also undergo rearrangements both within chromosomes and between chromosomes during evolution. An example of the differences between mouse and human chromosomes is shown in Figure 10.7. A much larger data collection from a variety of mammalian chromosomes suggests that each chromosome is a mosaic of a similar set of ancestral fragments (O'Brien et al. 1999). Similar studies with plant genomes have also indicated that they have a similar overall gene content but that many regional duplications and rearrangements have occurred during evolution (Bennetzen 1998, 2000; Bennetzen et al. 1998). The availability of genome sequences of plants and animals offers some exciting opportunities for determining the chromosomal changes that have occurred during evolution of the plant and animal kingdoms.

Computational Analysis of Gene Rearrangements

As genome-by-genome comparisons of the chromosomes of related species are made and the rearrangements are discovered, a further challenge to computational and evolutionary biologists is to estimate the number and types of rearrangements that have occurred and also to determine when they occurred. For example, a comparison of the mouse and human chromosomes reveals many rearrangements (Fig. 10.7). A computational approach to these questions is outlined in Figure 10.8. In aligning gene and protein sequences, one assumes a model in which no rearrangements have occurred so that lines can be drawn between the corresponding positions in the sequences and no lines will cross or intersect, as shown in Figure 10.8A. For comparing gene orders on chromosomes that have undergone rearrangements, lines joining the corresponding genes will intersect, as shown in Figure 10.8B, and the greater the amount of rearranging, the greater the number of intersects. In the random shuffling model, one tries to estimate the number of rearrangements that produces the observed number of intersections and to compare this number to one that would randomly shuffle the same fragments. The analysis shown in Figure 10.8C attempts to reconstruct the number and types of rearrangements (inversions, etc.) that have given rise to the observed variation in gene order between the chromosomes.

Clusters of Genes on Chromosomes Have a Metabolically Related Function

In a given organism or species, genes are found in a given order that is maintained on the chromosomes from one generation to the next. Genetic analysis has revealed that genes with a related function are frequently found to be clustered at one chromosomal location. Clustering of related genes presumably provides an evolutionary advantage to a species, but the underlying biological reason is not understood. One possibility is that there is genetic variation (alleles) within each gene in a cluster of a given species and that only certain allelic combinations of different genes are compatible. Another possibility is some kind of coordinated translation of the proteins that may aid their folding. In the model bacterial species *E. coli*, genes that act sequentially in a biochemical pathway are frequently found to be adjacent to each other at one chromosomal location. For example, the genes required for synthesis of the amino acid tryptophan (*trp* genes) are clustered together on the chromosome of *E. coli*, as illustrated in Figure 10.8, where their expression is coordi-

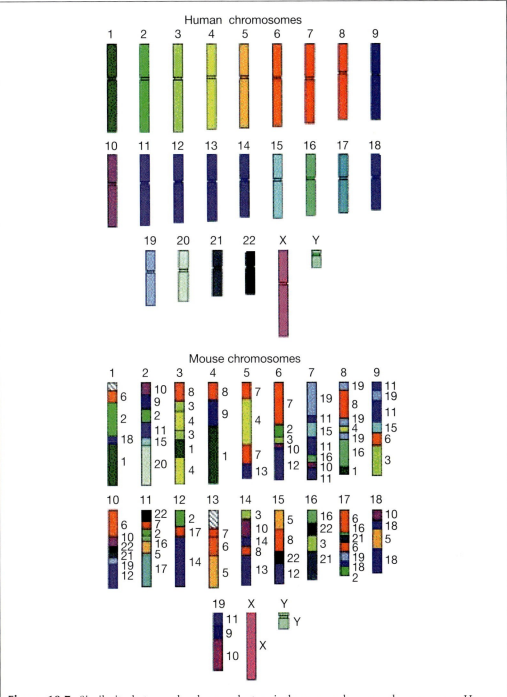

Figure 10.7. Similarity between local gene clusters in human and mouse chromosomes. Human chromosomes can be cut into >100 pieces and reassembled into a reasonable facsimile of the mouse chromosome. Only larger fragments are represented. Chromosomes of all mammals may share a similar relationship (O'Brien et al. 1999).

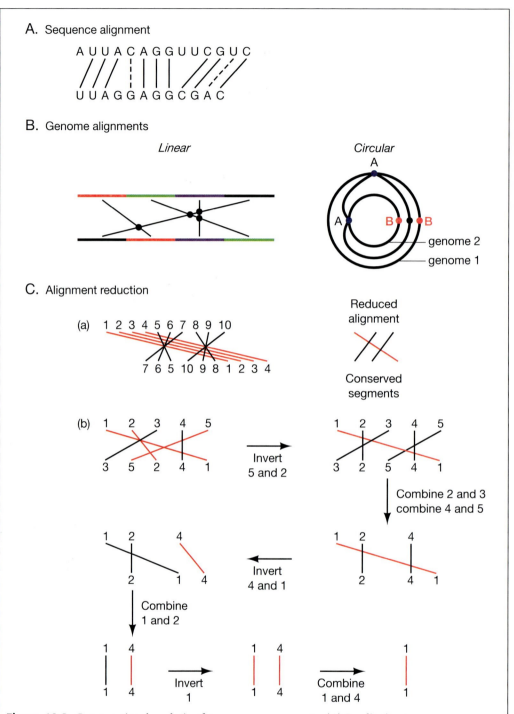

Figure 10.8. Computational analysis of genome arrangements. (*A*) In aligning two sequences, one sequence is written above the other and the highest number of consecutive matches between the sequences provides an optimal alignment as described in Chapter 3. The alignment includes matches (*solid lines*), mismatches (*dotted lines*), and insertions/deletions in order to produce an optimal number of matches. The matches are in a consecutive order in two sequences such that no rearrangements would be found. (*B*) Alignments of linear and circular chromosomes that have undergone rearrangements such as those found in mammalian chromosomes and mitochondria. In contrast to sequence alignment, lines indicating homologous positions in linear chromosomes (*left*) now cross, producing points of intersection. The more rearrangements there are, the more intersections will occur. For alignment of circular chromosomes (*right*), depending on how the chromosomes are aligned, there are two ways of showing a moved region. To go from A on the outer genome to A on the inner genome,

nately regulated by a common promoter. This coordination of expression avoids wasteful production of one enzyme when others in the same pathway are not available.

With the availability of other prokaryotic genome sequences, important metabolic genes such as *trp* can be identified in these species, and the chromosomal location of these genes can be compared with that of *E. coli*. Using the predicted tryptophan genes as an example (Fig. 10.9), the following observations were made: (1) At least some of the *trp* genes are also clustered together on the chromosomes of other species of Bacteria and Archaea; (2) the order of the genes within the cluster is conserved within the first four species listed in Figure 10.9, all of which are bacteria; (3) the order is much less conserved in the last three species, all of which are Archaea, and some of the genes have been moved to a more distant location; (4) there are multiple examples of gene fusions that give rise to a new protein that performs both biochemical functions of the single-gene, parent proteins. *trpC* has been fused independently with two other genes, *trpD* and *trpF*. Alternatively, a composite gene may produce two smaller single-component genes by fission of a parent composite gene. Fission events have only been observed in thermophiles among prokaryotes (Snel et al. 2000a). However, biochemical reasons have been presented that fission events may provide a mechanism for evolution of protein complexes (Marcotte et al. 1999b).

When a series of predicted genes in a known *E. coli* pathway is in the same order in another organism as in *E. coli*, e.g., *trpB-trpA* and *trpE-trpG* in the Archaea in Figure 10.9, then the same biochemical pathway is predicted also. Even if the genome annotation is based on a weak prediction of the biochemical function of two individual genes, the prediction is stronger if the two genes act in the same pathway and is strongest if the genes are clustered (Huynen et al. 2000). In the *trp* example shown in Figure 10.9, the presence of the genes in such a phylogenetically diverse group of organisms indicates that the pathway is an ancient one. Clustering of the genes further indicates that they probably originated as a group in the single chromosomal region of an ancient ancestor organism, assuming there has not been a driving force for repeated independent clustering events. What is also revealed in the *trp* example in Figure 10.9 is that some *trp* genes are found at a much more remote chromosomal location. The diverse location of the *trp* genes in *Methanococcus jannaschii* is an outstanding example. Apparently, rearrangements can break clusters and

The term clusters has been used in two different ways in the literature and in this chapter, and the two should not be confused. One use is to represent groups of genes in one or several organisms that share a significant degree of sequence similarity. An example is Figure 10.4C. A second use of clusters is to represent a physical clustering of genes on the same chromosome. An example is the arrangement of the trp *genes in Figure 10.8.*

the line joining them can go clockwise or counterclockwise and, as a result, there will be either 0 or 1 intersections with the line joining B. The complexity of alignments of circular chromosomes is reduced by limiting the joining lines to 180 degrees of relative genome positions. Sankoff and Goldstein (1989) devised a shuffling model for estimating the number of rearrangements when the number of intersections is known. The method is analogous to shuffling an ordered deck of cards and then predicting how much order remains. Eventually, after *n* log *n* shuffles, where *n* is the number of cards, the order becomes random. Given an observed remaining order, how many shuffles have occurred? The number of observed intersections is compared to the number expected for completely shuffled genomes (Sankoff et al. 1993). (*C*) Another method for determining numbers of rearrangements is to assume that they have occurred by a number of transposition or recombination events. The object of this analysis is to try to identify the rearrangements that occurred and then to undo (or derange) the alignments accordingly. The goal is to minimize the number of rearrangements, this number then representing a genetic distance between the sequences. (*a*) Alignments of genes 1–10 in two genomes where some genes are in the same order (*red lines*) and others are inverted (*blue lines*). Groups of genes such as the two joined by the blue lines may be combined into a single unit representing a conserved segment since no recombination event would be required. (*b*) Alignment that can be accounted for by these inversion events. The program DERANGEII is available from the authors and FTP from ftp.ebi.ac.uk/pub/software/unix/derange2.tar.Z. These methods have been used to analyze rearrangements in mitochondrial and bacterial genomes (Sankoff et al. 1992; Blanchette et al. 1996; Sankoff and Nadeau 1996) and additional algorithms have also been developed (Kececioglu and Sankoff 1995; Kececioglu and Gusfield 1998). (Adapted from Sankoff et al. 1992, 1993.)

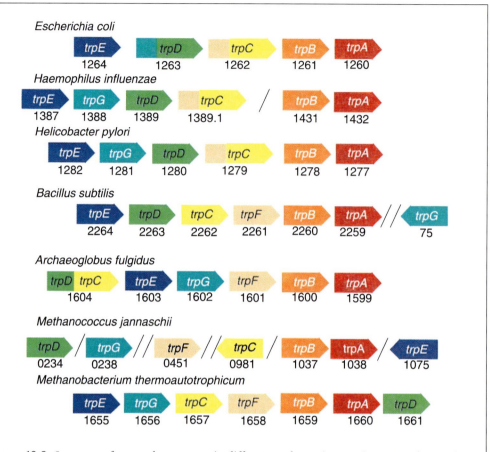

Figure 10.9. Structure of tryptophan operon in different prokaryotic organisms. Numbers indicate gene number in genome; arrows indicate direction of transcription; double lines indicate a separation of more than 50 genes due to dispersion of the operon. Shown also are examples of gene fusion so examples of domain fusions (e.g., *trpD* and *trpG*) are fused in *E. coli*. Note that only the *trpA* and *trpB* genes are genetically linked and separate genes in all of the species. (Reprinted, with permission, from Dandekar et al. 1998 [copyright Elsevier Science].)

move genes to other locations, although another possibility is that the dispersed arrangement is a more ancestral state.

Two methods have been described for identifying clusters or coordinately regulated genes. In one study with three separate groups of three distantly related prokaryotes (Dandekar et al. 1998), approximately 100 genes were found to be conserved as a cluster of two pairs. (Looking for a pair in three species avoided possible complications from horizontal transfer.) The direction of transcription was the same for all genes, implying a regulatory relationship as in an operon. For approximately 75% of the genes, a physical interaction between the genes had previously been demonstrated and could be predicted for almost all proteins based on additional sequence comparisons. These conserved proteins have core biological functions such as transcription, translation, and cell division.

In a second method (Overbeek et al. 1999), a full reciprocal search like that used in Figure 10.4 for comparing yeast, worm, and fly genomes and for making COGs was performed between the proteomes of two prokaryotes: Each protein of one proteome was used to search the proteome of the second. Protein pairs that gave a best hit with the other genome and that had an E value of less than 10^{-5} were identified, called a bidirectional best hit (BBH). Pairs of close bidirectional best hits (PCBBH) that are within 300 bp of each other on the chromosomes of the respective organisms and that are transcribed from the same strand, i.e., are in a "typical" operon, were then identified. A score for these pairs was formulated that is higher when the number of organisms in which the pair is observed is greater and the phylogenetic distance between the organisms is larger. Forty percent of a set of higher-scoring pairs corresponded to proteins that are known to act in a common metabolic pathway, as defined in metabolic function databases (see Table 10.1D). Hence, a significant proportion of the pairs of PCBBH correspond to genes that have a related function and lie on the same pathway. This same approach could play an important role in assigning a function to uncharacterized genes in genomes based on proximity to other genes of known function.

Composite Genes with a Multiple Set of Domains Predict Physical Interactions and Functional Relationships between Protein Pairs That Share the Same Domains

As illustrated in Figure 10.9, single *trp* genes can be fused into larger composite genes. Observation of such evolutionary events provided a major step forward in understanding relationships among the proteins of diverse organisms (Enright et al. 1999; Marcotte et al. 1999b). The occurrence of a fused or composite gene in one organism is called a "Rosetta Stone sequence" because it provides evidence that the single component genes in a separate organism encode proteins that physically interact (Marcotte et al. 1999b). For example, if a composite human gene has two domains A and B, the analysis assumes that A and B physically interact within the protein. If two separate genes in other organisms (yeast or *E. coli*) make two proteins, one with domain A and a second with domain B, then these two proteins are assumed to interact because A and B interact. These sequence relationships may be found by sequence alignment of the composite AB protein with each of the single-component A and B proteins. However, A and B will not align with each other. If A and B do not interact in composite proteins, the prediction is a false-positive result. However, these proteins are still predicted to have related functions based on the gene fusion result.

Composite proteins were found by searching SwissProt for statistically significant matches to domains in the ProDom domain database (see Table 9.5, p. 430). Six percent of the Rosetta Stone proteins were found to be represented in the DIP database of interacting proteins (see Table 9.5). Rosetta Stone predictions of interacting proteins were compared to predictions by another method for predicting related proteins, the phylogenetic profile method (Pelligrini et al. 1999; see also "bag of genes" concept in Huynen and Bork 1998). This method is based on the assumption that proteins that function together in a biochemical pathway should evolve in a correlated fashion. Databases are searched for significant matches to two proteins A and B. If A and B have related functions, they should be found together in a large proportion of genomes, whereas if they do not, they will be found to have a random association in genomes.

Enright et al. (1999) used reciprocal searches among three complete prokaryotic proteomes, as described above in Figure 10.4, and identified related proteins that have the expected alignments for composite (AB) and component (A or B) proteins. These proteins interact functionally, act in the same biochemical pathway, or are coregulated. Predictions are stronger when component proteins (A and B) have few paralogs, since the interacting pair can be more readily identified. Conversely, the presence of paralogs of the composite proteins increases the strength of the prediction because the number of possible interactions is increased (Enright et al. 1999).

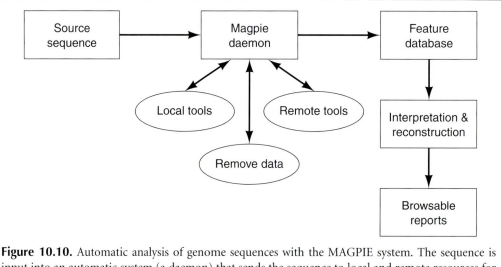

Figure 10.10. Automatic analysis of genome sequences with the MAGPIE system. The sequence is input into an automatic system (a daemon) that sends the sequence to local and remote resources for analysis (BLAST search, PROSITE search, etc.). The information retrieved is stored in a feature database, and the data are interpreted by a set of rules and placed in Web-browsable reports. (Redrawn, with permission, from Gasterland and Sensen 1998 [copyright Kluwer Academic/Plenum].)

Resources for Genome Analysis

The above types of analyses depend on a labor-intensive annotation of the genome and functional analysis of the predicted proteins. Computational tools have been made available to automate some of these steps. Examples are MAGPIE and GeneQuiz, listed in Table 10.1F.

MAGPIE analyzes the genome using a set of automated processes that are illustrated in Figure 10.10. Designed for high-throughput genome sequence analysis, MAGPIE automatically annotates genomic sequence data and maintains a daily up-to-date record in response to user queries about one or more genomes. The system also uses a set of rules in logic programming to make decisions that may be used to interpret information from various sources. MAGPIE has been used to locate potential promoters, terminators, start codons, Shine-Dalgarno sites, DNA motif sites, co-transcription units, and putative operons in microbial genomes. These sites are shown on a map display of the genome that may be edited.

GeneQuiz is an integrated system for large-scale biological sequence analysis that uses a variety of search and analysis methods using current sequence databases. By applying expert rules to the results of the different methods, GeneQuiz creates a compact summary of findings. It focuses on deriving a predicted protein function, based on a variety of available evidence, including the evaluation of the similarity to the closest homolog in a database.

GLOBAL GENE REGULATION

One way to obtain useful information about a genome is to determine which genes are induced or repressed in response to a phase of the cell cycle, a developmental phase, or a response to the environment, such as treatment with a hormone. Sets of genes whose expression rises and falls under the same condition are likely to have a related function. In addition, a pattern of gene expression may also be an indicator of abnormal cellular regulation and is a useful tool in cancer diagnosis (see, e.g., Golub et al. 1999; Perou et al. 1999). Because genomes, especially eukaryotic genomes, are so large, a new technology has been developed for studying the regulation of thousands of genes on a microscope slide.

Microarray (or microchip) analysis is a new technology in which all of the genes of an organism are represented by oligonucleotide sequences spread out in an 80×80 array on microscope slides, but can also be synthesized directly on the slide at densities of up to one million per square centimeter. The oligonucleotides are collectively hybridized to a labeled cDNA library prepared by reverse-transcribing mRNA from cells. The amount of label binding to each oligonucleotide spot reflects the amount of mRNA in the cell. The analysis of the data collected in this type of experiment is depicted and described in Figure 10.11. Genes that are responding the same way to an environmental signal, in this case the addition of serum to serum-starved skin cells, are clustered together in a display. From this analysis, a set of genes that responds in an identical manner may be identified. Automatic methods for clustering related sets of genes have been devised, and three representative methods are shown and described in Figure 10.12. The first of these methods, hierarchical clustering (Eisen et al. 1998), is commonly used, but the other two methods are better designed to detect differences in patterns over a set of time points or samples. The derivation of clustering algorithms for microarray analysis has become an active area of bioinformatics.

Once a set of genes that are coregulated has been found, the promoter regions of these genes may be analyzed for conserved patterns that represent sites of interaction with specific transcription factors. This type of analysis is described in detail in Chapter 8 (Table 8.6, p. 370), and additional resources are given in Table 10.1E.

Microarray analysis is designed to detect global changes in transcription in a genome but does not provide information about the levels of protein products of the genes, which may also be subject to translational regulation. Labeled protein samples may also be extracted from treated cells and separated by two-dimensional gel electrophoresis. The proteins are first separated in a column on the basis of size and then across a second dimension on a slab on the basis of charge. The amount of protein in each spot is then determined. This method also can resolve thousands of proteins based on size and charge. There are databases of the patterns found in different organisms; these are listed in Table 10.1E. The technology can also be extended to further purification and microsequencing of the protein spots or of proteins in complexes so that the genes encoding the protein may be identified by proteome similarity searches.

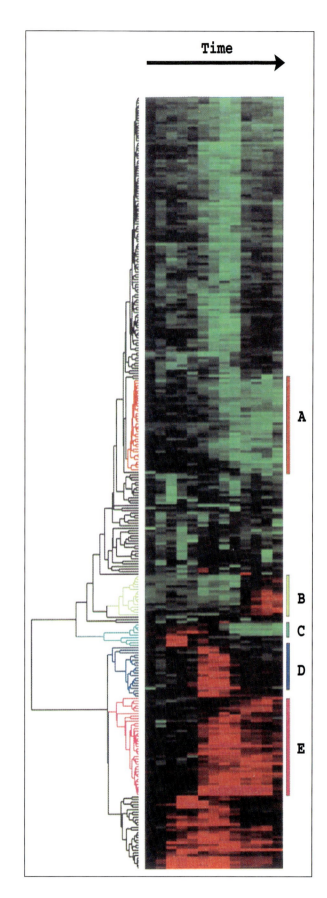

Figure 10.11. Example of cluster analysis of microarray data. Rows represent changes in an individual hybridization signal for a single gene on a cDNA microarray display system. Columns show changes in the expression of a selected 9800 human cDNA set. These genes change their level of expression in human skin fibroblasts that have been deprived of serum growth factors and serum then added back (time 0). The time points vary from 0 to 24 hours, left to right, and the last column is a control. RNA was removed from cells and the amount was measured by quantitative reverse transcription in the presence of the fluorescent dye Cy5. A reference time 0 sample was labeled in parallel with the green fluorescent dye Cy3 and mixed with samples taken at a later time. The labeled cDNA preparations were then hybridized to the cDNA microarray and the Cy5/Cy3 fluorescence ratio of each spot was measured. Each ratio is expressed as a log odds ratio to the base 2. Thus, a value of $+4$ at time t indicates 16 times more mRNA at time t than at time 0; 0 means no change and -4 means 16 times less RNA at time t than at time 0. Tables of these raw data are kept (see http://rana.stanford.edu/clustering). The color display in the figure varies from saturated green (log odds -3.0) to saturated red (log odds $+3.0$) with black as the intermediate color (log odds 0). The dendrogram on the right of the color display was made by a hierarchical clustering algorithm that is similar to the single-linkage cluster analysis described in Fig. 10.4. The object of clustering is to identify genes that respond the same way to the environmental treatment. Each gene is compared to every other gene and a gene similarity score (metric) is produced. If X_i is the log odds value for gene X at time i, then for two genes X and Y and N observations, a similarity score is calculated. (Reprinted, with permission, from Eisen et al. 1998 [copyright National Academy of Sciences].)

$$S(X, Y) = 1/N \sum_{i = 1,N} \left(\frac{X_i - X_{\text{offset}}}{\Phi_X} \right) \left(\frac{Y_i - Y_{\text{offset}}}{Q_Y} \right)$$

$$\text{where } \Phi_X = \sqrt{ \sum_{i = 1,N} \frac{(X_i - X_{\text{offset}})^2}{N} }$$

$$\text{and } Q_Y = \sqrt{ \sum_{i = 1,N} \frac{(Y_i - Y_{\text{offset}})^2}{N} }$$

$S(X,Y)$ is also known as the Pearson correlation coefficent. X_{offset} and Y_{offset} can be the mean of the observations on X or Y, respectively, in which case Φ is the standard deviation, or else X_{offset} and Y_{offset} can be set to zero when a reference state is used (as in the present example). After values of $S(X,Y)$ have been calculated for all gene combinations, the most closely related pairs are identified in an above-diagonal scoring matrix. A node is created between the highest-scoring pair, and the gene-expressed profiles of these two genes are averaged and the joined elements are weighted by the number of elements they contain. The matrix is then updated replacing the two joined elements by the node. For n genes, the process is repeated $n - 1$ times until a single element remains. In the final dendrogram, the order of genes within a cluster is determined by simple weighting schemes, e.g., average dendrogram level (Eisen et al. 1998). The software availability is given in Table 10.1E, microarray guide. This image is available at http://rana.stanford.edu/clustering/serum.html. On the left side of the color display are letters A–E which identify clusters of genes that show clearly distinct responses to the treatment.

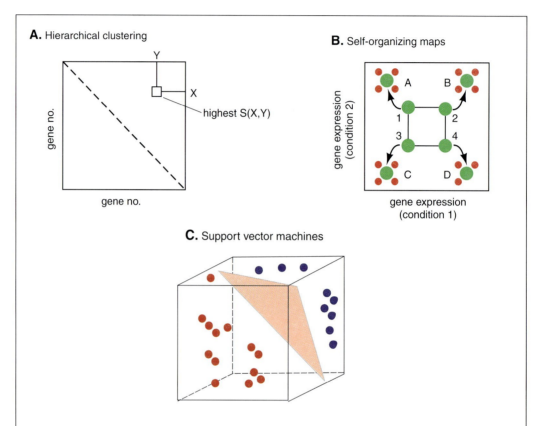

A. Hierarchical clustering

gene no.

gene no.

Y

X

highest S(X,Y)

B. Self-organizing maps

gene expression
(condition 2)

A   B

1   2

3   4

C   D

gene expression
(condition 1)

C. Support vector machines

Figure 10.12. Examples of methods of cluster analysis for microarray data. (*A*) Hierarchical cluster-ing (Eisen et al. 1998). (*B*) Self-organizing maps (Tamayo et al. 1999). (*C*) Support vector machines (Brown et al. 2000). (*A*) The hierarchical clustering method is described in detail in Fig. 10.11. Basi-cally, the method generates a similarity score [*S(X,Y)*] for all gene combinations, places the scores in a matrix, joins those genes that have the highest score, and then continues to join progressively less similar pairs. The disadvantage of this method is that it fails to discriminate between different patterns of variation. For example, a gene expression pattern for which a high value is found at an intermedi-ate time point will be clustered with another for which a high value is found at a late time point in the experiment. These variations have to be separated in a subsequent step (see Fig. 10.11). Methods B and C below are able to discriminate such differences. (*B*) In the SOM method, a choice is made of a number of clusters by which to organize the data. Shown is a 2 × 2 SOM comprising nodes 1–4 that assumes the presence of four clusters. Only two data points are shown in the example, but more data can be included by adding more dimensions to the analysis. The object is to move each node to the center of a cluster of data points. At each iteration, a data point P is selected, and the node closest to that point is identified. The location of that node is then moved slightly toward the point. Thus, node 1 in the example will gradually migrate to the center of cluster A, node 2 to cluster B, and so on. In practice, the size of the SOM is gradually increased until clearly different sets of expression patterns are identified. The computer program that performs the SOM analysis is GENECLUSTER (Table 10.1E). (*C*) SVMs are a binary classification method to discriminate one set of data points from anoth-er. They are similar to the types of discriminant analyses described for gene prediction in Chapter 9. For microarray analysis, sets of genes are identified that represent a target pattern of gene expression. The SVM is then trained to discriminate between the data points for that pattern and other data points

PREDICTION OF GENE FUNCTION BASED ON A COMPOSITE ANALYSIS

When two proteins share a considerable degree of sequence identity throughout the sequence alignment, they are likely to share the same function. A considerable fraction of a genome may encode proteins whose function may not be identified in this manner because the proteins are not related to another of known function. In the above sections, other types of evidence for a relationship between two genes are also given that are not dependent in sequence similarity. These include (1) genes are closely linked on the same chromosomes and transcribed from the same DNA strand, implying coordinated regulation in an operon-like structure; (2) gene fusions are observed between otherwise separate genes (suggests the encoded proteins are physically associated in a common complex); and (3) phylogenetic profiles reveal the genes are both commonly present in many organisms (implying they have interdependent metabolic functions). Three additional types of data have been used as evidence for gene relatedness: (1) the encoded proteins each have homologs in another organism that operate in a common metabolic pathway, (2) experimental data suggest an interaction between the proteins (stored in databases of interacting proteins; Table 9.5, p. 430), and (3) patterns of mRNA expressions are found to be correlated in microarray data. The results of using the above tests for the identification of a group of related genes in yeast are shown in Figure 10.13. In an examination of the entire yeast proteome, proteins that share a relationship with the yeast Sup35 protein based on one or more of the above tests are shown as points in a two-dimensional cluster where the distances between the points are proportional to the weight of the evidence for a relationship between the protein pair and the strength of the connection is proportional to the amount of evidence for a relationship. These types of predictions can be an important basis for hypotheses that can be tested experimentally.

that do not show the pattern. Shown in the diagram are two sets of data points (*red* and *blue*) in a three-dimensional plot that illustrate these two classes of data points. As the SVM learns to discriminate between the data sets, a hyper-plane (*pink*) is drawn between the sets. The hyper-plane is then used as a basis for classifying unknown data points. Only three dimensions are shown for illustrative purposes, but additional ones can be included, adding more dimensions to the analysis. SVMs were used to categorize genes based on 79 different sets of data points from studies of the yeast cell cycle and are particularly useful for such complex data sets. Data points are log-transformed and normalized as in method A, where for N observations of a gene i, the log transform X_i of the expression level E_i and reference level R_i is:

$$X_i = \frac{\log (E_i/R_i)}{\sqrt{\sum_{j = 1,N} \log_z (E_j/R_j)}}$$

so that X_i is positive if the gene is more strongly expressed than in the reference condition, and negative if expression is reduced. Gene combinations averaged over all experimental conditions are then examined by a multidimensional analysis (see http://www.cse.ucsc.edu/research/compbio/genex). A tutorial on SVMs is available through http://www-ai.cs.uni-dortmund.de/SOFTWARE/SVM_LIGHT/svm_light.eng.html. (Adapted, with permission, from Gaasterland and Bekinanov 2000 [copyright Nature Publishing].)

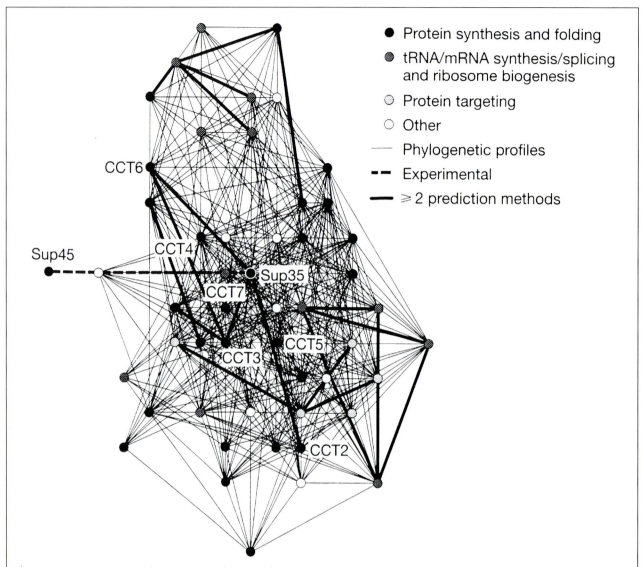

Figure 10.13. Genome-wide prediction of protein functions by a combinatorial method (Enright et al. 1999; Marcotte et al. 1999a). This figure shows the network of yeast proteins that are linked to the yeast prion and translation factor Sup35 (double circle in center of network). Each point represents a yeast protein, and branches between proteins indicate a relationship by one of several criteria indicated in the legend. Branch lengths are shorter for closely related proteins and thicker when two or more prediction methods indicate a relationship. Related to Sup35 protein are proteins involved in protein folding and targeting. The links are based on experimental data, proteins whose homologs are known to operate sequentially in metabolic pathways, proteins that evolved in a correlated fashion as evidenced by presence in fully sequenced genomes (see Snel et al. 1999), proteins whose homologs are fused into a single protein in another organism, and proteins whose mRNA expression profiles are similar under a range of cellular and environmental conditions. (Reprinted, with permission, from Marcotte et al. 1999a [copyright Macmillan].)

FUNCTIONAL GENOMICS

Genome analysis depends to a large extent on sequence analysis methods that identify gene function based on similarity between proteins of unknown function and proteins of known function. Known functions are derived from experimental evidence in molecular biology and genetic studies with model organisms. Orthologous genes between biologically distinct species (for example, yeast and fruit flies) can be identified, and the high sequence similarity between them is strong evidence for a related function. However, given the more complex multicellular biology of flies, the fly gene could have an additional function that is not predictable by the yeast model. In other cases, the occurrence of families of paralogous genes that share common domains can make a precise guess of function of one of these proteins more difficult because all match a model protein to some degree. Sequence-based methods of gene prediction can be augmented by the types of genome comparisons described above that are designed to identify related genes based on common patterns of expression, evolutionary profiles, chromosomal locations, and other features. However, all of the above methods can fail to provide a precise determination of gene function. Hence, methods have been devised for directing mutations into specific genes that inactivate or modify the gene function, and the effect is then analyzed in the mutant organism.

Two general types of approaches illustrated in Figure 10.14 are used—one in which a genetic construct is made that interferes with the expression of a particular gene (and sometimes a set of related genes) and a second in which a large number of random mutations are generated in a population of organisms. The individual with a mutation in a particular gene is then identified. Once mutants are obtained, the effect of the mutant genes on phenotype is determined. The gene function may then be predicted on the basis of the observed alterations. Because such extreme genetic experiments cannot be performed with humans, the mouse model for the human genome serves the same purpose. Web sites that compare the mouse and human genomes listed in Table 10.1C provide an important basis for analyzing the human genome. An orthologous gene is identified in the mouse genome, the sequence or expression of the gene is disrupted in some fashion, and a transgenic mouse homozygous for the mutant gene is then produced. Using this technology, one can systematically go through genes that regulate cell division, for example, and determine the significance of these genes in normal versus abnormal (tumor) growth.

PUTTING TOGETHER ALL OF THE INFORMATION INTO A GENOME DATABASE

A genome database may also be interfaced with other types of data, such as clinical data. This type of organization, termed data warehousing, can facilitate the search for novel relationships among the data by data-mining methods. These methods include genetic algorithms, neuronetworks, and others described elsewhere in this text.

The ultimate step in genome analysis is to collect the information found on gene and protein sequences, alignments, gene function and location, protein families and domains, relationships of genes to those in other organisms, chromosomal rearrangements, and so on, into a comprehensive database. This database should be logically organized so that all types of information are readily accessible and easily retrievable by users who have widely divergent knowledge of the organism. This goal is best achieved by using controlled vocabularies that can identify the same genetic or biochemical function in different organisms without ambiguity. Examples of groups that are developing systematic ways of defining terms and of collecting and organizing data are given in Table 10.1E. Other examples of database tools used to express biological information are given in Chapter 2 (page 44). The genome sites of model organisms listed in Table 10.1B, especially SGD and Flybase, provide examples for further study. In addition to the care needed in organizing genome databases, a

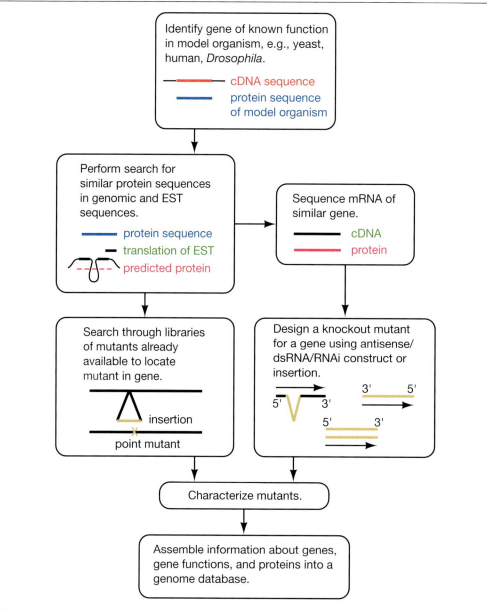

Figure 10.14. Reverse-genetics analysis of gene function. Steps for identification of gene function in an organism are identified. Even though a particular gene may be a highly predicted ortholog of a gene of known function in another organism, that gene may be acquired by a novel function. For example, a defect in a plant or animal gene that is a homolog of a yeast gene may have an effect on a developmental process or other biologically unique function of multicellular organisms. Information on knockout mutants in model organisms is available through the genome Web sites given in Table 10.1C. Directed gene knockout and mutagenesis methods are described in Fire et al. (1998) and McCallum et al. (2000), respectively.

great deal of human input is needed to annotate the genome manually with information about individual genes and proteins, effects of mutations in these genes, and other types of genome variations that cannot be readily incorporated into the database by automated methods. For the human genome, this activity will occupy the time of many scientists for many years to come.

REFERENCES

Adams M.D., Celniker S.E., Holt R.A., Evans C.A., Gocayne J.D., Amanatides P.G., Scherer S.E., Li P.W., Hoskins R.A., Galle R.F., et al. 2000. The genome sequence of *Drosophila melanogaster*. *Science* **287:** 2185–2195.

Andrade M.A., Brown N.P., Leroy C., Hoersch S., de Daruvar A., Reich C., Franchini A., Tamames J., Valencia A., Ouzounis C., and Sander C. 1999. Automated genome sequence analysis and annotation. *Bioinformatics* **15:** 391–412.

Bailey L.C., Jr., Fischer S., Schug J., Crabtree J., Gibson M., and Overton G.C. 1998. GAIA: Framework annotation of genomic sequence. *Genome Res.* **8:** 234–250.

Baker P.G., Goble C.A., Bechhofer S., Paton N.W., Stevens R., and Brass A. 1999. An ontology for bioinformatics applications. *Bioinformatics* **15:** 510–520.

Bansal A.K. 1999. An automated comparative analysis of 17 complete microbial genomes. *Bioinformatics* **15:** 900–908.

Bennetzen J.L. 1998. The structure and evolution of angiosperm nuclear genomes. *Curr. Opin. Plant Biol.* **1:** 103–108.

–––––––. 2000. Comparative sequence analysis of plant nuclear genomes: Microcolinearity and its many exceptions. *Plant Cell* **12:** 1021–1030.

Bennetzen J.L., SanMiguel P., Chen M., Tikhonov A., Francki M., and Avramova Z. 1998. Grass genomes. *Proc. Natl. Acad. Sci.* **95:** 1975–1978.

Bernardi G. 1995. The human genome: Organization and evolutionary history. *Annu. Rev. Genet.* **29:** 445–476.

Blanchette M., Kunisawa T., and Sankoff D. 1996. Parametric genome rearrangements. *Gene* **172:** GC11–17.

Blattner F.R., Plunkett G., III, Bloch C.A., Perna N.T., Burland V., Riley M., Collado-Vides J., Glasner J.D., Rode C.K., Mayhew G.F., Gregor J., Davis N.W., Kirkpatrick H.A., Goeden M.A., Rose D.J., Mau B., and Shao Y. 1997. The complete genome sequence of *Escherichia coli* K-12. *Science* **277:** 1453–1462.

Bork P. 1999. Powers and pitfalls in sequence analysis: The 70% hurdle. *Genome Res.* **10:** 398–400.

Brown M.P., Grundy W.N., Lin D., Cristianini N., Sugnet C.W., Furey T.S., Ares M., Jr., and Haussler D. 2000. Knowledge-based analysis of microarray gene expression data by using support vector machines. *Proc. Natl. Acad. Sci.* **97:** 262–267.

Brown T.A. 1999. *Genomes*. Wiley-Liss, New York.

Bult C.J., White O., Olsen G.J., Zhou L., Fleischmann R.D., Sutton G.G., Blake J.A., FitzGerald L.M., Clayton R.A., Gocayne J.D., Kerlavage A.R., Dougherty B.A., Tomb J.F., Adams M.D., Reich C.I., Overbeek R., Kirkness E.F., Weinstock K.G., Merrick J.M., Glodek A., Scott J.L., Geoghagen N.S.M., and Venter J.C. 1996. Complete genome sequence of the methanogenic archaeon, *Methanococcus jannaschii*. *Science* **273:** 1058–1073.

Cawthon R.M., O'Connell P., Buchberg A.M., Viskochil D., Weiss R.B., Culver M., Stevens J., Jenkins N.A., Copeland N.G., and White R.1990. Identification and characterization of transcripts from the neurofibromatosis 1 region: The sequence and genomic structure of EVI2 and mapping of other transcripts. *Genomics* **7:** 555–565.

Cherry J.M., Ball C., Weng S., Juvik G., Schmidt R., Adler C., Dunn B., Dwight S., Riles L., Mortimer R.K., and Botstein D. 1997. Genetic and physical maps of *Saccharomyces cerevisiae*. *Nature* (suppl. 6632) **387:** 67–73.

Chervitz S.A., Aravind L., Sherlock G., Ball C.A., Koonin E.V., Dwight S.S., Harris M.A., Dolinski K., Mohr S., Smith T., Weng S., Cherry J.M., and Botstein D. 1998. Comparison of the complete protein sets of worm and yeast: Orthology and divergence. *Science* **282:** 2022–2028.

Chung S.-Y. and Wong L. 1999. Kleisli: A new tool for data integration in biology. *Trends Biotechnol.* **17:** 351–355.

Copertino D.W. and Hallick R.B. 1993. Group II and group III introns of twintrons: Potential relationships with nuclear pre-mRNA introns. *Trends Biochem. Sci.* **18:** 467–471.

Dandekar T., Snel B., Huynen M., and Bork P. 1998. Conservation of gene order: A fingerprint of proteins that physically interact. *Trends Biochem. Sci.* **23:** 324–328.

de Bruijn F.J., Lupski J.R., and Weinstock G.M., Eds. 1998. *Bacterial genomes: Physical structure and analysis.* Chapman and Hall, New York.

Deckert G., Warren P.V., Gaasterland T., Young W.G., Lenox A.L., Graham D.E., Overbeek R., Snead M.A., Keller M., Aujay M., Huber R., Feldman R.A., Short J.M., Olsen G.J., and Swanson R.V. 1998. The complete genome of the hyperthermophilic bacterium *Aquifex aeolicus. Nature* **392:** 353–358.

Devos K.M. and Gale M.D. 2000. Genome relationships. The grass model in current research. *Plant Cell* **12:** 637–646.

Duran B.S. and Odell P.L. 1974. Cluster analysis: A survey. In *Lecture notes in economics and mathematical systems* (ed. M. Beckmann and H.P. Künzi). Springer-Verlag, New York.

Eisen M.B., Spellman P.T., Brown P.O., and Botstein D. 1998. Cluster analysis and display of genome-wide expression patterns. *Proc. Natl. Acad. Sci.* **95:** 14863–14868.

Enright A.J. and Ouzounis C.A. 2000. GeneRAGE: A robust algorithm for sequence clustering and domain detection. *Bioinformatics* **16:** 451–457.

Enright A.J., Iliopoulos I., Kyrpides N.C., and Ouzounis C.A. 1999. Protein interaction maps for complete genomes based on gene fusion events. *Nature* **402:** 86–90.

Fire A., Xu S., Montgomery M.K., Kostas S.A., Driver S.E., and Mello C.C. 1998. Potent and specific genetic interference by double-stranded RNA in *Caenorhabditis elegans. Nature* **391:** 806–811.

Fitch W.M. 1970. Distinguishing homologous from analogous proteins. *Syst. Zool.* **19:** 99–113.

Fleischmann R.D., Adams M.D., White O., Clayton R.A., Kirkness E.F., Kerlavage A.R., Bult C.J., Tomb J.-F., Dougherty B.A., Merrick J.M., et al. 1995. Whole-genome random sequencing and assembly of *Haemophilus influenzae* Rd. *Science* **269:** 496–512.

Force A., Lynch M., Pickett F.B., Amores A., Yan Y.L., and Postlethwait J. 1999. Nucleotide preservation of duplicate genes by complementary, degenerate mutations. *Genetics* **151:** 1531–1545.

Gaasterland T. and Bekiranov S. 2000. Making the most of microarray data. *Nat. Genet.* **24:** 204–206.

Gaasterland T. and Sensen C.W. 1998. MAGPIE: A multipurpose automated genome project investigation environment for ongoing sequencing projects. In *Bacterial genomes: Physical structure and analysis* (ed. F.J. de Bruijn et al.), pp. 559–582. Chapman and Hall, New York.

Gilbert W., de Souza S.J., and Long M. 1997. Origin of genes. *Proc. Natl. Acad. Sci.* **94:** 7698–7703.

Golub T.R., Slonim D.K., Tamayo P., Huard C., Gaasenbeek M., Mesirov J.P., Coller H., Loh M.L., Downing J.R., Caligiuri M.A., Bloomfield C.D., and Lander E.S. 1999. Molecular classification of cancer: Class discovery and class prediction by gene expression monitoring. *Science* **286:** 531–537.

Green P., Lipman D., Hillier L., Waterston R., States D., and Claverie J.-M. 1993. Ancient conserved regions in new gene sequences and the protein databases. *Science* **259:** 1711–1716.

Henikoff S., Greene E.A., Pietrokovski S., Bork P., Attwood T.K., and Hood L. 1997. Gene families: The taxonomy of protein paralogs and chimeras. *Science* **278:** 609–614.

Himmelreich R., Plagens H., Hilbert H., Reiner B., and Herrmann R. 1997. Comparative analysis of the genomes of the bacteria *Mycoplasma pneumoniae* and *Mycoplasma genitalium. Nucleic Acids Res.* **25:** 701–712.

Himmelreich R., Hilbert H., Plagens H., Pirkl E., Li B.C., and Herrmann R. 1996. Complete sequence of the genome of the bacterium *Mycoplasma pneumoniae. Nucleic Acids Res.* **24:** 4420–4449.

Hoersch S., Leroy C., Brown N.P., Andrade M.A., and Sander C. 2000. The GeneQuiz web server: Protein functional analysis through the Web. *Trends Biochem. Sci.* **25:** 33–35.

Hoogland C., Sanchez J.C., Tonella L., Binz P.A., Bairoch A., Hochstrasser D.F., and Appel R.D. 2000. The 1999 SWISS-2DPAGE database update. *Nucleic Acids Res.* **28:** 286–288.

Hughes J.D., Estep P.W., Tavazoie S., and Church G.M. 2000. Computational identification of *cis*-regulatory elements associated with groups of functionally related genes in *Saccharomyces cerevisiae. J. Mol. Biol.* **296:** 1205–1214.

Huttley G.A., Easteal S., Southey M.C., Tesoriero A., Giles G.G., McCredie M.R., Hopper J.L., and Venter D.J. 2000. Adaptive evolution of the tumour suppressor BRCA1 in humans and chimpanzees. Australian breast cancer family study. *Nat. Genet.* **25:** 410–413.

Huynen M.A. and Bork P. 1998. Measuring genome evolution. *Proc. Natl. Acad. Sci.* **95:** 5449–5856.

Huynen M., Snel B., Lathe W., III, and Bork P. 2000. Exploitation of gene context. *Curr. Opin. Struct. Biol.* **10:** 366–370.

Kanehisa M. and Goto S. 2000. KEGG: Kyoto encyclopedia of genes and genomes. *Nucleic Acids Res.* **28:** 27–30.

Kaneko T., Katoh T., Sato S., Nakamura Y., Asamizu E., Kotani H., Miyajima N., and Tabata S. 1999. Structural analysis of *Arabidopsis thaliana* chromosome 5. IX. Sequence features of the regions of 1,011,550 bp covered by seventeen P1 and TAC clones. *DNA Res.* **6:** 183–195.

Kaneko T., Sato S., Kotani H., Tanaka A., Asamizu E., Nakamura Y., Miyajima N., Hirosawa M., Sugiura M., Sasamoto S., Kimura T., Hosouchi T., Matsuno A., Muraki A., Nakazaki N., Naruo K., Okumura S., Shimpo S., Takeuchi C., Wada T., Watanabe A., Yamada M., Yasuda M., and Tabata S. 1996a. Sequence analysis of the genome of the unicellular cyanobacterium *Synechocystis* sp. strain PCC6803. II. Sequence determination of the entire genome and assignment of potential protein-coding regions. *DNA Res.* **3:** 109–136.

———. 1996b. Sequence analysis of the genome of the unicellular cyanobacterium *Synechocystis* sp. strain PCC6803. II. Sequence determination of the entire genome and assignment of potential protein-coding regions. *DNA Res.* (suppl.) **3:** 185–209.

Karlin S., Campbell A.M., and Mrázek J. 1998. Comparative DNA analysis across diverse genomes. *Annu. Rev. Genet.* **32:** 185–225.

Karp P.D., Riley M., Saier M., Paulsen I.T., Paley S., and Pellegrini-Toole A. 2000. The EcoCyc and Meta-Cyc databases. *Nucleic Acids Res.* **28:** 56–59.

Kawarabayasi Y., Sawada M., Horikawa H., Haikawa Y., Hino Y., Yamamoto S., Sekine M., Baba S., Kosugi H., Hosoyama A., Nagai Y., Sakai M., Ogura K., Otsuka R., Nakazawa H., Takamiya M., Ohfuku Y., Funahashi T., Tanaka T., Kudoh Y., Yamazaki J., Kushida N., Oguchi A., Aoki K., and Kikuchi H. 1998. Complete sequence and gene organization of the genome of a hyper-thermophilic archaebacterium, *Pyrococcus horikoshii* OT3 (supplement). *DNA Res.* **5:** 147–155.

Kececioglu J. and Gusfield D. 1998. Reconstructing a history of recombinations from a set of sequences. *Discrete Appl. Math.* **88:** 239–260.

Kececioglu J. and Sankoff D. 1995. Exact and approximation algorithms for sorting by reversals, with application to genome rearrangement. *Algorithmica* **13:** 180–210.

Kidwell M.G. and Lisch D.R. 1997. Transposable elements as sources of variation in plants and animals. *Proc. Natl. Acad. Sci.* **94:** 7704–7711.

———. 2000. Transposable elements and host genome evolution. *Trends Ecol. Evol.* **15:** 95–99.

Koonin E.V. and Galperin M.Y. 1997. Prokaryotic genomes: The emerging paradigm of genome-based microbiology. *Curr. Opin. Genet. Dev.* **7:** 757–763.

Koonin E.V., Tatusov R.L., and Galperin M.Y. 1998. Beyond complete genomes: From sequence to structure and function. *Curr. Opin. Struct. Biol.* **8:** 355–363.

Kunst F., Ogasawara N., Moszer I., Albertini A.M., Alloni G., Azevedo V., Bertero M.G., Bessieres P., Bolotin A., Borchert S., et al. 1997. The complete genome sequence of the gram-positive bacterium *Bacillus subtilis*. *Nature* **390:** 249–256.

Kyrpides N.C. 1999. Genomes OnLine database (GOLD 1.0): A monitor of complete and ongoing genome projects world-wide. *Bioinformatics* **15:** 773–774.

Kyrpides N.C. and Ouzounis C.A. 1999. Whole-genome sequence annotation: "Going wrong with confidence." *Mol. Microbiol.* **32:** 886–887.

Labedan B. and Riley M. 1995. Gene products of *Escherichia coli*: Sequence comparisons and common ancestries. *Mol. Biol. Evol.* **12:** 980–987.

Li W.-H. 1997. *Molecular evolution.* Sinauer Associates, Sunderland, Massachusetts.

Liang F., Holt I., Pertea G., Karamycheva S., Salzberg S.L., and Quackenbush J. 2000. Gene index analysis of the human genome estimates approximately 120,000 genes. *Nat. Genet.* **25:** 239–240.

Lin X., Kaul S., Rounsley S., Shea T.P., Benito M.I., Town C.D., Fujii C.Y., Mason T., Bowman C.L., Barnstead M., et al. 1999. Sequence and analysis of chromosome 2 of the plant *Arabidopsis thaliana*. *Nature* **402:** 761–768.

Link A.J., Eng J., Schieltz D.M., Carmack E., Mize G.J., Morris D.R., Garvik B.M., and Yates J.R., III. 1999. Direct analysis of protein complexes using mass spectrometry. *Nat. Biotechnol.* **17:** 676–682.

Lupski J.R. 1998. Genomic disorders: Structural features of the genome can lead to DNA rearrangements and human disease traits. *Genetics* **14:** 417–422.

Marcotte E.M., Pellegrini M., Thompson M.J., Yeates T.O., and Eisenberg D. 1999a. A combined algorithm for genome-wide prediction of protein function. *Nature* **402:** 83–86.

Marcotte E.M., Pellegrini M., Ng H., Rice W.D., Yeates T.O., and Eisenberg D. 1999b. Detecting protein function and protein–protein interactions from genome sequences. *Science* **285:** 751–753.

McCallum C.M., Comai L., Greene E.A., and Henikoff S. 2000. Targeting induced local lesions in genomes (TILLING) for plant functional genomics. *Plant Physiol.* **123:** 439–442.

McGuire A.M., Hughes J.D., and Church G.M. 2000. Conservation of DNA regulatory motifs and discovery of new motifs in microbial genomes. *Genome Res.* **10:** 744–757.

Myers E.W., Sutton G.G., Delcher A.L., Dew I.M., Fasulo D.P., Flanigan M.J., Kravitz S.A., Mobarry C.M., Reinert K.H.J., Remington K.A., et al. 2000. A whole-genome assembly of *Drosophila*. *Science* **287:** 2196–2204.

O'Brien S.J., Menotti-Raymond M., Murphy W.J., Nash W.G., Wienberg J., Stanyon R., Copeland N.G., Jenkins N.A., Womack J.E., and Marshall Graves J.A. 1999. The promise of comparative genomics in mammals. *Science* **286:** 458–462, 479–481.

Ochmann H., Lawrence J.G., and Groisman E.A. 2000. Lateral gene transfer and the nature of bacterial innovation. *Nature* **405:** 299–304.

Oliver S.G., van der Aart Q.J., Agostoni-Carbone M.L., Aigle M., Alberghina L., Alexandraki D., Antoine G., Anwar R., Ballesta J.P., Benit P., et al. 1992. The complete DNA sequence of yeast chromosome III. *Nature* **357:** 38–46.

Ouzounis C., Casari G., Sander C., Tamames J., Valencia A. 1996. Computational comparisons of genomes. *Trends Biotechnol.* **14:** 280–285.

Overbeek R., Fonstein M., D'Souza M., Pusch G.D., and Maltsev N. 1999. The use of gene clusters to infer functional coupling. *Proc. Natl. Acad. Sci.* **96:** 2896–2901.

Overbeek R., Larsen N., Pusch G.D., D'Souza M., Selkov E., Jr., Kyrpides N., Fonstein M., Maltsev N., and Selkov E. 2000. WIT: Integrated system for high-throughput genome sequence analysis and metabolic reconstruction. *Nucleic Acids. Res.* **28:** 123–125.

Paterson A.H., Lin Y.-R., Li Z., Schertz K.F., Doebley J.F., Pinson S.R.M., Liu S.-C., Stansel J.W., and Irvine J.E. 1995. Convergent domestication of cereal crops by independent mutations at corresponding genetic loci. *Science* **269:** 1714–1718.

Pellegrini M., Marcotte E.M., Thompson J.M., Eisenberg D., and Yeates T.O. 1999. Assigning protein functions by comparative genome analysis: Protein phylogenetic profiles. *Proc. Natl. Acad. Sci.* **96:** 4285–4288.

Perou C.M., Jeffrey S.S., van de Rijn M., Rees C.A., Eisen M.B., Ross D.T., Pergamenschikov A., Williams C.F., Zhu S.X., Lee J.C., Lashkari D., Shalon D., Brown P.O., and Botstein D. 1999. Distinctive gene expression patterns in human mammary epithelial cells and breast cancers. *Proc. Natl. Acad. Sci.* **96:** 9212–9217.

Pruitt K.D., Katz K.S., Sicotte H., and Maglott D.R. 2000. Introducing RefSeq and LocusLink: Curated human genome resources at the NCBI. *Trends Genet.* **16:** 44–47.

Retief J.D., Lynch K.R., and Pearson W.R. 1999. Panning for genes: A visual strategy for identifying novel gene orthologs and paralogs. *Genome Res.* **9:** 373–382.

Riley M. 1993. Functions of the gene products of *Escherichia coli*. *Microbiol. Rev.* **57:** 862–952.

———. 1998. *E. coli* genes: Ancestry and map locations. In *Bacterial genomes: Physical structure and analysis* (ed. F.J. de Bruijn et al.), pp. 187–195. Chapman and Hall, New York.

Roth F.P., Hughes J.D., Estep P.W., and Church G.M. 1998. Finding DNA regulatory motifs within unaligned noncoding sequences clustered by whole-genome mRNA quantitation. *Nat. Biotechnol.* **16:** 939–945.

Rowen L., Koop B.F., and Hood L. 1996. The complete 685-kilobase DNA sequence of the human beta T cell receptor locus. *Science* **272:** 1755–1762.

Rubin G.M., Yandell M.D., Wortman J.R., Gabor Miklos G.L., Nelson C.R., Hariharan I.K., Fortini M.E., Li P.W., Apweiler R., Fleischmann W., et al. 2000. Comparative genomics of the eukaryotes. *Science* **287:** 2204–2215.

Ruiz M., Giudicelli V., Ginestoux C., Stoehr P., Robinson J., Bodmer J., Marsh S.G., Bontrop R., Lemaitre M., Lefranc G., Chaume D., and Lefranc M.P. 2000. IMGT, the international ImMunoGeneTics database. *Nucleic Acids Res.* **28:** 219–221.

Sanger F., Coulson A.R., Hong G.F., Hill D.F., and Petersen G.B. 1982. Nucleotide sequence of bacteriophage lambda DNA. *J. Mol. Biol.* **162:** 729–773.

Sankoff D. and Goldstein M. 1989. Probabilistic models of genome shuffling. *Bull. Math. Biol.* **51:** 117–124.

Sankoff D. and Nadeau J.H. 1996. Conserved synteny as a measure of genomic distance. *Discrete Appl. Math.* **71:** 247–257.

Sankoff D., Cedergren R., and Abel Y. 1993. Genome divergence through gene rearrangement. *Methods Enzymol.* **183:** 428–438.

Sankoff D., Leduc G., Antoine N., Paquin B., Lang B.F., and Cedergren R. 1992. Gene order comparisons for phylogenetic inference: Evolution of the mitochondrial genome. *Proc. Natl. Acad. Sci.* **89:** 6575–6579.

SanMiguel P., Gaut B.S., Tikhonov A., Nakajima Y., and Bennetzen J.L. 1998. The paleontology of intergene retrotransposons of maize. *Nat. Genet.* **20:** 43–45.

SanMiguel P., Tikhonov A., Jin Y.K., Motchoulskaia N., Zakharov D., Melake-Berhan A., Springer P.S., Edwards K.J., Lee M., Avramova Z., and Bennetzen J.L. 1996. Nested retrotransposons in the intergenic regions of the maize genome. *Science* **274:** 765–768.

Schultz J., Copley R., Doerks T., Ponting C.P., and Bork P. 2000. SMART: A web-based tool for the study of genetically mobile domains. *Nucleic Acids Res.* **28:** 231–234.

Smit A.F. 1996. The origin of interdispersed repeats in the human genome. *Curr. Opin. Genet. Dev.* **6:** 743–748.

Snel B., Bork P., and Huynen M. 1999. Genome phylogeny based on gene content. *Nat. Genet.* **21:** 108–110.

———. 2000a. Genome evolution: Gene fusion versus gene fission. *Trends Genet.* **16:** 9–11.

Snel B., Lehmann G., Bork P., and Huynen M.A. 2000b. STRING: A web-server to retrieve and display the repeatedly occurring neighbourhood of a gene. *Nucleic Acids Res.* (in press).

Stoeckert C.J., Jr., Salas F., Brunk B., and Overton G.C. 1999. EpoDB: A prototype database for the analysis of genes expressed during vertebrate erythropoiesis. *Nucleic Acids Res.* **27:** 200–203.

Tamayo P., Slonim D., Mesirov J., Zhu Q., Kitareewan S., Dmitrovsky E., Lander E.S., and Golub T.R. 1999. Interpreting patterns of gene expression with self-organizing maps: Methods and application to hematopoietic differentiation. *Proc. Natl. Acad. Sci.* **96:** 2907–2912.

Tatusov R.L., Koonin E.V., and Lipman D.J. 1997. A genomic perspective on protein families. *Science* **278:** 631–637.

Tatusov R.L., Galperin M.Y., Natale D.A., and Koonin E.V. 2000. The COG database: A tool for genome-scale analysis of protein functions and evolution. *Nucleic Acids Res.* **8:** 33–36.

Tatusov R.L., Mushegian A.R., Bork P., Brown N.P., Hayes W.S., Borodovsky M., Rudd K.E., and Koonin E.V. 1996. Metabolism and evolution of *Haemophilus influenzae* deduced from a whole-genome comparison with *Escherichia coli. Curr. Biol.* **6:** 279–291.

Tekaia F., Lazcano A., and Dujon B. 1999. The genomic tree as revealed from whole genome comparisons. *Genome Res.* **9:** 550–557.

Tipton K. and Boyce S. 2000. History of the enzyme nomenclature system. *Bioinformatics* **16:** 34–40.

Tomita M., Hashimoto K., Takahashi K., Shimizu T.S., Matsuzaki Y., Miyoshi F., Saito K., Tanida S., Yugi K., Venter J.C., and Hutchison C.A., III. 1999. E-CELL: Software environment for whole-cell simulation. *Bioinformatics* **15:** 72–84.

Volff J.N. and Altenbuchner J. 2000. A new beginning with new ends: Linearisation of circular chromosomes during bacterial evolution. *FEMS Microbiol Lett.* **86:** 143–150.

Webb E.C. 1992. *Enzyme nomenclature 1992: Recommendations of the Nomenclature Committee of the International Union of Biochemistry and Molecular Biology on the Nomenclature and Classification of Enzymes.* Academic Press, San Diego, California.

Weiner A.M. 2000. Do all SINEs lead to LINEs? *Nat. Genet.* **24:** 332–333.

Glossary

Abstract Syntax Notation (ASN.1)

A language that is used to describe structured data types formally. Within bioinformatics, it has been used by the National Center for Biotechnology Information to encode sequences, maps, taxonomic information, molecular structures, and biographical information in such a way that it can be easily accessed and exchanged by computer software.

Accession number

A unique identifier that is assigned to a single database entry for a DNA or protein sequence.

Affine gap penalty

A gap penalty score that is a linear function of gap length, consisting of a gap opening penalty and a gap extension penalty multiplied by the length of the gap. Using this penalty scheme greatly enhances the performance of dynamic programming methods for sequence alignment. See also Gap penalty.

Algorithm

A systematic procedure for solving a problem in a finite number of steps, typically involving a repetition of operations. Once specified, an algorithm can be written in a computer language and run as a program.

Alignment

Refers to the procedure of comparing two or more sequences by looking for a series of individual characters or character patterns that are in the same order in the sequences. Of the two types of alignment, local and global, a local alignment is generally the most useful. See also Local and Global alignments.

Alignment score

An algorithmically computed score based on the number of matches, substitutions, insertions, and deletions (gaps) within an alignment. Scores for matches and substitutions are derived from a scoring matrix such as the BLOSUM and PAM matrices for proteins, and affine gap penalties suitable for the matrix are chosen. Alignment scores are in log odds units, often bit units (log to the base 2). Higher scores denote better alignments. See also Similarity score, Distance in sequence analysis.

Alphabet

The total number of symbols in a sequence—4 for DNA sequences and 20 for protein sequences.

Annotation

The prediction of genes in a genome, including the location of protein-encoding genes, the sequence of the encoded proteins, any significant matches to other proteins of known function, and the location of RNA-encoding genes. Predictions are based on gene models; e.g., hidden Markov models of introns and exons in proteins encoding genes, and models of secondary structure in RNA.

Anonymous FTP

When a FTP service allows anyone to log in, it is said to provide anonymous FTP service. A user can log in to an anonymous FTP server by typing *anonymous* as the user name and his E-mail address as a password. Most Web browsers now negotiate anonymous FTP logon without asking the user for a user name and password. See also FTP.

ASCII

The American Standard Code for Information Interchange (ASCII) encodes unaccented letters a–z, A–Z, the numbers 0–9, most punctuation marks, space, and a set of control characters such as carriage return and tab. ASCII specifies 128 characters that are mapped to the values 0–127. ASCII files are commonly called "plain text," meaning that they only encode text without extra markup.

Back-propagation

When training feed-forward neural networks, a back-propagation algorithm can be used to modify the network weights. After each training input pattern is fed through the network, the network's output is compared with the desired output and the amount of error is calculated. This error is back-propagated through the network by using an error function to correct the network weights. See also Feed-forward neural network.

Baum-Welch algorithm

An expectation maximization algorithm that is used to train hidden Markov models.

Bayes' rule

Forms the basis of conditional probability by calculating the likelihood of an event occurring based on the history of the event and relevant background information. In terms of two parameters A and B, the theorem is stated in an equation: The conditional probability of A, given B, $P(A|B)$, is equal to the probability of A, $P(A)$, times the conditional probability of B, given A, $P(B|A)$, divided by the probability of B, $P(B)$. $P(A)$ is the historical or prior distribution value of A, $P(B|A)$ is a new prediction for B for a particular value of A, and $P(B)$ is the sum of the newly predicted values for B. $P(A|B)$ is a posterior probability, representing a new prediction for A given the prior knowledge of A and the newly discovered relationships between A and B.

Bayesian analysis

A statistical procedure used to estimate parameters of an underlying distribution based on an observed distribution. See also Bayes' rule.

Biochips

Miniaturized arrays of large numbers of molecular substrates, often oligonucleotides, in a defined pattern. They are also called DNA microarrays and microchips.

Bioinformatics

An interdisciplinary field involving biology, computer science, mathematics, and statistics to analyze biological sequence data, genome content, and arrangement, and to predict the function and structure of macromolecules.

Bit units

From information theory, a bit denotes the amount of information required to distinguish between two equally likely possibilities. The number of bits of information, N, required to convey a message that has M possibilities is $\log_2 M = N$ bits.

Block

Conserved ungapped patterns approximately 3–60 amino acids in length in a set of related proteins.

BLOSUM matrices

An alternative to PAM tables, BLOSUM tables were derived using local multiple alignments of more distantly related sequences than were used for the PAM matrix. These are used to assess the similarity of sequences when performing alignments.

Boltzmann distribution

Describes the number of molecules that have energies above a certain level, based on the Boltzmann gas constant and the absolute temperature.

Boltzmann probability function

See Boltzmann distribution.

Bootstrap analysis

A method for testing how well a particular data set fits a model. For example, the validity of the branch arrangement in a predicted phylogenetic tree can be tested by resampling columns in a multiple sequence alignment to create many new alignments. The appearance of a particular branch in trees generated from these resampled sequences can then be measured. Alternatively, a sequence may be left out of an analysis to determine how much the sequence influences the results of an analysis.

Branch length

In sequence analysis, the number of sequence changes along a particular branch of a phylogenetic tree.

Chebyshev's inequality

The probability that a random variable exceeds its mean is less than or equal to the square of 1 over the number of standard deviations from the mean.

Cluster analysis

A method for grouping together a set of objects that are most similar from a larger group of related objects. The relationships are based on some criterion of similarity or difference. For sequences, a similarity or distance score or a statistical evaluation of those scores is used.

Cobbler

A single sequence that represents the most conserved regions in a multiple sequence alignment. The BLOCKS server uses the cobbler sequence to perform a database similarity search as a way to reach sequences that are more divergent than would be found using the single sequences in the alignment for searches.

Coding system (neural networks)

Regarding neural networks, a coding system needs to be designed for representing input and output. The level of success found when training the model will be partially dependent on the quality of the coding system chosen.

Codon usage

Analysis of the codons used in a particular gene or organism.

COG

Clusters of orthologous groups in a set of groups of related sequences in microorganisms and yeast (*S. cerevisiae*). These groups are found by whole proteome comparisons and include orthologs and paralogs. See also Orthologs and Paralogs.

Comparative genomics

A comparison of gene numbers, gene locations, and biological functions of genes in the genomes of diverse organisms, one objective being to identify groups of genes that play a unique biological role in a particular organism.

Complexity (of an algorithm)

Describes the number of steps required by the algorithm to solve a problem as a function of the amount of data; for example, the length of sequences to be aligned.

Conditional probability

The probability of a particular result (or of a particular value of a variable) given one or more events or conditions (or values of other variables).

Consensus

A single sequence that represents, at each subsequent position, the variation found within corresponding columns of a multiple sequence alignment.

Context-free grammars

A recursive set of production rules for generating patterns of strings. These consist of a set of terminal characters that are used to create strings, a set of nonterminal symbols that correspond to rules and act as placeholders for patterns that can be generated using terminal characters, a set of rules for replacing nonterminal symbols with terminal characters, and a start symbol.

Contig

A set of clones that can be assembled into a linear order.

CORBA

The Common Object Request Broker Architecture (CORBA) is an open industry standard for working with distributed objects, developed by the Object Management Group. CORBA allows the interconnection of objects and applications regardless of computer language, machine architecture, or geographic location of the computers.

Correlation coefficient

A numerical measure, falling between −1 and 1, of the degree of the linear relationship between two variables. A positive value indicates a direct relationship, a negative value indicates an inverse relationship, and the distance of the value away from zero indicates the strength of the relationship. A value near zero indicates no relationship between the variables.

Covariation (in sequences)

Coincident change at two or more sequence positions in related sequences that may influence the secondary structures of RNA or protein molecules.

Database

A computerized storehouse of data that provides a standardized way for locating, adding, removing, and changing data. See also Object-oriented database, Relational database.

Dendogram

A form of a tree that lists the compared objects (e.g., sequences or genes in a microarray analysis) in a vertical order and joins related ones by levels of branches extending to one side of the list.

Dirichlet mixtures

Defined as the conjugational prior of a multinomial distribution. One use is for predicting the expected pattern of amino acid variation found in the match state of a hidden Markov model (representing one column of a multiple sequence alignment of proteins), based on prior distributions found in conserved protein domains (blocks).

Distance in sequence analysis

The number of observed changes in an optimal alignment of two sequences, usually not counting gaps.

Dot matrix

Dot matrix diagrams provide a graphical method for comparing two sequences. One sequence is written horizontally across the top of the graph and the other along the left-hand side. Dots are placed within the graph at the intersection of the same letter appearing in both sequences. A series of diagonal lines in the graph indicate regions of alignment. The matrix may be filtered to reveal the most-alike regions by scoring a minimal threshold number of matches within a sequence window.

Dynamic programming

A dynamic programming algorithm solves a problem by combining solutions to sub-problems that are computed once and saved in a table or matrix. Dynamic programming is typically used when a problem has many possible solutions and an optimal one needs to be found. This algorithm is used for producing sequence alignments, given a scoring system for sequence comparisons.

Entropy

From information theory, a measure of the unpredictable nature of a set of possible elements. The higher the level of variation within the set, the higher the entropy.

Erdos and Renyi law

In a toss of a "fair" coin, the number of heads in a row that can be expected is the logarithm of the number of tosses to the base 2. The law may be generalized for more than two possible outcomes by changing the base of the logarithm to the number of outcomes. This law was used to analyze the number of matches and mismatches that can be expected between random sequences as a basis for scoring the statistical significance of a sequence alignment.

Expect value (*E*)

In a database similarity search, the probability that an alignment score as good as the one found between a query sequence and a database sequence would be found in as many comparisons between random sequences as was done to find the matching sequence. In other types of sequence analysis, *E* has a similar meaning.

Expectation maximization (sequence analysis)

An algorithm for locating similar sequence patterns in a set of sequences. A guessed alignment of the sequences is first used to generate an expected scoring matrix representing the distribution of sequence characters in each column of the alignment, this pattern is matched to each sequence, and the scoring matrix values are then updated to maximize the alignment of the matrix to the sequences. The procedure is repeated until there is no further improvement.

Extreme value distribution

Some measurements are found to follow a distribution that has a long tail which decays at high values much more slowly than that found in a normal distribution. This slow-falling type is called the extreme value distribution. The alignment scores between unrelated or random sequences are an example. These scores can reach very high values, particularly when a large number of comparisons are made, as in a database similarity search. The probability of a particular score may be accurately predicted by the extreme value distribution, which follows a double negative exponential function after Gumbel.

False negative

A negative data point collected in a data set that was incorrectly reported due to a failure of the test in avoiding negative results.

False positive

A positive data point collected in a data set that was incorrectly reported due to a failure of the test. If the test had correctly measured the data point, the data would have been recorded as negative.

Feed-forward neural network

Organizes nodes into sequence layers in which the nodes in each layer are fully connected with the nodes in the next layer, except for the final output layer. Input is fed from the input layer through the layers in sequence in a "feed-forward" direction, resulting in output at the final layer. See also Neural network.

Filtering (window size)

During pair-wise sequence alignment using the dot matrix method, random matches can be filtered out by using a sliding window to compare the two sequences. Rather than comparing a single sequence position at a time, a window of adjacent positions in the two sequences is compared and a dot, indicating a match, is generated only if a certain minimal number of matches occur.

Fourier analysis

Studies the approximations and decomposition of functions using trigonometric polynomials.

Format (file)

Different programs require that information be specified to them in a formal manner, using particular keywords and ordering. This specification is a file format.

Forward-backward algorithm

Used to train a hidden Markov model by aligning the model with training sequences. The algorithm then refines the model to reduce the error when fitted to the given data using a gradient descent approach.

FTP (File Transfer Protocol)

Allows a person to transfer files from one computer to another across a network using an FTP-capable client program. The FTP client program can only communicate with machines that run an FTP server. The server, in turn, will make a specific portion of its file system available for FTP access, providing that the client is able to supply a recognized user name and password to the server.

Functional genomics

Assessment of the function of genes identified by between-genome comparisons. The function of a newly identified gene is tested by introducing mutations into the gene and then examining the resultant mutant organism for an altered phenotype.

Gap

Mismatch in the alignment of two sequences caused by either an insertion in one sequence or a deletion in the other.

Gap penalty

A numeric score used in sequence alignment programs to penalize the presence of gaps within an alignment. The value of a gap penalty affects how often gaps appear in alignments produced by the algorithm. Most alignment programs suggest gap penalties that are appropriate for particular scoring matrices.

Genetic algorithm

A kind of search algorithm that was inspired by the principles of evolution. A population of initial solutions is encoded and the algorithm searches through these by applying a pre-defined fitness measurement to each solution, selecting those with the highest fitness for reproduction. New solutions can be generated during this phase by crossover and mutation operations, defined in the encoded solutions.

Genome

The genetic material of an organism, contained in one haploid set of chromosomes.

Gibbs sampling method

An algorithm for finding conserved patterns within a set of related sequences. A guessed alignment of all but one sequence is made and used to generate a scoring matrix that represents the alignment. The matrix is then matched to the left-out sequence, and a probable location of the corresponding pattern is found. This prediction is then input into a new alignment and another scoring matrix is produced and tested on a new left-out sequence. The process is repeated until there is no further improvement in the matrix.

Global alignment

Attempts to match as many characters as possible, from end to end, in a set of two or more sequences.

Graph theory

A branch of mathematics which deals with problems that involve a graph or network structure. A graph is defined by a set of nodes (or points) and a set of arcs (lines or edges) joining the nodes. In sequence and genome analysis, graph theory is used for sequence alignments and clustering alike genes.

Half-bits

Some scoring matrices are in half-bit units. These units are logarithms to the base 2 of odds scores times 2.

Heuristic

A procedure that progresses along empirical lines by using rules of thumb to reach a solution. The solution is not guaranteed to be optimal.

Hexadecimal system

The base 16 counting system that uses the digits 0–9 followed by the letters A–F.

Hidden Markov Models (HMM)

In sequence analysis, a HMM is usually a probabilistic model of a multiple sequence alignment, but can also be a model of periodic patterns in a single sequence, representing, for example, patterns found in the exons of a gene. In a model of multiple sequence alignments, each column of symbols in the alignment is represented by a frequency distribution of the symbols called a state, and insertions and deletions by other states. One then moves through the model along a particular path from state to state trying to match a given sequence. The next matching symbol is chosen from each state, recording its probability (frequency) and also the probability of going to that particular state from a previous one (the transition probability). State and transition probabilities are then multiplied to obtain a probability of the given sequence. Generally speaking, a HMM is a statistical model for an ordered sequence of symbols, acting as a stochastic state machine that generates a symbol each time a transition is made from one state to the next. Transitions between states are specified by transition probabilities.

Hidden layer

An inner layer within a neural network that receives its input and sends its output to other layers within the network. One function of the hidden layer is to detect covariation within the input data, such as patterns of amino acid covariation that are associated with a particular type of secondary structure in proteins.

Hierarchical clustering

The clustering or grouping of objects based on some single criterion of similarity or difference. An example is the clustering of genes in a microarray experiment based on the correlation between their expression patterns. The distance method used in phylogenetic analysis is another example.

Hill climbing

A nonoptimal search algorithm that selects the singular best possible solution at a given state or step. The solution may result in a locally best solution that is not a globally best solution.

Homolog

A similar component in two organisms (e.g., genes with strongly similar sequences) that can be attributed to a common ancestor of the two organisms during evolution.

Horizontal transfer

The transfer of genetic material between two distinct species that do not ordinarily exchange genetic material. The transferred DNA becomes established in the recipient genome and can be detected by a novel phylogenetic history and codon content compared to the rest of the genome.

HTML

The Hyper-Text Markup Language (HTML) provides a structural description of a document using a specified tag set. HTML currently serves as the Internet lingua franca for describing hypertext Web page documents.

Hyperplane

A generalization of the two-dimensional plane to N dimensions.

Hypercube

A generalization of the three-dimensional cube to N dimensions.

Indel

An insertion or deletion in a sequence alignment.

Information content (of a scoring matrix)

A representation of the degree of sequence conservation in a column of a scoring matrix representing an alignment of related sequences. It is also the number of questions that must be asked to match the column to a position in a test sequence. For bases, the maximum possible number is 2, and for proteins, 4.32 (logarithm to the base 2 of the number of possible sequence characters).

Information theory

A branch of mathematics that measures information in terms of bits, the minimal amount of structural complexity needed to encode a given piece of information.

Input layer

The initial layer in a feed-forward neural net. This layer encodes input information that will be fed through the network model.

Interface definition language

Used to define an interface to an object model in a programming language neutral form, where an interface is an abstraction of a service defined only by the operations that can be performed on it.

Internet

The network infrastructure, consisting of cables interconnected by routers, that provides global connectivity for individual computers and private networks of computers. A second sense of the word "internet" is the collective computer resources available over this global network.

Interpolated Markov model

A type of Markov model of sequences that examines sequences for patterns of variable length in order to discriminate best between genes and non-gene sequences.

Iterative

A sequence of operations in a procedure that is performed repeatedly.

K-tuple

Identical short stretches of sequences, also called word*s*.

Likelihood

The hypothetical probability that an event which has already occurred would yield a specific outcome. Unlike probability, which refers to future events, likelihood refers to past events.

Linear discriminant analysis

An analysis in which a straight line is located on a graph between two sets of data points in a location that best separates the data points into two groups.

Local alignment

Attempts to align regions of sequences with the highest density of matches. In doing so, one or more islands of subalignments are created in the aligned sequences.

Log odds score

The logarithm of an odds score. See also Odds score.

Machine learning

The training of a computational model of a process or classification scheme to distinguish between alternative possibilities.

Markov chain

Describes a process that can be in one of a number of states at any given time. The Markov chain is defined by probabilities for each transition occurring; that is, probabilities of the occurrence of state s_j given that the current state is s_i. Substitutions in nucleic acid and protein sequences are generally assumed to follow a Markov chain in that each site changes independently of the previous history of the site. With this model, the number and types of substitutions observed over a relatively short period of evolutionary time can be extrapolated to longer periods of time. In performing sequence alignments and calculating the statistical significance of alignment scores, sequences are assumed to be Markov chains in which the choice of one sequence position is not influenced by another.

Maximum likelihood (phylogeny, alignment)

The most likely outcome (tree or alignment), given a probabilistic model of evolutionary change in DNA sequences.

Maximum parsimony

The minimum number of evolutionary steps required to generate the observed variation in a set of sequences, as found by comparison of the number of steps in all possible phylogenetic trees.

Method of moments

The mean or expected value of a variable is the first moment of the values of the variable around the mean, defined as that number from which the sum of deviations to all values is zero. The standard deviation is the second moment of the values about the mean, and so on.

Minimum spanning tree

Given a set of related objects classified by some similarity or difference score, the minimum spanning tree joins the most-alike objects on adjacent outer branches of a tree and then sequentially joins less-alike objects by more inward branches. The tree branch lengths are calculated by the same neighbor-joining algorithm that is used to build phylogenetic trees of sequences from a distance matrix. The sum of the resulting branch lengths between each pair of objects will be approximately that found by the classification scheme.

Molecular clock hypothesis

The hypothesis that sequences change at the same rate in the branches of an evolutionary tree.

Monte Carlo

A method that samples possible solutions to a complex problem as a way to estimate a more general solution.

Mutation data matrix

A scoring matrix compiled from the observation of point mutations between aligned sequences. Also refers to a Dayhoff PAM matrix in which the scores are given as log odds scores.

Nats (natural logarithm)

A number expressed in units of the natural logarithm.

Needleman-Wunsch algorithm

Uses dynamic programming to find global alignments between sequences.

Neighbor-joining method

Clusters together alike pairs within a group of related objects (e.g., genes with similar sequences) to create a tree whose branches reflect the degrees of difference among the objects.

Neural network

From artificial intelligence algorithms, techniques that involve a set of many simple units that hold symbolic data, which are interconnected by a network of links associated with numeric weights. Units operate only on their symbolic data and on the inputs that they receive through their connections. Most neural networks use a training algorithm (see Back-propagation) to adjust connection weights, allowing the network to learn associations between various input and output patterns. See also Feed-forward neural network.

Noise

In sequence analysis, a small amount of randomly generated variation in sequences that is added to a model of the sequences; e.g., a hidden Markov model or scoring matrix, in order to avoid the model overfitting the sequences. See also Overfitting.

Normal distribution

The distribution found for many types of data such as body weight, size, and exam scores. The distribution is a bell-shaped curve that is described by a mean and standard deviation of the mean. Local sequence alignment scores between unrelated or random sequences do not follow this distribution but instead the extreme value distribution which has a much extended tail for higher scores. See also Extreme value distribution.

Object Management Group (OMG)

A not-for-profit corporation that was formed to promote component-based software by introducing standardized object software. The OMG establishes industry guidelines and detailed object management specifications in order to provide a common framework for application development. Within OMG is a Life Sciences Research group, a consortium representing pharmaceutical companies, academic institutions, software vendors, and hardware vendors who are working together to improve communication and interoperability among computational resources in life sciences research.

Object-oriented database

Unlike relational databases (see entry), which use a tabular structure, object-oriented databases attempt to model the structure of a given data set as closely as possible. In doing so, object-oriented databases tend to reduce the appearance of duplicated data and the complexity of query structure often found in relational databases.

Odds score

The ratio of the likelihoods of two events or outcomes. In sequence alignments and scoring matrices, the odds score for matching two sequence characters is the ratio of the frequency with which the characters are aligned in related sequences divided by the frequency with which those same two characters align by chance alone, given the frequency of occurrence of each in the sequences. Odds scores for a set of individually aligned positions are obtained by multiplying the odds scores for each position. Odds scores are often converted to logarithms to create log odds scores that can be added to obtain the log odds score of a sequence alignment.

Optimal alignment

The highest-scoring alignment found by an algorithm capable of producing multiple solutions. This is the best possible alignment that can be found, given any parameters supplied by the user to the sequence alignment program.

Orthologs

A pair of genes found in two species are orthologous when the encoded proteins are 60–80% identical in an alignment. The proteins almost certainly have the same three-dimensional structure, domain structure, and biological function, and the encoding genes have originated from a common ancestor gene at an earlier evolutionary time. Two orthologs I and II in genomes A and B, respectively, may be identified when the complete genomes of two species are available: (1) in a database similarity search of all of the proteome of B using I as a query, II is the best hit found, and (2) I is the best hit when II is used as a query of the proteome of B. The best hit is the database sequence with the highest expect value (*E*). Orthology is also predicted by a very close phylogenetic relationship between sequences or by a cluster analysis. Compare to Paralogs. See also Cluster analysis.

Output layer

The final layer of a neural network in which signals from lower levels in the network are input into output states where they are weighted and summed to give an output signal. For example, the output signal might be the prediction of one type of protein secondary structure for the central amino acid in a sequence window.

Overfitting

Can occur when using a learning algorithm to train a model such as a neural net or hidden Markov model. Overfitting refers to the model becoming too highly representative of the training data and thus no longer representative of the overall range of data that is supposed to be modeled.

Pair-wise sequence alignment

An alignment performed between two sequences.

PAM scoring matrices

Percent Accepted Mutation or PAM matrices describe the probability that one base or amino acid has changed during the course of evolution. Amino acid PAM matrices are derived from families of closely related sequences and are used to assess the similarity of sequences when performing alignments.

Paralogs

Genes that are related through gene duplication events. These events may lead to the production of a family of related proteins with similar biological functions within a species. Paralogous gene families within a species are identified by using an individual protein as a query in a database similarity search of the entire proteome of an organism. The process is repeated for the entire proteome and the resulting sets of related proteins are then searched for clusters that are most likely to have a conserved domain structure and should represent a paralogous gene family.

Parametric sequence alignment

An algorithm that finds a range of possible alignments based on varying the parameters of the scoring system for matches, mismatches, and gap penalties. An example is the Bayes block aligner.

Pearson correlation coefficent

A measure of the correlation between two variables that reflects the degree to which the two variables are related. For example, the coefficient is used as a measure of similarity of gene expression in a microarray experiment. See also Correlation coefficient.

Percent identity

The percentage of the columns in an alignment of two sequences that includes identical amino acids. Columns in the alignment that include gaps are not scored in the calculation.

Percent similarity

The percentage of the columns in an alignment of two sequences that includes either identical amino acids or amino acids that are frequently found substituted for each other in sequences of related proteins (conservative substitutions). These substitutions may be found in an amino acid substitution matrix such as the Dayhoff PAM and

Henikoff BLOSUM matrices. Columns in the alignment that include gaps are not scored in the calculation.

Perceptron

A neural network in which input and output states are directly connected without intervening hidden layers.

Poisson distribution

Used to predict the occurrence of infrequent events over a long period of time or when there are a large number of trials. In sequence analysis, it is used to calculate the chance that one pair of a large number of pairs of unrelated sequences may give a high local alignment score.

Position-specific scoring matrix (PSSM)

Represents the variation found in the columns of an alignment of a set of related sequences. Each subsequent matrix column corresponds to the next column in the alignment and each row corresponds to a particular sequence character (one of four bases in DNA sequences or 20 amino acids in protein sequences). Matrix values are log odds scores obtained by dividing the counts of the residue in the alignment, dividing by the expected number of counts based on sequence composition, and converting the ratio to a log score. The matrix is moved along sequences to find similar regions by adding the matching log odds scores and looking for high values. There is no allowance for gaps. Also called a weight matrix or scoring matrix.

Posterior (Bayesian analysis)

A conditional probability based on prior knowledge and newly evaluated relationships among variables using Bayes' rule. See also Bayes' rule.

Prior (Bayesian analysis)

The expected distribution of a variable based on previous data.

Profile

A matrix representation of a conserved region in a multiple sequence alignment that allows for gaps in the alignment. The rows include scores for matching sequential columns of the alignment to a test sequence. The columns include substitution scores for amino acids and gap penalties.

Profile hidden Markov model

A hidden Markov model of a conserved region in a multiple sequence alignment that includes gaps and may be used to search new sequences for similarity to the aligned sequences.

Proteome

The entire collection of proteins that are encoded by the genome of an organism. Initially the proteome is estimated by gene prediction and annotation methods but eventually will be revised as more information on the sequence of the expressed genes is obtained.

Pseudocounts

Small number of counts that is added to the columns of a scoring matrix to increase the variability either to avoid zero counts or to add more variation than was found in the sequences used to produce the matrix.

Receiver operator characteristic

The receiver operator characteristic (ROC) curve describes the probability that a test will correctly declare the condition present against the probability that the test will declare the condition present when actually absent. This is shown through a graph of the test's sensitivity against one minus the test's specificity for different possible threshold values.

Regular expressions

This computational tool provides a method for expressing the variations found in a set of related sequences including a range of choices at one position, insertions, repeats, and so on. For example, these expressions are used to characterize variations found in protein domains in the PROSITE catalog.

Regularization

A set of techniques for reducing data overfitting when training a model. See also Overfitting.

Relational database

Organizes information into tables where each column represents the fields of information that can be stored in a single record. Each row in the table corresponds to a single record. A single database can have many tables and a query language is used to access the data. See also Object-oriented database.

Scoring matrix

See Position-specific scoring matrix.

Selectivity (in database similarity searches)

The ability of a search method to locate members of a protein family without making a false-positive classification of members of other families.

Sensitivity (in database similarity searches)

The ability of a search method to locate as many members of a protein family as possible, including distant members of limited sequence similarity.

Significance

A significant result is one that has not simply occurred by chance, and therefore is probably true. Significance levels show how likely a result is due to chance, expressed as a probability. In sequence analysis, the significance of an alignment score may be calculated as the chance that such a score would be found between random or unrelated sequences. See Expect value.

Similarity score (sequence alignment)

The sum of the number of identical matches and conservative (high scoring) substitutions in a sequence alignment divided by the total number of aligned sequence characters. Gaps are usually ignored.

Simulated annealing

A search algorithm that attempts to solve the problem of finding global extrema. The algorithm was inspired by the physical cooling process of metals and the freezing process in liquids where atoms slow down in movement and line up to form a crystal. The algorithm traverses the energy levels of a function, always accepting energy levels that are smaller than previous ones, but sometimes accepting energy levels that are greater, according to the Boltzmann probability distribution.

Single-linkage cluster analysis

An analysis of a group of related objects, e.g., similar proteins in different genomes to identify both close and more distant relationships, represented on a tree or dendogram. The method joins the most closely related pairs by the neighbor-joining algorithm by representing these pairs as outer branches on the tree. More distant objects are then progressively added to lower tree branches. The method is also used to predict phylogenetic relationships by distance methods. See also Hierarchical clustering, Neighbor-joining method.

Smith-Waterman algorithm

Uses dynamic programming to find local alignments between sequences. The key feature is that all negative scores calculated in the dynamic programming matrix are changed to zero in order to avoid extending poorly scoring alignments and to assist in identifying local alignments starting and stopping anywhere with the matrix.

Space or time complexity

An algorithm's complexity is the maximum amount of computer memory or time required for the number of algorithmic steps to solve a problem.

Specificity (in database similarity searches)

The ability of a search method to locate members of one protein family, including distantly related members.

Stochastic context-free grammar

A formal representation of groups of symbols in different parts of a sequence; i.e., not in the same context. An example is complementary regions in RNA that will form secondary structures. The stochastic feature introduces variability into such regions.

Stringency

Refers to the minimum number of matches required within a window. See also Filtering.

Sum of pairs method

Sums the substitution scores of all possible pair-wise combinations of sequence characters in one column of a multiple sequence alignment.

Synteny

The presence of a set of homologous genes in the same order on two genomes.

Threading

In protein structure prediction, the aligning of the sequence of a protein of unknown structure with a known three-dimensional structure to determine whether the amino acid sequence is spatially and chemically compatible with that structure.

Uncertainty

From information theory, a logarithmic measure of the average number of choices that must be made for identification purposes. See also Information content.

Unified Modeling Language (UML)

A standard sanctioned by the Object Management Group that provides a formal notation for describing object-oriented design.

Viterbi algorithm

Calculates the optimal path of a sequence through a hidden Markov model of sequences using a dynamic programming algorithm.

Weight matrix

See Position-specific scoring matrix.

Index

Page numbers followed by *f* indicate a figure and by *t* indicate a table.